Springer Monographs in Mathematics

This series publishes advanced monographs giving well-written presentations of the "state-of-the-art" in fields of mathematical research that have acquired the maturity needed for such a treatment. They are sufficiently self-contained to be accessible to more than just the intimate specialists of the subject, and sufficiently comprehensive to remain valuable references for many years. Besides the current state of knowledge in its field, an SMM volume should ideally describe its relevance to and interaction with neighbouring fields of mathematics, and give pointers to future directions of research.

Shigeki Matsutani

The Weierstrass Sigma Function in Higher Genus and Applications to Integrable Equations

Springer

Shigeki Matsutani
Institute of Science and Engineering
Kanazawa University
Kanazawa, Ishikawa, Japan

ISSN 1439-7382 ISSN 2196-9922 (electronic)
Springer Monographs in Mathematics
ISBN 978-981-97-8162-1 ISBN 978-981-97-8163-8 (eBook)
https://doi.org/10.1007/978-981-97-8163-8

Mathematics Subject Classification: 14H40, 14H42, 14H50, 14H05, 14H70, 14-03, 32-03, 01A55, 14H81, 37J70, 35Q51, 35Q53, 34-03, 01A45, 01A50, 14H45, 37K10, 30-03, 74-03

This Springer imprint is published by the registered company Springer Nature Singapore Pte Ltd.
The registered company address is: 152 Beach Road, #21-01/04 Gateway East, Singapore 189721, Singapore

Dedicated to Emma Previato

Preface

This book provides a reconstruction of the Abelian function theory of Weierstrass based on the modern knowledge of algebraic geometry, the results of its related studies, and its applications to integrable equations.

Weierstrass' elliptic function theory starts with Weierstrass' standard equation $f_E = y^2 - 4x^3 - g_2x - g_3 = 0$, which determines the curve E, and constructs the Weierstrass sigma function, σ, on $\mathbb{C}$ as the covering space of its Jacobian $J_E := \mathbb{C}/\Gamma_E$. It is obtained that $\left(\wp(u) = -\frac{d^2}{du^2}\log\sigma(u), \frac{d\wp(u)}{du}\right)$ is identical to a point (x, y) of the curve E.

The algebraic structures associated with $R_E := \mathbb{C}[x, y]/(f_E)$ determine the algebraic properties of the meromorphic functions $\mathscr{M}_E$ on E, whereas the sigma function associated with the J_E provides the transcendental properties of the meromorphic functions $\mathscr{M}_{J_E}$ on J_E. The identification yields the relations among these algebraic and transcendental properties. The equivalence between the algebraic side associated with R_E and the transcendental side related to $\mathscr{M}_{J_E}$, based on the identification, plays a crucial role in Weierstrass' elliptic function theory and affects several fields in mathematics, science, and technology. It displays a picture to unify the algebraic objects and the transcendental objects in theory.

Weierstrass envisioned himself extending the picture to general algebraic curves, but his attempt failed due to difficulties and obstacles. The developments of algebraic geometry in the 20th century were to overcome these, and after these problems were settled, it was natural to extend Weierstrass' program to construct the Abelian function theory in the manner of his elliptic function theory.

We have considered the generalization of the picture to algebraic curves with higher genera in a series of studies as an extension of Mumford's studies for the hyperelliptic curves based on the modern algebraic geometry of the 1980s.

This book shows the results of these studies and gives the straightforward generalization of Weierstrass' elliptic function theory or the theory of the elliptic sigma function to the general algebraic curves.

Chapter 1 describes the history and overview of the sigma function theory from Jacob Bernoulli's elastica to the contemporary sigma function theory through the studies in the 18th and the 19th centuries. It is not familiar that the origin of the elliptic function was the elastica problems studied by Jacob, Daniel Bernoulli, and Euler. We first mention the fascinating relationship between the elastica (thin elastic curve) and the elliptic function, which is also the origin of the sigma function. In contrast, it is shown that the lemniscate played a subsidiary role in certain earlier stages. Even Lagrange and Gauss followed their study on elastica and investigated the elliptic function. Gauss's approach to the elliptic function, probably through the study of the elastica, stands for the theory of the sigma function, in a sense. He extracted the nonlinear differential equations in the elliptic integral to find the theta series using the bilinear equations. Taking over Gauss's approach, probably through Gudermann, Weierstrass discovered the original version of the sigma function, the Al function, of the elliptic curves and attempted to extend them to the Abelian function theory after investigating the hyperelliptic function theory. Though Riemann introduced the compact Riemann surfaces and the Riemann theta function to breathe new life into the studies on the Abelian function theory, Weierstrass insisted on the style in which we started from algebraic curves. It seems that Weierstrass attempted to construct the Abelian function theory, which is comparable to and compatible with his elliptic function theory. However, Weierstrass and his successors faced several serious problems, and these studies failed at the end of the 19th century. It is in contrast to Riemann's approach, which was taken over and developed in modern mathematics.

However, the attempt by Weierstrass and his successor, especially Baker, includes quite exciting views. After settling the problems thanks to the development of mathematics, including the modern Abelian function theory, these studies were reborn in the 1990s in a movement that aims to reconstruct the Abelian function theory so that we obtain the theory comparable to and compatible with his elliptic function theory. We give the overview of the contemporary sigma function as a result of this movement.

Chapter 2 is devoted to Weierstrass curves, or W-curves for short, and their affine rings R_X, with Sect. 2.1 listing the conventions and notations used in this book. We have studied further generalization of the picture in Weierstrass' elliptic function theory. As Weierstrass restricted the standard form f_E of the elliptic curves, we also introduce the Weierstrass canonical form to express general algebraic curves, whose origin was also in Weierstrass' Abelian function theory based on Abel's last paper.

The Weierstrass curve X is a normalized curve of the curve given by the Weierstrass canonical form, $y^r + A_1(x)y^{r-1} + A_2(x)y^{r-2} + \cdots + A_{r-1}(x)y + A_r(x) = 0$ where each A_j is a polynomial in x of degree $\le js/r$ for certain coprime positive integers r and s ($r < s$); the Weierstrass non-gap sequence at $\infty \in X$ is given by the numerical semigroup H_X whose generators of H_X contain r and s. Every compact Riemann surface has a Weierstrass curve X which is birational to the surface. It provides the projection $\varpi_r : X \to \mathbb{P}$, (e.g., $(x, y) \mapsto x$) as a covering space. Let $R_X := \mathbf{H}^0(X, \mathscr{O}_X(*\infty))$ and $R_{\mathbb{P}} := \mathbf{H}^0(\mathbb{P}, \mathscr{O}_{\mathbb{P}}(*\infty))$. We have the explicit description of the complementary module $R_X^{\mathfrak{c}}$ of $R_{\mathbb{P}}$-module R_X in Sect. 2.2 and the trace structure. They lead to the explicit expressions of the holomorphic one-form except for ∞, $\mathbf{H}^0(\mathbb{P}, \mathscr{A}_{\mathbb{P}}(*\infty))$ and the fundamental 2-form of the second kind in Sect. 2.4, whereas Sect. 2.3 demonstrates the examples of the W-curves. Section 2.5 introduces the differential operators, which play a crucial role in Chap. 4. Further, Sect. 2.6 reviews the addition structure arising from linear equivalence in R_X, and the division polynomials in R_X. Section 2.7 gives a short review of some examples of the period integrals of W-curves. Section 2.8 mentions covering spaces of typical W-curves which are related to al-functions which are used in Chaps. 3 and 4. As an example of the degenerating family of W-curves, Sect. 2.9 gives an explanation of a particular degeneration process of the cyclic (3,4)-curve. Appendix 2.10 recalls the normalization of the curves. Appendix 2.11 reviews the complementary modules of the extension of the ring. Appendix 2.12 provides a short summary of the deformation of curves.

Chapter 3 is on the sigma function for W-curves. The elliptic sigma function was developed on Weierstrass' elliptic Al function by himself after profound investigations. Klein refined Weierstrass' hyperelliptic Al function to define the generalized sigma function for the hyperelliptic Jacobians by using only the data of compact Riemann surfaces transcendentally. Baker reformulated Klein's sigma function from algebraic viewpoints by following Weierstrass' approach, which he started from hyperelliptic curves and showed several crucial results. Recently, D. Korotkin and V. Shramchenko, and A. Nakayashiki defined the sigma function of every compact Riemann surface following Klein's approach, transcendentally. Every compact Riemann surface can be characterized by a W-curve, as mentioned in Chap. 2. In this book, we provide the connection between the W-curves in Chap. 2 and the sigma function from an algebraic viewpoint, as Baker did for the hyperelliptic sigma function, and show the properties of the sigma function.

Section 3.1 reviews the properties of the Riemann theta function, significantly shifting the Riemann constant due to the spin structure, and the precise version of the Riemann–Kempf theorem to define the sigma function. Section 3.2 is devoted to defining the sigma function for W-curves and showing their primitive properties. In Sect. 3.3, we show the Jacobi inversion formula, which we have illustrated in Fig. 1.9; we can handle the meromorphic functions on both W-curves and their Jacobian equivalently without loss of concreteness.

Using them, we show several properties of the sigma function in Sects. 3.4–3.7; Sect. 3.4 provides the addition theorems of the sigma function for the particular curves. Section 3.5 gives an algebraic expression of prime form in terms of the sigma function. Section 3.6 considers the al function, which is a generalization of the Jacobi sn, cn, dn functions. Using the al function, we demonstrate that we can handle the behavior of the sigma function for the degenerating family of curves in Sect. 3.7. Appendix 3.8 is for basic results of the theta function and complicated computations associated with theta function. In Appendix 3.9, we review the related topics associated with the universal Grassmannian manifolds, and in Appendix 3.10, we show the differential properties of the Schur function. Appendix 3.11 provides a short review of the heat equations of the sigma function.

Chapter 4 covers the application of the sigma function theory to the integrable systems. Section 4.1 shows the KdV hierarchy based on Baker's 1903 study and related topics, including so-called soliton solutions. Section 4.2 is on the integrable system related to the al-function, i.e., the modified KdV equation, the generalized Lamé equation, the sine-Gordon equation, and the Neumann system. Section 4.3 is on the differential relations on the integrable system associated with the addition theorems of the sigma function, i.e., Toda equation. Section 4.4 is on the difference relations related to the addition theorem. Section 4.5 provides the dispersionless KP hierarchy. Section 4.6 reviews the generalized elastica problem and shows its shapes associated with higher genera beyond elastica of Euler and the Bernoullis.

Here we remark that there are already excellent reviews and books on the algebro-geometric solutions of the integral system, [Mum1981, Mum1984], [TanD1979, BelBEIM1994, GeH2003, GeHMT2008] and on the sigma function [BelE2001, BelE2002, BucEL1997b, BucEL2012, On2013], which complement this book by describing much of what is not mentioned here. This book follows the ideas of Weierstrass on his Abelian function theory, which has long been out of the limelight, as mentioned in Chap. 1. We first present the ideas of Weierstrass with historical background in Chap. 1 and demonstrate the recent achievements on related topics along his line of thought. We believe that it is time to publish them as a book from that viewpoint.

Kanazawa, Japan Shigeki Matsutani
June 2024

Acknowledgments

The writing of this book was initiated by an offer from Springer in 2018. I asked my collaborator Emma Previato to write this book with me. She agreed, and we wrote a proposal for this book to Springer. Following the style of our collaborations, we thought that I would write the manuscript first, and then she would revise it. So I had been writing this manuscript since 2018.

However, her health became gradually too poor especially since 2020. We have had several studies published [MatP2022, KomMP2022a, KomMP2022b] directly related to this book. Writing a book seemed to be a difficult task for her. Her direct contributions to these studies became less and less, especially after our DNA paper [MatP2022]. She always pointed me in the right direction for these papers [MatP2022, KomMP2022a, KomMP2022b] as I explain below, but unfortunately she had no strength left to make further contributions to this book. She had encouraged the writing of this book, as I had sent her the latest version (pdf-file and bound draft) via e-mail and air-mail every six months since 2019, and we had discussed ways to revise it in the future.

However, she could not revise anything, as she passed away in June 2022.

In other words, this book was written by me alone except for a few (not so crucial) statements in Chap. 1, albeit consequently. I state here that had she lived or had we found her heirs, this book should have been published as a co-authored work. Due to copyright issues (though I wanted a different solution if possible), I eventually deleted and rewrote the parts she wrote (pp. 2, 3 and pp. 10–12) and made it a single-author book for publication according to her legacy and will.

Thus, I emphasize that this book is the culmination of 18 years of my collaboration with Emma Previato, even though it is a significant departure from the first plan she and I presented in our proposal.

I am indebted to numerous researchers who have assisted me in the completion of this book. I extend my sincerest gratitude to each of them. I am particularly grateful to Chris Eilbeck, Victor Enolskii, Yuri Fedorov, John Gibbons, Yuji Kodama, Jiryo Komeda, Yoshihiro Ônishi, and Kazuhiko Aomoto for their invaluable contributions to several collaborative projects, including crucial discussions and valuable comments. I would also like to thank Victor M. Buchstaber, Takeo

Ohsawa, Atsushi Nakayashiki, Takao Kato, Chris Athorne, Julia Bernatska, Tony Shaska, and Takanori Ayano for their valuable and helpful comments during our research on the sigma function. I am also indebted to Takeo Ohsawa for the references [Hej1972, Ste1986] and invaluable insights, and to Da-jun Zhang for [New1985]. I am grateful to John McKay, Tadashi Ashikaga, Hajime Kaji, Yohei Komori, Yoshinori Machida, Katsuyuki Takashima, Tetsuji Tokihiro, Satoru Saito, Hideo Tsuru, Hideo Mitsuhashi, and Norio Konno for their fruitful discussions, helpful comments and support during the course of my long-term research activity. I would like to express my particular gratitude to Kenichi Tamano for his comprehensive guidance over the course of 27 years. I am supported by the Grant-in-Aid for Scientific Research (C) of Japan Society for the Promotion of Science Grant, No. 16K05187 and No. 21K0328.

I would like to express my sincerest gratitude to the anonymous reviewers for their invaluable comments. I am indebted to Glenn Stevens, Ron Donagi, Takeo Ohsawa, and Masayuki Nakamura and Francesca Bonadei at Springer Inc. for their assistance in resolving issues related to the publication of this book.

Special Acknowledgments

I would like to record here how the book was written, taking into account some more emotional aspects of the book. It is about my very beautiful collaboration with Emma Previato.

After I had almost finished this book, including some parts of the Preface above, I received an e-mail from Boston University informing me that Emma Previato had passed away on June 29, 2022.

Our collaboration started in Montreal in May 2004. John McKay invited us to a private meeting at Concordia University. He had found my preprint on the generalized elastica (KdV flow) and the hyperelliptic sigma function [MatO2003a] and had written an e-mail to me in 2000 in which the study must have been related to his replicable function and Monster group, and he invited me to Montreal. As Emma Previato investigated the mKdV flow in [Pr1993] and published the papers of the hyperelliptic sigma function with Chris Eilbeck, and Victor Enolskii [EilEP2001, EilEP2003], I asked him to invite her. Then our collaboration started, and we have 18 published papers. In our collaboration, John McKay's insight turned out tremendously profound; our studies seem to be closely connected with his research on the monstrous moonshine problem as in [KomMP2013, MatP2016, MatP2022, McH2022]. He also passed away, on April 19, 2022.

Our latest projects with Jiryo Komeda [KomMP2022a, KomMP2022b] are on a problem on algebraic construction of the sigma function, which is stated in a celebration paper for her 65th birthday [KomM2020, DoS2020] at her suggestion. At the same time, we started the plan for this book in 2018. We discussed the contents, and she summarized the first plan of this book. I have been writing for four years, according to our plan, which is very different from our first plan, but which preserves the spirit in which we first conceived of this book. Our goal in this book is to show the unification between the algebraic curves and their Jacobi varieties in terms of the sigma function following Weierstrass' thoughts, one of which is in quoted Weierstrass' letter to Schwartz in Subsect. 1.6.1. When I quoted the letter at the end of [KomMP2022b], she told me it was incredible and set the stage (for our subjects).

After losing her mother in early 2020, who was long bedridden in hospital and her only family, her health deteriorated and she gradually became too weak. Writing a book appeared to be a difficult task for her. Thus her direct contributions became less gradual, particularly after our DNA paper [MatP2022] mentioned in our final section in this book, which corresponds to the topic in the 2004 meeting in Montreal. However, she always pointed me in the right direction through these papers and short messages in e-mails.

I might think that Emma Previato anticipated the situation that I would have to write this book as a single-author book; she may have known it was time to leave us soon. She strongly recommended me to publish the new contents of the book as papers, which are [KomMP2022a, KomMP2022b]. In fact, she prepared the journal in order to publish one of them as an editor and strongly suggested to me, in the case of *MDPI Mathematics* of [KomMP2022b], a few months before she passed away. By following her suggestions, I was able to make this decision because I can mention all except for a few (not so important) statements of her contributions to the new content by citing these papers. (Without her suggestion to publish these papers, this publication would have been dead in the water.)

For a final brushing-up, we waited a month and decided on the submission date of two papers [KomMP2022a, KomMP2022b]. However, after she wrote me that she could not read e-mails for a while, in the middle of June, she ceased communication. I thought that she was hospitalized and would return to work since she had returned to work many times, but it was not realized.

I completed this book alone. As I stated above, had she lived, or had we found her heirs, this book would have been published as a co-authored work with her. This book is the culmination of 18 years of our collaboration, and I have decided to publish it as a single-author book, following her legacy and strong will.

Thus, this book is dedicated to Emma Previato, an outstanding mathematician, my collaborator, my master in algebraic geometry, and my thoughtful friend.

Kanazawa, Japan
June 2024

Shigeki Matsutani

Contents

Chapter 1
Overview of Work on Sigma Function from Historical Viewpoint

This chapter aims to put the sigma function into a historical perspective. In searching for the origin of periodic real or complex functions, we were inevitably led to problems of engineering and the art of craft. The interplay of transcendental and algebraic curves, which will be one of the main themes of this book, was understood incrementally. We also found several little-known calculations, forerunners of the "integrable hierarchies" that emerged in the 1970s.

1.1 Brief History of Elliptic Integrals from Elastica

1.1.1 Elastica from Instrumentation to Mathematics: Euler, the Bernoulli Family, Lagrange, Gauss

The philosopher Edmund G. A. Husserl (1859–1938), who studied under Leopold Kronecker and Karl Weierstrass at the Frederick William University of Berlin, and after obtaining his Ph.D. worked as an assistant to Karl Weierstrass, wrote in his famous essay "the origin of geometry" [Hus1936],

> *We can always presuppose some measuring technique, whether of a lower or higher type, in the essential forward development of culture, [as well as] the growth of such a technique, thus also including the art of design for buildings, of surveying fields, pathways, etc.; such a technique is always already there, already abundantly developed and regimen to the philosopher who did not yet know geometry but who should be conceivable as its inventor.*

In that book [Hus1936], he stated that the origin of geometry was not in nature but in instrumentation. As we will show in this section, the elliptic functions, the simplest Abelian functions, originated from engineering problems. As a student of Weierstrass, Husserl studied Abelian function theory, but his background was

S. Matsutani, *The Weierstrass Sigma Function in Higher Genus and Applications to Integrable Equations*, Springer Monographs in Mathematics,
https://doi.org/10.1007/978-981-97-8163-8_1

also celestial mechanics and optics. His thesis, supervised by Leo Königsberger (a former student of Weierstrass) was about calculus of variations, which is related to both Abelian function theory and applied mathematics. This combination brought together his original scientific interests and is the spirit of our monograph.

The purpose of this book is dual: on the one hand, we show recent progress on the Abelian function theory, as it originated with Weierstrass. On the other, we pursue applications of Abelian functions to physics and other sciences, particularly integrable systems, which are connected with engineering and the art of craft (technology). As Husserl stated, mathematical theories were motivated by applications.

In this section, we show that the origin of elliptic function theory was a problem in architecture, namely the bending of beams under stress [Tr1960, Tr1968, Tr1983, Mat2010]. The study of the bending of the beam can be found in the notebooks of Leonardo da Vinci (1452–1519) together with his sketches; Galileo Galilei (1564–1642) also investigated it.

Jacob Bernoulli (1655–1705) in 1691 proposed the problem of how to express the bending of a thin elastic beam on the plane. He called it elastica, signifying an ideal thin elastic rod with non-stretching constraint, which is to say, an analytic isometric immersion of $\mathbb{R}$ into $\mathbb{C} = \mathbb{R}^2$. An example of elastica is a string of a guitar. As we show later in Sect. 4.6, the statistical mechanical treatment of elastica is related to hyperelliptic curves, but this connection does not use the curve in the same way as the string (a compact Riemann surface) in the string theory of elementary particle physics. Curves, or Riemann surfaces, are instrumental to different areas of modern physics. However, even at the classical level, the behavior of elastica is attractive for modern mathematics, due to the isometric deformation.

The shape of an elastica can be realized as the shape of paper or carpet in real life. It also appears in the shape of cable or paper in a printer (e.g. [HonJJPP2007]) and in the soft robot arm [ArDNB2017]. Even now, the shape of an elastica is concerned with industrial problems.

Jacob discussed with Gottfried W. Leibniz (1646–1716) how to apply Leibniz's new analytic approach, differential and integral analysis, to the problems in nature and the art of craft. Jacob taught the novel method to his brother, Johann (1667–1748), to investigate the catenary.

As Johann settled the catenary problem in 1691, Jacob proposed the problem of elastica the same year; Jacob regarded the elastica problem as the sequel to the catenary, but the problem of elastica had been regarded as a more difficult one than the catenary. Though Christiaan Huygens (1629–1695) criticized the elastica problem, Jacob devoted himself to it [Tr1960].

Jacob had investigated the elastica problem for three years and found that the moment of force of the elastica was proportional to the curvature k of the elastica and a typical shape of an elastica drawn in June 1694 is shown in the left-hand side of Fig. 1.1 [Ber1694a].

For the shape of the elastica illustrated in Fig. 1.1, which is now generally known as the rectangular elastica, the arclength s and the height Y are given by

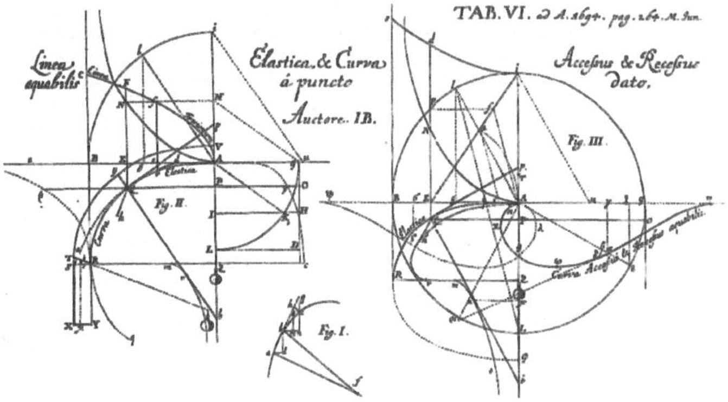

Fig. 1.1 Shape of an elastica by Jacob Bernoulli [Ber1694a]

$$s = \int_0^X \frac{dx}{\sqrt{1 - X^4}}, \quad Y = \int_0^X \frac{X^2 dx}{\sqrt{1 - X^4}}. \tag{1.1}$$

In order to obtain the quadrature of (1.1), Bernoulli investigated the geometric equivalent problem [Bos1993, 2],

$$\sqrt{ar} = \int_0^u \frac{a^2 du}{\sqrt{a^4 - u^4}},$$

which is of the paracentric isochrone posed by Leibniz [Bos1993]. The paracentric curve is a curve given by the following: 1) Fix a point O in the plane parallel to the uniform gravity and consider an orbit of a particle depending on time t in the plane. 2) Consider that the particle descends by the pull of gravity by restricting that its distance from the point O is proportional to the t. Bos stated that Jacob mainly studied the arc length differential,

$$\frac{d(ar)}{2\sqrt{ar}} = \frac{a^2 du}{\sqrt{a^4 - u^4}}.$$

He recognized this as the arc length differential of the elastica, and concluded that this enabled him to give a construction by rectification.

After finding the integrals of the rectangular elastica, Jacob discovered the lemniscate displayed in Fig. 1.2 which is given by

$$(X^2 - Y^2) = (X^2 + Y^2)^2 \tag{1.2}$$

so that its arclength is also written as

$$s = \int_0^X \frac{dX}{\sqrt{1 - X^4}}$$

as a toy model of the integral (October 1694 [Ber1694b]). After obtaining the integral, they discussed whether the lemniscate integral could be written as the trigonometric function by some transformation.

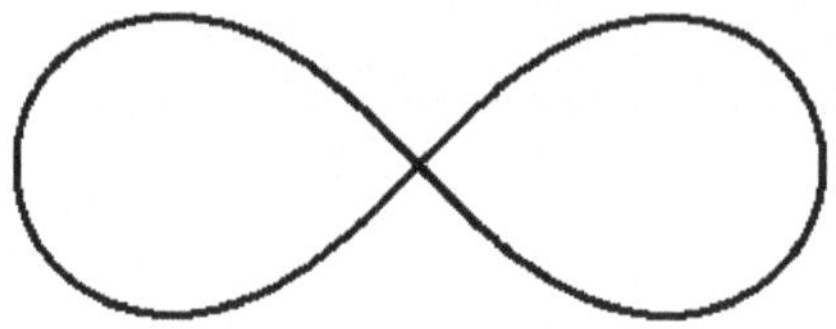

Fig. 1.2 Lemniscate

It is remarkable that Jacob's construction also connected the lemniscate and Leibniz's problem. We quote again [Bos1993, 2] "And this was not merely a curious idiosyncrasy of one mathematician. Shortly afterwards three further articles appeared, by Leibniz, Johann Bernoulli (Jakob's brother), and Jakob himself, each containing reduction of the formula (...) to the arc length of a curve. It was the *Lemniscate*, whose origin, therefore, lies in a preference for rectifications over quadratures in the representation of transcendental curves."

It should also be remarked that, as in Fig. 1.3, Euler classified the shape of an elastica completely by numerical computations and there is a figure of eight, which is different from the lemniscate. As pointed out in [Mat1995], there is a transformation from the rectangular elastica in Fig. 1.1 to the figure of eight, though it is not realized as an elastica: for a tangential angle ϕ of the rectangular elastica, we consider $3\phi/2$, and then we obtain the lemniscate curve. The rectification in Bernoulli's construction via the paracentric isochrone was the geometrical approach to this transformation. It could be obtained from the difference of the integrals of the rectangular elastica,

$$(s \pm Y)(X) = \int_0^X \sqrt{\frac{1 \pm X^2}{1 \mp X^2}} dx.$$

Since Huygens criticized Jacob's approach for not being able to represent the variety of shapes of elastica and gave the question of how to proceed to the moduli of elastica, it is almost clear that one of Jacob's motivations for presenting this figure of eight was to avoid Huygens' criticism.

Daniel Bernoulli (1700–1782) and Leonhard Euler (1707–1783) took over Jacob's investigation on elastica. Daniel was a son of Johann, and Euler's father was an acquaintance of Johann. They studied mathematics under Johann.

Daniel Bernoulli discovered the fact that the shape of an elastica was realized as the minimal point of the energy,

$$\int k^2 ds, \tag{1.3}$$

where k is the curvature of the elastica, and wrote this in a letter to Euler in 1738.

This minimal principle was obtained beyond the idea of Fermat's principle of optics, in which is the minimal optical length. Daniel's discovery extends the minimal principle in optics into mechanics. It is the birth of analytical mechanics.

This energy can be considered the simplest and oldest harmonic map from a modern viewpoint. For the U(1) bundle, $\mathrm{U}(1)_{\mathbb{R}}$, over a subset of the real one-dimensional space $\mathbb{R}$, the harmonic map is obtained by

$$\int (g^{-1}(s)dg(s)) * (g^{-1}(s)dg(s)) = \int \left(\frac{d\phi}{ds}\right)^2 ds = \int k^2 ds,$$

where g is the section of the bundle, $g(s) = \mathrm{e}^{\sqrt{-1}\phi(s)} \in \mathrm{U}(1)_s$, which can be regarded as Daniel's energy, as Bryant and Griffiths investigated [BryG1986].

Based on Daniel's discovery, Euler wrote a book, *Method* in 1744 [Eu1744], in which the variational method, isoperimetric method, was introduced. Euler applied his variational method to the elastica problem under the isometric restriction in the appendix of the book. He found that the arclength s and the height Y were given by the horizontal position X;

$$s = \int^X \frac{a^2 dX}{\sqrt{a^4 - (\alpha + \beta X + \gamma X^2)^2}},$$

$$Y = \int^X \frac{(\alpha + \beta X + \gamma X^2) dX}{\sqrt{a^4 - (\alpha + \beta X + \gamma X^2)^2}}, \tag{1.4}$$

where a, α, β, γ are real constant numbers. They are the elliptic integrals of the first kind and the second kind. The inverse function of the first equation is the elliptic function.[1]

He showed all the possible forms of the elastica and classified them as in Fig. 1.3. He settled the problem in its entirety. In the computations, he developed his theory of the elliptic integrals and moduli of elliptic curves. It could be interpreted that his study means the beginning of the Abelian function theory because periodic and quasi-periodic functions appeared naturally.

The study by Euler and the Bernoullis is quite interesting. There appear many investigations using their study of elastica even now, and similar shapes of the curve to Euler's result appear.

---

[1] Scarpello and Ritelli wrote [ScaR2011] "The subject attracted also C. Maclaurin, who realized the elastica had to be connected to elliptic integrals, see *A treatise on fluxions*, 1742, at §927: The construction of the elastic curve, and of other figures, by the rectification of the conic sections" [Mac1742, §927]. However Bistafa pointed out that Euler derived (1.4) in 1732 based on Daniel's investigation under the isoperimetric condition [Bi2023]. Further, although Euler's derivation in 1744 was so complicated and has not been well interpreted, it was clarified that he used Neother's method found by Neother 1918 [Noe1918] to obtain (1.4) [Mat2024b].

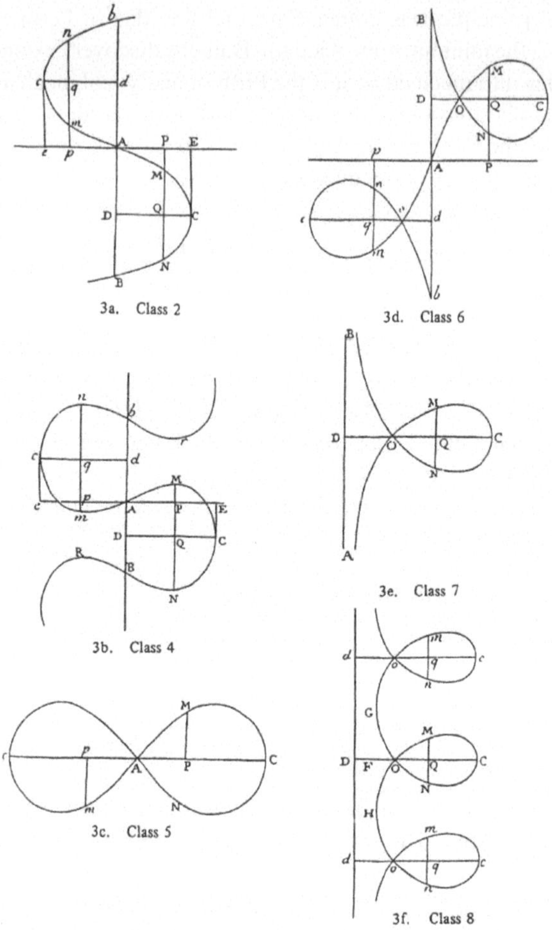

Fig. 1.3 Shapes of an elastica (except for the circle, Fig. 1.1, line)

In terms of the Weierstrass ζ function, Euler's result is represented by [Mat2010]

$$X(s) + \sqrt{-1}Y(s) = \zeta(s + u_0) - e_1 s, \tag{1.5}$$

where e_1 and u_0 are certain constants. Its X component is also expressed by an elliptic function expressed by the Weierstrass $\wp$ functions,

$$X(s) = \frac{\wp_s(s + u_0)}{4\sqrt{-1}(\wp(s + u_0) - e_1)} + X_0$$

for a certain constant X_0. Due to the reality condition of an elastica, the equality between the elliptic function and the real part of $\zeta(s)$ holds.

On the illustration of an elastica in Fig. 1.3, if we consider (X, Y) as a function of the arclength s, X is the elliptic function, and Y is the imaginary part of the elliptic ζ function; X behaves periodically, whereas the behavior of Y is not periodic but quasi-periodic, which comes from the quasi-periodic property of the theta function; the shapes of the elastica show the behavior of the theta functions.

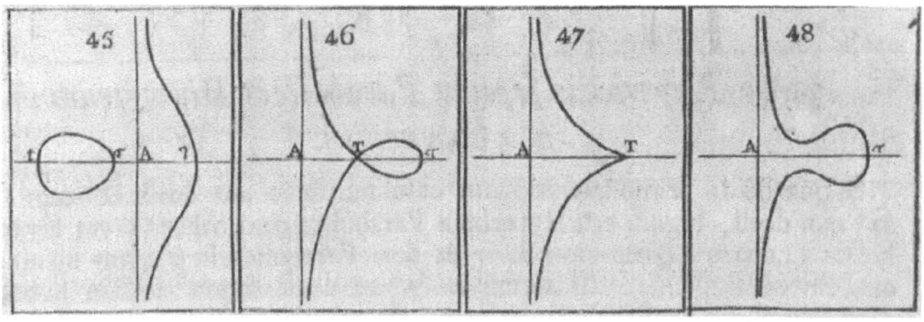

Fig. 1.4 Enumeratio linearum tertii ordinis" [New1711]

Further, Euler substantially investigated the dependence of the elliptic integrals on a, α, β, γ. He implicitly found that due to the reality condition of an elastica, the moduli space is given by a real one-dimensional space. It means the investigation of the moduli space of the elliptic curves. Fraser stated that Euler's own classification was inspired by Newton's "Enumeratio linearum tertii ordinis" (1711) illustrated in Fig. 1.4 [Fra1991]; Newton studied the moduli space of curves $ay^2 + by = cx^3 + dx^2 + ex + f$ [New1711].

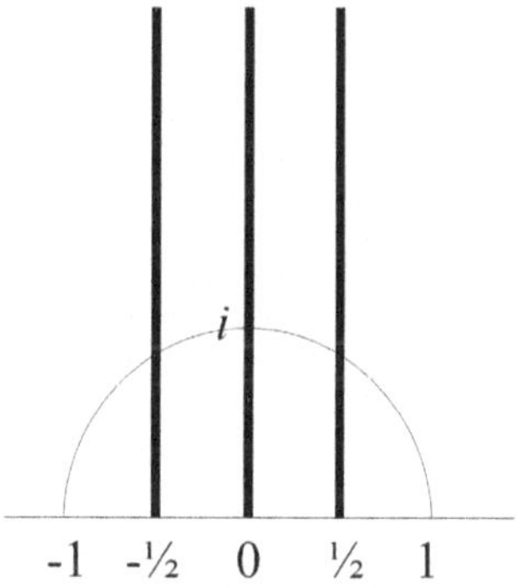

Fig. 1.5 Moduli of an elastica

In the paper [Mum1993], D. Mumford determined the moduli of an elastica from a very simple consideration of the theta function as illustrated in Fig. 1.5. It also shows the profound property of an elastica.

Further, Euler computed the relation

$$\int_0^1 \frac{dx}{\sqrt{1-X^4}} \int_0^1 \frac{X^2 dX}{\sqrt{1-X^4}} = \frac{\pi}{4}, \tag{1.6}$$

for the rectangular elastica of Fig. 1.1 and sent it to Johann. This relation is a special case of the Legendre relation of the elliptic curve.

An elastica has a fascinating symmetry. For example, the curvature k is proportional to a difference of position X from a certain point X_0:

$$X(s) - X_0 = ak(s),$$

where a is a constant factor. This differential equation generates the properties of an elastica and elliptic function.

According to Scarpello and Ritelli [ScaR2011], Euler studied this elastica problem again and wrote a paper, *Sur la force des colonnes* (1757, published 1759), concerning the buckling of columns, where the simplified expression for the elastica's curvature and the differential equations were derived. Following this study, in 1770, Joseph-Louis Lagrange (1736–1813) published the paper *Sur la figure des colonnes* [Lag1869, tome 2, p. 123], unifying Euler's results, and found the differential equations

$$ds = \frac{dy}{\sqrt{2a(f^2 - y^2) - a^2(f^2 - y^2)^2}}, \quad ds = \frac{d\varphi}{\sqrt{2a - a^2 f^2 \cos^2 \varphi}}, \tag{1.7}$$

where a and f are certain constants for the shapes of elastica displayed in Fig. 1.6.

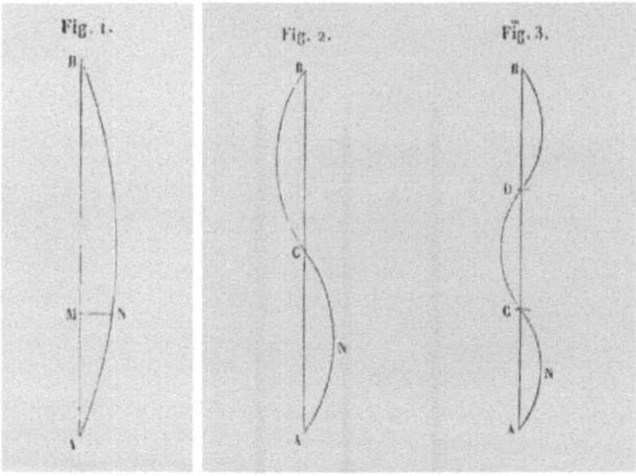

Fig. 1.6 Shapes of an elastica by Lagrange [Lag1869, tome 2, p. 123]

After an investigation of the elastica based on the elliptic integrals, Lagrange published the paper "Nouvelle Méthode de Calcul Intégral" 1784 [Lag1869, tome 2, p. 251], in which the integrals

$$\frac{Z(z)dz}{\sqrt{(1 - r^2z^2)(1 - s^2z^2)}}, \quad \frac{(\frac{c}{r} + \frac{d}{r^3} \sin^2 \varphi)d\varphi}{\sqrt{1 - \frac{s^2}{r^2} \sin^2 \varphi)}}, \tag{1.8}$$

by setting $Z = c + dz^2$ and $rz = \sin \varphi$, were studied. Though (1.8) were, sometime, recognized as the first general elliptic integral in history, they are the same as (1.4) and (1.7).

On the other hand, Johann Carl Friedrich Gauss (1777–1855) also studied the minimal problem of the Euler–Bernoulli energy (1.3) in [Ga1929, Bd. 12, pp. 51–52] and derived the static sine-Gordon equation (pendulum equation),

$$\frac{d^2}{ds^2}\phi = A\sin\phi + B\cos\phi, \tag{1.9}$$

as the determination of the shapes of the elastica, which was written in a unpublished single note; according to Schlesinger, the single note of the study on the elastica was written in 1825. Equation (1.9) was derived by the variational equations for the tangential angle ϕ of the elastica with Lagrange multipliers A and B,

$$\frac{\delta\left[\int_0^L \left(\frac{1}{2}\frac{d\phi}{ds}\right)^2 ds - A\left(X_0 - \int_0^L \cos\phi\, ds\right) - B\left(Y_0 - \int_0^L \sin\phi\, ds\right)\right]}{\delta\phi} = 0. \tag{1.10}$$

It is known that the elastica with isometric condition obeys the static modified Korteweg–de Vries equation but also the sine-Gordon equation; they have the same elliptic solutions. Probably Lagrange derived the second Eq. (1.7) by a similar consideration.

According to Houzel [Hou1978], following Lagrange's paper of 1784, Gauss investigated the relation between the arithmetic-geometric mean $M(a, b)$ and the complete elliptic integral by 1800 [Ga1929, Bd 3, S. 33]. Gauss also studied Euler's elastica in the 1790s, and wrote in his diary of his decision on January 8, 1797 to study the rectangular elastica [Kl1903]. As Schlesinger pointed out [Ga1929, Bd 10b, Abh. 2], his first goal was to elucidate the lemniscate integral through Euler's discovery (1.6), which was known as the arithmetic-geometric mean $M(a, b)$. However, he seemed to want to understand the study of Euler's elastica completely and brush it up. In 1799, Gauss investigated the inverse of the elliptic integral, $\displaystyle\int \frac{du}{\sqrt{1-k^2\sin^2 u}}$, and studied this differential equation,

$$\frac{d^2}{du^2}\log x = \frac{1}{x^2} + \mu x^2, \tag{1.11}$$

for the elliptic integral,

$$u = \int^{(x,y)} \frac{dx}{\sqrt{(1-x^2)(1-\mu x^2)}}, \tag{1.12}$$

to obtain the theta series [Ga1929, Bd. 8 p. 96] in 1800. By setting $\phi = \log(\sqrt{-\mu}x)$ and $u = s$, (1.11) is reduced to (1.9) with certain A and B. After he set $x = P(u)/Q(u)$, Gauss derived the bilinear equations (cf. Definition 4.22 and Proposition 4.23),

$$PP'' - P'P' = -Q^2, \quad QQ'' - Q'Q' = -\mu P^2, \tag{1.13}$$

to obtain the several terms of power expansions of the θ series P and Q. These P and Q are prototypes of the theta and the sigma functions. It is worthwhile noting that Gauss determined their power series and properties using the bilinear expression of the nonlinear differential equation. Further, it is remarkable that Gauss discovered his theta functions from the nonlinear differential equation (1.11), besides the differential equation

$$\frac{dx}{du}(u) = \sqrt{(1-x^2)(1-\mu x^2)} \tag{1.14}$$

directly obtained from (1.12).

In 1827 (after being confronted with Abel's and Jacobi's studies on elliptic functions) he improved this study by introducing his new theta functions R and S besides P, Q. He investigated the moduli of the elliptic curves, as he might have traced Euler's study of elastica. We note that the pictures similar to Fig. 1.5 appeared in Gauss' note in 1827 [Ga1929, Bd. 3, p.478].[2]

It is unclear how Gauss was motivated to study elastica, but it is clear that there was a tendency based on his diary of January 8, 1797, and the single note on 1825, the study of $M(a, b)$ based on (1.6), and the study of the elliptic integrals (1.12) including the moduli study, associated with elastica.

From the history of the elastica, it is most likely that the mathematics of elliptic functions came from the technological problem (or the art of craft) as Leonardo da Vinci wrote, "Mechanics is the paradise of the mathematical sciences because by means of it, one comes to the fruits of mathematics" [dV1888, 1155]. Husserl regarded that the origin of mathematics was in technology or the art of craft. It is valid for the elliptic function theory, which his teacher, Weierstrass, passionately studied. However, it is not correct, of course, that the origin of all mathematics is technology or the art of craft. The value of mathematics is not determined by its usefulness, and its usefulness does not enhance or diminish the value of mathematics. One of the goals of this book is to reconstruct some of the results of modern Abelian function theory in a more useful style, while paying tribute to these historical achievements.

1.1.2 Lemniscate: Algebraic Versus Transcendental

To consider algebra and curves, we return to the days of ancient Greek mathematics. Dieudonné stated [Di1972] "the Greeks invented geometry as a deductive science, they never (contrary to popular beliefs) made any attempt to divorce it from algebra. On the contrary, one of their main trends was to use geometry to solve algebraic problems, and this is best exemplified in the invention of the conics, the first curves which they thoroughly studied after straight lines and circles." "The Greeks knew simple geometric constructions for the root of the equation $x^2 = ab$, a and b being given as lengths of segments, and the unknown x being considered as the side of

[2] On Gauss' researches on elliptic functions, see [Ga1929, Bd. 3, pp. 404–477].

a square; they usually wrote the equation as a 'proportion' $a/x = x/b$. The 'Delic problem' called for construction of a length x of given cube, $x^3 = a^2b$; this was transformed by Hippocrates of Chio (around 420 B.C.) into a 'double proportion' $a/x = x/y = y/b$ for two unknown lengths x, y. Menechmus (ca. 350 B.C.) had the idea of considering the loci given by the two equations $ay = x^2$ and $xy = ab$, whose intersection has as coordinates x, y a solution of the problem." These recall analytic geometry, but their analytic geometry was primitive.

It was necessary to wait for René Descartes (Renatus Cartesius) (1596–1650) to arrive at modern mathematical descriptions of and thoughts about curves. It was Descartes who pushed forward algebraic and analytic geometry. He began to use symbols to describe mathematical quantities and curves in the Cartesian coordinate system in the chapter, "La géométrie" in his book, [Des1637].

The historian Goodman pointed out [Goo1974] that Descartes regarded arithmetic and geometry as the only disciplines that had acquired the certainty he sought by citing Descartes's statements, "Of the sciences now known only arithmetic and geometry are free from falsity and uncertainty, and we must carefully inquire why this is so." "My thoughts having led me from the particular study of arithmetic and geometry to a general study of mathematics, I first sought the exact meaning of this word, and why not only the two sciences already mentioned but also music, optics, mechanics etc., were regarded as a part of mathematics...." Descartes had strong faith in geometry and algebra as tools for obtaining truth, and he believed that there was truth in them.

However, Descartes' statements were too strong. According to Bos, "Descartes had introduced the techniques of analytic geometry through which curves could be represented by equations. He had also forcefully voiced his opinion about what was acceptable in geometry and what was not, asserting that curves were acceptable only if their equations were algebraic, that is, if these equations involved no other algebraic operations than addition, multiplication, division, and subtraction."

This led to hesitation in dealing with new transcendental curves by Huygens. Bos stated [Bos1993, 1] "We conclude that Huygens's tractrix did not serve as a mathematical model to describe tractional motion, but conversely tractional motion so to speak helped Huygens mathematically to lay his hands on the tractrix, and thereby on the quadrature of the hyperbola[3] and on all other geometrical problems that depended on it. Why was that necessary? Because apparently the quadrature of the hyperbola (we would say the logarithm function) was not considered truly known, and because a curve as the tractrix could not just simply be taken to serve as a solution of that problem. Now who decided these things? In this case it was Descartes." Some fifty years earlier Descartes stated that the acceptable curves were given only by algebraic equations. "The tractrix does not have such an equation; it is a transcendental, that is, a nonalgebraic curve."

Through them we can see the archetype at the birth of the transcendental curves obtained by the elastica mentioned above. Descartes' spell of "being algebraic" created certain obstacles in the discovery of Huygens, and his thought led to the

[3] The quadrature of the hyperbola means $\log z = \displaystyle\int_1^z \frac{dx}{x}$.

discovery of algebraic properties of transcendental curves and functions, such as the $\log(x + 2\pi\sqrt{-1}) = \log(x)$ algebraic properties of logarithmic functions. The algebraic properties of the curves obtained from an elastica were the driving force behind the move from elliptic integral theory to elliptic function theory, i.e. the studies of the lemniscate integrals (1.1).[4]

The "transcendental" properties of elliptic integrals had been demonstrated in the study of the elastica, but the algebraic properties were hidden in the integral by the study of the lemniscate integral. In the discovery of the algebraic curve of the lemniscate curve, (1.2) from the transcendental curve of the rectangular elastica, we should note Descartes' thought behind it. They may have studied the integral of the algebraic curve in the context of Descartes' thought.

Due to the lemniscate curve of Jacob Bernoulli and its integral, Giulio Carlo Fagnano dei Toschi (1682–1766) found additional relations in the lemniscate integrals in 1718.[5] Fagnano solved the differential equation [Ric2008]

$$\frac{dt}{\sqrt{1-t^4}} = \frac{2dx}{\sqrt{1-x^4}},$$

and obtained its solution,

$$t = \frac{2x\sqrt{1-x^4}}{1+x^4}.$$

It means that

$$2\int_0^x \frac{dt}{\sqrt{1-t^4}} = \int_0^{\frac{2x\sqrt{1-x^4}}{1+x^4}} \frac{dt}{\sqrt{1-t^4}}.$$

The study stimulated Euler. Weil wrote in [Wei1983] that according to Jacobi, the birth of elliptic function theory occurred between December 23, 1751 and January 27, 1752. The first date was the day that Euler received Fagnano's collected works on the lemniscate and the second was when Euler submitted his first work on the lemniscate to the Berlin Academy.

Surely, Euler had studied the properties of the integrals by investigating the rectangular elastica. However, he recognized the algebraic meaning of the study of Fagnano immediately. Beginning with the general differential equation,

$$\frac{mdx}{\sqrt{1-x^4}} = \frac{ndy}{\sqrt{1-y^4}},$$

for positive integers m and n, he found that u, z and w satisfying

---

4 In footnote 11, (1.42), and Theorem 1.1 5 one may notice the continuity from tractrix, elastica, elliptic functions, to the book's theory of general algebraic curves. Further, we note that the algebraic properties are built into the equation of Theorem 1.1 5.

5 One of the motivations for his research seems to have been to generalize the transformation from the rectangular elastica to the lemniscate and to obtain transformations to Huygens' critique of the elastica problem.

$$w = \frac{z\sqrt{1-u^4} + u\sqrt{1-z^4}}{1+uuzz}$$

implies that

$$\int_0^w \frac{dt}{\sqrt{1-t^4}} = \int_0^u \frac{dt}{\sqrt{1-t^4}} + \int_0^z \frac{dt}{\sqrt{1-t^4}}.$$

It became known as the addition theorem of the lemniscate function. These should be expressed as the inverse function of the elliptic integral and show the algebraic properties of the integrals. Jacobi recognized the birth of the elliptic function as mentioned above.

However, we add a remark that when a beam of building material is bent and used in construction, it is important in application to know where the point is at half its length, or what shape the point will be when one arc length is added to another arc length. Euler must have understood immediately about the importance of its application as well, e.g., the paper of 1757. It is important to emphasize, however, that in the history of conventional elliptic function theory, the lack of perspective on the importance of the role of the elastica was a major motivation for studying the "transcendentality–algebraicity" relationship that they require to describe the world (or, more clearly, technology). It was never only for the beauty of the mathematics.

Here we note that there are two aspects of correspondences between "transcendental" and "algebraic" properties: 1) from curves $y = x$ and $y^2 = (1 - x^4)$ for $y = dx/dz$ to transcendental curves such as the elastica and the tractrix obtained by some integrals, and 2) in the integrals, we found algebraic properties again.[6] The algebraic properties and the inverse functions of the elliptic integrals, in the second sense, have been the central problems in the elliptic function theory and Abelian function theory.

Further, we are interested in the applications of these functions to the real world. It should be noted that these, which are our theme in this book, appeared in the studies of the elastica and lemniscate.

1.2 Elliptic Functions of Abel and Jacobi Based on Legendre's Results

In this section, we review the elliptic integrals and functions established by Adrien-Marie Legendre (1752–1833), Niels Henrik Abel (1802–1829), and Carl Gustav Jacob Jacobi (1804–1851) based on [Mar1996, Hou1978, Hou2002]. Behind the integrals, they considered the elliptic curves $\widehat{E}$ given by

$$y^2 = (1 - x^2)(1 - k^2x^2). \tag{1.15}$$

[6] Later, the first corresponds to the Abelian integral, the second to the algebraic properties in Jacobi varieties. They are our main subject illustrated in Figs. 1.7 and 1.8.

It is a very interesting point that Abel and Jacobi found their theta functions, in contrast to Gauss's approach, through the differential equations (1.11) and (1.13).

Following Euler, Fagnano and Lagrange, Legendre investigated the elliptic integrals; he called the elliptic integral the *elliptic function* and classified the integrals into three kinds [Hou1978]:[7]

$$\text{The first kind: } F(\varphi) := \int_0^{\varphi} \frac{d\varphi}{\Delta(\varphi)}, \quad \text{the second kind: } E(\varphi) := \int_0^{\varphi} \Delta(\varphi)\, d\varphi,$$

$$\text{the third kind: } \Pi(\varphi) := \int_0^{\varphi} \frac{d\varphi}{(1+n\sin^2\varphi)\Delta(\varphi)}, \tag{1.16}$$

where $\Delta(\varphi) := \sqrt{1-k^2\sin^2\varphi}$.

Building on Legendre's work, Abel considered elliptic integrals [Hou2002],

$$u = \int_0 \frac{dx}{\sqrt{(1-c^2x^2)(1+e^2x^2)}}, \tag{1.17}$$

and finally arrived at its inverse, i.e., his elliptic function $\lambda(u) = x$, where c and e are real numbers. He put $f(u) = \sqrt{1-c^2x^2}$ and $F(u) = \sqrt{1+e^2x^2}$, and found the addition theorem of these functions, e.g.,

$$\begin{aligned} \lambda(u+v) &= \frac{\lambda(u)f(v)F(v) + \lambda(v)f(u)F(u)}{1+e^2c^2\lambda^2(u)\lambda^2(v)}, \\ f(u+v) &= \frac{f(u)f(v) - c^2\lambda(u)\lambda(v)F(u)F(v)}{1+e^2c^2\lambda^2(u)\lambda^2(v)}, \\ F(u+v) &= \frac{F(u)F(v) + e^2\lambda(u)\lambda(v)f(u)f(v)}{1+e^2c^2\lambda^2(u)\lambda^2(v)}. \end{aligned} \tag{1.18}$$

Using the addition theorem, he obtained his theta function, e.g., φ, and ψ (original one f); in the letter to Legendre for $e^2 = -1$ and by letting $\lambda(u) = \dfrac{\varphi(u)}{\psi(u)}$ [Abl1830], Abel wrote the relations

$$\begin{aligned} \varphi(u+v)\varphi(u-v) &= (\varphi(u)\psi(v))^2 - (\varphi(v)\psi(u))^2, \\ \psi(u+v)\psi(u-v) &= (\psi(u)\psi(v))^2 - c^2(\varphi(v)\varphi(u))^2. \end{aligned} \tag{1.19}$$

---

[7] Legendre envisioned the construction of tables of integrals to separate the mathematics of elliptic integrals from their applications and make them available to applied mathematicians. To do this, it was necessary to consider the classification of elliptic integrals as functions. It should be noted that both in Euler's sketch and in Lagrange's calculations, X and the tangent angles φ are functions of the arc length s, and it was natural for anyone except Legendre to consider the inverse function of an elliptic integral as a function even then, although the concept of function was not defined well.

Here ϖ and ψ correspond to σ and σ_1 in Weierstrass elliptic function theory respectively.

Jacobi proposed that the name *elliptic function* was for the inverse function of the elliptic integral of the first kind, since Abel and Jacobi had investigated the inverse function.[8]

Let c be denoted by k. Jacobi introduced the amplitude function am such that [Mar1996]

$$x(\alpha) = \sin \mathrm{am}(\alpha), \quad \cos \mathrm{am}(\alpha) := \sqrt{1 - \sin^2 \mathrm{am} u}, \quad \Delta \mathrm{am}(\alpha) := \sqrt{1 - k^2 \sin^2 \mathrm{am} u}. \tag{1.20}$$

It is also noteworthy that Jacobi did not refer to these elliptic functions as sn, cn, dn; Christof Gudermann (1798–1852) gave such abbreviated names to Jacobi's functions. Jacobi also discovered the theta function, which became its acknowledged discovery since both Gauss and Abel found them privately, whereas Jacobi made them public. In Jacobi's elliptic function theory, the integral of the third kind plays a critical role. By investigating it, he obtained his theta function and connected it with the integral of the third kind [Jac1829]

$$\Pi(u, a) = uZ(a) + \frac{1}{2} \log \frac{\Theta(u + v)}{\Theta(u - v)}. \tag{1.21}$$

He also showed the relation between the integral of the third kind and the Legendre relation (using Legendre symbols, i.e., E and F),

$$\int_0^{\pi/2} \frac{k^2 \sin \alpha \, \cos \alpha \, \Delta\alpha \, \sin^2 \varphi \, d\varphi}{(1 - k^2 \sin^2 \alpha \sin^2 \varphi) \Delta\varphi} = F^I E(\alpha) - E^I F(\alpha), \tag{1.22}$$

where $\varphi := \mathrm{am} u$, $\alpha := \mathrm{am} a$.

We display three different approaches to the discovery of the theta functions by Gauss, Abel, and Jacobi. Their results exhibit that it is a concern to clarify the direct relationship between algebraic and transcendental properties in elliptic functions and elliptic integrals. This book shows that all of them are essential and are generalized to general algebraic curves in determining their sigma function.

1.3 Weierstrass Elliptic Function Theory

In this book, since we show the generalization of Weierstrass' theory of elliptic functions, we will focus upon Weierstrass' work and brief review the Weierstrass elliptic function theory.

---

[8] Jacobi was also involved in the compilation of Euler's complete works and was aware of Euler's developments during 1746–1766 but does not seem to have been aware of elastica. Therefore, he only described the above events of 1751.

1.3.1 Birth of Weierstrass Elliptic Function Theory Based on Gauss' Approach

Karl Theodor Wilhelm Weierstrass (1815–1897) was born in Ostenfelde, part of Ennigerloh, Province of Westphalia [Po1910, Kl1926]. After leaving the University of Bonn, at the Academy of Münster in 1839, he encountered the elliptic functions and works of Abel. At the Academy, he learned mathematics from Gudermann, who studied at the University of Göttingen; his academic advisor seemed to be Gauss. Gudermann considered that the power series expansion of the functions was the nature of functions [Jah2003]. In the paper [Wei1840], Weierstrass studied the Jacobi elliptic functions and his theta functions, Al functions,

$$\mathrm{sn}(u) = \frac{\mathrm{Al}_1(u)}{\mathrm{Al}_0(u)}, \quad \mathrm{cn}(u) = \frac{\mathrm{Al}_2(u)}{\mathrm{Al}_0(u)}, \quad \mathrm{dn}(u) = \frac{\mathrm{Al}_3(u)}{\mathrm{Al}_0(u)} \tag{1.23}$$

by citing Abel's two papers including [Abl1830] which contains (1.19). Here the Al functions are the origin of his sigma functions, and the name was in honor of Abel according to Klein [Kl1926].[9]

For the investigation of these Al functions or determination of their power expansions, Weierstrass started with a differential relation,

$$\frac{d^2}{du^2} \log \mathrm{sn}(u) = \mathrm{sn}(u)^2 - k^2 \frac{1}{\mathrm{sn}^2(u)}, \tag{1.24}$$

which was the same as Gauss' study of the differential equation, the pendulum, or the static sine-Gordon equation (1.11) [Ga1929, Bd. 8 p. 96, Bd. 3 p. 473] of 1800 and 1827, whose origin might be his elastica theory. Weierstrass used the same symbols P, Q, R and S of Gauss' and the bilinear equations to investigate the elliptic functions, and then, renamed them Al functions [Wei1840].

In other words, while showing respect for Abel, Weierstrass, in fact, succeeded to Gudermann's philosophy that the function should be regarded as a power series, and to Gauss's approach, probably, through Gudermann.

This approach characterizes Weierstrass' elliptic function theory and Abelian function theory. It is in contrast to the elliptic function theory of Abel and Jacobi. Abel and Jacobi found and studied the theta series using the addition theorem and the elliptic integrals of the third kind. The third differential plays an essential role under the influence of Legendre there. In Weierstrass' elliptic function theory, the role of the differential equations, e.g. (1.24) and (1.31), whose origin might be Gauss's elastica theory, occupy a special position, though the third derivative and the additive theorem are, of course, significant.

---

[9] According to Klein, the names sn, cn, dn functions for Jacobi's sin am, cos am, δam functions were owing to Gudermann; sn means sinus amplitudinis as initial symbol and final consonant. This nomenclature was inherited by his student, Weierstrass as Al corresponded to Abel [Kl1926].

After obtaining his university degree, he became an assistant teacher of mathematics and physics at Pro-Gymnasium at Deutsch-Krone in 1842. He was promoted to an ordinary teacher in 1845 with additional subjects, including gymnastics. He was moved to the Gymnasium at Braunsberg. His theme moved from the elliptic functions to hyperelliptic functions, thereby extending Gauss' approaches. Through the study he was to embody his respect for Abel.

After Weierstrass moved to Berlin by taking a chair at the University of Berlin, he went on to study the elliptic functions. As Poincaré described [Po1910]:

1. To deepen the general theory of functions, first that of the functions of a variable, then that of the functions of two variables; this was the base on which the whole pyramid was to rise.
2. The Abelian functions being the natural extension of elliptic functions, it was necessary to perfect the theory of these transcendent ones and to put it in a form in which generalization became easy.
3. Finally, it remained to attack the Abelian functions themselves.

Refining the function theory, he gave lectures on elliptic functions, which we called Weierstrass' theory on the elliptic functions;[10] His lecture note on the elliptic function theory was published in [Wei1882].

Considering $\frac{d^2}{du^2}\log \mathrm{Al}(u)$ might have made Weierstrass reach his standard form of the elliptic curve E in the following subsection as a differential identity; he regarded that E is more fundamental than the curve $\widehat{E}$ of (1.15) for the theory.

1.3.2 Short Review of Weierstrass Elliptic Function Theory

Though the following is not a historical description, we give a short review of Weierstrass theory [Wei1875b].

In Weierstrass' theory of the elliptic functions, Weierstrass restricted himself to handling curves with the canonical form;

$$E := \left\{ (x, y) \;\middle|\; \begin{array}{l} y^2 = x^3 + \lambda_2 x^2 + \lambda_1 x + \lambda_0 \\ \quad = (x - e_1)(x - e_2)(x - e_3) \end{array} \right\} \cup \{\infty\}, \tag{1.25}$$

where e_1, e_2 and e_3 are mutually disjoint complex numbers. We denote its affine ring by $R_E := \mathbb{C}[x, y]/(y^2 - x^3 - \lambda_2 x^2 - \lambda_1 x - \lambda_0)$. We note $R_E = \mathbf{H}^0(E, \mathcal{O}_E(*\infty))$.

[10] Weierstrass's lectures were repeated every half to two years, covering the theory of analytic functions, the theory of elliptic functions, the application of elliptic functions to physics, mechanics, and geometry, and the theory of Abelian functions. As the refinement of "elliptic functions" through the applications is in the tradition of Bernoulli, Euler, Lagrange, and Gauss, he polished his theory of elliptic functions through the applications. It was not mere mathematics for mathematics's sake, but he constructed his theory by thinking through and confirming what was crucial in the theory. Therefore, these definitions are steadfast and extremely sharp in their applications. We believe it is the prototype in mathematics of the Japanese term "yoh-no-bi" (the beauty of use).

For $\overline{y} := 2y$ and $\overline{x}; = x - \lambda_2/3$, E recovers the Weierstrass standard form

$$\overline{y}^2 = 4\overline{x}^3 + g_2\overline{x} + g_3, \tag{1.26}$$

where $g_2 = 4\left(\lambda_1 - \frac{1}{3}\lambda_2\right)$ and $g_3 = 4\left(\lambda_0 - \frac{1}{3}\lambda_2\lambda_1 + \frac{2}{27}\lambda_2^3\right)$.

The holomorphic one-form of E, the differential of the first kind, is given by

$$du := \nu^{\mathrm{I}} := \frac{dx}{2y},$$

whereas the differential of the second kind is

$$\nu^{\mathrm{II}} := \frac{xdx}{2y}.$$

When we consider the incomplete elliptic integral rigorously, we require the Abelian covering $\widetilde{E}$ of E, $\kappa_E : \widetilde{E} \to E$, which is obtained through the Abelianization of the homotopy equivalent quotient of the path space of the curve E; it has also a natural embedding $\iota_E : E \to \widetilde{E}$. The elliptic integral (EI) is given as a map,

$$\widetilde{w} : \widetilde{E} \to \mathbb{C}, \quad \left(u = \widetilde{w}(x, y) = \int_{\gamma_{(x,y),\infty}} \nu^{\mathrm{I}}, \ \nu^{\mathrm{I}} = du = \frac{dx}{2y}\right),$$

where $\gamma_{(x,y),\infty}$ is a path from ∞ to (x, y) in E, and an element in $\widetilde{E}$. Since it is generalized to the (W-normalized) Abelian integral, we also call this the *Abelian integral* (AI) here. We refer to the convention as *W-normalized*; it is normalized as $du = -(1+$ higher order terms of $t)dt$ for $t = x^2/y$ at ∞.

The complete elliptic integrals of the first kind and the second kind are

$$\omega_i := \int_{\infty}^{(e_i,0)} \nu^{\mathrm{I}}, \quad \eta_i := -\int_{\infty}^{(e_i,0)} \nu^{\mathrm{II}}, \quad \nu^{\mathrm{I}} := \frac{dx}{2y}, \quad (i = 1, 2, 3).$$

Let standard homology bases be α and β, i.e., $\mathbf{H}_1(E, \mathbb{Z}) = \mathbb{Z}\alpha \oplus \mathbb{Z}\beta$, and then the half periods and related integrals are given by

$$\omega' = \frac{1}{2}\oint_{\alpha} \nu^{\mathrm{I}}, \quad \omega'' = \frac{1}{2}\oint_{\beta} \nu^{\mathrm{I}}, \quad \eta' = -\frac{1}{2}\oint_{\alpha} \nu^{\mathrm{II}}, \quad \eta'' = -\frac{1}{2}\oint_{\beta} \nu^{\mathrm{II}}. \tag{1.27}$$

There are correspondences among them; $\omega' := \omega_1, \omega'' := \omega_3, \eta' := \eta_1$, and $\eta'' := \eta_3$. The W-normalized and the normalized lattices are given by

$$\Gamma_E = 2\mathbb{Z}\omega_1 + 2\mathbb{Z}\omega_3 \subset \mathbb{C}. \quad \Gamma_E^{\circ} = \mathbb{Z} + \mathbb{Z}\tau, \quad (\tau := \omega_3/\omega_1).$$

The former one, in this book, we call W-normalized because it plays an important role in the Weierstrass theory of elliptic functions. The W-normalized and the normalized Jacobians (Jacobi varieties) are defined by

$$\kappa_J : \mathbb{C} \to J_E = \mathbb{C}/\Gamma_E, \quad \kappa_J^\circ : \mathbb{C} \to J_E^\circ = \mathbb{C}/\Gamma_E^\circ.$$

We have the W-normalized and the normalized Abel–Jacobi maps (AJM) as birational maps,

$$w : E \to J_E, \quad (w := \kappa_J \circ \widetilde{w} \circ \iota_E), \quad w^\circ : E \to J_E^\circ, \quad (w^\circ := \kappa_J^\circ \circ \widetilde{w}^\circ \circ \iota_E),$$

Then we have the following diagrams since there are $\mathrm{SL}(2, \mathbb{Z})$ actions on Γ_E and $\mathbb{Z}\alpha \oplus \mathbb{Z}\beta$ as the loop space of E with the base point ∞ up to homotopy equivalence, i.e., $LE(\infty)$.

Algebraic space

Covering space $\widetilde{E}$ =
Abelianization of
quotient space of
path space of E
$\kappa_E : \widetilde{E} \to E$
$\widetilde{E} \supset LE(\infty) \cong \mathbb{Z}\alpha \oplus \mathbb{Z}\beta$
SL(2,$\mathbb{Z}$) $\curvearrowright$ this expression.

Transcendental space

Covering space $\mathbb{C}$ =
Abelianization of
quotient space of
path space of J_E
$\kappa_J : \mathbb{C} \to J_E = \mathbb{C}/\Gamma_E$
$\Gamma_E = \mathbb{Z}(2\omega') + \mathbb{Z}(2\omega'')$
SL(2,$\mathbb{Z}$) $\curvearrowright \Gamma_E, J_E$

We are concerned with the meromorphic functions on E and J_E, which are not related to the actions of $\mathrm{SL}(2, \mathbb{Z})$. It implies that the entire functions on $\mathbb{C}$ should be invariant for the action. Weierstrass' sigma function is defined by [HurC1893]

$$\sigma(u) = 2\omega_1 \mathrm{e}^{\left(\frac{\eta_1 u^2}{2\omega_1}\right)} \frac{\theta_1(\frac{u}{2\omega_1})}{\theta_1'^0} = \left(\frac{2\pi}{\omega' \Delta^{1/4}}\right)^{1/2} \mathrm{e}^{\left(\frac{\eta_1 u^2}{2\omega_1}\right)} \theta_1\left(\frac{u}{2\omega'}\middle|\frac{\omega''}{\omega'}\right),$$

$$\theta_1(z, \tau) = -\sum_{n=-\infty}^{\infty} \mathrm{e}^{\sqrt{-1}\pi\tau(n+\frac{1}{2})^2 + \sqrt{-1}\pi(2n+1)(z+\frac{1}{2})}, \tag{1.28}$$

where $\Delta = (g_2^3 - 27g_3^2)$, and it has the following properties:

1. σ is an entire function over $\mathbb{C}$,
2. for $\ell := 2\ell_1\omega' + 2\ell_2\omega''$ of $\ell_1, \ell_2 \in \mathbb{Z}$, the translational formula is obtained as

$$\sigma(u + \ell) = (-1)^{\ell_1+\ell_2+\ell_1\ell_2} \exp\left((\ell_1\eta_1 + \ell_2\eta_3)(2u + \ell)\right)\sigma(u),$$

3. the zero of σ is given by {zeros of σ} $\equiv 0$ modulo Γ_E,

4. the leading term of the expansion of σ at the zero point,

$$\sigma(u) = u + \text{higher order terms of } u, \quad \text{and}$$

5. it is modular invariant for SL(2, $\mathbb{Z}$)-action.

We should regard σ as the base of line bundle $\mathcal{L}_{J_E}$ on the Jacobian J_E, $\mathcal{L}_{J_E} = \mathbb{C}\sigma$, and by using the Abel–Jacobi map $w : E \to J_E$, we also consider it as the base of line bundle $\mathcal{L}_E$ on E, $\mathcal{L}_E = \mathbb{C}\sigma$.

Using the sigma function, we have the elliptic Weierstrass' ζ and $\wp$ functions:

$$\zeta(u) = \frac{d}{du}\log\sigma(u), \quad \wp(u) = -\frac{d^2}{du^2}\log\sigma(u). \tag{1.29}$$

By letting $f_u := \dfrac{d}{du} f$ for a function f of u, since we have

$$\wp(u) = \frac{\sigma_u(u)^2 - \sigma_{uu}(u)\sigma(u)}{\sigma(u)^2},$$

the properties of the translational formula of the sigma function show that $\wp(u)$ is a meromorphic function over J_E and it belongs to $\mathbf{H}^1(J_E, \mathcal{O}_{J_E}(2\{0\}))$, i.e.,

$$\wp(u + 2\ell'\omega' + 2\ell'\omega'') = \wp(u), \quad \zeta(u + 2\ell'\omega' + 2\ell'\omega'') = \zeta(u) + 2\ell'\eta' + 2\ell'\eta''$$

for integers ℓ' and ℓ''. For $u = w(x, y)$, we have the "Jacobi inversion formula",

$$(x, y) = (\wp(u), \wp_u(u)/2), \quad \wp_u(u) := \frac{d}{du}\wp(u). \tag{1.30}$$

This relation means the nonlinear ordinary differential equation,

$$\wp_u^2/4 = \wp^3 + \lambda_2\wp^2 + \lambda_1\wp + \lambda_0, \tag{1.31}$$

which recovers the data of the curve E. The relation of $\wp$ in (1.30) is called the Jacobi inversion formula in this book.

By differentiating (1.31), we obtain $\dfrac{\partial^2\wp}{\partial u^2} = 6\wp^3 - \dfrac{1}{2}g_2$, which is known as the static Korteweg–de Vries equation. Noting that the $\wp$ function is also a function of g_2 and g_3, we differentiate (1.31) in g_2 and g_3 to have [Wei1882, Wei1875b]

$$2\frac{\partial\wp}{\partial u}\frac{\partial^2\wp}{\partial u\partial g_2} = 2\frac{\partial^2\wp}{\partial u^2}\frac{\partial\wp}{\partial g_2} - \wp,$$

$$2\frac{\partial\wp}{\partial u}\frac{\partial^2\wp}{\partial u\partial g_3} = 2\frac{\partial^2\wp}{\partial u^2}\frac{\partial\wp}{\partial g_3} - 1.$$

Roughly speaking, we integrate them by u twice, and they are reduced to the heat equations of σ [EilO2020]:

$$\begin{aligned}\frac{\partial^2\sigma}{\partial u^2} - 12g_3\frac{\partial\sigma}{\partial g_2} - \frac{2}{3}g_2^2\frac{\partial\sigma}{\partial g_3} + \frac{1}{12}g_2u^2\sigma &= 0,\\ u\frac{\partial\sigma}{\partial u} - 4g_2\frac{\partial\sigma}{\partial g_2} - 6g_3\frac{\partial\sigma}{\partial g_3} - \sigma &= 0.\end{aligned} \tag{1.32}$$

Using them, we have the recursion relations of the coefficients of the power expansion of σ at $u=0$,

$$\sigma(u) = \sum_{m,n} a_{m,n}\left(\frac{1}{2}g_2\right)^m (2g_3)^n \frac{u^{4m+6n+1}}{(4m+6n+1)!}, \tag{1.33}$$

where

$$a_{m,n} = 3(m+1)a_{m+1,n-1} - \frac{16}{3}a_{m-2,n+1} + \frac{1}{3}(2m+3n-1)(4m+6n-1)a_{m-1,n}.$$

Consequently, we explicitly obtain

$$\begin{aligned}\sigma(u) = u &- \frac{1}{240}g_2u^5 - \frac{1}{840}g_3u^7 - \frac{1}{161280}g_2^2u^9 - \frac{1}{2217600}g_2g_3u^{11}\\ &+ \left(\frac{23g_2^3}{16605388800} - \frac{g_3^2}{28828800}\right)u^{13}\cdots.\end{aligned}$$

Weierstrass also investigated the expansions of $\sigma_i(u) = \dfrac{e^{-\eta_i u}\sigma(u+\omega_i)}{\sigma(\omega_i)}$.

These elliptic integral (EI), Abel–Jacobi map (AJM) and the Jacobi inversion formula (JIF) are illustrated in Fig. 1.7 below.

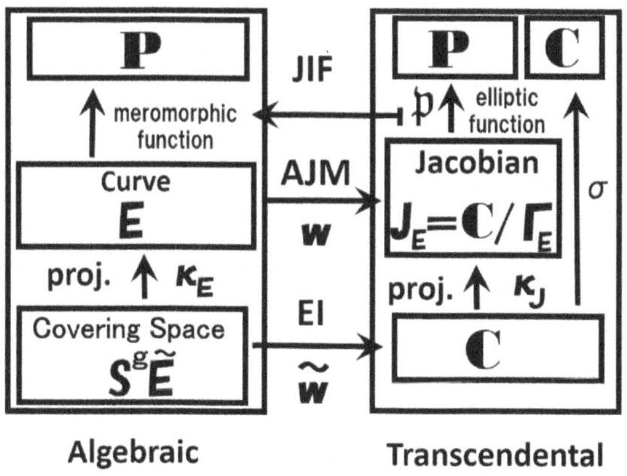

Fig. 1.7 Algebraic versus transcendental in $g=1$

These Abelian integral, Abel–Jacobi map and the Jacobi inversion formula enable us to handle both the algebraic system and transcendental system equivalently. Using the natural properties of a one-sided system, we sometimes determine the properties of the other system. At the dawn of elliptic function theory by Abel, Jacobi, and so on, the relation between algebraic and transcendental properties had been the most important theme.

We also point out that the shape of an elastica is classified on the transcendental side because it is represented by the elliptic ζ function (1.5) and by the integrals (1.4), while the elliptic curve E is characterized on the algebraic side. It is also noted that the additive structure in the curve E is determined by the natural addition in the transcendental side; for two arclengths $s_1, s_2, s_1 + s_2$ is naturally defined, which is also regarded as the addition in the path spaces, or, the fundamental groupoid [BotT1993, Sp1966]. We claim, here, that such a dichotomy was built-in from its inception.

The addition theorem is given by

$$x(u) - x(v) = \wp(u) - \wp(v) = -\frac{\sigma(u+v)\sigma(u-v)}{\sigma(u)^2\sigma(v)^2}. \tag{1.34}$$

As Weierstrass cited [Abl1830] in his earliest work [Wei1840], Abel found this relation in 1828 in [Abl1830, p. 76]. Further, Hermite, Frobenius and Stickelberger showed the following interesting results [WhW1927, p. 458]: for $u^{(i)} := w(x_i, y_i)$ $(i = 1, \dots, n)$,

$$\frac{\sigma(u^{(0)} + u^{(1)} + \cdots + u^{(n)}) \prod_{i<j} \sigma(u^{(i)} - u^{(j)})}{\sigma^{n+1}(u^{(0)})\sigma^{n+1}(u^{(1)}) \cdots \sigma^{n+1}(u^{(n)})} = c_n \begin{vmatrix} 1 & \phi_1(x_0, y_0) & \cdots & \phi_n(x_0, y_0) \\ 1 & \phi_1(x_1, y_1) & \cdots & \phi_n(x_1, y_1) \\ \vdots & \vdots & \ddots & \vdots \\ 1 & \phi_1(x_n, y_n) & \cdots & \phi_n(x_n, y_n) \end{vmatrix}, \tag{1.35}$$

where c_n is a constant factor and $\phi_j(x, y) = d^{j-1}x/du^{j-1}, \phi_1(x, y) = x, \phi_2(x, y) = 2y$, $\phi_3(x, y) = 3x^2 + 2\lambda_2 x + \lambda_1$, $\phi_4(x, y) = 2(6xy + 2\lambda_2)y$, $\cdots$. Using the relation, Kiepert and Brioschi showed [Ki1873, BriH1864] that the division polynomial,

$$\psi_n(u) := \frac{\sigma(nu)}{\sigma(u)^{n^2}}, \tag{1.36}$$

satisfies the relation [WhW1927, p. 460]

$$\psi_n(u) = c_n' \begin{vmatrix} \phi_2(x, y) & \phi_3(x, y) & \cdots & \phi_n(x, y) \\ \phi_3(x, y) & \phi_4(x, y) & \cdots & \phi_{n+1}(x, y) \\ \vdots & \vdots & \ddots & \vdots \\ \phi_n(x, y) & \phi_{n+1}(x, y) & \cdots & \phi_{2n-2}(x, y) \end{vmatrix}, \tag{1.37}$$

where c'_n is a constant factor. The zero u of ψ_n is the point such that $nu = 0$ modulo Γ_E. These relations are generalized as in (1.57) in Sect. 1.9.1 (cf. Sect. 3.4).

1.3.3 Jacobi's Elliptic Function Theory in Weierstrass' Framework

Jacobi discovered his theta function by considering the elliptic, differentials, and integrals of the third kind, which were essential in his elliptic function theory; they were generalized to the higher genus case by Galois and Riemann, as we will mention later.

Thus in the framework of Weierstrass elliptic function theory, we give descriptions of the differentials of the third kind following Jacobi's treatments (1.21) and (1.22), and its related matters, which are also generalized to the higher genus case in Sect. 2.4.3. We also show the al functions associated with Jacobi's elliptic functions.

For $P := (x_P, y_P)$, $Q := (x_Q, y_Q) \in X$, let us define

$$\Sigma(P, Q) := \left[\frac{y_P + y_Q}{(x_P - x_Q)2y_P}\right] dx_P, \tag{1.38}$$

which has the properties as a function of P;

1) it is holomorphic except at infinity and Q (even for $x_P = x_Q$ and $y_P = -y_Q$, it is holomorphic), and
2) at Q, it behaves like $\dfrac{dt_Q}{t_Q} + \cdots$, where t_Q is the local parameter at Q.

The third differential $\nu^{\mathrm{III}}_{Q_1\text{-}Q_2}(P)$ is given by

$$\nu^{\mathrm{III}}_{Q_1\text{-}Q_2}(P) := \Sigma(P, Q_1) - \Sigma(P, Q_2), \tag{1.39}$$

which is holomorphic except at Q_1 and Q_2 and at Q_a it behaves like $\dfrac{\pm dt}{t} + \cdots$ for a local parameter t at Q_a. We thus define

$$\Omega(P, Q) := \left(\frac{\partial}{\partial x_Q}\Sigma(P, Q)\right) dx_Q + \nu^{\mathrm{I}}(P)\nu^{\mathrm{II}}(Q), \tag{1.40}$$

which is equal to $\left(\dfrac{\partial}{\partial x_Q}\left[\dfrac{y_P + y_Q}{(x_P - x_Q)2y_P}\right] + \dfrac{1}{2y_P}\dfrac{x_Q}{2y_Q}\right) dx_P dx_Q$. Then due to the addition theorem of σ, we identify it with

$$\wp(u - v)du dv = \frac{1}{2}\frac{\partial}{\partial v}\left[\frac{\wp'(u) + \wp'(v)}{\wp(u) - \wp(v)}\right] du dv + \wp(v) du dv \tag{1.41}$$

for certain u and v. $\Omega(P, Q)$ has the properties:

1) $\Omega(P, Q) = \Omega(Q, P)$,
2) as a function of $P(Q)$ it is holomorphic except at $Q(P)$, and
3) at Q it behaves like $\dfrac{dt_P dt_Q}{t_Q^2} + \cdots$, where t_Q is the local parameter at Q.

Since $\Omega(P, Q)$ is given by $\dfrac{\partial}{\partial u}\dfrac{\partial}{\partial v} \log \sigma du dv$, we obtain $\sigma(u - v)$ by integrating $\exp\left(\int^u \int^v \wp(u - v) du dv\right)$. It means that (1.41) corresponds to Jacobi's formula (1.21), i.e.,

$$\int_{u'}^{u} \int_{v'}^{v} \wp(u - v) du dv = \log\left(\frac{\sigma(u - u')\sigma(v - v')}{\sigma(v - u')\sigma(u - v')}\right). \tag{1.42}$$

Since $\Omega(P, Q)$ is holomorphic over X except at Q as a function of P and at Q and behaves like $\dfrac{dt_P dt_Q}{(t_P - t_Q)^2} + \cdots$, this analogue on the Riemann sphere $\mathbb{P}$ is reduced to a simple relation, $\displaystyle\int^t \int^s \frac{dt' ds'}{(t' - s')^2} = \log(t - s)$[11], and then its integral means $e^{\int^t \int^s \frac{dt'ds'}{(t'-s')^2}} = t - s$; it is reduced to the quadrature of the hyperbola by Huygens mentioned in Sect. 1.1.2. Jacobi, Abel, and Riemann owned this approach to finding the theta series using this correspondence.

Further, we touch upon Jacobi's sn, cn, and dn functions in Weierstrass' framework. Their history is more extended than the $\wp$ function. Gauss and Abel considered the elliptic integral,

$$u(x) = \int^x \frac{dx}{\sqrt{(1 - x^2)(1 - k^2 x^2)}}.$$

Abel found that the inverse function of $u(x)$ is more important than the integral itself, though Gauss also studied the lemniscate function for the $k = \sqrt{-1}$ case, and (1.12). For the elliptic curve E of $y^2 = (x - e_1)(x - e_2)(x - e_3)$, let us consider a parameter z satisfying $z^2 = (x - e_1)/(e_2 - e_1)$,

$$y^2 = (e_1 - e_2)^2 (e_3 - e_1) z^2 (1 - z^2)\left(1 - \frac{e_1 - e_2}{e_1 - e_3} z^2\right).$$

Since the curve is singular due to the point $z = 0$, we further introduce $w = \dfrac{y}{z(e_3 - e_1)^{1/2}(e_1 - e_2)}$ and $k^2 = \dfrac{e_1 - e_2}{e_1 - e_3}$, and then find a natural curve $\widehat{E}$,

$$w^2 = (1 - z^2)(1 - k^2 z^2).$$

[11] It reminds us of Huygens's tractrix. We might interpret it as the origin of the elliptic function theory as mentioned in the footnote 4.

Recalling $\wp(u) = x(u)$, we define $\mathrm{al}_1(u) := z(u)$ [HurC1893, II Chap. 2, §18] by

$$\mathrm{al}_1(u) = \sqrt{x(u) - e_1} = \sqrt{\wp(u) - e_1} = \frac{\mathrm{e}^{-\eta_1 u}\sigma(u+\omega_1)}{\sigma(u)}. \tag{1.43}$$

Similarly, it is natural to consider

$$\mathrm{al}_2(u) = \sqrt{\wp(u) - e_2} = \frac{\mathrm{e}^{-\eta_2 u}\sigma(u+\omega_2)}{\sigma(u)}, \quad \mathrm{al}_3(u) = \sqrt{\wp(u) - e_3} = \frac{\mathrm{e}^{-\eta_3 u}\sigma(u+\omega_3)}{\sigma(u)}.$$

They are elliptic al functions which were initially introduced by Weierstrass based on the connections between $\wp$ function, for hyperelliptic curves [Wei1854] and Jacobi's elliptic functions,

$$\mathrm{sn}(u) = \sqrt{\frac{e_1 - e_3}{\wp(u) - e_3}}, \quad \mathrm{cn}(u) = \sqrt{\frac{\wp(u) - e_1}{\wp(u) - e_3}}, \quad \mathrm{dn}(u) = \sqrt{\frac{\wp(u) - e_2}{\wp(u) - e_3}},$$

$$\mathrm{sn}(u) = (e_1 - e_3)^{1/2}\frac{\sigma(u)\sigma(\omega_3)}{\mathrm{e}^{-\eta_3 u}\sigma(u+\omega_3)}, \quad \mathrm{cn}(u) = \mathrm{e}^{(\eta_3-\eta_1)u}\frac{\sigma(u+\omega_1)\sigma(\omega_3)}{\sigma(u+\omega_3)\sigma(\omega_1)},$$

$$\mathrm{dn}(u) = \mathrm{e}^{(\eta_3-\eta_2)u}\frac{\sigma(u+\omega_2)\sigma(\omega_3)}{\sigma(u+\omega_3)\sigma(\omega_2)}.$$

Then they are related to the sine-Gordon equation (1.9) and the ellipses,

$$\mathrm{sn}^2(u) + \mathrm{cn}^2(u) = 1, \quad k^2\mathrm{sn}^2(u) + \mathrm{dn}^2(u) = 1.$$

As introduced and discussed by Weierstrass, the al function plays an important role in discussing the behavior at a finite number of branching points, as will be mentioned in Sects. 2.8 and 3.6. We note Abel's f and F in (1.18) also could be regarded as al functions.

1.3.4 Summary of the Weierstrass Elliptic Function Theory

Here we summarize the Weierstrass elliptic function theory:

1. We consider an elliptic curve E of standard form $y^2 = x^3 + g_2 x + g_3$, which represents an equivalent class as its representative element.
2. We define the differentials of the first kind, the second kind and the third kind.
3. Using the data of the integrals of these differentials, we define the Jacobian and the σ function.
4. We obtain the Jacobi inversion formula $(x, y) = (-\frac{d^2}{du^2}\log\sigma, -\frac{1}{2}\frac{d^3}{du^3}\log\sigma)$.

5. Using the relation $y^2 = x^3 + g_2 x + g_3$, we obtain the differential equation, $\wp_u^2/4 = \wp^3 + g_2\wp + g_3$, where $\wp = -\frac{d^2}{du^2}\log\sigma$.
6. Investigating the differential equation, we determine the power expansion of the σ function.
7. With the help of σ, we consider the structure of the algebraic curves, $\wp$'s, and the Jacobian.

1.4 Dawn of the Abelian Function Theory of Higher-Genus Case

1.4.1 Abel, Jacobi, and Galois

Abel studied integrals of one-forms of a hyperelliptic curve and their algebraic properties beyond his elliptic function theory 1828 [Abl1828].

Further, in the two-page paper of 1829 [Abl1829], which is a summary of his Parisian memoir of 1826 [Abl1841], he also investigated an algebraic curve given by

$$y^n + p_{n-1}y^{n-1} + \cdots + p_3 y^3 + p_2 y^2 + p_1 y + p_0 = 0, \tag{1.44}$$

where p_i is an entire function of x, and its general Abelian integrals. Then by considering an equation,

$$q_{n-1}y^{n-1} + q_{n-2}y^{n-2} + \cdots + q_3 y^3 + q_2 y^2 + q_1 y + q_0 = 0, \tag{1.45}$$

where q_i is an entire function of x, he investigated a sum of contour integrals (Abel integrals) associated with roots of (1.45) over the curve (1.44). Then Abel found a theorem which, later, was known as the Abel theorem. Equation (1.44) corresponds to (1.48), and (1.45) to (1.58).

In those days, Abel also investigated the algebraic properties of an algebraic equation $y^n + a_{n-1}y^{n-1} + \cdots + a_3 y^3 + a_2 y^2 + a_1 y + a_0 = 0$ for $a_i \in \mathbb{Q}$ and the group of their roots to show the impossibility of solving the quintic equation in terms of radicals. His investigation above should also be understood in terms of the Galois theory of algebraic equations from modern viewpoints. Thus he must have been concerned with the algebraic (group-theoretic) properties of the Eq. (1.44), which is the algebraic equation over the function field $\mathbb{C}(x)$. It implies that Abel's considered objects are the following:

1. generalization of the relations between transcendental objects of the elliptic integrals and the algebraic properties on the elliptic curves to more general algebraic curves, and
2. generalization of the algebraic properties of the algebraic curves over $\mathbb{Q}$ to the function field $\mathbb{C}(x)$ (noting that $p_i \in \mathbb{C}[x]$ in (1.44)).

Since he knew that the degree of the polynomial is significant to the algebraic equation, it was expected that the degrees of (1.44) also are crucial, and the algebraic properties did not mean only the additive ones and might include Galois groups associated with (1.44). We should regard that his considered problem is to reveal the algebraic and transcendental properties of (1.44) and (1.45) or to construct the theory to describe the algebraic equations over the function field $\mathbb{C}(x)$. We call it Abel's intention.

On the other hand, following Abel's study on hyperelliptic integrals, Jacobi investigated the integrals, which were called the Abelian integrals later [Jac1835, Jac1846]. The Abelian integral is a generalization of the elliptic integral. Since the inverse function of the elliptic integral is the elliptic function, Jacobi investigated the "inverse" function of the Abel integral for an algebraic curve of genus two in 1835 [Jac1835]. Jacobi investigated the curve of genus two,

$$y^2 = x(1-x)(1-\kappa^2 x)(1-\lambda^2 x)(1-\mu^2 x),$$

and its Abelian integrals. He found quartic periodicity and concluded that the image of the integrals should be of higher dimensions rather than one–dimensional space. By considering the integrals

$$\int_a^{(x,y)} \frac{(\alpha+\beta x)dx}{y} + \int_b^{(x',y')} \frac{(\alpha+\beta x)dx}{y} = u,$$

$$\int_a^{(x,y)} \frac{(\alpha'+\beta' x)dx}{y} + \int_b^{(x',y')} \frac{(\alpha'+\beta' x)dx}{y} = u',$$

where a, b, α, β, α', and β' are constants, he showed that U, U', and U'' satisfying

$$Ux^2 - U'x + U'' = 0$$

are functions of u and u'. By sending a and b to ∞, they turn out to be the elementary symmetric functions, $U'/U = x + x'$ and $U''/U = xx'$. This study was the starting point of the function theory of the several variables [Jac1832a, Jac1832b, Jac1835]. It means that Jacobi posed the so-called "Jacobi inversion problem" [Jac1846, Tak1998, Tak2012]. Jacobi considered x and x' as functions of u and u', i.e., $x = x(u, u')$ and $x' = x'(u, u')$ and called them "Abelian functions", which differs from the more recent name [Jac1846].[12]

Though Évariste Galois (1811–1832) did not consider the function theory of several variables, in his last letter to his friend, Chiveral, six days before his death

[12] The Abelian functions, in modern terminology, mean the meromorphic functions defined on the Abelian variety $\mathbb{C}^g/\Gamma$; they satisfy $\varphi(z+\ell) = \varphi(z)$ for $z \in \mathbb{C}^g$ and $\ell \in \Gamma$. We also note that by our sigma function theory, we obtain the expression of x and x' as a meromorphic function as in Corollary 3.62. It implies that our sigma function theory of the algebraic curves may also realize Jacobi's original inversion formula and Abelian function.

in 1832, he showed the Legendre relation or the symplectic structure of a curve by investigating the integrals of the first, the second, and the third kind for a curve [Ga1846, Tak1998], i.e.,

$$EF' - E'F = \frac{\sqrt{-1}}{2}\pi. \tag{1.46}$$

It might be a myth of history; the investigation of Galois must be based on the fundamental consideration of the meromorphic functions on curves. His derivation in the letter resembles Weierstrass' one mentioned in Sect. 1.6.2 and ours in Sects. 1.9.3 and 2.4. See also Theorem 2.148.

1.4.2 Weierstrass' Paper of 1854

In his time at the Gymnasium, Weierstrass studied Abel's papers. After constructing the elliptic Al function, he attempted to solve the Jacobi inversion problem for a hyperelliptic curve of general genus g and to build concrete expressions for Abel's theory in his paper of 1829. Though the genus two cases were settled by Johann Georg Rosenhain (1816–1887) and Adolph Göpel (1812–1847), Weierstrass considered the general genus case. In the paper published in Bulletin of the Gymnasium of 1848-49, he generalized Jacobi's sn, cn, and dn functions to hyperelliptic versions. He also studied the Jacobi inversion problem using his theta function, Al. However, the paper probably never even reached the attention of mathematicians because the Bulletin of the Gymnasium was only a local publication.

After then, he published its sequel in Crelle's Journal in 1854, which was widely respected. He started with a curve

$$X_g := \{(x, y) \mid y^2 = (x - b_1) \cdots (x - b_{2g+1})\} \cup \{\infty\} \tag{1.47}$$

He also called the meromorphic function on the gth symmetric product $S^g X_g$ of X_g, or the Abelian function, al, in honor of Abel and he investigated his problem, i.e., al is expressed by a ratio of Al functions; roughly speaking, we show

$$\mathrm{al}_r(v) = \gamma_r \sqrt{(x_1 - b_r) \cdots (x_g - b_r)},$$

where γ_r is a certain constant factor, $v \in \mathbb{C}^g$ is given by the Abelian integrals,

$$v = (v_i), \quad v_i = \sum_{i=1}^{g} \widehat{w}((x_i, y_i)), \quad \widehat{w}((x_i, y_i)) = \int_{\infty}^{(x_i, y_i)} dv_i,$$

for the points $(x_i, y_i) \in E$ $(i = 1, \ldots, g)$, and the twisted W-normalized holomorphic one-form dv_i (cf. Sect. 4.2.3). Then $\mathrm{al}_r(v)$ is expressed by a ratio of the entire Al functions on $\mathbb{C}^g$, i.e., $\mathrm{al}_r(v) = \dfrac{\mathrm{Al}_r(v)}{\mathrm{Al}(v)}$ as in (1.23) and (1.43) for the genus one case.

Investigating the differential equation (1.11) was also a model even for this case. In the determination of Al functions, he regarded (1.24) as a prototype, whose origin must be Gauss (1.11). (As mentioned in Sect. 4.2.3, these al_r functions satisfy the sine-Gordon equation (cf. Sect. 4.2.3).)

1.5 Riemann's Abelian Function Theory and Subsequent Developments

1.5.1 Riemann's Abelian Function Theory

In 1857, Georg Friedrich Bernhard Riemann (1826–1866) published the paper [Rie1857], in which he constructed the Riemann theta function for general compact Riemann surfaces and investigated its Abelian functions [Tak2016]. He defined the compact Riemann surfaces based on his study of the complex function theory, which was the theme in his thesis of 1851, and on his mathematical intuition from the mathematical investigation of physical phenomena, e.g., electric systems and fluid dynamics. We state that from the beginning, his work was complete. His approach entirely differed from the traditional approaches in the 19th century; for example, Abel, Jacobi, and Weierstrass started from the equations (1.44) and (1.47) and investigated the naturally obtained affine rings, the integrals, and their inverse functions. The curve given by an equation has several ambiguities and should be regarded as a representative element in the quotient set of algebraic curves divided by the birational equivalence. Riemann defined a compact Riemann surface directly as an element of quotient space, which is very natural and elegant as an extension of the one-variable complex holomorphic function theory. The Jacobian (Jacobi variety) as the target space of the Abelian integral of the differentials of the affine ring is generalized to the Abelian variety. Basically, he developed Jacobi's approaches (1.21) and presented relations between meromorphic functions on the compact Riemann surfaces and the Abelian varieties. Thus the Jacobi inversion formulae were obtained as the relations between the theta functions $\vartheta^{(a)}$ and the integrals of the third kind $\pi^{(\mu)}$ like [Rie1857, p. 151].

$$\log \vartheta^{(2)} - \log \vartheta^{(1)} = \sum_{1}^{g} \pi^{(\mu)}(\varepsilon_1, \varepsilon_2).$$

However, we note that it implies a break with the concreteness of the known elliptic function theory. (This relation is the original version of the so-called Jacobi–Riemann fundamental relation, which induces the Jacobi inversion formulae for the sigma function in Sects. 3.1 and 3.2; cf. Theorem 3.37.)

Though Riemann's study might deflate Weierstrass' results, Weierstrass' approach had "remarkable unity of thought, persisting through the scope and a variety of his works. From the beginning, he set himself a definite goal; he created methods to reach it" [Po1910].

After Riemann passed away, Weierstrass published a paper [Wei1915, II, pp. 49–54], in which Riemann's study had a gap because Riemann used Dirichlet's principle, i.e., the minimal principle, for the existence of the Abelian functions over a compact Riemann surface [Bot1986, Bot2003]. Dirichlet's principle needs the limit operation in the function space, but in mathematics in those days, there were no words describing such a limit precisely. There was a period of stagnation as Riemann's approach was inaccurate. However, sensing the potential of Riemann's approach, attempts were made, in various directions, to guarantee its legitimacy. Starting from particular curves, there were moves to utilize the theory of compact Riemann surfaces, for example, while guaranteeing the existence of functions from the standpoint of curves. Then what was treated as a flaw was justified by Hilbert in 1900. Thereby, Riemann's approach became more common than Weierstrass' in modern algebraic geometry and complex function theory for several variables.

1.5.2 Dedekind and Weber, Klein: The Successors of Riemann

The application of Riemann's results on compact Riemann surfaces to the study of algebraic curves was carried out by Karl H. A. Schwarz (1843–1921) 1864, Gustav Roch (1839–1866) 1865, and Alfred Clebsch (1833–1872) 1873 according to Houzel [Hou1978]. Schwarz, a student of Weierstrass, conducted the research under the direction of Weierstrass, whereas Roch and Clebsch were encouraged in their study by direct influence from Riemann. Schwartz proposed a method related to Dirichlet's principle, the gap of Riemann's paper, which inspired Weierstrass to write the paper of 1870.

Felix C. Klein (1849–1925) was one of those struck by Riemann's approach. In 1886, late in Weierstrass' life, Klein investigated Weierstrass' hyperelliptic Al function and refined his Al function to define the hyperelliptic sigma function in terms of data of compact Riemann surfaces of hyperelliptic curves. His approach was to redefine the hyperelliptic sigma function by refining the theta function so that it held a modular invariant using the data of compact Riemann surfaces and Jacobians with properties of hyperelliptic curves rather than curves themselves. Klein considered that the compact Riemann surfaces and their Jacobians are essential rather than the algebraic curves themselves and probably thought that Weierstrass' approach might be old-fashioned. Indeed, even for the elliptic curves, it is well-known that the $\wp$ function can be defined by the data of the Jacobian,

$$\wp(u) = \frac{1}{u^2} + \sum_{n^2+m^2\neq 0} \left[\frac{1}{(u + m\omega_1 + n\omega_3)^2} - \frac{1}{(m\omega_1 + n\omega_3)^2} \right],$$

and then we obtain the sigma function by integrating it twice. For $u := (2\omega_1)z$ and $\frac{\eta_1}{\omega_1}u^2 = \omega_1\eta_1 z^2 = \frac{1}{12}\frac{\theta_1'''}{\theta_1'}z^2$, we have the sigma function only by using the data of the Jacobian and θ_1 function.

Klein wrote about his hyperelliptic sigma function in his book [Kl1926, p. 293], "In my work on hyperelliptic and Abelian functions, Math. Ann. Vol. 27 to 36 1) (1886–89), I transferred the idea of σ to the corresponding higher cases and, in particular, gave, in my opinion, the final expression of the Weierstrass thought on the decomposition of the algebraic functions of a Riemann surface into prime factors and units." The papers [Kl1886, Kl1888, Kl1890] could be regarded as implicitly showing the old-fashioned manner developed in the middle of the 19th century, including Weierstrass's approach, to prepare for the end. He defined his sigma function of genus two in terms of data of the compact Riemann surface of genus two,

$$\sigma(v_1, v_2) = \mathrm{e}^{-\frac{1}{20}\left(\sum_1^{10}\frac{\vartheta_{11}}{\vartheta}\cdot v_1^2 + \sum_1^{10}\frac{\vartheta_{12}}{\vartheta}\cdot v_1 v_2 + \sum_1^{10}\frac{\vartheta_{22}}{\vartheta}\cdot v_2^2\right)} \cdot \frac{\vartheta(v_1, v_2)}{\vartheta},$$

where $\vartheta_{ij} := 4\pi\sqrt{-1}\frac{\partial\vartheta}{\partial\tau_{ij}}$ and the sum is over the even theta characteristics. Following this approach, he extended his sigma functions to more general curves, or so-called trigonal curves [Kl1890].

The definition of the Kleinian sigma function formed the direction of algebraic geometry in the 20th century. In modern algebraic geometry, due to Torelli's theorem, the importance of the curves is less than before.

Julius W. R. Dedekind (1831–1916) developed the algebraic tools to reconstruct Riemann's results algebraically with Heinrich M. Weber (1842–1913) for their respected friend Riemann. They also championed Riemann. In the paper [DedW1882], Dedekind and Weber demonstrated the differentials and the "points" on the compact Riemann surfaces were expressed algebraically. This paper presented tools as the algebraic embodiment of the ideas and methods pioneered by Riemann; they showed consistency between local and global considerations and extracted the essentials of the entire objects explicitly and abstractly.[13] (We also use their concept, the complementary basis of the module in Sect. 2.2.5 [KomMP2022a].)

[13] Their work could also be regarded as an arithmetization under the influence of Weierstrass and Kronecker [DedW1882, Translator's Introduction]. Further, it means that this paper could be interpreted as a realization of Abel's intention [Ch1951]. In fact, Dedekind and Weber started with the Weierstrass polynomial $y^r + b_1 y^{r-1} + b_2 y^{r-2} + \cdots + b_r = 0$, where $b_i \in \mathbb{C}(x)$, as (1.44) with $p_i \in \mathbb{C}(x)$ instead of $\mathbb{C}[x]$ in §1: the field of algebraic functions [DedW1882]. However, they are concerned with the local properties; it might be regarded as a polynomial appearing in the Weierstrass preparation theorem. The approach from local to global data was one of the ideas of Weierstrass, which appeared in the analytic continuation [Bot1986]. Its origin was Gudermann, and it was inherited in the foundational part of the sheaf theory. However, what we address in this book is what Weierstrass tried to construct more directly, taking over Abel's intention.

In other words, the philosophy of Weierstrass that a nice representative element represents a particular equivalent class seemed to be outdated. Around that time the modern idea that equivalent classes should be expressed by a particular set, module, ideal, etc.began to emerge. A compact Riemann surface represents a class of algebraic curves; thus, the concept of the Riemann surfaces is more natural in modern algebraic geometry. Hermann K. H. Weyl (1885–1955) showed the uniformization between Riemann's concept based on the compact Riemann surfaces and Weierstrass' concept based on the algebraic curves [Wey1955]. Weyl stated, in his book about the Jacobi inversion problem: "The principal significance of the inversion problem to us today lies primarily, not in its intrinsic value, but in the splendid developments created by Riemann and Weierstrass in their efforts to solve the problem". It means that in the theory of compact Riemann surfaces, the Jacobi inversion formula studied by Riemann and Weierstrass might have only historical value rather than a mathematical one.

1.6 Weierstrass' Abelian Function Theory and Subsequent Developments

1.6.1 Weierstrass: The Successor of Abel After 1854

In contrast to Riemann and his successors, Weierstrass trod a different path, standing on his own philosophy on the Abelian functions. We show this here.

After taking a chair at the University of Berlin, Weierstrass started to construct the theory of rigorous analysis to develop his Abelian function theory. He basically repeated the four lectures over a period of about a year on 1) the foundation of analysis, 2) the elliptic function theory, 3) the application of the elliptic function theory to geometry and physics, and 4) the hyperelliptic function theory or the Abelian function theory [Wei1915, III pp. 355–360].

He inherited Abel's program on relations between the transcendental and the algebraic objects.

Weierstrass stated in the Academic Inquires held at the public session of the Berlin Academy on July 9, 1857 [Wei1915, I pp. 223–226]: "Abel, who was used to taking the highest point of view everywhere, had put forward a theorem that encompassed all transcendents arising from the integration of algebraic differentials and had the same meaning for them as the one Euler saw for the elliptic ones. Carried away in the prime of his life, Abel had not been able to pursue his great discovery; but Jacobi had succeeded in linking a no less important one to it by proving the existence of periodic functions of several arguments, the fundamental properties of which are founded in the Abelian theorem, whereby at the same time the true meaning and the actual essence of it were opened up. To really represent these quantities of a completely new kind, for which calculus had no example yet, and to fathom their properties more closely, became from now on one of the main tasks of mathematics, in which I also decided to try myself, as soon as I had clearly recognized the sense and the meaning of it. Of course, it would have been foolish to even think about solving such a problem if I had not prepared myself by thoroughly studying the available tools and by dealing with less difficult tasks."

Weierstrass prepared carefully. He laid the foundations of analysis from its very substructure. He has promoted the construction of the Abelian function theory and refined his elliptic function theory to perfection without a single flaw.

As Weierstrass wrote in 1875 in a letter to Schwarz: "The more I think about the principles of function theory—and I do continuously—the more I am convinced that this must be built on the foundations of algebraic truths, and that it is consequently not correct to resort on the contrary to the 'transcendent', to express myself briefly, as the basis of simple and fundamental algebraic propositions. This view seems so attractive at first sight, in that through it Riemann was able to discover so many important properties of algebraic functions. (It is self-evident that, as long as he is working, the researcher must be allowed to follow every path he wishes; it is only a matter of systematic foundations.)"[Jah2003][14][Wei1915, II p. 235]. This phrase was also cited by Poincaré [Po1910]. Bottazzini refered this opposition as to "algebraic truths versus geometric fantasies" [Bot2003].

As mentioned in Sect. 1.5.2, Riemann's approaches were very attractive and had many supporters even in those days. It is no exaggeration to say that modern algebraic geometry is based on Riemann's thoughts rather than Weierstrass'. However, Weierstrass chose a different path. He attempted to build the Abelian function theory using the algebraic, not the transcendental, as its cornerstone, but failed due to several following obstacles. This book shows the results as conducted research according to Weierstrass's thought because we find it essential and valuable.

In order to achieve the program to realize his Abelian function theory, Weierstrass ordered his students, including Schwarz, Ludwig Stickelberger (1850–1936), and Ferdinand G. Frobenius (1849–1917) [Haw2013]. For a long time, the most important parts of Weierstrass' works remained unpublished, whereas they were in his oral teaching, which we can read in his *Werke* [Wei1875a, Wei1875b]. However, his spirit inspired not only those who had had his words, but those who had received only an indirect talk [Po1910]. His lecture notes also influenced the researchers including Klein, Oskar Bolza (1857–1942), and Henry F. Baker (1866–1956) [Hod1956]. Bolza and Baker learned his theory of the Abelian function theory and championed

[14] "Je mehr ich über die Principien der Functionentheorie nachdenke—und ich thue dies unablässig –, um so fester wird meine Überzeugung, dass diese auf dem Fundamente algebraischer Wahrheiten aufgebaut werden muss, und dass es deshalb nicht der richtige Weg ist, wenn umgekehrt zur Begründung einfacher und fundamentaler algebraischer Sätze das "Transcendente", um mich kurz auszudrücken, in Anspruch genommen wird—so bestechend auch auf den ersten Anblick z. B. die Betrachtungen sein mögen, durch welche Riemann so viele der wichtigsten Eigenschaften algebraischer Functionen entdeckt hat. (Dass dem Forscher, so lange er sucht, jeder Weg gestattet sein muss, versteht sich von selbst; es handelt sich nur um die systematische Begründung.)"

Weierstrass; in fact, they reformulated Klein's sigma function following Weierstrass' thought and developed it, though Klein orchestrated it.[15]

After the construction of rigorous analysis, he revised his elliptic function in terms of the sigma function by refining the elliptic Al function[16] as mentioned above. Further, the rigorous treatments of analysis of the several variables had been developed as mentioned above following Poincaré [Po1910]. For example, for a given Abelian function, Stickelberger found the existence theorem of the entire functions whose ratio is equal to it. Further, after finding the gap in Riemann's approach, Weierstrass discussed with his student Schwarz to prove the existence theorem of meromorphic functions over a compact Riemann surface without using the Dirichlet principle. (His student Husserl's thesis was on the variational principle).

However, he seemed to insist on the algebraic curves and his philosophy that a nice representative element represents a particular equivalent class. His philosophy in his elliptic function theory surely succeeded. His theory has influenced various fields in mathematics, science, and technology. It contrasts with the modern theory of the algebraic curves with higher genera based on Riemann's approach, which, while its importance is unquestionable in pure mathematics, only affects some abstract theories that cannot be directed to experiments, such as elementary particle physics, except in mathematics. In the 21st century, advances in other areas of mathematics and technological advances have led to demands to handle the Abelian functions for higher genera as an extension of Weierstrass' elliptic functions. Such research is ongoing, and this book is written against that background.

1.6.2 Abel's Intention and Weierstrass Canonical Form

Weierstrass had schemed to realize Abel's intention; he aimed to generalize the Abelian function of a hyperelliptic curve to a more general algebraic curve X defined by a standard form (1.44) with some restrictions, to which in this book, we refer as the Weierstrass canonical form [Wei1867].

Even after the publication of Riemann's studies, Weierstrass was concerned with the importance of the curves themselves rather than compact Riemann surfaces. In the construction of the theory, he investigated their algebraic properties like algebraic integers in the number theory, which is regarded as his arithmetization of analysis

---

[15] At the time, Klein was initiating various actions to balance the development of mathematics. Both the completion of his sigma function and the development in the direction of Weierstrass, as opposed to his own philosophy, seem to have been intended by him. The preface of Baker's book [Ba1897] notes that his work was initiated on Klein's advice. Klein also influenced an American mathematician, William F. Osgood (1864–1943), so that he wrote a related book [Os1912] which was described from the point of view of the several complex variables and the kernel functions; it is cited in [Hej1972].

[16] According to Takeo Ohsawa [Ohs2017], his student pointed out that $\wp$ might come from the inverse of "al". If so, σ is related to the symbol $\mathfrak{a}$, and Abel lives on in the function name of Weierstrass.

[Po1910]. Though Klein considered his thought as a decomposition into prime factors and units and carried out its extensions transcendently, and Dedekind and Weber implemented the arithmetization in the framework of their module theory, because the Riemann surface does not represent a single curve, these do not seem to show properties of Weierstrass' elliptic function theory even for the genus one, and, we consider, differ from his final goal.

In [Wei1867], he stated that the Weierstrass points exist in the curves X and showed that the numerical semigroup expressed the Weierstrass non-gap sequence.

As mentioned above, Weierstrass introduced a canonical form of curves and their gap sequences at its single ∞, which was given by the numerical semigroup; we call it the *Weierstrass semigroup* [Del2008]. Baker also inherited Weierstrass' intention and gave the review of Weierstrass' canonical form [Ba1897]; for coprime positive integers $r, s, (r < s)$,

$$f_X(x, y) := y^r + A_1(x)y^{r-1} + A_2(x)y^{r-2} + \cdots + A_{r-1}(x)y + A_r(x) = 0, \tag{1.48}$$

where the $A_i(x)$'s are polynomials in x, $A_i = \sum_{j=0}^{\lfloor is/r \rfloor} \lambda_{i,j} x^j$, and $\lambda_{i,j} \in \mathbb{C}$, $\lambda_{r,r} = 1$. This formula is a refined version of Abel's (1.44). Equation (1.48) uniquely determines a non-singular curve up to certain equivalences, which is denoted by X. Equation (1.48) is constructed so that X has a single ∞.

Then the monomial $x^\ell y^{\ell'}$ diverges at ∞ if $\ell, \ell' > 0$. It is a meromorphic function over X which has singularity only at ∞. Due to Weierstrass' investigation, the ∞ is sometimes the Weierstrass point. By a local parameter t at ∞, x and y has the local parameterization

$$x = \frac{1}{t^r}, \quad y = \frac{1}{t^s}(1 + (\text{holomorphic with respect to } t)).$$

In [Wei1867], Weierstrass investigated the behavior of the singularity at ∞ and found that 1) there are the gaps, the so-called Weierstrass gap and 2) the non-gap sequence obeys the numerical semigroup H_X. The numerical semigroup H is a submonoid of the monoid $\mathbb{N}_0 = \{0, 1, 2, \ldots\}$ generated by the minimal set of positive integers $b_1, b_2, \ldots, b_m$, $(b_1 < b_2 < \cdots < b_m)$ so that its complement is a finite set. We refer to it as $H = \langle b_1, b_2, \ldots, b_m \rangle$. The cardinality of the complement $H^c := \mathbb{N}_0 \setminus H$ of H is called the genus, $g = g(H)$. Weierstrass demonstrated the cases of $g = 1, 2$ and 3, though he wrote that the method was applicable for $g = 4$ but would be complicated for $g > 4$. For $g = 1$, he derived his standard form (1.26), and for $g = 2$ he obtained $y^2 = 4x^5 - g_1x^3 - g_2x^2 - g_3x - g_4$.[17] The $g = 3$ case is realized only by three H_X^c's $\{1, 3, 5\}$, $\{1, 2, 5\}$, and $\{1, 2, 4\}$, which correspond to $\langle 2, 5 \rangle$, $\langle 3, 4 \rangle$ and $\langle 3, 5, 7 \rangle$, and

[17] In this book, we refer to (1.48) as the Weierstrass canonical form, but his investigation contained the study on the moduli parameters. From the standpoint of historical rigor, it is clear that we should call the Eq. (2.150) shown in Appendix 2.12 the Weierstrass canonical form.

$$
\begin{aligned}
f_{\langle 2,5\rangle}(x,y) &= y^2 + (\lambda_{1,2}x^2 + \lambda_{1,1}x + \lambda_{1,0})y - (x^5 + \lambda_{2,4}x^4 + \cdots + \lambda_{2,1}x + \lambda_{2,0}),\\
f_{\langle 3,4\rangle}(x,y) &= y^3 + (\lambda_{1,1}x + \lambda_{1,0})y^2(\lambda_{2,2}x^2 + \lambda_{2,1}x + \lambda_{2,0})y\\
&\qquad -(x^4 + \lambda_{3,3}x^3 + \lambda_{3,2}x^2 + \lambda_{3,1}x + \lambda_{3,0}),\\
f_{\langle 3,5,7\rangle}(x,y) &= y^3 + \alpha(x)y^2\varepsilon(x)\beta(x)y + \varepsilon^2(x)\gamma(x),
\end{aligned}
\tag{1.49}
$$

where α, β, γ, ε are coprime polynomials in $\mathbb{C}[x]$, $\alpha(x) = \alpha_1 x + \alpha_0$, $\beta(x) = \beta_2 x^2 + \beta_1 x + \beta_0$ $\gamma(x) = x^3 + \gamma_2 x^2 + \gamma_1 x + \gamma_0$, and $\varepsilon(x) = x + \varepsilon_0$. Further, these coefficients λ's are given so that $f_{\langle 2,5\rangle}$ and $f_{\langle 3,4\rangle}$ are non-singular curves and the affine rings $R_{\langle 2,5\rangle} = \mathbb{C}[x,y]/(f_{\langle 2,5\rangle})$ and $R_{\langle 3,4\rangle} = \mathbb{C}[x,y]/(f_{\langle 3,4\rangle})$ are integral rings. On the other hand, $f_{\langle 3,5,7\rangle}$ corresponds to a singular curve and thus the curve X is regarded as its normalized one; there exist three curves that are non-singular by normalizing the ring and mutually transferable by a holomorphic map. By regarding the transformation as an equivalence, the normalized curve is unique up to the equivalence. The affine ring R_X of X is given by

$$
R_{\langle 3,5,7\rangle} = \mathbb{C}[x,y,z]/(y^2 + \varepsilon z + \alpha y + \varepsilon\beta,\ yz - \varepsilon\gamma,\ z^2 + \beta z + \gamma y + \alpha\gamma).
$$

Table 1.1 Weierstrass non-gap sequence

	0	1	2	3	4	5	6	7	8	9	...
$\langle 2,5\rangle$	1	–	x	–	x^2	y	x^3	xy	x^4	x^2y	...
$\langle 3,4\rangle$	1	–	–	x	y	–	x^2	xy	y^2	x^3	...
$\langle 3,5,7\rangle$	1	–	–	x	–	y	x^2	z	xy	x^3	...

Table 1.1 is known as the Weierstrass non-gap sequence. Even for the normalized curve, there is a single ∞ as a feature of the Weierstrass canonical form. Thus in this book, we call these non-singular curves given by the Weierstrass canonical form (1.48) *Weierstrass curves* or *W-curves*.

Further, the numerical semigroup H is connected with the Young diagram Λ_H. For its complement $H^c = \{N^c(0), \ldots, N^c(g-1)\}$, $(N^c(i) < N^c(i+1))$, Λ_H is defined by $\Lambda_H = (\Lambda_1, \ldots, \Lambda_g)$, $\Lambda_i = N^c(g-i) - g + i - 1$, $i = 1, 2, \ldots, g$. We say that the numerical semigroup H is symmetric if the transpose of the Young diagram associated with Λ_H is self-adjoint, and otherwise we say that it is not symmetric. The semigroup $\langle 3,5,7\rangle$ is not symmetric, whereas $\langle 2,5\rangle$ and $\langle 3,4\rangle$ are symmetric. The corresponding Young diagrams are given as follows:

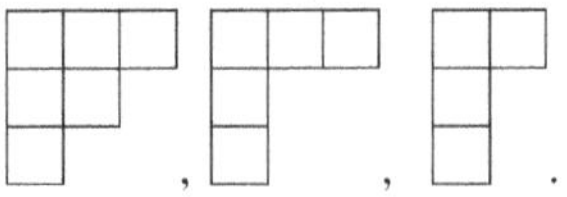

$\langle 2,5\rangle$ $\quad\langle 3,4\rangle$ $\quad\langle 3,5,7\rangle$

In *Werke IV*, Weierstrass generalized Σ (1.38) and Ω (1.40), which are denoted by H and G in his conventions, and showed Jacobi's result (1.21) (cf. Theorem 3.37). We thus consider the plane curve X represented by Weierstrass canonical form $f_X(x, y) = 0$. Then Weierstrass took particular note of

$$h_X(x, y, y') = \frac{f_X(x, y) - f_X(x, y')}{y - y'}, \tag{1.50}$$

and considered [Wei1875a, p. 78, p. 241]

$$H(x, y, x', y') := \frac{h(x, y, y')dx}{\partial f_X(x, y)/\partial y}, \tag{1.51}$$

which behaves like $H(x, y, x', y') = -\dfrac{dt}{t} + d_{>0}(t)$ for the local parameter t at (x', y') as a function of (x, y), and holomorphic except at the point (x', y') and ∞. He defined $G(x, y, x', y')$ which behaves like $G(x, y, x', y') = -\dfrac{dt}{t^2} + d_{>0}(t)$ in terms of H. For the case of the hyperelliptic curve, $y^2 = f(x) = x^{2g+1} + \lambda_{2g}x^{2g} + \cdots + \lambda_1 x + \lambda_0$ and the affine ring $R_X = \mathbb{C}[x, y]/(y^2 - f(x))$, he defined

$$f(x, x') := \sum_{i=0}^{g} \left(\lambda_{2i} x^i x'^i + \frac{1}{2}\lambda_{2i+1}(x + x')x^i x'^i \right),$$

which satisfies [Wei1875a, p. 273]

$$f(x, x) = f(x), \quad \left.\frac{\partial f(x, x')}{\partial x}\right|_{x=x'} = \frac{1}{2} f'(x), \quad f(x, x') = f(x', x),$$

and found [Wei1875a, p. 274]

$$G(x, y, x', y') = -\frac{yy' + f(x, x')}{2(x - x')^2 yy'} dx dx'. \tag{1.52}$$

Obviously H and G are higher genus analogues of Σ in (1.38) and Ω in (1.40) and correspond to them in Proposition 2.84 and Theorem 2.84. Using them, Weierstrass derived the generalized Legendre relations, which contains Galois' one (1.46) and are completely the same as Theorem 2.148 [Wei1875a, p. 330]. It is noted that (1.46) means that Galois essentially found these tools.

1.6.3 Weierstrass' Program of the Abelian Function Theory

Weierstrass seemed to have his own program of his Abelian function theory as follows:

1. Consider an algebraic curve X with a canonical form.
2. Define the differentials of the first kind, the second kind and the third kind on X.
3. In terms of the data of the integrals of these differentials, define the Jacobian and the sigma function, Al or σ.
4. Obtain the Jacobi inversion formula using Jacobi's method (1.21).
5. Using the data of the algebraic curve X or its affine ring R_X, obtain differential equations.
6. Analyze the differential equation to determine the power expansion of the σ function.
7. With the help of σ and $\wp$'s, investigate the algebraic, geometric and transcendental structures of the algebraic curves X and the Jacobian J_X.

Various problems had to be solved in order to execute these programs, which he did one by one. For example, one of them is the problem of the uniqueness of expressions of meromorphic functions defined on an Abelian variety by the related theta function. Following his study, Frobenius precisely investigated the Abelian function and theta function on the Abelian variety [Haw2013].

As a side note, we point out that the leading terms of the sigma function are determined by the Schur functions, and the hyperelliptic function cases were essentially obtained by Weierstrass himself. The name of these functions is in honor of Schur. Looking at the profound correspondence between the Schur functions and the Abelian functions shows that Issai Schur (1875–1941) himself seemed to be influenced by Weierstrass through his supervisor Frobenius.

1.6.4 Bolza and Baker: The Successors of Weierstrass

At the end of the 19th century, Weierstrass' approach encountered several problems. However, Bolza and Baker proceeded with Weierstrass' program in Sect. 1.6.3 using the Kleinian sigma function of hyperelliptic curves. Bolza, who studied under both Weierstrass and Klein, understood both their thoughts. He followed to Weierstrass' approach to investigate Klein's sigma function. He developed Klein's sigma function and connected them with the algebraic curves in particular for genus two [Bol1895, Bol1899a, Bol1899b]. Following Bolza's works, Baker investigated the properties of the sigma function and provided construction of the sigma function using the extensions of Weierstrass' H in (1.51), G in (1.52) and the differentials of the third kind [Ba1897]. As mentioned above, Klein probably orchestrated these movements. Baker connected the properties of the sigma function with the data of algebraic curves rather than compact Riemann surfaces, particularly, for general hyperelliptic curves.

As we mention below, he was very close to completing Weierstrass' program for the hyperelliptic curves. In parallel, for their generalization, he also studied curves given by the Weierstrass canonical forms (1.48). Indeed, it is remarkable that Baker started the book [Ba1897] by presenting the related equation to (1.48). "1. This book is concerned with a particular development of the theory of the algebraic irrationality arising when a quantity y is defined in terms of a quantity x by means of an equation of the form

$$a_0 y^n + a_1 y^{n-1} + \cdots + a_{n-1} y + a_n = 0,$$

wherein $a_0, a_1, \cdots, a_n$ are rational integral polynomials in x. The equation is supposed to be irreducible; that is, the left-hand side cannot be written as the product of other expressions of the same rational form."

After publishing the book on the Abelian function theory [Ba1897],[18] Baker challenged the fifth and the sixth steps in Weierstrass' program of Sect. 1.6.3. He found the series of the differential equations (1.53) corresponding to (1.31) of genus two in 1898 [Ba1898b], which are so-called the KdV hierarchy nowadays. As mentioned in Sect. 1.3.3, in the paper [Ba1898b], Baker seemed to consider the integration of the differential equations in the Jacobian. He demonstrated the series of the differential equations of genus two and considered the first integrals, which showed the embedding of the Jacobian as a quartic surface.

He also published a paper [Ba1898a] in which he handled the extensions of the addition theorem of the sigma function of genus one, (1.19) or (1.34) by considering the behavior of the al functions. Then probably, he found a formula associated with Weierstrass' G (1.52), which is identical to the formula known as Fay's formula nowadays (cf. Lemma 3.79 and Proposition 3.87).

Further in the paper [Ba1903] (cf. Sect. 4.1) following [Ba1898b], he presented the generating relations of the differential equations for every genus g with derivations by starting the formula associated with Weierstrass' G (1.52), cf. (4.1), and explicitly showed the series of the differential equations of genus three (1.53). He pursued his own program, cited in the paper [Ba1898b]. However, due to the lack of sufficient knowledge of algebraic geometry, and the complex analysis of several variables, Baker came to a dead end.

We mention the papers [Ba1898b] and [Ba1903] more precisely (cf. Sect. 4.1). He introduced the hyperelliptic $\wp$ function by

$$\wp_{\mu,\nu}(u) := -\frac{\partial^2}{\partial u_\mu \partial u_\nu} \log \sigma(u)$$

for $u \in \mathbb{C}^g$; the Abel–Jacobi map $w : S^g X \to J_X = \mathbb{C}^g / \Gamma_X$ is given by $u_\mu = \sum_{i=1}^{g} w_\mu((x_i, y_i))$, and $w_\mu((x_i, y_i)) = \int_\infty^{(x_i, y_i)} \frac{x^{\mu-1} dx}{2y}$ with $\kappa_J : \mathbb{C}^g \to J_X$. He rewrote

[18] In the book, he formulated Abel's theorem [Abl1829] explicitly on p. 213 but later, he called the Jacobi–Riemann fundamental formula on p. 275 Abel's theorem (cf. Sect. 3.1.5).

the relation by defining the bilinear differential operator, $\Delta_\lambda = \dfrac{\partial}{\partial u_\lambda} - \dfrac{\partial}{\partial u'_\lambda}$; he described it by $\wp_{\mu,\nu} = -\sigma(u)^{-2}\Delta_\mu\Delta_\nu\sigma(u)\sigma(u')$ for replacing u' with u after acting on them.[19]

After rewriting the differential equations by the bilinear operator corresponding to (1.13), he stated "This form shews, by equating terms of the same dimension on the two sides, that the equations can be used to obtain the expansion of the σ function in integral powers of the arguments" [Ba1898b]; he aimed to trace the path Weierstrass took; the starting point was Gauss via Gudermann, and could be traced back to Euler's elastica. Baker continued "as a practical example, I have obtained the expansion of the function $\sigma_{12}(u_1, u_2)$ up to terms of the ninth order, assuming only the terms of the first order to be u_1. It is probably better, however, that the linear partial differential equations, which for any specified function are more convenient than these, should first be deduced from them" [Ba1898b]. We remark that the motivation for studying the differential equations with matrix coefficients, known as the Baker–Akhiezer theory, was the expansion of the sigma function. In developing the theory of the linear differential equations, he also found the results known as the Baker–Campbell–Hausdorff formula.

1903, he published the paper in which he showed the generating relations among the hyperelliptic $\wp$ functions of genus g (cf. Theorem 4.3) with derivations following Weierstrass' thought, and explicit formulae as series of the following differential equations of genus three, $y^2 = \lambda_7 x^7 + \lambda_6 x^6 + \cdots + \lambda_1 x + \lambda_0$ [Ba1903]:

$$
\begin{aligned}
&\wp_{3333} - 6\wp_{33}^2 = 2\lambda_5\lambda_7 + 4\lambda_6\wp_{33} + 4\lambda_7\wp_{32},\\
&\wp_{3332} - 6\wp_{33}\wp_{32} = 4\lambda_6\wp_{32} + 2\lambda_7(3\wp_{31} - \wp_{22}),\\
&\wp_{3331} - 6\wp_{31}\wp_{33} = 4\lambda_6\wp_{31} - 2\lambda_7\wp_{21},\\
&\wp_{3322} - 4\wp_{32}^2 - 2\wp_{33}\wp_{22} = 2\lambda_5\wp_{32} + 4\lambda_6\wp_{31} - 2\lambda_7\wp_{21},\\
&\wp_{3321} - 2\wp_{33}\wp_{21} - 4\wp_{32}\wp_{31} = 2\lambda_5\wp_{31},\\
&\wp_{3311} - 4\wp_{31}^2 - 2\wp_{33}\wp_{11} = 2\Delta,\\
&\wp_{3222} - 6\wp_{32}\wp_{22} = -4\lambda_2\lambda_7 - 2\lambda_3\wp_{33} + 4\lambda_4\wp_{32} + 4\lambda_5\wp_{31} - 6\lambda_7\wp_{11},\\
&\wp_{3221} - 4\wp_{32}\wp_{21} - 2\wp_{31}\wp_{22} = -2\lambda_1\lambda_7 + 4\lambda_4\wp_{31} - 2\Delta,\\
&\wp_{3211} - 4\wp_{31}\wp_{21} - 2\wp_{32}\wp_{11} = -4\lambda_0\lambda_7 + 2\lambda_3\wp_{31},\\
&\wp_{3111} - 6\wp_{31}\wp_{11} = 4\lambda_0\wp_{33} - 2\lambda_1\wp_{32} + 4\lambda_2\wp_{31},\\
&\wp_{2222} - 6\wp_{22}^2 = -8\lambda_2\lambda_6 + 2\lambda_3\lambda_5\\
&\qquad - 6\lambda_1\lambda_7 - 12\lambda_2\wp_{33} + 4\lambda_3\wp_{32} + 4\lambda_4\wp_{22} + 4\lambda_5\wp_{21} - 12\lambda_6\wp_{11} + 12\Delta,\\
&\wp_{2221} - 6\wp_{22}\wp_{21} = -4\lambda_1\lambda_6 - 8\lambda_0\lambda_7 - 6\lambda_1\wp_{33} + 4\lambda_3\wp_{31} + 4\lambda_4\wp_{21} - 2\lambda_5\wp_{11},
\end{aligned}
$$

[19] This operator is known as Hirota's bilinear operator due to Hirota's direct method on the nonlinear equation [Hir1971, Hir1976]. However, it was defined by Baker for analyzing the KdV hierarchy for characterizing the sigma function in 1898. Further, it is also remarkable that Guass already considered the bilinear equations to define his theta series from the static sine-Gordon equation as in (1.13), and Weierstrass took over the method in [Wei1840]. Baker also took it over.

$$\begin{aligned}
\wp_{2211} - 4\wp_{21}^2 - 2\wp_{22}\wp_{11} &= -8\lambda_0\lambda_6 - 8\lambda_0\wp_{33} - 2\lambda_1\wp_{32} + 4\lambda_2\wp_{31} + 2\lambda_3\wp_{21},\\
\wp_{2111} - 6\wp_{21}\wp_{11} &= -4\lambda_0\lambda_5 - 8\lambda_0\wp_{32} + 2\lambda_1(3\wp_{31} - \wp_{22}) + 4\lambda_2\wp_{21},\\
\wp_{1111} - 6\wp_{11}^2 &= -8\lambda_0\lambda_4 + 2\lambda_1\lambda_3 + 4\lambda_0(4\wp_{31} - 3\wp_{22}) + 4\lambda_1\wp_{21} + 4\lambda_2\wp_{11},
\end{aligned} \tag{1.53}$$

where

$$\Delta = \wp_{32}\wp_{21} - \wp_{31}\wp_{22} + \wp_{31}^2 - \wp_{33}\wp_{11}.$$

These could be regarded as the KdV-hierarchy, a part of the KP-hierarchy if we use modern language [BucEL1997a, Mat2002c] (cf. Sect. 4.1). Baker also posed a problem:

> to obtain a theory of differential equations which shall shew from them why, if we assume $\wp_{\mu,\nu} = -\dfrac{\partial^2}{\partial u_\mu \partial u_\nu} \log \sigma(u)$ the function $\sigma(u)$ has the properties which a priori we know it to possess, and how far the forms of the equations are essential to these properties.

In this paper, Baker proceeded with Weierstrass' program in Sect. 1.6.3, and considered the sigma function in terms of the differential equations.

1.6.5 Obstacles in Weierstrass' Program

(1.50) is defined on the enveloping algebra $R_X \otimes_{\mathbb{C}[x]} R_X$ for the affine algebra $R_X = \mathbb{C}[x, y]/(f_X(x, y))$, whereas (1.51) and (1.52) are defined on the quotient field $\mathcal{Q}(R_X \otimes_{\mathbb{C}} R_X)$. The difference in these algebras caused a serious problem; even for hyperelliptic curves, G has several different expressions; in other words, it is not uniquely determined. The problem is settled by the cohomological meaning, or quasi-isomorphism as mentioned in Chap. 2 (cf. Proposition 2.122 4). It might not be a coincidence that Hodge, who had regular informal contact with Baker, studied the cohomology [At1976].

Further, the study of the Abelian function theory in the 19th century faced difficulty due to a lack of tools and of rigorousness; in those days, there was no Emmy Neother's theory, Zariski's topology, Oka's work on sheaves, nor Grotherndick's scheme theory.

The paper of Baker in 1903 could be interpreted as the end of the theory of Abelian functions in the 19th century. After then, Baker started to study the linear differential equation theory, which is known as the Baker–Akhiezer theory.

After Baker published a summary of his series of work in 1907, restricting it to genus two, he essentially did not touch the work related to the sigma function. However, when J. L. Burchnall and T. W. Chaundy published the series of papers on the commutative ordinary differential operators [BurC1922, BurC1928, BurC1931], Baker wrote a note in [Ba1928] that their problem was considered in the context of his study of the sigma function and connected with the algebraic curves he considered.

1.7 Weierstrass' Semigroup Based on Riemann Surfaces

As mentioned in Sect. 1.6.1, Weierstrass found the Weierstrass canonical form (1.48), by which a curve is fixed so that its Weierstrass non-gap sequence at ∞ obeys the semigroup.

For a numerical semigroup $H = \langle b_1, b_2, \ldots, b_m \rangle$, we call the number which is the least integer x such that $x + n \in H$ for all $n \in \mathbb{N}_0$ the *conductor* of H, denoted by $c(H)$, and $c(H) - 1$ the *Frobenius number* of H, denoted by $F(H)$. The Frobenius problem is to find the formulae of $g(H)$ and $F(H)$ in terms of generators $\{b_1, b_2, \ldots, b_m\}$ of H. Brauer stated that Frobenius mentioned it occasionally in his lecture [Bra1942].

When the number m of the generators is two, b_1 and b_2, Sylvester showed that the genus is equal to $(b_1 - 1)(b_2 - 1)/2$ and $F(S) = b_1 b_2 - b_1 - b_2$ [Sy1884]. However, the Frobenius problem remains open for numerical semigroups with $m > 2$. Brauer found a result by the discussion with Schur [Bra1942, BraS1962] and the problems have been studied by several authors after it.

Further, Weierstrass investigated the complex multiplication problems as the relation between algebraic and transcendental properties on the algebraic curves. He studied it with Kronecker, Weber, Frobenius following Hermite's [Haw2013]; For $J = \begin{pmatrix} 0 & I_g \\ -I_g & 0 \end{pmatrix}$, where I_g is the $g \times g$ unit matrix. if a matrix A satisfies ${}^t A J A = nJ$, we say that n is the order of A, i.e., $\mathrm{ord}(A) = n$ and we call such a matrix an Abelian matrix. Hermite observed when $g = 2$ that if A has order m and B has order n, ${}^t(AB)J(AB) = nmJ$, which means the order is abelian, i.e., $\mathrm{ord}(AB) = \mathrm{ord}(BA) = nm$. It is obvious that the Abelian matrix is connected with the complex multiplication of the Abelian functions.

Adolf Hurwitz (1859–1919) studied the elliptic modular and Abelian functions under Klein after receiving guidance from Hermann Schubert (1848–1911). It is well-known that Klein recognized that geometry should be described and dominated by a group. In a paper [Hur1888] in 1888, Hurwitz considered the group actions or the transformation of the meromorphic functions on the compact Riemann surfaces based on the results of Dedekind and Weber [DedW1882]. He investigated the problem to determine all irreducible algebraic equations, $f(s, z) = 0$, which can be transformed into themselves by a rational uniquely reversible transformation, $s' = \varphi(s, z)$ and $z = \psi(s, z)$ as the connection with the problem to indicate all those compact Riemann surfaces on which $(s, z; s', z')$ exists. In other words, he argued the group actions (or the transformation) on the meromorphic functions on the Riemann surfaces, which is also related to the complex multiplication [Haw2013] and the covering of the Riemann surfaces, which is known as the Riemann–Hurwitz theorem. In 1891, he also investigated the behavior of the leaves for the covering of the compact Riemann surfaces on $\mathbb{P}$, $\varpi_x : X \to \mathbb{P}$ [Hur1891]; at each ramification point, the connection among the leaves is expressed by a (monodromy) group action. Each group action must be consistent with the global (Galois) group associated with ϖ_x [Hur1891].

He studied the consistency there. It is now known as the Hurwitz action in the braid group problem.

Further, in 1893 Hurwitz argued the construction of a compact Riemann surface governed by a finite group, as a unique transformation [Hur1893]. In the paper, he handled the local properties of the differential forms (which are very similar to 2.2.6.5 in this book). He saw through the fact that the origin of the group action is the numerical semigroup. It means that Hurwitz suggested the problem for a given numerical semigroup to construct a non-singular algebraic curve whose Weierstrass non-gap sequence agrees with it. This is known as the Hurwitz problem on the numerical semigroup. It is expressed by the problem whether, for a given numerical semigroup H, there exists a non-singular algebraic curve that has H as the Weierstrass non-gap sequence at a certain point,

Due to these studies by Hurwitz and several other results, including [DedW1882], the problem has been studied on the compact Riemann surfaces instead of explicit curves given by the Weierstrass normal form.

Let $\mathcal{M}_{g,P}^{H}$ be the moduli space of non-singular algebraic curves whose genus is g, and P is a point with the non-gap sequence given by the numerical semigroup H of genus g. We refer to such a semigroup as the Weierstrass semigroup or Weierstrass if $\mathcal{M}_{g,P}^{H}$ is not empty. In modern terms, the Hurwitz problem means whether $\mathcal{M}_{g,P}^{H}$ is not empty for any numerical semigroup H. In other words, the Hurwitz problem is closely related to the moduli space of the algebraic curves.

According to [EisH1987], after the problem was posed by Hurwitz, almost immediately, M. Haure incorrectly gave further restrictions in 1896, and K. Hensel and G. Landsberg incorrectly showed there was no such empty set $\mathcal{M}_{g,P}^{H}$ in 1902.

R. Apéry investigated the properties of semigroup in 1946 [Ape1946]. Clifford and Preston also studied the algebraic properties of the semigroup following the group theory in 1961 and 1967 [CliP1967]. Under E. Kunz, J. Herzog investigated the semigroup rings, and the monomial rings associated with the numerical semigroups [Ku1970, Her1970]. H. C, Pinkham studied the deformations from the monomial curves to the algebraic curves, in order to settle the Hurwitz problem by clarifying the properties of the moduli space of the curves [Pin1974, MorP1986].

The answer to the Hurwitz problem was obtained by R.-O. Buchweitz in 1979 negatively. Its first counterexample is the numerical semigroup generated by $\{13, 14, 15, 16, 17, 18, 20, 22, 23\}$, though recently infinitely many counterexamples have been obtained.

On the other hand, T. Kato showed that every 3-numerical semigroup is a Weierstrass semigroup [Kat1979] and following his study, J. Komeda also showed that every 4-numerical semigroup is Weierstrass [Kom1983]. We offer the recent results in Sect. 2.2.

For a semigroup H, the so-called Schubert index α_H, which is essentially the same as the Young diagram, is determined (cf. (2.3)) and vice versa. The Schubert index plays a crucial role in embedding the affine ring R_X into the universal Grassmannian manifold (cf. Appendix 3.9).

1.8 Completely Integrable Systems: from Differential Equations to Theta Series and Curves

The method of determining theta functions from the nonlinear differential equations whose origin must be Gauss had revived since the 1950s.

Diederik J. Korteweg (1848–1941) received a Ph.D. from the University of Amsterdam in 1878 on the study whose title is "On the Propagation of Waves in Elastic Tubes". In 1881, he joined the University of Amsterdam as a Professor of Mathematics, Mechanics and Astronomy. In 1895 he published a notable paper [KorV1885] on the equation

$$v_t + 6vv_x + v_{xxx} = 0, \tag{1.54}$$

which was independently of Euler's results, movements of Weierstrass and Baker. In the paper [KorV1885], they investigated its solutions in terms of hyperbolic functions as meromorphic functions of the degenerated curves.

In 1970, B. B. Kadomtsev and V. I. Petviashvili discovered the equation [KadP1970],

$$\partial_x(\partial_t u + u\partial_x u + \varepsilon^2 \partial_x^3 u) + \lambda \partial_y^2 u = 0, \tag{1.55}$$

where $\lambda = \pm 1$ as a result of studies on plasma physics. In the paper, the studies of Baker were not mentioned, of course.

We mention here that both (1.54) and (1.55) appear in the papers of Baker in 1898 and in 1903 [Ba1898b, Ba1903] (cf. Sect. 4.1). (1.54) is called the *Korteweg–de Vries (KdV) equation*, whereas (1.55) is known as the *Kadomtsev–Petviashvili (KP) equation*.

1.8.1 Mid-20th Century: from Differential Equations to Curves (Integrable System)

Baker derived the nonlinear differential equations from the theory of the Abelian function in 1898 to find the properties of curves and sigma function in terms of the differential equations following Weierstrass' program, whose origin is Gauss. In other words, he aimed at the inverse direction, and then, using the nonlinear differential equations, he would characterize the sigma function, i.e., from differential equations to general algebraic curves:

19th c.	Alg. Curve	$\rightarrow$	Diff. Eq.
20th c.	Alg. Curve	$\leftarrow$	Diff. Eq.

Weierstrass obtained both directions for elliptic curves following Gauss, whose origin must have been Euler's elastica.

However, even though the right arrow was formidable, its inverse was much more difficult. After the middle of the 20th century, the inverse direction was studied in the framework of the nonlinear integrable system.

The modern study of nonlinear integrable equations started in the 1950s. The computer experiment by E. Fermi, J. R. Pasta and S. M. Ulam in the 1950s provided a problem in the nonlinear system [FePU1955]. There exists a class of nonlinear equations which has a remarkable property, a completely integrable property. Studying such completely integrable nonlinear equations means going from differential equations to algebraic curves. This method was achieved by many distinguished researchers, such as M. D. Kruskal, N. J. Zabusky, R. M. Miura, C. S. Gardner, J. M. Greene, P, D. Lax, M. J. Ablowitz, H. Segur, V. E. Zakharov, A. B. Shabat, L. D. Faddeev, I. M. Gel'fand, L. A. Dickey, S. P. Novikov, B. Dubrovin, I. Krichever, D. Mumford, M. Toda, R. Hirota, J. Satsuma, M. Wadati, E. Date, S. Tanaka, M. Sato, Y. Sato, M. Jimbo, M. Kashiwara, T. Miwa, M. Mulase, G. Segal, G. Wilson, T. Shiota, and so on [New1985, TanD1979, To1989, Pr1996]. It is a well-known fact and a significant event in mathematics as a nonlinear integrable theory. Thus, there is much literature; therefore, we will not give details here, though we give a short review on the theory of KP-hierarchy by Dubrovin, Krichever, Mumford and Shiota, and its relation to the universal Grassmannian manifold in the Sato theory in Appendix 3.9.

Nonlinear differential equations played essential roles in the study of the elliptic functions by Euler, Gauss, and Weierstrass. However, it was not until the 20th century that the path from differential equations to Abelian functions seemed realistically constructible. It required rigorous mathematical tools, e.g., modern algebraic geometry, sheaf theory, several complex variables, and algebraic analysis. In other words, the problem proposed by Baker was too complicated in his day, and we had to wait for three-quarters of a century.

In the study, it is interesting that Baker–Akhiezer theory plays a central role, as in the papers [Ba1898a] and [Ba1903] of Baker, which already alluded to the necessity of the study. In the derivation [Ba1903], he used bilinear operators and the Pfaffian to obtain his hierarchy. Further, we also note the genealogy, Weierstrass, Frobenius, and Schur. It is ironic that the Schur polynomial plays a vital role in the Sato theory and in the sigma function for curves given by the Weierstrass canonical forms (1.48).

Even though in the theory of the integrable system, the studies of Baker or Weierstrass were not mentioned, they realized the path from the nonlinear differential equations to the theta series as the origin of the sigma functions; Gauss found this approach, and Weierstrass and Baker adopted it. Looking back at the history of Euler, Lagrange, Gauss, and Weierstrass, we might say that the path they took was already carefully planned.

1.8.2 Mixing up Theories in the 19th and the 20th Centuries

Just as the theory of elliptic functions played a vital role in theoretical physics, number theory, and engineering applications, attempts to apply Abelian function theory to experimentally verifiable physics and number theory problems reveal that the theory is not well established. We can reconstruct the Abelian function theory of algebraic curves at the same level as the elliptic function theory by unifying the advantages of the 19th and 20th centuries.

19th c. Concrete	→	20th c. Abstract	→	21th c. Concrete +Abstract

The prototype of the study is *Tata Lecture of Theta I, II* by D. Mumford [Mum1981, Mum1984]. By reconsidering the theory of Abelian functions in the 19th century based on modern algebraic geometry, he found the hyperelliptic solutions of the KdV equation, the Neumann equation, and the sine-Gordon equation. He introduced the nice meromorphic functions and the Abelian functions, known as UVW representation for the study. By connecting them to the Riemann theta function, he clarified algebraic and differential structures of the Abelian functions and theta functions. Then he found several hyperelliptic solutions of these nonlinear differential equations.

We address his approach in the study as Mumford's program: "For the reconstruction of the Abelian functions over hyperelliptic curves, we study the hyperelliptic curves by following steps.
(1): Evaluate Jacobi's inversion formula of hyperelliptic curves (UVW representation).
(2): Clarify the algebraic and differential structures of the Abelian functions and theta functions.
(3): Apply the results of (1) and (2) to the nonlinear differential equations and feed back their results to the problems (1) and (2)."

This book aims to reconstruct the Abelian function theory of a more general algebraic curve following Mumford's program following the earliest works [Pr1985, Pr1986, Pr1987]. Instead of the theta function, we use the Weierstrass–Klein sigma function as in the title of [Pr1987], "Generalized Weierstrass $\wp$-functions and KP flows in affine space". One of the properties of the sigma function is the modular invariant. It is the most crucial property in reconstructing the theory.

One of our motivations for these studies described in this book is to investigate the generalization of Euler's elastica theory as will be mentioned in the final Sect. 4.6 following [Pr1993, Mat1998, MatP2016, MatP2022].

1.9 Contemporary Sigma Since 1990

This section reviews the works in the recent study of the generalization of Weierstrass' sigma function since 1990.[20] Baker's book was reissued in 1995, in which I. Krichever wrote the foreword. The 1990s was the period of the reappearance of classical sigma.

[20] The late Vector Z. Enolskii played a central role in these movements for three decades. In the article [PrBMEBKL] for him, the recent movements related to him are reported.

Then the Abelian function theory of the 19th century was re-evaluated. The study of the sigma function is to construct the Abelian function theory of higher genus that is comparable to, and compatible with, the Weierstrass elliptic function theory for applying the Abelain functions to several fields in mathematics, science, and technology since Weierstrass' elliptic function theory is applicable to many fields.

Such movements had been concurrently started mainly in Russia, Ukraine, the UK, and Japan by C. Athorne, T. Ayano, E. D. Belokolos, J. Bernatska, H. W. Braden, S. Baldwin, V. M. Buchstaber, E. Y. Bunkova, C. Eilbeck, M. England, V. Z. Enolskii, Y. N. Fedorov, J. Gibbons, D. Korotkin, D. V. Leykin, A. Nakayashiki, Y. Ônishi, V. Shramchenko, K. Yori, and so on. The author of this book also has worked with E. Previato, J. Komeda and Y. Kodama to investigate these problems. As the book *Multi-dimensional sigma-functions* [BucEL2012] by V. M. Buchstaber, V. Z. Enolskii, and D. V. Leykin is an excellent book on these researches, we show the movements from a different aspect.

1.9.1 Revaluation of the Sigma Function for Hyperelliptic Curves

The first step was done by D. Grant [Gra1990], though Yusuke Hagihara (1897–1979) reviewed the work of Baker in the celestial mechanics [Hag1970][21] and B. Crespi, S.-J. Chang and K.-J. Shi [CrCS1993] used Baker's work in their Poncelet's theorems in n-dimensional space as a billiard system.

From a number theoretic viewpoint, Grant re-evaluated Baker's work and gave the modern view of the (1.53) of genus two, $y^2 = x^5 + \lambda_4 x^4 + \cdots + \lambda_0$, by the linear equivalence, though he did not do further work on the sigma function. Let the dimension of $\mathcal{L}_{J_X}(D) := \mathbf{H}^0(J_X, \mathcal{O}_{J_X}(D))$ denoted by $\ell_{J_X}(D)$, and Θ be the divisor of σ acted by $\kappa : \mathbb{C}^g \to J_X$. Since Θ is an ample divisor with a self-intersection number of two, $\ell_{J_X}(n\Theta) = n^2$ of $n \geq 1$ from the Riemann–Roch theorem [Lan1982, p. 99, Theorem 4.1]. The linear bases of the spaces related to Θ are given by

$$\mathcal{L}_{J_X}(\Theta) = \mathbb{C},$$

$$\mathcal{L}_{J_X}(2\Theta)/\mathcal{L}_{J_X}(\Theta) = \mathbb{C}\wp_{11} \oplus \mathbb{C}\wp_{12} \oplus \mathbb{C}\wp_{22},$$

$$\mathcal{L}_{J_X}(3\Theta)/\mathcal{L}_{J_X}(2\Theta) = \mathbb{C}\wp_{111} \oplus \mathbb{C}\wp_{112} \oplus \mathbb{C}\wp_{122} \oplus \mathbb{C}\wp_{222} \oplus \mathbb{C}\wp,$$

$$\mathcal{L}_{J_X}(4\Theta)/\mathcal{L}_{J_X}(3\Theta) = \mathbb{C}\wp_{11}^2 \oplus \mathbb{C}\wp_{11}\wp_{12} \oplus \mathbb{C}\wp_{12}\wp_{22} \oplus \mathbb{C}\wp_{22}^2 \oplus \mathbb{C}\frac{\partial\wp}{\partial u_1} \oplus \mathbb{C}\frac{\partial\wp}{\partial u_2},$$

where $\wp := \wp_{11}\wp_{12} - \wp_{22}^2$. On the other hand, there are too many functions belonging to $\mathcal{L}_{J_X}(2\Theta)/\mathcal{L}_{J_X}(\Theta)$. The KdV hierarchy (1.53) obtained by Baker means the linear

[21] The relation between a problem in celestial mechanics and the hyperelliptic sigma function theory was revived by Enolskii and coworkers in [EnHKKLS2012] for an orbit of a particle around a black hole [Hag1970, 2.15–2.20]; Hagihara studied differential equations under Baker.

dependence among the bases. This estimation cannot be applicable to higher genus cases; for example, we encounter obstacles even in genus three of the hyperelliptic type as in Proposition 4.24. However, this investigation provides the modernized interpretation of (1.53); it should be essentially regarded as the Plücker embedding in the Sato theory [SatS1981, SatS1982].

Similarly from a number theoretic viewpoint, Y. Ônishi showed a natural extension of the Frobenius–Stickelberger relations for genera two and three [On1998, On2002], and a general genus of hyperelliptic curve [On2005].

In the investigation, the explicit representation of the Riemann–Kempf theory (due to G. Kempf [Ke1973]), which we show in Sects. 3.1.7 and 3.2.10 more systematically was obtained: Let $\natural_n$ be the multi-indices of $\{1, 2, \dots, g\}$ such that

$$\natural_n = \begin{cases} n+1, n+3, \dots, g-1 & \text{for } g-n \equiv 0 \text{ modulo } 2, \\ n+1, n+3, \dots, g & \text{for } g-n \equiv 1 \text{ modulo } 2, \end{cases}$$

whose examples are listed in Table 1.2, and

$$\sigma_{\natural_n} := \left(\prod_{i \in \natural_n} \frac{\partial}{\partial u_i} \right) \sigma(u). \tag{1.56}$$

Table 1.2 Examples of $\sigma_{\natural_i}$

Genus	$\sigma_{\natural_1}$	$\sigma_{\natural_2}$	$\sigma_{\natural_3}$	$\sigma_{\natural_4}$	$\sigma_{\natural_5}$	$\sigma_{\natural_6}$	$\sigma_{\natural_7}$	$\sigma_{\natural_8}$	$\cdots$
1	σ	σ	σ	σ	σ	σ	σ	σ	$\cdots$
2	σ_2	σ	σ	σ	σ	σ	σ	σ	$\cdots$
3	σ_2	σ_3	σ	σ	σ	σ	σ	σ	$\cdots$
4	σ_{24}	σ_3	σ_4	σ	σ	σ	σ	σ	$\cdots$
5	σ_{24}	σ_{35}	σ_4	σ_5	σ	σ	σ	σ	$\cdots$
6	σ_{246}	σ_{35}	σ_{46}	σ_5	σ_6	σ	σ	σ	$\cdots$
7	σ_{246}	σ_{357}	σ_{46}	σ_{57}	σ_6	σ_7	σ	σ	$\cdots$
8	σ_{2468}	σ_{357}	σ_{468}	σ_{57}	σ_{68}	σ_7	σ_8	σ	$\cdots$
$\vdots$	$\vdots$	$\vdots$	$\vdots$	$\vdots$	$\vdots$	$\vdots$	$\vdots$	$\vdots$	$\ddots$

The Riemann–Kempf theorem for hyperelliptic curves is refined as follows [On2005]:

1. For $u \in \kappa_J^{-1} w(S^n X)$ and every multi-index $(a_1, a_2, \dots, a_\ell)$ of $a_i \in \{1, 2, \dots, g\}$, $(i = 1, 2, \dots, \ell)$ of $\ell < |\natural_n|$

$$\frac{\partial^\ell}{\partial u_{a_1} \partial u_{a_2} \cdots \partial u_{a_\ell}} \sigma(u) = 0.$$

2. There is $u \in \kappa_J^{-1} w(S^n X)$ satisfying

$$\sigma_{\natural_n}(u) \neq 0.$$

Using the results, Ônishi also generalized the Hermite–Frobenius–Stickelberger relation of elliptic curve to a general hyperelliptic curve of genus g as an addition theorem [On1998, On2002, On2005] (cf. Theorem 3.69).

For positive integers (m, n), points $(x_i, y_i) \in X_g$ $(i = 1, \ldots, m)$, and $u^{(i)} = \int_\infty^{(x_i, y_i)} \nu^{\mathrm{I}}$,

$$\frac{\sigma_{\natural^n}(\sum_{i=1}^n u^{(i)}) \prod_{i<j} \sigma_{\natural_2}(u^{(i)} - u^{(j)})}{\prod_{i=1}^n \sigma_{\natural_1}(u^{(i)})^n} = \epsilon_n \psi_n((x_1, y_1), \ldots, (x_n, y_n)), \tag{1.57}$$

where $\epsilon_n = (-1)^{-g^2 n/2 + (n-2)g/2 + 1}$ for $n \leq g$ and $\epsilon_n = (-1)^{-g^2 n/2 + ((n-2)g)/2}$ for $n \geq g + 1$, and $\psi_n((x_1, y_1), \ldots, (x_n, y_n))$ is given by

$$\psi_n(P_1, \ldots, P_n) = \psi_n((x_1, y_1), \ldots, (x_n, y_n))$$
$$:= \begin{vmatrix} 1 & \phi_1(x_1, y_1) & \cdots & \phi_{n-2}(x_1, y_1) & \phi_{n-1}(x_1, y_1) \\ 1 & \phi_1(x_2, y_2) & \cdots & \phi_{n-2}(x_2, y_2) & \phi_{n-1}(x_2, y_2) \\ \vdots & \vdots & \ddots & \vdots & \vdots \\ 1 & \phi_1(x_{n-1}, y_{n-1}) & \cdots & \phi_{n-2}(x_{n-1}, y_{n-1}) & \phi_{n-1}(x_{n-1}, y_{n-1}) \\ 1 & \phi_1(x_n, y_n) & \cdots & \phi_{n-2}(x_n, y_n) & \phi_{n-1}(x_n, y_n) \end{vmatrix}.$$

Here $\phi_i(P) = \phi_i(x, y)$ is a monomial of the Weierstrass non-gap sequence, e.g., $\phi_0 = 1, \phi_1 = x, \cdots$.

$\psi_n((x_1, y_1), \ldots, (x_n, y_n))$ is referred to as the Frobenius–Stickelberger determinant and plays a crucial role in algebraic properties of σ [MatP2008].

Using this, we could find the addition formulae as mentioned in Sect. 3.4.

In the book *Multi-dimensional sigma-functions* [BucEL2012], we find the crucial and interesting results after V. M. Buchstaber, V. Z. Enolskii, D. V. Leykin and their coauthors further developed Baker's investigations and results, especially hyperelliptic curves.

(1) They showed how to get the full expression of $\wp_{ij}$ of the hyperelliptic curves of $(i, j = 1, \cdots, 9)$.
(2) They investigated the relations between the zeta function and the hyperelliptic integrals of the second kind.
(3) They generalized the addition formulae.

Enolskii and his coauthors applied them to physical models, including monopoles and Young–Mills excitation fields [BraDE2011, EnPR2003]. The author of this book also applied Baker's results to integrable systems [Mat2002b, Mat2002d, Mat2008a, MatP2009b, MatP2022].

1.9.2 Extension of Sigma Function to General Curves

As mentioned, Weierstrass investigated general algebraic curves, and Klein also defined the sigma function for so-called trigonal curves. Thus, it is natural to extend the theory of hyperelliptic sigma function to more general curves. Buchstaber, Enolskii and Leykin [BucEL1999], and Eilbeck, Enolskii and Leykin [EilEL2000] generalized the sigma function to every (n, s)-curve,[22] which is based on Baker's investigation of Weierstrass' canonical form. (Its origin was the two-page paper of Abel in 1829 and Weierstrass' Abelian function theory.) They studied a non-singular curve given by (1.48).

Baker's construction of the sigma function was inherited and revived by Eilbeck, Enolskii, and Leykin [EilEL2000] for a general plane curve in 2000 and by the author of this book with E. Previato and J. Komeda [KomM2020, KomMP2019] for certain classes of space curves. The authors in [EilEL2000] provided a straightforward procedure on how to define the differentials of the first, the second, and the third kinds of the (n, s)-curve, based on Baker's approach [BucEL1999]; we sometimes refer to it the EEL construction. (It could not be done in the 19th century.) It implies that we could basically define the sigma function and show the Jacobi inversion formulae like the elliptic curve case. Further, Buchstaber, Enolskii and Leykin demonstrated that the Schur polynomials give the leading term of the sigma function for (n, s)-curves [BucEL1999].

A. Nakayashiki clarified the relation between the sigma function of (n, s)-curves and the theta function in Fay's approach [Nak2010a]. Nakayashiki also connected, there, the sigma function of (n, s)-curves and the tau function in the Sato theory of the universal Grassmannian manifold [SatN1984, SeW1985] and proved that the Schur polynomials give the leading term of the sigma function. Further, by investigating the sigma function in the framework of the Sato theory, he displayed the expansions of the sigma function of every (n, s)-curve in terms of Schur polynomials [Nak2010b]. It is noted that Enolskii, Eilbeck and Gibbons had also attempted such a study for a decade [EilEG2010].

On the other hand, we have handled a natural meromorphic function, μ_ℓ, of such curves (cf. Definition 2.156). For the case (r, s)-curve

$$\mu_\ell = y^{r-1} + \widehat{q}_{r-2} y^{r-2} + \cdots + \widehat{q}_1 y + \widehat{q}_0, \tag{1.58}$$

where $\widehat{q}_j$ $(j = 1, \cdots, r-2)$ is a rational function of x and points (x_i, y_i) $i = 1, 2, \cdots, \ell$ corresponding to (1.45) with $\widehat{q}_j = q_j/q_{n-1}$ for the $n = r = \ell$ case. The equation $\mu_\ell = 0$ as a meromorphic function of the curve naturally provided the addition structure of the principal divisor. In other words, the Abelian integral of the zeros of μ_ℓ is equal to a lattice point associated with its Jacobian. This μ_ℓ could be regarded as a generalization of the U function of Mumford's UVW-triplet, and as a refined version of Abel's (1.45).

---

[22] We find the use of the convention (n, s) in the paper by Coppens on the Weierstrass points with two prescribed non-gaps [Cop1988]. However, we don't know the origin of the name (n, s).

We extend the pictures which we have for the elliptic curve to a general curve X:

Algebraic space

Covering space $\widetilde{X}$ =
Abelianization of
quotient space of
path space of X
$\kappa_X : \widetilde{X} \to X$
$\widetilde{X} \supset LX(\infty) \cong \langle \alpha, \beta \rangle_{\mathbb{Z}}$
$\mathrm{Sp}(2g, \mathbb{Z}) \curvearrowright$ this expression.

Transcendental space

Covering space $\mathbb{C}^g$ =
Abelianization of
quotient space of
path space of J_X
$\kappa_J : \mathbb{C} \to J_X = \mathbb{C}/\Gamma_X$
$\Gamma_X = \langle 2\omega', 2\omega'' \rangle_{\mathbb{Z}}$
$\mathrm{Sp}(2g, \mathbb{Z}) \curvearrowright \Gamma_X, J_X$

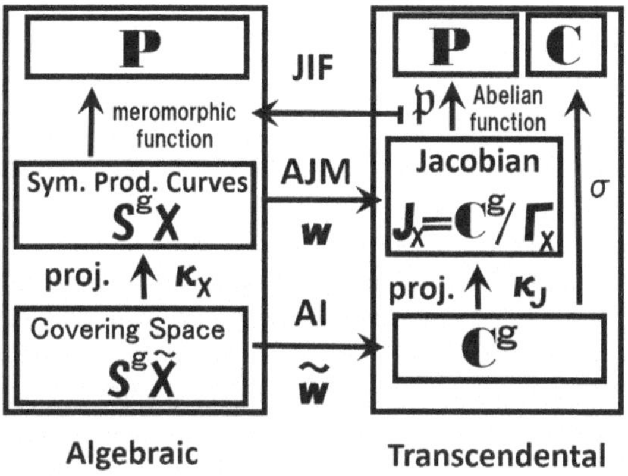

Fig. 1.8 Algebraic versus transcendental for general g

The Abelian function theory consists of the algebraic space, the left-hand side, and the transcendent space, the right-hand side in Fig. 1.8. The arrows connect both spaces; the arrows from left to right are the Abel–Jacobi map and the Abelian integrals, whereas the arrow from right to left is the Jacobi inversion formula. The Abel–Jacobi map is a map from the curves to the Jacobian, and the Abelian integrals are from the covering space to the complex vector space.

The Abelian function is a function over the Jacobian as a multiple periodic function, which is expressed by sigma function but is identical to a meromorphic function over the symmetric product of the curves.

The covering space of the symmetric product of an algebraic curve is a path space of the symmetric product of an algebraic curve. Thus there is a natural projection from the covering space to the base space. There is an ambiguity in its lifting, which corresponds to the modular transformation. Thus in the left space, the meromorphic function does not depend on the ambiguity.

Even on the right-hand side, the meromorphic function on J_X must be free from the ambiguity. The modular invariance of the sigma function guarantees the natural correspondence among both spaces; the sigma function is connected with the coordinate of x and y for an affine curve, e.g., $y^3 = x^5 - 1$, through the Jacobi inversion formula.

D. Korotkin and V. Shramchenko extended Klein's definition of sigma function to the general compact Riemann surfaces [KorS2012]. Further, Nakayashiki provided another definition based on the universal Grassmannian manifold [Nak2016]. They defined the sigma functions by using only the data of the theta function, the compact Riemann surfaces, and the Jacobians. In other words, their sigma functions are not directly related to certain algebraic curves following Klein's thought.

It implies that we should rewrite their sigma functions in terms of certain algebraic curves given by the Weierstrass canonical form following the Weierstrass program.

1.9.3 Weierstrass' Dream in the 21st Century

Weierstrass' dream was based on Abel's thoughts, and his approach differs from Riemann's. Even though the curve given by the Weierstrass canonical form (1.48) is a mere representative element of a set of compact Riemann surfaces, the definitive treatment provided by the Weierstrass canonical form has the advantage if we consider the applications of the Abelian function theory to a particular field.

Of course, the Weierstrass canonical form survives in several kinds of literature in the modern algebraic curves, sometimes called Weierstrass polynomials. However, due to the obstacles mentioned in Sect. 1.6.5, it is used only for the local treatment of an affine part of the curves since every algebraic curve is birational to a curve representative by a Weierstrass plane curve locally (cf. Proposition 2.23). Modern algebraic geometry has many tools, in the sheaf theory and cohomology theory, to find the global structure by connecting the local data. For example, Dedekind and Weber also constructed the module theory in the algebraic curves for local senses [DedW1882] and C. Chevalley followed the local treatment of the algebraic variety [Ch1951]. Chevalley also implicitly used the Weierstrass polynomial (or Abel's (1.44)) to investigate algebraic functions of one variable because at the locus, the algebraic element in terms of x and y shows the covering structure sufficiently for the analysis which we require. Though the role of the Weierstrass polynomial deviates from the original meaning, it is satisfied from the viewpoints of the investigations of the algebraic curves themselves since the treatment of both local and global data by the connections of the local data provides the geometrical and algebraic structure of curves. The treatment is consistent with the philosophy of modern algebraic geometry. On the other hand, the idea of Weierstrass has become archaic and such treatment has been avoided for the past century.

However, when we consider some applications of the algebraic curves to other fields of science, we should revive Weierstrass' approach, in which we first determine the curves globally by certain standard equations. By using the tools in modern algebraic geometry and commutative algebra, we can revive it [Miu1998, Sti2008].

Kato gave a natural proof that the Weierstrass canonical form represents a set of the compact Riemann surfaces [Kat1980, CopK2014], though Baker gave a classical proof in [Ba1897]. It is possible in principle to construct the sigma function for the curve given in Weierstrass canonical form corresponding to any compact Riemann surface.

In this book, we refer to the normalized curve given by the Weierstrass canonical form as the *Weierstrass curve* or *W-curve*. The curve with a non-symmetric Weierstrass semigroup H_X is complicated in principle, but we have heuristically obtained the descriptions of the sigma function of certain space curves, e.g. $\langle 3, p, q\rangle$ curves [MatK2013, KomMP2013, KomMP2019]. By analyzing them, the author of this book with Previato and Komeda provided an algebraic construction of the sigma function for every W-curve [KomMP2022a, KomMP2022b]. We will sketch it but first, we mention the degenerating family of the curves.

For a given curve whose Weierstrass gap is related to a numerical semigroup $\langle a_1, a_2, \dots, a_\ell\rangle$, we may regard the curve as a degeneration of a curve for $\langle a_1, a_2, \dots, a_{\ell-1}\rangle$, e.g., $\langle 3, 4, 5\rangle \leftarrow \langle 3, 4\rangle$. Thus we find a partial ordering structure in the numerical semigroup. It implies that a similar structure exists in the set of curves. The degeneration sometimes plays a valuable role as in Newton's pictures in Fig. 1.4 and the generalized elastica problem (related to the shape of DNA) in Fig. 4.8.

In [FeKMPA2022], the author of this book with Previato, Fedorov, Komeda, and Aomoto handled the degenerating family of the curves $\langle 3, 4, 5\rangle \leftarrow \langle 3, 4\rangle$, In this book, we demonstrate it (cf. Sects. 2.9 and 3.7). We consider a curve X given by $y^3 = x(x-s)(x-b_1)(x-b_2)$ and its limit of $s \to 0$. It is the singular curve $y^3 = x^2 k_2(x)$. $k_2(x) = (x-b_1)(x-b_2)$. By normalizing it, we exclude the singularity, and then we have the variable w such that $w^3 = xk_2^2(x)$, and the governing equation,

$$y^2 = xz, \quad xz = xk_2(x), \quad z^2 = k_2(x)y.$$

In other words, we have a space curve $\widehat{X} := \{(x, y, z) \mid \text{above relations}\} \cup \{\infty\}$. The behavior of the monomial at ∞ is given by the relation, $y \sim 1/t^4, z \sim 1/t^5, x \sim 1/t^3$ where t is the local parameter at ∞, and its non-gap sequence forms the numerical semigroup $H_{\widehat{X}}$ generated by $\{3, 4, 5\}$. It is shown in the following table (Table 1.3):

Table 1.3 Non-gap sequence of $\widehat{X}$

	0	1	2	3	4	5	6	7	8	9	...
$\widehat{X}$	1	–	–	x	y	w	x^2	xy	xw	x^3	...

$H_{\widehat{X}}$ is connected with the Young diagram; the gap corresponds to $\begin{smallmatrix}\square\\\square\end{smallmatrix}$. The $\widehat{X}$ is the simplest curve whose Weierstrass non-gap sequence at ∞ is the non-symmetric numerical semigroup $H_{\widehat{X}}$, and the infinity at $\widehat{X}$ is not the Weierstrass point of the curve. As in [MatK2013], we have the sigma function of $\widehat{X}$. By comparing the sigma function for $\widehat{X}$ and for the plane (3, 4)-curve, we can investigate the behavior of sigma of the limit of $s \to 0$ of curve $y^3 = x(x-s)(x-b_1)(x-b_2)$ [FeKMPA2022] (cf. Sect. 3.7).

Since every (n, s)-curve generates certain space curves by such a degenerating process, we should construct the sigma function theory of more general curves given by the Weierstrass canonical form rather than plane (n, s)-curves.

Here we briefly present the recent progress in the algebraic construction of the sigma function of the Weierstrass curves [KomMP2022a, KomMP2022b].

The set of W-curves represents the set of compact Riemann surfaces. It provides the projection $\varpi_r : X \to \mathbb{P} := \mathbb{P}^1$ $(e.g., (x, y) \mapsto x)$ as a covering space. For $R_X = \mathbf{H}^0(X, \mathcal{O}_X(*\infty))$ and $R_{\mathbb{P}} = \mathbf{H}^0(\mathbb{P}, \mathcal{O}_{\mathbb{P}}(*\infty))$, let these quotient fields be $\mathcal{Q}(R_X)$ and $\mathcal{Q}(R_{\mathbb{P}})$. Then $\mathcal{Q}(R_X)$ is a field extension of $\mathcal{Q}(\mathbb{P})$, or X is the r-sheeted holomorphic covering on $\mathbb{P}$, and R_X is regarded as an $R_{\mathbb{P}}$-module.

In the general theory of algebraic curves [DedW1882, Ku2004, Sti2008], the algebraic curves with a covering are studied well, and it turns out that the complementary module plays a critical role and is connected with the meromorphic one-forms. The origin of the complementary modules is in the study by Dedekind and Weber [DedW1882]; Weierstrass already stated some results in [Wei1875a]. Kunz showed the explicit description of the complementary module for every plane W-curve [Ku2004, Theorem 15.1], i.e., $R_X^{\mathfrak{c}} = \dfrac{1}{\partial f_X(x, y)/\partial y} R_X$, though he did not call it the Weierstrass curve. Further, it can be connected with the R_X [KomMP2022a, KomMP2022b]. The algebraic properties of R_X as an $R_{\mathbb{P}}$-module and the complementary module $R_X^{\mathfrak{c}}$ are revealed. Then we have $\mathbf{H}^0(X, \mathcal{A}_X(*\infty)) = R_X^{\mathfrak{c}} dx$. Here we assume that the sheaf holomorphic one-form on X is denoted by $\mathcal{A}_X$.

By investigating them, we found the essentials of the general constructions of $R_X^{\mathfrak{c}}$ in terms of data of R_X for general W-curves with the Weierstrass semigroup $H_X = \langle r = r_1, r_2, \ldots, r_{m_X} \rangle$ as follows. Since the W-curves X are characterized by the Weierstrass semigroups, H_X [Wei1867], the monomial curves $X_H = \operatorname{Spec} R_H$ and the numerical semigroup rings $R_H = \mathbb{C}[t^{r_1}, t^{r_2}, \ldots, t^{r_{m_X}}]$ associated with H_X determine the behavior of R_X at $\infty \in X$. Following such studies on the monomial curves by Kunz [Ku1970], Herzog [Her1970], and Pinkham [Pin1974], we use the multiple groups $\mathbf{G}_{\mathrm{m}}$ action on the monomial curve X_H and the monomial ring R_H^Z. As we assume that the generators of the Weierstrass semigroup H_X include the minimal number $r = r_1$, the cyclic group $\mathfrak{C}_r$ of the order r gives the standard basis of H_X with respect to r.

The standard basis $\{\mathfrak{e}_i\}$ of H_X, i.e., $H_X = \bigoplus_{i=0}^{r-1} r\mathbb{N}_0 \mathfrak{e}_i$ as a monoid, governs the monomial ring and induces the standard basis of R_X as the $R_{\mathbb{P}}$-module globally,

$$R_X = \bigoplus_{i=0}^{r-1} R_{\mathbb{P}} \mathfrak{y}_{\mathfrak{e}_i}, \quad \mathfrak{y}_{\mathfrak{e}_0} = 1, \quad \mathfrak{y}_{\mathfrak{e}_i} = \frac{1}{t^{\mathfrak{e}_i}}(1 + d_{>0}(t)), \text{ at } \infty \in X.$$

The standard basis of R_X enables us to investigate the $R_{\mathbb{P}}$-module structure of R_X. We find $m_X - 1$ identities $f_X^{(i)}(x, y_{r_i}) = 0$ in R_X as Weierstrass polynomials and thus we find a subring $R_X^{(i)} = \mathbb{C}[x, y_{r_i}]/(f_X^{(i)}(x, y_{r_i}))$ of R_X (cf. Proposition 2.23), which induces the injection $\iota_{r,r_j} : R_X^{(i)} \to R_X$ and the projection $\varpi_{r,r_i} : X \to X_i = \operatorname{Spec} R_X^{(i)}$, which is regarded as a singular plane curve. Roughly speaking, by aiding with the structure of $R_X^{(i)}$ as an $R_{\mathbb{P}}$-module and its analogue in R_H, we construct the finding of the dual basis $\widehat{\mathfrak{y}}_{\mathfrak{e}_i}$ in R_X with respect to the trace $\tau_{\widetilde{h}}$ such that $\tau_{\widetilde{h}}(\widehat{\mathfrak{y}}_{\mathfrak{e}_i}\mathfrak{y}_{\mathfrak{e}_j}) = \delta_{ij}$ (cf. Sect. 2.2.6). The complementary module $R_X^{\mathfrak{c}}$ is determined by a fractional ideal of R_X,

$$R_X^{\mathfrak{c}} = \frac{\langle \widehat{\mathfrak{y}}_{\mathfrak{e}_1}, \widehat{\mathfrak{y}}_{\mathfrak{e}_2}, \ldots, \widehat{\mathfrak{y}}_{\mathfrak{e}_{r-1}} \rangle_{R_X}}{h_X(x, y_{r_1}, \ldots, y_{r_{m_X}})},$$

where h_X is a certain element of R_X. Then we describe the holomorphic one-form except at ∞, which is given by

$$\mathbf{H}^0(X, \mathcal{A}_X(*\infty)) = R_X^{\mathfrak{c}} dx = \bigoplus_{i=0, j=0}^{r-1, \infty} \frac{x^j \widehat{\mathfrak{y}}_{\mathfrak{e}_i} dx}{h_X(x, y_{r_1}, \ldots, y_{r_{m_X}})}.$$

Further, the holomorphic one-forms, or the Abelian differentials of the first kind $\{\nu_i^{\mathrm{I}}\}_{i=1,\ldots,g}$, are given by its subset.

The trace $\tau_{\widetilde{h}}$ defined in $R_X \otimes_{R_{\mathbb{P}}} R_X$ can be extended to the trace operator $p_\varpi \in \mathcal{Q}(R_X \otimes_{\mathbb{C}} R_X)$, the quotient field of $R_X \otimes_{\mathbb{C}} R_X$; p_ϖ is holomorphic for $(X \setminus \{$ branch points$\}) \times (X \setminus \{\infty\})$ such that $p_\varpi(P, Q) = 1$ for $P = Q$, $= 0$ for $P \neq Q$ and $\varpi(P) = \varpi(Q)$, which was a crucial obstacle at the end of the 19th century. However, it is turned out that the uniqueness of the extension is justified in the meaning of the cohomology.

Then Weierstrass' H (1.51) and (1.38) are generalized to every W-curve (cf. Proposition 2.120),

$$\Sigma(P, Q) = \frac{p_\varpi(P, Q) dx_P}{(x_P - x_Q) h_X(P)},$$

for $P, Q \in X$. By investigating its differential $d_Q \Sigma(P, Q)$, we have the generalization of Weierstrass' G (1.51) and (1.40),

$$\Omega(P, Q) := d_Q \Sigma(P, Q) + \sum_{i=1}^{g} \nu_i^{\mathrm{I}}(P) \otimes \nu_i^{\mathrm{II}}(Q),$$

where $\nu_i^{\mathrm{II}} \in \mathbf{H}^0(X, \mathcal{A}_X(*\infty))$ is the Abelian differentials of the second kind satisfying

$$\oint_\infty \left(\int_\infty^P \nu_i^{\mathrm{I}}(Q) \right) \nu_j^{\mathrm{II}}(P) = \delta_{ij}.$$

We also introduce the Abelian integral of the third kind, $\Pi_{Q_1, Q_2}^{P_1, P_2} := \int_{P_2}^{P_1} \int_{Q_2}^{Q_1} \Omega(P, Q)$, which is a generalization of (1.21). By integrating ν^{I} and ν^{II} over the homology basis of X, we have the period matrices (ω', ω'') and (η', η''); it is easy to find that they satisfy the generalized Legendre relation (cf. Proposition 2.148).

On the other hand, since the canonical divisor K_X of a space W-curve of genus g is not $(2g-2)\infty$ in general and for the case where the Riemann constant does not correspond to a point of a half lattice of $\Gamma_X := \langle \omega', \omega'' \rangle_{\mathbb{Z}}$. It can be redressed by shifting the Riemann constant ξ_X and the Abelian integrals. For the abelian covering of $\widetilde{X}$ of X and the Abelian integral $\widetilde{w} : S^k\widetilde{X} \to \mathbb{C}^g$ $(k = 1, 2, \ldots, g)$, we introduce the shifted Abelian integral $\widetilde{w}_{\mathfrak{s}}(\cdot) := \widetilde{w}(\cdot) + \widetilde{w}(\mathfrak{K}_{\mathfrak{s}})$ where $K_X = 2\mathfrak{K}_{\mathfrak{s}} + (2g - 2 - \deg(\mathfrak{K}_{\mathfrak{s}}))\infty$, which corresponds to the spin connection on X [At1971, KorS2012], and the shifted Riemann constant $\xi_{\mathfrak{s}} = \xi_X - \widetilde{w}(\mathfrak{K}_{\mathfrak{s}})$, which corresponds to a theta characteristic δ in the half lattice $\frac{1}{2}\Gamma_X$ [KomMP2016].

Following Ônishi's refinement of the Riemann–Kempf theory [On2005] and using the Sato theory on the universal Grassmannian manifold, Nakayashiki defined the sigma function of a pointed compact Riemann surface whose Weierstrass non-gap sequence is given by H_X. We modify Nakayashi's sigma function to define σ as an entire function of (a column-vector) $u = {}^t(u_1, u_2, \ldots, u_g) \in \mathbb{C}^g$ for δ so that it holds the spin structure proposed by Korotkin and Shramchenko [KorS2012];

$$\sigma(u) = \sigma(u; M) := \frac{\varepsilon_{\Lambda,\natural_{X,0}} \exp(-\frac{1}{2}\,{}^t u\eta'\omega'^{-1}u)\vartheta\begin{bmatrix}\delta''\\ \delta'\end{bmatrix}(\frac{1}{2}\omega'^{-1}u;\ \omega'^{-1}\omega'')}{\partial_{\natural_{X,0}}\vartheta\begin{bmatrix}\delta''\\ \delta'\end{bmatrix}(\frac{1}{2}\omega'^{-1}u;\ \omega'^{-1}\omega'')\Big|_{u=\widetilde{w}(\iota_X\mathfrak{K}_{\mathfrak{s}})}}$$
$$= \frac{\varepsilon_{\Lambda,\natural_{X,0}} \exp(-\frac{1}{2}\,{}^t u\eta'\omega'^{-1}\,u)}{\partial_{\natural_{X,0}}\vartheta\begin{bmatrix}\delta''\\ \delta'\end{bmatrix}(\frac{1}{2}\omega'^{-1}u;\ \omega'^{-1}\omega'')|_{u=\widetilde{w}(\iota_X\mathfrak{K}_{\mathfrak{s}})}}$$
$$\times \sum_{n\in\mathbb{Z}^g} \exp\left[\pi\sqrt{-1}\{\,{}^t(n+\delta'')\omega'^{-1}\omega''(n+\delta'') + {}^t(n+\delta'')(\omega'^{-1}u + 2\delta')\}\right], \tag{1.59}$$

where $\varepsilon_{\Lambda,\natural_{X,0}}$ is defined in (3.16) and $\natural_{X,0}$ is defined in Definition 3.18.

It is shown that $\sigma(u)$ has the following properties (cf. Chap. 3).

Theorem 1.1 *The sigma function has the following properties:*

1. *It is modular invariant.*
2. *It obeys the translational formula; for* $u, v \in \mathbb{C}^g$, *and* $\ell\ (= 2\omega'\ell' + 2\omega''\ell'') \in \Gamma_X$), *if we define* $L(u, v) := 2\,{}^t u(\eta' v' + \eta'' v'')$, $\chi(\ell) := \exp[\pi\sqrt{-1}(2({}^t\ell'\delta'' - {}^t\ell''\delta') + {}^t\ell'\ell'')]$, *the following holds*
$$\sigma(u + \ell) = \sigma(u)\exp(L(u + \frac{1}{2}\ell, \ell))\chi(\ell). \tag{1.60}$$
3. *Its divisor is* $\kappa_J^{-1}\Theta_{X,\mathfrak{s}} \subset \mathbb{C}^g$, *where* $\Theta_{X,\mathfrak{s}} := \kappa_J\widetilde{w}_{\mathfrak{s}}(S^{g-1}\widetilde{X}) \subset J_X$.
4. *It satisfies the Jacobi–Riemann fundamental relation; for* $(P, Q, P_i, P_i') \in X^2 \times (S^g(X) \setminus S_1^g(X)) \times (S^g(X) \setminus S_1^g(X))$,
$$u := \widetilde{w}_{\mathfrak{s}}(P_1, \ldots, P_g), \quad v := \widetilde{w}_{\mathfrak{s}}(P_1', \ldots, P_g'),$$

$$\exp\left(\sum_{i,j=1}^{g} \Pi_{P_i,P'_j}^{P,Q}\right) = \frac{\sigma(\widetilde{w}(P)-u)\sigma(\widetilde{w}(Q)-v)}{\sigma(\widetilde{w}(Q)-u)\sigma(\widetilde{w}(P)-v)}$$
$$= \frac{\sigma(\widetilde{w}(P)-\widetilde{w}_{\mathfrak{S}}(P_1,\cdots,P_g))\sigma(\widetilde{w}(Q)-\widetilde{w}_{\mathfrak{S}}(P'_1,\cdots,P'_g))}{\sigma((\widetilde{w}(Q)-\widetilde{w}_{\mathfrak{S}}(P_1,\cdots,P_g))\sigma(\widetilde{w}(P)-\widetilde{w}_{\mathfrak{S}}(P'_1,\cdots,P'_g))},$$

which generates the Jacobi inversion formulae for $S^k X$.[23]

5. *The leading term in the Taylor expansion of the* σ *function associated with* X *is expressed by a Schur function,*

$$\sigma(u+\widetilde{w}(\iota_X\mathfrak{K}_{\mathfrak{S}})) = S_\Lambda(T)|_{T_{\Lambda_i+g-i}=u_i} + \sum_{|\mathrm{w}_g(\alpha)|>|\Lambda_X|} a_\alpha u^\alpha,$$

where $a_\alpha \in \mathbb{Q}[b_1,\cdots,b_{s+r}]$, $\alpha := (\alpha_1,\ldots,\alpha_g)$, $u^\alpha := u_1^{\alpha_1}\cdots u_g^{\alpha_g}$, *and* $\mathrm{w}_g(\alpha) = \sum \alpha_i \mathrm{wt}(u_i)$, *(*$S_\Lambda(T)$ *is the Schur polynomial associated with the numerical semigroup* H_X, *and the lowest-order term in a certain degree of the* u_i; $\sigma(u)$ *is homogeneous* $|\Lambda|$ *with respect to the other certain degree).*

6. $\sigma(-u) = \pm\sigma(u)$.
7. *If the Galois group* G_X *on* X, $\widehat{\zeta} \in G_X$, *satisfies* $\widehat{\zeta}\Gamma_X = \Gamma_X$, $\widehat{\zeta}^\ell = \mathrm{id}$, *and* $\underset{\frown}{\zeta} w(\mathfrak{K}_{\mathfrak{S}}) = w(\mathfrak{K}_{\mathfrak{S}})$, *the action gives the one-dimensional representation such that*

$$\widehat{\zeta}\sigma(u) = \rho_{\widehat{\zeta}}\sigma(u),$$

where $\rho_{\widehat{\zeta}}^\ell = 1$.

This is a natural extension of Weierstrass' sigma function of genus one and Baker's reformulation of the Klein sigma function [Ba1897]. Though Klein's approach for the sigma does not use the data of algebraic curves, Weierstrass–Baker's approach gives the connection between the algebraic and the transcendental. Since Korotkin and Shramchenko, and Nakayashiki generalized Klein's sigma function, we should add the algebraic meaning to their sigma functions as Baker interpreted Klein's sigma function in an algebraic framework following Weierstrass' manner. As the author of this book with Previato and Komeda provided the algebraic interpretations of the sigma functions [KomMP2022a, KomMP2022b], the role of Weierstrass' sigma function of general W-curves can be illustrated in Fig. 1.9.

As mentioned in Remark 2.136, in this framework, we are concerned with the first de Rham cohomology $\mathbf{H}^1_{\mathrm{DR}}(X \setminus \{\infty\})$ rather than $\mathbf{H}^1_{\mathrm{DR}}(X)$. The former is generated by $\{\nu_i^{\mathrm{I}}, \nu_i^{\mathrm{II}}\}_{i=1,\ldots,g}$. In contrast, the latter is by $\{\nu_i^{\mathrm{I}}, \overline{\nu_i^{\mathrm{I}}}\}_{i=1,\ldots,g}$ and is the base of Hodge's theory on the compact Riemann surfaces refining Riemann's approach [Gri1989]. Both are isomorphic as $\mathbb{C}$ vector spaces [Del1989]. As the Hurwitz problem, it is critical whether the curve X given by (1.48) is birational to a compact Riemann

[23] This formula is a direct generalization of (1.42). Further, footnotes 4 and 11 show that this origin is Huygens's tractrix.

surface.[24] If so, the curve X can be projectivized and embedded in a projective space (cf. Examples 2.241 and 2.243). However, the affine part plays a crucial role in Weierstrass' approach as in his elliptic function theory; the projective curve actively does not take advantage of that property in the theory.

Contrasting to Hodge's theory in which the Hermitian form is focused, in this framework, the symplectic structure given by the Legendre relation is critical; it recovers the Hermitian structure (cf. Remark 3.184). This symplectic structure is reduced to $\left[\frac{d}{dx}, x\right] = \frac{d}{dx}x - x\frac{d}{dx} = 1$, which is the fundamental relation in Sato's theory of the universal Grassmannian manifold (cf. Appendix 3.9).

Adopting Weierstrass' approach instead of Riemann's allows for affine ring aggressive data handling and provides the scheme shown in Fig. 1.9. "Global" use of data from the affine ring characterizes Weierstrass' approach and shows its advantage. In its applications, the use allows for an algebraic, minute, and elaborate handling of the transcendent functions if necessary. The sigma function guarantees the existence of both arrows connecting both spaces.

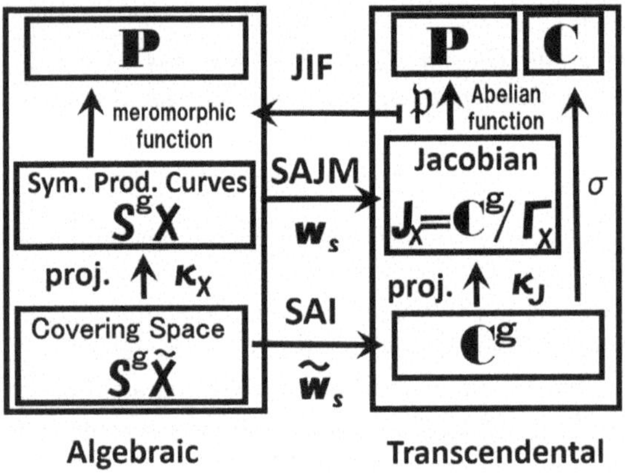

Fig. 1.9 Algebraic versus transcendental for general W curves

As Weierstrass criticized Riemann's approach and insisted on the algebraic ways associated with Weierstrass curves, we emphasize that the above properties of the sigma function imply completing his program in a certain sense. Poincaré finalized his essay on Weierstrass [Po1910] by citing Weierstrass' words in the letter mentioned in Sect. 1.6.

In this book, we demonstrate the Abelian function theory following Weierstrass' intention, including his sigma function for Weierstrass curves that correspond to

[24] Since data of $X \setminus \{\infty\}$ cannot recover data of X in general, it is undoubtedly the reason why Hodge, who Baker influenced, excluded Weierstrass' approach to construct the general theory. However, when the affine data recovers the global data of X, it is natural to employ the vision and investigate the properties of individual affine curves $X \setminus \{\infty\}$ and their meromorphic functions.

certain compact Riemann surfaces in terms of the data of the curves; it might trace an aspect of Baker's roles for the hyperelliptic sigma function. Chapter 2 is devoted to the investigations of the meromorphic function and differentials of the Weierstrass curves, and Chap. 3 is on the sigma function. We remark that based on the modern algebraic theory, we will construct them. Thus, we use the facts, including the results obtained by the transcendental investigations, from a rigorous Weierstrass viewpoint. It implies that it might not be the complete realization of Weierstrass' goal. We say that, at last, we stand at the starting line.

However, as mentioned in the beginning of Sect. 1.9, we regard that the recent movements on the sigma function aim to construct the Abelian function theory of higher genus that is comparable to, and compatible with, the Weierstrass elliptic function theory. We believe that it is time to summarize the recent achievements following Weierstrass' intention, though there are already excellent reviews [BelE2001, BelE2002, BucEL1997b, BucEL2012, On2013].

For this reason, we have presented his intention in this chapter.

Chapter 4 displays the applications of the sigma function theory to the integrable systems. As mentioned, the Abelian function theory started from the problem of elastica as a problem of the building material. One of the modern problems related to elastica is to find the shape of DNA, which is closely related to the problem in biophysics and modern medical technology. We will discuss the problem of the shape of DNA at the end of Chap. 4. To settle the problem beyond Euler's results, the reconstruction of the Abelian function theory following the Weierstrass program is required, which is the book's central theme. The author of this book with Emma Previato planned this research project in 2004 and have been working on it for two decades; as they took a step forward in 2022, we report on it there [MatP2016, MatP2022, DAGJMMRT2023, Mat2024a].

Chapter 2
Curves in Weierstrass Canonical Form (W-Curves)

This chapter introduces the Weierstrass curves and demonstrates their properties, which are higher genus analogue of the curves of Weierstrass standard form in the Weierstrass elliptic function theory.

Section 2.1 lists the conventions and notations used in this book.

The Weierstrass curve X is a smooth curve given by the Weierstrass canonical form, $y^r + A_1(x)y^{r-1} + A_2(x)y^{r-2} + \cdots + A_{r-1}(x)y + A_r(x) = 0$ where each A_j is a polynomial in x of a certain degree and a positive integer r so that it has a single ∞ and the Weierstrass non-gap sequence at $\infty \in X$ is given by the numerical semigroup H_X whose minimal generator is r. It has a natural projection $\varpi_r : X \to \mathbb{P}$ as a covering space. Let $R_X := \mathbf{H}^0(X, \mathscr{O}_X(*\infty))$ and $R_{\mathbb{P}} := \mathbf{H}^0(\mathbb{P}, \mathscr{O}_{\mathbb{P}}(*\infty))$. We have the explicit description of the complementary module $R_X^{\mathfrak{c}}$ of the $R_{\mathbb{P}}$-module R_X and the trace structure in Sect. 2.2. They supply the explicit expressions of the holomorphic one-form except at ∞, $\mathbf{H}^0(\mathbb{P}, \mathscr{A}_{\mathbb{P}}(*\infty))$ and the fundamental 2-form of the second kind in Sect. 2.4. Section 2.3 shows some examples of the W-curves. Section 2.5 introduces the differential operators, which are crucial in Chap. 4. Further, Sect. 2.6 describes the addition structure based on the linear equivalence in R_X and the division polynomials in R_X. Section 2.7 mentions the period integrals of W-curves. Section 2.8 provides covering spaces of typical W-curves which are related to al functions used in Chaps. 3 and 4. As an example of the degenerating family of W-curves, we discuss a particular degeneration process of the cyclic (3,4)-curve in Sect. 2.9. Appendix 2.10 recalls the normalization of the curves, Appendix 2.11 reviews the complementary modules of the extension of the ring, and Appendix 2.12 shows a short summary of the deformation of curves.

S. Matsutani, *The Weierstrass Sigma Function in Higher Genus and Applications to Integrable Equations*, Springer Monographs in Mathematics,
https://doi.org/10.1007/978-981-97-8163-8_2

2.1 Conventions and Notations

Before prepare to introduce the Weierstrass curves, we describe our conventions in this book.

In this book, we loosely deal with smooth connected complex projective algebraic curves X as complex analytic spaces and as algebraic varieties based on the Zariski topology, following the GAGA principle [Ser1989, Theorem 6.1.1].[1]

We use the symbols $\mathbb{N}_0$ as the set of non-negative integers and $\mathbb{N}$ as positive integers; further, $\mathbb{C}$ denotes the complex number field, $\mathbb{R}$, the real number field, and $\mathbb{Z}$, the ring of (rational) integers. $\mathbb{P}$ and $\mathbb{P}^n$ denote the projection line and the n-dimensional projective space, respectively, i.e., $\mathbb{P} = \mathbb{P}^1$ but $\mathbb{P}^n \neq (\mathbb{P}^1)^n$.

The set of the $\ell \times m$ matrices over k is denoted by $\mathrm{Mat}_k(\ell, m)$; $\mathrm{Mat}_k(\ell) := \mathrm{Mat}_k(\ell, \ell)$.

For a ring R, we let $\mathscr{Q}(R)$ be its quotient field.

Let $\mathrm{Div}(X)$ be a formal finite additive group over X defined by the divisor D [Mir1995],

$$D = \sum_{P \in X} a_P \cdot P, \quad a_P \in \mathbb{Z}$$

such that $|\{a_P \mid a_P \neq 0\}| < \infty$. The degree of $\mathrm{Div}(X)$ is given as

$$\deg : \mathrm{Div}(X) \to \mathbb{Z}, \quad \deg\left(\sum_{P \in X} a_P \cdot P\right) = \sum_{P \in X} a_P.$$

The kernel of deg, a subgroup of $\mathrm{Div}(X)$, is denoted by $\mathrm{Div}^0(X)$. Let $\mathscr{M}_X(X)$ be the set of meromorphic functions over X and $\mathscr{M}_X(X)^\times := \mathscr{M}_X(X) \setminus \{0\}$. For a meromorphic function $f \in \mathscr{M}_X(X)^\times$, we let $\deg_{P,0}(f)$ is the degree of zero at P such that if P is not a zero of f, we set $\deg_{P,0} f = 0$, and $\deg_{P,\infty} f$ is the degree of singularity at P such that if P is not a singular point of f, we set $\deg_{P,\infty}(f) = 0$. We define the following:

$$\mathrm{div}_0(f) = (f)_0 := \sum_{P \in X} \deg_{P,0}(f) \cdot P,$$

$$\mathrm{div}_\infty(f) = (f)_\infty := \sum_{P \in X} \deg_{P,\infty}(f) \cdot P,$$

and

$$(f) := \mathrm{div}(f) := (f)_0 - (f)_\infty.$$

[1] The related matters associated with the algebraic curves as topics in this book, including the complex topology and Zariski topology, are reviewed well for the same purpose in "§0. Review of background in algebraic geometry" of Mumford's book [Mum1984, §0]. If necessary, the book should be referred to.

It is known that $\mathrm{div}(f)$ belongs to $\mathrm{Div}^0(X)$, referred to as the *principal* divisor. Let $\mathrm{PDiv}(X) := \{\mathrm{div}(f) \mid f \in \mathscr{M}_X^{\times}(X)\}$.

Two divisors D_1 and D_2 on X are said to be *linearly equivalent*, written $D_1 \sim D_2$ if their difference is a principal divisor, i.e., if their difference is the divisor of a meromorphic function on X. It induces the principal divisor class,

$$\mathrm{Pic}^0(X) := \mathrm{Div}^0(X)/\mathrm{PDiv}(X). \tag{2.1}$$

Further, K_X means the canonical divisor on X.

For an algebraic variety Z, the i-th homology over Z with the coefficients $\mathscr{K}$ is denoted by $\mathbf{H}_i(Z, \mathscr{K})$, whereas the cohomology is written as $\mathbf{H}^i(Z, \mathscr{K})$. We let $\mathscr{O}_Z$ denote the sheaf of the holomorphic functions over Z, and $\mathscr{A}_Z$, the sheaf of the holomorphic one-forms over Z. The sheaf cohomology over $U \subset Z$ of a sheaf $\mathscr{F}_Z$ is written as $\mathbf{H}^i(U, \mathscr{F}_Z)$; we set $\mathscr{F}_Z(U) := \mathbf{H}^0(U, \mathscr{F}_Z)$. We sometimes use the holomorphic line bundle of the degree d denoted by $\mathscr{L}_Z^d$. For two effective divisors D and D', let $\mathscr{L}_Z^d(D - D')$ denote the sheaf given by germs of meromorphic sections of $\mathscr{L}_Z^d$ whose poles at most at D and zeros are at least at D': $\mathscr{O}_Z$ is identified with $\mathscr{L}_Z^0(0)$. Further, for two divisors D and D' in X, we let $\mathscr{L}_Z^d(D)(D') := \mathscr{L}_Z^d(D + D')$. For a divisor D of X,

$$\mathbf{H}^0(U, \mathscr{O}_Z(D)) = \{f \in \mathscr{M}_Z(U) \mid (f) + D \geq 0\}.$$

Thus $\mathbf{H}^0(U, \mathscr{O}_Z(D)) = \mathbf{H}^0(U, \mathscr{L}_Z(D))$. Similarly, we have $\mathbf{H}^0(U, \mathscr{A}_Z(D))$.

Since the symbols h, ℓ, and L are used for some tasks, we express, for example, the Riemann–Roch theorem for a compact Riemann surface X as

$$\dim \mathbf{H}^0(X, \mathscr{O}_X(D)) - \dim \mathbf{H}^0(X, \mathscr{O}_X(K_X - D)) = \deg D - g + 1.$$

Further, for $P \in Z$, $\mathscr{L}_Z^d(*P)$ denotes the sheaf of germs of meromorphic sections of $\mathscr{L}_Z^d$ which are holomorphic except at P, and $\mathscr{O}_Z(*P) = \mathscr{L}_Z^0(*P)$: Since for $P \in Z$ and an open subset $U \subset Z$, $\mathbf{H}^0(U, \mathscr{O}_Z(nP))$ is a sub-space in $\mathbf{H}^0(U, \mathscr{O}_Z((n+1)P))$ as a $\mathbb{C}$-vector space, $\mathbf{H}^0(U, \mathscr{O}_Z(*P))$ is also regarded as the inductive limit, $\mathbf{H}^0(U, \mathscr{O}_Z(*P)) = \varinjlim \mathbf{H}^0(U, \mathscr{O}_Z(nP))$. Similarly, we have $\mathbf{H}^0(U, \mathscr{A}_Z(*P)) = \varinjlim \mathbf{H}^0(U, \mathscr{A}_Z(nP))$.

2.2 Weierstrass Canonical Form

In this section, we introduce natural generalizations of the Weierstrass standard form of the elliptic curves to the general compact Riemann surfaces, called W-curves, given by the Weierstrass canonical form and show their algebraic properties following [KomMP2022a].

2.2.1 Numerical and Weierstrass Semigroup

This subsection is on numerical, and Weierstrass semigroups based on [ADGS2016, KomMP2013, KomMP2016, KomMP2019, KomM2020, RosGS2009]. An additive sub-monoid of the monoid, non-negative integers $\mathbb{N}_0$ is called the *numerical semigroup* if its complement in $\mathbb{N}_0$ is a finite set. In this subsection, we review the numerical semigroups associated with algebraic curves.

Let X be a smooth (complex projective) curve of genus g. For a point $P \in X$,

$$H(X, P) := \{n \in \mathbb{N}_0 \mid \text{there exists } f \in \mathscr{M}_X(X) \text{ such that } (f)_\infty = nP\}$$

forms a numerical semigroup by the Riemann–Roch theorem, which is called the Weierstrass semigroup at the point P. If the Weierstrass gap sequence $H^{\mathfrak{c}}(X, P) := \mathbb{N}_0 \setminus H(X, P)$ differs from the set $\{1, 2, \cdots, g\}$, we say that P is a Weierstrass point of X [FarK1991]. (Later by fixing $P = \infty$, we denote them by H_X and $H_X^{\mathfrak{c}}$ respectively.)

In this book, we consider a *pointed curve*, a pair (X, P) with P a point of the curve X with the Weierstrass semigroup $H(X, P)$.

In general, a numerical semigroup H has a unique (finite) minimal set of generators, $M = M(H)$, $(H = \langle M \rangle)$ and the finite cardinality g of $H^{\mathfrak{c}} = \mathbb{N}_0 \setminus H$; g is the genus of H or $H^{\mathfrak{c}}$ and $H^{\mathfrak{c}}$ is called a gap-sequence in this book. For example,

$$\begin{aligned}
&H^{\mathfrak{c}} = \{1, 2, 4, 5\}, \ \text{for } H = \langle 3, 7, 8\rangle, \\
&H^{\mathfrak{c}} = \{1, 2, 3, 4, 6, 8, 9, 13\}, \ \text{for } H = \langle 5, 7, 11\rangle, \ \text{and} \\
&H^{\mathfrak{c}} = \{1, 2, 3, 4, 5, 7, 8, 9, 10, 11, 17, 23\}, \ \text{for } H = \langle 6, 13, 14, 15, 16\rangle.
\end{aligned} \tag{2.2}$$

We let $r_{\min}(H)$ be the smallest positive integer of $M(H)$, called *multiplicity* of H. We call the semigroup H an $r_{\min}(H)$*-semigroup*, so that $\langle 3, 7, 8\rangle$ is a 3-semigroup and $\langle 6, 13, 14, 15, 16\rangle$ is a 6-semigroup. Let $N(i)$ and $N^{\mathfrak{c}}(i)$ be the i-th ordered element of $H = \{N(i) \mid i \in \mathbb{N}_0\}$ and $H^{\mathfrak{c}} = \{N^{\mathfrak{c}}(i) \mid i = 0, 1, \ldots, g-1\}$ satisfying $N(i) < N(i+1)$, and $N^{\mathfrak{c}}(i) < N^{\mathfrak{c}}(i+1)$ respectively. The Schubert index of the set $H^{\mathfrak{c}}$ is

$$\mathfrak{a}(H) := \{\mathfrak{a}_0(H), \mathfrak{a}_1(H), \ldots, \mathfrak{a}_{g-1}(H)\}, \tag{2.3}$$

where $\mathfrak{a}_i(H) := N^{\mathfrak{c}}(i) - i - 1$, $(i = 1, 2, \ldots, g)$ [EisH1987]. Moreover,

$$\mathfrak{a}(\langle 3, 7, 8\rangle) = \{0^2, 1^2\} \quad \text{and} \quad \mathfrak{a}(\langle 6, 13, 14, 15, 16\rangle) = \{0^5, 1^5, 6, 11\}.$$

When an $\mathfrak{a}_i$ is repeated $j > 1$ times we write $\mathfrak{a}_i^j(H)$ in $\mathfrak{a}(H)$. We set $\mathrm{wt}_H(H^{\mathfrak{c}}) := \sum_{i=0}^{g-1} \mathfrak{a}_i(H)$, and refer to it as the *weight* of $H^{\mathfrak{c}}$.

Further, the conductor c_H of H is defined by the minimal natural number satisfying $c_H + \mathbb{N}_0 \subset H$ [RosGS2009]. The number $c_H - 1$ is known as the Frobenius number due to the study of Frobenius (cf. Sect. 1.7); the Frobenius number is the largest element of $H^{\mathfrak{c}}$, i.e., $c_H = N^{\mathfrak{c}}(g-1)+1$.

By letting the row lengths be $\Lambda_i = \mathfrak{a}_{g-i}(H)+1$, $(i=1,2,\cdots,g)$, we have the Young diagram of the semigroup, $\Lambda := (\Lambda_1,\dots,\Lambda_g)$, $(\Lambda_i \ge \Lambda_{i+1})$. The Young diagram Λ is a partition of $|\Lambda| := \sum_{i=1}^{g} \Lambda_i$. We say that such a Young diagram is associated with the numerical semigroup. If for a given Young diagram Λ, we cannot find any numerical semigroups H such that $\Lambda_i = \mathfrak{a}_{g-i}(H)+1$, we say that Λ is not associated with the numerical semigroup. It is obvious that in general, the Young diagrams are not associated with the numerical semigroups. We show the examples of the Young diagrams associated with numerical semigroup:

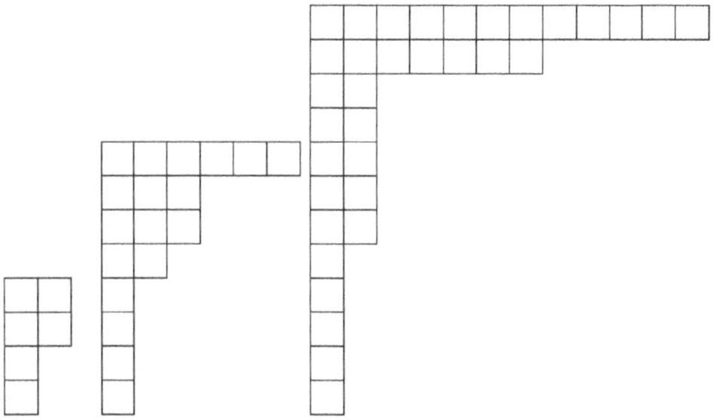

$\langle 3,7,8\rangle$, $\langle 5,7,11\rangle$, $\langle 6,13,14,15,16\rangle$. (2.4)

The Young diagram and the associated numerical semigroup are called symmetric if the Young diagram is invariant under reflection across the main diagonal. It is known that the numerical semigroup is symmetric if and only if $2g-1$ occurs in the gap sequence. It means that if $c_H = 2g$, H is symmetric.

We obviously have the following proposition:

Proposition 2.1 *The following hold:*

1. $N(n) - n \le g$ *for every* $n \in \mathbb{N}_0$,
2. $N(n) - n = g$ *for* $N(n) \ge c_X = N(g)$ *or* $n \ge g$,
3. $N(n) - n < g$ *for* $0 \le N(n) < c_X$ *or* $n < g$,
4. $\#\{n \mid N^{\mathfrak{c}}(n) \ge g\} = \#\{n \mid N(n) < g\}$,
5. *for* $N(i) < N^{\mathfrak{c}}(j)$, $N^{\mathfrak{c}}(j) - N(i) \in H^{\mathfrak{c}}$,
6. *when* H *is symmetric,* $c_H = N(g) = 2g$ *and* $c_H - N(i) - 1 = N^{\mathfrak{c}}(g-i-1)$ *for* $0 \le i \le g-1$, *and*
7. $N^{\mathfrak{c}}(0) = 1$.

Proof *1–3*, *5* and *7* are obvious. By noting $\#H^{\mathfrak{c}} = g$, *4* means that what is missing must be filled later for $H^{\mathfrak{c}}$. *6* is left to [ADGS2016]. ∎

The length r_Λ of the diagonal of the Young diagram Λ is called the rank of Λ. The number of boxes below and to the right of the i-th box of the diagonal from lower right to upper left are assumed $a_{r_\Lambda-i+1}$ and $b_{r_\Lambda-i+1}$ respectively. The Young diagram is represented by $(a_{r_\Lambda}, \dots, a_2, a_1; b_{r_\Lambda}, \dots, b_2, b_1)$, which is known as the Frobenius representation or characteristics of Λ. We employ a modified version of the original convention in [FuH1991, Sect. 4.1, p. 51] such that $a_i \geq a_{i+1}$ and $b_i \geq b_{i+1}$. Then $\ell_i := a_i + b_i + 1$ is called the hook length of the characteristics. For $\Lambda = (6, 3, 3, 2, 1, 1, 1, 1)$ associated with $H = \langle 5, 7, 11\rangle$, it is $(0, 2, 7; 0, 1, 5)$ and rank of Λ is three. For $\Lambda = (6, 3, 3, 3, 1, 1, 1, 1)$ associated with $H = \langle 5, 6, 14\rangle$, it is $(1, 2, 7; 0, 1, 5)$ and rank of Λ is three. We show their examples of the Young diagram:

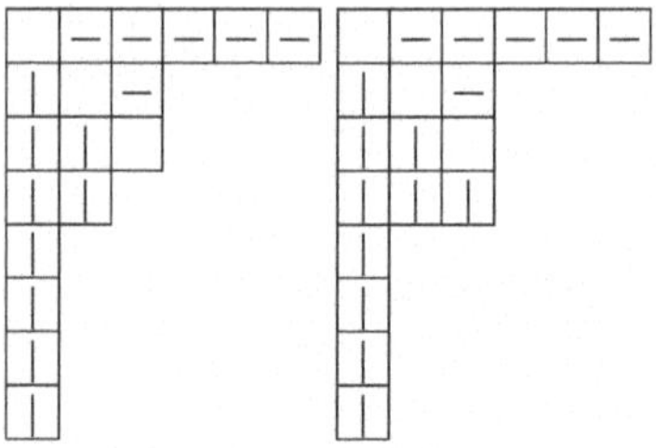

$(0, 2, 7; 0, 1, 5)$, $(1, 2, 7; 0, 1, 5)$,
$\langle 5, 7, 11\rangle$, $\langle 5, 6, 14\rangle$. (2.5)

Definition 2.2 For a given Young diagram $\Lambda = (\Lambda_1, \Lambda_2, \cdots, \Lambda_g)$, we define

$$\Lambda^{(k)} := (\Lambda_1, \Lambda_2, \cdots, \Lambda_k), \quad \Lambda^{[k]} := (\Lambda_{k+1}, \Lambda_{k+2}, \cdots, \Lambda_g).$$

Let g_Λ be the number of the rows of Λ, i.e., $g_{\Lambda^{(k)}} = k$ and $g_{\Lambda^{[k]}} = g - k$.

We show the properties of the Young diagram associated with the numerical semigroup H in the following lemma, which is geometrically obvious:

Lemma 2.3 *1. If we put the number on the boundary of the Young diagram Λ from the lower to upper right as in (2.6), each number in the right-hand side box of the i-th row corresponds to the gap number in $N^{\mathfrak{c}}(g-i) \in H^{\mathfrak{c}}$ or $\Lambda_i + g - i$, i.e., $H^{\mathfrak{c}} \ni N^{\mathfrak{c}}(g-i) = \Lambda_i + g - i$, for $0 < i \leq g$, and each number in the top numbered box of the i-th column corresponds to $N(i)$ for $0 < i < \Lambda_1$.*

2. The hook length $\ell_i = a_i + b_i + 1$ is given by $\ell_i = N^{\mathfrak{c}}(g-i) - N(i-1)$ $(i = 1, \dots, r_\Lambda)$, which belongs to $H^{\mathfrak{c}}$, and thus let $L_i \in \{0, 1, \dots, g-1\}$ such that $N^{\mathfrak{c}}(g - L_i) = \ell_i$. Then $\Lambda_{L_i} + g - L_i = \ell_i$, and $L_1 = 1$.

3. For $\Lambda := (\Lambda_1, \dots, \Lambda_g)$ associated with the numerical semigroup $H := \langle M\rangle$, the truncated Young diagram $\Lambda^{[k]} := (\Lambda_{k+1}, \dots, \Lambda_g)$ is also associated with a numerical semigroup H', i.e., H' is generated by

$$M \cup \{N^{\mathfrak{c}}(g-k), N^{\mathfrak{c}}(g-k+1), \dots, N^{\mathfrak{c}}(g-1)\},$$

though the generators are not minimal in general. Further, its complement H'^{c} is a subset of H^{c}, i.e., $H'^{\mathrm{c}} \subset H^{\mathrm{c}}$.

4. *For $\Lambda := (a_{r_\Lambda}, \dots, a_2, a_1; b_{r_\Lambda}, \dots, b_2, b_1)$ associated with the numerical semigroup H, $\Lambda[\ell] := (a_{r_\Lambda}, \dots, a_{\ell+1}, a_\ell; b_{r_\Lambda}, \dots, b_{\ell+1}, b_\ell)$, and then $\Lambda[\ell]$ is, in general, not associated with the numerical semigroup unless $\ell = 1$. Further, $g_{\Lambda[\ell+1]} \le g_{\Lambda[\ell]}$. Moreover, we have $|\Lambda| = \sum_{i=1}^{r_\Lambda}(a_i + b_i + 1) = \sum_{i=1}^{r_\Lambda} \ell_i$, and for a symmetric Λ, if r_Λ is odd (even) if and only if $|\Lambda|$ is odd (even).*

$\Lambda^{[1]}$ of the numerical semigroup $\langle 5, 7, 11\rangle$ is associated with the numerical semigroup $\langle 5, 7, 11, 13\rangle$. Further, $\{1, 2, \dots, c_H - 1\}$ are embedded in Λ as follows.

$$\underset{\langle 3,7,8\rangle}{\begin{array}{|c|c|}\hline & 5 \\ \hline 3 & 4 \\ \hline 2 \\ \cline{1-1} 1 \\ \cline{1-1}\end{array}}, \quad \underset{\langle 5,7,11,13\rangle}{\begin{array}{|c|c|c|}\hline & & 9 \\ \hline & 7 & 8 \\ \hline 5 & 6 \\ \cline{1-2} 4 \\ \cline{1-1} 3 \\ \cline{1-1} 2 \\ \cline{1-1} 1 \\ \cline{1-1}\end{array}}, \quad \underset{\langle 6,10,11,13,14,15\rangle}{\begin{array}{|c|c|}\hline & 9 \\ \hline & 8 \\ \hline 6 & 7 \\ \hline 5 \\ \cline{1-1} 4 \\ \cline{1-1} 3 \\ \cline{1-1} 2 \\ \cline{1-1} 1 \\ \cline{1-1}\end{array}}. \tag{2.6}$$

In this book, we mainly consider the r-numerical semigroup, H. We introduce the tools as follows:

Definition 2.4 1. Let $\mathbb{Z}_r := \{0, 1, 2, \dots, r-1\}$ and $\mathbb{Z}_r^\times := \mathbb{Z}_r \setminus \{0\}$.
2. Let $\widetilde{\mathfrak{e}}_i := \min\{h \in H \mid i \equiv h \text{ modulo } r\}$, $i \in \mathbb{Z}_r$.
3. Let $\widetilde{\mathfrak{E}}_H := \{\widetilde{\mathfrak{e}}_i \mid i \in \mathbb{Z}_r\}$ be the *standard basis* of H. Further, we define the ordered set $\mathfrak{E}_H := \{\mathfrak{e}_i \in \widetilde{\mathfrak{E}}_H \mid \mathfrak{e}_i < \mathfrak{e}_{i+1}\}$, and $\mathfrak{E}_H^\times := \mathfrak{E}_H \setminus \{0\}$, e.g., $\mathfrak{e}_0 = \widetilde{\mathfrak{e}}_0 = 0$.
4. Let $\mathfrak{e}_{\ell,i}^*$ be the element in $\mathfrak{E}_H$ such that $\mathfrak{e}_{\ell,i}^* = \mathfrak{e}_\ell - \mathfrak{e}_i$ modulo r.
5. Let $\overline{H}^{\mathrm{c}} := H^{\mathrm{c}} \bigcup(-\mathbb{N})$, where $\mathbb{N} := \mathbb{N}_0 \setminus \{0\}$.
6. The Apéry set $\mathrm{Ap}(H, n)$ (in honor of Apéry [Ape1946]) for a positive integer n is defined by
$$\mathrm{Ap}(H, n) := \{s \in H \mid s - n \notin H\}.$$

Since it is obvious that $\mathrm{Ap}(H, r) = \mathfrak{E}_H = \widetilde{\mathfrak{E}}_H$ as sets, the standard basis is sometimes defined by the Apéry set $\mathrm{Ap}(H, r)$.

We have the following elementary but essential results:

Lemma 2.5 *For $a \in \mathbb{N}_0$, we define*

$$[a]_r := \{a + kr \mid k \in \mathbb{N}_0\}, \quad \overline{[a]}_r^{\mathrm{c}} := \{a - kr \mid k \in \mathbb{N}\}, \quad [a]_r^{\mathrm{c}} := \overline{[a]}_r^{\mathrm{c}} \cap \mathbb{N}.$$

1. *We have the following decomposition;*

   a. $\mathbb{N}_0 = \bigoplus_{i \in \mathbb{Z}_r} [i]_r$,

b. $H = \bigoplus_{i \in \mathbb{Z}_r} [\mathfrak{e}_i]_r$,

c. $\overline{H}^{\mathrm{c}} = \bigoplus_{i \in \mathbb{Z}_r} \overline{[\mathfrak{e}_i]}_r^{\mathrm{c}}$, $H \bigcup \overline{H}^{\mathrm{c}} = \mathbb{Z}$,

d. $H^{\mathrm{c}} = \bigoplus_{i \in \mathbb{Z}_r} [\mathfrak{e}_i]_r^{\mathrm{c}} = \bigoplus_{i \in \mathbb{Z}_r^{\times}} [\mathfrak{e}_i]_r^{\mathrm{c}}$, $H \bigcup H^{\mathrm{c}} = \mathbb{N}_0$.

2. *For every* $x_i \in [\mathfrak{e}_i]_r$ $(i \in \mathbb{Z}_r)$,

$$\{x_i \text{ modulo } r \mid i \in \mathbb{Z}_r\} = \mathbb{Z}/r\mathbb{Z},$$

especially for $x \in [\widetilde{\mathfrak{e}}_i]_r$, $x = i$ *modulo* r.

3. *We have* $\mathfrak{e}^*_{\ell,\ell} = \mathfrak{e}_0 = 0$, *and* $\mathfrak{e}^*_{\ell,0} = \mathfrak{e}_\ell$.

Proof *1.a*, *2* and *3* are apparent. From the definition of $\mathfrak{E}_H$, $H = \{\mathfrak{e}_i + kr \mid i \in \mathbb{Z}_r,\ k \in \mathbb{N}_0\}$. For $i \neq j$, $[\mathfrak{e}_i]_r \cap [\mathfrak{e}_j]_r = \emptyset$ and thus we have the decomposition in *1.b*. Since $H^{\mathrm{c}} = \mathbb{N}_0 \setminus H$, we have *1.d* and *c* noting *1.a*. ■

The following is obvious:

Lemma 2.6 *For the generators* r *and* s *in the numerical semigroup* H, *there are positive integers* i_s *and* i_r *such that* $i_s s - i_r r = 1$. *We fix* (i_s, i_r) *as the minimal pair with respect to* $i_s + i_r$.

We remark that $\overline{H}^{\mathrm{c}}$ of the Weierstrass semigroup determines the structure of the differentials on a certain curve X in Lemma 2.109 and the embedding of the curve into the universal Grassmannian manifold as in Sect. 3.9.3.

A numerical semigroup H is said to be Weierstrass if there exists a pointed curve (X, P) such that $H = H(X, P)$, as mentioned above. Hurwitz posed the problem of whether any numerical semigroup H is Weierstrass. The question was revived in the 1980s. In [BucG1979, BucG1980], a counterexample was given for the semigroup H_B generated by 13, 14, 15, 16, 17, 18, 20, 22, and 23, whose genus is 16: this can be seen as follows. Assume that $H_B = H(X, P)$ for a pointed curve (X, P), then X would have a holomorphic differential vanishing to order $\ell - 1$ at P for any $\ell \in H_B^{\mathrm{c}}$. By letting $H^{\mathrm{c}\,2}$ be the set of all sums of two elements of H^{c} for a semigroup H, X would have a holomorphic quadratic differential vanishing to order $m - 2$ at P for any $m \in H_B^{\mathrm{c}\,2}$, which contradicts the fact that $\# H_B^{\mathrm{c}\,2} = 46 > 45 = (3 \times 16) - 3 = \dim \mathbf{H}^0(X, K_X^{\otimes 2})$. Note that we write $K_X^{\otimes 2}$ for the sheaf of regular quadratic differentials on X.

We mainly refer to the work of Komeda and co-authors for further information [Kom1983, Kom1993, Kom1999, Kom2004, KomO2004, KomO2008a, KomO2008b, EisH1987]. D. Eisenbud with J. Harris, C. Maclachlan and I. Morrison, jointly with H. Pinkham are other authors who wrote on this issue. We sketch a list of sufficient properties for H to be Weierstrass:

1. semigroup whose cardinality of the generators is less than 4,
2. semigroup whose genus is less than 9,
3. semigroup which is primitive, i.e., twice the smallest positive integer in $H >$ the largest integer in $H^{\mathfrak{c}}$, of genus 9,
4. semigroup whose $r_{\min}(H^{\mathfrak{c}}) = 2, 3, 4, 5$,
5. semigroup which is primitive and $\mathrm{wt}_H(H^{\mathfrak{c}}) \leqq g - 1$,
6. semigroup whose $\mathfrak{a}(H) = (0^{g-2}, m, n)$ for genus g and $\mathrm{wt}_H(H^{\mathfrak{c}}) = g$,
7. semigroup whose $\mathfrak{a}(H) = (0^{g-r}, m^r)$ for genus g, and
8. many other special cases of n-semigroups generated by 4 elements.

The following are well-known as typical solutions of the Frobenius problem:

Lemma 2.7 *1. For a numerical semigroup $H = \langle r, s\rangle$, $(r < s, (r, s) = 1)$,*

$$g = \frac{(r-1)(s-1)}{2}, \quad c_H = (r-1)(s-1), \quad |\Lambda_H| = \frac{(r^2-1)(s^2-1)}{24},$$

and particularly $|\Lambda_{\langle 2,2g+1\rangle}| = g(g+1)/2$.

2. *For a numerical semigroup $H = \langle r_1, \ldots, r_m\rangle$, $(r_i < r_{i+1}, (r_i, r_j) = 1)$, $m \geq 2$,*

$$g \leq \frac{(r_1-1)(r_j-1)}{2}, \quad c_H \leq (r_1-1)(r_j-1), \quad |\Lambda_H| \leq \frac{(r_1^2-1)(r_j^2-1)}{24}$$

for $j = 2, \ldots, m$.

Proof The estimation of g in 1 is proved in [Sy1884] by considering the expansion $(1 + z^s + z^{2s} + \cdots + z^{rs})(1 + z^r + z^{2r} + \cdots + z^{rs})$. The estimation of $|\Lambda_H|$ is also given by the polynomial; we follow the proof of Buchstaber and Leykin [BucL2004, p. 97]. Let us consider the formal series $H(z) := \sum_{i=0}^{\infty} z^{N_H(i)}$, which is realized in the inductive limit of the numerical semigroup ring R_H, and $H^{\mathfrak{c}}(z) := \sum_{i=0}^{g} z^{N_H^{\mathfrak{c}}(i)}$. Both are related due to

$$H^{\mathfrak{c}}(z) = \frac{1}{1-z} - H(z).$$

Hence we have the estimation of g and $|\Lambda_H|$ by

$$g = H^{\mathfrak{c}}(1), \quad |\Lambda_H| = \frac{dH^{\mathfrak{c}}(z)}{dz}\Big|_{z=1} - \frac{g(g-1)}{2}.$$

For the case of the numerical semigroup generated by two entities,

$$H(z) = \frac{1-z^{rs}}{(1-z^r)(1-z^s)},$$

and thus we obtain the formula of g and

$$|\Lambda_H| = \frac{(2rs-r-s-1)(r-1)(s-1)}{12} - \frac{g(g-1)}{2} = \frac{(r^2-1)(s^2-1)}{24}.$$

These evaluations are obtained by the expansion of $H^{\mathfrak{c}}(1+t)$ with respect to t. Then 2 is obvious. ■

2.2.2 *Weierstrass Canonical Form and Weierstrass Curves (W-Curves)*

In this subsection, we now review the "Weierstrass canonical form" ("Weierstrass normal form") based on [KomMP2016, KomMP2019, KomM2020, KomMP2022a], which is a generalization of Weierstrass' standard form for elliptic curves (1.26). This form originated from Abel's insight [Abl1829, Abl1830] and Weierstrass investigated its primitive property [Wei1867, Wei1875a] as mentioned in Sect. 1.6.2, (see also Definition 2.253 and its remark).

Baker [Ba1897, Ch. V, §§60-79] gave a complete review, proof, and examples of the theory. Here we refer to Kato [Kat1980, CopK2014], who also produces this representation with proof.

We consider a pointed smooth curve (X, P), i.e., a pointed compact Riemann surface, and its Weierstrass semigroup $H(X, P)$. Let $r = r_{\min}(H(X, P))$ and let s be the least integer in $H(X, P)$ which is prime to r. By denoting $P = \infty$, X can be viewed as an $r:1$ cover of $\mathbb{P}$ via an equation of the following form: $f_X(x, y) = y^r + A_1(x)y^{r-1} + \cdots + A_{r-1}(x)y + A_r(x) = 0$, where x is an affine coordinate on $\mathbb{P}$, the point ∞ on X is mapped to $x = \infty \in \mathbb{P}$, and each A_j is a polynomial in x of degree $\le js/r$, with equality being attained only for $j = r$. We call $f_X(x, y)$ the *Weierstrass canonical form*.[2]

The algebraic curve $\mathrm{Spec}\,\mathbb{C}[x, y]/f_X(x, y)$ is, in general, singular, and thus, we let X denote its unique normalization up to equivalent relation.

Proposition 2.8 [Kat1980, CopK2014] *For a pointed smooth curve (X, ∞) with Weierstrass semigroup $H_X := H(X, \infty)$ for which $r_{\min}(H_X) = r$, and $\mathfrak{e}_i \in \mathfrak{E}_{H_X}$, ($i \in \mathbb{Z}_r^\times$) in Definition 2.4, and we let $s := \min_{i \in \mathbb{Z}_r^\times}\{\mathfrak{e}_i \in \mathfrak{E}_{H_X} \mid (\mathfrak{e}_i, r) = 1\}$ and $s = \mathfrak{e}_{\ell_s}$. (X, ∞) is defined by an irreducible equation,*

---

2 Kunz also mentioned it as a plane curve with no name [Ku2004]. A similar form is argued in the books of Forster [Fo1981], Griffiths [Gri1989], and Hansen [Han1989], in which f_X is called Weierstrass' polynomial (see also footnote 13 in Chap. 1).

Further, we note several variants on the definition of the Weierstrass canonical form. For example, van Hoeij proposed an algorithm to transform a governing equation to the Weierstrass normal form, which is implemented in the software Maple as a package, algcurves [vH1983]. However, it differs from our Weierstrass canonical form. It allows that the obtained curves sometimes have several infinite points. Our Weierstrass canonical form provides a curve with a single infinity.

Moreover, the curve X can be regarded as a projective curve following the procedure in [Mum1984, 3.16] for the hyperelliptic curves. (cf. Sect. 2.8.2, and Examples 2.241 and 2.243.)

$$f_X(x, y) = 0, \tag{2.7}$$

for a polynomial $f_X \in \mathbb{C}[x, y]$ *of type,*

$$f_X(x, y) := y^r + A_1(x)y^{r-1} + A_2(x)y^{r-2} + \cdots + A_{r-1}(x)y + A_r(x), \tag{2.8}$$

where the $A_i(x)$*'s are polynomials in* x, $A_0 = 1$,

$$A_i = \sum_{j=0}^{\lfloor is/r \rfloor} \lambda_{i,j} x^j =: \prod_{j=1}^{\lfloor is/r \rfloor} a_{i,0}(x - a_{i,j}), \tag{2.9}$$

and $\lambda_{i,j} \in \mathbb{C}$, $\lambda_{r,s} = -1$.

Proof We let $\overline{\mathfrak{E}}_{H_X} := \{\mathfrak{e}_1, \mathfrak{e}_2, \ldots, \mathfrak{e}_{r-1}\} \setminus \{s\}$, $\overline{\mathfrak{E}}_{H_X} =: \{\overline{\mathfrak{e}}_2, \ldots, \overline{\mathfrak{e}}_{r-1}\}$, ($\overline{\mathfrak{e}}_i < \overline{\mathfrak{e}}_j$ for $i < j$), and $\overline{\mathfrak{e}}_1 = s$. Let $\widetilde{\mathfrak{y}}_{\overline{\mathfrak{e}}_i}$ be a meromorphic function on X whose only pole is ∞ with order $\overline{\mathfrak{e}}_i$, ($i \in \mathbb{Z}_r^\times$), taking $y_s := \widetilde{\mathfrak{y}}_s = \widetilde{\mathfrak{y}}_{\overline{\mathfrak{e}}_1}$ and $1 = \widetilde{\mathfrak{y}}_{\overline{\mathfrak{e}}_0}$. From the definition of X, we have, as $\mathbb{C}$-vector spaces,

$$\mathbf{H}^0(X, \mathscr{O}_X(*\infty)) = \sum_{i=0}^{r-1} \sum_{j=0} \mathbb{C} x^j \widetilde{\mathfrak{y}}_{\overline{\mathfrak{e}}_i}. \tag{2.10}$$

Let the affine ring R_X of X be defined by $\mathbf{H}^0(X, \mathscr{O}_X(*\infty))$, i.e., $R_X := \mathbf{H}^0(X, \mathscr{O}_X(*\infty))$. Thus for every $\overline{\mathfrak{e}}_j \in \overline{\mathfrak{E}}_{H_X}$ $(j = 2, 3, \ldots, r-1)$, we obtain the following equations:

$$\begin{cases} y_s \widetilde{\mathfrak{y}}_{\overline{\mathfrak{e}}_2} = A_{2,0} + A_{2,1} y_s + A_{2,2} \widetilde{\mathfrak{y}}_{\overline{\mathfrak{e}}_2} + \cdots + A_{2,r-2} \widetilde{\mathfrak{y}}_{\overline{\mathfrak{e}}_{r-2}} + A_{2,r-1} \widetilde{\mathfrak{y}}_{\overline{\mathfrak{e}}_{r-1}}, \\ y_s \widetilde{\mathfrak{y}}_{\overline{\mathfrak{e}}_3} = A_{3,0} + A_{3,1} y_s + A_{3,2} \widetilde{\mathfrak{y}}_{\overline{\mathfrak{e}}_2} + \cdots + A_{3,r-2} \widetilde{\mathfrak{y}}_{\overline{\mathfrak{e}}_{r-2}} + A_{3,r-1} \widetilde{\mathfrak{y}}_{\overline{\mathfrak{e}}_{r-1}}, \\ \quad \cdots \\ y_s \widetilde{\mathfrak{y}}_{\overline{\mathfrak{e}}_{r-1}} = A_{r-1,0} + A_{r-1,1} y_s + A_{r-1,2} \widetilde{\mathfrak{y}}_{\overline{\mathfrak{e}}_2} + \cdots + A_{r-1,r-2} \widetilde{\mathfrak{y}}_{\overline{\mathfrak{e}}_{r-2}} + A_{r-1,r-1} \widetilde{\mathfrak{y}}_{\overline{\mathfrak{e}}_{r-1}}, \end{cases} \tag{2.11}$$

$$y_s^2 = A_{1,0} + A_{1,1} y_s + A_{1,2} \widetilde{\mathfrak{y}}_{\overline{\mathfrak{e}}_2} + \cdots + A_{1,r-2} \widetilde{\mathfrak{y}}_{\overline{\mathfrak{e}}_{r-2}} + A_{1,r-1} \widetilde{\mathfrak{y}}_{\overline{\mathfrak{e}}_{r-1}}, \tag{2.12}$$

where $A_{i,j} \in \mathbb{C}[x]$.

When $r = 2$, (2.7) equals (2.12). We assume that $r > 2$ and then (2.11) is reduced to $(r-2)$ linear equations,

$$\begin{pmatrix} A_{2,1} - y_s & A_{2,2} & \cdots & A_{1,r-1} \\ A_{3,1} & A_{3,2} - y_s & \cdots & A_{3,r-1} \\ \vdots & \vdots & \ddots & \vdots \\ A_{r-1,1} & A_{r-1,2} & \cdots & A_{r-1,r-1} - y_s \end{pmatrix} \begin{pmatrix} \widetilde{\mathfrak{y}}_{\overline{\mathfrak{e}}_2} \\ \widetilde{\mathfrak{y}}_{\overline{\mathfrak{e}}_3} \\ \vdots \\ \widetilde{\mathfrak{y}}_{\overline{\mathfrak{e}}_{r-1}} \end{pmatrix} = - \begin{pmatrix} A_{2,0} + A_{2,1} y_s \\ A_{3,0} + A_{3,1} y_s \\ \vdots \\ A_{r-1,0} + A_{r-1,1} y_s \end{pmatrix}. \tag{2.13}$$

One can check that the determinant of the matrix on the left-hand side of (2.13) is not equal to zero by computing the order of pole at ∞ of the monomials $B_i y_s^{r-2-i}$ in the expression,

$$P(x, y_s) := \begin{vmatrix} A_{2,1} - y_s & A_{2,2} & \cdots & A_{2,r-1} \\ A_{3,1} & A_{3,2} - y_s & \cdots & A_{3,r-1} \\ \vdots & \vdots & \ddots & \vdots \\ A_{r-1,1} & A_{r-1,2} & \cdots & A_{r-1,r-1} - y_s \end{vmatrix}$$
$$= y_s^{r-2} + B_1 y_s^{r-3} + \cdots + B_{r-3} y_s + B_{r-4},$$

which is $s(r-2-i) + r \cdot \deg_x B_i$ by letting $\deg_x h(x)$ be the degree of h with respect to x. The fact that $(r, s) = 1$ shows that $s(r-2-i) + r \cdot \deg_x B_i \neq s(r-2-j) + r \cdot \deg_x B_j$ for $i \neq j$.

Hence by solving Eq. (2.13), we have

$$\widetilde{\mathfrak{y}}_{\mathfrak{e}_i} = \frac{Q_i(x, y_s)}{P(x, y_s)}, \tag{2.14}$$

where $Q_i(x, y_s) \in \mathbb{C}[x, y_s]$ and a polynomial of order at most $r-2$ in y_s. Note that the Eqs. (2.14) are not algebraically independent in general but in any case the function field of the curve can be generated by some of these $\widetilde{\mathfrak{y}}_{\mathfrak{e}_j}$'s, and its affine ring R_X can be given by a quotient ring of $\mathbb{C}[x, y_{r_2}, y_{r_3}, \ldots, y_{r_{m_X}}]$ for $i_j \in M_X := M(H_X)$, where $M_X = \{r_1, r_2, \ldots, r_{m_X}\} \subset \mathbb{N}$ with the conditions that mutually coprime, $(r_1, \ldots, r_{m_X}) = 1, r_1 = r, r_2 = s$, and $r_i < r_j$ for $2 < i < j$, is a minimal set of generators for H_X. Here the cardinality of the generator M_X of H_X is $m_X (< r)$.

By putting (2.14) into (2.12), we obtain (2.8) if it is irreducible. If it is reducible, $f_X(x, y_s)$ is decomposed to polynomials whose degree is less than r with respect to y_s. However, the relation $(r, s) = 1$ shows that there do not exist such monic polynomials due to the order of the singularity at ∞.

Further, the order of the singularity of $A_i y_s^{r-i}$ is given by $s(r-i) + r \cdot \deg_x A_i$. The cases satisfying that $s(r-i) + r \cdot \deg_x A_i = s(r-j) + r \cdot \deg_x A_j$ mean that $i = j$ or $(i, j) = (0, r), (r, 0)$ due to $(r, s) = 1$. Hence $\deg_x A_r = s$. For $i = 1, \ldots, r-1$, $s(r-i) + r \cdot \deg_x A_i < rs$ leads to $\deg_x A_i < si/r$. ■

Remark 2.9 Let us address the curve in Proposition 2.8 as a *Weierstrass curve* or a *W-curve* in this book. The Weierstrass canonical form characterizes the W-curve, which has only one infinity point ∞. The infinity point ∞ is a Weierstrass point if $H_X^{\mathfrak{c}} = H^{\mathfrak{c}}(X, \infty) = \{N^{\mathfrak{c}}(i)\}$ differs from $\{1, 2, \cdots, g\}$. Since every compact Riemann surface of the genus, $g(> 1)$, has a Weierstrass point whose Weierstrass gap sequence has genus g [ArCGH1985, FarK1991], it characterizes the behavior of the meromorphic functions at the point, and thus there is a W-curve which is birationally equivalent to the compact Riemann surface.

Further, Proposition 2.8 is also applicable to a pointed compact Riemann surface (Y, P) of genus g whose point P is a non-Weierstrass point rather than the Weierstrass point; its Weierstrass gap sequence at P is $H^{\mathfrak{c}}(Y, P) = \{1, 2, \ldots, g\}$. Even for this case, we find the Weierstrass canonical form f_X and the W-curve X with $H_X^{\mathfrak{c}} = \{1, 2, \ldots, g\}$ which is bi-rational to Y so that $P(\in Y)$ corresponds to $\infty \in X$.

We also remark that we can consider X as a projective curve since we can find the chart and its affine equation at ∞ by using the birational curve mentioned in Sect. 2.8.2 as in [Mum1984, 3.13], (cf. Examples 2.241 and 2.243 in Appendix 2.10).

Remark 2.10 [Ku2004, Fu1969] Let $R_{X^\circ}^\circ := \mathbb{C}[x, y]/(f_X(x, y))$ for (2.7) and its normalized ring be R_X° if $X^\circ :=$ Spec $R_{X^\circ}^\circ$ is singular. R_X° is the coordinate ring of the affine part of $X \setminus \{\infty\}$ and we identify R_X° with $R_X = \mathbf{H}^0(X, \mathscr{O}_X(*\infty))$. Then the quotient field $\mathbb{C}(X) := \mathscr{Q}(R_X)$ of R_X is considered as an algebraic function field on X over $\mathbb{C}$.

By introducing $R_{\mathbb{P}} := \mathbf{H}^0(\mathbb{P}, \mathscr{O}_{\mathbb{P}}(*\infty)) = \mathbb{C}[x]$ and its quotient field $\mathbb{C}(x) := \mathscr{Q}(R_{\mathbb{P}})$, $\mathscr{Q}(R_X)$ is considered a finite extension of $\mathscr{Q}(R_{\mathbb{P}})$. We regard R_X as a finite extended ring of $R_{\mathbb{P}}$ of rank r, e.g., $R_{X^\circ}^\circ = R_{\mathbb{P}}[y]/(f_X(x, y))$ as mentioned in Sect. 2.2.5 [Ku2004]. Further, as we will mention in Sect. 2.2.4, $\mathfrak{y}_{\widetilde{\mathfrak{e}}_i}$ in the proof of Proposition 2.8 is the standard basis of R_X as an $R_{\mathbb{P}}$-module, and the matrix in (2.13) are naturally interpreted as the $R_{\mathbb{P}}$-module.

For the local ring $R_{X,P}$ of R_X at $P \in X$, we have the ring homomorphism,

$$\varphi_P : R_X \to R_{X,P}, \quad R_X = \bigcap_{P \in X \setminus \{\infty\}} R_{X,P}.$$

We note that $R_{X,\infty}$ plays crucial roles in the Weierstrass canonical form.

We let $M_X = \{r_1, r_2, \ldots, r_{m_X}\}$ be the minimal generator of the numerical semigroup $H_X = H(X, \infty)$ appearing in the proof of Proposition 2.8. The W-curve admits a local cyclic $\mathfrak{C}_r = \mathbb{Z}/r\mathbb{Z}$-action at ∞ (cf. Sect. 2.2.3). The genus of X is denoted by g_X, for short, g and the conductor of $H_X = \{N_X(i)\}$ is denoted by $c_X := c_{H_X}$; the Frobenius number $c_X - 1$ is the maximal gap in H_X. The gap sequence is denoted by $H_X^{\mathfrak{c}} = \{N_X^{\mathfrak{c}}(i)\}$, and we also let the entities defined in Sect. 2.2.1 be similar, e.g., $\overline{H}_X^{\mathfrak{c}}$, the Young diagram associated with H_X, $\Lambda_X = (\Lambda_1, \ldots, \Lambda_g)$. However, sometimes we also drop the index X, e.g. just Λ, $N(i)$.

2.2.2.1 Uniqueness of the Weierstrass Curve

In general, the curve X° given by (2.7) is singular and thus its normalization X is unique up to isomorphism as in Theorem 2.239 [Gri1989]. There are isomorphic curves to X, i.e., there is a holomorphic map φ among them:

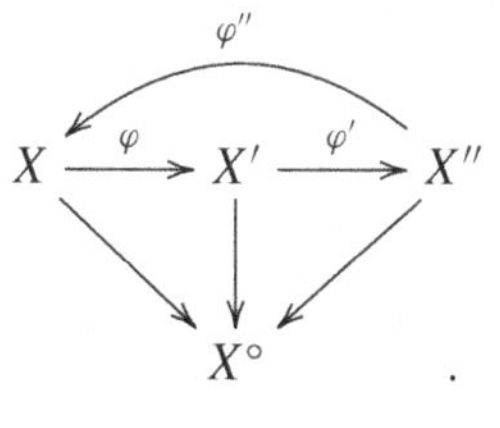

(2.15)

We fix one of them, which X denotes.

2.2.2.2 Projection from X to $\mathbb{P}$

For a pointed smooth curve (X, ∞) which satisfies the conditions in Proposition 2.8, we can find the Weierstrass canonical form and the generators $x, y_{r_2}, y_{r_3}, \ldots, y_{r_{m_X}}$ of the numerical semigroup H_X though the relations (2.14) in its proof, e.g. (2.63) and (2.72) (cf. Remark 2.15). It means that a point in a W-curve X is expressed by a point $(x, y_{r_2}, y_{r_3}, \ldots, y_{r_{m_X}})$ in the m_X dimensional affine space with the common infinite point ∞. In the following, we show how it is embedded in the space, but first, note its projective structure,

$$\varpi_{r_j} : X \to \mathbb{P}, \qquad (\varpi_{r_j}(x, y_{r_2}, \ldots, y_{r_{m_X}}) = y_{r_j}) \tag{2.16}$$

such that $\varpi_{r_j}(\infty) = \infty \in \mathbb{P}$. Let ϖ_r and ϖ_s be denoted by ϖ_x and ϖ_y for some cases.

Let $\{y_\bullet\} := \{y_s = y_{r_2}, y_{r_3}, \ldots, y_{r_{m_X}}\}$ and $\mathbb{C}[x, y_\bullet] := \mathbb{C}[x, y_s = y_{r_2}, y_{r_3}, \ldots, y_{r_{m_X}}]$.

2.2.2.3 Cyclic Weierstrass Curves

We, sometimes, consider the special case of the Weierstrass canonical form, the *cyclic Weierstrass form*,

$$y^r = x^s + \lambda_{s-1}x^{s-1} + \cdots + \lambda_1 x + \lambda_0, \tag{2.17}$$

whose normalized curve is called the *cyclic Weierstrass curve* or *cyclic W-curve* in this book. It is, sometimes, referred to as the *superelliptic curve* [Lan1978, p. 147][GaPS2000, Koo1999] or Z_n*-curves* [FarZ2011].

2.2.2.4 Symmetric and Non-symmetric Weierstrass Curves

We also investigate the Weierstrass curves whose Weierstrass semigroups H_X are symmetric and non-symmetric, which are called the *symmetric Weierstrass curve* (*symmetric W-curve*), and *non-symmetric Weierstrass curve* (*non-symmetric W-curve*) respectively.

2.2.3 The Monomial Curves and W-Curves

This subsection shows the monomial curves, and their relation to W-curves based on [KomMP2016, KomMP2019, KomM2020].

For a given W-curve X with the Weierstrass semigroup $H = H_X$, and its generator $M_X = \{r = r_1, r_2, \ldots, r_{m_X}\}$, the behavior of singularities of the elements in R_X at

∞ is described by a monomial curve X_H^Z. For the numerical semigroup $H = \langle M_X \rangle$, the *numerical semigroup ring* R_H is defined as $R_H := \mathbb{C}[z^{r_1}, z^{r_2}, \cdots, z^{r_{m_X}}]$.

Following a result of Herzog's [Her1970], we recall the well-known proposition for a polynomial ring $\mathbb{C}[Z] := \mathbb{C}[Z_{r_1}, Z_{r_2}, \cdots, Z_{r_{m_X}}]$.

Proposition 2.11 *For the $\mathbb{C}$-algebra homomorphism $\widetilde{\varphi}_H^Z : \mathbb{C}[Z] \to R_H$, the kernel of $\widetilde{\varphi}_H^Z$ is generated by certain binomials $f_i^H \in \mathbb{C}[Z]$ $(i = 1, 2, \ldots, k_X)$ for a positive integer k_X, $(m_X - 1 \le k_X < \infty)$, i.e., $\ker \widetilde{\varphi}_H^Z = (f_1^H, f_2^H, \ldots, f_{k_X}^H)$, and*

$$R_H \simeq \mathbb{C}[Z]/\ker \widetilde{\varphi}_H^Z =: R_H^Z.$$

Proof This follows from a result of Herzog's [Her1970]. There are the multiplicative group actions ($\mathbb{G}_{\mathrm{m}}$-actions) on Z_{r_i}'s, whereas R_H is invariant for the action. It means that the number of generators of $\ker \widetilde{\varphi}_H^Z$ is determined, i.e., k_X. The relation in the $\ker \widetilde{\varphi}_H^Z$ is reduced to a binary one. ■

We call $R_H^Z = \mathbb{C}[Z]/\ker \widetilde{\varphi}_H^Z$ a *monomial ring*. By sending Z_r to $1/x$ and Z_{r_i} to $1/y_{r_i}$, the monomial ring R_H^Z determines the structure of gap sequence of X at ∞ [Her1970, Pin1974]. Bresinsky showed that in general, k_X can be any finitely large number if $m_X > 3$ [Bre1975].

Let $X_H := \operatorname{Spec} R_H^Z$, which we call a *monomial curve*. We also define the ring isomorphism on R_H^Z induced from $\widetilde{\varphi}_H^Z$, which is denoted by φ_H^Z,

$$\varphi_H^Z : \mathbb{C}[Z]/\ker \widetilde{\varphi}_H^Z = R_H^Z \to R_H. \tag{2.18}$$

Further, we let $\{Z_\bullet\} := \{Z_{r_2}, Z_{r_3}, \ldots, Z_{r_{m_X}}\}$, and $\mathbb{C}[Z_r, Z_\bullet] := \mathbb{C}[Z_r, Z_{r_2}, \ldots, Z_{r_{m_X}}]$. A monomial curve is an irreducible affine curve with $\mathbb{G}_{\mathrm{m}}$-action, where $\mathbb{G}_{\mathrm{m}}$ is the multiplicative group of the complex numbers; $Z_a \mapsto g^a Z_a$ for $g \in \mathbb{G}_{\mathrm{m}}$, and it induces the action on the monomial ring R_H^Z.

The following action of the cyclic group of order r plays a crucial role in this book.

Lemma 2.12 *The cyclic group $\mathfrak{C}_r$ of order r acts on the monomial ring R_H^Z; the action of the generator $\widehat{\zeta_r} \in \mathfrak{C}_r$ on Z_a is defined by sending Z_a to $\zeta_r^a Z_a$, where ζ_r is a primitive r-th root of unity.*

By letting $\mathfrak{r}_i^* := (r, r_i)$, $\mathfrak{r}_i := r/\mathfrak{r}_i^*$, and $\bar{\mathfrak{r}}_i := r_i/\mathfrak{r}_i^*$, the orbit of Z_{r_i} forms $\mathfrak{C}_{\mathfrak{r}_i}$; especially for the case that $(r, r_i) = 1$, it recovers $\mathfrak{C}_r$.

Thus in R_H^Z, we have the following identities:

$$f_H^{(j)}(Z_r, Z_{r_j}) = 0, \quad f_H^{(j)} := Z_{r_j}^{\mathfrak{r}_j} - Z_r^{\bar{\mathfrak{r}}_j}, \quad (j = 2, \ldots, m_X). \tag{2.19}$$

For example, the case $M_X = \{3, 7, 8\}$ provides that these elements $\{f_1^H, f_2^H, f_3^H\}$ are given by the 2×2 minors of $\begin{vmatrix} Z_3^2 & Z_7 & Z_8 \\ Z_7 & Z_8 & Z_3^3 \end{vmatrix}$. There is a cyclic group $\mathfrak{C}_3 := \{\zeta_3^a \mid a = 0, 1, 2\}$ action on R_H as a $\mathbb{G}_{\mathrm{m}}$ action. Due to the relation, there are other possibilities which are given by the 2×2 minors of $\begin{vmatrix} Z_3^2 & \zeta_3^a Z_7 & \zeta_3^{2a} Z_8 \\ \zeta_3^a Z_7 & \zeta_3^{2a} Z_8 & Z_3^3 \end{vmatrix}$ for $a = 0, 1, 2$. It means that f_i^H is unique up to the $\mathbb{G}_{\mathrm{m}}$ action. On the other hand, $f_H^{(2)} = Z_7^3 - Z_3^7$ and $f_H^{(3)} = Z_8^3 - Z_3^8$ for (2.19).

There are non-negative integers $h_j^{(i\pm)}$ such that

$$f_i^H = \left(\prod_{j=2} Z_{r_j}^{h_j^{(i+)}}\right) - \left(\prod_{j=1} Z_{r_j}^{h_j^{(i-)}}\right), \tag{2.20}$$

where, in the first term, Z_{r_1} does not exist because $(r_1, r_2, \ldots, r_{m_X}) = 1$.

Recalling the standard basis of H_X in Definition 2.4, we find the monic monomial $\mathfrak{z}_{\mathfrak{e}_i} \in R_H^Z$ such that $\varphi_H^Z(\mathfrak{z}_{\mathfrak{e}_i}) = z^{\mathfrak{e}_i}$ and the standard basis $\{\mathfrak{z}_{\mathfrak{e}_i} \mid i \in \mathbb{Z}_r\}$; $\mathfrak{z}_{\mathfrak{e}_0} = 1$ of the monomial ring R_H^Z.

Lemma 2.13 *The $\mathbb{C}[Z_r]$-module R_H^Z is given by*

$$R_H^Z = \mathbb{C}[Z_r] \oplus \mathbb{C}[Z_r]\mathfrak{z}_{\mathfrak{e}_1} \oplus \cdots \oplus \mathbb{C}[Z_r]\mathfrak{z}_{\mathfrak{e}_{r-1}},$$

and thus $\mathfrak{z}_H := \{\mathfrak{z}_{\mathfrak{e}_0} = 1, \mathfrak{z}_{\mathfrak{e}_1}, \ldots, \mathfrak{z}_{\mathfrak{e}_{r-1}}\}$ is the basis of the $\mathbb{C}[Z_r]$-module R_H^Z. Then there is a monomial $b_{ijk} \in \mathbb{C}[Z_r]$ such that

$$\mathfrak{z}_{\mathfrak{e}_i}\mathfrak{z}_{\mathfrak{e}_j} = \sum_{k \in \mathbb{Z}_r} b_{ijk}\mathfrak{z}_{\mathfrak{e}_k}.$$

Further, there are elements $\mathfrak{z} \in R_H^Z$ and $\mathfrak{z}_{\mathfrak{e}_i}^* \in R_H^Z$, $(i \in \mathbb{Z}_r^\times)$ satisfying

$$\mathfrak{z}_{\mathfrak{e}_i}^*\mathfrak{z}_{\mathfrak{e}_i} = \mathfrak{z}, \quad \textit{for} \quad i \in \mathbb{Z}_r^\times,$$

and $\{Z_{r_j} \mid j = 2, \ldots, m_X\} \subset \mathfrak{z}_H$.

Moreover, the cyclic group of order r acts on these elements; the action of the generator $\widehat{\zeta}_r \in \mathfrak{C}_r$ on $\mathfrak{z}_{\mathfrak{e}_i}$ $(i = 0, \ldots, r-1)$ is defined by sending $\mathfrak{z}_{\mathfrak{e}_i}$ to $\zeta_r^{\mathfrak{e}_i}\mathfrak{z}_{\mathfrak{e}_i}$. For $f, g \in R_H^Z$, by letting $\widehat{\zeta}_r(fg) = \widehat{\zeta}_r(f)\widehat{\zeta}_r(g)$, $\widehat{\zeta}_r(\mathfrak{z}_{\mathfrak{e}_1}\mathfrak{z}_{\mathfrak{e}_2}\cdots\mathfrak{z}_{\mathfrak{e}_{r-1}})^2 = (\mathfrak{z}_{\mathfrak{e}_1}\mathfrak{z}_{\mathfrak{e}_2}\cdots\mathfrak{z}_{\mathfrak{e}_{r-1}})^2$.

Proof They are obtained from Definition 2.4 and Lemma 2.5. From the definition of M_X, every Z_{r_j} $(j = 2, \ldots, m_X)$ belongs to $\mathfrak{z}_H$. The group action is obvious from Lemma 2.12. ■

To construct our curve X, we could first consider the numerical semigroup H. We follow Pinkham's strategy [Pin1974]; namely, we start with an irreducible singular curve X_H with $\mathbb{G}_{\mathrm{m}}$ action, described parametrically by its numerical semigroup ring R_H (see Appendix 2.12). We deform the matrix, e.g., (2.13) or (b_{1ij}) in Lemma 2.13, of the relations. This deformation space $\bar{U}$ classifies the set of pairs consisting of a smooth and proper algebraic curve and X_H together with the semigroup $H(X_H, \infty)$. If $\bar{U}$ is non-void, then we have directly constructed a compactification of a moduli space for curves with points of semigroup H ([Pin1974], Introduction). For a given monomial curve, it is not trivial whether its associated W-curve exists, known as the Hurwitz problem [Hur1893] as mentioned in Sect. 2.2.1. Therefore, in this book, we start with the W-curve X and its numerical semigroup H_X to avoid the problem.

However, it is useful to look at the curves in Pinkham's picture. We will show the picture. In the picture we could consider that the coefficients $\{\lambda_{i,k}\}_{i,k}$ in $f_X(x, y)$ (2.8) are algebraically independent. For $a_{i,j}$ in (2.9), let $\mathbb{A}_{\mathbb{C}} := \mathbb{C}[a_{1,0}, \cdots, a_{r,s}]$ be the coefficient algebra so that $\lambda_{i,j} \in \mathbb{A}_{\mathbb{C}}$. Then f_X is regarded as an element in $\mathbb{A}_{\mathbb{C}}[x, y]$ as a higher dimensional algebraic object.

The picture is realized by the commutative diagram (cf. Appendix 2.12),

$$\begin{array}{ccc} \operatorname{Spec} R_H & \longrightarrow & \operatorname{Spec} R_X \\ \downarrow & & \downarrow \\ \operatorname{Spec} \mathbb{C} & \longrightarrow & \operatorname{Spec} \mathbb{A}_{\mathbb{C}} \end{array} \quad (2.21)$$

More precisely, for the dimension d of the moduli space with H, Pinkham considers the Weierstrass canonical form with coefficients given as elements of $\mathbb{A}_{\mathbb{C}}^{\circ} := \mathbb{C}[\widetilde{a}_1, \cdots, \widetilde{a}_d]$ and deformation of curves in the diagram. However, since the above diagram holds even for $\mathbb{A}_{\mathbb{C}}$, we use this picture in this book.

Let $\mathfrak{m}_{\mathbb{A}}$ be a maximal ideal in $\mathbb{A}_{\mathbb{C}}$ such that $\lambda_{i,j} \equiv 0$ modulo $\mathfrak{m}_{\mathbb{A}}$ except for certain (i, j), e.g., (r, s); $\mathbb{A}_{\mathbb{C}}/\mathfrak{m}_{\mathbb{A}} = \mathbb{C}$. Pinkham's investigations provide the following proposition ([Pin1974], see also [ADGS2016, p. 80]):

Proposition 2.14 *For a given W-curve X and its associated monomial ring, $R_H^Z = \mathbb{C}[Z]/(f_1^H, f_2^H, \dots, f_{k_X}^H)$, there are a surjective ring-homomorphism,*

$$\varphi_H^X : R_X \to R_H^Z$$

such that $R_X/\mathfrak{m}_{\mathbb{A}}$ is isomorphic to R_H^Z and $\varphi_H^X(y_{r_i}) = Z_{r_i}$, and a polynomial $\{f_i^X\}_{i=1,\dots,k_X} \in \mathbb{C}[x, y_\bullet]$ satisfying:

1. *$(f_i^X$ modulo $\mathfrak{m}_{\mathbb{A}}) = f_i^H$ for $(i = 1, \dots, k_X)$,*
2. *the affine part of R_X is given by $R_X = \mathbb{C}[x, y_\bullet]/(f_1^X, f_2^X, \dots, f_{k_X}^X)$, and*
3. *the rank of the matrix $\left(\dfrac{\partial f_i^X}{\partial y_{r_j}}\right)_{i=1,2,\dots,k_X, j=1,2,\dots,m_X}$ is $m_X - 1$ for every point P in X.*

Proof As we showed in the proof of Proposition 2.8, R_X is a quotient ring of $\mathbb{C}[x, y_\bullet]$. Since φ_H^X must be a surjective, there is a prime ideal $(f_1^X, f_2^X, \dots, f_{k_X}^X) \subset \mathbb{C}[x, y_\bullet]$ such that satisfying *1*. Thus we prove *2*. By noting $k_X \geq m_X$, Nagata's Jacobi criterion [Mat1986, Theorem 30.10] shows *3*. ∎

Remark 2.15 Nagata's Jacobi criterion is crucial. When the Weierstrass canonical form (2.8) is non-singular, $k_X = 1$, $m_X = 2$, $f_1^X = f_X$, and the rank of matrix

$\left(\frac{\partial f_X}{\partial x}, \frac{\partial f_X}{\partial y}\right)$ is equal to 1. On the other hand, for $m_X > 2$, there is, at least, one singular point at which both $\frac{\partial f_X}{\partial x}$ and $\frac{\partial f_X}{\partial y}$ vanish, and due to the normalization, the f_i^X $(i = 1, \ldots, k_X)$ compensates it.

Further, some of f_i^X is equal to $P(x, y_s)\widetilde{\mathfrak{y}}_{\mathfrak{e}_i} = Q(x, y_s)$ (2.14).

We introduce several tools for the investigation of W-curves, which were supplied by Ônishi [On2018]:

Definition 2.16 1. Recalling Lemma 2.6, we define the *arithmetic local parameter* at ∞ by

$$t = \frac{x^{i_r}}{y^{i_s}}.$$

2. The degree at $\mathscr{Q}(R_{X,\infty})$ as the order of the singularity with respect to t is naturally defined by

$$\mathrm{wt} = \deg_\infty : \mathscr{Q}(R_X) \to \mathbb{Z}, \tag{2.22}$$

which is called the *Sato–Weierstrass weight*.[3]

3. In the ring of the formal power series $\mathbb{C}[[t_1, \ldots, t_\ell]]$, we define the symbols $d_{>n}(t_1, \ldots, t_\ell)$ and $d_{\geq n}(t_1, \ldots, t_\ell)$, which express that they belong to the ideals

$$d_{>n}(t_1, \ldots, t_\ell) \in \{\sum a_{i_1,\ldots,i_\ell} t_1^{i_1} \cdots t_\ell^{i_\ell} \mid a_{i_1,\ldots,i_\ell} = 0 \text{ for } i_1 + \cdots + i_\ell \leq n\},$$

$$d_{\geq n}(t_1, \ldots, t_\ell) \in \{\sum a_{i_1,\ldots,i_\ell} t_1^{i_1} \cdots t_\ell^{i_\ell} \mid a_{i_1,\ldots,i_\ell} = 0 \text{ for } i_1 + \cdots + i_\ell < n\}.$$

The weight of y_{r_i} is given by

$$\mathrm{wt}(y_{r_i}) = -r_i, \quad (i = 1, 2, \ldots, m_X), \qquad y_{r_i} = \frac{1}{t^{r_i}}(1 + d_{>0}(t)).$$

Since R_X or R_X° is given by a quotient ring of $\mathbb{C}[x, y_\bullet]/(f_1^X, f_2^X, \ldots, f_{k_X}^X)$ in Proposition 2.14, we have the decomposition of R_X as a $\mathbb{C}$-vector space:

Lemma 2.17 *R_X is decomposed as a $\mathbb{C}$-vector space:*

$$R_X = \bigoplus_{i=0} \mathbb{C}\phi_i, \tag{2.23}$$

where ϕ_i is a monomial in R_X satisfying $\phi_i = t^{N(i)}(1 + d_{>0}(t))$ for the arithmetic parameter t at ∞. i.e., $\phi_0 = 1$, $\phi_1 = x$, ... We also have the inequalities $-\mathrm{wt}\,\phi_i < -\mathrm{wt}\,\phi_j$ for $i < j$,

[3] It is sometimes referred to as Sato-weight in honor of Sato in the Sato theory on the universal Grassmaniann manifold [SatS1982] because, in the theory, the weight plays a crucial role. However, since this weight essentially appeared in Weierstrass's paper [Wei1867], we refer to it so.

Further, by assigning a certain weight on each coefficient $\lambda_{i,j} \in \mathbb{A}_{\mathbb{C}}$ in (2.7) and (2.21) so that (2.7) is a homogeneous equation of weight rs, we also define another weight,

$$\mathrm{wt}_\lambda : R_X \to \mathbb{Z}. \tag{2.24}$$

Definition 2.18 We define the basis $S_X := \{\phi_i \mid i = 0, 1, 2 \ldots\}$ of R_X as a $\mathbb{C}$ vector space as in (2.23).

Lemma 2.19 *1.* $N(i) = -\mathrm{wt}(\phi_i)$, *i.e.*, $\{N(i) \mid i \in \mathbb{N}_0\} = H_X$.
2. Let t be the arithmetic local parameter at ∞ *of* R_X.

a. $R_{X,\infty}$ *is a subring of* $\mathbb{C}\left[\dfrac{1}{t}\right][[t]]$.

b. By the isomorphism $\varphi_{\mathrm{inv}} : z \mapsto \dfrac{1}{t}$, $\varphi_{\mathrm{inv}}(R_H)(\cong R_H^Z)$ *is a subring of* $\mathbb{C}\left[\dfrac{1}{t}\right]$.
We sometimes identify $\varphi_{\mathrm{inv}}(R_H)$ *with* R_H.

c. There is a surjection of ring $\varphi_\infty : R_X \to R_H(\cong R_H^Z)$*; for* $f \in R_X$*, there is* $g(z) \in R_H$ *such that*

$$(f)_\infty = g\left(\frac{1}{t}\right)(1 + d_{>0}(t)) \in R_{X,\infty},$$

where $(f)_\infty$ *means the germ at* ∞ *or* $(f)_\infty \in R_{X,\infty}$ *via* φ_H^X *in Proposition 2.14. It induces the surjection* $R_{X,\infty} \to R_H(\cong R_H^Z)$.

Proof By letting $g = \varphi_H^Z \circ \varphi_H^X(f)$, we find $g \in R_H$. The others are obvious. ∎

Lemma 2.19 plays a critical role to investigate the properties of the trace operator p_ϖ in Sect. 2.2.6, W-normalized differentials of ν^{I}, ν^{II}, ν^{III}, Σ and $d\Sigma$ in Sects. 2.4.2 and 2.4.3. Further, these correspondences also are essential in Appendix 3.9.4. Thus Lemma 2.19 is the key lemma in the Weierstrass Abelian function theory.

2.2.4 $R_{\mathbb{P}}$-Module R_X

This subsection is devoted to algebraic properties of the $R_{\mathbb{P}}$-module R_X. Its affine part is given by the quotient ring of $R_{\mathbb{P}}[y_\bullet]$. We recall Definition 2.4 and Lemma 2.13, and apply them to W-curves:

Proposition 2.20 *For* $\mathfrak{e}_i \in \mathfrak{E}_{H_X}$*, there exists a monic monomial* $\mathfrak{y}_{\mathfrak{e}_i}$ *in* R_X *whose weight is* $-\mathfrak{e}_i$*,* $(\mathfrak{y}_{\mathfrak{e}_0} = 1)$ *satisfying*

$$R_X = R_{\mathbb{P}} \oplus \bigoplus_{i=1}^{r-1} R_{\mathbb{P}}\mathfrak{y}_{\mathfrak{e}_i} = \bigoplus_{i=0}^{r-1} R_{\mathbb{P}}\mathfrak{y}_{\mathfrak{e}_i} = \langle \mathfrak{y}_{\mathfrak{e}_0}, \mathfrak{y}_{\mathfrak{e}_1}, \ldots, \mathfrak{y}_{\mathfrak{e}_{r-1}} \rangle_{R_{\mathbb{P}}}$$

with the relations

$$\mathfrak{y}_{\mathfrak{e}_i}\mathfrak{y}_{\mathfrak{e}_j} = \sum_{k=0}^{r-1} \mathfrak{a}_{ijk}\mathfrak{y}_{\mathfrak{e}_k}, \tag{2.25}$$

where $\mathfrak{a}_{ijk} \in R_{\mathbb{P}}$, $\mathfrak{a}_{ijk} = \mathfrak{a}_{jik}$, *especially* $\mathfrak{a}_{0jk} = \mathfrak{a}_{j0k} = \delta_{jk}$. *Further, we let* $\widehat{H}_{\mathfrak{e}} := \left\{ -\mathrm{wt}(f) \mid f \in \bigoplus_{\mathfrak{e}_i \in \mathfrak{E}_H^\times} R_{\mathbb{P}}\mathfrak{y}_{\mathfrak{e}_i} \right\}$ *and then* $\widehat{H}_{\mathfrak{e}} \subset H_X$.

Proof The generating formula is directly obtained from (2.10) by noting $\mathfrak{y}_{\mathfrak{e}_0} = 1$ and Lemma 2.13. $\widehat{H}_{\mathfrak{e}} \subset H_X$ is obvious. ■

The set $\mathfrak{Y}_X := \{\mathfrak{y}_0, \mathfrak{y}_{\mathfrak{e}_1}, \ldots, \mathfrak{y}_{\mathfrak{e}_{r-1}}\}$ is called the standard basis of R_X as an $R_{\mathbb{P}}$-module, which is essentially the same as $\widetilde{\mathfrak{y}}_{\overline{\mathfrak{e}}_i}$ in the proof of Proposition 2.8; $\mathfrak{Y}_X = \{\widetilde{\mathfrak{y}}_{\overline{\mathfrak{e}}_i}\} \cup \{1, y_s\}$, and thus we let ℓ_s be $y_s = \mathfrak{y}_{\mathfrak{e}_{\ell_s}}$. The $\mathfrak{a}_{\ell_s,k,\ell}$ in (2.25) corresponds to $A_{k\ell}$ in the proof of Proposition 2.8.

The following corollary is obvious:

Corollary 2.21 *In* $\mathbb{C}[x, y_\bullet]$, *the ideal generated by (2.25) is a subideal of* $\mathscr{I}_{R_X} := (f_1^X, f_2^X, \ldots, f_{k_X}^X)$ *in Proposition 2.14.*

As we regard an element in R_X as an element in an $R_{\mathbb{P}}$-module, we interpret a certain polynomial in R_X as an $R_{\mathbb{P}}$-monomial such that it is given by $\delta(x)\prod_{i=2}^{m_X} y_{r_i}^{h_i}$ for certain non-negative integers h_i and $\delta(x) \in R_{\mathbb{P}}$, e.g., $(x^2 - ax - b)y_{r_2}^2 y_{r_3}$. Since we also define the weight, wt_λ, on the $R_{\mathbb{P}}$-monomials, we can consider homogeneous polynomials as elements in the $R_{\mathbb{P}}$-module.

Remark 2.22 Corresponding to (2.20), the relation f_i^X in Proposition 2.14 is decomposed as

$$f_i^X = \left(\prod_{j=2} y_{r_j}^{h_j^{(i+)}}\right) - \delta_{f_i^X}(x)\left(\prod_{j=2} y_{r_j}^{h_j^{(i-)}}\right) + \text{lower weight terms with respect to } -\mathrm{wt} \tag{2.26}$$

as an $R_{\mathbb{P}}$-module, which is relevant to (2.25). Here $\delta_{f_i^X}(x)$ is an element of $R_{\mathbb{P}}$ whose degree is $h_1^{(i+)}$ in (2.20). The first and the second terms are homogeneous polynomials with respect to the Sato–Weierstrass weight wt.

2.2.4.1 Embedding of X into $(\mathbb{P}^2)^{m_X-1}$

The projection from X to $\mathbb{P}$ in Sect. 2.2.2.2 with the $R_{\mathbb{P}}$-module structure induces an embedding of X into $(\mathbb{P}^2)^{m_X-1}$ as follows (cf. Sect. 2.1). Besides $\{f_i^X \mid i = 1, \ldots, k_X\}$, we introduce a subset of polynomials $\{f_X^{(i)} \mid i = 2, \cdots, m_X\}$ of $\mathbb{C}[x, y_\bullet]$, which are R_X-analogue of $f_H^{(i)}$ in Lemma 2.12 such that $f_X^{(i)} = 0$ is an identity in R_X.

Proposition 2.23 *Let X be the W-curve in Proposition 2.8 with the affine ring $R_X = \mathbb{C}[x, y_\bullet]/(f_1^X, f_2^X, \ldots, f_{k_X}^X)$.*

1. *There are polynomials $A_i^{(j)} \in \mathbb{A}_{\mathbb{C}}[x]$ $(i = 2, 3, \ldots, \mathfrak{r}_j = r/(r, r_j), j = 2, \cdots, m_X)$ satisfying $A_i^{(j)} = \sum_{k=0}^{\lfloor i\bar{\mathfrak{r}}_j/\mathfrak{r}_j \rfloor} \lambda_{i,k}^{(j)} x^k$, where $\bar{\mathfrak{r}}_j = r_j/(r, r_j)$, $\lambda_{i,k}^{(j)} \in \mathbb{A}_{\mathbb{C}}$, $\lambda_{r,r_j}^{(j)} = 1$, and an irreducible polynomial,*
$$f_X^{(j)}(x, y_{r_j}) := y_{r_j}^{\mathfrak{r}_j} + A_1^{(j)} y_{r_j}^{\mathfrak{r}_j-1} + A_2^{(j)} y_{r_j}^{\mathfrak{r}_j-2} + \cdots + A_{\mathfrak{r}_j-1}^{(j)} y_{r_j} + A_{\mathfrak{r}_j}^{(j)}, \quad (2.27)$$
in $\mathbb{C}[x, y_{r_j}]$, especially $f_X^{(2)} = f_X$ in (2.8), so that $f_X^{(j)}$ satisfies the identity $f_X^{(j)}(x, y_{r_j}) = 0$ in R_X, and
2. *for the case of cyclic W-curve, $f_X^{(j)}(x, y_{r_j}) = 0$ is reduced to*
$$y_{r_j}^{\mathfrak{r}_j} = x^{\bar{\mathfrak{r}}_j} + \lambda_{\bar{\mathfrak{r}}_j-1}^{(r_j)} x^{\bar{\mathfrak{r}}_j-1} + \cdots + \lambda_1^{(r_j)} x + \lambda_0^{(r_j)}. \quad (2.28)$$

Proof The $m_X = 2$ case is trivial. Let us consider $m_X > 2$ case. Then Spec $(\mathbb{C}[x, y] /(f_X(x, y)))$ is singular and the commutative ring $\mathbb{C}[x, y]/(f_X(x, y))$ is not normal. We normalize it to obtain R_X, in which every element in $\mathcal{Q}(R_X)$ is expressed by a monic equation with coefficients in R_X. It means that y_{r_j} satisfies a certain relation $y_{r_j}^n + b_1 y_{r_j}^{n-1} + \cdots + b_{n-1} y_{r_j} + b_n = 0$ for certain positive integer n and $b_i \in R_X$. We show that when $n = r_j$, it is irreducible and $b_j \in R_{\mathbb{P}}$ as follows.

We remark that y_{r_j} is equal to an element of the standard basis $\mathfrak{Y}_X$ from the definition of y_{r_j} in the proof of Proposition 2.8; we take $\widetilde{j}$ such that $y_{r_j} = \mathfrak{y}_{\mathfrak{e}_{\widetilde{j}}}$. We apply the investigation of $A_{k\ell}$ for $\mathfrak{y}_{\mathfrak{e}_{\ell_s}} = y_s$ in the proof of Proposition 2.8 and $\mathfrak{a}_{\ell_s ik}$ in (2.25) to this $\mathfrak{y}_{\mathfrak{e}_{\widetilde{j}}} = y_{r_j}$ case. We introduce
$$\widetilde{A}_{i,k}^{(j)} := \mathfrak{a}_{\widetilde{j},i,k} - \mathfrak{y}_{\mathfrak{e}_{\widetilde{j}}}\delta_{i,k}, \quad \widetilde{b}_i := -\mathfrak{a}_{\widetilde{j},i,0} - \mathfrak{a}_{\widetilde{j},i,\ell_{\widetilde{j}}}\mathfrak{y}_{\mathfrak{e}_{\widetilde{j}}}, \quad (i, k \in \mathbb{Z}_r^\times \setminus \{\widetilde{j}\}), \quad (2.29)$$
and then we consider the $\mathfrak{y}_{\mathfrak{e}_{\widetilde{j}}} = y_{r_j}$ action on $\mathfrak{Y}_X$ in (2.25), which is described by
$$\sum_{\ell \in \mathbb{Z}_r^\times \setminus \{\widetilde{j}\}} \widetilde{A}_{k\ell}^{(j)} \mathfrak{y}_{\mathfrak{e}_\ell} = \widetilde{b}_k, \quad (k \in \mathbb{Z}_r^\times \setminus \{\widetilde{j}\}), \quad (2.30)$$
$$y_{r_j}^2 = \overline{A}_{\widetilde{j},0} + \overline{A}_{\widetilde{j}\widetilde{j}} y_{r_j} + \sum_{k \in \mathbb{Z}_r^\times \setminus \{\widetilde{j}\}} \overline{A}_{\widetilde{j},k} \mathfrak{y}_{e_k}, \quad (2.31)$$
where $\overline{A}_{\widetilde{j},\ell} := \mathfrak{a}_{\widetilde{j},\widetilde{j},\ell} \in \mathbb{C}[x]$.

If $(r_j, r) = 1$, the matrix $\widetilde{A}_{k\ell}^{(j)}$ is regular and we obtain (2.27) as in the proof of Proposition 2.8.

Thus we assume that r_j and r are not coprime. We recall $\mathfrak{r}_j^* = (r, r_j)$, $\mathfrak{r}_j = r/\mathfrak{r}_j^*$ and $\bar{\mathfrak{r}}_j = r_j/\mathfrak{r}_j^*$.

If the determinant of the $(r-2)\times(r-2)$ matrix $\widetilde{A}^{(j)}$ is not equal to zero, it is reduced to the above case. Then we obtain the formula f which is given by $y_{r_j}^r - x^{r_j} + \cdots$ and thus its image of φ_H^X is reduced to $Z_{r_j}^r - Z_r^{r_j}$, which is decomposed into $(Z_{r_j}^{\mathfrak{r}_j} - Z_r^{\overline{\mathfrak{r}}_j})(Z_{r_j}^{\mathfrak{r}_j(\mathfrak{r}_j^*-1)} + \cdots + Z_r^{\overline{\mathfrak{r}}_j(\mathfrak{r}_j^*-1)})$. If the formula f is irreducible, the infinity point ∞ in X must not be unique, which contradicts the uniqueness of ∞ in W-curve as mentioned in Remark 2.9. Thus the formula must be reduced to two formulae; one of them must be the Sato–Weierstrass weight, $-rr_j/\mathfrak{r}_j^*$ and contain the terms $y_{\mathfrak{r}_j}^r - x^{\overline{\mathfrak{r}}_j}$, which is equal to $f_X^{(j)}$ (2.27). If the other formula $h_X^{(j)}$ is algebraically independent to $f_X^{(j)}$, there are two projections, $\pi_{r,r_j} : X \to \mathrm{Spec}\,(\mathbb{C}[x, y_{r_j}]/(f_X^{(j)}))$ and $\pi'_{r,r_j} : X \to \mathrm{Spec}\,(\mathbb{C}[x, y_{r_j}]/(h_X^{(j)}))$. It also contradicts the uniqueness of the infinity ∞ at X in W-curves. Hence $f = (f_X^{(j)})^{\mathfrak{r}_j^*}$ up to a constant factor. Hence we have (2.27) and the identity $f_X^{(j)} = 0$.

Hence we further assume that the matrix $\widetilde{A}^{(j)}$ is singular, and its rank is $q(< r-2)$.

Now we consider the case that $q = 0$. Then in the relation (2.31), we show that $\overline{A}_{\widetilde{j},k}$ $(k \in \mathbb{Z}_r \setminus \{\widetilde{j}\})$ must vanish. If some of $\overline{A}_{\widetilde{j},k}$ does not vanish, there is an element $\widetilde{y} \in \bigoplus_{i\in\mathbb{Z}_r\setminus\{\widetilde{j}\}} R_{\mathbb{P}}\mathfrak{y}_{\mathfrak{e}_i}$ which is expressed by a meromorphic function of x and y_{r_j} from the relation. However, it means that there is a non-trivial relation in (2.30) and thus the rank q must not be zero. It contradicts the assumption. Hence $y_{r_j}^2$ is expressed as $\overline{A}_{\widetilde{j},0} + \overline{A}_{\widetilde{j}\widetilde{j}}y_{r_j}$; the order of $\overline{A}_{\widetilde{j},0}$ in x is $\overline{\mathfrak{r}}_j$. Accordingly, if $q = 0$, there exists an $R_{\mathbb{P}}$-submodule $\widetilde{R}' = R_{\mathbb{P}} \oplus R_{\mathbb{P}}y_{r_j}$, which forms a subring of R_X, $\mathbb{C}[x, y_{r_j}]/\,(y_{r_j}^2 - \overline{A}_{\widetilde{j},0} - \overline{A}_{\widetilde{j}\widetilde{j}}y_{r_j}) \subset R_X$. X is a covering of a hyperelliptic (or elliptic) curve, and $\mathrm{wt}(x)$ is divisible by two and $2|r$. It is obvious that $(y_{r_j}^2 - \overline{A}_{\widetilde{j},0} - \overline{A}_{\widetilde{j}\widetilde{j}}y_{r_j})$ is irreducible as an $R_{\mathbb{P}}$-module since X is not a covering of decomposed curve, and we obtain (2.27).

Hence we let $q \neq 0$ or $0 < q < r-2$. We introduce a subset $I := \{n_1, n_2, \ldots, n_q\}$ of $\mathbb{Z}_r^\times$ and a submodule $\widetilde{R}_X$ of the $R_{\mathbb{P}}$-module R_X defined by

$$\widetilde{R}_X = R_{\mathbb{P}} \oplus R_{\mathbb{P}}y_{r_j} \oplus \bigoplus_{i\in I} R_{\mathbb{P}}\mathfrak{y}_{\mathfrak{e}_i}.$$

We assume that $\widetilde{R}_X$ is closed for the y_{r_j} action on $\widetilde{R}_X$, i.e.,

$$y_{r_j}\widetilde{R}_X \subset \widetilde{R}_X. \tag{2.32}$$

The assumption enables us to find the regular submatrix $\widetilde{A}^I := (\widetilde{A}_{i,k}^{(j)})_{i,k\in I}$ of $\widetilde{A}^{(j)}$ satisfying

$$\sum_{k\in I} \widetilde{A}_{ik}^{(j)}\mathfrak{y}_{\mathfrak{e}_k} = \widetilde{b}_i, \quad i \in I. \tag{2.33}$$

We let $\overline{I} := I \cup \{0, r_j\}$, and $\mathfrak{Y}_X^{\overline{I}} := \{\mathfrak{y}_{\mathfrak{e}_i} \mid i \in \overline{I}\}$.

By considering the image of (2.33) under $\varphi_H^Z \circ \varphi_H^X$, the weight $-\mathrm{wt}$ of each component in $\widetilde{A}^I$ obviously leads to the fact that there is a sub-monoid $H' := \langle r, r_j, \mathfrak{e}_{n_1}, \dots, \mathfrak{e}_{n_q}\rangle$ such that $\langle r, r_j\rangle \subset H' \subset H_X$, and the set $\overline{I}$ is characterized by

$$\overline{I} = \{\ell \in \mathbb{Z}_r^\times \mid \mathfrak{e}_\ell \in H'\} \neq \varnothing, \quad \text{and} \quad \sum_{k\in\overline{I}} [\mathfrak{e}_k]_r = H'. \tag{2.34}$$

Then we find an expression $\mathfrak{y}_{e_k}$ of $k \in I$ as a meromorphic function of x and y_{r_j},

$$\mathfrak{y}_{\mathfrak{e}_k} = \frac{\widetilde{Q}_k(x, y_{r_j})}{|\widetilde{A}^I|(x, y_{r_j})}, \tag{2.35}$$

where $|\widetilde{A}^I|$ is a monic degree q polynomial of y_{r_j} with $R_{\mathbb{P}}$ coefficients, whereas the degree of $\widetilde{Q}_k$ in y_{r_j} is $q-1$. (2.35) means that $\mathfrak{y}_{\mathfrak{e}_\ell}$ satisfies the relation $|\widetilde{A}^I|\mathfrak{y}_{\mathfrak{e}_k} = \widetilde{Q}_k(x, y_{r_j})$.

Then by substituting (2.35) into $\mathfrak{y}_{\mathfrak{e}_k}$ in the relation $y_{r_j}^2 = \overline{A}_{\tilde{j},0} + \overline{A}_{\tilde{j}\tilde{j}} y_{r_j} + \sum_{k\in I} \overline{A}_{\tilde{j},k}$ $\mathfrak{y}_{e_k}$, we obtain

$$y_{r_j}^{q+2} + \overline{A}_1 y_{r_j}^{q+1} + \cdots + \overline{A}_{q+1} y_{r_j} + \overline{A}_{q+2} = 0,$$

where $\overline{A}_i$ is a certain element in $R_{\mathbb{P}}$. We state $(q+2)|r$ because $\varphi_H^X(y_{r_j}^{\mathfrak{r}_j} - x^{\bar{\mathfrak{r}}_j}) = 0$ and it must belong to $\varphi_H^{X-1}(\{0\})$. Thus we let $(q+2) = n\mathfrak{r}_j$ for an integer $n(\geq 1)$. However, due to $\varphi_H^X(y_{r_j}^{\mathfrak{r}_j} - x^{\bar{\mathfrak{r}}_j}) = 0$ again, $y_{r_j}^{\mathfrak{r}_j} - x^{\bar{\mathfrak{r}}_j}$ is equal to lower weight terms with respect to $-\mathrm{wt}$ because of the uniqueness of the ∞ in X. Thus the equation is reduced to multiply the same equation by order $\mathfrak{r}_j$ in y_{r_j}, due to the above arguments. It means that $H' \subset \langle r, r_j\rangle$, and thus $H' = \langle r, r_j\rangle$, i.e., $q = n\mathfrak{r}_j - 2$. Hence we obtain (2.27) as in the proof of Proposition 2.8, which is irreducible.

We now show (2.32). Assume that $y_{r_j}\widetilde{R}_X \setminus \widetilde{R}_X \neq \varnothing$ and let $\overline{I}^{\mathrm{c}} := \mathbb{Z}_r^\times \setminus (\overline{I})$. Obviously $y_{r_j}\mathfrak{y}_{\mathfrak{e}_i}$ belongs to $\widetilde{R}_X$ for $i \in I$ because of (2.33). Hence the assumption means

$$y_{r_j}^2 = \overline{A}_{\tilde{j},0} + \overline{A}_{\tilde{j}\tilde{j}} y_{r_j} + \sum_{i\in I} \overline{A}_{\tilde{j},i}\mathfrak{y}_{\mathfrak{e}_i} + \sum_{k\in\overline{I}^{\mathrm{c}}} \overline{A}_{\tilde{j},k}\mathfrak{y}_{\mathfrak{e}_k},$$

where there exists, at least, a non-vanishing $\overline{A}_{\tilde{j},\ell} \in R_{\mathbb{P}}$ for a certain $\ell \in \overline{I}^{\mathrm{c}}$, and $\overline{A}_{\tilde{j},i} \in R_{\mathbb{P}}$. By considering the Sato–Weierstrass weight of both sides, we have $-\mathrm{wt}(\overline{A}_{\tilde{j},\ell}\mathfrak{y}_{\mathfrak{e}_\ell}) = 2r_j \in H'$ and $\mathfrak{e}_\ell$ belongs to H'. It means $\ell \in \overline{I}$ and thus contradicts (2.34). Hence we show $y_{r_j}\widetilde{R}_X \subset \widetilde{R}_X$ or (2.32).

For the cyclic case, we have the global cyclic action on X and thus (2.27) is reduced to (2.9). ∎

Proposition 2.23 obviously leads us to the following observations. Using (2.27) in Proposition 2.23, we introduce $R_{X^{(i)}} := \mathbb{C}[x, y_{r_i}]/(f_X^{(i)})$, $(i = 2, \cdots, m_X)$ which are $R_{\mathbb{P}}$-modules and the unnormalized rings for $m_X > 2$, and their associated singular curves $X^{(i)}$ $(i = 2, \cdots, m_X)$ with the projection $\varpi_{X^{(i)}} : X^{(i)} \to \mathbb{P}$, $(\varpi_{X^{(i)}}(x, y_i) = x)$, $(i = 2, \ldots, m_X)$. Since $f_X^{(i)}$ is irreducible, $R_{X^{(i)}}$ is a subring of R_X, and R_X is the normalized ring of $R_{X^{(i)}}$. There are injective ring-homomorphisms $R_{\mathbb{P}} \xrightarrow{\iota_r^{(i)}} R_{X^{(i)}} \xrightarrow{\iota_{r,r_i}} R_X$; thus it induces the projections $\varpi_{r_i,r} : X \to X^{(i)}$ $((x, y_\bullet) \mapsto (x, y_{r_i}))$ and $\varpi_r^{(i)} : X^{(i)} \to \mathbb{P}$ $((x, y_{r_i}) \mapsto x)$; They satisfy the commutative diagrams

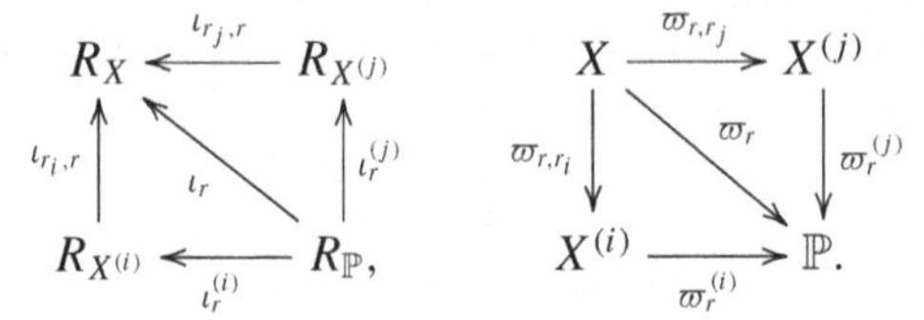

(2.36)

Further, we also define the tensor product of these rings $R_{X^{(2)}} \otimes_{R_{\mathbb{P}}} R_{X^{(3)}} \otimes_{R_{\mathbb{P}}} \cdots \otimes_{R_{\mathbb{P}}} R_{X^{(m_X)}}$, and its geometrical picture $X_{\mathbb{P}}^{[m_X-1]} := X^{(2)} \times_{\mathbb{P}} X^{(3)} \times_{\mathbb{P}} \cdots \times_{\mathbb{P}} X^{(m_X)}$. By identifying $\mathbb{C}[x, y_\bullet] / (f_X^{(2)}, f_X^{(3)}, \ldots, f_X^{(m_X)}) =: R_X^{\otimes[m_X-1]}$ with a ring $R_{X^{(2)}} \otimes_{R_{\mathbb{P}}} R_{X^{(3)}} \otimes_{R_{\mathbb{P}}} \cdots \otimes_{R_{\mathbb{P}}} R_{X^{(m_X)}}$, we have the natural projection $\varphi_{R_X^{\otimes[m_X-1]}} : R_X^{\otimes[m_X-1]} \to R_X$, i.e., $R_X = R_X^{\otimes[m_X-1]}/(f_1^X, \ldots, f_{k_X}^X)$, and the injection $\iota_{R_X^{\otimes[m_X-1]}} : R_{\mathbb{P}} \hookrightarrow R_X^{\otimes[m_X-1]}$. It induces the injection $\iota_{X_{\mathbb{P}}^{[m_X-1]}} : X \to X_{\mathbb{P}}^{[m_X-1]}$ and the projection $\prod_i \varpi_{X^{(i)}} : X_{\mathbb{P}}^{[m_X-1]} \to \mathbb{P}$.

Moreover, we also define the direct product of these rings $R_X^{[m_X-1]} := R_{X^{(2)}} \times R_{X^{(3)}} \times \cdots \times R_{X^{(m_X)}}$, and its geometrical picture $X^{[m_X-1]} := \prod X^{(i)} \subset (\mathbb{P}^2)^{m_X-1}$. Then we have the following proposition.

Proposition 2.24 *Expression (2.27) in Proposition 2.23 provides surjective and injective ring homomorphisms,*

$$\varphi_{R_X^{\otimes[m_X-1]}} : R_X^{\otimes[m_X-1]} \to R_X, \quad \iota_{R_X^{\otimes[m_X-1]}} : R_{\mathbb{P}} \hookrightarrow R_X^{\otimes[m_X-1]},$$

so that they satisfy the commutative diagrams

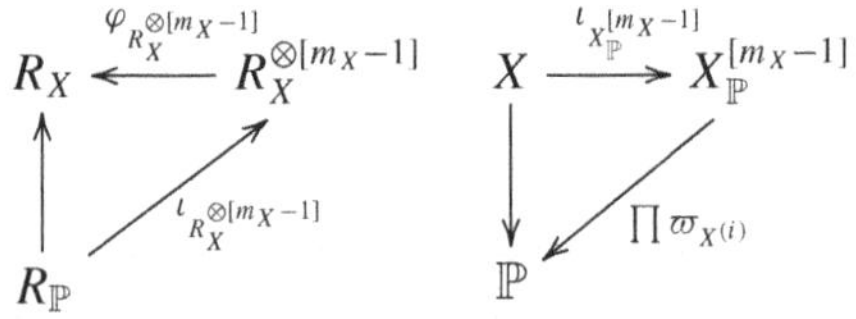

which is consistent with the projection $\varpi_{X_{\mathbb{P}}^{[m_X-1]}} : X_{\mathbb{P}}^{[m_X-1]} \to X$. They induce the homomorphism by the ideal $(x - x_2, x - x_3, \ldots, x - x_{m_X})$ and an embedding,

$$\varphi_{R_X^{[m_X-1]}} : R_X^{[m_X-1]} \to R_X, \quad \iota_{X^{[m_X-1]}} : X \hookrightarrow X^{[m_X-1]}(\subset (\mathbb{P}^2)^{m_X-1}).$$

Remark 2.25 Since the normalization of a ring is not unique in general, as mentioned in (2.15), the surjective ring homomorphism $\varphi_{R_X^{\otimes[m_X-1]}}$ is not injective except for $m_X = 2$. For example, in the (3, 4, 5) case, there is a surjective ring homomorphisms whose φ_H image is given by

$$\begin{aligned}&\mathbb{C}[Z_3, Z_4, Z_5]/(Z_4^3 - Z_3^4, Z_5^2 - Z_3^5)\\ &\to \mathbb{C}[Z_3, Z_3, Z_4]/(Z_4^2 - \zeta_3^a Z_3 Z_5, Z_4 Z_5 - Z_3^3, Z_5^2 - \zeta_3^{2a} Z_3^2 Z_4).\end{aligned}$$

Thus y and y_{r_j} are related via f_i^X's whose images of φ_H^X are binomial relations.

Direct computation gives the following relations based on (2.27).

Lemma 2.26 *1. For a point (x, y_{r_j}) satisfying $f_X^{(j)}(x, y) = 0$,*

$$f_{X,y}^{(j)}(x, y_{r_j})dy_{r_j} = -f_{X,x}^{(j)}(x, y_{r_j})dx, \tag{2.37}$$

where $f_{X,x}^{(j)} := \frac{\partial}{\partial x} f_X^{(j)}(x, y)$ and $f_{X,y}^{(j)} := \frac{\partial}{\partial y} f_X^{(j)}(x, y)$, and if $f_{X,y}^{(j)}$ does not vanish,

$$\frac{dy_{r_j}}{dx} = -\frac{f_{X,x}^{(j)}}{f_{X,y}^{(j)}}(x, y_{r_j}).$$

2. $f_X^{(j)}(x, y') = \sum_{\ell=0}^{\mathfrak{r}_j} \frac{1}{\ell!} f_{X,y^\ell}^{(j)}(x, y)(y' - y)^\ell$, where $f_{X,y^\ell}^{(j)} := \frac{\partial^\ell}{\partial y^\ell} f_X^{(j)}(x, y)$ and $f_{X,y^{\mathfrak{r}_j}}^{(j)} = \mathfrak{r}!$.

3. $(\delta_{Y,y} f_X^{(j)})(x, Y, y) := \dfrac{f_X^{(j)}(x, Y) - f_X^{(j)}(x, y)}{Y - y} = \sum_{\ell=0}^{\overline{\mathfrak{r}}_j-1} A_\ell^{(j)}(x) \sum_{i=0}^{\mathfrak{r}_j-\ell-1} Y^i y^{\mathfrak{r}_j-\ell-i-1}$.

2.2.5 *Covering Structure in W-Curves*

Here we follow [Gri1989, Ku2004, Sti2008] to investigate the covering structure in W-curves.

2.2.5.1 The r-Sheeted Holomorphic Covering

As mentioned in Remark 2.10, let us consider the Riemann sphere $\mathbb{P}$ and $R_{\mathbb{P}} = \mathbf{H}^0(\mathbb{P}, \mathscr{O}_{\mathbb{P}}(*\infty))$. We identify $R_{\mathbb{P}}$ with its affine part $R_{\mathbb{P}}^\circ = \mathbb{C}[x]$ and its quotient field is denoted by $\mathbb{C}(x) = \mathscr{Q}(R_{\mathbb{P}})$. The quotient field $\mathscr{Q}(R_X) = \mathbb{C}(X)$ of R_X is an extension of the field $\mathbb{C}(x)$.

In general, let M and L be fields such that M is a finite field extension of L. We recall that a finite field extension M/L is said to be a Galois extension if the order of the automorphism group $\mathrm{Aut}(M/L) = \{s : M \to M \mid s \text{ is an isomorphism with } s(x) = x \text{ for every } x \in L\}$ is equal to $[M : L]$, and then we write $\mathrm{Gal}(M/L) = \mathrm{Aut}(M/L)$, which is called the Galois group of M/L [Ku2004, Sti2008]. Further, the geometry related to the integral rings in M and L shows the Galois covering. Thus we can regard $\mathscr{Q}(R_X)/\mathscr{Q}(R_{\mathbb{P}})$ as the Galois extension for a certain X, though, in general, $\mathscr{Q}(R_X)/\mathscr{Q}(R_{\mathbb{P}})$ is not a Galois extension (Fig. 2.1).

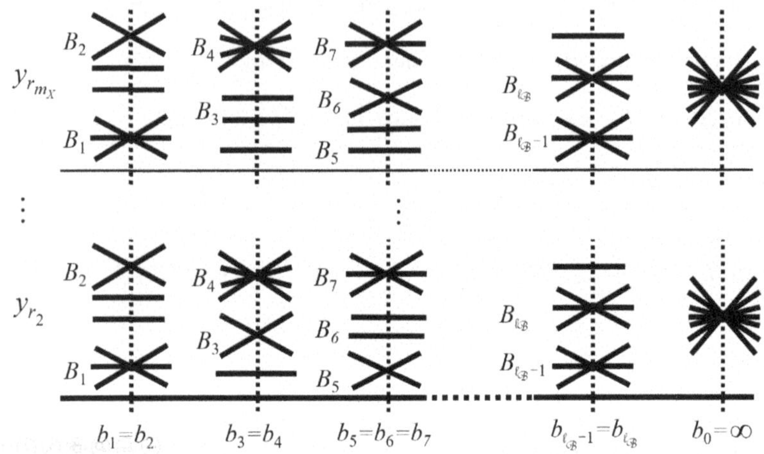

Fig. 2.1 Covering space

In order to show its geometrical meaning, we recall the Riemann existence theorem and its consequence [Har2003, Ser1989]: In general, let V be a smooth connected complex algebraic curve, which we also regard as a one-dimensional algebraic variety with the Zariski topology, as a complex analytic space and as a topological space with respect to the classical topology; it is the GAGA principle [Ser1989, Chap. 6]. Then the following categories are equivalent [Har2003];

1. finite étal covers of the variety V,
2. finite analytic covering maps of V, and
3. finite covering space of the topological space V.

Thus we regard $\mathscr{Q}(R_X)/\mathscr{Q}(R_{\mathbb{P}})$ as a field extension, and we consider the W-curve X as a (finite analytic) covering of $\mathbb{P}$ following Forster's book [Fo1981].

Let Z and Y be topological spaces. A continuous mapping $f : Y \to Z$ between two locally compact spaces is called *proper* if the preimage of every compact set is compact. A mapping $\varpi : Y \to Z$ is called a *covering map* if the following holds. Every point $z \in Z$ has an open neighborhood U and a discrete set J such that its preimage $\varpi^{-1}(U)$ can be represented as

$$\varpi^{-1}(U) = \bigcup_{j \in J} V_j,$$

where V_j $(j \in J)$ are disjoint open subsets of Y, and every mapping $\varpi|_{V_j} V_j \to U$ is a homeomorphism or ϖ is a local homeomorphism. For $z \in Z$, the set $\varpi^{-l}(z)$ is called the fiber of ϖ over z. If $y \in \varpi^{-l}(z)$, then one says that the point y lies over z.

In general, if Y and Z are locally compact spaces and the local homeomorphism $\varpi : Y \to Z$ is proper, ϖ is a covering map. For topological spaces Y' and Z with covering map $\varpi' : Y' \to Z$, the continuous map $f : Y \to Y'$ is fiber-preserving map if $\varpi' = \varpi \circ f$.

When Y is a Riemann surface and f is a meromorphic function, $f : X \to \mathbb{P}$ is a holomorphic mapping if $f(p)$ is defined by $\infty \in \mathbb{P}$ for an each pole p of f. When Y and Z are Riemann surfaces and $\varpi : Y \to Z$ is a non-constant holomorphic map, a point $y \in Y$ is called a ramification point, and its image $\varpi(y)$ is called a branch point[4] if there is no neighborhood V of y such that $\varpi|_V$ is injection. The holomorphic map ϖ is called an unbranched holomorphic map if it has no branch points, and otherwise, ϖ is called a branched holomorphic map.

Due to [Fo1981, Theorem 4.24], we have the fact:

Proposition 2.27 *Suppose Y and Z are Riemann surfaces and $\varpi : Y \to Z$ is a proper non-constant holomorphic map. Then there exists a natural number n such that ϖ takes value $c \in Z$, counting multiplicities, n times.*

Proof Let $B \subset Z$ be the set of branch points and let $Z' := Z \setminus B$, and $Y' := \varpi^{-1}(Z')$. Let n be the number of sheets of the unbranched covering. For $b \in B$, assume that $\varpi^{-1}(b) = \{y_1, \dots, y_\ell\}$. Then there are local parameters t_Y at y_i and t_Z at b such that $t_Z = c\, t_Y^{k_i}$, where $c \in \mathbb{C}^\times$, where k_i is called the multiplicity. In other words, there exist disjoint neighborhoods U_j of y_j and V_j of b such that for every $z \in V_i \setminus \{b\}$, the set $\varpi^{-1}(z) \cap U_j$ consists of exactly k_j points for $j = 1, \dots, \ell$. Since Y and Z are locally compact and ϖ is proper and discrete, there exist V_j' such that $\varpi^{-1}(V_j') \subset V'$ for the neighborhoods of $\varpi^{-1}(z)$, and the set of $\varpi^{-1}(z)$ is finite in general. So we can find a neighborhood $V \subset V_1 \cap \cdots \cap V_\ell$ of b such that $\varpi^{-1}(V) \subset U_1 \cup \cdots \cup U_\ell$. Then, for any point $z \in V \cap Z'$, we have that $\varpi^{-1}(z)$ consists of $k_1 + \cdots + k_\ell$ points. On the other hand, for $z \in Z'$, the cardinality of $\varpi^{-1}(z)$ is equal to n. So $n = k_1 + \cdots + k_\ell$. ∎

We refer to such a proper non-constant holomorphic map in Proposition 2.27 as an *n-sheeted holomorphic covering map* following [Fo1981, p. 30]. It is obvious that the n-sheeted holomorphic covering maps are allowed to have ramification points. We note that they are not covering maps at the ramification points, but we follow the conventions in the theory of compact Riemann surfaces; the Galois covering in the following definition should also be mentioned.

Definition 2.28 Suppose Z and Y are connected Hausdorff spaces and $\varpi : Y \to Z$ is a covering map. We call a fiber-preserving homeomorphism $f : Y \to Y$ *a covering transformation* of this covering. With the operation of the composition of mappings, the set of all covering transformations of $\varpi : Y \to Z$ forms a group, denoted by

[4] We follow the conventions on branch and ramification points of Namba [Nam1979].

$\mathrm{Aut}(Y/Z)$ or $\mathrm{Aut}(\varpi)$. If for every pair of points $y_0, y_1 \in Y$ with $\varpi(y_0) = \varpi(y_1)$, there exists a covering transformation $f : Y \to Y$ such that $f(y_0) = y_1$, we call it a *Galois covering*.

As mentioned above, for $B \subset Z$ be the set of branch points, let $Z' := Z \setminus B$, and $Y' := \varpi^{-1}(Z')$ for a non-constant holomorphic map $\varpi : Y \to Z$. Then $\varpi : Y \to X$ is called a Galois covering if the covering $\varpi : Y' \to Z'$ is Galois as connected Hausdorff spaces.

Following the above description, we consider the W-curve X. The covering $\varpi_r = \varpi_x : X \to \mathbb{P}$ is obviously the an r-sheeted holomorphic covering.

If there is an automorphism $\mathrm{Aut}(X/\mathbb{P})$, i.e., $\varpi_r : X \to \mathbb{P}$ is the Galois covering, we have the Galois group $\mathrm{Gal}(\mathcal{Q}(R_X)/\mathcal{Q}(R_{\mathbb{P}})) = \mathrm{Aut}(X/\mathbb{P}) = \mathrm{Aut}(\varpi_r)$, which is denoted by G_X.

Even when $\varpi_r : X \to \mathbb{P}$ is not a Galois covering, we find a group action for the fiber $\varpi_r^{-1}x$ at each $x \in \mathbb{P}$ since for a point $x \in \mathbb{P}$, the Weierstrass canonical form $f_X(x, y_r)$ (2.7) can be regarded as a monic polynomial of a number field $\mathbb{C}$; the coefficients $A_i(x)$ belong to $\mathbb{C}$ and we can apply the ordinary Galois theory to the roots of the algebraic equation over the number field $\mathbb{C}$. Some of them are trivial due to the degenerating or the ramification, but we find a group governing the roots of the equation $f_X(x, y_r) = 0$, which is denoted by $G_{X,x}$.[5] Further, as mentioned in Proposition 2.31 for the case of the Galois action G_X, each y_{r_j}, $(j = 3, \ldots, m_X)$ is also acted on by $G_{X,x}$ through the equation $f_X^{(j)}(x, y_j) = 0$ in Proposition 2.23 and the condition that $(x, y_\bullet)$ lies on X even for the action.

We may also consider the projections $\varpi_{r_j} : X \to \mathbb{P}$, $((x, y_\bullet) \mapsto y_{r_j})$. These ϖ_{r_j} are finite branched coverings. A ramification point of ϖ_{r_j} is defined as a point of such that it is not biholomorphic at the point. The image ϖ_{r_j} of the ramification point is called the branch point of ϖ_{r_j}. Let the set of these ramification points $B_{r_j,i}$ of $\varpi_{r_j} : X \to \mathbb{P}$ be denoted by $\mathfrak{B}_{X,r_j} := \{B_{r_j,i} \mid i = 0, 1, 2, \ldots, \ell_{\mathfrak{B},r_j}\}$ and their image by $\mathfrak{B}_{\mathbb{P},r_j} = \varpi_{r_j}(\mathfrak{B}_{X,r_j})$. Since $\mathfrak{B}_{X,r_j}$ contains $\infty \in X$, let $B_{r_j,0} = \infty \in X$.

We basically focus on the an r-sheeted holomorphic covering $\varpi_x = \varpi_r : X \to \mathbb{P}$:

Definition 2.29 Let $\mathfrak{B}_X := \mathfrak{B}_{X,r} = \{B_i\}_{i=0,\ldots,\ell_{\mathfrak{B}}}$ and $\mathfrak{B}_{\mathbb{P}} := \varpi_r(\mathfrak{B}_X) = \{b_i\}_{i=0,\ldots,\ell_{\mathfrak{B}}}$, where $\ell_{\mathfrak{B}} := \#\mathfrak{B}_X - 1$, $B_i := B_{r,i}$, $B_0 = \infty \in X$ and $b_i := \varpi_r(B_i)$.[6]

By considering (2.37), we have the following results.

Lemma 2.30 *Every element B_i in $\mathfrak{B}_X \setminus \{\infty\}$ is given by the point $(x, y_\bullet)$ at which there exists, at least, a certain j in $\{2 \le j \le m_X\}$ such that* $\dfrac{\partial f_X^{(j)}(x, y_{r_j})}{\partial y_{r_j}} = 0$ *and*

$$\frac{\partial f_X^{(j)}(x, y_{r_j})}{\partial x} \neq 0.$$

---

[5] Roughly speaking, when each $G_{X,x}$ does not depend on the position x, it can be globally defined and agrees with G_X, e.g., a cyclic group for the cyclic Weierstrass curve (2.17). It is difficult to find G_X even if $G_{X,x}$ of X has the property.

[6] In Sects. 2.7–2.9 and 3.6–3.7, we adopt a different convention of assigning B_0 to a finite branch point of interest.

Proof When $\dfrac{\partial f_X^{(i)}(x, y_{r_i})}{\partial x} = 0$ and $\dfrac{\partial f_X^{(i)}(x, y_{r_i})}{\partial y_{r_j}} = 0$ at a point $P \in X$, the point means the singular point as the plane curve given by $f_X^{(j)}(x, y_{r_j}) = 0$ at $\varpi_{r,r_i}(P)$. However, since X is not singular, there exists j satisfying the condition. Then at the point, dx is identically zero, and thus at the point, x is not a local parameter of X. Thus P must be an element in $\mathfrak{B}_X \setminus \{\infty\}$, $P = B_i$. It means $dx = d(t^{e_i})(c + d_{>0}(t))$ for a certain $c \in \mathbb{C}^\times$, and a positive integer $e_i > 1$ in terms of the local parameter t such that $t(B_i) = 0$, and there exists j such that $dy_{r_j} = dt$ and $dy_{r_i} = t^{f_i} dt(c_i + d_{>0}(t))$ $(i \neq j, f_i \geq 0)$, $c_i \in \mathbb{C}^\times$. ■

The e_i appearing in the proof for the ramification point B_i in Lemma 2.30 is called *the ramification index*, and denoted by e_{B_i}.

When X is the Galois coverings over $\mathbb{P}$, we can consider the Galois group action on $X^{(j)}$ for the finite branch coverings over $\varpi_r^{(j)} : X^{(j)} \to \mathbb{P}$, and we have the group $G_{r_j} = \mathrm{Aut}(\varpi_r^{(j)})$ action. The Galois group G_X could be thought of as consisting of G_{r_j}.

Let us consider the G_{r_j}. For fixing $x = \varpi_r^{(j)}((x, y_{r_j}))$ in $X^{(j)}$, we have

$$f_X^{(j)}(x, Y) = (Y - y_{r_j}^{(1)})(Y - y_{r_j}^{(2)}) \cdots (Y - y_{r_j}^{(\mathfrak{r}_j)}). \tag{2.38}$$

Every group action on $R_{X^{(j)}} = \mathbb{C}[x, y_{r_j}]/(f_X^{(j)}(x, y_{r_j}))$ is given as an element of a subgroup of the permutation group $\mathfrak{S}_{\mathfrak{r}_j} = \left\{ \tau = \begin{pmatrix} 1 & 2, \dots, \mathfrak{r}_j \\ i_1 & i_2, \dots, i_{\mathfrak{r}_j} \end{pmatrix} \right\}$,

$$G_{r_j} := \left\{ \widehat{\zeta_{\tau,j}} \;\middle|\; \widehat{\zeta_{\tau,j}} \in \mathrm{Aut}(\varpi_r^{(j)}) \right\} \subset \mathfrak{S}_{\mathfrak{r}_j}$$

so that its group action on $y_{r_j}^{(i_\ell)}$ is given by $y_{r_j}^{(i_\ell)} = \widehat{\zeta_{\tau,j}} y_{r_j}^{(\ell)}$ [Gri1989, II Lemma 5.6] and $\zeta_{i,i'}^{(j)} \infty = \infty$. In the case of hyperelliptic curves, m_X must be two and the group G_{2g+1} is the hyperelliptic involution.

We note that X is embedded in $\mathbb{P}^{m_X}$ by the projectivization but we focus on a geometrical structure, i.e., for a point $P \in X$, a unique component $(x, y_s^{(j_2)}, y_{r_3}^{(j_3)}, \dots, y_{r_{m_X}}^{(j_{m_X})}) = \prod_i^{m_X} \varpi_{r_i}(P) \in (\mathbb{P}^1)^{m_X}$, (recalling $s = r_2$). The Galois action $\widehat{\zeta}$ on X can be represented by the permutation of each component in G_{r_j}; For the action $\widehat{\zeta_{\tau,2}}$ on $X^{(2)}$ such that $y_s^{(j_2)}$ is replaced with $y_s^{(j_2')}$, the other components $y_{r_3}^{(j_3')}, \dots, y_{r_{m_X}}^{(j_{m_X}')}$ are also exchanged uniquely. It means that for a permutation $\tau \in \mathfrak{S}_n$ satisfying $\widehat{\zeta_{\tau,2}} \in G_s$, there exists a unique map $\iota_G : G_s \to \prod_j G_{r_j}$ $(\iota_G(\widehat{\zeta_{\tau,2}}) = (\widehat{\zeta_{\tau_3,3}}, \dots, \widehat{\zeta_{\tau_{m_X},m_X}}) \in \prod_j G_{r_j})$, which induces $(\widehat{\zeta_{\tau_3,3}}, \dots, \widehat{\zeta_{\tau_{m_X},m_X}}) \in \prod_{i=3}^{m_X} G_{r_i}$ so that $(x, \widehat{\zeta_{\tau,2}} y_s^{(j_2)}, \dots, \widehat{\zeta_{\tau_{m_X},m_X}} y_{r_{m_X}}^{(j_{m_X})})$ should be on the curve X and represents the automorphism

$$\widehat{\zeta_\tau}(x, y_s^{(j_2)}, \ldots, y_{r_{m_X}}^{(j_{m_X})}) := (x, \widehat{\zeta}_{\tau,2} y_s^{(j_2)}, \widehat{\zeta}_{\tau_3,3} y_s^{(j_3)}, \ldots, \widehat{\zeta}_{\tau_{m_X},m_X} y_{r_{m_X}}^{(j_{m_X})}). \tag{2.39}$$

This investigation shows the following proposition:

Proposition 2.31 *Let $\varpi_r : X \to \mathbb{P}$ be the Galois covering. The Galois group G_X for $\varpi_r : X \setminus \mathfrak{B}_X \to \mathbb{P}$ is realized as the group action on the component expression of the curve, i.e., $\widehat{\zeta_\tau}(x, y_\bullet)$ as in (2.39) for each $x \in X \setminus \mathfrak{B}_X$.*

Especially for the cyclic W-curve and $\widehat{\zeta}_r \in G_X$ ($\widehat{\zeta}_r \neq \mathrm{id}$, $\widehat{\zeta}_r^r = \mathrm{id}$),

$$\widehat{\zeta}_r(x, y, y_{r_3}, \ldots, y_{r_{m_X}}) = (x, \zeta_r y, \zeta_r^{i_3} y_{r_3}, \ldots, \zeta_r^{i_{m_X}} y_{r_{m_X}}),$$

where i_j is a certain element in $\mathbb{Z}_r^\times$ ($j = 3, \ldots, m_X$) and $\zeta_r := \mathrm{e}^{2\pi\sqrt{-1}/r}$.

Remark 2.32 We also note that $G_{X,x}$ for a point $x \in \mathbb{P}$ is similarly realized as mentioned in Proposition 2.31. We refer to $G_{X,x}$ as the *local Galois action* or simply the group action, which determines the behavior of the Abelian differentials.

Morrison and Pinkham investigated when the Galois group $G_H = \mathbb{Z}/r\mathbb{Z}$ acting on R_H via $\mathscr{Q}(R_H)$ is extended to X [MorP1986]. It is a quite an interesting problem and they call such a Weierstrass point a *Galois Weierstrass point* and its numerical semigroup *a Galois Weierstrass semigroup*. They showed that there are relatively few Galois Weierstrass semigroups in the Weierstrass semigroups.

2.2.5.2 Riemann–Hurwitz Theorem

Let us consider the behaviors of the covering $\varpi_r : X \to \mathbb{P}$, including the ramification points. The Riemann–Hurwitz theorem [Fu1969, Ku2004] shows the following:

Proposition 2.33 *We have the following relations for the covering $\varpi_r : X \to \mathbb{P}$:*

1. *Generally,*

$$2g - 2 = -2r + \sum_{i=1}^{\ell_{\mathfrak{B}}} (e_{B_i} - 1) + (r - 1) = \sum_{i=1}^{\ell_{\mathfrak{B}}} (e_{B_i} - 1) - (r + 1), \tag{2.40}$$

2. *For the cyclic W-curve case, $\ell_{\mathfrak{B}} = \ell_{\mathfrak{B}}^{(c)} := [\mathscr{Q}(R_X) : \mathscr{Q}(R_{\mathbb{P}})]$ and*

$$2g - 2 = \ell_{\mathfrak{B}}^{(c)}(r - 1) - (r + 1).$$

3. *For the cyclic plane W-curve ($m_X = 2$) case, $\ell_{\mathfrak{B}}^{(c)} = s$ and*

$$2g - 2 = s(r - 1) - (r + 1).$$

Lemma 2.34 *For $B_i \in \mathfrak{B}_X$ and $\varpi_{y_j}(B_i) =: b_{i,j}$,* $\mathrm{div}(y_{r_j} - b_{i,j})$ *is expressed by* $\mathrm{div}(y_{r_j} - b_{i,j}) = d_{i,j}B_i + E_{B_i}^{(j)} - r_j\infty$, *where $E_{B_i}^{(j)}$ is an effective divisor whose degree is $r_j - d_{i,j}$ and $d_{i,j}$ is a positive integer such that for every B_i, there is $j \in \{2, \ldots, m_X\}$ satisfying $d_{i,j} = 1$,*

Proof From Lemma 2.30, they are proved. ■

Due to Lemma 2.30, the Riemann–Hurwitz relation (2.40) in Proposition 2.33 corresponds to a divisor of dx:

Corollary 2.35

$$\mathrm{div}(dx) = \sum_{i=1}^{\ell_{\mathfrak{B}}} (e_{B_i} - 1)B_i - (r+1)\infty.$$

2.2.6 *Trace in the Covering Structure*

Since the trace property in the covering structure associated with W-curves X plays a crucial role, we mention it here. It is known that the Riemann–Hurwitz relation and the divisor of dx are obtained based on the Dedekind different [Ku2004, Theorem 15.11]. In this subsection, we consider the properties of the $R_{\mathbb{P}}$-module R_X, which are related to the trace, the complementary module, and the Dedekind different [Ku2004, Chap. 15] based on [KomMP2022a].

2.2.6.1 Trace Structure in $\mathbb{Q}[x]/\mathbb{Q}$

We start with recalling primitive results:

Lemma 2.36 *For an irreducible polynomial f in $\mathbb{Q}[x]$ given by $f(x) = x^n + \lambda_{n-1}x^{n-1} + \cdots + \lambda_0 = (x-a_1)^{\ell_1}\cdots(x-a_m)^{\ell_m}$, where a_i's are mutually disjoint elements in the algebraic closure $\overline{\mathbb{Q}}$ and*

$$h(x, x') := \frac{f(x) - f(x')}{x - x'}, \quad p(x, x') := \lim_{\substack{y \to x \\ z \to x'}} \frac{h(y, z)}{f'(y)} \in \mathbb{Q}(x, x'),$$

we have the results

$$h(x, x') \in \mathbb{Q}[x, x'], \quad p(a_i, a_j) = \begin{cases} 1 & \text{for } i = j, \\ 0 & \text{otherwise.} \end{cases}$$

Here the limit in the definition means that we employ l'Hopital's rule.

Proof Obviously $h(x, x') = \sum_{j=0}^{n-1} \lambda_{j+1} \sum_{i=0}^{j} x^{j-i} x'^i$ and thus $h(x, x')$ is a polynomial in $\mathbb{Q}[x, x']$. For $a_i \neq a_j$, it is obvious $p(a_i, a_j) = 0$, whereas we obtain

$$h(a_i + t, a_i + s) = \frac{t^{\ell_i} - s^{\ell_i}}{t - s} \prod_{j \neq i} (a_i - a_j)^{\ell_j} (1 + d_{>0}(t, s))$$

for every t, s and thus $p(a_i, a_i) = 1$. ■

Lemma 2.37 *For* $f(x, z) = x^{r-1} + x^{r-2}z + x^{r-3}z^2 + \cdots + xz^{r-2} + z^{r-1}$ *and* $\zeta_r = \mathrm{e}^{2\pi\sqrt{-1}/r}$, *we have the identity*

$$x^{r-i-1}z^i = \frac{1}{r} \sum_{j=0}^{r-1} f(\zeta_r^{-j} x, \zeta_r^{(r-i-1)j} z).$$

Lemma 2.38 *For* $f(x) = (x - a_1) \cdots (x - a_r)$ *and a non-negative integer* $k < r$, *where* a_i*'s are mutually disjoint elements in* $\mathbb{C}$,

$$\sum_{i=1}^{r} \frac{a_i^k}{f'(a_i)} = 0.$$

Proof It is proved from Lemma 4.8 1; $\chi_{k,g-1} = 1$ and $g = r$. ■

2.2.6.2 Trace in $R_X/R_{\mathbb{P}}$

We review the general results of R_X as a ring extension of $R_{\mathbb{P}}$ following [Ku2004] (see Appendix 2.11). Let us consider the covering structures of $\varpi_r : X \to \mathbb{P}$ to discriminate its lifted points, and the enveloping field $\mathscr{Q}(R_X)^e = \mathbb{C}(X)^e := \mathbb{C}(X) \otimes_{\mathbb{C}(x)} \mathbb{C}(X) = \mathscr{Q}(R_X) \otimes_{\mathscr{Q}(R_{\mathbb{P}})} \mathscr{Q}(R_X)$.

The field extension $\mathbb{C}(X)/\mathbb{C}(x)$ induces the extension ring R_X of $R_{\mathbb{P}}$. As mentioned in Appendix 2.11, we consider the dual of R_X with respect to $R_{\mathbb{P}}$,

$$\omega_{R_X/R_{\mathbb{P}}} := \mathrm{Hom}_{R_{\mathbb{P}}}(R_X, R_{\mathbb{P}}),$$

which is a free $R_{\mathbb{P}}$-module. For $R_X = \oplus_{i=0}^{r-1} R_{\mathbb{P}} \mathbf{y}_i$ (i.e., $\mathbf{y}_i = \mathfrak{y}_{\mathfrak{e}_i}$), it is expressed by the dual basis $\{\mathbf{y}_i^*\}_{i=0,\ldots,r-1}$ such that

$$\omega_{R_X/R_{\mathbb{P}}} = \bigoplus_{i=0}^{r-1} R_{\mathbb{P}} \mathbf{y}_i^*, \quad \text{and} \quad \mathbf{y}_i^* \mathbf{y}_j = \delta_{i,j}. \tag{2.41}$$

Here $\mathbf{y}_i^*$ and $\mathbf{y}_j$ correspond to the point in the fiber which lies over $x \in R_{\mathbb{P}}$. We call an element $\tau \in \omega_{R_X/R_{\mathbb{P}}}$ a *trace* if it satisfies

$$\omega_{R_X/R_{\mathbb{P}}} = R_X \cdot \tau.$$

(Here, we use the symbol τ for the trace only in this subsection and Appendix 2.11, whereas τ means the matrix in the Siegel upper half space in the other subsections.) Further, we can define the standard trace $\tau_{R_X/R_{\mathbb{P}}} := \sum_{i=0}^{r-1} \mathbf{y}_i \circ \mathbf{y}_i^*$; we note that $\tau_{R_X/R_{\mathbb{P}}}$ is one of the traces since $R_X/R_{\mathbb{P}}$ is separable.

For a trace τ, there is a dual basis $\{\widehat{\mathbf{y}}_i\}_{i=0,\dots,r-1}$ of R_X with respect to τ defined by $\tau(\widehat{\mathbf{y}}_i \mathbf{y}_j) = \delta_{ij}$.

For the enveloping algebra $R_X^e := R_X \otimes_{R_{\mathbb{P}}} R_X$, there is the canonical multiplication map,

$$\mu : R_X \otimes_{R_{\mathbb{P}}} R_X \to R_X, \quad \mu(a \otimes b) = a \cdot b.$$

We introduce the annihilator of the diagonal, the kernel $\mathrm{Ker}\mu$, of μ,

$$\mathrm{Ann}_{R_X^e}(\mathrm{Ker}\mu) = \{z \in R_X^e \mid z \cdot \mathrm{Ker}\mu = 0\}.$$

Since there are natural two embeddings,

$$R_X \to R_X^e \quad (a \to a \otimes 1) \quad \text{and} \quad R_X \to R_X^e \quad (a \to 1 \otimes a),$$

for every $a \in R_X$, we obtain

$$(a \otimes 1 - 1 \otimes a) \cdot \mathrm{Ann}_{R_X^e}(\mathrm{Ker}\mu) = 0.$$

Then we have the isomorphism as an R_X-module,

$$\varphi : \mathrm{Ann}_{R_X^e}(\mathrm{Ker}\mu) \longrightarrow \mathrm{Hom}_{R_X}(\omega_{R_X/R_{\mathbb{P}}}, R_X),$$

which induces the correspondence; for every trace $\tau \in \omega_{R_X/R_{\mathbb{P}}}$ satisfying $\sum_{i=0}^{r-1} \tau(\widehat{\mathbf{y}}_i)\mathbf{y}_i = 1$ or $\tau(\widehat{\mathbf{y}}_i \mathbf{y}_j) = \delta_{i,j}$, there is a unique element $\Delta_\tau = \sum_{i=0}^{r-1} \widehat{\mathbf{y}}_i \otimes \mathbf{y}_i \in \mathrm{Ann}_{R_X^e}(\mathrm{Ker}\mu)$. Then the standard trace in $R_X/R_{\mathbb{P}}$ given by $\tau_{R_X/R_{\mathbb{P}}} = \sum_{i=0}^{r-1} \mathbf{y}_i \circ \mathbf{y}_i^*$ such that $\tau_{R_X/R_{\mathbb{P}}} = \mu(\Delta_\tau) \cdot \tau$.

We can define the complementary module $R_X^{\mathfrak{c}}$ over $R_X/R_{\mathbb{P}}$ for a trace $\tau \in \omega_{R_X/R_{\mathbb{P}}}$. We fix the trace by the standard trace $\tau_{R_X/R_{\mathbb{P}}}$. For the standard trace $\tau_{R_X/R_{\mathbb{P}}} \in \omega_{R_X/R_{\mathbb{P}}}$, we define the complementary module $R_X^{\mathfrak{c}}$ over $R_X/R_{\mathbb{P}}$ with respect to the standard trace $\tau_{R_X/R_{\mathbb{P}}}$ by

$$R_X^{\mathfrak{c}} := \{z \in \mathscr{Q}(R_X) \mid \tau_{R_X/R_{\mathbb{P}}}(za) \in R_{\mathbb{P}}, \ \ \forall a \in R_X\}. \tag{2.42}$$

It is obvious that the localization $(R_X^{\mathfrak{c}})_P$ at $P \in X$ equals $R_{X,P}^{\mathfrak{c}}$ from the definition. Since for a point $P \in X$, $R_{X,P}$ is a principal ideal domain, every ideal is generated by a certain element in $R_{X,P}$. There is an element $\mathfrak{h}_{X,P} \in R_{X,P}$ such that $R_{X,P}^{\mathfrak{c}} = \mathfrak{h}_{X,P} R_{X,P}$ [Sti2008, Proposition 3.4.2]. Following [Sti2008, Definition 3.4.3], we define the different of $R_X/R_{\mathbb{P}}$.

Definition 2.39 The different $\operatorname{diff}(R_X/R_{\mathbb{P}})$ is a divisor defined by

$$\operatorname{diff}(R_X/R_{\mathbb{P}}) := \sum_{P \in X \setminus \{\infty\}} d_P \, P,$$

where the different divisor $d_P := -\deg_P(\mathfrak{h}_{X,P})$ for $R_{X,P}^{\mathfrak{c}} = \mathfrak{h}_{X,P} R_{X,P}$.

By Dedekind's different theorem, Theorem 2.251, we have the following [Ku2004, Theorem 15.11][Sti2008, Theorem 3.5.1]:

Proposition 2.40 $d_{B_i} = e_{B_i} - 1$ *for* $P \in \mathfrak{B}_X \setminus \{\infty\}$, *and the support of* $\operatorname{diff}(R_X/R_{\mathbb{P}})$ *is equal to* $\mathfrak{B}_X \setminus \{\infty\}$.

Proof See Theorem 2.251. ∎

Remark 2.41 Since the fiber product $X \times_{\mathbb{P}} X$ associated with R_X^{e} is given by the commutative diagram

$$\begin{array}{ccc} X \times_{\mathbb{P}} X & \longrightarrow & X \\ \downarrow & & \downarrow \\ X & \longrightarrow & \mathbb{P}, \end{array}$$

Proposition 2.40 could be regarded as the base-change.

2.2.6.3 Trace Operator for Plane W-Curves: $R_X = R_{X^\circ}$ ($m_X = 2$) Case

Following Kunz [Ku2004, Theorem 15.1], we review $R_X^{\mathfrak{c}}$ of the $m_X = 2$ case:

Proposition 2.42 [Ku2004, Theorem 15.1] *For the plane W-curve* ($m_X = 2$)*, we have the relation of the complementary module* $R_X^{\mathfrak{c}}$,

$$R_X^{\mathfrak{c}} = \frac{1}{f_{X,y}} \cdot R_X. \tag{2.43}$$

Proof Let us show the proof by Kunz. We note that the extension of the field $\mathscr{Q}(R_X)$ of $\mathscr{Q}(R_{\mathbb{P}})$ is separable and f_X is monic,

$$\mathscr{Q}(R_X) = \mathscr{Q}(R_{\mathbb{P}})[Y]/(f_X) = \bigoplus_{i=0}^{r-1} \mathscr{Q}(R_{\mathbb{P}})y^i,$$

and thus we note $y^i = \mathbf{y}_i = \mathfrak{y}_{\mathfrak{e}_i}$ in Proposition 2.20 and in Appendix 2.11, and

$$\mathscr{Q}(R_X)^e = \mathscr{Q}(R_X) \otimes_{\mathscr{Q}(R_{\mathbb{P}})} \mathscr{Q}(R_X) \cong \mathscr{Q}(R_X)[Y]/(f_X).$$

Using the ring-homomorphism $\mu : \mathscr{Q}(R_X)^e \to \mathscr{Q}(R_X)$ $(a \otimes b \mapsto ab)$, the standard trace $\tau_{R_X/R_{\mathbb{P}}} := \sum_{i=0}^{r-1} y^i \circ \mathbf{y}_i^*$ is obtained by the extension of a trace τ. Let us find the basis $\{\widehat{\mathbf{y}}_i\}$ of R_X with respect to τ, and $\Delta_\tau := \bigoplus_{i=0}^{r-1} \widehat{\mathbf{y}}_i \otimes y^i$ as an element of the $R_{\mathbb{P}}$-module $\mathrm{Ann}_{R_X^e}(\mathrm{Ker}\mu)$. If we find $\{\widehat{\mathbf{y}}_i\}$ and τ, using them, we obtain the standard trace $\tau_{R_X/R_{\mathbb{P}}} = \mu(\Delta_\tau) \circ \tau$. Accordingly, we construct the τ and $\{\widehat{\mathbf{y}}_i\}$ as follows.

For an element $\widetilde{h} \in \mathscr{Q}(R_X)[Y]$ such that $f_X(x, Y) = (Y - y) \cdot \widetilde{h}$, we can identify $\mathrm{Ann}_{R_X^e}(\mathrm{Ker}\mu)$ with the principal ideal $(\widetilde{h})/(f_X(x, Y))$ in $\mathscr{Q}(R_X)[Y]/(f_X(x, Y))$ since

$$f_X(x, Y) = f_X(x, Y) - f_X(x, y) = (Y - y)\frac{f_X(x, Y) - f_X(x, y)}{Y - y} = (Y - y)\widetilde{h} = 0.$$

It means that $\widetilde{h}(x, Y, y) = (\delta_{Y,y} f_X)(x, Y, y)$ belongs to $R_X \otimes_{R_{\mathbb{P}}} R_X$ noting Lemma 2.26 3. Thus we have

$$\Delta_{\widetilde{h}} := \sum_{\ell=0}^{r-1} A_\ell(x) \sum_{j=0}^{r-\ell-1} y^j \otimes y^{r-\ell-j-1} \in R_X^e = R_X \otimes_{R_{\mathbb{P}}} R_X,$$

which generates the ideal $\mathrm{Ann}_{R_X^e}(\mathrm{Ker}\mu)$ in R_X^e, i.e., $(a \otimes 1 - 1 \otimes a)\Delta_{\widetilde{h}} = 0$ for every $a \in R_X$. Further, from Proposition 2.249, it corresponds to the trace $\tau_{\widetilde{h}} \in \omega_{R_X/R_{\mathbb{P}}}$ by

$$1 = \sum_{\ell=0}^{r-1} A_\ell(x) \sum_{j=0}^{r-\ell-1} \tau_{\widetilde{h}}(y^j) y^{r-\ell-j-1}.$$

For example, for $r = 4$, it is

$$1 = \tau_{\widetilde{h}}(y^3 + A_1 y^2 + A_2 y + A_3) + \tau_{\widetilde{h}}(y^2 + A_1 y + A_2)y + \tau_{\widetilde{h}}(y + A_1)y^2 + \tau_{\widetilde{h}}(1)y^3,$$

which should be interpreted as $\tau_{\widetilde{h}}(1) = 0$, $\tau_{\widetilde{h}}(y) = 0$, $\tau_{\widetilde{h}}(y^2) = 0$, and thus $\tau_{\widetilde{h}}(y^3) = 1$. Similarly, since $A_0 = 1$ and $y^{r-\ell-j-1} = 1$ when $\ell = 0$ and $j = r - 1$, we compare both sides in the equation and obtain

$$\tau_{\widetilde{h}}(y^i) = \begin{cases} 1 & \text{for } i = r-1, \\ 0 & \text{otherwise.} \end{cases}$$

In other words, for $\mathbf{y}_i = y^i$, $(i \in \mathbb{Z}_r)$, we have

$$\widehat{\mathbf{y}}_i = \sum_{\ell=0}^{r-1-i} A_\ell(x) y^\ell$$

as the dual basis of the basis $\{y^i = \mathbf{y}_i = \mathfrak{y}_{\mathfrak{e}_i}\}_{i \in \mathbb{Z}_r}$ with respect to $\tau_{\widetilde{h}}$ in Appendix 2.11. Then $\tau_{\widetilde{h}}(R_X) \subset R_{\mathbb{P}}$. It is obvious that $\mu(\Delta_{\widetilde{h}}) = f_{X,y}$. Thus the standard trace $\tau_{R_X/R_{\mathbb{P}}}$ of $R_X/R_{\mathbb{P}}$ is given by

$$\tau_{R_X/R_{\mathbb{P}}} = \mu(\Delta_{\widetilde{h}}) \circ \tau_{\widetilde{h}} = f_{X,y} \circ \tau_{\widetilde{h}}.$$

We have

$$\tau_{R_X/R_{\mathbb{P}}}\left(\frac{y^i}{f_{X,y}(x,y)}\right) = \begin{cases} 1 & \text{for } i = r-1, \\ 0 & \text{otherwise.} \end{cases}$$

Let us consider an element in $R_X^{\mathfrak{c}}$ as in (2.42). We consider $u \in \mathcal{Q}(R_X)$ which is expressed by

$$u = \frac{1}{f_{X,y}(x,y)} \sum_{i=0}^{r-1} a_i y^i \in \mathcal{Q}(R_X),$$

and its conditions in (2.42). For every element $v = \sum_{j=0}^{r-1} b_j y^j \in R_X$, $(b_j \in R_{\mathbb{P}})$, u satisfies $\tau_{R_X/R_{\mathbb{P}}}(uv) \in R_{\mathbb{P}}$, i.e.,

$$\tau_{R_X/R_{\mathbb{P}}}(uv) = \sum_{i,j} \tau_{\widetilde{h}}(a_i b_j y^i y^j) = \sum_{i=0}^{r-1} a_i b_{r-1-i} \in R_{\mathbb{P}}.$$

It implies that $a_i \in R_{\mathbb{P}}$ for every $i = 0, 1, \ldots, r-1$. Thus we obtain the relation (2.43). ■

Remark 2.43 Here we note that instead of $\{\widehat{\mathbf{y}}_i\}$ in the proof, we define a simpler dual basis of $\{y^i = \mathbf{y}_i = \mathfrak{y}_{\mathfrak{e}_i}\}_{i \in \mathbb{Z}_r}$ with respect to $\tau_{\widetilde{h}}$ by

$$\{\widehat{\mathfrak{y}}_{\mathfrak{e}_i} := y^{r-i-1}\}_{i \in \mathbb{Z}_r},$$

because it is obvious that $\tau_{\widetilde{h}}(\widehat{\mathfrak{y}}_{\mathfrak{e}_i} \mathfrak{y}_{\mathfrak{e}_j}) = \delta_{ij}$ from the definition of $\tau_{\widetilde{h}}$.

From (2.43), the Dedekind different of the plane W-curve is given by

$$\mathrm{diff}(R_X/R_{\mathbb{P}}) = -\sum_{P \in X \setminus \{\infty\}} \deg_P(f_{X,y}) P.$$

By Dedekind's different theorem (Proposition 2.40), we have the following [Ku2004, Chap. 15]:

Proposition 2.44 $\deg_{B_i}(f_{X,y}) = e_{B_i} - 1$.

2.2.6.4 Trace in the Covering for R_X

We generalize the result of the plane curves to the space curves. The surjective ring homomorphism in Proposition 2.24 can be interpreted as follows. We implicitly introduce the dual modules $\omega_{R_{X^{(i)}}/R_{\mathbb{P}}}$ of $R_{X^{(i)}}$ and their tensor product $\omega_{R_X/R_{\mathbb{P}}}^{\otimes[m_X-1]} := \omega_{R_{X^{(2)}}/R_{\mathbb{P}}} \otimes_{R_{\mathbb{P}}} \cdots \otimes_{R_{\mathbb{P}}} \omega_{R_{X^{(m_X)}}/R_{\mathbb{P}}}$ to find $R_X^{\mathfrak{c}}$. More precisely, we implicitly construct the trace τ in R_X by using the data of $(\tau_2, \cdots, \tau_{m_X})$ of $R_X^{\otimes[m_X-1]}$ by regarding R_X as $R_X = R_X^{\otimes[m_X-1]}/(f_1^X, f_2^X, \ldots, f_{k_X}^X)$. By investigating them, we obtain $R_X^{\mathfrak{c}}$ for the $m_X > 2$ case.

2.2.6.5 Trace in the Covering at R_H

In order to generalize the investigation of the complementary module for the plane curves $m_X = 2$ to general W-curves, we deal with the trace structure at R_H since the monomial curve in Sect. 2.2.3 is crucial in W-curves. We assume $m_X \geq 2$.

We use the surjection $\varphi_\infty : R_X \to R_H$ in Lemma 2.19, and consider the behavior of the trace in R_H^Z. We investigate a "covering" structure in $\varpi_H : X_H = \operatorname{Spec} R_H^Z \to \operatorname{Spec} \mathbb{C}[Z_r]$. The cyclic group $\mathfrak{C}_r$ of order r acts on R_H^Z and X_H as the $\mathbb{G}_{\mathrm{m}}$ action. We regard it as the r-covering and consider the $\mathbb{C}[Z_r]$-module R_H^Z and R_H.

We introduce a meromorphic function on $\operatorname{Spec} R_H^Z \times_{\operatorname{Spec}\mathbb{C}[Z_r]} \operatorname{Spec} R_H^Z$ or an element p in the enveloping field $\mathcal{Q}(R_H^Z)^e := \mathcal{Q}(R_H^Z \otimes_{\mathbb{C}[Z_r]} R_H^Z)$, and an element h in its associated enveloping ring $R_H^{Z^e} := R_H^Z \otimes_{\mathbb{C}[Z_r]} R_H^Z$, i.e., $Z_{r_i}^r = Z_r^{r_i} = Z_{r_i}^{\prime r}$ for $i = 2, \ldots, m_X$. We extend the group action of $\mathfrak{C}_r$ to that on $\mathcal{Q}(R_H^Z \otimes_{\mathbb{C}[Z_r]} R_H^Z)$ such as $\widehat{\zeta} h(Z_r, Z_\bullet, Z'_\bullet) = h(Z_r, \widehat{\zeta} Z_\bullet, \widehat{\zeta} Z'_\bullet)$ for $\zeta \in \mathfrak{C}_r$. We define an element p_H in $\mathcal{Q}(R_H^Z \otimes_{\mathbb{C}[Z_r]} R_H^Z)$ by

$$\begin{aligned} p_H(Z_r, Z_\bullet, Z'_\bullet) &:= \prod_{i=2}^{m_X} p_{H,r_i}(Z_r, Z_{r_i}, Z'_{r_i}), \\ p_{H,r_i}(Z_r, Z_{r_i}, Z'_{r_i}) &:= \frac{Z_{r_i}^{\mathfrak{r}_i-1} + Z_{r_i}^{\mathfrak{r}_i-2} Z'_{r_i} + Z_{r_i}^{\mathfrak{r}_i-3} Z_{r_i}^{\prime 2} + \cdots + Z_{r_i}^{\prime \mathfrak{r}_i-1}}{r Z_{r_i}^{\mathfrak{r}_i-1}}. \end{aligned} \tag{2.44}$$

Noting $Z_{r_i}^{\mathfrak{r}_i} = Z_r^{\mathfrak{r}_i} = Z_{r_i}^{\prime\bar{\mathfrak{r}}_i}$ in Lemma 2.12, then each factor behaves like

$$p_{H,r_i}(Z_r, Z_{r_i}, Z'_{r_i}) = \begin{cases} 1 & \text{for } Z'_{r_i} = Z_{r_i}, \\ 0 & \text{otherwise,} \end{cases}$$

as the simplest case in Lemma 2.36. We have the trace property,

$$p_H(Z_r, Z_\bullet, Z'_\bullet) = \begin{cases} 1 & \text{for } Z'_\bullet = Z_\bullet, \\ 0 & \text{otherwise.} \end{cases} \tag{2.45}$$

Lemma 2.45 *There are polynomials* $\widetilde{h}_H(Z_r, Z_\bullet, Z'_\bullet) \in R_H^Z \otimes_{\mathbb{C}[Z_r]} R_H^Z$ *and* $h_H(Z) = h_H(Z_r, Z_\bullet) := \widetilde{h}_H(Z_r, Z_\bullet, Z_\bullet) \in R_H^Z$ *such that*

$$\varphi_H^Z(\widetilde{h}_H(Z_r, Z_\bullet, Z'_\bullet)) = z^{d_h} \sum_{i=0}^{r-1} z'^{\mathfrak{e}_i} z^{-\mathfrak{e}_i} = \sum_{i=0}^{r-1} z'^{\mathfrak{e}_i} z^{\widehat{\mathfrak{e}}_i} \in R_H \otimes_{\mathbb{C}[z^r]} R_H, \tag{2.46}$$

where 1) z *and* z' *are given by* $\varphi_H^Z(Z_{r_i}) = z^{r_i}$, $\varphi_H^Z(Z'_{r_i}) = z'^{r_i}$, *for* $i \in \mathbb{Z}_r$, *and* $z^r = z'^r$, *2)* $d_h, \widehat{\mathfrak{e}}_i, \delta_i$ $(i \in \mathbb{Z}_r)$ *and* ℓ *are non-negative integers satisfying the conditions that a)* $d_h \geq \mathfrak{e}_i$ *for every* $i \in \mathbb{Z}_r$, *b)* d_h *is determined by Lemma 2.55, c)* $d_h = \mathfrak{e}_\ell + \delta_0 r$ *such that* $\mathfrak{e}_\ell = \sum_{i=2}^{m_X} (\mathfrak{r}_i - 1) r_i$ *modulo* r, *and d)*

$$\widehat{\mathfrak{e}}_i := d_h - \mathfrak{e}_i = \mathfrak{e}^*_{\ell,i} + \delta_i r, \quad z^{\mathfrak{e}_i} z^{\widehat{\mathfrak{e}}_i} = z^{\mathfrak{e}_0} z^{\widehat{\mathfrak{e}}_0} = z^{d_h} \tag{2.47}$$

for every $i \in \mathbb{Z}_r$, *especially*

$$\widehat{\mathfrak{e}}_0 = d_h = \mathfrak{e}_\ell + \delta_0 r, \quad \widehat{\mathfrak{e}}_\ell = \delta_0 r, \quad (\delta_\ell = \delta_0), \tag{2.48}$$

and 3) $\widetilde{h}_H(Z_r, Z_\bullet, Z'_\bullet)$ *consists of* r *monomials corresponding to each term in (2.46) and satisfies*

$$p_H(Z_r, Z_\bullet, Z'_\bullet) = \frac{\widetilde{h}_H(Z_r, Z_\bullet, Z'_\bullet)}{h_H(Z)}. \tag{2.49}$$

Proof The cyclic group $\mathfrak{C}_r$ acts on $p_H(Z_r, Z_\bullet, Z'_\bullet)$ so that it is invariant. We consider R_H rather than R_H^Z. We note the following: 1) For $j = 2, \ldots, m_X$, r_i and $r_j \in M_X$, $\{r_j i \text{ modulo } r \mid i \in \mathbb{Z}_r\} = \mathbb{Z}_r$, 2) the numerator in each p_{H,r_i} in (2.44) is homogeneous, and 3) their product is also homogeneous. Therefore we see that there are non-negative integers, $\widetilde{d}_\Delta$ and Δ_i, $(\Delta_i < \Delta_{i+1})$, such that

$$\frac{z^{\widetilde{d}_\Delta} \sum_{i=0}^{r-1} z'^{\Delta_i} z^{-\Delta_i}}{r z^{\widetilde{d}_\Delta}}$$

has the property of the right-hand side of (2.45) after acting φ_H^Z on both sides in (2.45), and $\{\Delta_i \text{ modulo } r \mid i \in \mathbb{Z}_r\} = \mathbb{Z}_r$. $\{\Delta_i \text{ modulo } r \mid i \in \mathbb{Z}_r\}$ obviously equals $\mathbb{Z}_r$. It means that $\{\Delta_i \mid i \in \mathbb{Z}_r\}$ equals $\{\mathfrak{e}_i + n_i r \mid i \in \mathbb{Z}_r\}$ for a certain non-negative number $n_i \in \mathbb{N}_0$ due to Lemma 2.5, and $\Delta_0 = 0$.

Due to the isomorphism φ_H^Z, for sufficiently large n and m, we find an element in $R_H^Z \otimes_{\mathbb{C}[Z_r]} R_H^Z$ whose image of φ_H^Z is $z^n z'^m$. For ℓ_i satisfying $\sum_i r_i \ell_i \equiv 0$ modulo r, $\prod Z_{r_i}^{\ell_i} = \prod Z_{r_i}'^{\ell_i}$, and thus we can find d_h such that $z^{\widetilde{d}_\Delta} \sum_{i=0}^{r-1} z'^{\Delta_i} z^{-\Delta_i} =$

$z^{d_h}\sum_{i=0}^{r-1} z'^{\mathfrak{e}_i} z^{-\mathfrak{e}_i}$ by noting $z^r = z'^r$. There is a preimage $\widetilde{h}_H(Z_r, Z_\bullet, Z'_\bullet)$ as an element in $R_H^Z \otimes_{\mathbb{C}[Z_r]} R_H^Z$. It is obvious that $\widetilde{h}_H(Z_r, Z_\bullet, Z'_\bullet)$ consists of r monomials and satisfies (2.46) and (2.49) for $p_H(Z_r, Z_\bullet, Z'_\bullet)$.

We note that the determination of d_h has ambiguity up to r, and thus we set it such that it satisfies Lemma 2.55. Here $d_h = \widetilde{d}_h + n_h r$ so that $\widetilde{d}_h$ is the minimal element satisfying the above relations.

From the definition and Lemma 2.5 3, we obtain the relations (2.47) of $i = 0$ and $i = \ell$, and $\delta_0 = \delta_\ell$.

We consider the cases d_h modulo r.

1. $d_h = \mathfrak{e}_0 = 0$ modulo r case: We note that $\{\mathfrak{e}_i \text{ modulo } r \mid i \in \mathbb{Z}_r^\times\} = \mathbb{Z}_r^\times$, and $\mathfrak{e}_\ell < \mathfrak{e}_{r-1}$, whereas $\widehat{\mathfrak{e}}_\ell$ must be non-negative. Each $\mathfrak{e}_i$ ($i \in \mathbb{Z}_r^\times$) cannot be divided by r and thus $\mathfrak{e}_i$, especially, $\mathfrak{e}_{r-1}$ is not equal to d_h. From Definition 2.4. we find a non-negative integer δ_i such that $\widehat{\mathfrak{e}}_i = \mathfrak{e}_{0,i}^* + \delta_i r$, $i \in \mathbb{Z}_r^\times$. It is obvious that $d_h = \widehat{\mathfrak{e}}_0 = \delta_0 r > \mathfrak{e}_{r-1} > 0$. We also note $\widehat{\mathfrak{e}}_{r-1} > 0$ because of $\mathfrak{e}_{0,r-1}^* > 0$.
2. $d_h = \mathfrak{e}_\ell$ modulo r case ($\ell = 1, \cdots, r-2$): Similarly, since $d_h = \mathfrak{e}_\ell + \delta_0 r$ satisfies $d_h - \mathfrak{e}_i \ge 0$ for $i \in \mathbb{Z}_r$ (especially $d_h \ge \mathfrak{e}_{r-1}$), we find non-negative integers δ_i such that $\widehat{\mathfrak{e}}_i = \mathfrak{e}_{\ell,i}^* + \delta_i r$ for $i \in \mathbb{Z}_r$, and $\widehat{\mathfrak{e}}_\ell = \delta_\ell r = \delta_0 r > 0$, by noting $\mathfrak{e}_{\ell,\ell}^* = 0$. Then $\widehat{\mathfrak{e}}_{r-1} > 0$ because of $\mathfrak{e}_{\ell,r-1}^* > 0$.
3. $d_h = \mathfrak{e}_{r-1}$ modulo r case: Similarly, since $d_h = \mathfrak{e}_{r-1} + \delta_0 r$ satisfies $d_h - \mathfrak{e}_i \ge 0$, especially $d_h \ge \mathfrak{e}_{r-1}$, δ_0 is non-negative. We find non-negative integers δ_i such that $\widehat{\mathfrak{e}}_i = \mathfrak{e}_{r-1,i}^* + \delta_i r$ for $i \in \mathbb{Z}_r^\times$, and $\widehat{\mathfrak{e}}_{r-1} = \delta_{r-1} r \ge 0$ or $\delta_{r-1} = \delta_0$ because of $\mathfrak{e}_{r-1,r-1}^* = 0$.

These show the statements in the proposition. ■

Proposition 2.46 *The δ_0 in (2.48) equals zero if and only if $d_h = \mathfrak{e}_{r-1}$.*

The case $d_h = \mathfrak{e}_{r-1}$ or $\delta_0 = 0$, occurs if and only if H is symmetric, whereas the case $\delta_0 \neq 0$ occurs if and only if H_X is not symmetric.

Thus we say that if $\delta_0 = 0$ occurs, d_h is symmetric and otherwise, d_h is not symmetric.

Proof They are proved in Lemma 2.110 1. ■

Remark 2.47 We remark that R_H and R_H^Z are characterized by these parameters $(M_X = \{r_i\}, m_X, k_X, \{\mathfrak{e}_i\}, \{\widehat{\mathfrak{e}}_i\}, d_h, \{\delta_i\}, \ell)$. Especially, ℓ is a fixed number for a given X in this section.

Example 2.48 1. $H = \langle 4, 6, 7, 9\rangle$ (non-symmetric) case ($d_h = 4\delta_0 + 9,\ \delta_0 = 1,\ \ell = 3$):

$H = \{0, 4, 6, 7, 8, 9, 10, 11, \cdots\}$, $H^c = \{1, 2, 3, 5\}$.

$R_H = \mathbb{C}[Z_4, Z_6, Z_7, Z_9]/\sim$ and then

$$h_{R_Z^H}(Z, Z_\bullet, Z'_\bullet) = Z_4 Z_9 + Z_7 Z'_6 + Z_6 Z'_7 + Z_4 Z'_9$$

$$= (1, Z_6, Z_7, Z_9)\begin{pmatrix} & & & Z_4 \\ & & 1 & \\ & 1 & & \\ Z_4 & & & \end{pmatrix}\begin{pmatrix} 1 \\ Z'_6 \\ Z'_7 \\ Z'_9 \end{pmatrix}, \quad h_H(Z) = 4Z_4Z_9.$$

i	0	1	2	3
$\mathfrak{e}_i$	0	6	7	9
$\widehat{\mathfrak{e}_i}$	4+9	7	6	4

2. $H = \langle 5, 7, 11, 13\rangle$ (non-symmetric) case ($d_h = 5\delta_0,\ \delta_0 = 5,\ \ell = 0$):
   $H = \{0, 5, 7, 10, 11, 12, 13, 14, \cdots\}$, $H^{\mathfrak{c}} = \{1, 2, 3, 4, 6, 8, 9\}$.
   $R_H = \mathbb{C}[Z_5, Z_7, Z_{11}, Z_{13}]/\sim$ and then

$$h_{R_Z^H}(Z, Z_\bullet, Z'_\bullet) = Z_5^5 + Z_5 Z_{13} Z'_7 + Z_{14} Z'_{11} + Z_5 Z_7 Z'_{13} + Z_{11} Z'_{14}$$

$$= (1, Z_7, Z_{11}, Z_{13}, Z_{14}) \begin{pmatrix} Z_4^5 & & & & \\ & & & Z_5 & \\ & & & & 1 \\ & Z_5 & & & \\ & & 1 & & \end{pmatrix} \begin{pmatrix} 1 \\ Z'_7 \\ Z'_{11} \\ Z'_{13} \\ Z'_{14} \end{pmatrix}, \quad h_H(Z) = 5Z_5^5.$$

i	0	1	2	3	4
$\mathfrak{e}_i$	0	7	11	13	14
$\widehat{\mathfrak{e}_i}$	25	13+5	14	7+5	11

3. $H = \langle 6, 13, 14, 15, 16\rangle$ (symmetric) case ($d_h = \mathfrak{e}_{r-1},\ \delta_0 = 0,\ \ell = r - 1$):
   $H = \{0, 6, 12, 13, 14, 15, 16, 18, 19, 20, 21, 22, 24, \cdots\}$,
   $H^{\mathfrak{c}} = \{1, 2, 3, 4, 5, 7, 8, 9, 10, 11, 17, 23\}$.
   $R_H = \mathbb{C}[Z_6, Z_{13}, Z_{14}, Z_{15}, Z_{16}]/\sim$ due to $Z_{13}Z_{16} = Z_{14}Z_{15}$

$$h_{R_Z^H}(Z, Z_\bullet, Z'_\bullet) = Z_{13}Z_{16} + Z_{13}Z'_{16} + Z_{14}Z'_{15} + Z_{15}Z'_{14} + Z_{16}Z'_{13} + Z'_{13}Z'_{14}$$

$$= (1, Z_{13}, Z_{14}, Z_{15}, Z_{16}, Z_{13}Z_{16}) \begin{pmatrix} & & & & & 1 \\ & & & & 1 & \\ & & & 1 & & \\ & & 1 & & & \\ & 1 & & & & \\ 1 & & & & & \end{pmatrix} \begin{pmatrix} 1 \\ Z'_{13} \\ Z'_{14} \\ Z'_{15} \\ Z'_{16} \\ Z'_{13}Z'_{16} \end{pmatrix}, \quad h_H(Z) = 6Z_{13}Z_{16}.$$

i	0	1	2	3	4	5
$\mathfrak{e}_i$	0	13	14	15	16	29
$\widehat{\mathfrak{e}_i}$	29	16	15	14	13	0

By comparing the semigroup H_X with $\{\mathfrak{e}_i\}$, we have the following corollaries:

Corollary 2.49 *1.* $\widetilde{h}_H(Z_r, Z_\bullet, Z'_\bullet)$ *is a homogeneous polynomial whose degree is* $d_h = \mathfrak{e}_\ell + \delta_0 r$ *in (2.48),*

2. $\widetilde{h}_H(Z_r, Z_\bullet, Z'_\bullet) = \sum_{i=0}^{r-1} Z_r^{\delta_i} \mathfrak{Z}_{\mathfrak{e}^*_{\ell,i}} \mathfrak{Z}'_{\mathfrak{e}_i}$,
3. $Z_r^{\delta_0} \mathfrak{Z}_{\mathfrak{e}_\ell} = Z_r^{\delta_i} \mathfrak{Z}_{\mathfrak{e}^*_{\ell,i}} \mathfrak{Z}_{\mathfrak{e}_i}$, *and*

4. $h_H(Z) = rZ_r^{\delta_i}\mathfrak{Z}_{\mathfrak{e}_i^*}\mathfrak{Z}_{\mathfrak{e}_i} = rZ_r^{\delta_0}\mathfrak{Z}_{\mathfrak{e}_\ell}$ *for* $i \in \mathbb{Z}_r$.

Proof They are obvious from Definition 2.4, Lemma 2.5 3, Lemma 2.13 and Lemma 2.45. ∎

Corollary 2.50 $\mathfrak{e}_i - r - 1 \geq 0, \widehat{\mathfrak{e}}_i - r - 1 \geq 0$ *and* $d_h - r - 1 \geq 0$, *for* $i \in \mathbb{Z}_r^\times$.

Proof $\mathfrak{e}_i \geq \min_{j=2}^{m_X} r_j = r_2 \geq r+1$ because of $r + 1 \leq r_2$. ∎

Remark 2.51 Corollary 2.49 determines the $R_{\mathbb{P}}$-module structure of R_X in Proposition 2.20.

2.2.6.6 Trace Structure in Weierstrass Curves (W-Curves)

We use the $R_{\mathbb{P}}$-module structure of R_X in the previous subsubsection and the properties in R_H and R_H^Z, noting the surjection $\varphi_\infty : R_X \to R_H$, to define $p_X^{(j)}$ in the quotient field $\mathcal{Q}(R_X \otimes_{R_{\mathbb{P}}} R_X)$ for every affine ring $\mathbb{C}[x, y]/(f_X^{(j)})$:

Definition 2.52 For $f_X^{(j)} \in \mathbb{C}[x, y]$ and $(\delta_{y,y'} f_X^{(j)})$ in Lemma 2.26 3, we define

$$p_X^{(j)}(x, y, y') := \frac{(\delta_{y,y'} f_X^{(j)})(x, y, y')}{(f_{X,y}^{(j)})(x, y)}.$$

We regard $p_X^{(j)}(x, y, y')$ as an element in $\mathcal{Q}(\mathbb{C}[x, y]/(f_X^{(j)}) \otimes_{\mathbb{C}[x]} \mathbb{C}[x, y]/(f_X^{(j)}))$ associated with $\mathcal{Q}(R_X \otimes_{R_{\mathbb{P}}} R_X)$. We extend the group $G_{X,x}$ to the action on $R_X \otimes_{R_{\mathbb{P}}} R_X$, such that $\widehat{\zeta}(x, y_\bullet, y'_\bullet) = (x, \widehat{\zeta} y_\bullet, \widehat{\zeta} y'_\bullet)$.

Lemma 2.53 $p_X^{(j)}(x, y, y') = \begin{cases} 1 & \textit{for} \quad y = y', \\ 0 & \textit{for} \quad y \neq y', \end{cases}$ *and for a group action* $\widehat{\zeta} \in G_{X,x}$,

$$p_X^{(j)}(x, \widehat{\zeta} y, \widehat{\zeta} y') = p_X^{(j)}(x, y, y').$$

Let us consider $p_{R_X} := \prod_{j=2}^{m_X} p_X^{(j)}$ as an element of $\mathcal{Q}(R_X \otimes_{R_{\mathbb{P}}} R_X)$. The following is obvious:

Proposition 2.54 *For* $(P, Q) \in X \times_{\mathbb{P}} X$,

$$p_{R_X}(P, Q) = \begin{cases} 1 & \textit{for} \quad P = Q, \\ 0 & \textit{for} \quad P \neq Q. \end{cases}$$

However, some parts in its numerator and denominator are canceled because they belong to $R_{\mathbb{P}}$. Thus we introduce an element $h(x, y_\bullet, y'_\bullet) \in R_X \otimes_{R_{\mathbb{P}}} R_X$ such that $h(x, y_\bullet, y'_\bullet)/h(x, y_\bullet, y_\bullet)$ reproduces the product.

The ring homomorphism $\varphi^X_{R_X}$ in Proposition 2.14 is extended to the surjective ring homomorphism from $R_X \otimes_{R_{\mathbb{P}}} R_X$ to $R^Z_H \otimes_{\mathbb{C}[Z_r]} R^Z_H$.

Lemma 2.55 *For a point $(P = (x, y_\bullet), P' = (x, y'_\bullet)) \in X \times_{\mathbb{P}} X$, there is a polynomial $\widetilde{h}_{R_X}(x, y_\bullet, y'_\bullet) \in R_X \otimes_{R_{\mathbb{P}}} R_X$ such that:*

1. *by regarding the element a in R_X as $a \otimes 1$ in $R_X \otimes_{R_{\mathbb{P}}} R_X$, $\widetilde{h}_{R_X}(x, y_\bullet, y_\bullet)$ and $\widetilde{h}_{R_X}(x, y_\bullet, y'_\bullet)$ are coprime as elements in $R_X \otimes_{R_{\mathbb{P}}} R_X$,*
2. *for a group action $\widehat{\zeta} \in G_{X,x}$, $\dfrac{\widetilde{h}_{R_X}(x, \widehat{\zeta} y_\bullet, \widehat{\zeta} y'_\bullet)}{\widetilde{h}_{R_X}(x, \widehat{\zeta} y_\bullet, \widehat{\zeta} y_\bullet)} = \dfrac{\widetilde{h}_{R_X}(x, y_\bullet, y'_\bullet)}{\widetilde{h}_{R_X}(x, y_\bullet, y_\bullet)}$,*
3. *it satisfies $\dfrac{\widetilde{h}_{R_X}(x, y_\bullet, y'_\bullet)}{\widetilde{h}_{R_X}(x, y_\bullet, y_\bullet)} = p_{R_X}(x, y_\bullet, y'_\bullet)$, and*
4. *$\varphi^X_H(\widetilde{h}_{R_X}(x, y_\bullet, y'_\bullet)) = \widetilde{h}_H(Z_r, Z_\bullet, Z'_\bullet)$ $(\mathrm{wt}(\widetilde{h}_{R_X}(x, y_\bullet, y'_\bullet)) = d_h$ with respect to $(x, y))$ for $\widetilde{h}_H$ in Lemma 2.45 by letting $\varphi^X_H(y_i) = Z_i$, $\varphi^X_H(y'_i) = Z'_i$.*

Proof By considering the numerator and denominator in $\prod_{j=2}^{m_X} p^{(j)}_X\big(\varpi_r(P), \varpi_{r_j}(P), \varpi_{r_j}(P')\big)$ modulo $(f^X_i)_{i=1,\dots,k_X}$, they are reduced to $\widetilde{h}_{R_X}$. Let the numerator be denoted by $\widetilde{h}_{R_X}(x, y_\bullet, y'_\bullet) \in R_X \otimes_{R_{\mathbb{P}}} R_X$. It is obvious that $\dfrac{\widetilde{h}_{R_X}(x, y_\bullet, y'_\bullet)}{\widetilde{h}_{R_X}(x, y_\bullet, y_\bullet)}$ must be invariant for the group action $G_{X,x}$ on $R_X \otimes_{R_{\mathbb{P}}} R_X$. Due to the condition *1*, we have a unique $\widetilde{h}_{R_X}(x, y_\bullet, y_\bullet)$. Then we can find d_h in Lemma 2.45 such that the image $\varphi^X_H(\widetilde{h}_{R_X}(x, y_\bullet, y'_\bullet))$ is equal to $\widetilde{h}_H$ in Lemma 2.45 because it is invariant for the $\mathbb{G}_{\mathrm{m}}$ action; the reduction in *3* corresponds to the reduction in $R^Z_H \otimes_{\mathbb{C}[Z_r]} R^Z_H$ as in *4*. Then it is clearly that *1*, *2*, and *3* are satisfied. ■

Definition 2.56 Let $h_X(x, y_\bullet) := \widetilde{h}_{R_X}(x, y_\bullet, y_\bullet)$.

Noting Corollary 2.49, we have the expression of $\widetilde{h}_{R_X}(x, y_\bullet, y'_\bullet)$ in terms of the standard basis in Proposition 2.20 and its dual:

Proposition 2.57 *$\widetilde{h}_{R_X} \in R_X \otimes_{R_{\mathbb{P}}} R_X$ is expressed by*

$$\begin{aligned}\widetilde{h}_{R_X}(x, y_\bullet, y'_\bullet) &= \widehat{\Upsilon}_0 \cdot 1 + \widehat{\Upsilon}_1 \mathfrak{y}'_{\mathfrak{e}_1} + \cdots + \widehat{\Upsilon}_{r-1} \mathfrak{y}'_{\mathfrak{e}_{r-1}} \\ &= 1 \cdot \widehat{\Upsilon}'_0 + \mathfrak{y}_{\mathfrak{e}_1} \widehat{\Upsilon}'_1 + \cdots + \mathfrak{y}_{\mathfrak{e}_{r-1}} \widehat{\Upsilon}'_{r-1} \\ &= \mathring{\mathfrak{y}}_{\mathfrak{e}_0} \mathfrak{y}'_{\mathfrak{e}_0} + \mathring{\mathfrak{y}}_{\mathfrak{e}_1} \mathfrak{y}'_{\mathfrak{e}_1} + \cdots + \mathring{\mathfrak{y}}_{\mathfrak{e}_{r-1}} \mathfrak{y}'_{\mathfrak{e}_{r-1}} \\ &\qquad + \textit{lower weight terms with respect to } -\mathrm{wt}\end{aligned}$$

as an $R_{\mathbb{P}}$-module. Here $\mathfrak{y}'_{\mathfrak{e}_0} = 1$, and $\widehat{\Upsilon}_i$ and $\mathring{\mathfrak{y}}_{\mathfrak{e}_i}$ hold the following properties:

1. *$\widehat{\Upsilon}_i = \sum_{j=0}^{r-1} \mathfrak{b}_{i,j} \mathfrak{y}_{\mathfrak{e}_j}$, with certain $\mathfrak{b}_{i,j} \in \mathbb{C}[x]$.*
2. *$\widehat{\Upsilon}_i = \mathring{\mathfrak{y}}_{\mathfrak{e}_i} +$ lower weight terms with respect to $-\mathrm{wt}$, where $\mathring{\mathfrak{y}}_{\mathfrak{e}_i} = \widetilde{\delta}_i(x) \mathfrak{y}_{\mathfrak{e}^*_{\ell,i}}$ with a monic polynomial $\widetilde{\delta}_i(x) \in \mathbb{C}[x]$ whose weight is $-\delta_i r$, (especially, $\mathring{\mathfrak{y}}_{\mathfrak{e}_0} = \widetilde{\delta}_0(x) \mathfrak{y}_{\mathfrak{e}^*_{\ell,0}} = \widetilde{\delta}_0(x) \mathfrak{y}_{\mathfrak{e}_\ell}$) such that*

$$\mathring{\mathfrak{y}}_{\mathfrak{e}_0} = \mathring{\mathfrak{y}}_{\mathfrak{e}_i}\mathfrak{y}_{\mathfrak{e}_i} + \textit{lower weight terms with respect to} - \mathrm{wt}$$

for $i \in \mathbb{Z}_r$, $\mathrm{wt}(\mathring{\mathfrak{y}}_{\mathfrak{e}_i}) = -\widehat{\mathfrak{e}}_i = -(d_h - \mathfrak{e}_i)$ *in Lemma 2.45.*

3. *We have the relation among the matrices*

$$\begin{pmatrix} \widehat{\Upsilon}_0^{(0)} & \widehat{\Upsilon}_1^{(0)} & \cdots & \widehat{\Upsilon}_{r-1}^{(0)} \\ \widehat{\Upsilon}_0^{(1)} & \widehat{\Upsilon}_1^{(1)} & \cdots & \widehat{\Upsilon}_{r-1}^{(1)} \\ \vdots & \vdots & \ddots & \vdots \\ \widehat{\Upsilon}_0^{(r-1)} & \widehat{\Upsilon}_1^{(r-1)} & \cdots & \widehat{\Upsilon}_{r-1}^{(r-1)} \end{pmatrix} \begin{pmatrix} \mathfrak{y}_{\mathfrak{e}_0}^{(0)} & \mathfrak{y}_{\mathfrak{e}_0}^{(1)} & \cdots & \mathfrak{y}_{\mathfrak{e}_0}^{(r-1)} \\ \mathfrak{y}_{\mathfrak{e}_1}^{(0)} & \mathfrak{y}_{\mathfrak{e}_1}^{(1)} & \cdots & \mathfrak{y}_{\mathfrak{e}_1}^{(r-1)} \\ \vdots & \vdots & \ddots & \vdots \\ \mathfrak{y}_{\mathfrak{e}_{r-1}}^{(0)} & \mathfrak{y}_{\mathfrak{e}_{r-1}}^{(1)} & \cdots & \mathfrak{y}_{\mathfrak{e}_{r-1}}^{(r-1)} \end{pmatrix} = \begin{pmatrix} h_X^{(0)} & & & \\ & h_X^{(1)} & & \\ & & \ddots & \\ & & & h_X^{(r-1)} \end{pmatrix}, \tag{2.50}$$

which is denoted by $M_{\widehat{\Upsilon}} M_{\mathfrak{y}} = M_h$. *Then* $M_{\widehat{\Upsilon}}$ *is given by* $M_{\widehat{\Upsilon}} = {}^t M_{\mathfrak{y}} M_{\mathfrak{b}}$, *where the matrix* $M_{\mathfrak{b}} = (\mathfrak{b}_{i,j})$ *whose coefficients belong to* $R_{\mathbb{P}}$ *is symmetric, i.e.,* $M_{\mathfrak{b}} = {}^t M_{\mathfrak{b}}$. *Here, for the group action of* $G_{X,x}$, *we denote the* ℓ*th conjugate element of* g *by* $g^{(\ell)}$. *The inverse of the matrix* $M_{\mathfrak{y}}$ *is given by* $M_{\mathfrak{y}}^{-1} = M_h^{-1} M_{\widehat{\Upsilon}} = M_h^{-1}\, {}^t M_{\mathfrak{y}} M_{\mathfrak{b}}$.

Proof From the construction, $\widetilde{h}_{R_X}(x, y_\bullet, y'_\bullet) = \widetilde{h}_{R_X}(x, y'_\bullet, y_\bullet)$; $\widetilde{h}_{R_X}(x, y_\bullet, y'_\bullet)$ is invariant for the exchanging between $y_\bullet$ and $y'_\bullet$. From Proposition 2.20, $\widehat{\Upsilon}_i$'s are uniquely determined. Thus the matrix expression in 3 is obvious. Each $\widehat{\Upsilon}_i$ is an element in R_X and thus there is the expansion by the basis $\{\mathfrak{y}_i\}$ as an $R_{\mathbb{P}}$-module and thus $\mathfrak{b}_{ij}$ belongs to $\mathbb{C}[x] = R_{\mathbb{P}}$. However, by letting ℓ_i satisfy $\mathfrak{e}_{\ell_i} = \mathfrak{e}^*_{\ell,i}$, due to the Sato–Weierstrass weight of $\widehat{\Upsilon}_i$, we see that $-\mathrm{wt}(\mathfrak{b}_{i,j}\mathfrak{y}_{\mathfrak{e}_j}) < -\mathrm{wt}(\mathfrak{b}_{i,\ell_i}\mathfrak{y}_{\mathfrak{e}_{\ell_i}})$ and $\mathrm{wt}(\widehat{\Upsilon}_i) = \mathrm{wt}(\mathfrak{b}_{i,\ell_i}\mathfrak{y}_{\mathfrak{e}_{\ell_i}})$. Thus we let $\widetilde{\delta}_i := \mathfrak{b}_{i,\ell_i}$, and then we obtain $\mathring{\mathfrak{y}}_{\mathfrak{e}_i} = \widetilde{\delta}_i \mathfrak{y}_{\mathfrak{e}^*_{\ell,i}}$, in particular $\mathring{\mathfrak{y}}_{\mathfrak{e}_0} = \widetilde{\delta}_0 \mathfrak{y}_{\mathfrak{e}_\ell}$. $\mathring{\mathfrak{y}}_{\mathfrak{e}_i}$ is monic from Corollary 2.49. Hence $\widetilde{h}_{R_X}(x, y_\bullet, y'_\bullet)$ must have the form mentioned above.

Since $\mathring{\mathfrak{y}}_{\mathfrak{e}_0}$ and $\mathring{\mathfrak{y}}_{\mathfrak{e}_i}\mathfrak{y}_{\mathfrak{e}_i}$ are homogeneous with respect to $-\mathrm{wt}_\lambda$ and correspond to the relations in Corollary 2.49, $\mathring{\mathfrak{y}}_{\mathfrak{e}_0} - \mathring{\mathfrak{y}}_{\mathfrak{e}_i}\mathfrak{y}_{\mathfrak{e}_i}$ is given by the lower weight terms with respect to $-\mathrm{wt}$. Further, (2.25) shows $\widetilde{\delta}_0 \mathfrak{y}_{\mathfrak{e}_\ell}$.

$$\begin{aligned}\mathring{\mathfrak{y}}_{\mathfrak{e}_i}\mathfrak{y}_{\mathfrak{e}_i} &= \widetilde{\delta}_i \mathfrak{y}_{\mathfrak{e}_{\ell_i}}\mathfrak{y}_{\mathfrak{e}_i} = \sum_{j \in \mathbb{Z}_r} \widetilde{\delta}_i \mathfrak{a}_{i,\ell_i,j}\mathfrak{y}_{\mathfrak{e}_j} \\ &= \widetilde{\delta}_i \mathfrak{a}_{i,\ell_i,\ell}\mathfrak{y}_{\mathfrak{e}_\ell} + \text{lower weight terms with respect to } -\mathrm{wt}.\end{aligned}$$

Unless $\widetilde{\delta}_i \mathfrak{a}_{i,\ell_i,\ell}/\widetilde{\delta}_0$ belongs to $R_{\mathbb{P}}$, it must be expressed as α/β. where α and β are elements of $R_{\mathbb{P}}$ and their Sato–Weierstrass weights are the same, and thus we redefine

$$\mathring{\mathfrak{y}}_{\mathfrak{e}_i} := \beta\widetilde{\delta}_i \mathfrak{y}_{\mathfrak{e}^*_{\ell,i}}, \quad \mathring{\mathfrak{y}}_{\mathfrak{e}_0} := \alpha\widetilde{\delta}_0 \mathfrak{y}_{\mathfrak{e}_\ell}.$$

We repeat this operation for each i and then, for every i, we obtain

$$\mathring{\mathfrak{y}}_{\mathfrak{e}_i}\mathfrak{y}_{\mathfrak{e}_i} = \mathring{\mathfrak{y}}_{\mathfrak{e}_0} + \sum_{j \in \mathbb{Z}_r, j \neq \ell} \widetilde{\delta}_i \mathfrak{a}_{i,\ell_i,j}\mathfrak{y}_{\mathfrak{e}_j}.$$

However, if $\widetilde{h}_{R_X}(x, y_\bullet, y'_\bullet)$ and $\widetilde{h}_X(x, y_\bullet, y_\bullet) = h_X(x, y_\bullet)$ are not coprime, it means that there is a cofactor in $\widetilde{h}_{R_X}(x, y_\bullet, y'_\bullet)$ and we can divide $\widetilde{h}_{R_X}(x, y_\bullet, y_\bullet)$ by the factor.

On the matrix expression, $M_{\widehat{\Upsilon}}$ is given by [Tak1971]

$$M_{\widehat{\Upsilon}} = \begin{pmatrix} \mathfrak{y}^{(0)}_{\mathfrak{e}_0} & \mathfrak{y}^{(0)}_{\mathfrak{e}_1} & \cdots & \mathfrak{y}^{(0)}_{\mathfrak{e}_{r-1}} \\ \mathfrak{y}^{(1)}_{\mathfrak{e}_0} & \mathfrak{y}^{(1)}_{\mathfrak{e}_1} & \cdots & \mathfrak{y}^{(1)}_{\mathfrak{e}_{r-1}} \\ \vdots & \vdots & \ddots & \vdots \\ \mathfrak{y}^{(r-1)}_{\mathfrak{e}_0} & \mathfrak{y}^{((r-1)}_{\mathfrak{e}_1} & \cdots & \mathfrak{y}^{(r-1)}_{\mathfrak{e}_{r-1}} \end{pmatrix} \begin{pmatrix} \mathfrak{b}_{0,0} & \mathfrak{b}_{1,0} & \cdots & \mathfrak{b}_{r-1,0} \\ \mathfrak{b}_{0,1} & \mathfrak{b}_{1,1} & \cdots & \mathfrak{b}_{r-1,1} \\ \vdots & \vdots & \ddots & \vdots \\ \mathfrak{b}_{0,r-1} & \mathfrak{b}_{1,r-1} & \cdots & \mathfrak{b}_{r-1,r-1} \end{pmatrix}, \tag{2.51}$$

and thus it is equal to ${}^t M_{\mathfrak{y}} M_{\mathfrak{b}}$. The matrix $M_{\mathfrak{b}}$ is symmetric, i.e., $M_{\mathfrak{b}} = {}^t M_{\mathfrak{b}}$ because $M_{\widehat{\Upsilon}} M_{\mathfrak{y}} = {}^t M_{\mathfrak{y}} M_{\mathfrak{b}} M_{\mathfrak{y}} = M_h$ is symmetric. It provides the inverse of the matrix $M_{\mathfrak{y}}$. ■

For the case of $m_X = 2$, the second matrix in (2.51) is given by

$$\begin{pmatrix} A_{r-1} & A_{r-2} & \cdots & A_1 & 1 \\ A_{r-2} & A_{r-3} & \cdots & 1 & \\ \vdots & \vdots & \ddots & \vdots & \vdots \\ A_1 & 1 & \cdots & & \\ 1 & & \cdots & & \end{pmatrix}.$$

Recalling Proposition 2.46, we find that $\mathring{\mathfrak{y}}_{\mathfrak{e}_i}$ and $h_X(x, y_\bullet)$ are expressed as follows.

Corollary 2.58 *1. If d_h is symmetric, $\mathring{\mathfrak{y}}_{\mathfrak{e}_i} = \widetilde{\delta}_i(x)\mathfrak{y}_{\mathfrak{e}^*_{r-1,i}}$ for $i \in \mathbb{Z}_r$, $\widetilde{\delta}_0 = \widetilde{\delta}_{r-1} = 1$, $\mathring{\mathfrak{y}}_{\mathfrak{e}_0} = \mathfrak{y}_{\mathfrak{e}_{r-1}}$, and $\mathring{\mathfrak{y}}_{\mathfrak{e}_{r-1}} = \mathfrak{y}_{\mathfrak{e}_0} = 1$.*
2. If d_h is not symmetric, $\mathring{\mathfrak{y}}_{\mathfrak{e}_i} = \widetilde{\delta}_i(x)\mathfrak{y}_{\mathfrak{e}^*_{\ell,i}}$ for $i \in \mathbb{Z}_r^\times$, and $\widetilde{\delta}_0 \neq 1$.
3. $h_X(x, y_\bullet) = r\mathring{\mathfrak{y}}_{\mathfrak{e}_0} +$ lower weight terms with respect to $-$wt.

These matrices in the proof of Proposition 2.57 hold the following.

Corollary 2.59 *1. ${}^t M_{\mathfrak{y}} M_{\mathfrak{b}} M_{\mathfrak{y}} = M_h$, $M_{\mathfrak{b}} = {}^t M_{\mathfrak{b}}$, $M_{\mathfrak{y}}^{-1} = M_h^{-1} M_{\widehat{\Upsilon}} = M_h^{-1}\, {}^t M_{\mathfrak{y}} M_{\mathfrak{b}}$.*
2. Let $\widetilde{\mathfrak{a}}_{ij} := \sum_{\ell=0}^{r-1} \mathfrak{y}_i^{(\ell)} \mathfrak{y}_j^{(\ell)} = \sum_{k,\ell=0}^{r-1} \mathfrak{a}_{ijl} \mathfrak{y}_k^{(\ell)}$ and $M_{\widetilde{\mathfrak{a}}} := (\widetilde{\mathfrak{a}}_{ij})$. Then we obtain $M_{\mathfrak{y}}\, {}^t M_{\mathfrak{y}} = M_{\widetilde{\mathfrak{a}}}$, and $\det(M_{\widetilde{\mathfrak{a}}}) = (\det(M_{\mathfrak{y}}))^2$.
3. $M_{\mathfrak{y}} = M_{\widetilde{\mathfrak{a}}}\, {}^t M_{\widehat{\Upsilon}}\, {}^t M_h^{-1}$, i.e., $\mathfrak{y}_{\mathfrak{e}_i} = \sum_k \dfrac{\widetilde{\mathfrak{a}}_{ik}}{h_X} \widehat{\Upsilon}_k$.
4. $M_{\mathfrak{b}} = (\sum_k h_X^{(k)}) M_{\widetilde{\mathfrak{a}}}^{-1}$ and thus $M_{\mathfrak{b}}^{-1} = \dfrac{1}{\sum_k h_X^{(k)}} M_{\widetilde{\mathfrak{a}}}$.

Proof 3 is obtained by $M_{\mathfrak{y}} = M_{\mathfrak{y}}\, {}^t M_{\mathfrak{y}}\, {}^t M_{\mathfrak{y}}^{-1} = M_{\widetilde{\mathfrak{a}}}\, {}^t M_{\widehat{\Upsilon}}\, {}^t M_h^{-1}$. The others are obvious. ■

We extend the arguments for the $m_X = 2$ case to general m_X cases.

Obviously, due to Proposition 2.20, the dual $\omega_{R_X/R_{\mathbb{P}}}$ of R_X as an $R_{\mathbb{P}}$-module has the standard basis as a trace.

Lemma 2.60 *We define*

$$\Delta_{\widetilde{h}} := \widehat{\Upsilon}_0 \otimes 1 + \sum_{k\in\mathbb{Z}_r^\times} \widehat{\Upsilon}_k \otimes \mathfrak{y}_{\mathfrak{e}_k} = \sum_{i,k\in\mathbb{Z}_r} \mathfrak{y}_{\mathfrak{e}_i}\mathfrak{b}_{i,k} \otimes \mathfrak{y}_{\mathfrak{e}_k} = \sum_{i,k\in\mathbb{Z}_r} \mathfrak{y}_{\mathfrak{e}_i} \otimes \mathfrak{b}_{i,k}\mathfrak{y}_{\mathfrak{e}_k}.$$

Then for every $a \in R_X$, $(a \otimes 1 - 1 \otimes a)\Delta_{\widetilde{h}} = 0$, and $\Delta_{\widetilde{h}} R_X^e$ is equal to $\mathrm{Ann}_{R_X^e}$ $(\mathrm{Ker}\,\mu_{\mathscr{Q}})$.

Proof First, we consider $\mathfrak{y}_{\mathfrak{e}_j}\widetilde{h}_{R_X}(x, y_\bullet, y'_\bullet) = \sum_{j,k} \mathfrak{b}_{j,k}\mathfrak{y}_{\mathfrak{e}_i}\mathfrak{y}_{\mathfrak{e}_j}\mathfrak{y}'_{\mathfrak{e}_k} = \sum_{j,k,\ell} \mathfrak{b}_{j,k}\mathfrak{a}_{ij\ell}\mathfrak{y}_{\mathfrak{e}_\ell}$ $\mathfrak{y}'_{\mathfrak{e}_k}$ and $\mathfrak{y}'_{\mathfrak{e}_j}\widetilde{h}_{R_X}(x, y_\bullet, y'_\bullet) = \sum_{j,k,\ell} \mathfrak{b}_{j,k}\mathfrak{a}_{ik\ell}\mathfrak{y}_{\mathfrak{e}_j}\mathfrak{y}'_{\mathfrak{e}_\ell}$. The latter can be expressed by $\sum_{\ell,j,k} \mathfrak{b}_{\ell,j}\mathfrak{a}_{ijk}\mathfrak{y}_{\mathfrak{e}_\ell}\mathfrak{y}'_{\mathfrak{e}_k}$. Equation (2.50) shows that both cases agree:

$$\mathfrak{y}_{\mathfrak{e}_j}\widetilde{h}_{R_X}(x, y_\bullet, y'_\bullet) = \mathfrak{y}'_{\mathfrak{e}_j}\widetilde{h}_{R_X}(x, y_\bullet, y'_\bullet) = \begin{cases} \mathfrak{y}_{\mathfrak{e}_j} h_X & \text{for } y_\bullet = y'_\bullet, \\ 0 & \text{otherwise.} \end{cases}$$

It means that $\sum_j \mathfrak{b}_{k,j}\mathfrak{a}_{ij\ell} = \sum_j \mathfrak{b}_{\ell,j}\mathfrak{a}_{ijk}$. This relation shows $\sum_{i,k} \mathfrak{b}_{j,k}\mathfrak{y}_{\mathfrak{e}_i}\mathfrak{y}_{\mathfrak{e}_j} \otimes \mathfrak{y}_{\mathfrak{e}_k} = \sum_{i,k} \mathfrak{b}_{j,k}\mathfrak{y}_{\mathfrak{e}_j} \otimes \mathfrak{y}_{\mathfrak{e}_i}\mathfrak{y}_{\mathfrak{e}_k}$ and we obtain the equality $(a \otimes 1 - 1 \otimes a)\Delta_{\widetilde{h}} = 0$. ■

Further, from Proposition 2.249, $\Delta_{\widetilde{h}}$ in Lemma 2.60 provides the trace $\tau_{\widetilde{h}} \in \omega_{R_X/R_{\mathbb{P}}}$ with the dual basis $\{\widehat{\Upsilon}_i\}$ of R_X with respect to the trace $\tau_{\widetilde{h}}$.

Lemma 2.61 *$\{\widehat{\Upsilon}_i\}$ is the dual basis of R_X with respect to the trace $\tau_{\widetilde{h}}$, such that*

$$\tau_{\widetilde{h}}(\widehat{\Upsilon}_i) = \begin{cases} 1 & \textit{for } i = 0, \\ 0 & \textit{otherwise.} \end{cases}$$

Proof These $\{\widehat{\Upsilon}_i\}$ correspond to the dual basis $\{\widehat{\mathbf{y}}_i\}$ of R_X with respect to the trace $\tau_{\widetilde{h}}$. By considering

$$1 = \tau_{\widetilde{h}}(\widehat{\Upsilon}_0) \cdot 1 + \sum_{k\in\mathbb{Z}_r^\times} \tau_{\widetilde{h}}\left(\widehat{\Upsilon}_k\right) \mathfrak{y}'_{\mathfrak{e}_k},$$

we have the relation. ■

We introduce an $R_{\mathbb{P}}$-module, $R^*_{X,\tau_{\widetilde{h}}} := \langle \widehat{\Upsilon}_0, \widehat{\Upsilon}_1, \ldots, \widehat{\Upsilon}_{r-1} \rangle_{R_{\mathbb{P}}}$, and consider its structure as an R_X-module. Due to Proposition 2.249, $\omega_{R_X/R_{\mathbb{P}}} = R_X \circ \tau$, and thus $\omega_{R_X/R_{\mathbb{P}}} \cong R^*_{X,\tau_{\widetilde{h}}}$ as an R_X-module:

Lemma 2.62 *$R^*_{X,\tau_{\widetilde{h}}}$ is an ideal of R_X, especially $\mathfrak{y}_{\mathfrak{e}_i} R^*_{X,\tau_{\widetilde{h}}} \subset R^*_{X,\tau_{\widetilde{h}}}$, and*

$$R^*_{X,\tau_{\widetilde{h}}} = \langle \widehat{\Upsilon}_0, \widehat{\Upsilon}_1, \ldots, \widehat{\Upsilon}_{r-1} \rangle_{R_X} = \langle \widehat{\Upsilon}_1, \ldots, \widehat{\Upsilon}_{r-1} \rangle_{R_X}.$$

It means that as $R_{\mathbb{P}}$-modules,

$$\langle \widehat{\Upsilon}_0, \widehat{\Upsilon}_1, \ldots, \widehat{\Upsilon}_{r-1} \rangle_{R_{\mathbb{P}}} = \langle \widehat{\Upsilon}_1, \ldots, \widehat{\Upsilon}_{r-1} \rangle_{R_X}.$$

Proof Let us consider $\mathfrak{y}_{\mathfrak{e}_i} \widehat{\Upsilon}_j$, which is equal to

$$\mathfrak{y}_{\mathfrak{e}_i} \widehat{\Upsilon}_j = \sum_{k=0}^{r-1} \mathfrak{b}_{jk} \mathfrak{y}_{\mathfrak{e}_j} \mathfrak{y}_{\mathfrak{e}_k} = \sum_{k,\ell=0}^{r-1} \mathfrak{b}_{jk} \mathfrak{a}_{ik\ell} \mathfrak{y}_{\mathfrak{e}_\ell}.$$

However, from the proof in Lemma 2.60, $\sum_k \mathfrak{b}_{jk} \mathfrak{a}_{ik\ell} = \sum_k \mathfrak{b}_{\ell k} \mathfrak{a}_{ikj}$, and thus

$$\mathfrak{y}_{\mathfrak{e}_i} \widehat{\Upsilon}_j = \sum_{k=0}^{r-1} \mathfrak{b}_{k\ell} \mathfrak{a}_{ikj} \mathfrak{y}_{\mathfrak{e}_\ell} = \sum_{k=0}^{r-1} \mathfrak{a}_{ikj} \widehat{\Upsilon}_k.$$

Hence, $R^*_{X,\tau_{\widetilde{h}}}$ is closed for the action of $\mathfrak{y}_{\mathfrak{e}_i}$.

On the other hand, we take an integer i $(i \neq \ell)$ and then consider $\mathfrak{y}_{\mathfrak{e}_i} \widehat{\Upsilon}_i$ which is decomposed by $\widehat{\Upsilon}_i$'s but has the form

$$\mathfrak{y}_{\mathfrak{e}_i} \widehat{\Upsilon}_i = \widehat{\Upsilon}_0 + \sum_{j=1}^{r-1} \mathfrak{d}_{ij} \widehat{\Upsilon}_j, \tag{2.52}$$

where $\mathfrak{d}_{ij}$ belongs to $R_{\mathbb{P}}$ because the leading terms of both $\mathfrak{y}_{\mathfrak{e}_i} \widehat{\Upsilon}_i$ and $\widehat{\Upsilon}_0$ must agree in both sides. It means that $\widehat{\Upsilon}_0 \in \langle \widehat{\Upsilon}_1, \ldots, \widehat{\Upsilon}_{r-1} \rangle_{R_X}$. ■

This proof shows the following lemma:

Lemma 2.63 $\tau_{\widetilde{h}}(\mathfrak{y}_{\mathfrak{e}_i} \widehat{\Upsilon}_i) = 1$ *for every* $i \in \mathbb{Z}_r^\times$*, and thus* $\tau_{\widetilde{h}}(\mathfrak{y}_{\mathfrak{e}_i} \widehat{\Upsilon}_j) = \delta_{ij}$.

Proof (2.52) shows this. ■

Noting Remark 2.43 for the $m_X = 2$ case, we also introduce the more convenient basis $\{\widehat{\mathfrak{y}}_{\mathfrak{e}_i}\}_{i \in \mathbb{Z}_R}$ with respect to $\tau_{\widetilde{h}}$ instead of $\{\widehat{\Upsilon}_i\}_{i \in \mathbb{Z}_R}$:

Definition 2.64 For $i \in \mathbb{Z}_r$, we define a truncated polynomial $\widehat{\mathfrak{y}}_{\mathfrak{e}_i}$ of $\widehat{\Upsilon}_i$ such that the weight $-$wt of $\widehat{\Upsilon}_i - \widehat{\mathfrak{y}}_{\mathfrak{e}_i}$ is less than $-\mathrm{wt}(\mathring{\mathfrak{y}}_{\mathfrak{e}_i})$, i.e., $\widehat{\mathfrak{y}}_{\mathfrak{e}_i} = \mathring{\mathfrak{y}}_{\mathfrak{e}_i} +$ certain terms, and the number of the terms is minimal satisfying the relations as $R_{\mathbb{P}}$-modules,

$$\langle \widehat{\Upsilon}_1, \ldots, \widehat{\Upsilon}_{r-1} \rangle_{R_X} = \langle \widehat{\mathfrak{y}}_{\mathfrak{e}_1}, \ldots, \widehat{\mathfrak{y}}_{\mathfrak{e}_{r-1}} \rangle_{R_X} = \langle \widehat{\mathfrak{y}}_{\mathfrak{e}_0}, \ldots, \widehat{\mathfrak{y}}_{\mathfrak{e}_{r-1}} \rangle_{R_{\mathbb{P}}},$$

$$\tau_{\widetilde{h}}(\widehat{\mathfrak{y}}_{\mathfrak{e}_i}) = \begin{cases} 1 & \text{for } i = 0, \\ 0 & \text{otherwise.} \end{cases}$$

Lemma 2.65 *1. When* d_h *is symmetric,* $\widehat{\mathfrak{y}}_{\mathfrak{e}_i} = \mathfrak{y}_{\mathfrak{e}_{r-1-i}}$*, especially for* $m_X = 2$ *case,* $\widehat{\mathfrak{y}}_{\mathfrak{e}_i} = \mathring{\mathfrak{y}}_{\mathfrak{e}_i} = y_s^i$.
2. *When* X *is cyclic,* $\widehat{\mathfrak{y}}_{\mathfrak{e}_i} = \mathring{\mathfrak{y}}_{\mathfrak{e}_i}$.

Proof The $m_X = 2$ case is described in Sect. 2.2.6.3 and the statement is obvious. For the symmetric d_h case, under which the numerical semigroup H_X is symmetric due to Proposition 2.46, $\mathring{\mathfrak{y}}_{\mathfrak{e}_i} = \mathfrak{y}_{\mathfrak{e}_{r-1-i}}$, ($\mathring{\mathfrak{y}}_{\mathfrak{e}_{r-1}} = 1$) and thus it is obvious that

$$\langle \widehat{\Upsilon}_1, \widehat{\Upsilon}_2, \ldots, \widehat{\Upsilon}_{r-1} \rangle_{R_X} = \langle \mathfrak{y}_{\mathfrak{e}_0}, \mathfrak{y}_{\mathfrak{e}_1}, \ldots, \mathfrak{y}_{\mathfrak{e}_{r-2}} \rangle_{R_X} = R_X.$$

Further, when X is cyclic, $\widehat{\Upsilon}_i = \mathring{\mathfrak{y}}_{\mathfrak{e}_i}$ because of the invariance for the cyclic action. ■

Since $\mu(\Delta_{\widetilde{h}}) = h_X(x, y_\bullet)$, the trace $\tau_{R_X/R_{\mathbb{P}}}$ of $R_X/R_{\mathbb{P}}$ is given by

$$\tau_{R_X/R_{\mathbb{P}}} = h_X(x, y_\bullet) \circ \tau_{\widetilde{h}}.$$

Then we obviously have the following lemma:

Lemma 2.66 *For a monomial ϕ in R_X as an $R_{\mathbb{P}}$-module and $\widehat{\mathfrak{y}}_{\mathfrak{e}_i}$, we have the following:*

1. *When d_h is symmetric ($d_h = \mathfrak{e}_{r-1}, \delta_0 = 0$) in (2.48),*

$$\tau_{R_X/R_{\mathbb{P}}} \left(\frac{\widehat{\mathfrak{y}}_{\mathfrak{e}_i}}{h_X(x, y_\bullet)} \right) = \begin{cases} 1 & \text{for } i = 0, \\ 0 & \text{otherwise.} \end{cases}$$

2. *When d_h is not symmetric ($d_h = \mathfrak{e}_\ell + \delta_0 r, \delta_0 \neq 0$) in (2.48),*

$$\tau_{R_X/R_{\mathbb{P}}} \left(\frac{\phi}{h_X(x, y_\bullet)} \right) = \begin{cases} \phi/\widehat{\mathfrak{y}}_{\mathfrak{e}_0} & \text{for } \widehat{\mathfrak{y}}_{\mathfrak{e}_0} | \phi, \\ 0 & \text{otherwise.} \end{cases}$$

Here we note that the case $\widehat{\mathfrak{y}}_{\mathfrak{e}_0} | \phi$ means $(\mathfrak{y}_{\mathfrak{e}_\ell} \widetilde{\delta_0}) | \phi$, whereas the case $\widehat{\mathfrak{y}}_{\mathfrak{e}_0} \nmid \phi$ consists of two cases (1) $\phi = (f(x)\mathfrak{y}_{\mathfrak{e}_\ell})$, $\widetilde{\delta_0} \nmid f$ and (2) $\phi = (f(x)\mathfrak{y}_{\mathfrak{e}_i})$ $i \in \mathbb{Z}_r \setminus \{\ell\}$.

Let us consider elements in $R_X^{\mathfrak{c}}$ as in (2.42). We consider $u \in \mathscr{Q}(R_X)$ given by

$$u = \frac{1}{h_X(x, y_\bullet)} \sum_{i=0}^{r-1} a_i \widehat{\mathfrak{y}}_{\mathfrak{e}_i} \in \mathscr{Q}(R_X), \tag{2.53}$$

which satisfies the condition in (2.42). In fact, for any element $v = \sum_{j=0}^{r-1} b_j \mathfrak{y}_{\mathfrak{e}_j} \in R_X$, ($b_j \in R_{\mathbb{P}}$), u satisfies $\tau_{R_X/R_{\mathbb{P}}}(uv) \in R_{\mathbb{P}}$, i.e.

$$\tau_{R_X/R_{\mathbb{P}}}(uv) = \sum_{i,j} \tau_{\widetilde{h}}(a_i b_j \widehat{\mathfrak{y}}_{\mathfrak{e}_i} \mathfrak{y}_{\mathfrak{e}_j}) = \sum_{i=0}^{r-1} (a_i b_i) \in R_{\mathbb{P}}.$$

The condition $\tau_{R_X/R_{\mathbb{P}}}(uv) \in R_{\mathbb{P}}$ means that a_i belongs to $R_{\mathbb{P}}$ for every $i = 0, 1, \ldots, r-1$, and thus we obtain $R_X^{\mathfrak{c}}$.

Proposition 2.67 *The complementary module $R_X^{\mathfrak{c}}$ is given as a fractional ideal of R_X,*

$$R_X^{\mathfrak{c}} = \frac{\langle \widehat{\mathfrak{y}}_{\mathfrak{e}_1}, \ldots, \widehat{\mathfrak{y}}_{\mathfrak{e}_{r-1}} \rangle_{R_X}}{h_X(x, y_\bullet)}.$$

Proof It is obvious that due to the above arguments and the identity $\langle \widehat{\Upsilon}_0, \widehat{\Upsilon}_1, \ldots, \widehat{\Upsilon}_{r-1} \rangle_{R_X} = \langle \widehat{\mathfrak{y}}_{\mathfrak{e}_1}, \ldots, \widehat{\mathfrak{y}}_{\mathfrak{e}_{r-1}} \rangle_{R_X}$ due to Lemma 2.62 and Definition 2.64. ■

Let us identify the generator $\mathfrak{h}_{X,P}$ of the principal ideal $R_{X,P}^{\mathfrak{c}}$ locally.

Proposition 2.68 *1. Symmetric d_h case ($\delta_0 = 0$ in (2.48) or $\widetilde{\delta}_0 = 1$ and $d_h = \mathfrak{e}_{r-1}$): The complementary module (as a fractional ideal) is given by*

$$R_X^{\mathfrak{c}} = \frac{1}{h_X(x, y_\bullet)} R_X = \mathfrak{h}_X R_X, \tag{2.54}$$

and we define $\mathfrak{h}_X := \dfrac{1}{h_X(x, y_\bullet)}$.

2. Non-symmetric d_h case ($\delta_0 \neq 0$ in (2.48) or $\widetilde{\delta}_0 \neq 1$ and $d_h = \mathfrak{e}_\ell + \delta_0 r$): For complex numbers a_i (some of them $\neq 0$), e.g., $a_i = 1$, we define

$$\mathfrak{h}_X := \frac{\sum_{i=1}^{r-1} a_i \widehat{\mathfrak{y}}_{\mathfrak{e}_i}}{h_X(x, y_\bullet)}. \tag{2.55}$$

For both cases, we obtain the local expression of the complementary module at $P \in X$,

$$R_{X,P}^{\mathfrak{c}} = \mathfrak{h}_{X,P} R_{X,P}. \tag{2.56}$$

Proof Symmetric d_h case: The ideal $\mathscr{I}$ contains 1 because $\widehat{\mathfrak{y}}_{\mathfrak{e}_{r-1}} = 1$. It includes $m_X = 2$ case in Sect. 2.2.6.3. In other words, $\mathfrak{h}_{X,P}$ in Definition 2.39 is given by $\varphi_P(\mathfrak{h}_X) = \mathfrak{h}_{X,P}$ by using the homomorphism $\varphi_P : R_X \to R_{X,P}$ in Remark 2.10.

Non-symmetric d_h case: By noting

$$\deg_{P,0}\left(\sum_{i=1}^{r-1} a_i \widehat{\mathfrak{y}}_{\mathfrak{e}_i}\right) = \min(\deg_{P,0}(\widehat{\mathfrak{y}}_{\mathfrak{e}_i})),$$

we also identify $\mathfrak{h}_{X,P}$ in Definition 2.39 with $\varphi_P(\mathfrak{h}_X)$. ■

Using the complementary module for both cases, we have the Dedekind different,

$$\mathrm{diff}(R_X/R_{\mathbb{P}}) = - \sum_{P \in X \setminus \{\infty\}} \deg_P(\mathfrak{h}_X) P.$$

By Dedekind's different theorem (Proposition 2.40), we have the following.

Proposition 2.69 *$e_{B_i} - 1 = -\deg_{B_i}(\mathfrak{h}_X)$, for $B_i \in \mathfrak{B}_X \setminus \{\infty\}$, and the support of* $\mathrm{div}(\mathfrak{h}_X)$ *is equal to $\mathfrak{B}_X$.*

2.2.6.7 Trace Operator p_X for Weierstrass Curves (W-Curves)

As we have $\widetilde{h}_{R_X}$ in Lemma 2.55, we extend it in $R_X \otimes_{R_{\mathbb{P}}} R_X$ to an element in $R_X \otimes_{\mathbb{C}} R_X$, though the extension is not unique; there are two different $h(x, y_\bullet, x', y'_\bullet)$ and $h'(x, y_\bullet, x', y'_\bullet)$ in $R_X \otimes_{\mathbb{C}} R_X$ such that $h(x, y_\bullet, x, y'_\bullet) = h'(x, y_\bullet, x, y'_\bullet)$ in $R_X \otimes_{R_{\mathbb{P}}} R_X$. Since they are quasi-isomorphic, we select one of them, and thus, it is well-defined in the meaning of Propositions 2.120 4. and 2.122 4.

We first consider it in $\varphi_\infty(R_X \otimes_{\mathbb{C}} R_X)$.

Definition 2.70 We define a polynomial in $R_H^Z \otimes_{\mathbb{C}} R_H^Z$ such that

$$\widetilde{h}_H(Z, Z') := \sum_{i=0}^{r-1} Z_r^{\delta_i} \mathfrak{z}_{\mathfrak{e}^*_{\ell,i}} \mathfrak{z}'_{\mathfrak{e}_i},$$

and an element,

$$p_H(Z, Z') = \frac{\widetilde{h}_H(Z, Z')}{h_H(Z_r, Z_\bullet)} \in \mathcal{Q}(R_H^Z \otimes_{\mathbb{C}} R_H^Z).$$

This $\widetilde{h}_H(Z, Z')$ is uniquely defined in the meaning of Propositions 2.120 4. and 2.122 4.

Proposition 2.71 *When $Z_r = Z'_r$, $\widetilde{h}_H(Z, Z')$ and $p_H(Z, Z')$ agree with $\widetilde{h}_H(Z_r, Z_\bullet, Z'_\bullet)$ in Corollary 2.49 and $p_H(Z_r, Z_\bullet, Z'_\bullet)$ in Lemma 2.45 respectively.* $\varphi_H^Z(\widetilde{h}_H(Z, Z') = z^{d_h} \sum_{i \in \mathbb{Z}_r} \frac{z'^{\mathfrak{e}_i}}{z^{\mathfrak{e}_i}}$.

Definition 2.72 Using $\widehat{\Upsilon}_i$ in Proposition 2.57, for a point $(P = (x, y_\bullet), P' = (x', y'_\bullet)) \in X \times X$, we define a polynomial $\widetilde{h}_X(x, y_\bullet, y'_\bullet) \in R_X \otimes_{\mathbb{C}} R_X$ by

$$\widetilde{h}_X(x, y_\bullet, y'_\bullet) := \widehat{\Upsilon}_0 \cdot 1 + \widehat{\Upsilon}_1 \mathfrak{y}'_{\mathfrak{e}_1} + \cdots + \widehat{\Upsilon}_{r-1} \mathfrak{y}'_{\mathfrak{e}_{r-1}}.$$

Then the following lemma is evident from Lemma 2.55.

Lemma 2.73 *For a point $(P = (x, y_\bullet), P' = (x', y'_\bullet)) \in X \times X$, $\widetilde{h}_X(x, y_\bullet, y'_\bullet)$ satisfies*

1. *for a group action $\widehat{\zeta} \in G_{X,x}$,* $\dfrac{\widetilde{h}_X(x, \widehat{\zeta} y_\bullet, \widehat{\zeta} y'_\bullet)}{\widetilde{h}_X(x, \widehat{\zeta} y_\bullet, \widehat{\zeta} y_\bullet)} = \dfrac{\widetilde{h}_X(x, y_\bullet, y'_\bullet)}{\widetilde{h}_X(x, y_\bullet, y_\bullet)}$ *if $x = x'$,*
2. $\displaystyle\lim_{(x,y)\to\infty} \frac{\widetilde{h}_X(x, y_\bullet, y'_\bullet)}{\widetilde{h}_X(x, y_\bullet, y_\bullet)} = \frac{1}{r}$ *for $P' \in X \setminus \{\infty\}$,*

3. *when* $x = x'$ *or* $\widetilde{h}_X(x, y_\bullet, x', y'_\bullet)$ *belongs to* $R_X \otimes_{R_{\mathbb{P}}} R_X$, *it satisfies*

$$\widetilde{h}_X(x, y_\bullet, y'_\bullet) = \widetilde{h}_{R_X}(x, y_\bullet, y'_\bullet),$$

4. $h_X(x, y_\bullet) = \widetilde{h}_X(x, y_\bullet, y_\bullet)$, *and*
5. $\varphi_H^X(\widetilde{h}_X(x, y_\bullet, y'_\bullet)) = \widetilde{h}_H(Z_r, Z_\bullet, Z'_\bullet)$ $(\mathrm{wt}(\widetilde{h}_X(x, y_\bullet, y'_\bullet)) = d_h$.

Definition 2.74 Using $\widetilde{h}_X(x, y_\bullet, y'_\bullet)$ for a point $(P = (x, y_\bullet), P' = (x', y'_\bullet)) \in X \times X$, we define

$$p_\varpi(P, P') := \frac{\widetilde{h}_X(x, y_\bullet, y'_\bullet)}{\widetilde{h}_X(x, y_\bullet)}.$$

It is obvious that $p_\varpi(P, Q)$ belongs to $\mathfrak{h}_{X,P} R_{X,P}$ as a function of P at $P \in X$ and thus Dedekind's different theorem (Propositions 2.40 and 2.69) shows the proposition:

Proposition 2.75 $e_{B_i} - 1 = -\deg_{P=B_i}(p_\varpi(P, Q))$ *for* $Q \in X \setminus \{\infty\}$.

Further, direct computations provide the following proposition:

Proposition 2.76 $p_\varpi(P, Q)$ *is holomorphic on* $(X \setminus \mathfrak{B}_r) \times (X \setminus \{\infty\})$. *For* $Q \in X \setminus \{\infty\}$,

$$p_\varpi(P, Q) = \begin{cases} 1 & \text{for } P = Q, \\ 0 & \text{for } P \neq Q \text{ and } \varpi_r(P) = \varpi_r(Q), \end{cases}$$

$$\lim_{P \to \infty} p_\varpi(P, Q) = \frac{1}{r}, \quad \lim_{Q \to \infty} p_\varpi(P, Q) = \infty.$$

Remark 2.77 We should remark that $p(P, Q) \in \mathcal{Q}(R_X \otimes_{\mathbb{C}} R_X)$ which holds the relations in Proposition 2.76 is not unique, as mentioned in the introduction to this subsubsection. The problem comes from the fact that there are infinitely many different $p'(P, Q)$ from $p(P, Q)$ such that $p(P, Q) = p'(P, Q)$ for $\varpi_r(P) = \varpi_r(Q)$.

Proposition 2.78 *For a point* $Q \in X \setminus \mathfrak{B}_X$,

$$\mathrm{div}_P(p_\varpi(P, Q)) = \sum_{\zeta \in G_{X,\varpi_r(Q)}, \zeta \neq e} \zeta Q + E_Q - \sum_{i=1}^{\ell_{\mathfrak{B}}} (e_{B_i} - 1) B_i,$$

where E_Q *is an effective divisor such that* $\deg E_Q = \left(\sum_{i=1}^{\ell_{\mathfrak{B}}} (e_{B_i} - 1) - r + 1 \right) = 2g$.

Further, this relation is extended to the condition $Q \in \mathfrak{B}_X \setminus \{\infty\}$ *by considering the multiplicity of the action* $G_{X,\varpi_r(Q)}$.

Proof At the ramification point B_i of $\varpi_r : X \to \mathbb{P}$, Proposition 2.75 shows the third term. We note that $p_\varpi(\infty, Q) = 1/r$. From the Riemann–Hurwitz theorem (2.40), there exist the first and the second terms. ■

For an element $f \otimes f'$ in $R_X \otimes_{\mathbb{C}} R_X$, we define the weight $\mathrm{wt}(f \otimes f')$ by $\mathrm{wt}(f) + \mathrm{wt}(f')$.

Lemma 2.79 *In the sense of Lemma 2.45, $\widetilde{h}_X(x, y_\bullet, x', y'_\bullet)$ is a homogeneous element in $R_X \otimes_{\mathbb{C}} R_X$ whose weight is $-d_h$ or $-\mathrm{wt}\,\widetilde{h}_X(x, y_\bullet, x', y'_\bullet) = -\mathrm{wt}\,h_X(x, y_\bullet) = d_h$ and thus the weight of p_ϖ is zero in $\mathscr{Q}(R_X \otimes_{\mathbb{C}} R_X)$.*

Since some of $f_{X,y}^{(j)}(P) = 0$ at $P = B_i \in \mathfrak{B}_X \setminus \{\infty\}$, $h_X(x, y_\bullet) \in R_X$ vanishes only at the ramification point $B_i \in X$ from Proposition 2.75 and the construction of h_X, we have the following corollary:

Corollary 2.80

$$\mathrm{div}(h_X(x, y_\bullet)) = \sum_{B_i \in \mathfrak{B}_X \setminus \{\infty\}} d_{B_i}\, B_i - d_h \infty,$$

where $d_{B_i} := \deg_{B_i,0}(h_X) \geq (e_{B_i} - 1)$, and $d_h = \displaystyle\sum_{B_i \in \mathfrak{B}_X \setminus \{\infty\}} d_{B_i} = -\mathrm{wt}(h_X)$.

Definition 2.81 The effective divisor, $\displaystyle\sum_{B_i \in \mathfrak{B}_X \setminus \{\infty\}} (d_{B_i} - e_{B_i} + 1)\, B_i$, is denoted by $\mathfrak{K}_X$, i.e. $\mathfrak{K}_X > 0$ and let $\mathfrak{k}_X := \displaystyle\sum_{B_i \in \mathfrak{B}_X \setminus \{\infty\}} (d_{B_i} - e_{B_i} + 1) = \deg(\mathfrak{K}_X) \geq 0$.

Lemma 2.82 *The divisor of $\dfrac{dx}{h_X}$ is expressed by $(2g - 2 + \mathfrak{k}_X)\infty - \mathfrak{K}_X$, and $2g - 2 + \mathfrak{k}_X = d_h - r - 1$ or $\mathfrak{k}_X = d_h - 2g - r + 1$. $\varphi_\infty\left(\dfrac{dx}{h_X}\right) = t^{d_h - r - 1} dt$.*

Proof Since the meromorphic one-form, in general, has a degree of $2g - 2$, we obtain $\deg(dx/h_X) = 2g - 2$. From Definition 2.81, its divisor is expressed by $\mathrm{div}(dx/h_X) = (2g - 2 + \mathfrak{k}_X)\infty - \mathfrak{K}_X$. Further, since at the ∞, its degree $\deg_\infty(dx/h_X) = d_h - r - 1$, we have $2g - 2 + \mathfrak{k}_X = d_h - r - 1$. ■

From Corollary 2.35, we note that these $\mathfrak{K}_X$ and $\mathfrak{k}_X$ play crucial roles in the investigation of the differentials on X (e.g., Lemma 2.110).

Proposition 2.83 *$\mathfrak{k}_X$ is equal to zero if d_h is symmetric, whereas $\mathfrak{k}_X$ is not zero otherwise.*

Proof For the symmetric case, (2.54) and Proposition 2.69 show that $d_{B_i} = e_{B_i} - 1$, whereas for the non-symmetric case, (2.56) and Proposition 2.69 yield non-vanishing $\mathfrak{k}_X$. ■

We recall $\widehat{\mathfrak{e}}_i$ in (2.47), $\mathfrak{e}_i$ in Definition 2.4 for the standard basis in Lemma 2.13 and Proposition 2.20, and $\widehat{\mathfrak{y}}_{\mathfrak{e}_i}$ in Definition 2.64.

By applying Propositions 2.67 and 2.69 to differentials on X, we consider $\dfrac{x^k \widehat{\mathfrak{y}}_{\mathfrak{e}_i}(x, y_\bullet) dx}{h_X(x, y_\bullet)}$, for which holds the following proposition:

Proposition 2.84

$$\mathrm{div}\left(\frac{x^k\widehat{\mathfrak{y}}_{\mathfrak{e}_i}(x,y_\bullet)dx}{h_X(x,y_\bullet)}\right)=\mathrm{div}\left(\frac{x^k\widetilde{\delta}_i(x)\mathfrak{y}_{\mathfrak{e}_{\ell,i}}(x,y_\bullet)dx}{h_X(x,y_\bullet)}\right)$$

$$=k\,\mathrm{div}_0(x)+\mathrm{div}_0(\widehat{\mathfrak{y}}_{\mathfrak{e}_i})-\sum_{j=1}^{\ell_{\mathfrak{B}}}(d_{B_j}-e_{B_j}-1)B_j+(d_h-\widehat{\mathfrak{e}}_i-(k+1)r-1)\infty$$

$$=k\,\mathrm{div}_0(x)+\mathrm{div}_0(\widehat{\mathfrak{y}}_{\mathfrak{e}_i})-\mathfrak{K}_X+(\mathfrak{e}_i-(k+1)r-1)\infty,$$

where $\mathrm{div}_0(\widehat{\mathfrak{y}}_{\mathfrak{e}_i})-\mathfrak{K}_X\ge 0$. *We have*

$$\left\{\mathrm{wt}\left(\frac{x^k\widehat{\mathfrak{y}}_{\mathfrak{e}_i}(x,y_\bullet)dx}{h_X(x,y_\bullet)}\right)+1\;\middle|\;i\in\mathbb{Z}_r^\times,\ k\in\mathbb{N}_0,\ \mathfrak{e}_i-(k+1)r>0\right\}=H_X^{\mathfrak{c}},$$

and

$$\left\{\mathrm{wt}\left(\frac{x^k\widehat{\mathfrak{y}}_{\mathfrak{e}_i}(x,y_\bullet)dx}{h_X(x,y_\bullet)}\right)+1\;\middle|\;i\in\mathbb{Z}_r,\ k\in\mathbb{N}_0\right\}=\overline{H}_X^{\mathfrak{c}}.$$

Proof The former statement is asserted from the previous lemma and Proposition 2.69, whereas the latter two relations on $H_X^{\mathfrak{c}}$ and $\overline{H}_X^{\mathfrak{c}}$ are proved by Lemma 2.5. Here we note that $\mathrm{wt}(\widehat{\mathfrak{y}}_{\mathfrak{e}_i})=-\widehat{\mathfrak{e}}_i=-(d_h-\mathfrak{e}_i)$ due to (2.47) and $\mathfrak{e}_0=0$. ■

We mention the result in Proposition 2.84 in terms of the subsets in the numerical semigroup H_X:

Corollary 2.85 *Let* $\widehat{H}_X^{(g)}:=\{\widehat{\mathfrak{e}}_i+kr\mid i\in\mathbb{Z}_r^\times,\ k\in\mathbb{N}_0,\ \mathfrak{e}_i-(k+1)r>0\}$. *Then*

$$\widehat{H}_X^{(g)}\subset H_X,\quad \#\widehat{H}_X^{(g)}=g,\quad H_X^{\mathfrak{c}}=\{d_h-n-r\mid n\in\widehat{H}_X^{(g)}\}.$$

Proof $\{d_h-n-r\mid n\in\widehat{H}_X^{(g)}\}=\{d_h-\widehat{\mathfrak{e}}_i-kr-r\mid\ i\in\mathbb{Z}_r,\ k\in\mathbb{N}_0,\ \mathfrak{e}_i-(k+1)r>0\}$ means Lemma 2.5 *1*.c. Here we note $d_h-\widehat{\mathfrak{e}}_i=\mathfrak{e}_i$. ■

Lemma 2.86 $h_X(x,y_\bullet)$ *has the expression*

$$h_X(x,y_\bullet)=\frac{\prod_{j=2}^{m_X}f_{X,y}^{(j)}(x,y_\bullet)}{h_0(x)},$$

where $h_0(x)\in\mathbb{C}[x]$, *and the group action* $\widehat{\zeta}\in G_{X,x}$,

$$h_X(x,y_\bullet)=\frac{\prod_{j=2}^{m_X}\widehat{\zeta}^{(j)}f_{X,y}^{(j)}(x,y_\bullet)}{h_0(x)}.$$

Proof From Proposition 2.31 and Lemma 2.73, we have the group action $G_{X,x}$ on h_X at $x \in X$. h_0 must be invariant for the group action and thus it belongs to $R_{\mathbb{P}}$. ■

2.2.7 Universal Abelian Covering $\widetilde{X}$ of X and $\mathbf{H}_1(X, \mathbb{Z})$

As we handle the contour integrals on X in the Abelian function theory, the universal Abelian covering related to the path space of X is introduced in this subsection.

We regard X as a compact Riemann surface and consider paths in X. In other words, we are dealing with the path space $\mathrm{Path}(X)$ over X [BotT1993, Sp1966]. A continuous path in X from Q to P is denoted by $\gamma_{P,Q}$, i.e., a continuous map $\gamma_{P,Q}$: $[0, 1] \to X$ satisfies $\gamma_{P,Q}(0) = Q$ and $\gamma_{P,Q}(1) = P$; $\mathrm{Path}(X) := \{\gamma_{P,Q} \mid P,\ Q \in X\}$ over X.

Then we define the "addition" in $\mathrm{Path}(X)$; for $\gamma_{P,Q}, \gamma_{R,P} \in \mathrm{Path}(X)$,

$$\gamma_{R,Q} := \gamma_{R,P} \dot{+} \gamma_{P,Q} \in \mathrm{Path}(X). \tag{2.57}$$

This "addition" is essential in the additive structure in the Jacobi variety, which can be interpreted as groupoid. Here we note that the "addition" is not commutative in general; $\gamma \dot{+} \gamma' \neq \gamma' \dot{+} \gamma$, and both of them cannot be defined well in general. We also define $\gamma_{Q,P}^{-1} = \gamma_{P,Q} \in \mathrm{Path}(X)$ and thus

$$\gamma_{Q,P}^{-1} \dot{+} \gamma_{R,Q} = \gamma_{P,Q}.$$

Further, we consider the subspace $\mathrm{Path}_P(X) := \{\gamma_{Q,P} \mid Q \in X\}$ with the marked point $P \in X$. $\mathrm{Path}_P(X)$ has the fiber structure $p : \mathrm{Path}_P(X) \to X$ by $p(\gamma_{Q,P}) = Q$. Since $\gamma_{Q,P}$ is obviously homotopic to P, $\mathrm{Path}_P(X)$ is homotopic to $\{P\}$.

We consider the Abelianization of the quotient space of $\mathrm{Path}_P(X)$ by the homotopy equivalent, which is denoted by $\Gamma_P X$, and $\widetilde{X} := \Gamma_\infty X$. The surjection p induces the projection $\kappa_X : \widetilde{X} \to X$ and a natural embedding $\iota_X : X \to \widetilde{X}$ such that $\kappa_X \circ \iota_X = id$; ι_X is sometimes abbreviated as ι. The symbol $+$ and $-\gamma$ denote $\dot{+}$ and γ^{-1} after the Abelianization respectively.

For the k-symmetric product of $\widetilde{X}$, $S^k\widetilde{X}$, we also extend the projection and the embedding by $\kappa_X : S^k\widetilde{X} \to S^k X$ and $\iota_X : S^k X \to S^k\widetilde{X}$. The path space includes the loop space $\{\gamma \in \mathrm{Path}_P(X) \mid p(\gamma) = P\}$ and the additive structure is induced from the path space. The Abelian covering $\widetilde{X}$ also includes the Abelian loop space, which is governed by the homology $\mathbf{H}_1(X, \mathbb{Z})$ whose basis is given by loops α_i, β_i with the intersection

$$\langle \alpha_i, \beta_j \rangle = \delta_{i,j}, \quad \langle \alpha_i, \alpha_j \rangle = \langle \beta_i, \beta_j \rangle = 0. \tag{2.58}$$

We could regard $\mathbf{H}_1(X, \mathbb{Z})$ as a discrete subspace of $\widetilde{X}$.

The basis $\{\alpha_1, \ldots, \alpha_g, \beta_1, \ldots, \beta_g\}$ in $\mathbf{H}_1(X, \mathbb{Z})$ can be transformed by the symplectic group $\mathrm{Sp}(2g, \mathbb{Z})$ so that it satisfies (2.58).

Lemma 2.87 *For $\gamma = (\gamma_1, \gamma_2, \dots, \gamma_{2g}) = (\alpha_1, \dots, \alpha_g, \beta_1, \dots, \beta_g)$, every element ρ of the symplectic group*

$$\mathrm{Sp}(2g, \mathbb{Z}) = \left\{ \rho := \begin{pmatrix} a & b \\ c & d \end{pmatrix} \,\middle|\, {}^t\rho \begin{pmatrix} 0 & 1_g \\ -1_g & 0 \end{pmatrix} \rho = \begin{pmatrix} 0 & 1_g \\ -1_g & 0 \end{pmatrix}, a, b, c, d \in \mathrm{GL}(g, \mathbb{Z}) \right\}, \tag{2.59}$$

where 1_g is the unit matrix of $g \times g$, the transformation ${}^t\gamma' = \rho\,{}^t\gamma$, i.e., $\gamma' = \gamma\,{}^t\rho$, satisfies the relations

$$\langle \gamma_i', \gamma_{g+j}' \rangle = \delta_{i,j}, \quad \langle \gamma_i', \gamma_j' \rangle = \langle \gamma_{g+i}', \gamma_{g+j}' \rangle = 0,$$

where $i, j = 1, 2, \dots, g$.

Proof From the properties of $\mathrm{Sp}(2g, \mathbb{Z})$, it is obvious (see Sect. 3.8.1). ∎

Further, if there exists the Galois action $\widehat{\zeta} \in G_X$, ($\widehat{\zeta} : X \to X$), it also preserves the properties and thus the following is obvious (Fig. 2.2).

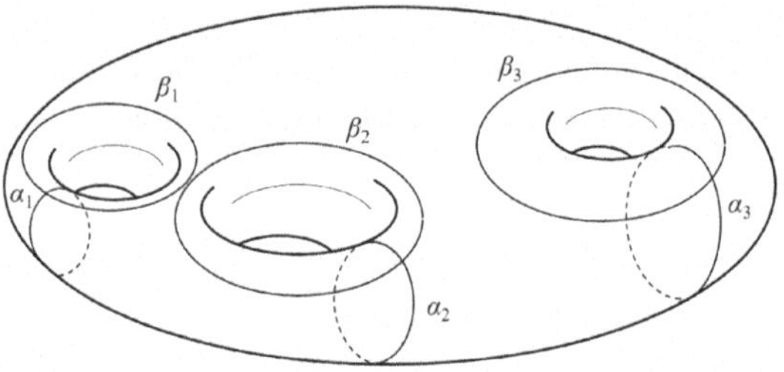

Fig. 2.2 Contours: α, β

Lemma 2.88 *Assume that there exists the Galois action $\widehat{\zeta} \in G_X$ on X. For the Galois action $\widehat{\zeta} \in G_X$, i.e., $\widehat{\zeta} : X \to X$, and for $\gamma = (\gamma_1, \gamma_2, \dots, \gamma_{2g}) = (\alpha_1, \dots, \alpha_g, \beta_1, \dots, \beta_g)$, the action $\gamma' = \widehat{\zeta}\gamma$ induces to an element $\rho_{\widehat{\zeta}} \in \mathrm{Sp}(2g, \mathbb{Z})$ of the symplectic group such that ${}^t(\widehat{\zeta}\gamma) = \rho_{\widehat{\zeta}}\,{}^t\gamma$, i.e., $\widehat{\zeta}\gamma = \gamma\,{}^t\rho_{\widehat{\zeta}}$.*

Since the loops can be deformed, we set the loops so that they are the boundary of the domain D_X as in Fig. 2.3 when we cut them, and it recovers the compact Riemann surface X, when we sew them together. We note that this operation can only be performed before the Abelianization They can be represented in terms of an extension of expressions in Path(X):

$$X = D_X / \partial D_X, \quad \partial D_X = \sum_{j=1}^{g}{}^{\boldsymbol{\cdot}} (\beta_j^{-1} \dot{+} \alpha_j^{-1} \dot{+} \beta_j \dot{+} \alpha_j), \tag{2.60}$$

where $\sum^{\boldsymbol{\cdot}}$ denotes the summation symbol associated with $\dot{+}$. The beginning point of α_j (β_j) is denoted by $\alpha_{j,0}$ ($\beta_{j,0}$) and the terminal point is by $\alpha_{j,1}$ ($\beta_{j,1}$).

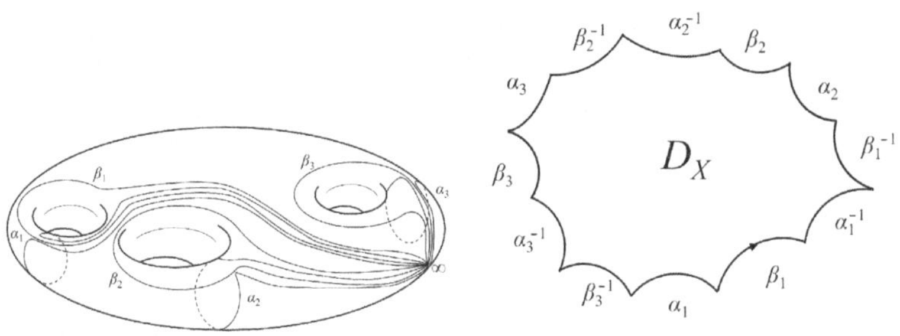

Fig. 2.3 D_X with ∂D_X

2.3 Examples of Weierstrass Curves (W-Curves)

This section enumerates examples of W-curves below. We clarify that the words "trigonal" and "pentagonal" are used here only as an indication of the fact that the pointed curve (X, ∞) has a Weierstrass semigroup of type three, five, respectively; a different convention requires that k-gonal curves not be j-gonal for $j < k$ (as in trigonal curves, which by definition are not hyperelliptic). For example, we say a non-singular curve $f_X(x, y) = 0$ in (2.8) with $(r = 3, s = 5)$, is a *trigonal curve of degree five*.

2.3.1 Plane W-Curves ((r, s)-Curves) and Hyperelliptic Curves

A plane Weierstrass curve (W-curve) X is a non-singular curve given by (2.8),

$$f_X(x, y) = y^r + A_1(x)y^{r-1} + A_2(x)y^{r-2} + \cdots + A_{r-1}(x)y + A_r(x) = 0. \tag{2.61}$$

The plane W-curve corresponds to the case $m_X = 2$. In this book, we also call it an (r, s)-*curve* following Buchstaber, Leykin and Enolskii [BucLE1999, BucEL2012, BucEL2020], and Eilbeck, Enolskii and Leykin [EilEL2000], though they refer to it as an (n, s)-curve, (cf. footnote in Sect. 1.9.2). Further, it is also called a C_{ab} curve for $a = r$ and $b = s$ by Miura's investigation [Miu1998] in 1998. We may refer to it as $X_{(r,s)}$, if necessary. For the case of $r = 2$, it is a hyperelliptic curve $X_{(2,2g+1)}$.

2.3.1.1 Plane (r, s)-Curves

We consider plane W-curves of (r, s)-type, or (r, s)-curves in this subsubsection. Its genus is given by the Sylvester formula, $g = \dfrac{(r-1)(s-1)}{2}$ as in Lemma 2.7 and Proposition 2.33 3. The related monomial curve is given by the relation, $Z_s^r = Z_r^s$.

For a (3, 4)-curve, (2.8) or (2.61) is reduced to

$$\begin{aligned} f_X(x,y) &= y^3 + A_1(x)y^2 + A_2(x)y + A_3(x), \\ A_1(x) &= \lambda_{1,1}x + \lambda_{1,0}, \quad A_2(x) = \lambda_{2,2}x^2 + \lambda_{2,1}x + \lambda_{2,0}, \\ A_3(x) &= -x^4 + \lambda_{3,3}x^3 + \lambda_{3,2}x^2 + \lambda_{3,1}x + \lambda_{3,0}. \end{aligned} \tag{2.62}$$

In the case of $r = 5$, (2.13) corresponds to

$$\begin{pmatrix} -y & 0 & 1 \\ 1 & -y & 0 \\ -A_3 & -A_2 & -A_1 - y \end{pmatrix} \begin{pmatrix} y^3 \\ y^2 \\ y^4 \end{pmatrix} = \begin{pmatrix} 0 \\ 0 \\ A_4 y + A_5 \end{pmatrix},$$

and thus (2.14) becomes

$$y^{2+i} = -\frac{(A_4 y + A_5)y^i}{y^3 + A_1 y^2 + A_2 y + A_3}. \tag{2.63}$$

The curve has a 5-semigroup at ∞ but is not necessarily cyclic. In this case ($m_X = 2$), f_X is equal to f_1^X ($m_X = 2$, $k_X = 1$). $R_X = \mathbb{C}[x, y]/(f_X(x, y))$.

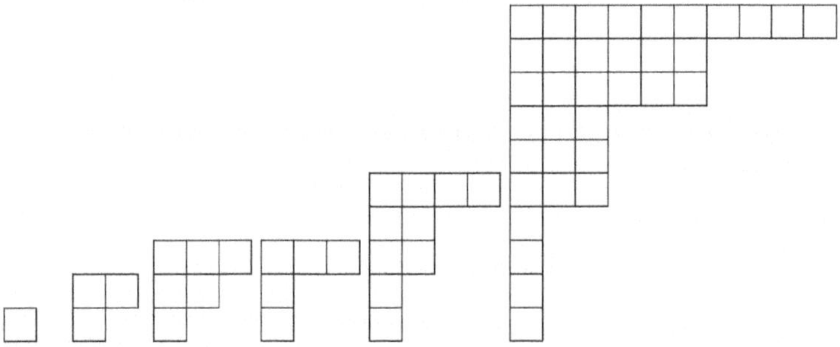

(2, 3), (2, 5), (2, 7), (3, 4), (3, 7), (5, 6).

The numerical semigroups $H_X = \langle r, s \rangle$ of X are symmetric.

$$p_\varpi((x,y),(x',y')) = \frac{(\delta_{y,y'} f_X)(x,y,y')}{f_{X,y}(x,y)}, \quad h_X(x,y) = f_{X,y}(x,y).$$

The basis of the affine ring $R_X = \mathbb{C}[x, y]/(f_X(x, y))$ as a $\mathbb{C}$ vector space, $R_X = \bigoplus_{i=0} \mathbb{C}\phi_i$ is given as the monomials whose order corresponds to the symmetric Weierstrass numerical semigroup (see Table 2.1).

Table 2.1 Examples of ϕ of (r, s)-curves, $c_X = 2g$

(r,s)	g	0	1	2	3	4	5	6	7	8	9	10	11	12	13	14	15	16	17	18	19	20	21	22
(2,3)	1	1	–	x	y	x^2	xy	x^3	x^2y	x^4	x^3y	x^5	x^4y	x^6	x^5y	x^7	x^6y	x^8	x^7y	x^9	x^8y	x^{10}	x^9y	x^{11}
(2,5)	2	1	–	x	–	x^2	y	x^3	xy	x^4	x^2y	x^5	x^3y	x^6	x^4y	x^7	x^5y	x^8	x^6y	x^9	x^7y	x^{10}	x^8y	x^{11}
(2,7)	3	1	–	x	–	x^2	–	x^3	y	x^4	xy	x^5	x^2y	x^6	x^3y	x^7	x^4y	x^8	x^5y	x^9	x^6y	x^{10}	x^7y	x^{11}
(3,4)	3	1	–	–	x	y	–	x^2	xy	y^2	x^3	x^2y	xy^2	x^4	x^3y	x^2y^2	x^5	x^4y	x^3y^2	x^6	x^5y	x^4y^2	x^7	x^6y
(3,5)	4	1	–	–	x	–	y	x^2	–	xy	x^3	y^2	x^2y	x^4	x^2y	xy^2	x^5	x^2y	x^2y^2	x^6	x^3y	x^2y^2	x^7	x^4y
(3,7)	6	1	–	–	x	–	–	x^2	y	–	x^3	xy	–	x^4	x^2y	y^2	x^5	x^3y	xy^2	x^6	x^4y	x^2y^2	x^7	x^6y
(5,6)	10	1	–	–	–	–	x	y	–	–	–	x^2	xy	y^2	–	–	x^3	x^2y	xy^2	y^3	–	x^4	x^3y	x^2y^2

2.3.1.2 Cyclic Plane (r, s)-Curves

In this book, we sometimes deal with simpler (r, s)-curves (2.61) which have the global cyclic group action as in Sect. 2.2.2.3. The cyclic plane W-curves, (r, s)-curves, are given by the Weierstrass canonical form,

$$y^r = x^s + \lambda_{s-1}x^{s-1} + \cdots + \lambda_1 x + \lambda_0, \tag{2.64}$$

which are studied in [GaPS2000, MatP2008]. These are sometimes called superelliptic curves [Koo1999]. The Weierstrass non-gap sequences are also displayed in Table 2.1.

The global cyclic Galois action $\widehat{\zeta_r} : X \to X$ is given by $(x, y) \mapsto (x, \zeta_r y)$, where $\zeta_r = \mathrm{e}^{2\pi\sqrt{-1}/r}$. The trace operator becomes very simple:

$$p_\varpi((x, y), (x', y')) = \frac{\sum_{i=1}^r y^{r-i} y'^{i-1}}{r y^{r-1}}, \quad h_X(x, y) = r y^{r-1}.$$

2.3.1.3 Hyperelliptic Curves

The $r = 2$ case means a hyperelliptic curve X. Though it has more general form, we mainly consider the cyclic hyperelliptic curve,

$$f_X(x, y) := y^2 - f(x), \quad f(x) := \lambda_{2g+1}x^{2g+1} + \lambda_{2g}x^{2g} + \cdots + \lambda_1 x + \lambda_0 = \prod_{i=0}^{2g}(x - b_i), \tag{2.65}$$

where complex numbers b_i's are mutually distinct, and $\lambda_{2g+1} = 1$. For later convenience, we introduce a decomposition of $f(x)$ by

$$f(x) = f_{\mathrm{I}}(x) f_{\mathrm{II}}(x), \quad f_{\mathrm{I}}(x) := \prod_{i=1}^{g}(x - a_i), \quad f_{\mathrm{II}}(x) := \prod_{i=0}^{g}(x - c_i), \tag{2.66}$$

where $a_i := b_{2i-1}$, $c_i := b_{2i}$ and $c := b_0$ for $i = 1, \ldots, g$. When $g = 1$, the cyclic curve corresponds to an elliptic curve denoted by E.

The automorphism $\widehat{\zeta}_2$ for every cyclic hyperelliptic curve is known as the hyperelliptic involution denoted by $\widehat{\zeta}_H$ ($\widehat{\zeta}_H(x, y) = (x, -y)$) because $\widehat{\zeta}_2^2 = \widehat{\zeta}_H^2$ is equal to the identity map.

The homology basis is illustrated in Fig. 2.4. (See also Fig. 2.6.)

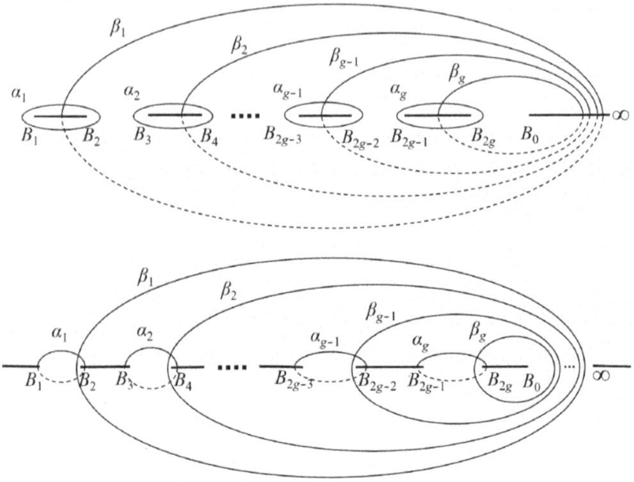

Fig. 2.4 Contours: α, β in dual expressions

For the hyperelliptic curves, the basis of the affine ring $R_X = \mathbb{C}[x, y]/(f_X(x, y))$ as a $\mathbb{C}$ vector space, $R_X = \bigoplus_{i=0} \mathbb{C}\phi_i$ is simply given by

$$\phi_i = \begin{cases} x^i & \text{for } i < g, \\ x^{\lfloor (i-g)/2 \rfloor} y & \text{for } i \geq g \ \&\ 2 \not| (i-g), \\ x^{\lfloor (i-g)/2 \rfloor + g} & \text{for } i \geq g \ \&\ 2 \mid (i-g). \end{cases}$$

2.3.1.4 Cyclic Trigonal Curves of Genus Three

A cyclic (3,4)-curve of genus three is given by the affine equation

$$\begin{aligned} y^3 &= x^4 + \lambda_3 x^3 + \lambda_2 x^2 \lambda_1 x + \lambda_0 \\ &= (x - b_0)(x - b_1)(x - b_2)(x - b_3), \end{aligned} \tag{2.67}$$

which is simpler than (2.62).

2.3.2 Telescopic Curves

Ayano studied the sigma function in telescopic curves [Ay2014, Ay2016]. Following [Ay2014, Ay2016], let us show the definition of telescopic curves given by Miura [Miu1998]. For an integer $m \geq 2$, we let $A_m := (a_1, \dots, a_m)$ be a sequence of positive integers such that these a_i are mutually coprime, $\gcd(a_1, \dots, a_m) = 1$, $a_i \geq 2$ $(i = 1, \dots, m)$, and for $2 \leq i \leq m$,

$$\frac{a_i}{d_i} \in \mathbb{N}_0 \frac{a_1}{d_{i-1}} + \cdots + \mathbb{N}_0 \frac{a_{i-1}}{d_{i-1}},$$

where $d_i := \gcd(a_1, \dots, a_{i-1})$.

Let $B(A_m) := \left\{ (\ell_1, \dots, \ell_m) \in \mathbb{N}_0^m \;\middle|\; 0 \leq i \leq \dfrac{d_{i-1}}{d_i} - 1, \quad (2 \leq i < m) \right\}$. In this situation, it is known that for every $a \in \langle a_1, \dots, a_m \rangle$, there is a unique element $(k_1, \dots, k_m) \in B(A_m)$ such that $\sum_{i=1}^m a_i k_i = a$. Using this fact, we also have a unique sequence $(\ell_{i,1}, \dots, \ell_{i,m}) \in B(A_m)$ such that $\sum_{j=1}^m a_j \ell_{i,j} = a_i \dfrac{d_{i-1}}{d_i}$. Then for $2 \leq i \leq m$, it turns out that $\ell_{i,j} = 0$ for $j > i$.

Using them, we consider polynomials in $\mathbb{C}[y_1, y_2, \dots, y_m]$,

$$f_i(y) = y_i^{d_{i-1}/d_i} - \prod_{j=1}^{i-1} y_j^{\ell_{i,j}} - \sum \lambda^{(i)}_{j_1,\dots,j_m} y_1^{j_1} \cdots y_m^{j_m},$$

for $2 \leq i \leq m$ and $\lambda^{(i)}_{j_1,\dots,j_m} \in \mathbb{C}$ such that $\sum_{k=1}^m a_k j_k < a_i \dfrac{d_{i-1}}{d_i}$.

Then let $R_X := \mathbb{C}[y_1, y_2, \dots, y_m]/(f_1, \dots, f_m)$ and its associated curve X, which is called the telescopic curve associated with $(a_1, \dots, a_m)$, when X is not singular. Then the genus g of X is given by $g = \dfrac{1}{2}\left(1 - a_1 + \sum_{i=2}^m \left(\dfrac{d_{i-1}}{d_i} - 1\right) a_i\right)$. The semigroup $H_X = \langle a_1, \dots, a_m \rangle$ is the Weierstrass semigroup of X.

Due to the conditions on $(a_1, \dots, a_m)$, the numerical semigroup is symmetric.

This telescopic curve includes the above (r, s)-curve as a particular case, and thus, Ayano studied this curve and its sigma function precisely.

2.3.3 Cyclic r Space Curves and Cyclic Trigonal Space Curves

The cyclic space W-curves

$$y^r = (x - b_1)^{d_1} \cdots (x - b_n)^{d_n},$$

where r, d_i, n are certain positive integers, are sometimes called superelliptic curves [Koo1999].

We deal with cyclic trigonal space curves, i.e., $r = 3$, particularly [KomMP2019]. For brevity, henceforth we let $\widehat{r} := 2\dot{r} + ds$ and $\widehat{s} := 2\dot{s} + \dot{r}$ so that for $\dot{r} < \dot{s}, \widehat{r} < \widehat{s}$. Unless $\dot{r} = 0$, we assume that $3, \widehat{r}, \widehat{s}$ are a minimal set of generators of the W-semigroup $H_X = H$; $H_X = \langle 3, \widehat{r}, \widehat{s}\rangle$, $m_X = 3$. When $\dot{s} > \dot{r} > 0$, the triple $(3, \widehat{r}, \widehat{s})$ is not telescopic in any order (because, as we demonstrate, the corresponding Young diagram is not symmetric), whereas a semigroup with two generators is clearly telescopic.

We give more detail on the monomial ring (defined in Sect. 2.2.3) of a 3-semigroup. Again by [Her1970, Th. 3.7],

Lemma 2.89 *The kernel of φ_H^Z is generated by*

$$f_{2\widehat{r}}^H = Z_{\widehat{r}}^2 - Z_3^{\dot{r}} Z_{\widehat{s}}, \quad f_{\widehat{s}+\widehat{r}}^H = Z_{\widehat{r}} Z_{\widehat{s}} - Z_3^{\dot{s}+\dot{r}}, \quad f_{2\widehat{s}}^H = Z_{\widehat{s}}^2 - Z_3^{\dot{s}} Z_{\widehat{r}}. \tag{2.68}$$

Let

$$k_{\dot{s}}(x) := (x - b_1) \cdots (x - b_{\dot{s}}) = x^{\dot{s}} + \sum_{i=1}^{\dot{s}} \lambda_i^{(\dot{s})} x^{\dot{s}-i},$$

$$k_{\dot{r}}(x) := (x - b_{\dot{s}+1}) \cdots (x - b_{\dot{s}+\dot{r}}) = x^{\dot{r}} + \sum_{i=1}^{\dot{r}} \lambda_i^{(\dot{r})} x^{\dot{r}-i},$$

$$k_{\dot{s}+\dot{r}}(x) := (x - b_1) \cdots (x - b_{\dot{s}}) \cdots (x - b_{\dot{s}+\dot{r}}) = x^{\dot{r}+\dot{s}} + \sum_{i=1}^{\dot{r}+\dot{s}} \lambda_i^{(\dot{s}+\dot{r})} x^{\dot{s}+\dot{r}-i},$$

and $\lambda_0^{(\dot{s})} = \lambda_0^{(\dot{r})} = \lambda_0^{(\dot{s}+\dot{r})} = 1$. We define $f_{2\widehat{r}}, f_{\widehat{s}+\widehat{r}}, f_{2\widehat{s}} \in \mathbb{C}[x, y_{\widehat{r}}, y_{\widehat{s}}]$ by

$$f_{2\widehat{r}}^X = y_{\widehat{r}}^2 - y_{\widehat{s}} k_{\dot{r}}(x), \quad f_{\widehat{s}+\widehat{r}}^X = y_{\widehat{r}} y_{\widehat{s}} - k_{\dot{r}}(x) k_{\dot{s}}(x), \quad f_{2\widehat{s}}^X = y_{\widehat{s}}^2 - y_{\widehat{r}} k_{\dot{s}}(x),$$

which are equal to the 2×2 minors of

$$\begin{vmatrix} k_{\dot{r}}(x) & y_{\widehat{r}} & y_{\widehat{s}} \\ y_{\widehat{r}} & y_{\widehat{s}} & k_{\dot{s}}(x) \end{vmatrix}. \tag{2.69}$$

We obtain cyclic Weierstrass canonical forms,

$$f_X = f_X^{(2)} = y_{\widehat{r}}^3 - k_{\dot{r}}^2(x) k_{\dot{s}}(x), \quad f_X^{(3)} = y_{\widehat{s}}^3 - k_{\dot{s}}^2(x) k_{\dot{r}}(x). \tag{2.70}$$

According to Pinkham [Pin1974], the monomial-ring action trivially extends to the ring

$$R_X^{\lambda} = \mathbb{C}[x, y_{\widehat{r}}, y_{\widehat{s}}, \lambda_1^{(\dot{s})}, \ldots, \lambda_{\dot{s}}^{(\dot{s})}, \lambda_1^{(\dot{r})}, \ldots, \lambda_{\dot{r}}^{(\dot{r})}]/(f_{2\widehat{r}}^X, f_{\widehat{s}+\widehat{r}}^X, f_{2\widehat{s}}^X),$$

graded by $\mathrm{wt}_\lambda : R_\lambda \to \mathbb{Z}$, $\lambda_i^{(\dot{s})} \mapsto 3i$ and $\lambda_i^{(\dot{r})} \mapsto 3i$, so that the equations that define the curve are homogeneous. This ring parameterizes the moduli (in a loose sense) of the curves over the rationals, and it can be used to address number-theoretic issues, although we do not do so in this book.

We consider the curve X as its affine part given by Spec R_X for $R_X = R_X^{\lambda}$ with certain fixed λ's, whose genus is $g = \dot{r} + \dot{s} - 1$. The curve X is a curve in affine 3-space with coordinates $(x, y_{\widehat{r}}, y_{\widehat{s}})$ whose ϕ in the Weierstrass non-gap sequence is given by Table 2.2. Equation (2.70) shows that there is a global $\mathbb{Z}/3\mathbb{Z}$ action on X given by:

$$\widehat{\zeta_3}(x, y_{\widehat{r}}, y_{\widehat{s}}) = (x, \zeta_3 y_{\widehat{r}}, \zeta_3^2 y_{\widehat{s}}),$$

where ζ_3 is a primitive third root of unity.

Table 2.2 Examples of ϕ of $(3, \widehat{r}, \widehat{s})$-curves

$(\dot{r}, \dot{s})$	$(\widehat{r}, \widehat{s})$	g	c_X	0	1	2	3	4	5	6	7	8	9	10	11	12	13	14
(1, 2)	(4, 5)	2	3	1	–	–	x	y_4	y_5	x^2	xy_4	xy_5	y_4y_5	x^2y_4	x^2y_5	xy_4y_5	$y_4^2y_5$	x^3y_5
(1, 3)	(5, 7)	3	5	1	–	–	x	–	y_5	x^2	y_7	xy_5	x^3	xy_7	x^2y_5	y_5y_7	x^2y_7	x^3y_5
(2, 3)	(7, 8)	4	6	1	–	–	x	–	–	x^2	y_7	y_8	x^3	xy_7	xy_8	x^4	x^2y_7	x^2y_8
(1, 5)	(7, 11)	5	9	1	–	–	x	–	–	x^2	y_7	–	x^3	xy_7	y_{11}	x^4	x^2y_7	xy_{11}
(2, 4)	(8, 10)	5	8	1	–	–	x	–	–	x^2	–	y_8	x^3	y_{10}	xy_8	x^4	xy_{10}	x^2y_8
(3, 4)	(10, 11)	6	9	1	–	–	x	–	–	x^2	–	–	x^3	y_{10}	y_{11}	x^4	xy_{10}	xy_{11}
(3, 5)	(11, 13)	7	11	1	–	–	x	–	–	x^2	–	–	x^3	–	y_{11}	x^4	y_{13}	xy_{11}

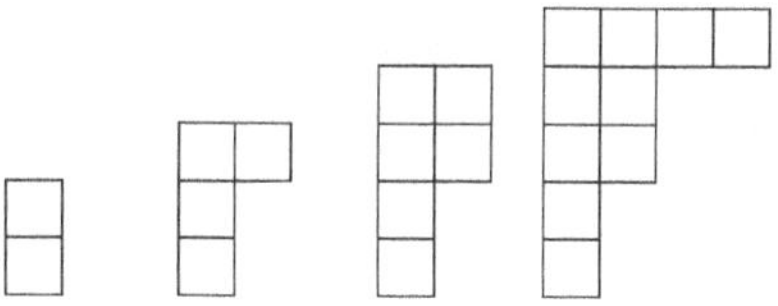

$(3, 4, 5)$, $(3, 5, 7)$, $(3, 7, 8)$, $(3, 7, 11)$.

The numerical semigroups H_X are non-symmetric, and the trace operator in Definition 2.74 is

$$p_{\varpi} = \frac{y_{\widehat{r}} y_{\widehat{s}} + y'_{\widehat{r}} y_{\widehat{s}} + y_{\widehat{r}} y'_{\widehat{s}}}{3(y_{\widehat{r}} y_{\widehat{s}})}, \quad \widetilde{h}_X(x, y_\bullet, y'_\bullet) = y_{\widehat{r}} y_{\widehat{s}} + y'_{\widehat{r}} y_{\widehat{s}} + y_{\widehat{r}} y'_{\widehat{s}},$$

$$\mathfrak{h}_X = \frac{y_{\widehat{r}} + y_{\widehat{s}}}{3(y_{\widehat{r}} y_{\widehat{s}})},$$

$h_X(x, y_\bullet) = 3(y_{\widehat{r}} y_{\widehat{s}}) = k_{\dot{s}}(x) k_{\dot{r}}(x) \in \mathbb{C}[x]$, and $d_h = \widehat{r} + \widehat{s} = 3(\dot{r} + \dot{s})$.

2.3.3.1 Cyclic (3, 4, 5)-Curves, and Uniqueness of Normalization

Let us consider the W-curves $X_s : y^3 = x(x-s)(x-b_1)(x-b_2)$ and its limit $s \to 0$, $X_0 : y^3 = x^2(x-b_1)(x-b_2)$ for mutually distinct $b_a (\neq 0) \in \mathbb{C}$, $(a = 1, 2)$. After its normalization, we have the space curve $X_{\widehat{0}} = X$, $R_X = \mathbb{C}[x, y, z]/(y^2 - xz, zy - x(x-b_1)(x-b_2), z^2 - (x-b_1)(x-b_2)y)$ whose Weierstrass semigroup is generated by $\{3, 4, 5\}$ and the Young diagram: $\Lambda_{(1,1)} = \boxminus$.

The affine model of the normalization is not unique, as the following lemma shows.

Lemma 2.90 *There is a group action* $(\widehat{\zeta_3^*})^a : z \to \zeta_3^a z$ $(a = 0, 1, 2)$ *on the prime ideals of* $\mathbb{C}[x, y, z]$, *i.e.,*

$$(y^2 - \zeta_3^a xz, \zeta_3^a zy - xk(x), \zeta_3^{2a} z^2 - k(x)y),$$

which induces the three different normalizations $R_{\widehat{0}}$, $R_{\widehat{0}}^*$, *and* $R_{\widehat{0}}^{**}$ *of* R_0 *with the* $\widehat{\zeta_3}$ *action*

$$\widehat{\zeta_3}(x, y, z) = (x, \zeta_3 y, \zeta_3^{2+a} z)$$

respectively, where $k(x) = (x-b_1)(x-b_2)$.

The normalization is unique up to the isomorphism given by the (biholomorphic) action $\widehat{\zeta_3^*}$. Corresponding to the relations among $R_{\widehat{0}}$, $R_{\widehat{0}}^*$, and $R_{\widehat{0}}^{**}$, the relations among the normalized curves $X_{\widehat{0}}$, $X_{\widehat{0}}^*$, and $X_{\widehat{0}}^{**}$ exist:

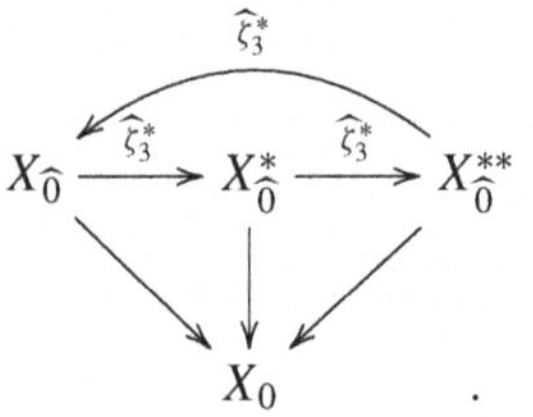

(2.71)

The isomorphism $\widehat{\zeta_3^*}$ also acts on the local parameter z at $B_0 = (x = 0, y = 0, z = 0)$.[7]

Remark 2.91 The action $\widehat{\zeta_3^*}$ can be induced on $\widehat{\pi}_3(X_{\widehat{0}}^{**}) = \mathbb{P}$ because it has an effect only on the third component z. The relation (2.71) among the three genus two curves $X_{\widehat{0}}$, $X_{\widehat{0}}^*$ and $X_{\widehat{0}}^{**}$ can be regarded as the relation among the genus two curve $X_{\widehat{0}}$, and two rational curves $\widehat{\pi}_3(X_{\widehat{0}}^*) = \mathbb{P}$ and $\widehat{\pi}_3(X_{\widehat{0}}^{**}) = \mathbb{P}$.

[7] Even though in Definition 2.29 we assign B_0 to ∞ in X, we employ another convention for the particular curves X_0 and $X_{\widehat{0}}$.

As we represent R_X by $\mathbb{C}[x, y, z]/(f_1^X, f_2^X, f_3^X)$, where $f_1^X = y^2 - xz$, $f_2^X = zy - xk(x)$, and $f_3^X = z^2 - k(x)y$, we check the rank of $\mathscr{M}_X := \left(\dfrac{\partial f_i^X}{\partial y_{r_j}}\right)_{i,j=1,2,3}$ in Proposition 2.14.

Lemma 2.92 *The rank of* $\mathscr{M}_X = \begin{pmatrix} z & 2y & -x \\ k(x) + xk'(x) & z & y \\ yk'(x) & -k(x) & 2z \end{pmatrix}$ *is two.*

Proof When y vanishes, and in consequence z vanishes, x must be one of $\{0, b_1, b_2\}$. For $x = 0$, $\mathscr{M}_X = \begin{pmatrix} 0 & 0 & 0 \\ k(0) & 0 & 0 \\ 0 & -k(0) & 0 \end{pmatrix}$ whose rank is obviously 2. Similarly for $x = b_a$ $a = 1, 2$. On the other hand, when y and z are nonzero,

$$\begin{pmatrix} 1 & & \\ 1 & -1 & 1 \\ & & 1 \end{pmatrix} \begin{pmatrix} z^2 & & \\ & yz & \\ & & y^2 \end{pmatrix} \mathscr{M}_X = \begin{pmatrix} 1 & & \\ 1 & -1 & 1 \\ & & 1 \end{pmatrix} \begin{pmatrix} z^3 & 2yz^2 & -z^2x \\ yz(xk(x))' & yz^2 & y^2z \\ y^3k'(x) & -y^2k(x) & 2y^2z \end{pmatrix}$$

shows that the rank is 2. ■

The infinity point of $X_{\widehat{0}}$ is not a Weierstrass point. Every genus-two curve is hyperelliptic, and thus the (3, 4, 5)-curve is birational to a hyperelliptic curve X_H. Since we have the global trigonal order cyclic action on the cyclic (3, 4, 5)-curve $X_{\widehat{0}}$, the hyperelliptic curve X_H also permits the group action. The birational map that sends $(x, y, z) \to (\xi_1 = y/x, \xi_2 = (x^2 - b_1b_2)/x)$ gives

$$\xi_2^2 = \xi_1^6 + 2(b_1 + b_2)\xi_1^3 + (b_1 - b_2)^2,$$

which is birationally equivalent to a hyperelliptic curve X_H given by (2.65) (cf. Sect. 2.8.2).[8]

[8] In general, for a hyperelliptic curve X_H of genus two given by $\widetilde{y}^2 = \widetilde{x}^5 + \widetilde{\lambda}_4\widetilde{x}^4 + \cdots + \widetilde{\lambda}_1\widetilde{x} + \widetilde{\lambda}_0 = \widetilde{f}(x)$, we consider a non-Weierstrass point $(\widetilde{x}_0, \widetilde{y}_0) \in X_H$. Then we can find

$$\xi_1 := \frac{(\widetilde{y} + \widetilde{y}_0) + \frac{\widetilde{f}'(x_0)}{2\widetilde{y}_0}(\widetilde{x} - \widetilde{x}_0) + \frac{1}{2}\frac{2\widetilde{f}(x_0)\widetilde{f}''(x_0) - \widetilde{f}'(x_0)^2}{4\widetilde{y}_0^2}(\widetilde{x} - \widetilde{x}_0)^2}{(\widetilde{x} - \widetilde{x}_0)^3 2\widetilde{y}_0} \in \mathbf{H}^0(X_H, \mathscr{O}_{X_H}(3(\widetilde{x}_0, \widetilde{y}_0))),$$

and similarly we may have $\xi_2 \in \mathbf{H}^0(X_H, \mathscr{O}_{X_H}(4(\widetilde{x}_0, \widetilde{y}_0)))$ and $\xi_3 \in \mathbf{H}^0(X_H, \mathscr{O}_{X_H}(5(\widetilde{x}_0, \widetilde{y}_0)))$ so that these ξ_1, ξ_2, and ξ_3 satisfies a Weierstrass canonical form of $\langle 3, 4, 5\rangle$.

2.3.4 *Special Curves: Pentagonal, Non-cyclic Trigonal, 6-Symmetric Curves*

We consider three special curves whose ϕ in the Weierstrass non-gap sequences are listed in Table 2.3 and whose associated Young diagrams are given as follows.

Table 2.3 Examples of ϕ of special curves

	0	1	2	3	4	5	6	7	8	9	10	11	12	13	14	15	16	17	18	19	20	21	22	23	24
I	1	–	–	x	–	–	x^2	y	w	x^3	xy	xw	x^4	x^2y	y^2	yw	w^2	xy^2	xyw	xw^2	x^2y^2	x^2yw	x^2w^2	x^3y	x^3w^2
II	1	–	–	–	–	x	–	y	–	–	x^2	w	xy	–	y^2	x^3	xw	x^2y	wy	xy^2	x^4	y^3	x^3y	xyw	x^2y^2
III	1	–	–	–	–	–	x	–	–	–	–	–	x^2	y_{13}	y_{14}	y_{15}	y_{16}	–	x^3	xy_{13}	xy_{14}	xy_{15}	xy_{16}	–	x^4

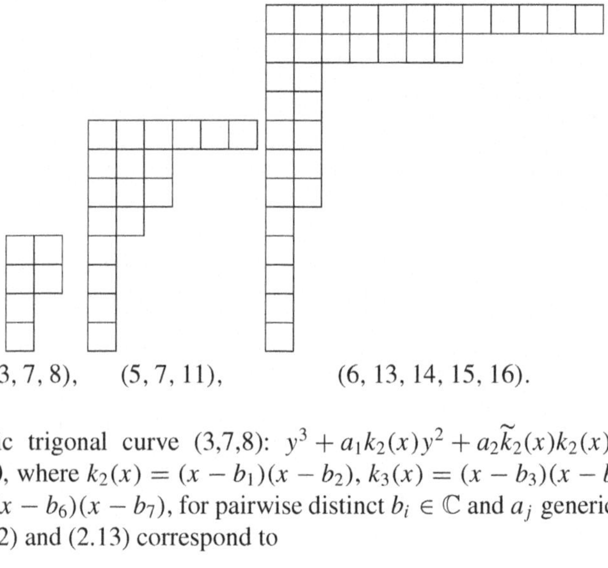

$(3, 7, 8)$, $(5, 7, 11)$, $(6, 13, 14, 15, 16)$.

I. Non-cyclic trigonal curve (3,7,8): $y^3 + a_1k_2(x)y^2 + a_2\widetilde{k}_2(x)k_2(x)y + k_2(x)^2 k_3(x) = 0$, where $k_2(x) = (x - b_1)(x - b_2)$, $k_3(x) = (x - b_3)(x - b_4)(x - b_5)$ $\widetilde{k}_2(x) = (x - b_6)(x - b_7)$, for pairwise distinct $b_i \in \mathbb{C}$ and a_j generic constants. Here (2.12) and (2.13) correspond to

$$y^2 = -a_1k_2y - k_2a_2\widetilde{k}_2 - k_2w, \quad yw = k_2k_3.$$

Multiplying the first equation by y and using the second equation give the curve's equation. Besides them, we have

$$w^2 = -(a_2\widetilde{k}_2w + a_1k_2k_3 + k_3y),$$

since multiplying the first equation by w^2 gives

$$w^3 + a_2\widetilde{k}_2w^2 + a_1k_2k_3w + k_2k_3^2 = 0.$$

This curve is trigonal with $H_X = \langle 3, 7, 8\rangle$ but not necessarily cyclic.

$$\begin{aligned}
\widetilde{h}_X(x, y, w, y', w') &= k_2 k_3^2 + k_3(yw' + y'w) \\
&+ \frac{1}{3} k_3 \left(2a_1 yy' + a_1 k_2(w + w') + (a_1^2 k_2 + a_2 \widetilde{k}_2)(y + y') + 2a_1 a_2 k_2 \widetilde{k}_2\right) \\
&+ \frac{1}{3} \widetilde{k}_2 \left(a_1 a_2 (yw' + y'w) + 2a_2 ww' + a_2^2 \widetilde{k}_2 (w + w')\right), \\
h_X(x, y, w) &= 3k_2 k_3^2.
\end{aligned}$$

II. Cyclic pentagonal curve (5,7,11): $y^5 = k_2(x)^2 k_3(x)$, where $k_2(x) = (x - b_1)(x - b_2)$, $k_3(x) = (x - b_3)(x - b_4)(x - b_5)$ for pairwise distinct $b_i \in \mathbb{C}$: (2.13) corresponds to

$$\begin{pmatrix} -y & 0 & 1 \\ k_2 & -y & 0 \\ 0 & 0 & -y \end{pmatrix} \begin{pmatrix} w \\ y^2 \\ wy \end{pmatrix} = \begin{pmatrix} 0 \\ 0 \\ -k_2^2 k_3 \end{pmatrix}.$$

The affine ring is $R_X^\circ = \mathbb{C}[x, y, w]/(y^3 - k_2 w, w^2 - k_3 y, y^2 w - k_2 k_3)$. Here (2.14) is reduced to

$$w = \frac{k_2 k_3}{y^2}, \quad yw = \frac{k_2 k_3}{y}, \quad y^2 = \frac{k_2 w}{y}. \tag{2.72}$$

This is a pentagonal cyclic curve (X, ∞) with $H_X = \langle 5, 7, 11 \rangle$.

$$\begin{aligned}
\widetilde{h}_X(x, y, w, y', w') &= y^2 w + ywy' + y^2 w' + wy'^2 + yw'y', \\
h_X(x, y, w) &= 5y^2 w = 5k_2 k_3.
\end{aligned}$$

These $\mathfrak{e}$'s and d_h in Lemma 2.45 are given as $\mathfrak{e}_0 = 0$, $\mathfrak{e}_1 = 7$, $\mathfrak{e}_2 = 11$, $\mathfrak{e}_3 = 14$, $\mathfrak{e}_4 = 18$, and $d_h = 25$.

III. 6-symmetric (6,13,14,15,16)-curve[9]:

We construct a non-singular curve X by giving an affine patch, an ideal in the ring $\mathbb{C}[x, y_{13}, y_{14}, y_{15}, y_{16}]$. For any complex numbers $\{b_i\}_{i=1,\dots,7}$ such that each is distinct from the others, we let

$$\begin{aligned}
k_3(x) &:= (x - b_1)(x - b_2)(x - b_3) = x^3 + \lambda_1^{(3)} x^2 + \lambda_2^{(3)} x + \lambda_3^{(3)}, \\
k_2(x) &:= (x - b_4)(x - b_5) = x^2 + \lambda_1^{(2)} x + \lambda_2^{(2)}. \\
\widehat{k}_2(x) &:= (x - b_6)(x - b_7) = x^2 + \widehat{\lambda}_1^{(2)} x + \widehat{\lambda}_0^{(2)}, \qquad \widehat{k}_5(x) := \widehat{k}_2(x) k_3(x), \\
k_{13}(x) &:= k_3(x) k_2(x)^2 \widehat{k}_2(x)^3, \qquad k_{14}(x) := k_7(x)^2 = k_3(x)^2 k_2(x)^4, \\
k_{15}(x) &:= \widehat{k}_5(x)^3, \qquad k_{16}(x) := k_8(x)^2 = k_3(x)^4 k_2(x)^2.
\end{aligned}$$

[9] This curve might be related to the Norton numbers {1, 2, 3, 4, 5, 7, 8, 9, 11, 17, 19, 23} for the representation of the Monster group [KomMP2013, McH2022] (cf. Table 2.3, except 10 and 19).

The Weierstrass canonical form is given by $y_{13}^6 = \widehat{k}_2^3 k_2^2 k_3$, which is normalized as follows.

Let the prime ideal $\mathscr{P}$ in $\mathbb{C}[x, y_{13}, y_{14}, y_{15}, y_{16}]$ be defined by

$$\mathscr{P} := (f_{12,1}, f_{12,2}, f_{12,3}, f_{12,4}, f_{12,5}, f_{12,6}, f_{12,7}, f_{12,8}, f_{12,9}),$$

where

$$\begin{aligned} f_{12,1} :=& y_{13}^2 - \widehat{k}_2(x) y_{14},\ f_{12,2} := y_{13}y_{14} - k_2(x) y_{15},\ f_{12,3} := \widehat{k}_2(x) y_{14}^2 - y_{13}y_{15}k_2(x),\\ f_{12,4} :=& y_{14}^2 - k_2(x) y_{16},\ f_{12,5} := y_{13}y_{16} - y_{14}y_{15},\ f_{12,6} := y_{15}^2 - \widehat{k}_2(x)k_3(x),\\ f_{12,7} :=& y_{14}y_{16} - k_2(x)k_3(x),\ f_{12,8} := y_{15}y_{16} - k_3(x)y_{13},\ f_{12,9} := y_{16}^2 - k_3(x)y_{14}, \end{aligned}$$

which are the 2×2 minors of

$$\begin{vmatrix} k_2(x) & y_{14} & y_{16} \\ y_{14} & y_{16} & k_3(x) \end{vmatrix}, \quad \begin{vmatrix} \widehat{k}_2(x) & y_{13} & y_{14}\widehat{k}_2(x) & y_{15} \\ y_{13} & y_{14} & y_{15}k_2(x) & y_{16} \end{vmatrix}; \tag{2.73}$$

again, the minor $y_{13}y_{15} - \widehat{k}_2(x) y_{16}$ is not in the list of $f_{i,j}$ and $f_{12,8}$ is not a minor, but they are compatible; the minor follows by combining $f_{12,8}$ with $f_{12,1}$ and $f_{12,9}$. Here the first matrix is the same as (2.69), and the latter is related to the double covering of a cyclic trigonal (3, 7, 8)-curve of genus four.

We define the $\mathbb{G}_{\mathrm{m}}$ action on x and y_a by $g^{-6}x$ and $g^{-a}y_a$ $(a = 13, 14, 15, 16)$ at $\infty \in X_{12}$.

Corresponding to Proposition 2.23, the affine ring is given by

$$R_X^\circ = \mathbb{C}[x, y_{13}, y_{14}, y_{15}, y_{16}]/\mathscr{P},$$

and the trace operator in Definition 2.74 is

$$p_\varpi = \frac{y_{13,1}y_{16,1} + y_{13,2}y_{16,1} + y_{13,1}y_{16,2} + y_{14,2}y_{15,1} + y_{14,1}y_{15,2} + y_{14,2}y_{15,2}}{6 y_{13,1} y_{16,1}}.$$

2.4 Abelian Differentials and Abelian Integrals in W-Curves

In this section, we mention the Abelian differentials on X following [KomMP2022a, KomMP2022b]. The Abelian differentials are the derivations of $\mathscr{Q}(R_X)$ or the meromorphic differential forms on X. The Abelian differential ν on X can be also naturally defined on D_X (2.60), $\mathrm{Path}_P(X)$ and $\widetilde{X}$ in Sect. 2.2.7. We recall that $\mathscr{A}_X$ is the sheaf of the holomorphic one-form on X, and let $\mathscr{A}_X(U) := \mathbf{H}^0(U, \mathscr{A}_X)$.

Lemma 2.93 *Let $P_1, \ldots, P_m, Q_1, \ldots, Q_n (\in D_X \setminus \partial D_X)$ be mutually disjoint. Let $\nu \in, \mathscr{A}_X(X \setminus \{P_1, \ldots, P_m\})$ and $\nu' \in, \mathscr{A}_X(X \setminus \{Q_1, \ldots, Q_n\})$ such that at every P_i,*

$\nu = \sum_{j=-N_i} a_j^{(i)} t^j dt$ $\nu' = \sum_{j=0} a_j'^{(i)} t^j dt$ *for* $N_i > 0$, *a local parameter* t *at* P_i *and* $a_j^{(i)}, a_j'^{(i)} \in \mathbb{C}$, *and at every* Q_i, $\nu = \sum_{j=0} b_j^{(i)} t^j dt$, $\nu' = \sum_{j=-N_i'} b_j'^{(i)} t^j dt$ *for* $N_i' > 0$, *a local parameter* t *at* Q_i *and* $b_j^{(i)}, b_j'^{(i)} \in \mathbb{C}$. *Then the following hold:*

$$\frac{1}{2\pi i}\sum_{i=1}^{g}\left(\int_{\alpha_i}\nu\int_{\beta_i}\nu' - \int_{\beta_i}\nu\int_{\alpha_i}\nu'\right) + \sum_{i=1}^{m}\left(a_{-1}^{(i)}\int_{\infty}^{P_i}\nu' + \sum_{k=0}^{\infty}\frac{a_{-k-2}^{(i)}a_k'^{(i)}}{k+1}\right)$$
$$-\sum_{j=1}^{n}\left(b_{-1}'^{(j)}\int_{\infty}^{Q_i}\nu + \sum_{k=0}^{\infty}\frac{b_{-k-2}'^{(j)}b_k^{(j)}}{k+1}\right) = 0,$$

and thus for any meromorphic differentials ν *and* ν',

$$\int_{\partial D_X}\left(\int_{\infty}^{P}\nu\right)\nu'(P) = \sum_{i=1}^{g}\left(\int_{\alpha_i}\nu\int_{\beta_i}\nu' - \int_{\beta_i}\nu\int_{\alpha_i}\nu'\right).$$

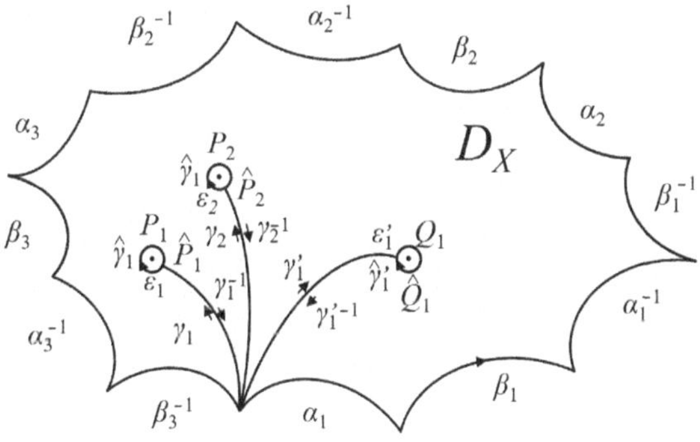

Fig. 2.5 ∂D_X and other contours

Proof From [Iw1993, Chap 5, p. 186, Theorem 5.5], we have these results. Here, we sketch them by referring to Fig. 2.5. We consider paths as elements in the quotient space of $\mathrm{Path}_\infty(X)$ by the homotopy equivalent before its Abelianization. We define a small circle $\widehat{\gamma}_i$ with the radius $\varepsilon_i (> 0)$ whose center is each P_i, and $\widehat{\gamma}_i'$ with the radius $\varepsilon_j' (> 0)$ for each Q_j. We take a point $\widehat{P}_i$ on each circle $\widehat{\gamma}_i$ and $\widehat{Q}_i$ on $\widehat{\gamma}_i'$; we regard $\widehat{\gamma}_i$ as a clockwise loop $\gamma_{\widehat{P}_i,\widehat{P}_i}$ and $\widehat{\gamma}_j$ as $\gamma_{\widehat{Q}_j,\widehat{Q}_j}$. Further, we consider the paths $\gamma_i := \gamma_{\widehat{P}_i,\infty}$ and $\gamma_j' := \gamma_{\widehat{Q}_j,\infty}$ so that they do not cross each other. We let the inverse of γ_i be denoted by γ_i^{-1}.

We deal with the closed contour $\gamma := \partial D_X \dot{+}\sum_i^{\cdot}(\gamma_i^{-1} \dot{+} \widehat{\gamma}_i \dot{+} \gamma_i) \dot{+}\sum_j^{\cdot}(\gamma_j^{\prime -1} \dot{+} \widehat{\gamma}_j^{\prime} \dot{+} \gamma_j^{\prime})$. Further, we let $h(\gamma_{P,Q}) := \int_{\gamma_{P,Q}} \nu$ and then we have the result, $\oint_\gamma h(\gamma_{P,\infty})\, \nu'(P) = 0$.

Thus we consider the integral related to $\partial D_X = \sum_i^{\cdot}(\beta_i^{-1} \dot{+} \alpha_i^{-1} \dot{+} \beta_i \dot{+} \alpha_i)$. In this argument, we treat the path space as one divided by the homotopy equivalence before the Abelianization; we also recognize that its elements are not commutative though the contour integrals over them are commutative. On the integral $I_i := \left[\int_{\alpha_i} h(\gamma_{P,\infty})\nu'(P) + \int_{\alpha_i^{-1}} h(\gamma_{P,\infty})\nu'(P)\right]$, we let the path $\gamma_{P,\infty}$ in the first integral be denoted by $\gamma_{P,\infty}^{\mathrm{I}}$, and then we note that $\gamma_{P,\infty}$ $(=: \gamma_{P,\infty}^{\mathrm{II}})$ in the second integral can be expressed by $\gamma_{P,\infty}^{\mathrm{II}} = \gamma_{P,\infty}^{\mathrm{I}} \dot{+} \beta_i$ because there is β_i between α_i and α_i^{-1} illustrated in Fig. 2.5. Further, since the directions of α_i and α_i^{-1} are inversed, the difference is represented by the sign of ν'. Hence we have

$$I_i = \left[\int_{\alpha_i} (h(\gamma_{P,\infty}^{\mathrm{I}}) - h(\gamma_{P,\infty}^{\mathrm{I}} \dot{+} \beta_i))\nu'(P)\right] = -\int_{\beta_i} \nu \int_{\alpha_i} \nu'.$$

Here we use $(h(\gamma_{P,\infty}^{\mathrm{I}}) - h(\gamma_{P,\infty}^{\mathrm{I}} \dot{+} \beta_i)) = -\int_{\beta_i} \nu$. By considering other contributions in ∂D_X, we obtain the first term in the first relation of Lemma 2.93.

Similarly, let $K_i := \int_{\gamma_i^{-1} \dot{+} \widehat{\gamma}_i \dot{+} \gamma_i} h(\gamma_{P,\infty})\nu'(P)$. Then by considering a homotopical deformation so that $\widehat{\gamma}_i$ belongs to $\mathrm{Path}_\infty(X)$, we obtain

$$\begin{aligned} K_i &= \int_{\gamma_i} [h(\gamma_{P,\infty}) - h(\gamma_{P,\infty} \dot{+} \widehat{\gamma}_i)]\nu'(P) + \int_{\widehat{\gamma}_i} h(\gamma_{P,\infty})\nu'(P) \\ &= 2\pi\sqrt{-1}\left[a_{-1}^{(i)} \int_{\gamma_i} \nu' + \sum_{k=0}^{\infty} \frac{a_{-k-2}^{(i)} a_k^{\prime (i)}}{k+1}\right], \end{aligned}$$

since we have $h(\gamma_{P,\infty}) - h(\gamma_{P,\infty} \dot{+} \widehat{\gamma}) = 2\pi\sqrt{-1}a_{-1}^{(i)}$ due to the clockwise direction at $\widehat{\gamma}_i$, and we use the following relation for the second term. Here noting the evaluation of $h(\gamma_{P,\infty}) = \int_{\gamma_{P,P_i}} \nu(P) + h(\gamma_{P_i,\infty})$, we have the expansion of $h(\gamma_{P,\infty})$ at P_i as

$$h(\gamma_{P,\infty}) = \int^P \sum_{j=-N_i} a_j^{(i)} t^j (dt) = \sum_{j=-N_i} \frac{a_j^{(i)}}{j+1} t^{j+1}.$$

Hence noting the clockwise integral again, we obtain the expression of the second term as

$$\int_{\widehat{\gamma}_i} h(\gamma_{P,\infty})\nu'(P) = -2\pi\sqrt{-1}\mathrm{Res}\left(\sum_{j=-N_i}\sum_{k=0}\frac{a_j^{(i)}a'^{(i)}_k}{j+1}t^{k+j+1}\right)$$
$$= 2\pi\sqrt{-1}\sum_{k=0}^{\infty}\frac{a^{(i)}_{-k-2}a'^{(i)}_k}{k+1}$$

due to $k+j+1=-1$ or $j=-k-2$. Further, we note that for every loop $\widetilde{\gamma}$ in X, we have the identity $\int_{\widetilde{\gamma}}(h(\gamma_{P,\infty})\nu'(P)+h'(\gamma_{P,\infty})\nu(P))=0$, where $h'(\gamma_{P,\infty}) := \int_{\gamma_{P,\infty}}\nu'$. Hence we can apply the method to other contours similarly. Then we obtain the above first relation by letting ε_i and ε'_j become zero. The second claim is evident due to the above argument. ■

Recalling Definition 2.16, we define the weight for the meromorphic differentials on X:

Definition 2.94 The degree of the meromorphic differentials ν as the order of the divisor is defined by

$$\mathrm{wt}(\nu) = \deg_{\infty}(\nu),$$

which is also called the Sato–Weierstrass weight.

2.4.1 *Normalized Abelian Differentials and Integrals*

As the normalized Abelian differentials play a crucial role in Riemann's Abelian function theory on the Riemann surfaces X, we introduce them here.

2.4.1.1 Normalized Abelian Differentials and Integrals of the First Kind

Let $\nu^{\mathrm{I}\circ}$ be the normalized Abelian differentials of the first kind on a W-curve X. The differential $\nu^{\mathrm{I}\circ}$ is defined as the column vector whose components are the normalized holomorphic one-forms $\nu_i^{\mathrm{I}\circ} \in \mathscr{A}_X(X)$ $(i=1,2,\ldots,g)$ of X so that $\int_{\alpha_i}\nu_j^{\mathrm{I}\circ}=\delta_{ij}$. Then the integrals $\tau_{X\,ij}$ (simply τ_{ij}) for β_i are given by

$$\tau_{X\,ij} := \int_{\beta_i}\nu_j^{\mathrm{I}\circ}. \tag{2.74}$$

Here the matrix $\tau_X \in \mathrm{Mat}_{\mathbb{C}}(g)$ (simply $\tau \in \mathrm{Mat}_{\mathbb{C}}(g)$) given by $\tau_X = \tau = (\tau_{X\,ij})$ determines a point in the Siegel upper half space $\mathfrak{H}_g$ as mentioned in Sect. 3.8.1,

Appendix 3.8. It is known that $\tau_X(=\tau)$ satisfies the Riemann bilinear relations [FarK1991, III,3]:

Proposition 2.95 *The matrix τ is the symmetric with positive definite imaginary part, i.e.,*

$$ {}^t\tau = \tau, \quad \mathrm{Im}(\tau) > 0. \tag{2.75}$$

Proof We obtain both by setting $\nu_i^{\mathrm{I}\circ}$ and $\nu_j^{\mathrm{I}\circ}$ into ν and ν' in Lemma 2.93, and $\nu_i^{\mathrm{I}\circ}$ and $\overline{\nu_j^{\mathrm{I}\circ}}$ to them. [FarK1991, III 2.8 Proposition]. ■

It consists of the Legendre relation in Theorem 2.148.

Using them, we define the normalized Abelian integral,

$$\widetilde{w}^\circ : \widetilde{X} \to \mathbb{C}^g \quad \text{and} \quad \widetilde{w}^\circ(\gamma_{P,\infty}) := \int_{\gamma_{P,\infty}} \nu^{\mathrm{I}\circ} \tag{2.76}$$

for a point $\gamma_{P,\infty} \in \widetilde{X}$. We extend it to the Abelian integral over $S^k\widetilde{X}$.

$$\widetilde{w}^\circ : S^k\widetilde{X} \to \mathbb{C}^g, \quad \widetilde{w}^\circ(\gamma_1, \cdots, \gamma_k) = \sum_{i=1}^k \widetilde{w}^\circ(\gamma_i) = \sum_{i=1}^k \int_{\gamma_i} \nu^{\mathrm{I}\circ}. \tag{2.77}$$

Since the $\mathbb{Z}$-module $\langle 1_g, \tau\rangle_{\mathbb{Z}}$ is the lattice denoted by Γ_X° as a subspace of $\mathbb{C}^g$, we introduce the normalized Jacobian (Jacobi variety) by a quotient space,

$$J_X^\circ := \mathbb{C}^g/\Gamma_X^\circ, \quad \kappa_J^\circ : \mathbb{C}^g \to J_X^\circ. \tag{2.78}$$

We also introduce the Abel–Jacobi map $w^\circ : X \to J_X^\circ$ by the Abel–Jacobi theorem such that $w^\circ := \kappa_J^\circ \circ \widetilde{w}^\circ \circ \iota_X$, and its extension $w^\circ : S^kX \to J_X$. Since the map w° embeds X into J_X°, we can generalize it to a map from the space of divisors of X into J_X° as $w^\circ(\sum_i n_i P_i) := \sum_i n_i w^\circ(P_i)$, $P_i \in X$, $n_i \in \mathbb{Z}$. Similarly, we will define $\widetilde{w}^\circ(\iota_X D)$ for a divisor D of X.

We recall the Abel and Jacobi theorems following [FarK1991]:

Theorem 2.96 Abel theorem [FarK1991, III 6.3 Theorem]: *The divisor $D \in \mathrm{Div}^0(X)$ is principal, i.e., $D \in \mathrm{PDiv}(X)$ if and only if D belongs to the kernel of the image of the Abel–Jacobi map, i.e., $w^\circ(D) = 0$ in J_X°.*

It means that the Abel–Jacobi map w is the injection as an additive group homomorphism from $\mathrm{Pic}^0(X)$ to J_X°.

Theorem 2.97 Jacobi theorem [FarK1991, III 6.6 Theorem]: *Every point in J_X° is given by the Abel–Jacobi map w° of $(k \geq g)$. It means that the Abel–Jacobi map w° is surjective.*

By unifying them, we have the Abel–Jacobi theorem:

Theorem 2.98 Abel–Jacobi theorem [FarK1991, III 6.6 Corollary]: $\mathrm{Pic}^0(X)$ *is isomorphic to* J_X° *as an additive group.*

They also show that the Abel integral $\widetilde{w} : \widetilde{X} \to \mathbb{C}^g$ is also isomorphic as an additive group.

Lemma 2.99 *For every* $\rho \in \mathrm{Sp}(2g, \mathbb{Z})$*, the lattice* Γ_X° *and* J_X° *are invariant via the action* $(1_g, \tau)\,{}^t\rho$*, i.e., for* $\tau' := \tau\,{}^t\rho$*,* $\langle 1_g, \tau\rangle_{\mathbb{Z}} = \langle 1_g, \tau'\rangle_{\mathbb{Z}}$.

Proof It is obvious. ■

2.4.1.2 Normalized Abelian Differentials of the Second Kind and the Third Kind

The non-holomorphic Abelian differentials with zero residues are said to be of the second kind. We recall the convention for the sheaf $\mathscr{A}_X(nP)$ in Sect. 2.1. The normalized differential of the second kind $\nu_{P,t,n}^{\mathrm{II}\circ} \in \mathbf{H}^0(X, \mathscr{A}_X(nP)) \setminus \mathbf{H}^0(X, \mathscr{A}_X(P))$ is defined such that

$$\int_{\alpha_i} \nu_{P,t,n}^{\mathrm{II}\circ} = 0$$

for every $i (= 1, \dots, g)$ and for a local parameter t at P, $\nu_{P,t,n}^{\mathrm{II}\circ} = \left(\dfrac{1-n}{t^n} + d_{\geq 0}(t)\right) dt$, $(n > 1)$.

Let us consider a differential $\nu_{P,Q}^{\mathrm{III}\circ} \in \mathscr{A}_X(X \setminus \{P, Q\})$ for a pair $(P, Q) \in X^2$ of points in X which is holomorphic on $X \setminus \{P, Q\}$ with

$$\deg_{P,\infty} \nu_{P,Q}^{\mathrm{III}\circ} = \deg_{Q,\infty} \nu_{P,Q}^{\mathrm{III}\circ} = 1, \qquad \mathrm{res}_P \nu_{P,Q}^{\mathrm{III}\circ} = 1, \quad \mathrm{res}_Q \nu_{P,Q}^{\mathrm{III}\circ} = -1,$$

and satisfies

$$\int_{\alpha_i} \nu_{P,Q}^{\mathrm{III}\circ} = 0$$

for every $i (= 1, \dots, g)$. $\mathrm{Res}_P \nu$ means the residue of ν at P. Such a $\nu_{P,Q}^{\mathrm{III}\circ}$ is called the normalized differential of the third kind.

We summarize their properties:

Lemma 2.100 *1. For* $P_1, P_2, Q_1, Q_2 \in X$, $\displaystyle\int_{P_1}^{P_2} \nu_{Q_1,Q_2}^{\mathrm{III}\circ} = \int_{Q_1}^{Q_2} \nu_{P_1,P_2}^{\mathrm{III}\circ}$.

2. *For* $P, P_0, Q_1, Q_2 \in X$*, and the local parameter* t *at* P*, i.e.,* $t(P) = 0$,
$$\frac{d}{dt} \int_{P_0}^{P} \nu_{Q_1,Q_2}^{\mathrm{III}\circ} \Big|_{t=0} = \int_{Q_2}^{Q_1} \nu_{P,t,2}^{\mathrm{II}\circ}, \text{ and } -\frac{2\pi\sqrt{-1}}{(n-1)!} \frac{d^{n-1}}{dt^{n-1}} \int_{P_0}^{P} \nu_j^{\mathrm{I}\circ} \Big|_{t=0} = \int_{\beta_j} \nu_{P,t,n}^{\mathrm{II}\circ}.$$

3. *For points* $P_1, P_2 \in X \setminus \{\bigcup_j \beta_j\}$, $2\pi\sqrt{-1}\int_{P_1}^{P_2} \nu_j^{\mathrm{I}\circ} = \int_{\beta_j} \nu_{P_1,P_2}^{\mathrm{III}\circ}$ *on* D_X.

Proof *1*: We use the formula in Lemma 2.93 by letting $m = n = 2$, $\nu = \nu_{P_1,P_2}^{\mathrm{III}\circ}$ and $\nu' = \nu_{Q_1,Q_2}^{\mathrm{III}\circ}$ and then we obtain the relation. In the derivation, we use the normalization condition.
2: We apply the formula in Lemma 2.93 by letting $m = 1, n = 2$, $\nu = \nu_{P,t,2}^{\mathrm{II}\circ}$ and $\nu' = \nu_{Q_1,Q_2}^{\mathrm{III}\circ}$ to the first relation. The normalization condition reduces the terms in the formula. Note that $a_{-k}^{(1)} = -\delta_{k,2}$ for a positive k, and $b_{-1}^{(1)} = -b_{-1}^{(2)} = 1$. Then we obtain the first relation. We also consider the case $m = 0, n = 1$, $\nu = \nu_j^{\mathrm{I}\circ}$ and $\nu' = \nu_{P,t,n}^{\mathrm{II}\circ}$ to obtain the second relation by noting $\int_{\alpha_i} \nu_j^{\mathrm{I}\circ} = \delta_{i,j}$ and the normalization condition of $\nu_{P,t,n}^{\mathrm{II}\circ}$.
3: We handle the case $m = 2, n = 0$, $\nu = \nu_{P_1,P_2}^{\mathrm{III}\circ}$ and $\nu' = \nu_j^{\mathrm{I}\circ}$ in Lemma 2.93, and then obtain the relation similarly. ■

These formulae are proved in [FarK1991, Chap. III] as the last equation on p. 68, (3.8.2) and (3.6.3).

2.4.1.3 Fundamental Normalized Differentials of the Second Kind

Let us consider the fundamental normalized differential of the second kind $\Omega^\circ(P_1, P_2)$ in [Fay1973, Corollary 2.6], which is given by the following proposition:

Proposition 2.101 *There exists a two-form* $\Omega^\circ(P_1, P_2)$ *for every* $(P_1, P_2) \in X \times X$, *which satisfies the following:*

1. *It is symmetric,*
$$\Omega^\circ(P_1, P_2) = \Omega^\circ(P_2, P_1). \tag{2.79}$$
2. *It is holomorphic everywhere except for a double pole along the diagonal of* $X \times X$, *in the vicinity of each point* (P_1, P_2) *with the local coordinates* t_1 *and* t_2,
$$\Omega^\circ(P_1, P_2) = \left(\frac{1}{(t_1 - t_2)^2} + d_{\geq 0}(t_1, t_2)\right) dt_1 dt_2.$$
3. *For fixing* $P_1 \in X$, *it satisfies*
$$\int_{\alpha_j} \Omega^\circ(P_1, P_2) = 0, \quad j = 1, \ldots, g.$$
4. *For every* $P_1, P_2, Q_1, Q_2 \in X$, *it holds the identity,*
$$\int_{P_2}^{P_1} \nu_{Q_1,Q_2}^{\mathrm{III}\circ} = \int_{P_2}^{P_1}\int_{Q_2}^{Q_1} \Omega^\circ(P, Q).$$

It results in the following identities:

Lemma 2.102 *For the arithmetic local parameters t_P, t_Q at $\infty \in X$, we expand $\Omega^\circ(P, Q)$ by*

$$\Omega^\circ(t_P, t_Q) - \frac{dt_P dt_Q}{(t_P - t_Q)^2} = \sum_{i,j \geq 1} q^\circ_{i,j} t_P^{i-1} t_Q^{j-1} dt_P dt_Q.$$

Then the normalized differential of the second kind $\mathscr{A}_X(X \setminus \infty)$ is expressed by

$$\nu^{\mathrm{II}\circ}_{\infty, t_P, n+1} = d\left[\frac{1}{t_P^n} - \sum_{j=1} \frac{q^\circ_{n,j}}{j} t_P^j\right].$$

Proof We combine the relations among $\nu^{\mathrm{III}\circ}$, Ω°, and $\nu^{\mathrm{II}\circ}$, and Lemma 2.93 to obtain the formula by extending Lemma 2.100 2 to

$$\frac{1}{(n-1)!} \frac{d^n}{dt^n} \int_{P_0}^{P} \nu^{\mathrm{III}\circ}_{Q_1, Q_2}\Big|_{t=0} = \int_{Q_1}^{Q_2} \nu^{\mathrm{II}\circ}_{P,t,n+1}. \qquad \blacksquare$$

We define $\Pi^{\circ}{}^{P_1, P_2}_{Q_1, Q_2} := \displaystyle\int_{P_2}^{P_1} \nu^{\mathrm{III}\circ}_{Q_1, Q_2}$, and then it is known that

$$\Pi^{\circ}{}^{P_1, P_2}_{Q_1, Q_2} = -\Pi^{\circ}{}^{P_2, P_1}_{Q_1, Q_2}, \quad \Pi^{\circ}{}^{P_1, P_2}_{Q_1, Q_2} = \Pi^{\circ}{}^{Q_1, Q_2}_{P_1, P_2}. \tag{2.80}$$

Remark 2.103 The existence and properties of $\Pi^{\circ}{}^{P_1, P_2}_{Q_1, Q_2}$ are discussed in detail by Schiffer and Spencer [SchS1951] and Hejhal [Hej1972]. Using $\Pi^{\circ}{}^{P_1, P_2}_{Q_1, Q_2}$, Hejhal showed the relations among the Bergman kernel, the Szegö kernel, the Green functions, and the Abelian differentials on compact Riemann surfaces.

Here we mention the relation between the kernel functions and Green's function. Roughly speaking, we note the fact $\dfrac{\sqrt{-1}}{2\pi} \dfrac{\partial^2 \log z}{\partial z \partial \overline{z}} = \delta(z)$, or $\dfrac{\sqrt{-1}}{2\pi} \dfrac{\partial}{\partial \overline{z}} \dfrac{1}{z} = \delta(z)$, where $\delta(z)$ is the Dirac distribution, and the origin $z = 0$ is also expressed by $\overline{z} = 0$. Thus for a Riemann surface, we assume that a point P expressed by the parameter p is also represented by an alternative parameter $\widetilde{p}$. The Green function is defined by

$$G(p, q) = \frac{1}{2}(\Pi^\circ_{q, \widetilde{q}}(p) - \Pi^\circ_{q, \widetilde{q}}(\widetilde{p})),$$

where $\Pi^\circ_{q_1, q_2}(p) := \Pi^{\circ}{}^{p, p_0}_{q_1, q_2}$ in (2.80) [SchS1951, (4.2.1)].

For a given Green function, the Bergman kernel is provided by the second derivative of the Green's function [SchS1951, SebF2007],

$$K(p, \widetilde{q}) = \frac{\partial^2 G(p, q)}{\partial p \partial \widetilde{q}}.$$

The adjoint kernel or the Schiffer kernel is defined by

$$L(p,q) = \frac{\partial^2 G(p,q)}{\partial p \partial q}.$$

The fundamental normalized differential of the second kind $\Omega^\circ(P_1, Q_1)$ could be identical to the Schiffer kernel. These are quite important for the bordered Riemann surfaces [Fay1973, Sect. VI]. In this book, we show the algebraic expression of the fundamental differential of the second kind in Theorem 2.140, (See Remark 2.136.)

2.4.1.4 Normalized Prime Form

As we show the explicit description of the normalized prime form of W-curves in Sect. 3.1.8, we introduce the prime form as a form of $\widetilde{X} \times \widetilde{X}$ following Mumford's expression in [Mum1984, 3.210]:

Definition 2.104 The normalized prime form E° is a holomorphic differential form of $(-1/2, -1/2)$ on $\widetilde{X} \times \widetilde{X}$ which satisfies the following relations:

1. $E^\circ(P_1, P_2)$ vanishes if and only if $\kappa_X(P_1) = \kappa_X(P_2)$.
2. $E^\circ(P_1, P_2)$ has a first-order zero along the diagonal $\kappa_X^* \Delta$ of $\Delta \subset X \times X$.
3. $E^\circ(P_1, P_2) = -E^\circ(P_2, P_1)$.
4. Let t_i be a local parameter at P_i of X such that $\zeta(P_i) = dt_i$, $(i = 1, 2)$,

$$E^\circ(P_1, P_2) = \frac{t_1 - t_2}{\sqrt{dt_1}\sqrt{dt_2}}(1 + d_> \left((t_1 - t_2)^0\right).$$

5. $E^\circ(P, Q)$ has the properties:
    1. $E^\circ(P, Q) = E^\circ(P + \alpha_i, Q)$ and $E^\circ(P, Q) = E^\circ(P, Q + \alpha_i)$,
    2. for $\gamma = \sum_{i=1}^{g} n_i \beta_i$,

$$E^\circ(P + \gamma, Q) = E^\circ(P, Q) \exp\left(2\pi\sqrt{-1}\left({}^t n\delta' - \frac{1}{2}{}^t n\tau n + {}^t n \int_P^Q \nu^{\mathrm{I}\circ}\right)\right),$$

$$E^\circ(P, Q + \gamma) = \pm E^\circ(P, Q) \exp\left(2\pi\sqrt{-1}\left({}^t n\delta' - \frac{1}{2}{}^t n\tau n - {}^t n \int_P^Q \nu^{\mathrm{I}\circ}\right)\right).$$

The normalized prime form is connected with the normalized differential of the third kind and the fundamental normalized differential of the second kind $\Omega^\circ(P_1, P_2)$.

Proposition 2.105 *1.* $\dfrac{E^\circ(P,Q)}{E^\circ(P,Q')} = \exp \displaystyle\int^P \nu^{\mathrm{III}\circ}_{Q,Q'}$, *and*

2. $\Omega^\circ(P,Q) = \dfrac{\partial^2}{\partial t_P \partial t_Q} \log E^\circ(P,Q)$.

Remark 2.106 In terms of the prime form, the theta function is described well as in [Fay1973, (44)]. Nakayashiki studied the sigma function in terms of the formula [Nak2010a].

Further, we refer to $E^\circ(P,Q)\sqrt{dt_P}|_{t_P=0}$ as $E^\circ(\infty, Q)$ for later convenience.

2.4.2 W-Normalized Abelian Differentials of the First Kind

Following Weierstrass [Wei1875a], Bolza [Bol1895, Bol1899a, Bol1899b], H. F. Baker [Ba1897], Buchstaber, Leykin and Enolskii [BucEL1999, BucLE1999], [BucEL2012], Eilbeck, Enolskii and Leykin [EilEL2000] and the author of this book with Previato and Komeda [KomMP2013, KomMP2019, KomM2020], we introduce different normalized Abelian differentials from the ones mentioned above, which we address as *W-normalized* here. We extend the W-normalized differentials for the hyperelliptic curves and plane Weierstrass curves (W-curves) to more general W-curves as in [KomMP2022a, KomMP2022b].

We start with the differentials of the first kind below based on Proposition 2.84 [KomMP2022a]. Due to the Riemann–Roch theorem, there is the i-th holomorphic one-form whose behavior at ∞ is given by

$$-\Big(t^{N^{\mathfrak{c}}(g-i)-1}(1 + d_{>0}(t))\Big)dt, \tag{2.81}$$

where $N^{\mathfrak{c}}(i) \in H^{\mathfrak{c}}_X$ $(i = 0, 1, 2, \ldots, g-1)$ satisfying $N^{\mathfrak{c}}(i) < N^{\mathfrak{c}}(i+1)$, and t is the arithmetic local parameter at ∞. This condition determines the basis of $\mathscr{A}_X(X)$ up to higher elements. We call this normalization the *W-normalization*.[10] Proposition 2.1 7 means that the g-th holomorphic one-form is given by $-dt(1 + d_{>0}(t))$.

2.4.2.1 W-Normalized Abelian Differentials of the First Kind

The W-normalized holomorphic one-forms are directly obtained from Proposition 2.84:

Lemma 2.107 *For* $x^k \widehat{\mathfrak{y}}_{\mathfrak{e}_i}$ *in Proposition 2.84, we have the relation,*

[10] There is a different convention in the sigma function theory for consistency with the so-called Sato theory, in which the minus sign is omitted, e.g., in [Nak2010a, Nak2010b, Nak2016, On2018]. In this book, emphasis is placed on consistency with Weierstrass' sigma function theory.

$$\left\langle \frac{x^k \widehat{\mathfrak{y}}_{\mathfrak{e}_i}}{h_X(x, y_\bullet)} dx \;\middle|\; i \in \mathbb{Z}_r^\times, k \in \mathbb{N}_0, \mathfrak{e}_i - (k+1)r > 0 \right\rangle_{\mathbb{C}} = \mathbf{H}^0(X, \mathscr{A}_X).$$

2.4.2.2 Canonical Basis $\widehat{S}_X$ in R_X

Using the Riemann–Roch theorem, we construct the basis of $\mathbf{H}^0(X, \mathscr{A}_X(*\infty))$ as a $\mathbb{C}$ vector space, which is given by $\left\{ \dfrac{x^k \widehat{\mathfrak{y}}_{\mathfrak{e}_j} dx}{h_X} \right\}$.

By re-ordering $x^k \widehat{\mathfrak{y}}_{\mathfrak{e}_j}$ with respect to the weight at ∞, we define the ordered set $\{\widehat{\phi}_i\}$:

Definition 2.108 1. Let us define the ordered subset $\widehat{S}_X$ of R_X by

$$\widehat{S}_X = \{\widehat{\phi}_i \mid i \in \mathbb{N}_0\}$$

such that $\widehat{\phi}_i$ is ordered by the Sato–Weierstrass weight, i.e., $-\mathrm{wt}\widehat{\phi}_i < -\mathrm{wt}\widehat{\phi}_j$ for $i < j$, and $\widehat{S}_X$ is equal to $\{x^k \widehat{\mathfrak{y}}_{\mathfrak{e}_i} \mid i \in \mathbb{Z}_r, \ k \in \mathbb{N}_0\}$ as a set up to constant factors.

2. Let $\widehat{R}_X$ be an R_X-module generated by $\widehat{S}_X$, i.e., $\widehat{R}_X := \langle \widehat{S}_X \rangle_{R_X} \subset R_X$.
3. Recalling $\mathfrak{K}_X$ and $\mathfrak{k}_X$ in Definition 2.81, we let $\widehat{N}_X(n) := -\mathrm{wt}\ (\widehat{\phi}_n) - \mathfrak{k}_X$ (i.e., $\mathrm{wt}(\widehat{\phi}_n) = -\widehat{N}_X(n) - \mathfrak{k}_X$), $\widehat{H}_X := \{-\mathrm{wt}\ (\widehat{\phi}_n) \mid n \in \mathbb{N}_0\}$, and we define the dual conductor $\widehat{c}_X$ as the minimal integer satisfying $\widehat{c}_X + \mathbb{N}_0 \subset \widehat{H}_X - \mathfrak{k}_X$. We sometimes use the short form $\widehat{N} = \widehat{N}_X$.
4. We define $\widehat{S}_X^{(g)} := \{\widehat{\phi}_0, \widehat{\phi}_1, \dots, \widehat{\phi}_{g-1}\}$ and the *W-normalized holomorphic one-form*, or the *W-normalized Abelian differentials of the first kind* ν_i^{I} as the canonical basis of X, satisfying (2.81), i.e.,

$$\left\langle \nu_i^{\mathrm{I}} := \frac{\widehat{\phi}_{i-1} dx}{h_X} \;\middle|\; \widehat{\phi}_{i-1} \in \widehat{S}_X^{(g)} \right\rangle_{\mathbb{C}} = \mathbf{H}^0(X, \mathscr{A}_X). \tag{2.82}$$

Lemma 2.109 *1. $\widehat{\phi}_i$ is not a monomial in R_X in general, and is given as $x^k \widehat{\mathfrak{y}}_{\mathfrak{e}_i} = x^k (\widetilde{\delta}_i(x) \mathfrak{y}_{\mathfrak{e}^*_{\ell,i}} + \cdots)$ in Definition 2.64.*

2. $\mathbf{H}^0(X, \mathscr{A}_X(*\infty)) = \displaystyle\bigoplus_{i=0} \mathbb{C} \frac{\widehat{\phi}_i dx}{h_X} = \frac{1}{h_X} \widehat{R}_X dx = R_X^{\mathfrak{c}} dx.$
3. *At ∞, ν_i^{I} behaves like $\nu_i^{\mathrm{I}} = (-t^{N^{\mathfrak{c}}(g-i-1)-1}(1 + d_{>0}(t))dt$ for the arithmetic local parameter t at ∞, and thus $\mathrm{wt}(\nu_i^{\mathrm{I}}) = N_X^{\mathfrak{c}}(g-i) - 1 = \widehat{c}_X - \widehat{N}_X(i-1) - 2$ for $(i = 1, 2, \dots, g)$, especially $\mathrm{wt}(\nu_g^{\mathrm{I}}) = 0$, or $N^{\mathfrak{c}}(0) = 1$.*
4. *$H_X^{(g)} = \{-\mathrm{wt}(\widehat{\phi}) \mid \widehat{\phi} \in \widehat{S}_X^{(g)}\} = \{\widehat{N}_X(i) + \mathfrak{k}_X \mid i = 0, 1, \dots, g-1\}$.*
5. *$\widehat{R}_X = \mathbf{H}^0(X, \mathscr{L}_X(-\mathfrak{K}_X)(*\infty))$.*
6. *$\widehat{R}_X$ is isomorphic to $\mathbf{H}^0(X, \mathscr{A}_X(*\infty))$ and $R_X^{\mathfrak{c}}$ as an R_X-module.*
7. *$\overline{H}_X^{\mathfrak{c}} = \left\{ \mathrm{wt}\left(\dfrac{\widehat{\phi}_i dx}{h_X}\right) + 1 \mid i \in \mathbb{N}_0 \right\} = \left\{ \mathrm{wt}(\widehat{\phi}_i) + d_h - r \mid i \in \mathbb{N}_0 \right\}$*

$$= \left\{ \mathrm{wt}(\widehat{\phi}_i) + 2g - 1 + \mathfrak{k}_X \mid i \in \mathbb{N}_0 \right\}.$$

8. $\widehat{N}_X(n) - \widehat{N}_X(n-1) = 1$ *for* $n > g$, *and* $\widehat{N}_X(g) - \widehat{N}_X(g-1) = 2$.

Proof *1*, *2*, and *4* are obviously obtained. The factor in *3* is determined by the ratio of $\widehat{y}_{\widehat{e}_i}$ and h_X or the trace in Definition 2.64. *6* is proved by Proposition 2.84, and Lemmas 2.5 and 2.82. Using *6*, we show *5* due to (2.54), (2.56), and Proposition 2.84. We recall the definition of $\overline{H}^{\mathfrak{c}}_X$ in Definition 2.4, and Proposition 2.84 yields *7*. Thus we obtain *8*. ■

The Riemann–Roch theorem shows that

$$\dim \mathbf{H}^0(X, \mathscr{O}_X(-n\infty)) - \dim \mathbf{H}^0(X, \mathscr{A}_X(n\infty)) - n = 1 - g. \tag{2.83}$$

Lemma 2.110 *1. If* $d_h = \mathfrak{e}_{r-1}$ *or* $\delta_0 = 0$ *in (2.48) (*d_h *is symmetric in Proposition 2.46),* H_X *is symmetric, otherwise (*$\delta_0 \neq 0$ *or* d_h *is not symmetric)* $\mathfrak{k}_X \neq 0$ *and* H_X *is not symmetric.*

2. $\{-\mathrm{wt}(\widehat{\phi}_i)\} = \{\widehat{\mathfrak{e}}_i + kr \mid i \in \mathbb{Z}_r,\ k \in \mathbb{N}_0\} = \{d_h - \mathfrak{e}_i + kr \mid i \in \mathbb{Z}_r,\ k \in \mathbb{N}_0\}$.

3. $\mathrm{div}(\widehat{\phi}_i) \geq (\mathfrak{K}_X - (2g - 2 + \mathfrak{k}_X)\infty)$ *for every* $\widehat{\phi}_i \in \widehat{S}_R^{(g)}$, $(i = 0, 1, 2, \ldots, g-1)$.

4. $\mathrm{div}(\widehat{\phi}_i) \geq (\mathfrak{K}_X - (g + \mathfrak{k}_X + i)\infty)$ *for every* $\widehat{\phi}_i \in \widehat{S}_R$, $(i \geq g)$.

Proof *2* is proved from Lemma 2.109 *6*. Let us consider *1*. If d_h is symmetric, $\mathfrak{k}_X = 0$ from Proposition 2.83. Thus if d_h is symmetric, $(dx/h_X) = (2g-2)\infty$, and due to the Riemann–Roch theorem, H_X is symmetric. It corresponds to $\delta_0 = 0$ and $d_h = \mathfrak{e}_{r-1}$ in Proposition 2.46 and Lemma 2.45. On the other hand, the case $\mathfrak{k}_X \neq 0$ means that H_X is not symmetric and $\delta_0 \neq 0$ in Lemma 2.45. Thus we prove *1*. The Riemann–Roch theorem shows *3* and *4*. ■

Proposition 2.111 *1. Assume* d_h *is symmetric or* $d_h = \mathfrak{e}_{r-1}$ $(\delta_0 = 0)$ *in (2.48). Then we have the following:*

a. $\mathfrak{k}_X = 0$ *in Definition 2.81 and* H_X *is symmetric,*

b. $\widehat{S}_X = S_X$ *in Definitions 2.18 and 2.108,* $\mathfrak{K}_X = 0$ *in Definition 2.81,* $\widehat{c}_X = c_X$, *and* $\widehat{R}_X = R_X$ *in Definition 2.108.*

2. In general, $\widehat{R}_X \neq R_X$ *and we have the equality if and only if* H_X *is symmetric.*

Proof We note Proposition 2.1 *6*. They are obvious. ■

2.4.2.3 Properties of W-Normalized Differentials of the First Kind

Due to the Abel–Jacobi theorem, Theorem 2.98, we introduce the W-normalized integrals of the first kind.

Definition 2.112 1. The matrices ω' and ω'' (W-normalized complete integrals of the first kind), $\omega := (\omega', \omega'')$, are defined by[11]

[11] See footnote 15 in this chapter.

$$\omega'_{ij} := (\omega'_{ij}), \quad \omega'' := (\omega''_{ij}), \quad \omega' := \frac{1}{2}\left(\int_{\alpha_j} \nu^{\mathrm{I}}_i\right), \quad \omega''_{ij} := \frac{1}{2}\left(\int_{\beta_j} \nu^{\mathrm{I}}_i\right).$$

2. The W-normalized (incomplete) Abelian integral is defined by

$$\widetilde{w} : S^k \widetilde{X} \to \mathbb{C}^g, \quad \widetilde{w}(\gamma_1, \cdots, \gamma_k) = \sum_{i=1}^{k} \widetilde{w}(\gamma_i) = \sum_{i=1}^{k} \int_{\gamma_i} \nu^{\mathrm{I}}.$$

3. We define the W-normalized lattice $\Gamma_X := \langle 2\omega', 2\omega'' \rangle_{\mathbb{Z}}$ and the Jacobian (Jacobi variety) $J_X := \mathbb{C}^g / \Gamma_X$ ($\kappa_J : \mathbb{C}^g \to J_X$).
4. The W-normalized Abel–Jacobi map is defined by $w := \kappa_J \circ \widetilde{w} \circ \iota_X : S^k X \to J_X$,

From the definition and Proposition 2.95, the following lemma is obvious:

Lemma 2.113 *The matrices ω' and ω'' are regular and satisfy the following relations:*

$$\nu^{\mathrm{I}} = 2\omega' \nu^{\mathrm{Io}}, \quad \textit{for} \quad \nu^{\mathrm{I}} := \begin{pmatrix} \nu_1 \\ \nu_2 \\ \vdots \\ \nu_g \end{pmatrix}, \quad \nu^{\mathrm{Io}} := \begin{pmatrix} \nu_1^{\mathrm{Io}} \\ \nu_2^{\mathrm{Io}} \\ \vdots \\ \nu_g^{\mathrm{Io}} \end{pmatrix},$$

and

$$\omega'^{-1} \omega = (1_g, \tau), \quad \tau = \omega'^{-1} \omega'', \quad {}^t\omega' \omega'' = {}^t\omega'' \omega', \quad \mathrm{Im}\left(\omega'^{-1} \omega''\right) > 0.$$

From Lemma 2.99, we address the following lemma:

Lemma 2.114 *For every $\rho \in \mathrm{Sp}(2g, \mathbb{Z})$, the lattices Γ_X and J_X are invariant via the action $(\omega', \omega'')\,{}^t\rho$, i.e., $(\widetilde{\omega}', \widetilde{\omega}'') := (\omega', \omega'')\,{}^t\rho$, $\langle 2\widetilde{\omega}', 2\widetilde{\omega}'' \rangle_{\mathbb{Z}} = \langle 2\omega', 2\omega'' \rangle_{\mathbb{Z}}$.*

Then we have the following lemma:

Lemma 2.115 *1.* $-\mathrm{wt}\, \widehat{\phi}_0 = \widehat{\mathfrak{e}}_{r-1} = d_h - \mathfrak{e}_{r-1} (= 0$ *if H_X is symmetric*$) = \widehat{c}_X + \mathfrak{k}_X - c_X = d_h - r - c_X + 1$,
2. $-\mathrm{wt}\ \widehat{\phi}_{g-1} = \widehat{c}_X + \mathfrak{k}_X - 2 = d_h - r - 1 = (2g-2) + \mathfrak{k}_X$, *i.e.,* $\widehat{N}(g-1) = 2g - 2$,
3. $\widehat{c}_X = 2g = d_h - \mathfrak{k}_X - r + 1$, $-\mathrm{wt}\,\widehat{\phi}_g = 2g + \mathfrak{k}_X = \widehat{c}_X + \mathfrak{k}_X = d_h - r + 1$,
4. $-\mathrm{wt}\,\widehat{\phi}_{g-1} + \mathrm{wt}\,\widehat{\phi}_0 = \mathfrak{e}_{r-1} - r - 1 = c_X - 2$, *and* $c_X = \mathfrak{e}_{r-1} - r + 1$.

Proof Lemma 2.110 2 means that $-\mathrm{wt}(\widehat{\phi}_0) = \widehat{\mathfrak{e}}_{r-1}$ due to the order of the weight. Then *1* is proved by the relations $-\mathrm{wt}\ \widehat{\phi}_0 = \widehat{N}(0) + \mathfrak{k}_X$ and $\widehat{N}(0) = \widehat{c}_X - c_X$. Let us consider *2*. Since from the Riemann–Roch theorem, wt $\nu^{\mathrm{I}}_g = 0$, whereas on $\nu^{\mathrm{I}}_g = \widehat{\phi}_{g-1} dx/h_X$, $\mathrm{wt}(dx/h_X) = (2g-2) + \mathfrak{k}_X = d_h - r - 1$, we have $-\mathrm{wt}\ \widehat{\phi}_{g-1} = (2g-2) + \mathfrak{k}_X$ or *2*. The Riemann–Roch theorem also shows that $-\mathrm{wt}\,\widehat{\phi}_g = -\mathrm{wt}\,\widehat{\phi}_{g-1} + 2 = \widehat{c}_X + \mathfrak{k}_X$, or *3*. We compare them and obtain *4*. ■

By the Abel–Jacobi theorem 2.98, $\mathfrak{K}_X$ in Definition 2.81 can be divided into two pieces, which is related to the spin structure in X.

Definition 2.116 Let $\mathfrak{K}_\mathfrak{s}$ and $\mathfrak{K}_X^\mathfrak{c}$ be the effective divisors which satisfy

$$\mathfrak{K}_X - \mathfrak{k}_X \infty \sim 2\mathfrak{K}_\mathfrak{s} - 2\mathfrak{k}_\mathfrak{s}\infty, \quad \mathfrak{K}_X + \mathfrak{K}_X^\mathfrak{c} - (\mathfrak{k}_X + \mathfrak{k}_X^\mathfrak{c})\infty \sim 0$$

as the linear equivalence, where $\mathfrak{k}_\mathfrak{s}$ and $\mathfrak{k}_X^\mathfrak{c}$ are the degree of $\mathfrak{K}_\mathfrak{s}$ and $\mathfrak{K}_X^\mathfrak{c}$ respectively. Further, we let the meromorphic functions associated with these linear equivalences denoted by $h_\mathfrak{s}$ and $h_X^\mathfrak{c}$ such that

$$(h_\mathfrak{s}) = \mathfrak{K}_X - 2\mathfrak{K}_\mathfrak{s} - (\mathfrak{k}_X - 2\mathfrak{k}_\mathfrak{s})\infty, \quad (h_X^\mathfrak{c}) = \mathfrak{K}_X + \mathfrak{K}_X^\mathfrak{c} - (\mathfrak{k}_X + \mathfrak{k}_X^\mathfrak{c})\infty.$$

We note that $h_\mathfrak{s}$ and $h_X^\mathfrak{c}$ belong to $\mathcal{Q}(R_X)$ and their behaviors at ∞ are given by

$$h_\mathfrak{s} = t^{2\mathfrak{k}_\mathfrak{s} - \mathfrak{k}_X}(1 + d_{>0}(t)), \quad h_X^\mathfrak{c} = t^{\mathfrak{k}_X + \mathfrak{k}_X^\mathfrak{c}}(1 + d_{>0}(t))$$

using the arithmetic parameter t in Definition 2.16.

Since the W-normalized holomorphic one-form is given by the basis (2.82), Definition 2.116 shows the canonical divisor:

Proposition 2.117 *The canonical divisor is given by*

$$\begin{aligned} K_X \sim (2g - 2 + \mathfrak{k}_X)\infty - \mathfrak{K}_X &\sim (2g - 2 + 2\mathfrak{k}_\mathfrak{s})\infty - 2\mathfrak{K}_\mathfrak{s} \\ &\sim (2g - 2 - \mathfrak{k}_X^\mathfrak{c})\infty + \mathfrak{K}_X^\mathfrak{c}. \end{aligned}$$

We may find the half-canonical divisor $(g - 1 + \mathfrak{k}_\mathfrak{s})\infty - \mathfrak{K}_\mathfrak{s}$, which plays a crucial role to define and the shifted Abelian integral (cf. Sect. 2.4.4) and the shifted Riemann constant (cf. Sect. 3.1.2).

2.4.2.4 Group Action on the Holomorphic One-Form and the Period Matrices

From Proposition 2.31, for an element of the group action $\widehat{\zeta} \in G_{X,x}$ for $(x, y_\bullet) \in X \setminus \mathfrak{B}_X$, we have

$$\widehat{\zeta}\left(\frac{\widehat{\phi}_i dx}{h_X(x, y_\bullet)}\right) = \frac{\widehat{\zeta}(\widehat{\phi}_i)dx}{\widehat{\zeta}(h_X(x, y_\bullet))}.$$

From Lemma 2.86, we have the following relation, which induces the "inverse" operation in the Jacobian:

Lemma 2.118

$$\sum_{\widehat{\zeta} \in G_{X,x}} \widehat{\zeta} \nu^{\mathrm{I}} = 0.$$

Proof It is proved essentially from Lemma 2.38 for $f_X(x, Y) = (Y - y)(Y - \widehat{\zeta} y) \cdots (Y - \widehat{\zeta}^{r-1} y)$ and $\widehat{\zeta} \in G_{X,x}$ when $m_X = 2$. In the general case, due to the construction of $h_X(x, y_\bullet)$, it is also reduced to Lemma 2.38. ■

When X is a Galois covering over $\mathbb{P}$, Lemma 2.88 is extended to the Galois action on the basis of the Homology $\mathbf{H}_1(Z, \mathbb{Z})$ shows the actions on these period matrices (ω', ω'') and Γ_X due to Lemma 2.114.

Lemma 2.119 *Assume that $\varpi_r : X \to \mathbb{P}$ is a Galois covering. For the Galois action $\widehat{\zeta} \in G_X$, i.e., $\widehat{\zeta} : X \to X$, its associated element $\rho_{\widehat{\zeta}}$ of* $\mathrm{Sp}(2g, \mathbb{Z})$ *acts on (ω', ω'') by*

$$\widehat{\zeta}(\omega', \omega'') = (\omega', \omega'')\,{}^t\rho_{\widehat{\zeta}},$$

and leaves the lattice Γ_X and J_X invariant,

$$\widehat{\zeta}\Gamma_X = \Gamma_X, \quad \widehat{\zeta} J_X = J_X,$$

i.e.,

$$\sum_{\widehat{\zeta} \in G_X} \widehat{\zeta} u = 0 \text{ modulo } \Gamma_X \text{ for } u \in \mathbb{C}^g.$$

Proof From Definition 2.112, its action is transpose of the matrix $\rho_{\widehat{\zeta}}$ from the right-hand side. ■

2.4.2.5 Examples of W-Normalized Abelian Differentials of the First Kind

1. Cyclic hyperelliptic curves ((2, 2g + 1)-curves of genus g) (2.66): $\mathfrak{k}_X = \mathfrak{k}_\mathfrak{s} = 0$, $c_X = \widehat{c}_X = 2g$.

$$\nu_i^{\mathrm{I}} = \frac{x^{i-1} dx}{2y}, \quad (i = 1, 2, \ldots, g), \quad h_X = 2y, \quad d_h = 2g + 1,$$

   and

$$\phi_i = \widehat{\phi}_i, \quad \widehat{N}(i) = N(i).$$

2. (3, 4)-curves (2.62): $\mathfrak{k}_X = \mathfrak{k}_\mathfrak{s} = 0$.

$$\nu_1^{\mathrm{I}} = \frac{dx}{f_{X,y}(x, y)}, \quad \nu_2^{\mathrm{I}} = \frac{x dx}{f_{X,y}(x, y)}, \quad \nu_3^{\mathrm{I}} = \frac{y dx}{f_{X,y}(x, y)}.$$

3. (r, s)-curves (2.61): $\mathfrak{k}_X = \mathfrak{k}_\mathfrak{s} = 0$, $c_X = \widehat{c}_X$. In terms of $\phi_i \in R_X = \widehat{R}_X$,

$$\nu_i^{\mathrm{I}} = \frac{\phi_{i-1} dx}{f_{X,y}(x, y)}, \quad (i = 1, \ldots, g), \quad \phi_i = \widehat{\phi}_i, \quad h_X(x, y) = f_{X,y}(x, y), \quad d_h = (r - 1)s,$$

and further, for the cyclic (r, s)-curves, $f_{X,y}(x, y) = ry^{r-1}$.

4. Cyclic $(3, \widehat{r}, \widehat{s})$-curves (2.68): $\mathfrak{K}_X = B_1 + \cdots + B_{\dot{r}+\dot{s}}$, $\mathfrak{K}_{\mathfrak{s}} = B_1 + \cdots + B_{\dot{r}}$ (or $= B_1 + \cdots + B_{\dot{s}}$), $\mathfrak{k}_X = \dot{r} + \dot{s}$, $\mathfrak{k}_{\mathfrak{s}} = \dot{r}$ (or $\mathfrak{k}_{\mathfrak{s}} = \dot{s}$), $\widehat{c}_X = c_X + \dot{r}$, $d_h = \widehat{r} + \widehat{s} = 3(\dot{r} + \dot{s})$, $\widehat{N}(0) = \dot{r}$, and
$$\nu_i^{\mathrm{I}} = \frac{\widehat{\phi}_{i-1}}{3 y_{\widehat{r}} y_{\widehat{s}}} dx, \quad (i = 1, 2, \ldots, g),$$
where $\widehat{\phi}_i$ is listed in Tables 2.4 and 2.5. It can be extended to more general cyclic space W-curves or space superelliptic curves [Koo1999].
5. Special curve cases in Sect. 2.3.4: As in Table 2.6, we have the holomorphic forms of the special curves as follows.

   a. I-type:
$$\nu_1^{\mathrm{I}} = \frac{y dx}{3k_2k_3^2}, \quad \nu_2^{\mathrm{I}} = \frac{w dx}{3k_2k_3^2}, \quad \nu_3^{\mathrm{I}} = \frac{x y dx}{3k_2k_3^2}, \quad \nu_4^{\mathrm{I}} = \frac{x w dx}{3k_2k_3^2}.$$

   b. II-type:
$$\nu_i^{\mathrm{I}} = \frac{\widehat{\phi}_{i-1} dx}{5k_2k_3}.$$

   c. III-type: $(i = 1, 2, \cdots, 12)$
$$\nu_i^{\mathrm{I}} = \frac{\phi_{i-1} dx}{6 y_{13} y_{16}} = \frac{\phi_{i-1} dx}{6 y_{14} y_{15}}.$$

Table 2.4 Examples of $\widehat{\phi}_i$ $(i = 0, 1, \ldots)$ of cyclic $(3, \widehat{r}, \widehat{s})$-curves with respect to $\mathrm{wt}(\widehat{\phi}_i)$

$(\dot{r}, \dot{s})$	$(\widehat{r}, \widehat{s})$	g	c_X	h_X	0	1	2	3	4	5	6	7	8	9	10	11	12	13	14	15	16
(1, 2)	(4, 5)	2	3	$3y_4y_5$	–	–	–	–	y_4	y_5	–	xy_4	xy_5	y_4y_5	x^2y_4	x^2y_5	xy_4y_5	$y_4^2y_5$	x^3y_5	y_5^3	x^4y_4
(1, 3)	(5, 7)	3	5	$3y_5y_7$	–	–	–	–	–	y_5	–	y_7	xy_5	–	xy_7	x^2y_5	y_5y_7	x^2y_7	x^3y_5	xy_5y_7	x^3y_7
(2, 3)	(7, 8)	4	6	$3y_7y_8$	–	–	–	–	–	–	–	y_7	y_8	–	xy_7	xy_8	–	x^2y_7	x^2y_8	y_7y_8	x^3y_7
(1, 5)	(7, 11)	5	9	$3y_7y_{11}$	–	–	–	–	–	–	–	y_7	–	–	xy_7	y_{11}	–	x^2y_7	xy_{11}	–	x^3y_7
(2, 4)	(8, 10)	5	8	$3y_8y_{10}$	–	–	–	–	–	–	–	–	y_8	–	y_{10}	xy_8	–	xy_{10}	x^2y_8	–	x^2y_{10}

Table 2.5 Examples of $\widehat{\phi}_i$ $(i = 0, 1, \ldots)$ of cyclic $(3, \widehat{r}, \widehat{s})$-curves with respect to $\widehat{N}(i)$

$(\dot{r}, \dot{s})$	$(\widehat{r}, \widehat{s})$	g	d_h	$\mathfrak{k}_X$	$\widehat{c}_X$	0	1	2	3	4	5	6	7	8	9	10	11	12	13
(1, 2)	(4, 5)	2	9	3	4	–	y_4	y_5	–	xy_4	xy_5	y_4y_5	x^2y_4	x^2y_5	xy_4y_5	$y_4^2y_5$	x^3y_5	y_5^3	x^4y_4
(1, 3)	(5, 7)	3	12	4	6	–	y_5	–	y_7	xy_5	–	xy_7	x^2y_5	y_5y_7	x^2y_7	x^3y_5	xy_5y_7	x^3y_7	x^4y_5
(2, 3)	(7, 8)	4	15	5	8	–	–	y_7	y_8	–	xy_7	xy_8	–	x^2y_7	x^2y_8	y_7y_8	x^3y_7	x^3y_8	xy_7y_8
(1, 5)	(7, 11)	5	18	6	10	–	y_7	–	–	xy_7	y_{11}	–	x^2y_7	xy_{11}	–	x^3y_7	x^2y_{11}	y_7y_{11}	x^4y_7
(2, 4)	(8, 10)	5	18	6	10	–	–	y_8	–	y_{10}	xy_8	–	xy_{10}	x^2y_8	–	x^2y_{10}	x^3y_8	y_8y_{10}	x^3y_{10}
(3, 4)	(10, 11)	6	21	7	12	–	–	–	y_{10}	y_{11}	–	xy_{10}	xy_{11}	–	x^2y_{10}	x^2y_{11}	–	x^3y_{10}	x^3y_{11}
(3, 5)	(11, 13)	7	24	8	14	–	–	–	y_{11}	–	y_{13}	xy_{11}	–	xy_{13}	x^2y_{11}	–	x^2y_{13}	x^3y_{11}	–

Table 2.6 Examples of $\widehat{\phi}$ in $\widehat{S}_X$ of special curves with respect to $\mathrm{wt}(\widehat{\phi}_i)$

	0	1	2	3	4	5	6	7	8	9	10	11	12	13	14	15	16	17	18	19	20	21	22	23	24
I	–	–	–	–	–	–	–	y	w	–	xy	xw	–	x^2y	y^2	yw	w^2	xy^2	xyw	xw^2	x^2y^2	x^2yw	x^2w^2	x^3y	x^3w^2
II	–	–	–	–	–	–	–	y	–	–	–	w	xy	–	y^2	–	xw	x^2y	wy	xy^2	–	y^3	x^3y	xyw	x^2y^2
III	1	–	–	–	–	–	x	–	–	–	–	–	x^2	y	z	v	w	–	x^3	xy	xz	xv	xw	–	x^4

2.4.3 W-Normalized Differentials of the Second and the Third Kinds

Based on the trace structure of the complementary module $R_X^{\mathfrak{c}}$ in Sect. 2.2.6.7, we construct the W-normalized differentials of the second and the third kinds in this subsection.

2.4.3.1 The One-Form Σ on X

We construct an algebraic representation of the fundamental normalized differential of the second kind in [Fay1973, Corollary 2.6], namely, a two-form $\Omega(P_1, P_2)$ on $X \times X$ which is symmetric and has quadratic singularity as in Theorem 2.140.

Following K. Weierstrass [Wei1875a], O. Bolza [Bol1895, Bol1899a, Bol1899b], H. F. Baker [Ba1897], V.M. Buchstaber, D.V. Leykin and V.Z. Enolskii [BucEL1999, BucLE1999, BucEL2012], J.C. Eilbeck, V.Z. Enolskii, and D.V. Leykin [EilEL2000], we construct the fundamental W-normalized differential of the second kind Ω and W-normalized differentials of the second and the third kind using a meromorphic one-form $\Sigma(P, Q)$ on $X \times X$. As we extend the approach for the hyperelliptic curves and plane Weierstrass curves (W-curves) to more general W-curves in [KomMP2022b], we show it here, to introduce $\Sigma(P, Q)$ on $X \times X$ based on Definition 2.74 and Proposition 2.76.

Proposition 2.120 *For a point $(P, Q) \in X \times X$,*

$$\Sigma(P, Q) := \frac{dx_P}{(x_P - x_Q)} p_\varpi(P, Q) = \frac{dx_P}{(x_P - x_Q)} \frac{\widetilde{h}_X(x_P, y_{\bullet P}, y_{\bullet Q})}{h_X(x_P, y_{\bullet P})}, \tag{2.84}$$

has the following properties:

1. *$\Sigma(P, Q)$ is holomorphic over X except Q and ∞ as a function of P such that*
   a. *at Q, in terms of the local parameters $t_Q(P) = 0$ and $t_P(Q) = 0$, it behaves like*
   $$\Sigma(P, Q) = \frac{dt_P}{t_P}(1 + d_{>0}(t_P, t_Q)),$$
   b. *at ∞, the local parameter t_P $(t_P(\infty) = 0)$, it behaves like*

$$\Sigma\big(P, Q\big) = -\frac{dt_P}{t_P}(1 + d_{>0}(t_P)).$$

2. *$\Sigma(P, Q)$ as a function of Q is singular at P and ∞ such that*

   a. *at P, in terms of the local parameters $t_Q(P) = 0$ and $t_P(Q) = 0$, it behaves like*

$$\Sigma\big(P, Q\big) = -\frac{dt_P}{t_Q}(1 + d_{>0}(t_P, t_Q)),$$

   b. *at ∞, the arithmetic local parameter t_Q ($t_Q(\infty) = 0$), it behaves like*

$$\Sigma\big(P, Q\big) = \frac{dx_P}{\mathfrak{y}_{\mathfrak{e}_{r-1}P}t_Q^{\mathfrak{e}_{r-1}-r}}(1 + d_{>0}(t_Q)) = \frac{dx_P}{\mathfrak{y}_{\mathfrak{e}_{r-1}P}t_Q^{c_X-1}}(1 + d_{>0}(t_Q)).$$

3. *Let $\widetilde{h}'_X(x_P, y_{\bullet P}, y_{\bullet Q})$ be an element in $R_X \otimes_{\mathbb{C}} R_X$ satisfying the conditions in Lemma 2.73, i.e., $\widetilde{h}'_X(x_P, y_{\bullet P}, y_{\bullet Q}) = \widetilde{h}_X(x_P, y_{\bullet P}, y_{\bullet Q})$ at $\varpi_r(P) = \varpi_r(Q)$, and let*

$$\Sigma'\big(P, Q\big) := \frac{dx_P}{(x_P - x_Q)} \frac{\widetilde{h}'_X(x_P, y_{\bullet P}, y_{\bullet Q})}{h'_X(x_P, y_{\bullet P})}.$$

   Then $\Sigma'\big(P, Q\big) - \Sigma\big(P, Q\big)$ belongs to the set

$$\mathbf{H}^0(X, \mathscr{A}_X) \otimes_{\mathbb{C}} \mathbf{H}^0(X, \mathscr{O}_X(*\infty)) = \mathbf{H}^0(X, \mathscr{A}_X) \otimes_{\mathbb{C}} R_X.$$

Proof Lemma 2.73 1 shows *1*. By noting the properties in Proposition 2.78 and Corollary 2.35, the numerator of Σ is zero with the first-order at the points which $P \neq Q$ and $P \in \varpi_r^{-1}(\varpi_r(Q))$ and thus, Σ behaves like finite one-form there. At $P = Q$, the numerator is equal to one and thus we have $\Sigma = dx_P/(x_P - x_Q)$, which means $\Sigma = dt_P/t_P(1 + d_{>0}(t_P))$ in *2.a* and *3.a*.

Recalling $\varphi_\infty : R_X \to R_H$ in Lemma 2.19, let t_P and t_Q be the local parameters at ∞ corresponding to P and Q respectively. Proposition 2.76 shows $\lim_{P\to\infty} p_\varpi(P, Q) = \frac{1}{r}$ whereas $\lim_{P\to\infty} \frac{dx_P}{(x_P - x_Q)} = -r(1 + d_{>0}(t_P))\frac{dt_P}{t_P}$, and thus we have *2.b*. On the other hand, the following Lemma 2.121 3 shows that $\Sigma(P, Q)$ behaves $\sum_{k=1}^{\infty}\sum_{i=0}^{r-1} \frac{t_P^{\mathfrak{e}_i - rk}}{t_Q^{\mathfrak{e}_i - rk}} \frac{dt_P}{t_P}(1 + d_{>0}(t_P))$. The maximum of $\{\mathfrak{e}_i - r\}_{i\in\mathbb{Z}_r}$ is $\mathfrak{e}_{r-1} - r$, which is equal to $c_X - 1$ due to Lemma 2.115 4, and thus $\varphi_\infty\left(\frac{dx}{\mathfrak{y}_{\mathfrak{e}_{r-1}}}\right) = \frac{t_P^{\mathfrak{e}_{r-1}}}{rt_P^{r+1}}dt_P = \frac{t_P^{c_X-2}dt_P}{r}$. We obtain *3.b*.

Let us consider *4*. Since both $\Sigma'\big(P, Q\big)$ and $\Sigma\big(P, Q\big)$ satisfy these properties *1–3*, their difference is holomorphic over X with respect to P, and over $X \setminus \{\infty\}$ with respect to Q. It shows *4*. ■

Further, we may consider the group action on $\Sigma(P, Q)$; there exists a group action $\widehat{\zeta} \in G_{X,\varpi_r(P)}$ such that $\Sigma(\widehat{\zeta} P, \widehat{\zeta} Q) = \Sigma(P, Q)$ if $\varpi_r(P) = \varpi_r(Q)$.

2.4.3.2 The One-Form Σ at ∞

Noting Lemma 2.19, we address a derivation in $\mathscr{Q}(R_H^Z \otimes R_H^Z)$ for the monomial curve $X_H \times X_H$ instead of $R_{X,\infty}$ using surjection φ_∞.

In order to investigate the behavior of Σ at ∞, we handle the differential in monomial curve $X_H \times X_H$ and element in $\mathscr{Q}(R_H^Z \otimes_{\mathbb{C}} R_H^Z)$ (rather than $\mathscr{Q}(R_H^Z \otimes_{\mathbb{C}[Z_r]} R_H^Z)$),

$$\frac{dZ_r}{(Z_r - Z_r')} p_H(Z_r, Z_\bullet, Z_\bullet'). \tag{2.85}$$

Noting $p_H(Z_r, Z_\bullet, Z_\bullet') = \dfrac{\widetilde{h}_H(Z_r, Z_\bullet, Z_\bullet')}{\widetilde{h}_H(Z_r, Z_\bullet, Z_\bullet)}$ and Lemmas 2.19, 2.45, and 2.79, we define

$$\Sigma_H\big(t_P, t_Q\big) := -\frac{t_Q^r}{t_Q^r - t_P^r} \sum_{i=0}^{r-1} \frac{t_P^{\mathfrak{e}_i}}{t_Q^{\mathfrak{e}_i}} \frac{dt_P}{t_P} \tag{2.86}$$

as an element in $\mathscr{Q}(R_H \otimes_{\mathbb{C}} R_H)$ using isomorphism φ_H^Z and the parameters $t_P = 1/z_P$ and $t_Q = 1/z_Q$; Σ_H is regard as a derivation in $\mathscr{Q}(R_{X,\infty} \otimes_{\mathbb{C}} R_{X,\infty})$.

Then the direct computations show the following results:

Lemma 2.121 *1.* $\varphi_\infty(\Sigma(P, Q)) = \Sigma_H\big(t_P, t_Q\big)$ *with the local parameters* t_P *and* t_Q *at* ∞, *and it shows the behavior at* ∞, *i.e.,*

$$\Sigma(P, Q) = \Sigma_H\big(t_P, t_Q\big)(1 + d_{>0}(t_P, t_Q))$$

for the local parameters at ∞, $t_P(\infty) = t_Q(\infty) = 0$.

2. *For the case* $|t_P| < |t_Q|$, $\Sigma_H\big(t_P, t_Q\big) = -\displaystyle\sum_{k=0}^{\infty}\sum_{i=0}^{r-1} \frac{t_P^{\mathfrak{e}_i+rk}}{t_Q^{\mathfrak{e}_i+rk}} \frac{dt_P}{t_P}$.

3. *For the case* $|t_Q| < |t_P|$, $\Sigma_H\big(t_P, t_Q\big) = \displaystyle\sum_{k=1}^{\infty}\sum_{i=0}^{r-1} \frac{t_P^{\mathfrak{e}_i-rk}}{t_Q^{\mathfrak{e}_i-rk}} \frac{dt_P}{t_P}$, *and* $|\mathfrak{e}_i - kr| \neq 0$, $k > 0$.

Proof *1* is obvious. The relations $\dfrac{t_Q^r}{t_Q^r - t_P^r} = \dfrac{1}{1 - (t_P/t_Q)^r} = \displaystyle\sum_{k=0}^{\infty} \left(\frac{t_P}{t_Q}\right)^{rk}$, and $\dfrac{t_Q^r}{t_Q^r - t_P^r} = -\dfrac{t_Q^r}{t_P^r} \dfrac{1}{1 - (t_Q/t_P)^r} = -\dfrac{t_Q^r}{t_P^r} \displaystyle\sum_{k=0}^{\infty} \left(\frac{t_Q}{t_P}\right)^{rk}$ show 2 and 3. ∎

2.4.3.3 Examples of Σ

1. Cyclic (r, s)-curves (2.64):

$$\Sigma(P, Q) = \frac{\sum_{i=0}^{r} y_P^{r-i} y_Q^{i}}{r(x_P - x_Q)y_P^{r-1}} dx_P.$$

2. Cyclic $((3, \widehat{r}, \widehat{s})$-curves (2.68):

$$\Sigma(P, Q) = \frac{y_{\widehat{r},P} y_{\widehat{s},P} + y_{\widehat{r},P} y_{\widehat{s},Q} + y_{\widehat{r},Q} y_{\widehat{s},P}}{3(x_P - x_Q) y_{\widehat{r},P} y_{\widehat{s},P}} dx_P.$$

3. Cyclic (6, 13, 14, 15, 16)-curves (2.69): $\Sigma\big(P, Q\big)$ is equal to

$$\frac{y_{13,P}y_{16,P} + y_{13,Q}y_{16,P} + y_{13,P}y_{16,Q} + y_{14,Q}y_{15,P} + y_{14,P}y_{15,2} + y_{14,Q}y_{15,Q}}{6(x_P - x_Q)y_{13,P}y_{16,P}} dx_P.$$

4. (r, s) plane curves (2.61): For $P = (x, y)$, $P' = (x', y') \in X$, Eilbeck, Enolskii and Leykin proposed an alternative function form $\Sigma^{\mathrm{EEL}}(P, P')$ by [EilEL2000]

$$\Sigma^{\mathrm{EEL}}(P, P') = \frac{(\delta_{y,y'} f_X)(x', y, y')dx}{(x - x')f_{X,y}(x, y)} = \frac{\sum_{\ell=0}^{s} A_\ell(x') \sum_{i=0}^{r-\ell-1} y^i y'^{r-\ell-i-1}}{(x - x')f_{X,y}(x, y)} dx.$$

We note that $\varphi_\infty(\Sigma^{\mathrm{EEL}}(P, P')) = \varphi_\infty(\Sigma(P, P'))$. From Lemma 2.26, the difference between $\Sigma^{\mathrm{EEL}}(P, P')$ and $\Sigma(P, P')$ is given by

$$\begin{aligned}\Sigma(P, P') - \Sigma^{\mathrm{EEL}}(P, P') &= \frac{dx}{f_{X,y}(x, y)} \sum_{\ell=1}^{s-1} \frac{A_\ell(x) - A_\ell(x')}{x - x'} \sum_{i=0}^{r-\ell-1} y^i y'^{r-\ell-i-1} \\ &= \frac{dx}{f_{X,y}(x, y)} \sum_{\ell=1}^{s-1} \sum_{k=0}^{\lfloor \ell s/r \rfloor -1} \lambda_{\ell,k} x^{\lfloor \ell s/r\rfloor - k-1} x'^k \sum_{i=0}^{r-\ell-1} y^{r-\ell-i} y'^i.\end{aligned}$$

For example, the case of (3,4)-curve (2.62), it is

$$\frac{dx}{f_{X,y}(x, y)} (\lambda_{1,1}(y + y') + \lambda_{2,1}(x + x') + \lambda_{2,1}).$$

For the cyclic curve plane curve, the difference disappears, i.e., $\Sigma(P, P') = \Sigma^{\mathrm{EEL}}(P, P')$. The difference affects only the lower order terms of x and $y_\bullet$ in the differentials of the second kind. Further, it turns out that the difference is not severe due to Proposition 2.151 and Corollary 2.152. Thus it does not have effects on Ω in (2.95) and (2.96) and the Jacobi inversion formulas in Sect. 3.3 because of Proposition 2.120 4.

2.4.3.4 The Differential $d\Sigma$ on X

In order to define Ω, we consider the derivative $d\Sigma$ in this subsubsection.

Proposition 2.122

$$d_Q\Sigma(P,Q) := dx_Q \frac{\partial}{\partial x_Q}\Sigma(P,Q) \tag{2.87}$$

has the following properties:

1. $d_Q\Sigma(P,Q)$ *as a function of* P *is holomorphic over* X *except at* Q *such that*
   a. *at* Q*, in terms of the local parameters* $t_P(Q)=0$ *and* $t_Q(P)=0$*, it behaves like*
   $$d_Q\Sigma(P,Q) = \frac{dt_P dt_Q}{t_P^2}(1+d_{>0}(t_P,t_Q)),$$
   b. *at* ∞*, in terms of the arithmetic parameters* t_P *and* t_Q *such that* $|t_Q|>|t_P|$*, it behaves like*
   $$d_Q\Sigma(P,Q) = r\frac{t_P^{r-1}}{t_Q^{r+1}}dt_P dt_Q(1+d_{>0}(t_P,t_Q)),$$
   and $d_Q\Sigma(P,Q)$ *is holomorphic at* ∞*.*
2. $d_Q\Sigma(P,Q)$ *as a function of* Q *is holomorphic over* X *except at* P *and* ∞ *such that*
   a. *at* P*, in terms of the local parameters* $t_Q(P)=0$ *and* $t_P(Q)=0$*, it behaves like*
   $$d_Q\Sigma(P,Q) = \frac{dt_P dt_Q}{t_Q^2}(1+d_{>0}(t_Q,t_P)),$$
   b. *at* ∞*, by the arithmetic parameter* $t_Q(\infty)=0$ *such that* $|t_P|>|t_Q|$*, it behaves like*
   $$d_Q\Sigma(P,Q) = \frac{1}{t_Q^{c_X}}dx_P dt_Q(1+d_{>0}(t_Q,t_P)).$$
3. *It holds the following relations:*
   a. $d_Q\Sigma(P,Q) - d_P\Sigma(Q,P) \in \mathbf{H}^0(X,\mathscr{A}_X(*\infty))^{2\otimes_{\mathbb{C}}}$,
   b. $\displaystyle\oint_{Q=\infty}\Sigma(P,Q)=0,$
   c. $\displaystyle\oint_{\infty}\left(\int^{Q} d_{Q'}\Sigma(P,Q)\right) = \oint_{P=\infty}\Sigma(P,Q) = 2\pi\sqrt{-1}.$
4. *For* $\Sigma'(P,Q)$ *defined in Proposition 2.120 4, we have*

$$d_Q\Sigma'(P,Q) - d_Q\Sigma(P,Q) \in \mathbf{H}^0(X,\mathscr{A}_X)\otimes_{\mathbb{C}} d\mathbf{H}^0(X,\mathscr{O}_X(*\infty))$$
$$= \mathbf{H}^0(X,\mathscr{A}_X)\otimes_{\mathbb{C}} dR_X.$$

Proof *1.a*, *2.a*, and finite part of *3.a* are directly obtained from Proposition 2.120. The properties of $d\Sigma$ on $X_H \times X_H$ determine the behavior at ∞ in Sect. 2.4.3.5 below.

We use the facts $dx_Q\dfrac{\partial}{\partial x_Q} = dt_Q\dfrac{\partial}{\partial t_Q}$, and $dt_Q\dfrac{\partial}{\partial t_Q}t_Q^\ell = \ell\, t_Q^{\ell-1}dt_Q$. Lemmas 2.124 and 2.121 show *1.b*, *2.b* and the behavior at ∞ of *3.a*. Lemma 2.121 3 shows *3.b*.

Lemma 2.128 shows *3.c*, and Proposition 2.120 4 shows *4*. ∎

We recall the expression of $h_X(x,y_\bullet)$ in Lemma 2.86.

Lemma 2.123 *$d_Q\Sigma(P,Q)/dx_Pdx_Q$ is equal to*

$$\frac{\widetilde{h}_X(x,y_\bullet,y'_\bullet)h_0(x')h_X(x',y'_\bullet) - (x-x')\sum_{k=2}^{m_X}\left(\frac{\partial}{\partial y'_{r_k}}\widetilde{h}_X(x,y_\bullet,y'_\bullet)\right)f_{X,x}^{(k)}(x',y')\left(\prod_{\ell=2,\neq k}^{r_\ell} f_{X,y}^{(\ell)}(x',y')\right)}{(x-x')^2h_X(x,y_\bullet)h_0(x')h_X(x',y'_\bullet)},$$

where the numerator is divided by $h_0(x')$ and further

$$(x_P - x_Q)^2 d_Q\Sigma(P,Q) \in \mathbf{H}^0(X,\mathscr{A}_X(*\infty))^{2\otimes_{\mathbb{C}}}.$$

Proof From Lemma 2.26, $\dfrac{d}{dx'}\left((\widetilde{h}_X(x,y_\bullet,y'_\bullet)\right) = \displaystyle\sum_{j=2}^{m_X}\frac{\partial}{\partial y'_{r_j}}\widetilde{h}_X(x,y_\bullet,y'_\bullet)$ $\dfrac{f_{X,x}^{(j)}(x',y')}{f_{X,y}^{(j)}(x',y')}$

$$=\prod_{j=2}^{m_X}\left(\frac{1}{f_{X,y}^{(j)}(x',y')}\right)\sum_{k=2}^{m_X}\left(\frac{\partial}{\partial y'_{r_k}}\widetilde{h}_X(x,y_\bullet,y'_\bullet)\right)f_{X,x}^{(k)}(x',y')\left(\prod_{\ell=2,\neq k}^{r_\ell} f_{X,y}^{(\ell)}(x',y')\right)$$

$$=\frac{\displaystyle\sum_{k=2}^{m_X}\left(\frac{\partial}{\partial y'_{r_k}}\widetilde{h}_X(x,y_\bullet,y'_\bullet)\right)f_{X,x}^{(k)}(x',y')\left(\prod_{\ell=2,\neq k}^{r_\ell} f_{X,y}^{(\ell)}(x',y')\right)}{h_0(x',y'_\bullet)h_X(x',y'_\bullet)},$$

where $h_0(x,y_\bullet) \in R_{\mathbb{P}}$. In general for $h \in \mathbf{H}^0(U,\mathscr{O}_U)$, its derivative by the local parameter t, i.e., dh/dt also belongs to $h \in \mathbf{H}^0(U,\mathscr{O}_U)$. Thus the last relation holds; it means that h_0 is reduced by the numerator. ∎

It means that there exist $F_{d\Sigma}(P,Q) \in R_X\otimes_{\mathbb{C}} R_X$ such that

$$d_Q\Sigma(P,Q) = \frac{F_{d\Sigma}(P,Q)dx_P\otimes dx_Q}{(x_P-x_Q)^2h_X(x_P,y_{\bullet P})h_X(x_Q,y_{\bullet Q})}. \tag{2.88}$$

2.4.3.5 The Differential $d\Sigma_H$ on X_H

Noting Lemma 2.19, we also consider $d\Sigma$ as a two-form in $\mathscr{Q}(R_H^Z \otimes_{\mathbb{C}} R_H^Z)$ and the monomial curve $X_H \times X_H$ instead of $R_{X,\infty}$.

We consider the differential of (2.85),

$$d\Sigma_H = dZ'_r \frac{\partial}{\partial Z'_r} \frac{dZ_r}{(Z_r - Z'_r)} \frac{\widetilde{h}_H(Z_r, Z_\bullet, Z'_\bullet)}{\widetilde{h}_H(Z_r, Z_\bullet, Z_\bullet)},$$

which is equal to

$$\frac{dZ_r}{(Z_r - Z'_r)^2} \frac{\widetilde{h}_H(Z_r, Z_\bullet, Z'_\bullet)}{\widetilde{h}_H(Z_r, Z_\bullet, Z_\bullet)} + \frac{dZ_r}{(Z_r - Z'_r)\widetilde{h}_H(Z_r, Z_\bullet, Z_\bullet)} \frac{\partial}{\partial Z'_r} \widetilde{h}_H(Z_r, Z_\bullet, Z'_\bullet).$$

We recall the correspondence between z of R_H and t of $R_{X,\infty}$ in Lemma 2.19. Using (2.86), we have

$$d_Q\Sigma_H(t_P, t_Q) = -dt_Q \frac{\partial}{\partial t_Q} \frac{t_Q^r}{t_Q^r - t_P^r} \sum_{i=0}^{r-1} \frac{t_P^{\mathfrak{e}_i}}{t_Q^{\mathfrak{e}_i}} \frac{dt_P}{t_P}.$$

Lemma 2.124 *1. We have the equality:*

$$d_Q\Sigma_H(t_P, t_Q) = \frac{dt_P dt_Q}{(t_Q^r - t_P^r)^2 t_P t_Q} \left[\sum_{i=0}^{r-1} \left(\mathfrak{e}_i \frac{t_P^{\mathfrak{e}_i}}{t_Q^{\mathfrak{e}_i - 2r}} - (\mathfrak{e}_i - r) \frac{t_P^{\mathfrak{e}_i + r}}{t_Q^{\mathfrak{e}_i - r}} \right) \right].$$

2. For the case $|t_Q| < |t_P|$, we have the expansion,

$$d_Q\Sigma_H(t_P, t_Q) = dt_P dt_Q \left[\sum_{i=0}^{r-1} \sum_{k=1}^{\infty} \left(-(\mathfrak{e}_i - kr) \frac{t_P^{\mathfrak{e}_i - rk - 1}}{t_Q^{\mathfrak{e}_i - rk + 1}} \right) \right]. \tag{2.89}$$

3. For the case $|t_P| < |t_Q|$, we have the expansion,

$$d_Q\Sigma_H(t_P, t_Q) = dt_P dt_Q \left[\sum_{i=0}^{r-1} \sum_{k=1}^{\infty} \left((\mathfrak{e}_i + kr) \frac{t_P^{\mathfrak{e}_i + rk - 1}}{t_Q^{\mathfrak{e}_i + rk + 1}} \right) \right]. \tag{2.90}$$

Proof *1*: Using the relation, $\displaystyle \frac{\partial}{\partial t_Q} \frac{t_Q^r}{t_Q^r - t_P^r} = \left(r - \frac{r t_Q^r}{t_Q^r - t_P^r} \right) \frac{t_Q^{r-1}}{t_Q^r - t_P^r} = -\frac{r t_Q^r t_P^r}{(t_Q^r - t_P^r)^2 t_Q}$, we have

$$d_Q\Sigma_H\big(t_P,t_Q\big)=\left[\left(\frac{rt_P^r}{t_Q^r-t_P^r}\right)\sum_{i=0}^{r-1}\frac{t_P^{\mathfrak{e}_i}}{t_Q^{\mathfrak{e}_i-r}}+\sum_{i=0}^{r-1}\mathfrak{e}_i\frac{t_P^{\mathfrak{e}_i}}{t_Q^{\mathfrak{e}_i-r}}\right]\frac{1}{t_Q^r-t_P^r}\frac{dt_Pdt_Q}{t_Pt_Q}.$$

The relation $rt_P^r+\mathfrak{e}_i(t_Q^r-t_P^r)=\mathfrak{e}_it_Q^r-(\mathfrak{e}_i-r)t_P^r$ enables us to obtain *1*.

2: Since $\dfrac{1}{(1-x)^2}=\dfrac{d}{dx}\dfrac{1}{(1-x)}=\sum_{k=1}kx^{k-1}$, for a polynomial $H(t_P,t_Q)\in\mathbb{C}[t_P,t_Q]$, we have

$$\frac{H(t_P,t_Q)}{(t_Q^r-t_P^r)^2}=\frac{H(t_P,t_Q)}{t_P^{2r}}\sum_{k=1}^{\infty}k\frac{t_Q^{r(k-1)}}{t_P^{r(k-1)}}=\frac{H(t_P,t_Q)}{t_P^{2r}}\left(\sum_{k=1}^{N}k\frac{t_Q^{r(k-1)}}{t_P^{r(k-1)}}+\sum_{k=N+1}^{\infty}k\frac{t_Q^{r(k-1)}}{t_P^{r(k-1)}}\right),$$

and

$$d_Q\Sigma_H\big(t_P,t_Q\big)=dt_Pdt_Q\left[\sum_{i=0}^{r-1}\sum_{k=1}^{\infty}\left(k\mathfrak{e}_i\frac{t_P^{\mathfrak{e}_i-r(k+1)-1}}{t_Q^{\mathfrak{e}_i-r(k+1)+1}}-k(\mathfrak{e}_i-r)\frac{t_P^{\mathfrak{e}_i-rk-1}}{t_Q^{\mathfrak{e}_i-rk+1}}\right)\right].$$

The relation $(k-1)\mathfrak{e}_i-k(\mathfrak{e}_i-r)=-(\mathfrak{e}_i-kr)$ shows *2*. Similarly, we prove *3*. ■

Lemma 2.125 *$d_Q\Sigma_H\big(t_P,t_Q\big)-d_P\Sigma_H\big(t_Q,t_P\big)$ is equal to*

$$dt_Pdt_Q\sum_{i=0}^{r-1}\frac{\mathfrak{e}_i(t_P^{2\mathfrak{e}_i-r}t_Q^r-t_Q^{2\mathfrak{e}_i-r}t_P^r)-(\mathfrak{e}_i-r)(t_P^{2\mathfrak{e}_i}-t_Q^{2\mathfrak{e}_i})}{(t_Q^r-t_P^r)t_P^{\mathfrak{e}_i-r+1}t_Q^{\mathfrak{e}_i-r+1}}.$$

When $t_P=t_Q+\varepsilon$, it vanishes for the limit $\varepsilon\to 0$.

Proof The direct computations show them. ■

For the case $|t_Q|<|t_P|$, due to the facts that $\varphi_\infty(h_X(P))=r\,t_P^{-d_h}$, and $\varphi_\infty(dx_P)=r\,t_P^{-r-1}dt_P$, (2.89) in Lemma 2.124 leads to the expression,

$$d_Q\Sigma_H\big(t_P,t_Q\big)=\frac{dt_P}{t_P^{r+1}}\frac{dt_Q}{t_Q^{r+1}}\left[\sum_{i=0}^{r-1}\sum_{k=1}^{\infty}\left(-(\mathfrak{e}_i-kr)t_P^{d_h}t_Q^{d_h}\frac{t_P^{\mathfrak{e}_i-d_h-r(k-1)}}{t_Q^{\mathfrak{e}_i+d_h-r(k+1)}}\right)\right].\qquad(2.91)$$

Lemma 2.126 *Recalling $\widehat{H}^{(g)}$ in Corollary 2.85, and letting $\widehat{H}_X^{(g)*}:=\{d_h+\mathfrak{e}_i-r(k+2)\mid\widehat{\mathfrak{e}}_i+rk\in\widehat{H}^{(g)}\}$, we have*

$$\widehat{H}_X^{(g)}\subset\widehat{H}_X,\quad \widehat{H}_X^{(g)*}\subset\widehat{H}_X,\quad \#\widehat{H}_X^{(g)}=g,\quad \#\widehat{H}_X^{(g)*}=g.$$

Proof Corollary 2.85 shows the relations of $\widehat{H}^{(g)}$. On the other hand, the condition in $\widehat{H}^{(g)*}$ means that $d_h+\mathfrak{e}_i-r(k+2)\geq(d_h-r+1)+(\mathfrak{e}_i-r)>\widehat{c}_X+\mathfrak{k}_X$ because of $\widehat{c}_X+\mathfrak{k}_X-2=d_h-r-1$ from Lemma 2.115. Further we note Lemma 2.109 7. Thus $\widehat{H}^{(g)*}\subset\widehat{H}_X$, and $\#\widehat{H}_X^{(g)*}=g$. ■

Since the order of the singularity of $\Sigma_H(t_P, t_Q)$ at $t_Q = \infty$ is $c_X - 1 = \mathfrak{e}_{r-1} - r$, we introduce the two-form,

$$\Omega_H^{(0)}(t_P, t_Q) := dt_P dt_Q \left[\sum_{i=0}^{r-1} \sum_{\substack{k=1 \\ \widehat{\mathfrak{e}}_i + rk \in \widehat{H}^{(g)}}} \left((\mathfrak{e}_i - kr) \frac{t_P^{\mathfrak{e}_i - rk - 1}}{t_Q^{\mathfrak{e}_i - rk + 1}} \right) \right].$$

Lemma 2.127 *$\Omega_H^{(1)}(t_P, t_Q) := d_Q \Sigma_H(t_P, t_Q) + \Omega_H^{(0)}(t_P, t_Q)$ has no singularity at $t_Q = \infty$.*

Lemma 2.128 *Each term in $\Omega_H^{(0)}(t_P, t_Q)$ has the property*

$$\oint_{t_Q=0} \left[\int_0^{t_Q} (\mathfrak{e}_i - kr) \frac{t_P^{\mathfrak{e}_i - kr - 1}}{t_Q^{\mathfrak{e}_i - kr + 1}} dt_P \right] dt_Q = 2\pi\sqrt{-1}.$$

Lemma 2.129 *For $i \in \mathbb{Z}_r$ and $k \in \mathbb{N}$ such that $\widehat{H}^{(g)} \ni \widehat{\mathfrak{e}}_i + rk$, there are $\widehat{\phi}_j$ $(j < g)$ and $\widehat{\phi}_{j'}$ $(j' \geq g)$ such that*

$$\varphi_\infty \left(\frac{\widehat{\phi}_j dx}{h_X} \right) = -t^{\mathfrak{e}_i - kr - 1} dt (1 + d_{>0}(t)), \quad \varphi_\infty \left(\frac{\widehat{\phi}_{j'} dx}{h_X} \right) = -\frac{dt}{t^{\mathfrak{e}_i - kr + 1}} (1 + d_{>0}(t)).$$

Proof Lemma 2.126 shows the existence. ∎

2.4.3.6 W-Normalized Differentials of the Second Kind

We introduce the W-normalized differentials of the second kind using this $d_Q \Sigma(P, Q)$.

Definition 2.130 We consider a sufficiently small simple closed contour C_ε at ∞. Let D_ε be the inner side of $C_\varepsilon = \partial D_\varepsilon$ including ∞ and ε' be a point in D_ε such that $\varepsilon' \neq \infty$. For differentials ν and ν' in $\mathbf{H}^0(X, \mathscr{A}_X(*\infty))$, we define a pairing:

$$\langle \nu, \nu' \rangle := \lim_{\varepsilon' \to \infty} \frac{1}{2\pi\sqrt{-1}} \oint_{C_\varepsilon} \left(\int_{\varepsilon'}^{P} \nu(Q) \right) \nu'(P).$$

Lemma 2.131 *$\langle \nu, \nu' \rangle = -\langle \nu', \nu \rangle$, and $\langle \nu, \nu' \rangle$ does not depend on ε.*

We note that since we consider only the elements in $\mathbf{H}^0(X, \mathscr{A}_X(*\infty))$ in $\langle\ ,\ \rangle$, the right-hand side in Definition 2.130 is written as the left-hand side in the second relation in Lemma 2.93.

Definition 2.132 We define the pre-normalized differentials $\widetilde{\nu}_i^{\mathrm{II}} \in \mathbf{H}^0(X, \mathscr{A}_X(*\infty))$ of the second kind ($i = 1, 2, \ldots, g$), which satisfies the relations (if they exist)

$$\langle \nu_i^{\mathrm{I}}, \widetilde{\nu}_j^{\mathrm{II}} \rangle = \delta_{ij}, \quad \langle \widetilde{\nu}_i^{\mathrm{II}}, \widetilde{\nu}_j^{\mathrm{II}} \rangle = 0. \tag{2.92}$$

It is obvious that from Lemma 2.109 2 and Lemma 2.129, $\widetilde{\nu}_j^{\mathrm{II}}$ with j' in Lemma 2.129 is expressed as

$$\widetilde{\nu}_j^{\mathrm{II}} = \frac{\sum_{i=0}^{j'} a_{j,i}\widehat{\phi}_i dx}{h_X(x, y_\bullet)}. \tag{2.93}$$

Noting Proposition 2.122 3.b, we have the following relations:

Corollary 2.133 $\nu_g^{\mathrm{II}} = \dfrac{\widehat{\phi}_g}{h_X(x, y_\bullet)} dx$, *wt* $\nu_g^{\mathrm{II}} = -2$, *wt* $\nu_1^{\mathrm{II}} = -c_X$, *and*

$$\nu_i^{\mathrm{II}} = \nu_{\infty, N_{g-i+1}^{\mathfrak{c}}+1, t}^{\mathrm{II}\circ} + d_{\ge 0}(t)dt$$

for the arithmetic parameter t.

Proof $-\mathrm{wt}(\widehat{\phi}_g) + \mathrm{wt}(\widehat{\phi}_{g-1}) = 2$ and thus

$$\frac{1}{2\pi\sqrt{-1}} \oint \left(\int^{(x,y_\bullet)} \frac{\widehat{\phi}_{g-1}}{h_X(x', y'_\bullet)} dx' \right) \frac{\widehat{\phi}_g}{h_X(x, y_\bullet)} dx = 1.$$

On the other hand, $-\mathrm{wt}(\widehat{\phi}_0) = d_h - r + 1 - c_X$ in Lemma 2.115 means

$$\mathrm{wt}\ \nu_1^{\mathrm{I}} = \mathrm{wt}\left(\frac{\widehat{\phi}_0 dx}{h_X(x, y_\bullet)} \right) = c_X - 2.$$

Hence $\mathrm{wt}\ \widetilde{\nu}_1^{\mathrm{II}} = -c_X$ and we obtain

$$\widetilde{\nu}_1^{\mathrm{II}} = \frac{\widehat{\phi}_0^* dx}{h_X}, \quad \widehat{\phi}_0^* = \frac{1}{t^{d_h + c_X - r + 1}}(1 + d_{>0}(t)), \quad dx = -r\frac{dt}{t^{r+1}}(1 + d_{>0}(t)).$$

Similarly, we have its relation to the normalized differential of the second kind. ■

Theorem 2.134 *There exist the pre-normalized differentials* ν_j^{II} $(j = 1, 2, \cdots, g)$ *of the second kind such that they have a simple pole at* ∞ *and satisfy the relation,*

$$d_Q\Sigma\big(P, Q\big) - d_P\Sigma\big(Q, P\big) = \sum_{i=1}^{g} \Big(\nu_i^{\mathrm{I}}(Q) \otimes \nu_i^{\mathrm{II}}(P) - \nu_i^{\mathrm{I}}(P) \otimes \nu_i^{\mathrm{II}}(Q) \Big), \tag{2.94}$$

where the set of differentials $\{\nu_1^{\mathrm{II}}, \nu_2^{\mathrm{II}}, \nu_3^{\mathrm{II}}, \cdots, \nu_g^{\mathrm{II}}\}$ *is determined modulo the linear space spanned by* $\langle \nu_j^{\mathrm{I}} \rangle_{j=1,\ldots,g}$ *and* dR_X.

We call these ν_i^{II}*'s* W-normalized differentials of the second kind.[12]

Proof From Proposition 2.137, Lemmas 2.127 and 2.128 show the fact. ■

Theorem 2.135

$$(\mathbf{H}^0(X, \mathscr{A}_X(*\infty))/d\mathbf{H}^0(X, \mathscr{O}_X(*\infty)) = \bigoplus_{i=1}^{g}(\mathbb{C}\nu_i^{\mathrm{I}} \oplus \mathbb{C}\nu_i^{\mathrm{II}})$$

and thus $\dim_{\mathbb{C}}(\mathbf{H}^0(X, \mathscr{A}_X(*\infty))/d\mathbf{H}^0(X, \mathscr{O}_X(*\infty)) = 2g$.

Proof It is obvious that $d\mathbf{H}^0(X, \mathscr{O}_X(*\infty)) \subset \mathbf{H}^0(X, \mathscr{A}_X(*\infty))$; $(j : dR_X \hookrightarrow \mathbf{H}^0(X, \mathscr{A}_X(*\infty)))$. We recall Lemma 2.109 2 and 7. $\mathbf{H}^0(X, \mathscr{A}_X(*\infty)) \setminus \mathbf{H}^0(X, \mathscr{A}_X)$ is the set of differentials of the second kind. In other words, there is no differential with the first-order singularity at ∞. The embedding j is realized by $j : d\phi_i \mapsto \dfrac{h_X d\phi_i}{h_X}$. Let $H_{dR_X} := \{\mathrm{wt}(j(d\phi_i)) \mid i \in \mathbb{N}_0\}$. Then $H_{dR_X} \subset -\mathbb{N}$ and $\#((-\mathbb{N}) \setminus (H_{dR_X} + 1))) = g$, whereas $\#H_X^{\mathfrak{c}} = g$. Thus $\#((\overline{H}_X^{\mathfrak{c}} - 1) \setminus H_{dR_X}) = 2g$. Due to the Riemann–Roch theorem, for every ℓ in $H_X^{\mathfrak{c}}$, there is an element k in $(-\mathbb{N}) \setminus (H_{dR_X} + 1)$ such that $\ell + k = 0$. It shows the relations. ■

Remark 2.136 $(\mathbf{H}^0(X, \mathscr{A}_X(*\infty))/d\mathbf{H}^0(X, \mathscr{O}_X(*\infty))$ is referred to as the first algebraic de Rham cohomology, $\mathbf{H}^1_{\mathrm{DR}}(X \setminus \{\infty\})$. Here $\mathbf{H}^k_{\mathrm{DR}}(U)$ denotes the k-th de Rham cohomology over U. As Deligne mentioned the relation between the algebraic de Rham cohomology and the classical theory for the elliptic curves [Del1989], it can be straightforwardly extended to these W-curves. Though we don't mention it precisely here, we comment on it.

The algebraic de Rham cohomology is an algebraic version of de Rham cohomology for smooth manifolds, which can be defined for algebraic variety with the Zariski topology. It is a dual space of the homology $\mathbf{H}_1(X, \mathbb{C}) := \mathbf{H}_1(X, \mathbb{Z}) \otimes \mathbb{C}$ with intersection relations of the cycles as mentioned in Theorem 2.148 and Corollary 2.149.

Theorem 2.135 states that $\mathbf{H}^1_{\mathrm{DR}}(X \setminus \{\infty\})$ is generated by the differentials of the first and the second kind. There is an isomorphism $j : \mathbf{H}^1_{\mathrm{DR}}(X) \to \mathbf{H}^1_{\mathrm{DR}}(X \setminus \{\infty\})$ as a $\mathbb{C}$ vector space. For a Riemann surface X as a two-dimensional real manifold, the first de Rham cohomology $\mathbf{H}^1_{\mathrm{DR}}(X)$ is generated by $2g$ smooth one-forms, which are given by ν_i^{I} and $\overline{\nu_i^{\mathrm{I}}}$, $(i = 1, \ldots, g)$ due to Hodge [Gri1989, Theorem 3.3]. It means that, in regard to the de Rham cohomology, the behavior of X is determined by $X \setminus \{\infty\}$. This fact is one of the reasons why the structure of the

[12] In the articles [KodMP2013, KomMP2013, Mat2002b, Mat2002c, Mat2008a, MatP2008, MatP2014, On2005, On2009, On2013], the definition of W-normalized differentials ν^{II} of the second kind differs from this definition by its sign. We also recall (2.81) and that there is another convention on ν^{I} as mentioned in footnote 10 in this chapter. The differences are not significant but affect the Legendre relation in Theorem 2.148 and the sign of the quadratic form in the definition of sigma function in Definition 3.28. However, this definition is consistent with the convention in Weierstrass' elliptic function theory [Wei1875b, p. 73], Baker [Ba1897, p. 197][Ba1907, p. 14] and Buchstaber, Enolskii and Leykin [BucEL1997b, BucEL2012]; $(\omega, \omega', \eta, \eta')$ reads $(\omega', \omega'', \eta', \eta'')$.

Riemann surface X or a projective curve X is determined by its affine part $X \setminus \{\infty\}$ or by the Weierstrass canonical form, as in Weierstrass' elliptic function theory. As it means that the projectivization is less critical, we have focused only on its affine part in the theory of W-curves. It also shows why we consider only the holomorphic part ν_i^{I} with ν_i^{II} on $(X \setminus \{\infty\})$ rather than with the anti-holomorphic part $\overline{\nu_i^{\mathrm{I}}}$ on X, $(i = 1, \ldots, g)$; it contrasts with the theory of Hodge on the compact Riemann surfaces as in [Gri1989][FarK1991, III.2.7]. (See also Remarks 2.103 and 3.184.) We also note that this view essentially causes the Hurwitz problem. This problem is also related to a problem considering the extension of regular functions on S^1 to a compact Riemann surface.

Proposition 2.137 *If by using $\Sigma'(P, Q)$ defined in Proposition 2.120 4, we define $\{\nu_i^{\mathrm{II}'}\}_{i=1,\ldots,g}$, these ν_i^{II} and $\nu_i^{\mathrm{II}'}$ give the same period matrices in Definition 2.145.*

Proof From the Proposition 2.122 4 and Theorem 2.134, it is obvious. ■

Remark 2.138 Since for a loop $\gamma \in \mathbf{H}_1(X, \mathbb{Z})$ and $f \in R_X$, we have $\int_\gamma df = 0$, we note that any complete Abelian integral for $\gamma \in \mathbf{H}_1(X, \mathbb{Z})$ can be expressed by the linear combination of the complete Abelian integrals of the first kind and the second kind [HurC1893].

Lemma 2.139 *$d_{P_2}\Sigma(P_1, P_2) + \sum_{i=1}^{g} \nu_i^{\mathrm{I}}(P_1) \otimes \nu_i^{\mathrm{II}}(P_2)$ is holomorphic over $X \setminus \{P_2\}$ as a function of P_1 and is holomorphic over $X \setminus \{P_1\}$ as a function of P_2.*

Proof From Proposition 2.122, $d_{P_2}\Sigma(P_1, P_2)$ is holomorphic over $X \setminus \{P_2\}$ as a function of P_1 whereas $d_{P_2}\Sigma(P_1, P_2)$ is holomorphic over $X \setminus (\{P_2\} \cup \{\infty\})$ as a function of P_2. The order of the singularity at $Q = \infty$ is $2g$ and thus, which can be canceled by $\nu_i^{\mathrm{II}}(P_2)$. Since the numerator of $d_{P_2}\Sigma(P_1, P_2)$ in (2.88) consists of the elements in $R_X \otimes_{\mathbb{C}} R_X dx_1 \otimes dx_2$, from Proposition 2.122, there is no term whose weight is -1 in $d_{P_2}\Sigma(P_1, P_2)$ as a function in Q. Noting the homogeneous property of $\widetilde{h}_X$ from Lemma 2.45, we have the result. ■

Theorem 2.140 *1. For $P_1, P_2 \in X \setminus \{\infty\}$ and $P \in X$, the one-form,*

$$\nu_{P_1,P_2}^{\mathrm{III}}(P) := \Sigma(P, P_1) - \Sigma(P, P_2),$$

is a differential of the third kind whose only (first-order) poles are $P = P_1$ and $P = P_2$ with residues $+1$ and -1 respectively.

2. For $P, Q \in X$, the symmetric two-form,

$$\Omega(P, Q) := d_Q\Sigma(P, Q) + \sum_{i=1}^{g} \nu_i^{\mathrm{I}}(P) \otimes \nu_i^{\mathrm{II}}(Q), \tag{2.95}$$

is the fundamental differential of the second kind, which has the following properties:

a. $\Omega(P, Q) = \Omega(Q, P)$,
b. $\Omega(P, Q)$ *is holomorphic except* Q *as a function of* P *and behaves like*

$$\Omega(P, Q) = \frac{dt_P dt_Q}{(t_P - t_Q)^2}(1 + d_{>0}(t_P, t_Q)), \qquad \textit{and}$$

c. for $P_a, Q_a \in X \setminus \{\infty\}$,

$$\int_{P_2}^{P_1} \nu^{\mathrm{III}}_{Q_1, Q_2} = \int_{P_2}^{P_1} \int_{Q_2}^{Q_1} \Omega(P, Q).$$

Proof *1* is directly obtained by Proposition 2.120. *2* is obvious from Theorem 2.134 and Proposition 2.122. ■

Further, we may consider the group action on $\Omega(P, Q)$; there exists a group action $\widehat{\zeta} \in G_{X, \varpi_r(P)}$ such that $\Omega(\widehat{\zeta} P, \widehat{\zeta} Q) = \Omega(P, Q)$ if $\varpi_r(P) = \varpi_r(Q)$.

We note that the W-normalized differentials of the first kind and the second kind are defined in (2.82) (2.88), (2.93), (2.94), and Definitions 2.108 and 2.132.

Definition 2.141 1. $\nu^{\mathrm{III}}_{P,Q}$ is called the *W-normalized differential of the third kind.*
2. $\Omega(P, Q)$ is the *W-normalized fundamental differential of the second kind* or the *fundamental W-normalized differential of the second kind*, and is expressed by

$$\Omega(P, Q) = \frac{F_\Omega(P, Q) dx_P \otimes dx_Q}{(x_P - x_Q)^2 h_X(x_P, y_{\bullet P}) h_X(x_Q, y_{\bullet Q})}, \tag{2.96}$$

where F_Ω is called *Klein's fundamental form* or *the Klein form* in $R_X \otimes_{\mathbb{C}} R_X$.

In [MatP2008], the author of this book with Previato investigated Klein's fundamental form for the cyclic plane W-curves, whereas Suzuki evaluated it for plane W-curves [Suz2017].

Lemma 2.142 *We have*

$$\lim_{P_1 \to \infty} \frac{F_\Omega(P_1, P_2)}{\widehat{\phi}_{g-1}(P_1)(x_1 - x_2)^2} = \widehat{\phi}_g(P_2). \tag{2.97}$$

Proof The expression (2.95) and Corollary 2.133 give the result. ■

In Chap. 3 we will use:

$$\begin{aligned} \Pi^{P_1, P_2}_{Q_1, Q_2} &:= \int_{P_2}^{P_1} \int_{Q_2}^{Q_1} \Omega(P, Q) \\ &= \int_{P_2}^{P_1} (\Sigma(P, Q_1) - \Sigma(P, Q_2)) + \sum_{i=1}^{g} \int_{P_2}^{P_1} \nu^{\mathrm{I}}_i(P) \int_{Q_2}^{Q_1} \nu^{\mathrm{II}}_i(P). \end{aligned} \tag{2.98}$$

Recalling (2.80), we have

Lemma 2.143 $\Pi_{Q_1,Q_2}^{P_1,P_2} = -\Pi_{Q_1,Q_2}^{P_2,P_1}, \quad \Pi_{Q_1,Q_2}^{P_1,P_2} = \Pi_{P_1,P_2}^{Q_1,Q_2}.$

Remark 2.144 We note that $\Pi_{Q_1,Q_2}^{P_1,P_2}$ is the W-normalized version of $\Pi^{\circ}{}_{Q_1,Q_2}^{P_1,P_2}$ in (2.80), which is realized as an integral consisting of the commutative ring R_X. As mentioned in Remark 2.103, $\Pi_{Q_1,Q_2}^{P_1,P_2}$ and $\Omega(P_1, P_2)$ are closely related to the Bergman kernel, the Schiffer kernel, and the Green functions.

It is known that the Bergman kernel is explicitly expressed by the elliptic $\wp$ functions and σ function for a particular, annulus case $r < |z| < 1$ $(r > 0)$. [Sui1972, SebF2007] for $(\omega' = -\log r, \omega'' = \pi\sqrt{-1})$. As Suita cited [CouH1953], in [CouH1953, Chap. V §15 6, 7], Green's functions for a rectangle and a circular ring are given; the former is the real part of (1.42). Ohsawa pointed out that according to Suita, this example had influenced his conjecture, known as the Suita conjecture related to the Bergman kernel, and the importance of this example [Ohs2018a, Ohs2018b].

We have the algebraic expressions of $\Omega(P_1, P_2)$ and Theorem 3.37 as a generalization of (1.42); they may have some implications for these studies, though the topic is far beyond the scope of this book and we will not cover it.

Definition 2.145 We define the complete Abelian integrals of the second kind,

$$\eta'_{ij} := -\frac{1}{2}\int_{\alpha_j} \nu_i^{\mathrm{II}}, \quad \eta''_{ij} := -\frac{1}{2}\int_{\beta_j} \nu_i^{\mathrm{II}}.$$

The following lemma corresponds to Corollary 2.6 (ii) in [Fay1973].

Lemma 2.146

$$\Pi_{Q_1,Q_2}^{P_1,P_2} = \int_{P_2}^{P_1} \nu_{Q_1,Q_2}^{\mathrm{III}\circ} - \sum_{i=1}^{g}\sum_{j=1}^{g} \gamma_{i,j} \int_{P_2}^{P_1} \nu_i^{\mathrm{I}} \int_{Q_2}^{Q_1} \nu_j^{\mathrm{I}},$$

where $\gamma = \eta'\omega'^{-1}$ *and* ${}^t\gamma = \gamma$.

Proof Since $\Pi_{Q_1,Q_2}^{P_1,P_2} - \int_{P_2}^{P_1} \nu_{Q_1,Q_2}^{\mathrm{III}\circ}$ is holomorphic, it is expressed by the linear combination of the holomorphic one-forms and thus we find a certain γ. Thus we show $\gamma = \eta'\omega'^{-1}$. We differentiate both sides using (2.98) by the local parameter t of P_1,

$$\Sigma(P, Q_1) - \Sigma(P, Q_2) + \sum_{i=1}^{g} \nu_i^{\mathrm{I}}(P) \int_{Q_2}^{Q_1} \nu_i^{\mathrm{II}} = \int_{Q_2}^{Q_1} \nu_{2,P}^{\mathrm{II}} - \sum_{i=1}^{g} \gamma_{i,j} \nu_i^{\mathrm{I}}(P) \int_{Q_2}^{Q_1} \nu_j^{\mathrm{I}}$$

using Lemma 2.100. By integrating the relation along α_j $(j = 1, \ldots, g)$,

$$-2\eta'\nu^{\mathrm{I}}(P) = -2\gamma\omega'\nu^{\mathrm{I}}(P)$$

and we obtain the relation on γ. The relation $\Pi^{P_1,P_2}_{Q_1,Q_2} = \Pi^{Q_1,Q_2}_{P_1,P_2}$ in Lemma 2.143 yields ${}^t\gamma = \gamma$. ■

The following is obvious:

Corollary 2.147

$$\Omega(P_1, P_2) = \Omega^\circ(P_1, P_2) + \sum_{i=1}^{g}\sum_{j=1}^{g} \gamma_{i,j} \nu_i^{\mathrm{I}}(P_1)\nu_j^{\mathrm{I}}(P_2),$$

where $\gamma = \eta'\omega'^{-1}$ *and* ${}^t\gamma = \gamma$.

We have the generalized Legendre relation given as follows[13]:

Theorem 2.148

$$\omega''\,{}^t\eta' - \omega'\,{}^t\eta'' = \frac{\pi}{2}\sqrt{-1}1_g,$$

where 1_g *is the unit* $g \times g$ *matrix.*

The following matrix satisfies the generalized Legendre relation:

$$M := \begin{bmatrix} 2\omega' & 2\omega'' \\ 2\eta' & 2\eta'' \end{bmatrix}, \quad M \begin{bmatrix} & 1_g \\ -1_g & \end{bmatrix} {}^tM = \frac{2\pi}{\sqrt{-1}} \begin{bmatrix} & 1_g \\ -1_g & \end{bmatrix}. \tag{2.99}$$

Proof The definition (2.92) and Lemma 2.93 show the generalized Legendre relation; we put ν_i^{I} and ν_j^{II} into ν and ν' in Lemma 2.93. Proposition 2.95 shows that $\omega'^{-1}\omega''$ is symmetric, i.e., ${}^t(\omega'^{-1}\omega'') = \omega'^{-1}\omega''$, and thus ${}^t\omega''\,{}^t\omega'^{-1} = \omega'^{-1}\omega''$, or $\omega'\,{}^t\omega'' = \omega''\,{}^t\omega'$.

With the symmetry of ${}^t\gamma = \gamma$ in Lemma 2.146, we have

$$\eta''\,{}^t\eta' = \eta'\,{}^t\eta'', \tag{2.100}$$

because $\eta''\,{}^t\eta' = \eta''\,{}^t\omega'\,{}^t\omega'^{-1}\,{}^t\eta' = (2\pi\sqrt{-1}1_g + \eta'\,{}^t\omega'')\omega'^{-1}\,{}^t\eta' = 2\pi\sqrt{-1}\omega'^{-1}\,{}^t\eta'$ $+\eta'({}^t\eta'' - 2\pi\sqrt{-1}\omega'^{-1})$ shows the relation. ■

From Definitions 2.130 and 2.132, we have the following corollary, which is the dual space of the homological relations (2.58) recalling Remark 2.136:

Corollary 2.149 $\langle \nu_i^{\mathrm{I}}, \nu_j^{\mathrm{I}}\rangle = \langle \nu_i^{\mathrm{II}}, \nu_j^{\mathrm{II}}\rangle = 0$, $\langle \nu_i^{\mathrm{I}}, \nu_j^{\mathrm{II}}\rangle = \delta_{ij}$ *for* $i, j = 1, 2, \ldots, g$.

Lemma 2.88 of the Galois action on the basis of the Homology $\mathbf{H}_1(Z, \mathbb{Z})$ shows the actions of these period matrices $2(\omega', \omega'')$:

Lemma 2.150 *Assume that* $\varpi_r : X \to \mathbb{P}$ *is a Galois covering. For the Galois action* $\widehat{\zeta} \in G_X$, *i.e.,* $\widehat{\zeta} : X \to X$, *its associated element* $\rho_{\widehat{\zeta}}$ *of* $\mathrm{Sp}(2g, \mathbb{Z})$ *acts on* (η', η'') *by*

---

[13] The sign of this relation depends on the sign of the W-normalized Abelian differentials of the second kind because there are variant definitions (cf. footnote 12) in this chapter.

$$\widehat{\zeta}(\eta', \eta'') = (\eta', \eta'')\, {}^t\rho_{\widehat{\zeta}},$$

and the generalized Legendre relation (2.99) is invariant for the action with Lemma 2.119.

Proof Due to the definition of $\mathrm{Sp}(2g, \mathbb{Z})$, we have ${}^t\rho_{\widehat{\zeta}} \begin{bmatrix} & 1_g \\ -1_g & \end{bmatrix} \rho_{\widehat{\zeta}} = \begin{bmatrix} & 1_g \\ -1_g & \end{bmatrix}$ and thus, (2.99) is invariant.

2.4.3.7 Example Ω

1. Cyclic $(2, 2g + 1)$-curves (2.66) [Bol1895, Bol1899a]:

$$\Omega(P, Q) = \frac{f_\Omega(x_P, x_Q) + 2y_P y_Q}{(x_P - x_Q)^2} \frac{dx_P}{2y_P} \frac{dx_Q}{2y_Q}, \tag{2.101}$$

where

$$f_\Omega(x, x') = \sum_{j=0}^{g} x^j x'^j (2\lambda_{2j} + \lambda_{2j+1}(x + x')), \tag{2.102}$$

which is called the Abelian form [Ba1897, p. 195] [Ba1903, p. 146].

Proposition 2.151 *For the plane W-curve ((r, s)-curve, $m_X = 2$), there exist $f_i(x, y) \in R_X$ ($i = 1, 2, \ldots, g$) such that*

$$dx' \frac{\partial}{\partial x'} \left(\Sigma(P, P') - \Sigma^{\mathrm{EEL}}(P, P')\right) = \sum_{i=1}^{g} \nu_i^{\mathrm{I}}(x, y) df_i(x', y').$$

Proof The difference from EEL-convention is given by

$$dx' \frac{\partial}{\partial x'} \left(\Sigma(P, P') - \Sigma^{\mathrm{EEL}}(P, P')\right) = \frac{dx}{f_{X,y}(x, y)} \sum_{i=1}^{g} h_i(x, y) dx' \frac{\partial}{\partial x'} f_i(x', y').$$

In the formula, the factor with respect to $P = (x, y)$ in each monomial in the bracket with $\dfrac{dx}{f_{X,y}(x, y)}$ turns out to be a holomorphic form. On the other hand, the factor with respect to $P' = (x', y')$ belongs to the derivative of R_X, since it is explicitly given by

$$dx'\frac{\partial}{\partial x'}\left(\Sigma(P,P') - \Sigma^{\mathrm{EEL}}(P,P')\right) = \frac{dx}{f_{X,y}(x,y)}\frac{dx'}{f_{X,y}(x',y')}$$
$$\sum_{\ell=1}^{s-1}\Big(\sum_{k=0}^{\lfloor \ell s/r\rfloor -1} k\,\lambda_{\ell,k}x^{\lfloor \ell s/r\rfloor -k-1}x'^{k-1}f_{X,y}(x',y')\sum_{i=0}^{r-\ell-1} y^{r-\ell-i}y'^{i}$$
$$-\sum_{k=0}^{\lfloor \ell s/r\rfloor -1}\lambda_{\ell,k}x^{\lfloor \ell s/r\rfloor -k-1}x'^{k}\sum_{i=0}^{r-\ell-1} i y^{r-\ell-i}y'^{i-1}f_{X,x}(x',y')\Big). \qquad \blacksquare$$

Corollary 2.152 *The difference from the EEL-convention contributes only the difference of the differentials of the second kind, which does not contribute to the integrals of the second kind, i.e., for every loop γ in X like*

$$\oint_\gamma df_i = 0.$$

Both are quasi-isomorphic as the de-Rham cohomology $\mathbf{H}^1(X,\mathbb{C})$.

2.4.3.8 Examples of ν^{II}

If we employ a different definition of ν^{II} in the following, the minus sign "−" appears on the right-hand side, as, e.g., in [On2005, EilEMOP2007].

1. Cyclic hyperelliptic $(2, 2g+1)$-curves (2.66):

$$\nu_j^{\mathrm{II}} = \frac{1}{2y}\sum_{k=j}^{2g-j}(k+1-j)\lambda_{k+1+j}x^k dx, \quad (j=1,\dots,g),$$

which is given as follows: Let $\nu^{\mathrm{II}} := {}^t(\nu_1^{\mathrm{II}}, \nu_2^{\mathrm{II}}, \cdots, \nu_g^{\mathrm{II}})$ and $\widetilde{\nu}_i^{\mathrm{II}} := \dfrac{x^{g+i-1}dx}{2y}$ $(i=1,\dots,g)$, and then we have

$$\nu^{\mathrm{II}} = \Lambda\begin{pmatrix}\nu^{\mathrm{I}}\\ \widetilde{\nu}^{\mathrm{II}}\end{pmatrix},$$

where Λ is a $g\times 2g$-matrix given by

$$\begin{bmatrix}
0 & \lambda_3 & 2\lambda_4 & 3\lambda_5 & \cdots & (g-1)\lambda_{g+1} & g\lambda_{g+2} & (g+1)\lambda_{g+3} & \cdots & (2g-2)\lambda_{2g} & (2g-1)\lambda_{2g+1}\\
 & 0 & \lambda_5 & 2\lambda_6 & \cdots & (g-2)\lambda_{g+2} & (g-1)\lambda_{g+3} & g\lambda_{g+4} & \cdots & (2g-3)\lambda_{2g+1} & 0\\
 & & 0 & \lambda_7 & \cdots & (g-3)\lambda_{g+3} & (g-2)\lambda_{g+4} & (g-1)\lambda_{g+5} & \cdots & 0 & \\
 & & & & \ddots & \vdots & \vdots & \vdots & \cdots & & \\
 & & 0 & & & \lambda_{2g-1} & 2\lambda_{2g} & 3\lambda_{2g+1} & & & 0\\
 & & & & & 0 & \lambda_{2g+1} & 0 & & &
\end{bmatrix}.$$

2. (3, 4)-curves (2.62) [EilEMOP2007]:

$$\nu_3^{\mathrm{II}} = x^2 \frac{dx}{f_{X,y}},$$

$$\nu_2^{\mathrm{II}} = \left(2xy - \lambda_{1,1}x^2\right) \frac{dx}{f_{X,y}},$$

$$\begin{aligned}\nu_1^{\mathrm{II}} =& \Big((5x^2 + (\lambda_{1,1}\lambda_{2,2} - 3\lambda_{3,3})x + \lambda_{2,2}\lambda_{1,0} + \lambda_{3,1})y + \lambda_{2,2}y^2 + 3\lambda_{1,1}x^3 \\ &- (\lambda_{2,2}^2 + 2\lambda_{3,3}\lambda_{1,1} - 2\lambda_{1,0})x^2 - (\lambda_{2,1}\lambda_{2,2} + \lambda_{3,2}\lambda_{1,1} + \lambda_{3,3}\lambda_{1,0})x \\ &- \frac{1}{4}\lambda_{1,1} f_{X,x}(x,y) - (\frac{4}{3}\lambda_{2,2} - \frac{1}{4}\lambda_{1,1}^2) f_{X,y}(x,y)\Big) \frac{dx}{f_{X,y}}.\end{aligned}$$

In the EEL-convention, the last line should read $\frac{3}{4}\lambda_{1,1} f_{X,x}(x,y) - (\frac{1}{3}\lambda_{2,2} - \frac{1}{4}\lambda_{1,1}^2) \times f_{X,y}(x,y)$ due to the difference, $\lambda_{1,1}y + \lambda_{2,2}x$.

3. Cyclic $(3, \widehat{r}, \widehat{s})$-curves (2.68): We define $\widehat{R}_{\widehat{r}} := \{x^i y_{\widehat{r}} \mid i = 0, \ldots, g_{\widehat{r}} - 1\}$ and $\widehat{R}_{\widehat{s}} := \{x^i y_{\widehat{s}} \mid i = 0, \ldots, g_{\widehat{s}} - 1\}$ by letting $g_{\widehat{r}} := \left\lfloor \frac{\widehat{s} - 1}{3} \right\rfloor$ and $g_{\widehat{s}} := \left\lfloor \frac{\widehat{r} - 1}{3} \right\rfloor$. For $f \in \mathbb{C}[x_P, x_Q, y_{\widehat{s},P} y_{\widehat{r},P}, y_{\widehat{s},Q} y_{\widehat{r},Q}]$, let $\widehat{\pi}_{Q,i}(f)$ be the $y_{\widehat{r},Q}\widehat{\phi}_i(Q)$-coefficient of f if $\widehat{\phi}_i \in \widehat{R}$ or the $y_{\widehat{s},Q}\widehat{\phi}_i(Q)$-coefficient if $\widehat{\phi}_i \in \widehat{R}$.

$$\nu_i^{\mathrm{II}}(P) = \begin{cases} \dfrac{(\widehat{\pi}_{Q,i}(\widehat{D}_{\dot{r},\dot{s}}(P,Q))/y_{\widehat{r},P})dx_P}{3y_{\widehat{s},P}} & \text{if } \widehat{\phi}_i \in \widehat{R}_{\widehat{r}}, \\ \dfrac{(\widehat{\pi}_{Q,i}(\widehat{D}_{\dot{r},\dot{s}}(P,Q))/y_{\widehat{s},P}))dx_P}{3y_{\widehat{r},P}} & \text{if } \widehat{\phi}_i \in \widehat{R}_{\widehat{s}}, \end{cases}$$

where $\widehat{D}_{\dot{r},s}(P,Q) := y_{\widehat{s},P} y_{\widehat{r},Q} D_{\dot{s},\dot{r}}^{(+)}(P,Q) + y_{\widehat{s},Q} y_{\widehat{r},P} D_{\dot{s},\dot{r}}^{(-)}(P,Q)$,

$$\begin{aligned} D_{\dot{r},ds}^{(+)}(P,Q) :=& \sum_{j=0}^{\dot{s}+\dot{r}-2} \sum_{i=0}^{\dot{s}+\dot{r}-j-2} (i+1)\lambda_j^{(\dot{s}+\dot{r})} x_P^{\dot{r}+\dot{s}-j-i-2} x_Q^i \\ &+ \sum_{j=0}^{\dot{s}-2} \sum_{i=0}^{\dot{s}-j-2} \sum_{k=0}^{\dot{r}} (i+1)\lambda_j^{(\dot{s})} \lambda_{\dot{r}-k}^{(\dot{r})} x_P^{\dot{s}-j-i-2} x_Q^{k+i}, \\ D_{\dot{r},\dot{s}}^{(-)}(P,Q) :=& \sum_{j=0}^{\dot{s}+\dot{r}-2} \sum_{i=0}^{\dot{s}+\dot{r}-j-2} (i+1)\lambda_j^{(\dot{s}+\dot{r})} x_Q^{\dot{r}+\dot{s}-j-i-2} x_P^i \\ &+ \sum_{j=0}^{\dot{r}-2} \sum_{i=0}^{\dot{r}-j-2} \sum_{k=0}^{\dot{s}} (i+1)\lambda_j^{(\dot{r})} \lambda_{\dot{s}-k}^{(\dot{s})} x_Q^{\dot{r}-j-i-2} x_P^{k+i}. \end{aligned}$$

4. Cyclic (3, 4) and (3, 4, 5)-curves in (2.67) and Sect. 2.3.3.1:

(i) $(3,4) : y^3 = x(x-s)(x-b_1)(x-b_2) = x^4 + \lambda_3 x^3 + \lambda_2 x^2 + \lambda_1 x + \lambda_0$:

The W-normalized differentials of the second kind (holomorphic except at ∞) are

$$\nu^{\mathrm{II}} := \begin{pmatrix} \nu_1^{\mathrm{II}} \\ \nu_2^{\mathrm{II}} \\ \nu_3^{\mathrm{II}} \end{pmatrix}, \quad \nu_1^{\mathrm{II}} := \frac{(5x^2 + 2\lambda_3 x + \lambda_2)dx}{3y},$$

$$\nu_2^{\mathrm{II}} := \frac{2xdx}{3y}, \quad \nu_3^{\mathrm{II}} := \frac{x^2dx}{3y^2}.$$

(ii) $(3, 4, 5) : y^3 = x^2(x - b_1)(x - b_2)$:

$$\nu_1^{\mathrm{II}} = \frac{2xdx}{3y}, \quad \nu_2^{\mathrm{II}} = \frac{xdx}{3z}.$$

2.4.4 W-Normalized Shifted Abelian Integrals

In this subsection, we introduce the shifted Abelian integrals for the non-symmetric W-curves for later convenience. The non-symmetric W-curves in Sect. 2.2.2.4 have the non-trivial R_X-module, $\widehat{R}_X$ in Definition 2.108, and the canonical divisor has a non-trivial divisor rather than $(2g-2)\infty$ as in Proposition 2.117. In the Abel–Jacobi map in Theorem 2.98 for $\widehat{R}_X$, $w(\mathfrak{K}_\mathfrak{s})$ and $w(\mathfrak{K}_X)$ naturally appear corresponding to Proposition 2.173. The form with $w(\mathfrak{K}_\mathfrak{s})$ is found to be natural, although it will be confirmed later. Thus recalling Definitions 2.81, 2.112, and 2.116, we introduce the shifted Abelian integral $\widetilde{w}_\mathfrak{s}$ and the Abel–Jacobi map $w_\mathfrak{s}$ as an extension of the W-normalized Abelian integral and the Abel–Jacobi map in Definition 2.112. which is critical in Sect. 2.6 and Chap. 3:

Definition 2.153 We define $\widetilde{w}_\mathfrak{s}$ and $w_\mathfrak{s}$ by

$$\widetilde{w}_\mathfrak{s} : S^k\widetilde{X} \to \mathbb{C}^g, \quad \widetilde{w}_\mathfrak{s}(\gamma_1, \cdots, \gamma_k) := \left[\sum_{i=1}^{k} \widetilde{w}(\gamma_i)\right] + \widetilde{w}(\iota_X \mathfrak{K}_\mathfrak{s}) = \sum_{i=1}^{k} \int_{\gamma_i} \nu^{\mathrm{I}} + \int_{\iota_X \mathfrak{K}_\mathfrak{s}} \nu^{\mathrm{I}},$$

$$w_\mathfrak{s} := \kappa_J \circ \widetilde{w}_\mathfrak{s} \circ \iota_X : S^k X \to J_X,$$

and $\widetilde{w}_\mathfrak{s}^\circ$ and $w_\mathfrak{s}^\circ$ by

$$\widetilde{w}_\mathfrak{s}^\circ := \frac{1}{2}\omega'^{-1}\widetilde{w}_\mathfrak{s} : S^k\widetilde{X} \to \mathbb{C}^g, \quad w_\mathfrak{s}^\circ := \frac{1}{2}\omega'^{-1}w_\mathfrak{s} : S^k X \to J_X^\circ.$$

For symmetric numerical semigroup case H_X, $\widetilde{w}_\mathfrak{s} = \widetilde{w}$, $\widetilde{w}_\mathfrak{s}^\circ = \widetilde{w}^\circ$, $w_\mathfrak{s} = w$, and $w_\mathfrak{s}^\circ = w^\circ$.

For given a divisor D, let $|D|$ be the set of effective divisors linearly equivalent to D and identified with $\mathbf{H}^0(X, \mathscr{O}_X(D))$.

Definition 2.154 We define the subvarieties by

$$W_X^k := w(S^k X),\ W_{X,\mathfrak{s}}^k := w_{\mathfrak{s}}(S^k X) \subset J_X,$$
$$W_X^{\circ k} := w^\circ(S^k X),\ W_{X,\mathfrak{s}}^{\circ k} := w_{\mathfrak{s}}^\circ(S^k X) \subset J_X^\circ,$$

and their strata,

$$W_X^{k,1} := w(S_1^k X),\quad W_{X,\mathfrak{s}}^{k,1} := w_{\mathfrak{s}}(S_1^k X),\quad W_X^{\circ k,1} := w^\circ(S_1^k X),\quad W_{X,\mathfrak{s}}^{\circ k,1} := w_{\mathfrak{s}}^\circ(S_1^k X),$$

where $S_m^n X := \{D \in S^n X \mid \dim |D| \geq m\}$.

We might like to refer to these W's as the Wirtinger varieties, though the Wirtinger variety was named after another variety [Gun1984].

In Chaps. 3 and 4, we shall encounter several results which are obtained through the embedding $\iota_X : S^k X \to S^k \widetilde{X}$. For such cases, we sometimes omit ι_X for maps $\widetilde{w} \circ \iota_X$ and $\widetilde{w}_{\mathfrak{s}} \circ \iota_X$, e.g., for $\widetilde{w} \circ \iota_X : S^k X \to \mathbb{C}^g$, we simply write $\widetilde{w}(P_1, P_2, \ldots, P_k)$ rather than $\widetilde{w}(\iota_X(P_1, P_2, \ldots, P_k))$.

2.5 Symmetric Affine Algebra, and Vector Fields

For later convenience, we introduce several algebraic tools associated with the affine rings R_X on W-curves X in this section.

2.5.1 Symmetric Affine Algebra

Since the affine ring R_X has a natural grade with respect to the Sato–Weierstrass weight (2.22) or (2.24) and is a graded $\mathbb{C}$ vector space with the basis $\{\phi_i\}$, we consider R_X as a graded ring.

Further, we introduce the symmetric affine algebra $S^k R_X$ in this subsection. We focus on a symmetric action τ of the k-symmetric group $\mathfrak{S}_k$ on every element of the k tensor product in $R_X^{k\otimes}$ over $\mathbb{C}$,

$$\tau \sum_{i_1,\ldots,i_k} \alpha_{i_1,\ldots,i_k} \phi_{i_1,1} \otimes \phi_{i_2,2} \otimes \cdots \otimes \phi_{i_k,k} = \sum_{i_1,\ldots,i_k} \alpha_{i_1,\ldots,i_k} \phi_{i_1,\tau(1)} \otimes \phi_{i_2,\tau(2)} \otimes \cdots \otimes \phi_{i_k,\tau(k)},$$

where $\alpha_{i_1,\ldots,i_k} \in \mathbb{C}$. Then let $S^k R_X$ be the invariant subring of $R_X^{k\otimes}$ with respect to the action $\mathfrak{S}_k$; every element $\mu \in S^k R_X$ satisfies $\tau(\mu) = \mu$ for any $\tau \in \mathfrak{S}_k$. In the following sections, we often consider the symmetric algebra $S^k R_X$ corresponding to the Abelian integrals for $S^k X$, and its modules, $S^k \widehat{R}_X$ and the regular module $S^k R_X$. We also consider its rational field, $\mathcal{Q}(S^k R_X)$ and the fractional ideals $\mathcal{Q}(S^k \widehat{R}_X)$ as $S^k R_X$-modules. Using the elements of these modules, we consider the addition structure in R_X in Sect. 2.6. Further, we introduce homogeneous elements in $R_X \otimes \mathcal{Q}(S^k R_X)$ and $\widehat{R}_X \otimes \mathcal{Q}(S^k \widehat{R}_X)$ with respect to the Sato–Weierstrass weight wt.

Similarly, we also define the alternation affine algebra $A^k R_X$ in $R_X^{k\otimes}$ such that for $\tau \in \mathfrak{S}_k$, $\tau'(\mu) = (-1)^\tau \mu$, where the sign is minus for odd permutation and plus for the even permutation.

2.5.2 *Meromorphic Functions in* $R_X \otimes \mathscr{Q}(S^g R_X)$ *and* $\widehat{R}_X \otimes \mathscr{Q}(S^g \widehat{R}_X)$

This subsection is on the natural meromorphic functions on $R_X \otimes \mathscr{Q}(S^k R_X)$ and $\widehat{R}_X \otimes \mathscr{Q}(S^k \widehat{R}_X)$, representing the algebraic structures of X as an algebraic aspect. We introduce the Frobenius–Stickelberger (FS) matrix with entries in R_X: let n be a positive integer and $P_1, \ldots, P_n$ be in $X \setminus \{\infty\}$,

$$\Psi_n^{(\ell)}(P_1, P_2, \ldots, P_n) := \begin{pmatrix} \phi_0(P_1) & \phi_1(P_1) & \cdots & \check{\phi}_\ell(P_1) & \cdots & \phi_n(P_1) \\ \phi_0(P_2) & \phi_1(P_2) & \cdots & \check{\phi}_\ell(P_2) & \cdots & \phi_n(P_2) \\ \phi_0(P_3) & \phi_1(P_3) & \cdots & \check{\phi}_\ell(P_3) & \cdots & \phi_n(P_3) \\ \vdots & \vdots & \ddots & \vdots & \ddots & \vdots \\ \phi_0(P_n) & \phi_1(P_n) & \cdots & \check{\phi}_\ell(P_n) & \cdots & \phi_n(P_n) \end{pmatrix}, \qquad (2.103)$$

and the Frobenius–Stickelberger (FS) matrix of $\widehat{R}_X$,

$$\widehat{\Psi}_n^{(\ell)}(P_1, P_2, \ldots, P_n) := \begin{pmatrix} \widehat{\phi}_0(P_1) & \widehat{\phi}_1(P_1) & \cdots & \check{\widehat{\phi}}_\ell(P_1) & \cdots & \widehat{\phi}_n(P_1) \\ \widehat{\phi}_0(P_2) & \widehat{\phi}_1(P_2) & \cdots & \check{\widehat{\phi}}_\ell(P_2) & \cdots & \widehat{\phi}_n(P_2) \\ \widehat{\phi}_0(P_3) & \widehat{\phi}_1(P_3) & \cdots & \check{\widehat{\phi}}_\ell(P_3) & \cdots & \widehat{\phi}_n(P_3) \\ \vdots & \vdots & \ddots & \vdots & \ddots & \vdots \\ \widehat{\phi}_0(P_n) & \widehat{\phi}_1(P_n) & \cdots & \check{\widehat{\phi}}_\ell(P_n) & \cdots & \widehat{\phi}_n(P_n) \end{pmatrix}, \qquad (2.104)$$

where a check on top of a letter signifies deletion. Further, we introduce their abbreviations,

$$\begin{aligned} \Psi_n(P_1, P_2, \ldots, P_n) &:= \Psi_n^{(n)}(P_1, P_2, \ldots, P_n), \\ \widehat{\Psi}_n(P_1, P_2, \ldots, P_n) &:= \widehat{\Psi}_n^{(n)}(P_1, P_2, \ldots, P_n). \end{aligned}$$

The *Frobenius–Stickelberger (FS) determinants* are

$$\begin{aligned} \psi_n^{(\ell)}(P_1, \ldots, P_n) &:= \det(\Psi_n^{(\ell)}(P_1, \ldots, P_n)), \\ \widehat{\psi}_n^{(\ell)}(P_1, \ldots, P_n) &:= \det(\widehat{\Psi}_n^{(\ell)}(P_1, \ldots, P_n)), \end{aligned}$$

$$\psi_n(P_1, \ldots, P_n) := \det(\Psi_n(P_1, \ldots, P_n)), \; \widehat{\psi}_n(P_1, \ldots, P_n) := \det(\widehat{\Psi}_n(P_1, \ldots, P_n)). \qquad (2.105)$$

Lemma 2.155 *Let* $\gamma_{P,\infty} \in \widetilde{X}$, $u = \kappa_J \widetilde{w}_{\mathfrak{s}}(\gamma_{P,\infty}) \in \mathbb{C}^g$, *and* $P := \kappa_X \gamma_{P,\infty}$.

1. u_i *is a function of* u_g *whose expansion is given by* $u_i = -\dfrac{1}{N^{\mathfrak{c}}(g-i+1)} \times (-u_g)^{N^{\mathfrak{c}}(g-i+1)} + d_{>N^{\mathfrak{c}}(g-i+1)}(u_g)$,
2. $\phi_i(P) = \dfrac{1}{(-u_g)^{N(i)}}(1 + d_{>1}(u_g))$, *and*
3. $\psi_n(P_1, \ldots, P_{n-1}, P)$ *as a function of* P *is expanded as* $\dfrac{\psi_{n-1}(P_1, \ldots, P_{n-1})}{(-u_g)^{N(n)}} + d_{>N(n)}(u_g)$.

Proof (2.81) shows that $u_g = -t + d_{>1}(t)$ for the arithmetic parameter t of P, and thus we naturally obtain them noting Lemma 2.17. ■

Definition 2.156 For $P, P_1, \ldots, P_n \in (X \setminus \{\infty\}) \times S^n(X \setminus \{\infty\})$, we define $\mu_n(P)$ by

$$\mu_n(P) := \mu_n(P; P_1, \ldots, P_n) := \lim_{P_i' \to P_i} \frac{1}{\psi_n(P_1', \ldots, P_n')} \psi_{n+1}(P_1', \ldots, P_n', P),$$

$$\widehat{\mu}_n(P) := \widehat{\mu}_n(P; P_1, \ldots, P_n) := \lim_{P_i' \to P_i} \frac{1}{\widehat{\psi}_n(P_1', \ldots, P_n')} \widehat{\psi}_{n+1}(P_1', \ldots, P_n', P),$$

where the P_i' are generic, the limit is taken (irrespective of the order) for each i. We also define $\mu_{n,k}(P_1, \ldots, P_n)$ and $\widehat{\mu}_{n,k}(P_1, \ldots, P_n)$ by

$$\begin{aligned} \mu_n(P) &= \phi_n(P) + \sum_{k=0}^{n-1} (-1)^{n-k} \mu_{n,k}(P_1, \ldots, P_n) \phi_k(P), \\ \widehat{\mu}_n(P) &= \widehat{\phi}_n(P) + \sum_{k=0}^{n-1} (-1)^{n-k} \widehat{\mu}_{n,k}(P_1, \ldots, P_n) \widehat{\phi}_k(P), \end{aligned} \tag{2.106}$$

with the convention $\mu_{n,n}(P_1, \ldots, P_n) = \widehat{\mu}_{n,n}(P_1, \ldots, P_n) = 1$.

Corollary 2.157 *For the hyperelliptic curves (2.65) of genus g case, if* $n \le g$, $\mu_n = \widehat{\mu}_n$ *is reduced to*

$$\mu_n(P) = F_n(x) = (x - x_1)(x - x_n) \cdots (x - x_n),$$

where each $P_i = (x_i, y_i)$ *and* $P = (x, y)$, *which is known as U function of the Mumford triplet UVW [Mum1984]. We sometimes abbreviate* F_g *to* F.

Apparently, μ_n and $\widehat{\mu}_n$ are the homogeneous elements in $R_X \otimes \mathcal{Q}(S^n R_X)$ and $\widehat{R}_X \otimes \mathcal{Q}(S^n \widehat{R}_X)$ with respect to the weight wt_λ respectively. Thus $\mu_{n,i}$ and $\widehat{\mu}_{n,i}$ belong to $\mathcal{Q}(S^n R_X)$ and $\mathcal{Q}(S^n \widehat{R}_X)$ respectively.

2.5.3 Vector Fields on $S^k X \setminus S_1^k X$

We could identify the quotient field $\mathcal{Q}(S^g R_X)$ of $S^g R_X$ with the meromorphic functions on J_X. We show the vector field structure on $S^k X \setminus S_1^k X$ in this subsection. For $P_1, \dots, P_k$ in X, we modify the Frobenius–Stickelberger (FS) matrix (2.104) for the vector fields as

$$\widehat{\Psi}_k^D := \widehat{\Psi}_k^D(P_1, P_2, \dots, P_k) := \begin{pmatrix} \widehat{\phi}_{g-k}(P_1) & \widehat{\phi}_{g-k-1}(P_1) & \cdots & \widehat{\phi}_{g-1}(P_1) \\ \widehat{\phi}_{g-k}(P_2) & \widehat{\phi}_{g-k-1}(P_2) & \cdots & \widehat{\phi}_{g-1}(P_2) \\ \vdots & \vdots & \ddots & \vdots \\ \widehat{\phi}_{g-k}(P_k) & \widehat{\phi}_{g-k-1}(P_k) & \cdots & \widehat{\phi}_{g-1}(P_k), \end{pmatrix}, \qquad (2.107)$$

and its associated $k \times k$ matrix,

$$\mathcal{H}_k := \begin{pmatrix} h_X(P_1) & & & \\ & h_X(P_2) & & \\ & & \ddots & \\ & & & h_X(P_k) \end{pmatrix}.$$

Lemma 2.158 *Let $d\mathbf{x}_k := {}^t(dx_1, dx_2, \cdots, dx_k)$, and $d\mathbf{x} := d\mathbf{x}_g$. Then we have*

$$\begin{pmatrix} \nu^{\mathrm{I}}_{g-k+1} \\ \vdots \\ \nu^{\mathrm{I}}_{g-1} \\ \nu^{\mathrm{I}}_{g} \end{pmatrix} = \mathcal{H}_k^{-1}\, {}^t\widehat{\Psi}_k^D \begin{pmatrix} dx_1 \\ dx_2 \\ \vdots \\ dx_k \end{pmatrix}.$$

When $k = g$, $\nu^{\mathrm{I}} = \mathcal{H}_g^{-1}\, {}^t\widehat{\Psi}_k^D\, d\mathbf{x}$, and thus, $\left(\left(\dfrac{\partial u_i}{\partial x_j}\right)_{ij}\right) = \mathcal{H}_g^{-1}\, {}^t\widehat{\Psi}_g^D.$

Corollary 2.159 $d\mathbf{x} = {}^t\widehat{\Psi}_g^{D\,-1} \mathcal{H}_g\, \nu^{\mathrm{I}}$, *and* $\left(\left(\dfrac{\partial x_i}{\partial u_j}\right)_{ij}\right) = {}^t\widehat{\Psi}_g^{D\,-1} \mathcal{H}_g.$

Lemma 2.160 *Let $\partial_{u_i} := \dfrac{\partial}{\partial u_i}$ and $\partial_{x_i} := \dfrac{\partial}{\partial x_i}$. We have*

$$\begin{pmatrix} \partial_{u_1} \\ \partial_{u_2} \\ \vdots \\ \partial_{u_g} \end{pmatrix} = {}^t\mathcal{H}_g \widehat{\Psi}_g^{D\,-1} \begin{pmatrix} \partial_{x_1} \\ \partial_{x_2} \\ \vdots \\ \partial_{x_g} \end{pmatrix}, \quad \begin{pmatrix} \partial_{x_1} \\ \partial_{x_2} \\ \vdots \\ \partial_{x_g} \end{pmatrix} = \widehat{\Psi}_g^{D}\, {}^t\mathcal{H}_g^{-1} \begin{pmatrix} \partial_{u_1} \\ \partial_{u_2} \\ \vdots \\ \partial_{u_g} \end{pmatrix}.$$

Proof $\left(\left(\frac{\partial x_i}{\partial u_j}\right)_{ji}\right) = {}^t\mathscr{H}_g \widehat{\Psi}_g^{D\,-1}$ and $\left(\left(\frac{\partial u_i}{\partial x_j}\right)_{ji}\right) = \widehat{\Psi}_g^{D}\,{}^t\mathscr{H}_g^{-1}$. Even if $\widehat{\Psi}_g^D$ is singular, ${}^t\mathscr{H}_g \widehat{\Psi}_g^{D\,-1}$ is defined by Lemma 2.158 and Theorem 2.98. ∎

We note that Lemma 2.160 and Corollary 2.159 hold for $k < g$ by restricting the locus in $\mathbb{C}^g$ as in Lemma 2.158.

Lemma 2.161 *For* $P = (x, y_\bullet) \in X$, $\sum_{j=1} \widehat{\phi}_{j-1}(P)\frac{\partial}{\partial w_j(P)} = h_X(P)\frac{\partial}{\partial x}$.

Proposition 2.162 *The following holds for any* i *and* j:

$$\left[\frac{\partial}{\partial u_i}, \frac{\partial}{\partial u_j}\right] = \frac{\partial}{\partial u_i}\frac{\partial}{\partial u_j} - \frac{\partial}{\partial u_j}\frac{\partial}{\partial u_i} = 0.$$

Proof It is obvious from the fact that u_i and u_j are the independent components in $\mathbb{C}^g$. ∎

Remark 2.163 The relation in Proposition 2.162 is known as the compatibility condition in the integrable system. It can be also proved by the relations among the differential operators in terms of elements $\mathscr{Q}(R_X^g)\left[\frac{\partial}{\partial x_1}, \ldots, \frac{\partial}{\partial x_g}\right]$ for each W-curve X and R_X. Weierstrass essentially showed the relation in [Wei1854], which is the foundation of Frobenius integrability condition [Haw2013, Chap. 6, The problem of Pfaff].

Lemma 2.164 *We let the right-hand side on* ∂_{u_i} *in Lemma 2.160 denoted by* $\widetilde{\partial}_{u_i,x} \in \mathscr{Q}(S^g R_X)\left[\frac{\partial}{\partial x_1}, \ldots, \frac{\partial}{\partial x_g}\right]$. *Then the action of* $\widetilde{\partial}_{u_i,x}$ *on* $\mathscr{Q}(S^g R_X)$ *is closed, i.e.,* $\widetilde{\partial}_{u_i,x} : \mathscr{Q}(S^g R_X) \to \mathscr{Q}(S^g R_X)$, *and* $[\widetilde{\partial}_{u_i,x}, \widetilde{\partial}_{u_j,x}] = 0$.

Proof $\widetilde{\partial}_{u_i,x}$ is invariant for the exchanging x_i with x_j, and the action of $\frac{\partial}{\partial x}$ on R_X is closed. Proposition 2.162 shows that they are obvious. ∎

Definition 2.165 We define the differential rings,

$$\mathscr{D}_X := \mathscr{Q}(S^g R_X)\left[\widetilde{\partial}_{u_1,x}, \ldots, \widetilde{\partial}_{u_g,x}\right], \qquad \mathscr{D}_J := \mathscr{M}_{J_X}\left[\frac{\partial}{\partial u_1}, \ldots, \frac{\partial}{\partial u_g}\right],$$

where $\mathscr{M}_{J_X}$ is the field of the meromorphic functions on J_X.

The differential rings $\mathscr{D}_X$ and $\mathscr{D}_J$ as the $\mathscr{Q}(S^g R_X)$-module and the $\mathscr{M}_J$-module are, respectively, decomposed by

$$\mathscr{D}_X = \bigoplus_\alpha \mathscr{Q}(S^g R_X)\,\widetilde{\partial}_{u,x}^\alpha, \quad \mathscr{D}_J = \bigoplus_\alpha \mathscr{M}_J\,\partial_u^\alpha, \tag{2.108}$$

where the multi-index $\alpha = (\alpha_1, \ldots, \alpha_g)$, $\widetilde{\partial}_{u,x}^\alpha := \widetilde{\partial}_{u_1,x}^{\alpha_1} \cdots \widetilde{\partial}_{u_g,x}^{\alpha_g}$, and $\partial_u^\alpha := \partial_{u_1}^{\alpha_1} \cdots \partial_{u_g}^{\alpha_g}$.

Remark 2.166 The Abel–Jacobi Theorem 2.98 and above lemmas show that the differential rings $\mathscr{D}_X$ and $\mathscr{D}_J$ are isomorphic in a certain sense. As in Theorem 3.66, we show that both left modules have correspondences; the equivalence provides critical tools in Chap. 4, and guarantees the diagram illustrated in Fig. 1.9 as the "Weierstrass" Abelian function theory.

Following [Bj1992, Chap. 1], we consider the $\mathscr{D}_X$-modules. We also note that the W-normalized differentials of the first kind are characterized by the basis of the holomorphic one-form $\mathbf{H}^0(X, \mathscr{A}_X)$ as $\mathbb{C}$ vector space such that each Sato–Weierstrass weight is equal to $N^c(g-i)-1$. However, there are infinitely many bases of $\mathbf{H}^0(X, \mathscr{A}_X)$, and there are transformation matrices among the bases; there exists an element of $\mathrm{GL}(g, \mathbb{C})$ for the two bases.

For simplicity we consider a basis $\{\nu_1^{\mathrm{I}'}, \ldots, \nu_g^{\mathrm{I}'}\}$ such that $M\,{}^t(\nu_1^{\mathrm{I}}, \ldots, \nu_g^{\mathrm{I}}) = {}^t(\nu_1^{\mathrm{I}'}, \ldots, \nu_g^{\mathrm{I}'})$. In fact, we show the twisted W-normalized differentials of the first kind in (4.51) which were introduced by Weierstrass [Wei1854]. By considering their Abelian integrals, we see that the matrix M induces the action on the coordinate in J_X, $M : u \to u' = Mu$, and thus its differential operators ∂'_{u_i} such that $\langle \partial'_{u_i} \nu_j^{\mathrm{I}'} \rangle = \delta_{ij}$. It also induces the action on $\mathscr{D}_X$ and $\mathscr{D}_J$ from the right via M^{-1}.

Lemma 2.167 *For* $(\widetilde{\partial}'_{u_1,x}, \ldots, \widetilde{\partial}'_{u_g,x}) = (\widetilde{\partial}_{u_1,x}, \ldots, \widetilde{\partial}_{u_g,x})M^{-1}$, $M \in \mathrm{GL}(g, \mathbb{C})$, *and* $\mathscr{D}'_X := \mathscr{Q}(S^g R_X)\left[\widetilde{\partial}'_{u_1,x}, \ldots, \widetilde{\partial}'_{u_g,x}\right]$, $\mathscr{D}'_X$ *is isomorphic to* $\mathscr{D}_X$ *as a ring.*

$\mathscr{Q}(S^g R_X)$ *is a left* $\mathscr{D}_X$*-module, and thus* $\mathrm{Hom}_{\mathscr{D}_X}(\mathscr{Q}(S^g R_X), \mathscr{Q}(S^g R_X)) = \mathbb{C}$.

Proof The first statement is obvious, and the second is obtained by considering $\mathscr{Q}(S^g R_X)$ as $\mathscr{D}_X/(\bigoplus_{i=1}^{g} \mathscr{Q}(S^g R_X)\,\widetilde{\partial}_{u_i,x})$. ∎

Lemma 2.168 *For* $P = \sum p_\alpha \widetilde{\partial}^\alpha_{u,x} \in \mathscr{D}_X$, *the transformation* $\varphi_M : \widetilde{\partial}_{u,x} \to \widetilde{\partial}'_{u,x} = \widetilde{\partial}_{u,x} M^{-1}$ *induces the transformation* $\varphi_M : \mathscr{D}_X \to \mathscr{D}_X$, $(p_\alpha \mapsto p'_\alpha = \sum_\beta m_{\alpha,\beta} p_\beta)$ *such that* $\varphi_M(P) = P$. *For every* $M \in \mathrm{GL}(g, \mathbb{C})$, $\mathscr{Q}(S^g R_X)$ *is invariant for* φ_M.

Remark 2.169 We consider the set of the left $\mathscr{D}_X$-module $\mathrm{Mod}(\mathscr{D}_X)$ as a category by fixing X; the set of the objects are $\mathscr{Q}(S^g R_X)$-modules, $\bigoplus_\alpha \mathscr{Q}(S^g R_X)\,\widehat{\partial}^\alpha_{u,x}$, where $\widehat{\partial}^\alpha_{u,x}$ forms a certain basis associated with $\widetilde{\partial}^\alpha_{u,x}$, and the morphisms are φ_M induced from the element M of $\mathrm{GL}(g, \mathbb{C})$, which obviously satisfies $\varphi_M \partial_u = \partial'_u \varphi_M$. In other words, we consider the case that every element of $\mathrm{Hom}_{\mathrm{Mod}(\mathscr{D}_X)}(\mathscr{D}_X, \mathscr{D}_X)$ is given by φ_M.

Remark 2.170 Further, for a certain locus of J_X, we will also consider the differential rings $\mathscr{Q}(S^k R_X)\left[\widetilde{\partial}_{u_{g-k+1},x}, \ \ldots, \widetilde{\partial}_{u_g,x}\right]$ and $\mathscr{M}_{J_X}\left[\dfrac{\partial}{\partial u_{g-k+1}}, \ \ldots, \dfrac{\partial}{\partial u_g}\right]$ in Sect. 4.1.

2.6 Addition Structures in W-Curves

This section is devoted to the explanations of the additional structures in W-curves X based on the affine rings R_X.

2.6.1 Meromorphic Functions and Linear Equivalence

As the addition structures in W-curves X, we consider the linear equivalence of the meromorphic functions μ_n and $\widehat{\mu}_n$ in this subsection. In the following, we will discuss them for points in $S^n X$ but note that some may not be well defined for points in $S_1^n X$. Therefore, in order to limit the situation to a strict definition, there are several places where $S^n(X \setminus \{\infty\})$ in the following is read as $S^n(X \setminus \{\infty\}) \setminus S_1^n(X)$. However, as it depends on local group actions and algebraic relations to encounter ill-posed situations, as in the proof of Lemma 2.160, we state them here without limitation from a generic standpoint.

The meromorphic functions μ_n and $\widehat{\mu}_n$ introduced in Definition 2.156 enable us to express the addition structure of Pic X (2.1) in terms of FS-determinants in (2.105). We developed this theory for every W-curve (X, P), including non-symmetric Weierstrass semigroup cases.

Proposition 2.171 *For n points $(P_i)_{i=1,\ldots,n} \in X \setminus \{\infty\}$, we find an element of R_X associated with any point $P = (x, y)$ in $(X \setminus \{\infty\})$, $\alpha_n(P) := \alpha_n(P; P_1, \ldots, P_n) = \sum_{i=0}^n a_i \phi_i(P)$, $a_i \in \mathbb{C}$ and $a_n = 1$, which has a zero at each point P_i (with multiplicity, if the P_i are repeated) and has the smallest possible order of pole at ∞ with this property. Then α_n is identified with $\mu_n(P; P_1, \ldots, P_n)$ in Definition 2.156.*

Proof μ_n has the same properties as α_n. Assume that $\alpha_n - \mu_n \neq 0$. Since both belong to R_X, which is given by the linear combination of basis $\{\phi_i(P)\}$, the difference is expressed by $\sum_{i=0}^n c_i \phi_i(P) \in R_X$. It must vanish at each point P_i and has singularity with order $N(\max\{j \mid c_j \neq 0\})$. However, $\max\{j \mid c_j \neq 0\} < n$ because of the monic properties. There is no such element in R_X. ■

We have the following proposition.

Proposition 2.172 *Let n be a positive integer. For $(P_i)_{i=1,\ldots,n} \in S^n(X \setminus \{\infty\})$, the function $\alpha_n = \mu_n$ over X induces the map (which we call by the same name):*

$$\alpha_n : S^n(X \setminus \{\infty\}) \to S^{N(n)-n}(X),$$

i.e., to $(P_i)_{i=1,\ldots,n} \in S^n(X \setminus \{\infty\})$ there corresponds an element $(Q_i)_{i=1,\ldots,N(n)-n} \in S^{N(n)-n}(X)$, such that

$$\sum_{i=1}^{n} P_i - n\infty \sim - \sum_{i=1}^{N(n)-n} Q_i + (N(n) - n)\infty.$$

Proof This is just a dimension count. By definition, there are exactly n independent functions with poles of order $\le N(n)$ at ∞, and passage through the points $P_1, \cdots, P_n$ imposes in general n linear conditions. For a concrete formula, consider the linear combinations $\alpha = a_0 + \sum_i^n a_i\phi_i$, $a_i \in \mathbb{C}$. The definition of the ϕ_i gives that $-\deg_\infty \alpha = N(n)$. If the P_i are generic points, the following system of linear equations has a $(n+1) \times n$ matrix of rank n

$$\begin{pmatrix} 1 & \phi_1(P_1) & \phi_2(P_1) & \cdots & \phi_n(P_1) \\ 1 & \phi_1(P_2) & \phi_2(P_2) & \cdots & \phi_n(P_2) \\ \vdots & \vdots & \vdots & \ddots & \vdots \\ 1 & \phi_1(P_n) & \phi_2(P_n) & \cdots & \phi_n(P_n) \end{pmatrix} \begin{pmatrix} a_0 \\ a_1 \\ \vdots \\ a_{n-1} \\ a_n \end{pmatrix} = 0.$$

So the system determines the a_i's and α up to a constant factor, and this polynomial will have $\mathrm{div}(\alpha) = \sum_{i=1}^n P_i + \sum_{i=1}^{N(n)-n} Q_i - N(n)\infty \sim 0$. When the matrix is singular, e.g., $P_1 = P_2$ for the map $\varpi_r : X \to \mathbb{P}$ in (2.16), of the monomial ϕ_i, and the other points are in general positions, we consider the matrix

$$\begin{pmatrix} 1 & \phi_1(P_1) & \phi_2(P_1) & \cdots & \phi_n(P_1) \\ 0 & \frac{\partial\phi_1}{\partial t}(P_1) & \frac{\partial\phi_2}{\partial t}(P_1) & \cdots & \frac{\partial\phi_n}{\partial t}(P_1) \\ \vdots & \vdots & \vdots & \ddots & \vdots \\ 1 & \phi_1(P_n) & \phi_2(P_n) & \cdots & \phi_n(P_n) \end{pmatrix} \begin{pmatrix} a_0 \\ a_1 \\ \vdots \\ a_{n-1} \\ a_n \end{pmatrix} = 0$$

instead, for it has rank n (here t is a local parameter at P_1, obtained by taking the appropriate root of a function that vanishes simultaneously at P_1 and P_2). ■

We consider an extension in Proposition 2.171 to $\mathbf{H}^0(X, \mathscr{A}_X(*\infty))$. Generally, the one-form $\widehat{\alpha}_n := \dfrac{\widehat{\mu}_n dx}{h_X}$ does not have a singularity over $X \setminus \{\infty\}$, whereas $\dfrac{\mu_n dx}{h_X}$ does. The divisor of $\widehat{\mu}_n(x)$ contains $\mathfrak{K}_X - \mathfrak{k}_X\infty$ for any positive n, which is linearly equivalent to $2\mathfrak{K}_{\mathfrak{s}} - 2\mathfrak{k}_{\mathfrak{s}}\infty$ as in Definition 2.116. We have the following proposition.

Proposition 2.173 *Let n be a positive integer. For $(P_i)_{i=1,\ldots,n} \in S^n(X \setminus \{\infty\})$, the one-form $\widehat{\alpha}_n = \dfrac{\widehat{\mu}_n dx}{h_X}$ over X induces the map (which we also call by the same name):*

$$\widehat{\alpha}_n : S^n(X \setminus \{\infty\}) \to S^{\widehat{N}(n)-n}(X),$$

i.e., to $(P_i)_{i=1,\ldots,n} \in S^n(X \setminus \{\infty\})$ there corresponds an element $(Q_i)_{i=1,\ldots,\widehat{N}(n)-n-g-1}$ $\in S^{\widehat{N}(n)-n}(X)$, such that

$$\sum_{i=1}^{n} P_i + \mathfrak{K}_{\mathfrak{s}} - (n + \mathfrak{k}_{\mathfrak{s}})\infty \sim -\left(\sum_{i=1}^{\widehat{N}(n)-n} Q_i + \mathfrak{K}_{\mathfrak{s}} - (\widehat{N}(n) - n - \mathfrak{k}_{\mathfrak{s}})\infty\right).$$

For $D = \sum_{i=1}^{n} P_i$ *and* $D' = \sum_{i=1}^{\widehat{N}(n)-n} Q_i$*, we let* D' *denoted by* $D' + \mathfrak{K}_{\mathfrak{s}} = [-1](D + \mathfrak{K}_{\mathfrak{s}})$ *because of* $w_{\mathfrak{s}}(D') = -w_{\mathfrak{s}}(D)$.

Proof Since we have

$$\sum_{i=1}^{n} P_i + \mathfrak{K}_X + \sum_{i=1}^{\widehat{N}(n)-n} Q_i - (\widehat{N}(n) + \mathfrak{k}_X)\infty$$
$$\sim \sum_{i=1}^{n} P_i + \sum_{i=1}^{\widehat{N}(n)-n} Q_i + 2\mathfrak{K}_{\mathfrak{s}} - (\widehat{N}(n) + 2\mathfrak{k}_{\mathfrak{s}})\infty \sim 0,$$

the linear equivalence holds. ■

Recalling Lemma 2.115, for $n = g - 1$ case, we have

$$\sum_{i=1}^{g-1} P_i + \sum_{i=1}^{g-1} Q_i + 2\mathfrak{K}_{\mathfrak{s}} - 2(g - 1 + \mathfrak{k}_{\mathfrak{s}})\infty \sim 0,$$

where Q_i's are certain points of X.

Remark 2.174 We note that α_m and $\widehat{\alpha}_m$, or μ_m and $\widehat{\mu}_m$, are defined on $X \times S^m X$. Corollary 2.179 shows that Proposition 2.173 is a natural extension of the addition on the elliptic curves.

Proposition 2.173 implies that the image of the Abel–Jacobi map has the symmetry of the Jacobian using the shifted Abel–Jacobi map in Definition 2.153:

Theorem 2.175

$$-w_{\mathfrak{s}}(S^{g-1}X) = w_{\mathfrak{s}}(S^{g-1}X), \quad -w_{\mathfrak{s}}(S^g X) = w_{\mathfrak{s}}(S^g X).$$

Proof Lemma 2.115 means $w_{\mathfrak{s}}(P_1, \ldots, P_{g-1}) = -w_{\mathfrak{s}}(Q_1, \ldots, Q_{g-1})$ and thus $-w_{\mathfrak{s}}(S^{g-1}X) \subset w_{\mathfrak{s}}(S^{g-1}X)$. It leads to the relation. ■

Remark 2.176 The first relation in Theorem 2.175 is known as the Serre duality, whereas the second in Theorem 2.175 is related to the symmetry in the Jacobian J_X. We sometimes refer to the operations in Proposition 2.173 for the Abel–Jacobi map image as $[-1]$, which is identified with the minus operator in $\mathbb{C}^g$ due to Theorem 2.175. The Abel–Jacobi theorem 2.98 and Theorem 2.175 mean that $W^g_{X.\mathfrak{s}} = J_X$ and we have the following theorem. We note that in general, $[-1]W^i_{X.\mathfrak{s}}$ does not equal to $W^i_{X.\mathfrak{s}}$ for $i < g - 1$ except in the case of the hyperelliptic curves.

From Remark 2.176, the following theorem is obvious.

Theorem 2.177 *$[-1]\Gamma_X = \Gamma_X$ and $[-1]J_X = J_X$.*
$[-1]W^i_{X.\mathfrak{s}} = W^i_{X.\mathfrak{s}}$ for $i = g-1, g$.

Proposition 2.173 yields the addition structure in X:

Theorem 2.178 *Let $\widehat{N}^{(1)}(\ell) := \widehat{N}(\ell) - \ell$, and $\widehat{N}^{(2)}(\ell) := \widehat{N}^{(1)} \circ \widehat{N}^{(1)}(\ell)$. For $(P_1, \dots, P_n) \in S^n X$ and $(P'_1, \dots, P'_m) \in S^m X$, there are some points $(Q_1, \dots, Q_{\widehat{N}^{(1)}(n+m)}) \in S^{\widehat{N}^{(1)}(n+m)} X$ and $(Q'_1, \dots, Q'_{\widehat{N}^{(2)}(n+m)}) \in S^{\widehat{N}^{(2)}(n+m)} X$ satisfying the relation*

$$\begin{aligned} &P_1 + \cdots + P_n + P'_1 + \cdots P'_m + \mathfrak{K}_\mathfrak{s} - (n + m + \mathfrak{k}_\mathfrak{s})\infty \\ &\sim -(Q_1 + \cdots + Q_{\widehat{N}^{(1)}(n+m)} + \mathfrak{K}_\mathfrak{s} - (\widehat{N}^{(1)}(n+m) - \mathfrak{k}_\mathfrak{s})\infty \\ &\sim Q'_1 + \cdots + Q'_{\widehat{N}^{(2)}(n+m)} + \mathfrak{K}_\mathfrak{s} - (\widehat{N}^{(2)}(n+m) + \mathfrak{k}_\mathfrak{s})\infty, \end{aligned}$$

i.e.,

$$\begin{aligned} w(P_1, \dots, P_n) + w_\mathfrak{s}(P'_1, \dots, P'_m) &= w_\mathfrak{s}(P_1, \dots, P_n) + w(P'_1, \dots, P'_m) \\ &= -w_\mathfrak{s}(Q_1, \dots, Q_{\widehat{N}^{(1)}(n+m)}) \\ &= w_\mathfrak{s}(Q'_1, \dots, Q'_{\widehat{N}^{(2)}(n+m)}). \end{aligned}$$

Note that due to the Abel–Jacobi theorem 2.98 for every $u \in J_X$, we find the point $(\widetilde{P}_1, \dots, \widetilde{P}_g) \in S^g X$ using the correspondence by the Abel–Jacobi map w. We regard it as $(u - w(\mathfrak{K}_\mathfrak{s})) + w(\mathfrak{K}_\mathfrak{s})$ and thus $w_\mathfrak{s}(\widetilde{P}_1 + \cdots + \widetilde{P}_g - \mathfrak{K}_\mathfrak{s}) = u$.

Corollary 2.179 *1. For the hyperelliptic curve X of (2.65) and $(x, y) \in X$, $-((x, y) - \infty) = (x, -y) - \infty$, or the minus operation is equivalent to hyperelliptic involution.*
(It is not true for a general curve $y^2 + A_1(x)y + A_2(x) = 0$ with $A_1 \neq 0$.)
2. For the elliptic curve E, and $P_i = (x_i, y_i)$ and $P = (x, y)$ in E, $\mu_2(P, P_1, P_2) = y - \dfrac{y_2 - y_1}{x_2 - x_1} x + \dfrac{x_1 y_2 - x_2 y_1}{x_1 - x_2}$. Thus its zero corresponds to the point $-Q = (x_0, -y_0)$ that exists on the line passing through P_1 and P_2; The addition by μ_2 is identical to the well-known addition in $P_1 + P_2 = Q = (x_0, y_0)$.

Proof 1. Since $(\mu_1(P, P_1)) = (x - x_1) = (x_1, y_1) + (x_1, -y_1) - 2\infty$, we obtain it. We directly prove 2 [Har1977, IV, Prop. 4.8]. ■

Remark 2.180 The inclusion $i_n : S^n(X) \to S^{N(n)}(X), \big((P_1, ..., P_n) \mapsto (P_1, ..., P_n, \widehat{\alpha}_n(P_1, ..., P_n))\big)$ maps the stratification of symmetric products of (r, s)-curves,

$$S^0(X) \subset S^1(X) \subset \ldots \subset S^{g-2}(X) \subset S^{g-1}(X) \subset S^g(X),$$

to the stratification,

$$S^0(X) \subset S^{N(1)}(X) \subset \ldots \subset S^{N(g-2)}(X) \subset S^{N(g-2)}(X) \subset S^{2g}(X),$$

and

$$S^0(X) \subset S^{\widehat{N}(1)}(X) \subset \ldots \subset S^{\widehat{N}(g-2)}(X) \subset S^{\widehat{N}(g-2)}(X) \subset S^{2g}(X).$$

For the symmetric W-curve, we recall that $N(i) = \widehat{N}(i)$ and the fact that $2g-1$ is a gap is equivalent to $(2g-2)\infty \sim K_X$ [ArCGH1985, Ch. 1 Exercises, E-2 (ii)].

We consider the Galois action $\widehat{\zeta} \in G_X = \mathrm{Aut}(\varpi_r)$ of X if it exists, which manifestly has $\widehat{\zeta}(\infty) = \infty$. We have the following theorem.

Theorem 2.181 *Assume that $\varpi_r : X \to \mathbb{P}$ is a Galois covering. For any $\widehat{\zeta} \in G_X$ and $k = 0, 1, \ldots, g$, $\widehat{\zeta} w(S^k X) = w(S^k X)$ and $\widehat{\zeta}\widetilde{w}(S^k \widetilde{X}) = \widetilde{w}(S^k \widetilde{X})$. Especially, $\widehat{\zeta} J_X = J_X$.*

If $\widehat{\zeta} w(\mathfrak{k}_{\mathfrak{s}}) = w(\mathfrak{k}_{\mathfrak{s}})$, $\widehat{\zeta} w_{\mathfrak{s}}(S^k X) = w_{\mathfrak{s}}(S^k X)$, and $\widehat{\zeta}\widetilde{w}_{\mathfrak{s}}(S^k \widetilde{X}) = \widetilde{w}_{\mathfrak{s}}(S^k \widetilde{X})$ for $k = 0, 1, \ldots, g$.

Proof Let us consider $\gamma_{P,\infty} \in \widetilde{X}$ and then, in general, $\widehat{\zeta}\gamma_{P,\infty} \neq \gamma_{P,\infty}$. However, we can find $\gamma_{Q,\infty} \in \widetilde{X}$ such that $\widehat{\zeta}\gamma_{Q,\infty} = \gamma_{P,\infty}$. Then the others are obviously obtained. ■

For later use in Sect. 3.3.1, we prove the following relations.

Proposition 2.182 *For every $(P_1, \ldots, P_{k-1}; P_k) \in S^{k-1}(X \setminus \{\infty\}) \times X$ $(k = 1, \ldots, g)$, the following relations hold:*

$$\lim_{P_k \to \infty} \frac{\widehat{\phi}_{k-1}(P_k)}{\widehat{\phi}_k(P_k)} \widehat{\mu}_{k,\ell}(P_1, P_2, \ldots, P_k) = \begin{cases} -\widehat{\mu}_{k-1,\ell}(P_1, P_2, \ldots, P_{k-1}) & \text{for } 0 \le \ell \le k-2, \\ -1 & \text{for } \ell = k-1, \\ 0 & \text{for } \ell = k. \end{cases}$$

Proof Noting (2.106), for the $\ell = k-1$ case, we see that

$$\begin{aligned}\frac{\widehat{\phi}_{k-1}(P_k)}{\widehat{\phi}_k(P_k)} \widehat{\mu}_{k,k-1} &= \frac{-\widehat{\psi}_{k-1}^{(\check{k-1})}(P_1, \ldots, P_{k-1})\phi_k(P_k) + d_{\ge}(t_\infty^{-N(k-2)})}{\widehat{\psi}_{k-1}^{(\check{k-1})}(P_1, \ldots, P_{k-1})\phi_{k-1}(P_k) + d_{\ge}(t_\infty^{-N(k-2)})} \cdot \frac{\phi_{k-1}(P_k)}{\phi_k(P_k)} \\ &\to -1, \quad (P_k \to \infty).\end{aligned}$$

For the $\ell \le k-2$ case,

$$\begin{aligned}\frac{\phi_{k-1}(P_k)}{\phi_k(P_k)} \mu_{k,\ell} &= \frac{-\psi_{k-1}^{(\check{\ell})}(P_1, \ldots, P_{k-1})\phi_k(P_k) + d_{\ge}(t_\infty^{-N(k-2)})}{\psi_{k-1}^{(\check{k-1})}(P_1, \ldots, P_{k-1})\phi_{k-1}(P_k) + d_{\ge}(t_\infty^{-N(k-2)})} \cdot \frac{\phi_{k-1}(P_k)}{\phi_k(P_k)} \\ &\to -\mu_{k-1,\ell}, \quad (P_k \to \infty).\end{aligned}$$

■

2.6.2 Kiepert Formula for Symmetric W-Curves X

In this subsection, we show the functions whose zero points are related to the n-division points in the Jacobian J_X following [MatP2009a]. It is the generalization of the division polynomials $\psi_n(u) = \sigma(nu)/\sigma^{n^2}(u)$ of the elliptic curves based on Frobenius–Stickelberger [FroS1877], Kiepert [Ki1873] and Brioschi and Hermite [BriH1864].

The generalization of ψ_n was studied by Ônishi for hyperelliptic curves [On2002, On2005], for (3, 4)- and (3, 5)-curves [On2009, On2011] and for a more general curves [On2010]. We mention it based on [MatP2009a].

We recall that a symmetric W-curve X is a Weierstrass curve whose Weierstrass gap at ∞ contains $2g-1$ as in Proposition 2.1 and Sect. 2.2.2.4. In this subsection, we consider only symmetric W-curves X and thus $\mathfrak{K}_X = 0$, $\mathfrak{k}_X = 0$, and $\widehat{\phi}_i = \phi_i \in R_X = \widehat{R}_X$ due to Proposition 2.111.

In order to find the $(n+1)$-division points, we extend the argument in the proof in Proposition 2.172. Noting the identification between α_{n+1} in Proposition 2.172 and μ_{n+1} in Definition 2.156, we go on to consider ψ_{n+1} in $\mu_{n+1}(P, P_1, \cdots, P_{n+1})$ of Definition 2.156. We note that for any point Q in X, $\mu_{n+1}(P, P_1, \cdots, P_{n+1})$ with $P_1 = P_2 = \cdots = P_{n+1} = Q$ shows the existence of $Q_1, \ldots, Q_{N(n+1)-n-1}$ such that $(n+1)Q - (n+1)\infty \sim -\left(\sum_{i=1}^{N(n+1)-n-n} Q_i - (N(n+1)-n-1)\infty\right)$. However, what we will argue in this subsection is what point P could be an $(n+1)$-division point; above μ_{n+1} is for any point Q in X but we will consider specific points P in X; for $g=1$, we will find the specific point P such that it satisfies $(n+1)P - (n+1)\infty \sim 0$, and its higher genus analogue. Thus we check the behavior ψ_{n+1} in (2.105) to evaluate the coefficients of its Taylor expansion at such a point P. In other words, we find where the coefficients up to the n-th order vanish.

Definition 2.183 For a point $P \in X$, $\phi_i := \phi_i(P)$ and $u_g := w_g(P)$ $(du_g(P) = \nu^{\mathrm{I}}_g(P))$, we define

$$W^{(n)}_\infty(P) := \begin{pmatrix} \frac{\partial \phi_1}{\partial u_g} & \frac{\partial \phi_2}{\partial u_g} & \cdots & \frac{\partial \phi_n}{\partial u_g} \\ \frac{\partial^2 \phi_1}{\partial u_g^2} & \frac{\partial^2 \phi_2}{\partial u_g^2} & \cdots & \frac{\partial^2 \phi_n}{\partial u_g^2} \\ \vdots & \vdots & \ddots & \vdots \\ \frac{\partial^n \phi_1}{\partial u_g^n} & \frac{\partial^n \phi_2}{\partial u_g^n} & \cdots & \frac{\partial^n \phi_n}{\partial u_g^n} \end{pmatrix},$$

$$\mathfrak{w}^{(n)}_\infty := \det W^{(n)}_\infty \cdot \nu^{\mathrm{I}\,\otimes n(n+1)/2}_g.$$

The differential $\mathfrak{w}_\infty^{(n)}$ can be viewed as a section in $\mathbf{H}^0(X, \mathscr{A}_X^{\otimes n(n+1)/2}(*\infty))$. The statement is checked by a change of variables, see [ArCGH1985, I E-8], and [FarK1991, pp. 84–87].

Theorem 2.184 *Let $n \geq g-1$ be a positive integer. For a symmetric W-curve X, the vanishing of $\mathfrak{w}_\infty^{(n)}$ at $P \in X$ is a necessary and sufficient condition for $w((n+1)\cdot P)$ to belong to W_X^{g-1}.*

The assumption $n \geq g-1$ can be weakened to $n > 0$ for the sufficiency in the following sense: if P is a zero of $\mathfrak{w}_\infty^{(n)}$ then $w((n+1)\cdot P)$ belongs to W_X^m, where $m = N(N(n)-n-1)-(N(n)-n-1) < g$.

This was proved for the case of hyperelliptic curves in [Can1994, On2005], certain trigonal curves in [On2009], and symmetric W-curves [MatP2009a].

As Theorem 2.184 determines inflectionary points for the canonical divisor, and according to [BiV2003], the vanishing of the theta function along the curve at the image of the point P when the Abel–Jacobi map is based at ∞.

Proof We first prove the condition is sufficient, including the $n < g$ case. Assume that the point P is a zero of $\mathfrak{w}_\infty^{(n)}$. This implies that there is an element α in $\mathbf{H}^0(X, \mathscr{O}_X(N(n)\infty))$ such that

$$\alpha = \sum_{i=1}^{n} a_i \phi_i,$$

where $a_i \in \mathbb{C}$ and

$$\left.\frac{\partial^j \alpha}{\partial z^j}\right|_P = 0, \quad \text{for } 1 \leq j \leq n. \tag{2.109}$$

Since α is expanded as $\alpha = \sum_{j=0}^{\infty} \left.\frac{\partial^j \alpha}{\partial z^j}\right|_P z^j$ at P, $\alpha - \alpha_P$ is expanded as

$$\alpha - \alpha_P = \sum_{j=n+1}^{\infty} \left.\frac{\partial^j \alpha}{\partial z^j}\right|_P z^j,$$

where $\alpha_P := \alpha(P)$. In other words, there exist points $(Q_j)_{j=1,\cdots,N(n)-n} \in X$ such that

$$\operatorname{div}(\alpha - \alpha_P) = (n+1)P + \sum_{j=1}^{N(n)-n-1} Q_j - N(n)\infty.$$

From Proposition 2.1, $N(n) - n \leq g$. Let $\ell_n := N(n) - n - 1$. By letting $Q_j = \infty$ for $j = \ell_n, \cdots, g-1$ if necessary, we see that $(n+1)P - (n+1)\infty$ is linearly equivalent to $-\left(\sum_{j=1}^{g-1} Q_j - (g-1)\infty\right)$.

$$(n+1)P - (n+1)\infty \sim -\left(\sum_{j=1}^{g-1} Q_j - (g-1)\infty\right).$$

Using the Serre involution as mentioned in Theorem 2.175, we have

$$(n+1)P - (n+1)\infty \sim \sum_{j=1}^{g-1} Q'_j - (g-1)\infty,$$

for appropriate points Q'_i, $(i = 1, \cdots, g-1)$. More precisely, using Theorem 2.178, when $\ell_n < g-1$ or $n < g$, we see that there exist Q'_i in X $(i = 1, \cdots, N(\ell_n) - \ell_n)$ such that

$$(n+1)P - (n+1)\infty \sim \sum_{j=1}^{N(\ell_n)-\ell_n} Q'_j - (N(\ell_n) - \ell_n)\infty,$$

and $N(\ell_n) - \ell_n \le g-1$. Thus we have proved the sufficiency for any $n > 0$.

We now prove the necessity for $n \ge g$. Let us consider a point P in X such that there exist points $Q_j (\in X \setminus \{\infty\})$, $j = 1, \cdots, \ell \le g-1$,

$$(n+1)P - (n+1)\infty \sim \sum_{j}^{\ell} Q_j - \ell\infty. \tag{2.110}$$

By Proposition 2.172, there is a divisor $\sum_j^{N(\ell)-\ell} Q'_j - (N(\ell) - \ell)\infty$ satisfying the relation

$$(n+1)P + \sum_j^{N(\ell)-\ell} Q'_j - (N(\ell) - \ell + n + 1)\infty \sim 0. \tag{2.111}$$

Thus there is a polynomial function α such that

$$\mathrm{div}(\alpha) = (n+1)P + \sum_j^{N(\ell)-\ell} Q'_j - (N(\ell) - \ell + n + 1)\infty.$$

This implies that there is an integer m such that

$$N(m) = N(\ell) - \ell + n + 1.$$

From the assumption that $\ell \le g-1$, and Proposition 2.1 3, we have $N(\ell) - \ell \le g-1$, and

$$0 < N(m) - n \le g.$$

On the other hand, $m \leq n$ is guaranteed by the assumption $n > g$. For the $m = n$ case, α should be written as

$$\alpha = a_0 + \sum_{i=1}^{n} a_i \phi_i, \quad (a_n = 1),$$

and P is a zero of $\mathfrak{w}_\infty^{(n)}$ when we expand it at P as in (2.109). It is the desired result.

Now we consider $m < n$: for the $m < n$ case, we let $\alpha = a_0 + \sum_{i=1}^{m} a_i \phi_i$, $(a_m = 1)$; again, the assumption on $\mathrm{div}(\alpha)$ gives, using the local parameter z at P,

$$\begin{pmatrix} \frac{\partial \phi_1}{\partial z} & \frac{\partial \phi_2}{\partial z} & \cdots & \frac{\partial \phi_m}{\partial z} \\ \frac{\partial^2 \phi_1}{\partial z^2} & \frac{\partial^2 \phi_2}{\partial z^2} & \cdots & \frac{\partial^2 \phi_m}{\partial z^2} \\ \vdots & \vdots & \ddots & \vdots \\ \frac{\partial^n \phi_1}{\partial z^n} & \frac{\partial^n \phi_2}{\partial z^n} & \cdots & \frac{\partial^n \phi_m}{\partial z^n} \end{pmatrix} \begin{pmatrix} a_1 \\ \vdots \\ a_{m-1} \\ 1 \end{pmatrix} = 0.$$

So the rank must be smaller than m, and the determinant of any $m \times m$-submatrix should vanish at P. The Laplace expansion of the determinant of the matrix $W_\infty^{(n)}$ then shows that $\mathfrak{w}_\infty^{(n)}$ must vanish.

Lastly, for $n = g - 1$, ℓ in (2.110) is $g - 1$ and thus $(N(\ell) - \ell + n + 1)$ in (2.111) is equal to $2g - 1$; we note that $N(g) = 2g = c_X$ is the conductor of H_X, and $c_X - 1 = 2g - 1$ is the Frobenius number in Sect. 2.2.1. However, $2g - 1$ is the gap in the Weierstrass semigroup H_X and thus Q'_{g-1} (one of Q') in (2.111) must be ∞. There is an element α of R_X such that

$$\mathrm{div}(\alpha) \sim gP + \sum_{j=1}^{g-2} Q' - (2g-2)\infty,$$

which means that $\mathfrak{w}_\infty^{(g-1)}$ vanishes at P. ■

Remark 2.185 We do not claim the necessity of the condition of Theorem 2.184 for the case $n < g - 1$. It holds for hyperelliptic [Can1994, On2005] and certain trigonal curves in [On2009], more generally, we suspect it holds for symmetric W-curves. Applying the Galois action, $\zeta : X \to X$, to the relation if it exists,

$$(n+1)P - (n+1)\infty - \sum_{i}^{g-1} Q_i + (g-1)\infty \sim 0,$$

may allow us to translate the vanishing of suitable holomorphic sections of smaller multiples of the canonical bundle into symmetries of the theta divisor. However, in general, our (global) proof does not go through, and more precise knowledge of the Weierstrass semigroup at ∞ may be needed to conclude that belonging to smaller strata is linked to the property of being an inflection for a divisor which is a fraction of the canonical.

We refine the statement of Theorem 2.184:

Theorem 2.186 *Let $n(> g)$, $k(< g-1)$ and $\ell := g-k-1$ be non-negative integers. For a symmetric W-curve X, the vanishing of $\mathfrak{w}_\infty^{(n+\ell)}$, $\mathfrak{w}_\infty^{(n+\ell-1)}$, ..., $\mathfrak{w}_\infty^{(n+1)}$, $\mathfrak{w}_\infty^{(n)}$, at a point P of X is a necessary and sufficient condition for the image of the Abel–Jacobi map w of $(n+1)\cdot P$ to belong to W_X^k.*

Proof We prove the condition is necessary. Let us assume

$$(n+1)P-(n+1)\infty \sim \sum_{j=1}^{k} Q_j^{(0)} - k\infty, \tag{2.112}$$

for $(Q_j^{(0)})_{j=1,\cdots,k} \in S^k(X)$. Then for an positive integer m such that $k+m \le g-1$, we obtain

$$(n+m+1)P-(n+m+1)\infty \sim mP + \sum_{j=1}^{k} Q_j^{(0)} - (k+m)\infty,$$

and the w-image of the right-hand-side belongs to W_X^{g-1} so we can use Theorem 2.184.

For the sufficiency, we apply Theorem 2.184 to the vanishing assumptions. We have $(Q_j^{(i)})_{j=1,\cdots,g-1} \in S^{g-1}(X)$, $(i=0,\cdots,\ell)$ satisfying

$$(n+\ell+1)P-(n+\ell+1)\infty \sim \sum_{j=1}^{g-1} Q_j^{(\ell)} - (g-1)\infty, \tag{2.113}$$

$$(n+\ell)P-(n+\ell)\infty \sim \sum_{j=1}^{g-1} Q_j^{(\ell-1)} - (g-1)\infty, \tag{2.114}$$

$$\cdots\cdots\cdots\cdots\cdots\cdots$$

$$(n+2)P-(n+2)\infty \sim \sum_{j=1}^{g-1} Q_j^{(1)} - (g-1)\infty, \tag{2.115}$$

$$(n+1)P-(n+1)\infty \sim \sum_{j=1}^{g-1} Q_j^{(0)} - (g-1)\infty. \tag{2.116}$$

(2.113)–(2.116) show that a degree-ℓ divisor made up with elements of $\{Q_j^{(0)}\}$ must be $\ell \cdot P$ up to linear equivalence and thus we obtain (2.112). ■

If we assume that the curve is hyperelliptic, $N(m) = 2m$ if $m < g$ and there is a symmetry such that $Q - \infty \sim -Q + \infty$ for $Q \in X$. In this case, we have the relation for $0 \le m \le \ell$,

$$\begin{aligned}(n-m+1)P - (n-m+1)\infty &\sim \sum_{i=1}^{k} Q_i^{(0)} - mP - (k-m)\infty \\ &\sim \sum_{i=1}^{k} Q_i^{(0)} + mP - (k+m)\infty,\end{aligned} \tag{2.117}$$

and thus the left-hand-side of (2.117) belongs to W_X^{m+k}. Accordingly, we have the following result.

Corollary 2.187 [Can1994] *Let $n(> g)$, $k(< g)$ and $\ell := g - k - 1$ be non-negative integers. For a hyperelliptic curve X branched at ∞, the vanishing of $\mathfrak{w}_\infty^{(n+\ell)}$, $\mathfrak{w}_\infty^{(n+\ell-1)}, \dots, \mathfrak{w}_\infty^{(n+1)}, \mathfrak{w}_\infty^{(n)}, \mathfrak{w}_\infty^{(n-1)}, \dots, \mathfrak{w}_\infty^{(n-\ell+1)}, \mathfrak{w}_\infty^{(n-\ell)}$, at a point P of X is a necessary and sufficient condition for $\omega((n+1)\cdot P)$ to belong to W_X^k.*

2.6.3 *Kiepert and Brioschi–Hermite Formulae for Cyclic (r, s)-Curves*

This subsection is for the Kiepert and Brioschi–Hermite formulae for cyclic (r, s)-curve.

Let X be a cyclic plane W-curve, (r, s)-curve, in Sect. 2.3.1.2 [GaPS2000, MatP2009a]. Following [MatP2009a], we investigate the explicit expression of the function $\mathfrak{w}_\infty^{(n)}$. We have a differential in R_X given by

$$D := r y^{r-1} \frac{d}{dx} : R_X \to R_X.$$

Since $\nu_1^{\mathrm{I}} = \dfrac{dx}{r y^{r-1}}$ is a holomorphic one-form, D is the dual of ν_1^{I}, i.e., $\langle D, du_1\rangle = 1$, Further, for smallest positive integers p and q satisfying the relation $ps - qr = 1$, $\nu_g^{\mathrm{I}} = \dfrac{y^{r-p-1}x^{q-1}dx}{r y^{r-1}}$ is a holomorphic one-form and thus its dual $\dfrac{\partial}{\partial u_g}$ is equal to $\dfrac{\partial}{\partial u_g} = \dfrac{1}{y^{r-p-1}x^{q-1}} D$. We consider the base change by $\mathfrak{w}_\infty^{(n)} = \psi_\infty^{(n+1)} \nu_1^{\mathrm{I} \otimes n(n+1)/2}$.

Proposition 2.188 *Under the assumption in this subsection, $R_X[D]$ is a differential ring.*

Definition 2.189 For an integer $n > 1$, we define

$$\psi_\infty^{(n+1)} := \det \begin{pmatrix} D\phi_1 & D\phi_2 & \cdots & D\phi_n \\ D^2\phi_1 & D^2\phi_2 & \cdots & D^2\phi_n \\ \vdots & \vdots & \ddots & \vdots \\ D^n\phi_1 & D^n\phi_2 & \cdots & D^n\phi_n \end{pmatrix}.$$

It follows easily from the assumptions that $\psi_\infty^{(n)}$ belongs to $R_X = \mathbf{H}^0(X, \mathscr{O}_X(*\infty))$. Hence we have the following theorem.

Theorem 2.190 *For a positive integer $n \geq g$, $\psi_\infty^{(n+1)}$ is an element in R_X whose vanishing at P on the curve X is a necessary and sufficient condition for the normalized image of $(n+1) \cdot P$ in the Picard group to belong to W_X^{g-1}.*

Proof Essentially the same as Appendix in [Mat2002a] ([On2005]). By the coordinate transformation in [[FarK1991], p. 87], the zeros of $\psi_\infty^{(n+1)} du_1^{\otimes n(n+1)/2}$ coincide with zeros of $\mathfrak{w}_\infty^{(n)}$. However, since du_1 is an element of $\mathbf{H}^0(X, \mathscr{A}_X)$, the measure $du_1^{\otimes n(n+1)/2}$ does not vanish over X. ■

Remark 2.191 $\psi_\infty^{(n+1)}$ is unique up to a constant factor. If we replace du_1 with another one-form in $\mathbf{H}^0(X, \mathscr{A}_X)$, the coefficient of the measure belongs to $\mathscr{Q}(R_X)$ but not in R_X.[14]

Let us consider a Brioschi–Hermite type expression of $\psi_\infty^{(n+1)}$ as follows. By letting $D_x = d/dx$, we notice that

$$\begin{pmatrix} D_x x & D_x x^2 & \cdots & D_x x^\ell \\ D_x^2 x & D_x^2 x^2 & \cdots & D_x^2 x^\ell \\ D_x^3 x & D_x^3 x^2 & \cdots & D_x^3 x^\ell \\ \vdots & \vdots & \ddots & \vdots \\ \vdots & \vdots & & \vdots \\ D_x^{n-1} x & D_x^{n-1} x^2 & \cdots & D_x^{n-1} x^\ell \\ D_x^n x & D_x^n x^2 & \cdots & D_x^n x^\ell \end{pmatrix} = \begin{pmatrix} 1 & 2x & 3x^2 & \cdots & \ell x^{\ell-1} \\ & 2 & 6x & \cdots & \vdots \\ & & 6 & \cdots & \vdots \\ & & & \ddots & \vdots \\ & & & & \ell! \\ & & & & \\ & & & & \end{pmatrix}.$$

We define $\eta_j \in R_X$, $\ell(n)$, and $L(n)$ so that: (1) $\phi_1 \wedge \phi_2 \wedge \cdots \wedge \phi_n = \epsilon_n x \wedge x^2 \wedge \cdots \wedge x^{\ell(n)} \wedge \eta_1 \wedge \cdots \wedge \eta_{L(n)}$, where ϵ_n is a sign, (2) η_j cannot be expressed by finite linear combination of x^k's $(k = 1, \cdots)$, (3) $\eta_j \neq \eta_i$ if $j \neq i$, (4) there uniquely exists $m(j)$ for j satisfying $\phi_{m(j)} = \eta_j$ and when $m(j)$ is between m and j, $m(i) > m(j)$ for $i > j$, and (5) $\ell(n) + L(n) = n$.

---

[14] For example, by letting $d\widetilde{u} := \dfrac{x dx}{r y^{r-1}}$ and $\mathfrak{w}_\infty^{(n)} = \widetilde{\psi}_\infty^{(n+1)} d\widetilde{u}^{\otimes n(n+1)/2}$, then we obtain $\widetilde{\psi}_\infty^{(n+1)} = x^{-n(n+1)/2} \psi_\infty^{(n+1)}$, which is not an element in R_X in general.

Proposition 2.192 *For a fixed integer $n > 1$ and an integer i, let $M(i) := m(i) + \ell(n)$. Then the $\psi_\infty^{(n+1)}$ is equal to*

$$\epsilon_n \left(\prod_{j=1}^{\ell(n)} j!\right) (ry^{r-1})^{n(n+1)/2} \det \begin{pmatrix} D_x^{M(1)}\eta_1 & D_x^{M(1)}\eta_2 & \cdots & D_x^{M(1)}\eta_{L(n)} \\ D_x^{M(2)}\eta_1 & D_x^{M(2)}\eta_2 & \cdots & D_x^{M(2)}\eta_{L(n)} \\ \vdots & \vdots & \ddots & \vdots \\ D_x^{M(L(n))}\eta_1 & D_x^{M(L(n))}\eta_2 & \cdots & D_x^{M(L(n))}\eta_{L(n)} \end{pmatrix}.$$

This is a generalization of the Brioschi–Hermite formula for the elliptic curve [BriH1864] and Cantor's formulas for hyperelliptic curves [Can1994]. For a hyperelliptic curve $y^2 = f(x)$, η_i is simply given by $\eta_i = x^{j-1}y$ for a certain integer j and thus using $D_x x^{j-1} y = x^{j-1} D_x y + \cdots$, we obtain a simpler expression [Can1994, Mat2002a, On2005].

Corollary 2.193 *For a hyperelliptic curve, $y^2 = f(x)$ and $n \geq g + 3$,*

$$\psi_\infty^{(n)} = \begin{cases} (2y)^{g(g+1)/2}\alpha_n(x) \text{ for } n+g \text{ odd,} \\ (2y)^{g(g-1)/2}\alpha_n(x) \text{ for } n+g \text{ even,} \end{cases}$$

where α_n belongs to $\mathbb{C}[x]$ which is coprime of y in general.

Proof Let $p_n = \lfloor (n+g-1)/2 \rfloor$ which is the maximal number of x^{p_n} in $\{1, x, \phi_2, \ldots, \phi_n\}$ and $q_n = \lfloor (n-g-2)/2 \rfloor$ which is the maximal number of $x^{q_n}y$ there. ψ_n is explicitly given by

$$\psi_\infty^{(n)} = \varepsilon_n (2y)^{n(n-1)/2} \begin{vmatrix} y^{[p_n+1]} & y^{[p_n]} & \cdots & y^{[p_n-q_n+1]} \\ y^{[p_n+2]} & y^{[p_n+1]} & \cdots & y^{[p_n+q_n]} \\ \vdots & \vdots & \ddots & \vdots \\ y^{[n-1]} & y^{[n]} & \cdots & y^{[p_n+1]} \end{vmatrix},$$

where $y^{[n]} = \dfrac{1}{n!}\dfrac{d^n}{dx^n} y$ and ε_n is a sign ± 1.

We consider the case where $n+g$ is odd and where $n+g$ is even. Assume that $n+g$ is odd, i.e., $n-g$ is also odd. Then $p_n = \dfrac{n+g-1}{2}$ and $q_n = \dfrac{n-g-3}{2}$. $p_n + q_n + 1 = n - 1$, whereas $p_n - q_n = g + 1$. Here $\dfrac{d}{dx} y = \dfrac{1}{2y}\dfrac{df}{dx}$, and thus $y^{2\ell-1} \cdot \dfrac{d^\ell}{dx^\ell} y$ belongs to $\mathbb{C}[x]$. Thus $y^{(2(p_n+2)-1)(q_n+1)}$ times the above determinant belongs to $\mathbb{C}[x]$, and thus $y^{n(n-1)/2} y^{-(2(p_n+2)-1)(q_n+1)}$ is the front factor.

$$-(2(p_n+2)-1)(q_n+1) = \frac{1}{2}n(n-1) - \frac{1}{2}g(g+1).$$

We obtain the above statement of the case that $g+n$ is odd.

We consider the case that $n+g$ is even, or $n-g$ is also even. Then $p_n = \dfrac{n+g-2}{2}$ and $q_n = \dfrac{n-g-2}{2}$. $p_n + q_n + 1 = n-1$, whereas $p_n - q_n = g$.

$$-(2(p_n+2)-1)(q_n+1) = \frac{1}{2}n(n-1) - \frac{1}{2}g(g-1).$$

We also prove the statement of the even case. ∎

2.7 Period Matrices on W-Curves

In this section, we show the particular properties of the period integrals ω_i' and ω_i'' on W-curves $i = 1, \dots, g$ defined in Definition 2.112. It is evident that the period integrals or their contours (the basis of $\mathbf{H}_1(X, \mathbb{Z})$) consist of the paths between the ramification points on X. Since the infinity on X is one of the ramification points as a particular property of W-curves, we can choose one of the endpoints of the paths as the base point. The additive structure in path space (as a groupoid) allows the base changes. Further, the projective structure of W-curves (2.16) reduces the other endpoints to branch point on $\mathbb{P}$; $\varpi_r : \mathfrak{B}_X \to \mathfrak{B}_{\mathbb{P}}$. In other words, the paths are given by lifting the path on $\mathbb{P}$.

Even though the integrand as a meromorphic one-form is defined on X, the contour of the period integral consists of the paths defined on $\mathbb{P}$, from ∞ to the finite branch point $b_r = \varpi_r(B_r)$.

In other words, we normalize the period matrices by the integrals from ∞ to the finite branch point $\gamma_{B_i,\infty}$. This manner appeared in the Weierstrass elliptic function theory, and Picard's investigation of the trigonal curve [Pic1883].

However, it is noted that the integral ω_{B_r} is usually chosen by a lift of the path from $\gamma_{b_r,\infty}$ on $\mathbb{P}$ (choice is made by the integrand), and thus we have an ambiguity regarding which B_r is chosen in $\varpi_r^{-1}b_r$. In other words, we must care about the global group actions, e.g., the Galois action if it exists.

2.7.1 Period Matrices and Branch-Point Integrals on $\gamma_{B_i,\infty}$

This subsection is for the period matrices and branch-point integrals on $\gamma_{B_i,\infty}$.

Let us take up the ramification point $B_i \in \mathfrak{B}_X$, $(i = 1, 2, \dots, \ell_{\mathfrak{B}})$, and its branch point $b_i = \varpi_r(B_i) \in \mathfrak{B}_{\mathbb{P}}$. We consider the path $\gamma_{B_i,\infty}$ in X such that there exists a path $\gamma_{b_i,\infty}$ in $\mathbb{P}$ given by the continuous bijection $\varpi_r : \gamma_{B_i,\infty} \to \gamma_{b_i,\infty}$. Following Weierstrass' conventions, we have an integral

$$\omega_{B_i} := \int_\infty^{B_i} \nu^{\mathrm{I}} := \int_{\gamma_{B_i,\infty}} \nu^{\mathrm{I}} \quad \left(\omega_{B_i,j} = \int_\infty^{B_i} \nu_j^{\mathrm{I}}\right), \tag{2.118}$$

which we call a *branch-point integral* (or *canonical integral*) since the integral can be computed in $\mathbb{P}$, i.e,

$$\omega_{B_i} = \int_{\gamma_{b_i,\infty}} \varpi_r(\nu^{\mathrm{I}}) \tag{2.119}$$

by formally regarding $(\varpi_r \nu_i^{\mathrm{I}})(P)$ at $P \in \varpi_r(\gamma_{B_i,\infty}) \in \mathbb{P}$ as a one-form in $\varpi_r(P)$. In terms of the branch-point integrals, we basically express the period matrices, $2\omega'$, and $2\omega''$. We show examples of the connections.

2.7.2 Elliptic Curve Case $y^2 = (x - e_1)(x - e_2)(x - e_3)$

By letting $\widehat{y} = y/2$, we have the standard Weierstrass form (1.26) of the elliptic curve E to consider the elliptic integrals in this subsection. It is known that the period matrix of the elliptic curve is given by the twice of the branch-point integrals,[15]

$$2\omega' = 2\omega_{B_1} = 2\int_\infty^{B_1} \nu^{\mathrm{I}}, \quad 2\omega'' = 2\omega_{B_3} = 2\int_\infty^{B_3} \nu^{\mathrm{I}},$$

where $B_i := (e_i, 0)$ because the homology basis $\{\alpha, \beta\}$ is illustrated in Fig. 2.6; due to the hyperelliptic involution $\widehat{\zeta}_H \nu^{\mathrm{I}} = -\nu^{\mathrm{I}}$, $\oint_\alpha \nu^{\mathrm{I}} = \int_\infty^{B_1} \nu^{\mathrm{I}} - \int_\infty^{B_1} \widehat{\zeta}_H \nu^{\mathrm{I}}$ asserts the equality and same as $2\omega''$ for β. Conventionally, ω_{B_i} is denoted by ω_i. Since Fig. 2.6 shows that $\alpha + \alpha_0 = 0$ homologically and $\oint_{\alpha_0} \nu^{\mathrm{I}} = 2(\omega_{B_3} - \omega_{B_2})$, we have the identity

$$\omega_{B_1} + \omega_{B_3} = \omega_{B_2}. \tag{2.120}$$

We remark that the origin of the identity (2.120) is geometrical or homological property of E. When we describe the period matrices in terms of the branch-point integrals, such an identity should be noticed.

---

[15] We mention, here, our convention regarding the period matrices:

The periods of the elliptic curves are conventionally denoted by (ω, ω') following Weierstrass [Wei1875b] and the alternative one is employed (ω_1, ω_2), e.g., Weber [Web1908]. Thus Baker employed the notations of the period matrices by (ω, ω') even for curves of higher genus [Ba1897, e.g., p. 197], and Buchstaber, Enolskii, and Leykin [BucEL2020] succeed Baker's convention. It is an excellent choice.

However, we sometimes want to represent it by ω. Further, we later express a vector z in $\mathbb{C}^g$ denoted by $z = \omega z' + \omega' z''$ or $z = z' + \tau z''$ e.g., $\delta = \delta' + \delta'' \tau$. They might look like mixed of two conventions. Baker introduced the notation $\Omega_P = \omega P + \omega' P'$ [Ba1897, p. 283] but it may be a little too exaggerated.

Accordingly, we have employed the notation $\omega = (\omega', \omega'')$ in this book. As we sometimes refer to the branch-point integral as $\omega_i = \omega_{B_i}$, it may also avoid some confusion.

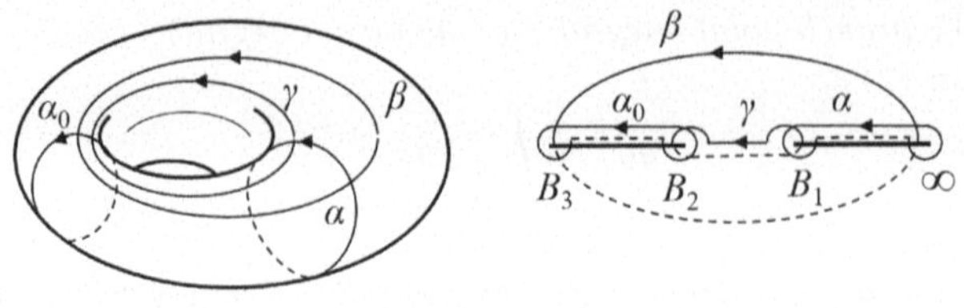

Fig. 2.6 The basis of $\mathbf{H}_1(E, \mathbb{Z})$

2.7.3 *Hyperelliptic Curve Case* $y^2 = \prod_{i=0}^{2g}(x - b_i)$

We consider the hyperelliptic curves X, $y^2 = \prod_{i=0}^{2g}(x - b_i)$, given by (2.65) in this subsection. In terms of the branch-point integrals for the ramification point $B_i = (b_i, 0)$,[16] the elements in the period matrix for the contours in Fig. 2.4 are expressed in the following lemma:

Lemma 2.194

$$2\omega_i'' = 2\omega_{B_{2i}}, \quad 2\omega_i' = 2(\omega_{B_{2i}} - \omega_{B_{2i+1}}).$$

Proof $2\omega_i' = \oint_{\alpha_i} \nu^{\mathrm{I}} = \int_{B_{2i-1}}^{B_{2i}} \nu^{\mathrm{I}} + \int_{B_{2i}}^{B_{2i-1}} \widehat{\zeta}_H \nu^{\mathrm{I}}$, and $2\omega_i'' = \oint_{\beta_i} \nu^{\mathrm{I}} = \int_{\infty}^{B_{2i}} \nu^{\mathrm{I}} + \int_{B_{2i}}^{\infty} \widehat{\zeta}_H \nu^{\mathrm{I}}$. ■

We investigate these integrals more precisely to represent the branch-point integral ω_{B_j} in terms of the period integrals, ω_i' and ω_i'' [FarK1991, pp. 322–3][EnR2008, BucEL2012].

We introduce the dual expressions of Fig. 2.4 as in Fig. 2.7 ($g = 3$ case). We find the identities among the integrals. Homologically we have the following identities:

$$\alpha_0 + \alpha_1 + \cdots + \alpha_g = 0, \quad \beta_g = \gamma_g + \alpha_0, \quad \beta_i = \beta_{i+1} + \gamma_i, \ (i = 1, 2, \ldots, g-1),$$

where γ's are loops as illustrated in Fig. 2.7. On the other hand, we have

$$\oint_{\alpha_0} \nu^{\mathrm{I}} = 2(\omega_{B_0}), \quad \oint_{\gamma_g} \nu^{\mathrm{I}} = 2(\omega_{B_{2g}} - \omega_{B_0}), \quad \oint_{\gamma_i} \nu^{\mathrm{I}} = 2(\omega_{B_{2i}} - \omega_{B_{2i+1}}),$$

$(i = 1, 2, \ldots, g-1)$.

Hence we have the identities corresponding to (2.120) so that the branch-point integrals are expressed by the period integrals:

---

[16] Even though in Definition 2.29, we assign B_0 to ∞ in X, we employ another convention for the particular curves.

Lemma 2.195

$$\begin{aligned}
\omega_{B_0} &= -\sum_{i=1}^{g} \omega_i', \\
\omega_{B_{2i}} &= \omega_{B_0} + \sum_{j=i+1}^{g} \omega_i' + \frac{1}{2}\sum_{j=i}^{g} \oint_{\gamma_i} \nu^{\mathrm{I}} = -\sum_{j=1}^{i} \omega_j' + \omega_i'', \quad (i = 1, \dots, g), \\
\omega_{B_{2i-1}} &= -\sum_{j=1}^{i-1} \omega_j' + \omega_i'', \quad (i = 1, \dots, g).
\end{aligned} \tag{2.121}$$

Proof They are proved by induction from $i = g$. ■

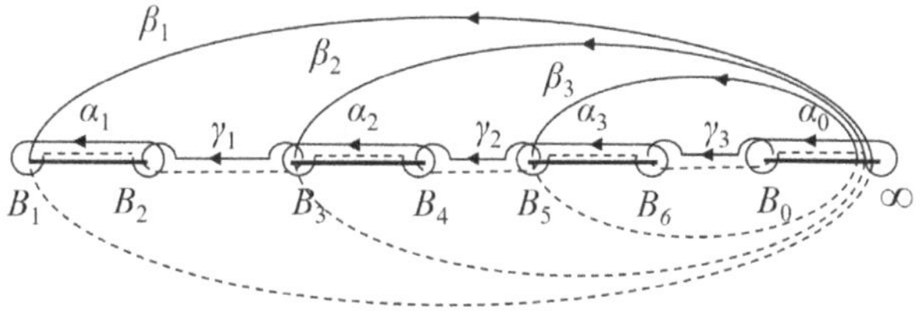

Fig. 2.7 The basis of $\mathbf{H}_1(X, \mathbb{Z})$

2.7.4 Cyclic (3,4)-Curve Case $y^3 = x(x - s)(x - b_1)(x - b_2)$

In this subsection, we handle the case of the (3,4)-curve (2.67), in which we let $b_0 = 0$ and $b_3 = s$. Let us refer to the curve as X_s.[17] This problem to find the period matrices of this curve was studied by Picard [Pic1883] and Shiga [Shi1988]. However, for later convenience, we employ different contours from them.

The standard homology basis $(\alpha_1, \alpha_2, \alpha_3, \beta_1, \beta_2, \beta_3)$ satisfying the relations, $\langle \alpha_i, \beta_j \rangle = \delta_{ij}$, $\langle \alpha_i, \alpha_j \rangle = 0$, and $\langle \beta_i, \beta_j \rangle = 0$, in (2.58) is illustrated in Fig. 2.8.

Let us consider the complete Abelian integrals ω' and ω'' of the first kind and the complete Abelian integrals η' and η'' of the second kind. Since we have the integral along the contour $\gamma_{B_a,\infty}$ from ∞ to the ramification point $B_a = (b_a, 0)$ $(a = 0, 1, 2, 3)$ [MatP2015], i.e.,

$$\omega_a := \int_{\infty}^{B_a} \nu^{\mathrm{I}} = \int_{\gamma_{B_a,\infty}} \nu^{\mathrm{I}}, \tag{2.122}$$

[17] Even though in Definition 2.29, we assign B_0 to ∞ in X, we employ another convention for the particular curves; B_0 denotes a finite ramification point of $b_0 = 0$.

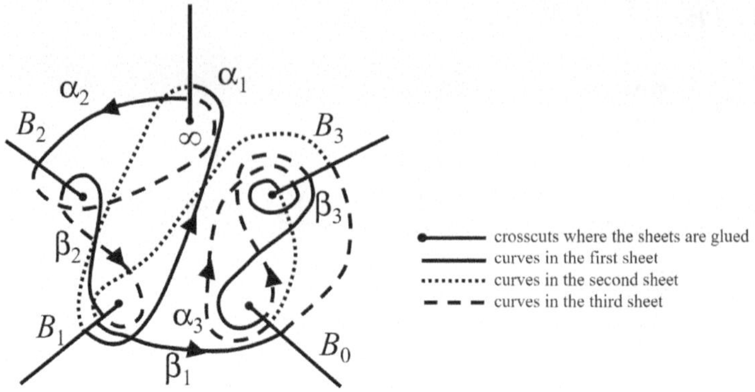

Fig. 2.8 The basis of $\mathbf{H}_1(X_s, \mathbb{Z})$

the period matrices $\omega' = (\omega_1', \omega_2', \omega_3')$ and $\omega'' = (\omega_1'', \omega_2'', \omega_3'')$ are described in terms of ω_{B_a} as follows.

Lemma 2.196

$$^t(\omega_1', \omega_2', \omega_3', \omega_1'', \omega_2'', \omega_3'') = {}^t((\omega_{B_0}, \omega_{B_1}, \omega_{B_2}, \omega_{B_3})W_{X_s}),$$

where W_{X_s} is a rank three matrix given by

$$W_{X_s} := \frac{1}{2}\begin{pmatrix} 0 & 0 & \widehat{\zeta}_3(1-\widehat{\zeta}_3) & 1-\widehat{\zeta}_3^2 & 0 & 1-\widehat{\zeta}_3^2 \\ \widehat{\zeta}_3 - 1 & 0 & 0 & \widehat{\zeta}_3 - 1 & \widehat{\zeta}_3^2 - 1 & 0 \\ 0 & 1-\widehat{\zeta}_3^2 & 0 & 0 & 1-\widehat{\zeta}_3^2 & 0 \\ 0 & 0 & \widehat{\zeta}_3^2 - \widehat{\zeta}_3 & \widehat{\zeta}_3^2 - \widehat{\zeta}_3 & 0 & \widehat{\zeta}_3^2 - 1 \end{pmatrix}.$$

The transpose is introduced because the action of $\widehat{\zeta}_3$ on the integrals is left-to-right. Since the integral over a path which is homotopic to a point, e.g.,

$$(1-\widehat{\zeta}_3^2)\omega_{B_2} + \widehat{\zeta}_3^2(1-\widehat{\zeta}_3^2)\omega_{B_1} + \widehat{\zeta}_3(1-\widehat{\zeta}_3^2)\omega_{B_0} + (1-\widehat{\zeta}_3^2)\omega_{B_3} = 0, \qquad (2.123)$$

is equal to zero, which corresponds to (2.120), ω_{B_a} $(a = 0, 1, 2, 3)$ are described in terms of $\omega' = (\omega_1', \omega_2', \omega_3')$:

Lemma 2.197

$$^t(\omega_{B_0}, \omega_{B_1}, \omega_{B_2}, \omega_{B_3}) = {}^t((\omega_1', \omega_2', \omega_3')V_{X_s}),$$

where

$$V_{X_s} := \frac{2}{3}\begin{pmatrix} \widehat{\zeta}_3^2 - 1 & \widehat{\zeta}_3^2 - 1 & 0 & \widehat{\zeta}_3^2 - 1 \\ (1-\widehat{\zeta}_3)\widehat{\zeta}_3 & 0 & 1-\widehat{\zeta}_3 & (1-\widehat{\zeta}_3)\widehat{\zeta}_3 \\ \widehat{\zeta}_3^2 - 1 & 0 & 0 & \widehat{\zeta}_3 - 1 \end{pmatrix}.$$

Proof By considering (2.123) and the relation

$$ {}^t(\omega'_1, \omega'_2, \omega'_3, 0) = {}^t((\omega_{B_0}, \omega_{B_1}, \omega_{B_2}, \omega_{B_3})V) $$

for a certain matrix V, the inverse matrix of V is reduced to V_{X_s}. ∎

Lemma 2.197 can be regarded as a generalization of Lemma 2.209 which holds for the genus one case.

Recalling (2.134) and (2.135) in the genus one case, we have the following lemma:

Lemma 2.198

$$ \begin{pmatrix} \omega''_{11} & \omega''_{12} & \omega''_{13} \\ \omega''_{21} & \omega''_{22} & \omega''_{23} \\ \omega''_{31} & \omega''_{32} & \omega''_{33} \end{pmatrix} = \begin{pmatrix} -\zeta_3^2\omega'_{12} & -\zeta_3^2\omega'_{11} + \omega'_{12} & -\zeta_3\omega'_{13} \\ -\zeta_3^2\omega'_{22} & -\zeta_3^2\omega'_{21} + \omega'_{22} & -\zeta_3\omega'_{23} \\ -\zeta_3\omega'_{32} & -\zeta_3\omega'_{31} + \omega'_{32} & -\zeta_3^2\omega'_{33} \end{pmatrix}, $$

$$ \begin{pmatrix} \eta''_{11} & \eta''_{12} & \eta''_{13} \\ \eta''_{21} & \eta''_{22} & \eta''_{23} \\ \eta''_{31} & \eta''_{32} & \eta''_{33} \end{pmatrix} = \begin{pmatrix} -\zeta_3\eta'_{12} & -\zeta_3\eta'_{11} + \eta'_{12} & -\zeta_3^2\eta'_{13} \\ -\zeta_3\eta'_{22} & -\zeta_3\eta'_{21} + \eta'_{22} & -\zeta_3^2\eta'_{23} \\ -\zeta_3^2\eta'_{32} & -\zeta_3^2\eta'_{31} + \eta'_{32} & -\zeta_3\eta'_{33} \end{pmatrix}. $$

2.7.5 *Periods of $X_{\hat{0}}$, (3, 4, 5)-Type*

In this subsection, for the case of $X_{\hat{0}}$ (3, 4, 5)-type, we specifically provide a basis of the homology, α's and β's, illustrated in Fig. 2.9.

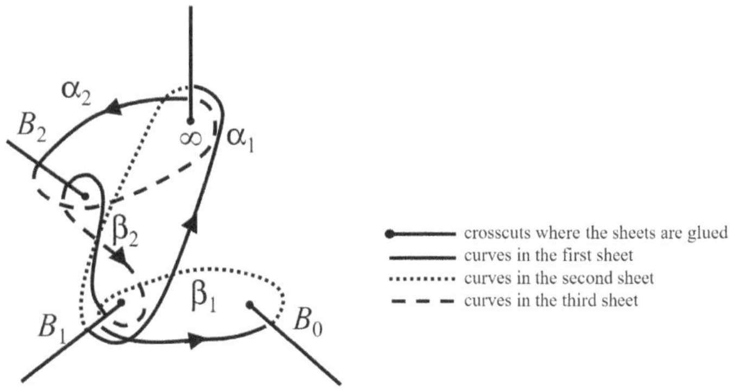

Fig. 2.9 The basis of $\mathbf{H}_1(X_{\hat{0}}, \mathbb{Z})$: We note the fact that B_0 is the singular point

Obviously, the period matrices, $\omega' = (\omega'_1, \omega'_2)$ and $\omega'' = (\omega''_1, \omega''_2)$ are described in terms of ω_{B_a}:

Lemma 2.199

$$ {}^t(\omega_1', \omega_2', \omega_1'', \omega_2'') = {}^t((\omega_{B_0}, \omega_{B_1}, \omega_{B_2}) W_{X_{\widehat{0}}}), $$

where

$$ W_{X_{\widehat{0}}} := \frac{1}{2} \begin{pmatrix} 0 & 0 & 1-\widehat{\zeta_3} & 0 \\ \widehat{\zeta_3}-1 & 0 & \widehat{\zeta_3}-1 & \widehat{\zeta_3^2}-1 \\ 0 & 1-\widehat{\zeta_3^2} & 0 & 1-\widehat{\zeta_3^2} \end{pmatrix}. $$

Graphically we also have the identity among ω_{B_a}:

Lemma 2.200

$$ (\widehat{\zeta_3} - \widehat{\zeta_3^2})\omega_{B_0} + (\widehat{\zeta_3^2} - \widehat{\zeta_3})\omega_{B_1} + (1 - \widehat{\zeta_3^2})\omega_{B_2} = 0. $$

We have the relations between ω' and ω'' (η' and η''):

Lemma 2.201

$$ \begin{pmatrix} \omega_{11}'' & \omega_{12}'' \\ \omega_{21}'' & \omega_{22}'' \end{pmatrix} = \begin{pmatrix} -\zeta_3^2\omega_{12}' & -\zeta_3^2\omega_{11}' + \omega_{12}' \\ -\zeta_3\omega_{22}' & -\zeta_3\omega_{21}' + \omega_{22}' \end{pmatrix}, $$

$$ \begin{pmatrix} \eta_{11}'' & \eta_{12}'' \\ \eta_{21}'' & \eta_{22}'' \end{pmatrix} = \begin{pmatrix} -\zeta_3\eta_{12}' & -\zeta_3\eta_{11}' + \eta_{12}' \\ -\zeta_3^2\eta_{22}' & -\zeta_3^2\eta_{21}' + \eta_{22}' \end{pmatrix}. $$

Proof It is proved similar to Lemma 2.198. ∎

2.8 Birational Curves and Finite Coverings of W-Curves

Though the Weierstrass canonical form is too restricted, we can handle more general curves in the framework of the theory of W-curves by considering the birational transformation of W-curves X, and thus in this section. we show them.

Since the W-curve X has the natural projection $\varpi_r : X \to \mathbb{P}$, it is easy to consider a birational transformation from $\mathbb{P}$ to $\mathbb{P}$ by preserving the fiber structure. We may take the projective linear transformation, PSL(2, $\mathbb{C}$) for $\mathbb{P}$; for $x \in \mathbb{P}$ and $g = \begin{pmatrix} a & b \\ c & d \end{pmatrix} \in \mathrm{SL}(2, \mathbb{C})$, the action is defined by $g(x) = \dfrac{ax+b}{cx+d}$. The action induces a transformation $g : X \to X'$ where X' is a non-singular algebraic curve; in general, X' is not a W-curve. We refer to X' as a *birational curve* in this book.

As mentioned in Remark 2.9 and Example 2.241 in Appendix 2.10, it also provides the affine system of the W-curve at ∞ as a projective curve; it means the *projectivization* of X. Further, at the branch point $B \in X$, we may consider a covering of $\mathbb{P}$, and then, it also induces a covering of X. In the following section, we describe the transformations for particular curves. We start with the double covering of the elliptic curves related to the Jacobi sn, cn, and dn functions.

2.8.1 Double Covering of Elliptic Curves

We treat the Jacobi elliptic functions in our framework. For the genus-one case, the Weierstrass $\wp(u)$ function represents a curve $E : y^2 = (x - e_1)(x - e_2)(x - e_3)$, whereas the Jacobi $z = \mathrm{sn}(u)$ functions is defined for the curve $\widehat{E}$ as a target space,

$$w^2 = (1 - z^2)(1 - k^2 z^2), \tag{2.124}$$

where $k^2 = (e_2 - e_1)/(e_3 - e_1)$. Both E and $\widehat{E}$ are connected as follows.

1. Let $z := \sqrt{(x - e_1)/(e_2 - e_1)}$ for $(x, y) \in E$, which is the local parameter at the ramification point $(e_1, 0)$, and then we obtain a singular curve $\widehat{E}' : y^2 = (e_2 - e_1)^3 z^2 (z^2 - 1)(z^2 - k^{-2})$ where $k^2 := (e_2 - e_1)/(e_3 - e_1)$.
2. We normalize $\widehat{E}'$ by introducing $w = yk/z\sqrt{(e_2 - e_1)^2}$ to obtain $\widehat{E}$.
3. $\widehat{E}$ is the double covering of E, $\varpi_{\widehat{E}} : \widehat{E} \to E$ whose local behavior is given by

$$du = \frac{dx}{2y} = 2\sqrt{e_2 - e_1}\frac{dz}{2w},$$

4. The sn function is defined so that it has the following properties:

    a. We note that z can be regarded as the local parameter at the ramification point $(e_1, 0)$ in E and as a global meromorphic function on $\widehat{E}$ (classically a global double-valued meromorphic function (a section of a line bundle) on E, $\varpi_{\widehat{E}}(z) = z^2 = (x - e_1)/(e_2 - e_1)$), and the Jacobian $J_{\widehat{E}}$ of $\widehat{E}$ is a double covering of the Jacobians J_E of E,
    b. there are two Galois action $\widehat{\zeta}'$ and $\widehat{\zeta}''$ on $\widehat{E}$; 1) $\widehat{\zeta}'$ is induced from $\widehat{\zeta}_2 = \widehat{\zeta}_H$ (hyperelliptic involution) on E, $\widehat{\zeta}'(z, wz) = (-z, -wz)$ on $\widehat{E}$, and 2) $\widehat{\zeta}''$ is induced from $\varpi_{\widehat{E}}$ such that $\widehat{\zeta}''(z, wz) = (z, -wz)$,
    c. the z as a function of Jacobian $J_{\widehat{E}}$, which is, here, denoted by $\mathrm{al}_1(u)$ (al_r is sometimes denoted by $\wp_r$ (co-pei) function), is given by the ratio of the sigma function of E such that $\mathrm{al}_1(u) = \dfrac{\mathrm{e}^{-\eta_1 u}\sigma(u + \omega_1)}{\sigma(\omega_1)\sigma(u)}$ and related to the sn function, $\mathrm{sn}(u) = \dfrac{1}{\mathrm{al}_1(u)}$; $\mathrm{al}_1(u + 2\omega') = -\mathrm{al}_1(u)$, $\mathrm{al}_1(u + 4m\omega' + 2n\omega'') = \mathrm{al}_1(u)$, and
    d. $\mathrm{sn}^2(u) + k^2\mathrm{cn}^2(u) = 1$.

We extend these properties to more general W-curves as follows.

2.8.2 Birational Curves

As mentioned in Remark 2.9 and Example 2.241 in Appendix 2.10, the birational curve provides the properties of W-curves as projective curves. This procedure is essentially the same as that in [Mum1984, 3.16] so that the affine hyperelliptic curve X can be regarded as a projective curve. In other words, the procedure is a projectivization of X. Though we don't touch the properties precisely in this book, we demonstrate the birational curves of the hyperelliptic curves as a relation between the $(2, 2g+1)$-curves and $(2, 2g+2)$-curves and a certain class of cyclic (r, s)-curves, and comment the projectivization of X in this subsection.

The degree of x of the defining equation of E is odd or 3, in contrast to that of $\widehat{E}$ is even or 4. We show how the $(2, 2g+1)$-curve of X is connected with a $(2, 2g+2)$-curve $X^\flat$ in this subsection.

Let us consider the rational functions,

$$\xi = \frac{\beta}{x} + \beta_{2g+2}, \quad \beta_i = \frac{\beta}{b_i} + \beta_{2g+2}, \quad \eta = \sqrt{\frac{\beta^{2g+1}}{-\prod_i^{2g-1} b_j} \frac{y}{x^{g+1}}} \tag{2.125}$$

for a nonvanishing parameter $\beta \in \mathbb{C} \setminus \{0\}$ and certain $\beta_{2g+2} \in \mathbb{C}$. They give the birational map from X to a hyperelliptic curve $X^\flat$:

$$\begin{aligned} \eta^2 &= \widehat{f}(\xi) \\ &= \bar{\lambda}_{2g+2}\xi^{2g+2} + \widehat{\lambda}_{2g+1}\xi^{2g+1} + \widehat{\lambda}_1 x \cdots + \bar{\lambda}_2\xi^2 + +\bar{\lambda}_1\xi + \bar{\lambda}_0 \\ &= (\xi - \beta_1)\cdots(\xi - \beta_g)(x - \beta_{g+1})\cdots(\xi - \beta_{2g})(\xi - \beta_{2g+1})(\xi - \beta_{2g+2}), \end{aligned} \tag{2.126}$$

where $\widehat{\lambda}_{2g+2} = 1$ and $\widehat{\lambda}_j$'s are defined by β_j's. Here β_{2g+2} is determined such that β_j's are mutually disjoint complex values.

In this transformation, we note the process; in the first step, by letting $\xi(x)$, we have a singular curve given by $f(x(\xi), y) = 0$, and in the second step, we normalize it by introducing η to obtain the above curve $X^\flat$ as mentioned in examples of Appendix 2.10. The ∞ in X is transposed to $(\beta_{2g+2}, 0)$ in $X^\flat$. In $X^\flat$, whereas we have two infinite-points, $(\infty, \pm\infty)$. Mumford introduced $X^\flat$ so that X can be regarded as a projective curve [Mum1984, 3.16]; we interpret that X consists of the affine covering (x, y) and (ξ, η), though we do not actively adopt that depiction in this book. (See Example 2.241.)

The W-normalized differentials of the first kind in $X^\flat$ are defined by

$$\nu_1^{\mathrm{I}} = \frac{d\xi}{2\eta}, \quad \nu_2^{\mathrm{I}} = \frac{\xi d\xi}{2\eta}, \quad \ldots, \quad \nu_g^{\mathrm{I}} = \frac{\xi^{g-1} d\xi}{2\eta}.$$

The differentials of the first kind ν_i^{I} are covariant, in contrast to that the differentials of the first kind ν_i^{II} are not; they are obtained by substituting $x(\xi)$ and $y(\eta, \xi)$ into $\nu_i^{\mathrm{II}}(x, y)$.

Similarly for a cyclic (r, s)-curve, especially $(r, \ell r - 1)$-curve X, $(\ell > 1)$,

$$y^r = \prod_{i=1}^{\ell r - 1} (x - b_i),$$

we consider a rational map,

$$\xi := \frac{\beta}{x} + \beta_{\ell r}, \quad \beta_i = \frac{\beta}{b_i} + \beta_{\ell r}, \quad \eta = \sqrt{\frac{\beta^{\ell r - 1}}{-\prod_i^{\ell r - 1} b_j} \frac{y}{x^\ell}},$$

to obtain the curve

$$\eta^r = \prod_{i=1}^{\ell r} (\xi - \beta_i).$$

We note that (ξ, η) and (x, y) cover the curve X as a projective curve as in [Mum1984, 3.16] (cf. Example 2.241). Extending this procedure to a general cyclic plane W-curve (r, s) is not tricky though we require a normalization of the ring in general. By using the projection $\varpi_r : X \to \mathbb{P}$, we can extend it to the general plane W-curve. Furthermore, since the W-curve is given by the polynomials f_i^X, $(i = 1, \ldots, k_X)$ as in Proposition 2.14, which can be naturally extended as homogeneous ones, we may also basically apply the procedure to X (cf. Example 2.243). Then if needed, we can consider X as a projective curve by the projectivization, though we do not mention it in detail in this book.

2.8.3 *Double Covering of Hyperelliptic Curves*

In this subsection, we consider a generalization of the Jacobi sn, cn, dn functions to the hyperelliptic curves, known as the al functions defined and investigated by Weierstrass. More precisely, we generalize the picture in Sect. 2.8.1 for the genus one to the general genus.

We handle the hyperelliptic curve X given by (2.65) with the hyperelliptic involution $\widehat{\zeta}_H$ as the Galois action $\widehat{\zeta}_2$ as mentioned in Sect. 2.2.5.1.

As in Corollary 2.157, we consider the meromorphic function F_k on $X \times S^k X$. For $(P = (x, y), (P_i = (x_i, y_i))_{i=1,\cdots,k}) \in X \times S^k X$, we deal with

$$F_k(x) = \mu_k(P; P_1, \cdots, P_k) = (x - x_1) \cdots (x - x_k), \tag{2.127}$$

which is known as U of Mumford's triple UVW for the hyperelliptic curves [Mum1981] and is equal to μ_k in (2.156) for $k \le g$; we follow the convention of Baker [Ba1897, Ba1898a, Ba1903]. We sometimes abbreviate F_g to F. We rewrite $\varpi_r : X \to \mathbb{P}$ by ϖ_x for the hyperelliptic curves.

Using the F function, we consider the al function roughly defined by

$$\mathrm{al}_r(u) = \sqrt{F(b_r)} \tag{2.128}$$

for a ramification point $B_r = (b_r, 0)$ in X. We precisely define it in Definition 2.202 and Theorem 2.205. In other words, we consider a particular line bundle on $S^g X$ and its section al_r.

The al functions have the properties:

1. It is associated with the finite ramification point B_i of X and is locally equal to its local parameter at B_i.
2. It is defined as a function of $S^g \widehat{X}$, where $\widehat{X}$ is a double covering of X, $\varpi_{\widehat{X}} : \widehat{X} \to X$ such that its affine coordinate (z, w) of the affine equation $f_{\widehat{X}}(z, w) = 0$ of $\widehat{X}$ satisfies the relations that $\varpi_{\widehat{X}}(z, w) = (z^2 + b_r, wz) \in X$, $z^2 = (x - b_r)$, and the Galois action $\widehat{\zeta} : \widehat{X} \to \widehat{X}$ $(\widehat{\zeta}(z) = -z)$.
3. It is expressed by the ratio of the sigma functions associated with B_r as mentioned in Sect. 3.6.1.

To obtain the al_r function is a generalization of the elliptic al function in Sect. 2.8.1, we consider a genus $2g - 1$ curve $\widehat{X}$ whose affine part is given by

$$w^2 = \prod_{i=0,\neq r}^{2g} (z^2 - \widehat{a}_i), \tag{2.129}$$

where $\widehat{a}_i = b_i - b_r$, $z^2 = x - b_r$ $(z = \sqrt{x - b_r})$, and $w = y/z$. Here w is introduced due to the normalization because the (z, y)-curve is singular.

Since the inverse image of $\varpi_{\widehat{X}}$ of the ramification point $(b_i, 0) = B_i \in X$ $(i \neq r)$ is also ramification points for $c_i^2 := \widehat{a}_i$, we let $(\pm c_i, 0) \in \widehat{X}$ be $\widehat{B}_i^{\pm}$ as finite ramification points in $\widehat{X}$.

There is an involution $\widehat{\zeta_{\mathrm{al}}} : (z, w) \mapsto (-z, w)$ as well as the hyperelliptic involution $\widehat{\zeta_{\widehat{H}}} : (z, w) \mapsto (z, -w)$ and $\widehat{\zeta_H} : (x, y) \mapsto (x, -y)$.

At the point $\pm\infty$ in $\widehat{X}$, we identify the actions $\widehat{\zeta_{\widehat{H}}}$ and $\widehat{\zeta_{\mathrm{al}}}$, i.e.,

$$\widehat{\zeta_{\widehat{H}}} : \pm\infty \mapsto \mp\infty, \quad \widehat{\zeta_{\mathrm{al}}} : \pm\infty \mapsto \mp\infty.$$

On the other hand, $(0, 0) \in \widehat{X}$, which corresponds to $B_r \in X$, is the fixed point of $\widehat{\zeta_{\widehat{H}}}$ and $\widehat{\zeta_{\mathrm{al}}}$.

Assume that $r = 1$ without loss of generality. Then there is a *double covering*:

$$\varpi_{\widehat{X}} : \widehat{X} \to X, \quad (\widehat{P} = (z, w) \mapsto P = (z^2 + b_r, wz)),$$

and

$$(0, 0) \mapsto B_1, \quad \widehat{B}_i^{\pm} \to B_i, \quad (i = 0, 2, 3, \cdots, 2g).$$

We illustrate this projection in Fig. 2.10, which is essentially the same as the picture in [ArCGH1985, p. 296]. Further, we note that there is a natural projection $\kappa_{\widehat{X}} : \widetilde{X} \to \widehat{X}$, satisfying the commutative diagram

$$\begin{array}{ccc} \widetilde{X} & \xrightarrow{\kappa_{\widehat{X}}} & \widehat{X} \\ & \searrow^{\kappa_X} & \downarrow{\scriptstyle \varpi_{\widehat{X}}} \\ & & X, \end{array} \tag{2.130}$$

i.e., $\kappa_X(\gamma_{P,\infty}) = \varpi_{\widehat{X}} \circ \kappa_{\widehat{X}}(\gamma_{P,\infty})$. We note $\varpi_{\widehat{X}}(0,0) \to B_1 \in X$. Further, let $\varpi_z : \widehat{X} \to \mathbb{P}$ by $\varpi_z(z,w) = z$.

For the local chart U at $(0,0) \in \widehat{X}$, $\varpi_z(U)$ is holomorphic to a certain local chart V at $B_1 \in X$. For two points $\gamma_{P',\infty}, \gamma_{P,\infty} \in \widetilde{X}$ at $\kappa_X^{-1}(B_1)$ such that $\varpi_x(P) = \varpi_x(P')$, we introduce the winding number $\varepsilon_1(\gamma_{P',P})$, $(\gamma_{P',P} = \gamma_{P',\infty} - \gamma_{P,\infty})$, at b_1 in $\varpi_x \kappa_X \widetilde{X}$. Then it can be extended to $\varepsilon_1(\gamma_{P,\infty})$, and z can be expressed by $(-1)^{\varepsilon_1(\gamma_{(x,y),\infty})}\sqrt{x - b_1}$.

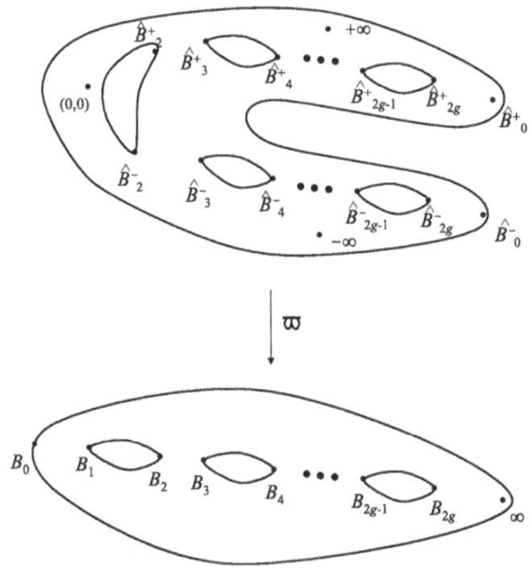

Fig. 2.10 $\varpi_{\widehat{X}} : \widehat{X} \to X$

Let us extend $\varpi_{\widehat{X}}$ to the projection $S^g\widehat{X} \to S^g X$ naturally, and let it be denoted by the same symbol $\varpi_{\widehat{X}}$, i.e., $\varpi_{\widehat{X}}(P_1, \ldots, P_g) = (\varpi_{\widehat{X}}(P_1), \ldots, \varpi_{\widehat{X}}(P_g))$.

Definition 2.202 With respect to the ramification point $(b_1, 0) \in X$, for $(P_1, \ldots, P_g) \in S^g\widehat{X}$ and $P_i = (z_i, w_i)$, we define

$$\mathrm{al}_1(P_1, \ldots, P_g) := (z_1 z_2 \cdots z_g).$$

This definition is a refinement of the rough definition (2.128). From the definition, we have the properties of the al_1 function as in Theorem 2.205.

Let us consider $\kappa_{\widehat{X}}^* \mathrm{al}_1$ as a function on $\mathbb{C}^g$ as the inverse of the Abelian integrals.

The W-normalized basis of holomorphic one-forms over $\widehat{X}$ is denoted by

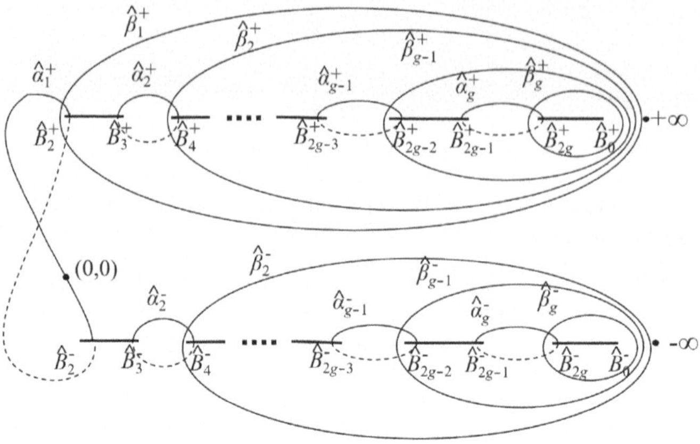

Fig. 2.11 $\varpi_{\widehat{X}} : \widehat{X} \to X$

$$\widehat{\nu}^{\mathrm{I}} := \begin{pmatrix} \widehat{\nu}_1^{\mathrm{I}} \\ \widehat{\nu}_2^{\mathrm{I}} \\ \vdots \\ \widehat{\nu}_{2g-1}^{\mathrm{I}} \end{pmatrix}, \quad \widehat{\nu}_j^{\mathrm{I}} = \frac{z^{j-1}dz}{w}, \quad (j = 1, 2, \ldots, 2g-1).$$

Here we have removed the factor $1/2$ for later convenience. For $P \in \widehat{X}$, let $\varGamma_P \widehat{X}$ be the path space of $\widehat{X}$ with the fixed point P. Here we regard $\varGamma_{-\infty}\widehat{X}$ as $\widetilde{X}$ due to (2.130). The Abelian integral is given by

$$\widehat{w} : S^k\widetilde{X} \longrightarrow \mathbb{C}^{2g-1}, \quad \left(\widehat{w}(\gamma_{(x_1,y_1),-\infty}, \ldots, \gamma_{(x_k,y_k),-\infty}) := \sum_{i=1}^k \int_{\gamma_{(x_i,y_i),-\infty}} \widehat{\nu}^{\mathrm{I}}\right).$$

As the contours in Fig. 2.11 illustrate, the associated periodic matrices $(\widehat{\omega}', \widehat{\omega}'')$ are given as,

$$\frac{1}{2}\left(\left(\int_{\widehat{\alpha}_1} \widehat{\nu}^{\mathrm{I}}, \left(\int_{\widehat{\alpha}_i^+} \widehat{\nu}^{\mathrm{I}}, \int_{\widehat{\alpha}_i^-} \widehat{\nu}^{\mathrm{I}}\right)_{i=2,\ldots,g}\right), \left(\int_{\widehat{\beta}_1} \widehat{\nu}^{\mathrm{I}}, \left(\int_{\widehat{\beta}_i^+} \widehat{\nu}^{\mathrm{I}}, \int_{\widehat{\beta}_i^-} \widehat{\nu}^{\mathrm{I}}\right)_{i=2,\ldots,g}\right)\right).$$

The lattice associated with the curve $\widehat{X}$ is denoted by $\varGamma_{\widehat{X}}$ and its Jacobian by $\widehat{J}_{\widehat{X}} = \mathbb{C}^{2g-1}/\varGamma_{\widehat{X}}$.

Direct computations show the following facts:

Proposition 2.203 *1. Let* $\varPi_{\widehat{X}} := \begin{pmatrix} 1\,0\,0 \cdots 0\,0\,0 \\ 0\,0\,1 \cdots 0\,0\,0 \\ \vdots \quad \ddots \quad \vdots \\ 0\,0\,0 \cdots 1\,0\,0 \\ 0\,0\,0 \cdots 0\,0\,1 \end{pmatrix} : \mathbb{C}^{2g-1} \to \mathbb{C}^g,$

$$\frac{z^{2i-2}dz}{w} = \frac{x^{i-1}dx}{2y}, \quad (i = 1, \ldots, g), \quad \varpi_{\widehat{X}}^{*} \nu^{\mathrm{I}} = \Pi_{\widehat{X}} \widehat{\nu}^{\mathrm{I}} = \begin{pmatrix} \widehat{\nu}_1^{\mathrm{I}} \\ \widehat{\nu}_3^{\mathrm{I}} \\ \vdots \\ \widehat{\nu}_{2g-1}^{\mathrm{I}} \end{pmatrix}.$$

2.

$$\widehat{\zeta}_{\widehat{H}} \varpi_{\widehat{X}}^{*} \nu^{\mathrm{I}} = \varpi_{\widehat{X}}^{*} \widehat{\zeta}_{H} \nu^{\mathrm{I}}.$$

3. *By defining*

$$\left(\sum_i^g \int_{\gamma_{(x_i, y_i), -\infty}} \varpi_{\widehat{X}}^{*} \nu^{\mathrm{I}} \right),$$

$\widehat{w}_{\varpi_{\widehat{X}}^{*}\nu^{\mathrm{I}}} : S^g \widetilde{X} \to \mathbb{C}^g$ *is a surjection.*

Figure 2.11 shows that as half of β_1 consists of the path from ∞ to B_1, the path from $\pm\infty$ to $(0, 0)$ in $\widehat{X}$ corresponds to a quarter of $\widehat{\beta}_1$. Each $\widehat{\beta}_a^{\pm}$ $(a = 2, \ldots, g)$ consists of a contour from $\pm\infty$ to $\widehat{B}_{2a-1}^{\pm}$. Similarly we have $\widehat{\alpha}_a^{\pm}$ $(a = 1, \ldots, g)$.

Noting that the path $\gamma_{B_2 B_1}$ lifts to $\widehat{\gamma}_{\widehat{X}, \widehat{B}_2^{(\pm)}, (0.0)}$, we find that

$$\int_{B_1}^{B_2} \nu^{\mathrm{I}} = \frac{1}{2} \int_{(0,0)}^{\widehat{B}_2} \varpi_{\widehat{X}}^{*} \nu^{\mathrm{I}}$$

is a half-period in $\widehat{X}$. The $(g \times 2(2g-1))$ matrix $(\widehat{\omega}', \widehat{\omega}'')|_{\varpi_{\widehat{X}}^{*}\nu^{\mathrm{I}}} = \Pi_{\widehat{X}}(\widehat{\omega}', \widehat{\omega}'')$ is given by

$$(\omega_1', \omega_2', \omega_2', \ldots, \omega_g', \omega_g', 2\omega_1'', \omega_2'', \omega_2'', \ldots, \omega_g'', \omega_g'').$$

Let $\widehat{\Gamma}_{\widehat{X}} := \left\langle 2\Pi_{\widehat{X}}(\widehat{\omega}', \widehat{\omega}'') \begin{pmatrix} {}^t\Pi_{\widehat{X}} & 0 \\ 0 & {}^t\Pi_{\widehat{X}} \end{pmatrix} \right\rangle_{\mathbb{Z}} \subset \mathbb{C}^g.$

Proposition 2.204 *Let*

$$\widehat{J}_X^{\mathrm{al},(1)} := \mathbb{C}^g / \widehat{\Gamma}_{\widehat{X}}.$$

Then $\widehat{J}_X^{\mathrm{al},(1)}$ *is a double covering of* J_X*;* $\varpi_{\widehat{J}} : \widehat{J}_X^{\mathrm{al},(1)} \to J_X$.

The al_1 function is characterized by the following theorem:

Theorem 2.205 *1. For* $(P_i) \in S^g \widehat{X}$, $\mathrm{al}_1^2(P_1, \ldots, P_g) = \varpi_{\widehat{X}}^{*}(F(b_r))$,

2. $\mathrm{al}_1(\widehat{\zeta}_{\mathrm{al}}(P_1), \ldots, P_g) = -\mathrm{al}_1(P_1, \ldots, P_g)$,
3. *at each* $\varpi_{\widehat{X}}(P_i) = B_1 = \varpi_{\widehat{X}}(0)$ *and for other generic* P_j $(j \neq i)$, *it agrees with the local parameter at* B_1,
4. al_1 *can be regarded as a function of* $S^g \widetilde{X}$, *as* $\kappa_{\widehat{X}}^{*}\mathrm{al}_1$, *simply denoted by* al_1 *again, and it can be regarded as*

$$\mathrm{al}_1(\gamma_1, \ldots, \gamma_g) = (-1)^{\varepsilon_1'(\gamma_1) + \cdots + \varepsilon_1'(\gamma_g)} \sqrt{F(b_r)},$$

where $x_i = \varpi_x(\kappa_X(\gamma_i))$, *and*

5. *by identifying* $\widetilde{w}(S^g \widetilde{X}) = \mathbb{C}^g$ *and* $\widehat{w}_{\varpi^*\nu^{\mathrm{I}}}(S^g \widetilde{X}) = \mathbb{C}^g$, *the* al_1 *function is expressed as a function on* $\widehat{J}_X^{\mathrm{al},(1)}$,

$$\mathrm{al}_1(u) = (z_1 z_2 \cdots z_g)(u),$$

i.e., for $\ell \in \widehat{\Gamma}_{\widehat{X}}$ *and* $u \in \mathbb{C}^g$, $\mathrm{al}_1(u) = \mathrm{al}_1(u + \ell)$.

Definition 2.206 By exchanging B_1 for the al_1 in Definition 2.202 with B_r, we define al_r. For b_r belonging to c_i of $f_{\mathrm{II}}(x)$ in (2.65), by regarding $\mathrm{al}_r(u)$ as a function over $\mathbb{C}^g$ in the sense of Theorem 2.205 5, we define

$$\mathfrak{al}_r(u) = \frac{1}{\sqrt{f_{\mathrm{II}}'(c_r)}} \mathrm{al}_r(u).$$

Proposition 2.207 *Let* t *be the local parameter at* $B_1 \in X$. *We consider* $U := \{(0,0)\} \times S^{g-1}(\widehat{X} \setminus \{\pm\infty\})$ *whose* $\varpi_{\widehat{X}}$*-image is* $B_1 \times S^{g-1}(X \setminus \{\infty\})$ *by fixing an element* $(P_2, \cdots, P_g) \in S^{g-1}(\widehat{X} \setminus \{\pm\infty\})$. *Let* $(P_1, \cdots, P_g) \in U$. *We consider* t *as a function of* $\varpi_{\widehat{X}}(P_1)$. *Then for* $\widehat{w}(P_1, \cdots, P_g) = u$ *and certain* $c_i \in \mathbb{C}$, $\sum_{i=1}^g c_i u_i = t(1 + d_{>0}(t))$ *as a function of* P_1 *and the* $\mathrm{al}_1(u)$ *function is proportional to* t *as a function of* $\varpi_{\widehat{X}}(P_1)$.

Proposition 2.208 *For* $u \in \mathbb{C}^g$, *we have*

$$\sum_{i=1}^{g+1} \mathfrak{al}_r^2(u) = 1.$$

We remark that this proposition is a natural generalization of the relation, $\mathrm{sn}^2(u) + k^2\mathrm{cn}^2(u) = 1$ as mentioned in Sect. 2.8.1.

Proof The left-hand side is given by

$$\sum_{i=0}^{g} \frac{F(c_i)}{f_{\mathrm{II}}'(c_i)} = \frac{1}{2} \sum_{i=0}^{g} \mathrm{res}_{(c_i,0)} \frac{F(x)}{f_{\mathrm{II}}(x)},$$

since at the finite ramified point $(c_i, 0)$ of the curve X, we have a local parameter $t^2 = (x - c_i)$ and

$$\mathrm{res}_{(c_i,0)} \frac{F(x)}{f_{\mathrm{II}}(x)} dx = \mathrm{res}_{(c_i,0)} \frac{2F(t^2 + c_i)t dt}{(t^2 + c_i - c_0) \cdots t^2 \cdots (t^2 + c_i - c_g)}.$$

Let us consider an integral over a boundary of polygon expression D_X of X,

$$\oint_{\partial D_X} \frac{F(x)}{f_{\mathrm{II}}(x)} dx = 0,$$

which gives the relation

$$\sum_{i=0}^{g} \mathrm{res}_{(c_i,0)} \frac{F(x)}{f_{\mathrm{II}}(x)} dx = -\mathrm{res}_{\infty} \frac{F(x)}{f_{\mathrm{II}}(x)} dx.$$

At ∞, the local parameter t is given by $x = 1/t^2$:

$$\mathrm{res}_{\infty} \frac{F(x)}{f_{\mathrm{II}}(x)} dx = \mathrm{res}_{\infty} \frac{\frac{1}{t^{2g}}(1 - x_1 t^2)\cdots(1 - x_g t^2)}{\frac{1}{t^{2g+2}}(1 - c_0 t^2)\cdots(1 - c_g t^2)} \frac{-2}{t^3} dt = -2.$$

Hence it is proved. ■

2.8.4 Triple Covering of Elliptic Curves

In this subsection, we consider the triple covering al function of genus one based on Appendix C in [FeKMPA2022].

Let us consider the elliptic curve E_s :

$$y(y - s) = x^3 \tag{2.131}$$

with the trigonal symmetry $\widehat{\zeta_3}(x, y) = (\zeta_3, y)$ where $\zeta_3 = \mathrm{e}^{2\pi\sqrt{-1}/3}$. In this subsection, we will define the trigonal al function in Proposition 2.210, which is roughly given by

$$\mathrm{al}_s(u) = \sqrt[3]{y - s}. \tag{2.132}$$

The curve corresponds to the Weierstrass standard form,

$$(\wp')^2 = 4\wp^3 - g_3 = 4(\wp - e_1)(\wp - e_2)(\wp - e_3),$$

where $g_3 = -4s^2$ and $\wp(u)' = 2y + s$, i.e.,

$$\left(y - \frac{s}{2}\right)^2 = \left(x + \sqrt[3]{\frac{s^2}{4}}\right)\left(x + \zeta_3\sqrt[3]{\frac{s^2}{4}}\right)\left(x + \zeta_3^2\sqrt[3]{\frac{s^2}{4}}\right).$$

Its limit $s \to 0$ corresponds to the type IV of the degeneration in Kodaira's classification. Here $e_j = -\zeta_3^{1-j}\sqrt[3]{\dfrac{s^2}{4}}$ $(j = 1, 2, 3)$.

We let the integral from the point at infinity to $(x, y) = (0, s)$ be denoted by ω_{B_s}, and that to $(0, 0)$ by ω_{B_0},

$$\omega_{B_s} = \int_{\infty}^{(0,s)} du, \quad \omega_{B_0} = \int_{\infty}^{(0,0)} du, \quad du = \frac{dx}{2y-s} = \frac{dy}{3x^2}. \tag{2.133}$$

Further, the standard half-period integrals are given by [On1998],

$$\omega' = \omega_1 = \int_{\infty}^{e_1} du, \quad \omega_2 = \zeta_3^2\omega', \quad \omega'' = \omega_3 = \zeta_3\omega', \tag{2.134}$$

and

$$\eta_i = -\int_{\infty}^{e_i} x du, \quad \eta' = \eta_1, \quad \eta'' = \eta_3 = \zeta_3^2\eta'. \tag{2.135}$$

They satisfy the following:

$$\omega_1 + \omega_2 + \omega_3 = 0, \quad \eta_1 + \eta_2 + \eta_3 = 0, \quad \eta'\omega'' - \eta''\omega' = \frac{\pi\sqrt{-1}}{2},$$

and [On1998]

$$\eta'\omega' = \frac{\pi\sqrt{-1}}{2(\zeta_3 - \zeta_3^2)} = \frac{\pi}{2\sqrt{3}} \in \mathbb{R}. \tag{2.136}$$

Further, due to [FeKMPA2022, Appendix C], we obtain

$$\omega_{B_s} = \frac{1-\zeta_3^2}{3}2\omega', \quad \omega_{B_0} = -\omega_{B_s},$$

$$\omega' = \frac{3}{2}\frac{1}{\zeta_3 - \zeta_3^2}\frac{\Gamma\left(\frac{1}{3}\right)^2}{s^{1/3}\Gamma\left(\frac{2}{3}\right)}, \quad \eta' = \frac{\pi\sqrt{-1}}{3\sqrt{3}}\frac{s^{1/3}\Gamma\left(\frac{2}{3}\right)}{\Gamma\left(\frac{1}{3}\right)^2}.$$

Using these, we define the lattice $\Gamma_s := \mathbb{Z}2\omega' + \mathbb{Z}2\omega''$ [EilMO2011]. Further, the affine coordinate (x, y) of the curve E_s is written in terms of the $\wp$ function due to the results in [EilMO2011]: Γ_s is illustrated in Fig. 2.12.

Lemma 2.209 *1.* $\wp(u) = x(u), \quad \wp'(u) = 2y(u) - s, \quad y(u) = \frac{1}{2}(\wp'(u) + s),$

2. $x(u) = x(-u), \quad y(-u) = -y(u) + s, \quad (y(-u) - s) = -y(u),$

3. $\wp(\zeta_3 u) = \zeta_3\wp(u), \quad \wp'(\zeta_3 u) = \wp'(u)$ *for* $u \in \mathbb{C}$,

4. $x(\zeta_3^r\omega_{B_s}) = \wp(\zeta_3^r\omega_{B_s}) = 0, (r = 0, 1, 2), \quad y(\omega_{B_0}) = 0, \quad y(\omega_{B_s}) = s$, *and,*

5. $\omega_{B_0} = \frac{1-\zeta_3}{3}2\omega', \quad \omega_{B_s} = \frac{1-\zeta_3^2}{3}2\omega' = \frac{2+\zeta_3}{3}2\omega' = \frac{2}{3}(2\omega' + \omega'').$

Let $z^3 = y - s$. It gives the singular curve $z^3(z^3 + s) = x^3$. We normalize the curve by $w = x/z$ and obtain the elliptic curve $\widehat{E}_s$ defined by

$$z^3 + s = w^3$$

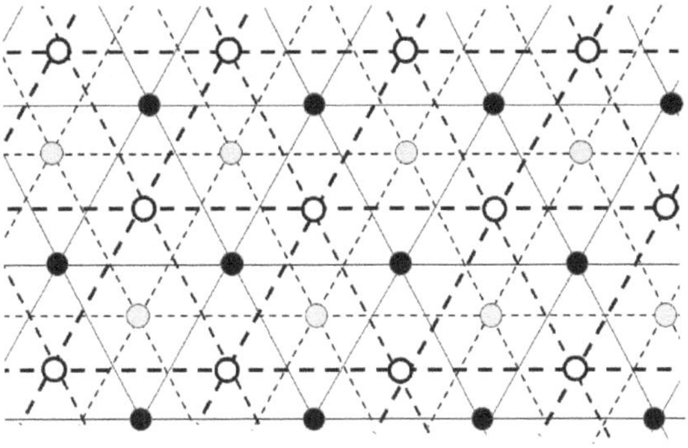

Fig. 2.12 The lattice points of Γ_s are denoted by the black dots, ω_{B_0} and ω_{B_s} with Γ_s translations are denoted by gray dots and white dots, respectively

as the triple covering curve $\varpi_{\widehat{E}_s} : \widehat{E}_s \to E_s$. This w^3 in the right-hand side agrees with $y = z^3 + s = w^3$. The point $w = 0$ corresponds to the point $\varpi_{\widehat{E}_s}^{-1}(x = 0, y = 0)$ and thus there is a symmetry group $G_{\widehat{E}_s}$,

$$(z, w) \mapsto (-w, -z),$$

which comes from the hyperelliptic involution $(x, y - s/2) \to (x, -y + s/2)$. Further, we have the two trigonal cyclic group actions $\widehat{\zeta_3}$ and $\widehat{\zeta_3'}$ on $\widehat{E}_s$

$$\widehat{\zeta_3}(z, w) = (z, \zeta_3 w), \quad \widehat{\zeta_3'}(z, w) = (\zeta_3 z, \zeta_3^{-1} w),$$

where $\widehat{\zeta_3}$ is induced from that of E_s. Since ∞ is the fixed point of the action $\widehat{\zeta_3}$, we note that $\widehat{E}_s$ has three different infinite points

$$(\zeta_3^p \infty, \zeta_3^{-p} \infty), \quad (p = 0, 1, 2) \tag{2.137}$$

as $\varpi_{\widehat{E}_s}^{-1}\infty$. Further, by noting $w^2 dw = z^2 dz$, we have the differential of the first kind (the holomorphic one-form) of $\widehat{E}_s$ by the relation

$$\frac{dx}{2y - s} = \frac{d(wz)}{2z^3 + s} = \frac{dz}{w^2}.$$

Proposition 2.210 *There is a function z of $u \in \mathbb{C}$ satisfying $z^3 = y - s$ and let*

$$\mathrm{al}_s^{(r)}(u) := \zeta_3^r z(u).$$

This definition is a refinement of the primitive definition (2.132).

Proof In order to use the results of the Weierstrass elliptic function theory for $\widehat{E}_s$, we introduce another curve $\widehat{E}'_s$ expressed by the Weierstrass canonical form. The $\widehat{E}_s$ is birational to the curve $\widehat{E}'_s$ defined by

$$W^2 - 3\sqrt{-3}sw = Z^3,$$

$$\left(W - \frac{3\sqrt{-3}}{2}s\right)^2 = Z^3 - \frac{27}{4}s^2 = \prod_{i=0}^{2}\left(Z + 3\sqrt[3]{\frac{s^2}{4}}\zeta_6^{1+2i}\right),$$

where $\zeta_6 := \mathrm{e}^{2\pi\sqrt{-1}/6} = 1 + \zeta_3$,

$$Z := \frac{3s}{z - w}, \quad W := 3\frac{(z+w)Z + (1+2\zeta_3)}{2} = 9s\frac{(z+w) + (1+2\zeta_3)(z-w)}{2(z-w)},$$

or

$$W - \frac{3\sqrt{-3}}{2}s = 9s\frac{(z+w)}{2(z-w)}, \quad w = \frac{W - 3(2+\zeta_3)}{3Z}, \quad z = \frac{W + 3(2+\zeta_3^2)}{6Z}.$$

$\widehat{E}'_s$ is also a trigonal covering of $\varpi_{\widehat{E}'_s} : \widehat{E}'_s \to E_s$ as the above sense. Let $\widehat{e}_i = 3\sqrt[3]{\frac{s^2}{4}}\zeta_6^{1-2i}$. Here $Z = 0$ means two infinity points in $\widehat{E}_s$, $(\zeta_3^p\infty, \zeta_3^{-p}\infty)$, $(p = 1, 2)$, whereas $Z = W = \infty$ corresponds to $(w, z) = (\infty, \infty)$ of $\widehat{E}_s$. The point $W = \frac{3\sqrt{-3}}{2}s$ in $\widehat{E}'_s$ corresponds to the point $z = -w$ in $\widehat{E}_s$, i.e., $w = -z = \zeta_3^r\sqrt[3]{\frac{s}{2}}$ $(r = 0, 1, 2)$. Thus these points $W = \frac{3\sqrt{-3}}{2}s$ and $w = -z = \zeta_3^r\sqrt[3]{\frac{s}{2}}$ $(r = 0, 1, 2)$ also correspond to the ramification points $x = zw = e_i$ $(i = 0, 1, 2)$ of E_s and $\widehat{e}_i$'s of $\widehat{E}'_s$.

Since the differential of the first kind $d\widehat{u}$ of $\widehat{E}'_s$ is given by $d\widehat{u} = \frac{dZ}{2W - 3\sqrt{-3}s} = -\frac{dz}{3w^2}$, we have the relation between the differentials of the first kind of E_s and $\widehat{E}'_s$,

$$du = -3d\widehat{u}.$$

Let us consider the half-period integrals of $\widehat{E}'_s$,

$$\Omega_i := \int_{\infty}^{\widehat{e}_i} d\widehat{u},$$

and then we obviously have the relation

$$\Omega_i = 3\zeta_6\omega_i, \quad (i = 0, 1, 2)$$

because it can be obtained by the variable change $Z = 3x\zeta_6$ in the integral. Hence the Jacobian $J_{\widehat{E}'_s}$ of $\widehat{E}'_s$ is given by

$$J_{\widehat{E}'_s} := \mathbb{C}/(6\zeta_6\omega'\mathbb{Z} \times 6\zeta_6\omega''\mathbb{Z}).$$

Then z is a well-defined function of the Jacobian $J_{\widehat{E}'_s}$, which contains the nine points which correspond to the infinite points of $\widehat{E}_s$,

$$(z, w) = (\zeta_3^p\infty, \zeta_3^{q-p}\infty), \quad (p, q = 0, 1, 2)$$

due to the actions $\widehat{\zeta_3}$ and $\widehat{\zeta'_3}$. However, due to (2.137), q should be fixed under the action of $\widehat{\zeta_3}$.

We note that $\mathrm{al}_s^{(r)}$ and z have the same poles in $\mathbb{C}$. The Jacobian $J_{\widehat{E}'_s}$ contains nine $z = 0$ points, which corresponds to

$$(\omega_{B_s} + 2m\omega' + 2n\omega'') \quad \text{modulo} \quad J_{\widehat{E}'_s}.$$

The involution $(z, w) \mapsto (-w, -z)$ in E_s is related to $\sigma(-u) = -\sigma(u)$. As in Lemma 2.209, z vanishes at ω_{B_s}, $-\omega_{B_0}$, $\zeta_3\omega_{B_s}$, $-\zeta_3\omega_{B_0}$, $\zeta_3^2\omega_{B_s}$, and $-\zeta_3^2\omega_{B_0}$ modulo $6\zeta_6\omega'\mathbb{Z} \times 6\zeta_6\omega''\mathbb{Z}$.

Therefore the fundamental domain J_z of $z(u)$ is smaller than $J_{\widehat{E}_s}$ and J_z has six zeros of z and then the cardinality of $\varpi_z^{-1}\infty$ is six for the covering $\varpi_z : J_z \to J_{E_s}$.

We should define the difference among $\mathrm{al}_s^{(r)}$'s, which are only the phase factor. ■

The following characterize the $\mathrm{al}_s^{(r)}$ functions:

Theorem 2.211 *1. For $P \in \widehat{E}_s$, $(\mathrm{al}_s^{(r)})^3(P) = \varpi_{\widehat{E}}^*(y - s)$,*
2. *$\mathrm{al}_s^{(r)}(\widehat{\zeta'_3}P) = \zeta_3\mathrm{al}_s^{(r)}(P)$,*
3. *at $\varpi_{\widehat{X}}(P) = (0, s) = \varpi_{\widehat{X}}(0)$, it agrees with the local parameter at $(0, s)$,*
4. *$\mathrm{al}_s^{(r)}(P)$ can be regarded as a function of the Abelian universal covering $\widetilde{E}_s$ of E_s,*

$$\mathrm{al}_s^{(r)}(\gamma) = (-1)^{\text{phase factor}}\sqrt[3]{y - s},$$

where $y = \varpi_y(\kappa_{E_s}(\gamma))$, and
5. *due to the proof in the previous proposition, $\mathrm{al}_s^{(r)}(P)$ function is also a function on J_z.*

The following is obvious:

Proposition 2.212 *By letting $\mathrm{al}_0^{(r)} := \zeta_3^r\sqrt[3]{y}$, we have*

$$\frac{\mathrm{al}_s^{(r)3}(u)}{s} + \frac{\mathrm{al}_0^{(r)3}(u)}{-s} = 1, \qquad \frac{\prod_{r=0}^2 \mathrm{al}_s^{(r)}(u)}{s} + \frac{\prod_{r=0}^2 \mathrm{al}_0^{(r)}(u)}{-s} = 1.$$

2.8.5 *Triple Covering of Cyclic (3,4)-Curves*

In this subsection, we deal with a cyclic triple covering of a (3, 4)-curve to define the trigonal al function following [MatP2015].

Let us consider the cyclic W-curve X, $y^3 = (x - b_0)(x - b_1)(x - b_2)(x - b_3)$ (2.67), where we sometimes assume $(b_0 = b_4 = 0, b_3 = s)$. Here we will precisely define the trigonal al function, which is roughly defined by

$$\mathrm{al}_a^{(c)}(u) = \zeta_3^c \frac{\mu_3(B_a)}{\sqrt[3]{F(B_a)}},$$

where B_a is the ramification point in X, $w((x_1, y_1), \ldots, (x_3, y_3)) = u \in \mathbb{C}^3$, and $F(B_a) = (x_1 - b_a)(x_2 - b_a)(x_3 - b_a)$, which locally agrees with the local parameter at $B_a = (b_a, 0)$. In other words, we consider a particular line bundle on $S^3 X$ and its section $\mathrm{al}_{\underline{a}}^{(c)}$, as for the hyperelliptic al functions.

Since ζ_3 is an automorphism of the curve X, which admits the Galois action, $\widehat{\zeta_3^c}(2\omega'_{X,a})$ $(a = 1, 2, 3, c = 0, 1, 2)$ in Lemma 2.197 is a point in the lattice Γ_s, and thus Lemma 2.197 is reduced to the following lemma:

Lemma 2.213 ([MatP2015, Prop. 2.2]) *The set* $(\widehat{\zeta_3^c}\omega_{X,a})$ $(a = 0, 1, 2, 3, c = 0, 1, 2)$ *is a subset of* Γ_s. *Thus there are integers* $h_{a,b}^{(c)\prime}$ *and* $h_{a,b}^{(c)\prime\prime}$ *satisfying the relation*

$$3\widehat{\zeta_3^c}\omega_{X,a} = \sum_{b=1}^{3} \left(h_{a,b}^{(c)\prime}(2\omega'_{X,a}) + h_{a,b}^{(c)\prime\prime}(2\omega''_{X,a}) \right) \textit{ modulo } \Gamma_s.$$

This corresponds to (2.134) and (2.135) of the genus-one case. The factor 3 means that $\widehat{\zeta_3^c}\omega_{X,a}$ is a 3-division point in the lattice Γ_s.

The Legendre relation (2.99) also determines the symplectic structure of the Jacobian J_X.

When we consider the preimage of $J_X^{(a;c)}$, we use a parameterization $z = \sqrt[3]{x - b_a}$ or $z^3 = x - b_a$. As in the standard construction of a Galois cover of a curve with $\mathbb{Z}_p$ action at a given point P, cf. [Cor1987], namely by normalizing the curve obtained by taking the inverse image of the 1-section under $\mathscr{L} \to \mathscr{L}^p$, in the total space of a line bundle $\mathscr{L}$ such that $\mathscr{L}^p = \mathscr{O}(P)$, we consider a curve $\widehat{X}$ whose affine representation can be given by

$$w^3 = \prod_{i=0,\neq a}^{3} (z^3 - a_i), \tag{2.138}$$

where $a_i = b_i - b_a$, $z^3 = x - b_a$. The birational change of variables $w = y/z$ transforms the plane curve X to a plane curve.[18]

---

[18] More precisely, we consider an extended ring $R' := R[z]/(z^3 - x + b_a)$ but since $\mathrm{Spec}(R')$ is a singular curve, we normalize it to $\widehat{R} = \mathbb{C}[w, z]/(w^3 - \prod(z^3 - a_i))$. $\widehat{X}$ is the projectivization of $\mathrm{Spec}(\widehat{R})$. As in (2.139), we extend the $\mathbb{Z}_3$-action consistently with the Galois action on R.

Then as illustrated in Fig. 2.14, there is a *trigonal covering*: $\widehat{X} := \widehat{X}^{(a)}$

$$\varpi_{\widehat{X}} : \widehat{X} \to X, \quad (\widehat{P} = (z, w) \mapsto P = (z^3 + b_a, wz)),$$

and

$$(0, 0) \mapsto B_a, \quad \widehat{B}_i^{(d)} \mapsto B_i, \quad (i \in \{1, 2, 3, 4\} \setminus \{a\}).$$

On the other hand, for $c_i^3 := a_i$, let $(\zeta_3^d c_i, 0) \in \widehat{X}$, $(d = 0, 1, 2)$, be a finite ramification point $\widehat{B}_i^{(d)}$ and $(\zeta_3^d z, w) \in \widehat{X}$, $(d = 0, 1, 2)$ a generic point $\widehat{P}^{(d)}$.

There is an automorphism,

$$\widehat{\zeta_3} : \widehat{X} \to \widehat{X}, \quad ((z, w) \mapsto (\zeta_3 z, w)),$$

whereas there are the trigonal automorphisms

$$\widehat{\zeta_3} : \widehat{X} \to \widehat{X} \text{ and } \widehat{\zeta_3} : X \to X, \quad ((z, w) \mapsto (z, \zeta_3 w), \quad (x, y) \mapsto (x, \zeta_3 y)).$$

These automorphisms generate distinct subgroups except at infinity.

The point at infinity of $\widehat{X}$ is resolved into three points $\infty^{(c)}$, $(c = 0, 1, 2)$. At each $\infty^{(c)}$ of $\widehat{X}$, $\widehat{\zeta_3}$ and ι_{ζ_3} are identified, i.e.,

$$\iota_{\zeta_3} : \infty^{(c)} \mapsto \infty^{(c+1 \bmod 3)}, \quad \widehat{\zeta_3} : \infty^{(c)} \mapsto \infty^{(c+1 \bmod 3)}. \tag{2.139}$$

Also, $(0, 0) \in \widehat{X}$ which corresponds to $B_a \in X$ is the fixed point of ι_{ζ_3} and $\widehat{\zeta_3}$.

As in the hyperelliptic curve case, we note that there is a natural projection $\kappa_{\widehat{X}} : \widetilde{X} \to \widehat{X}$, satisfying the commutative diagram

$$\begin{array}{ccc} \widetilde{X} & \xrightarrow{\kappa_{\widehat{X}}} & \widehat{X} \\ & \searrow^{\kappa_X} & \downarrow \varpi_{\widehat{X}} \\ & & X, \end{array} \tag{2.140}$$

i.e., $\kappa_X(\gamma_{P,\infty}) = \varpi_{\widehat{X}} \circ \kappa_{\widehat{X}}(\gamma_{P,\infty})$. We note $\varpi_{\widehat{X}}(0, 0) \to B_a \in X$. Further, let $\varpi_z : \widehat{X} \to \mathbb{P}$ by $\varpi_z(z, w) = z$.

For the local chart U at $(0, 0) \in \widehat{X}$, $\varpi_z(U)$ is holomorphic to a certain local chart V at $B_a \in X$. For two points $\gamma_{P',\infty}, \gamma_{P,\infty} \in \widetilde{X}$ at $\kappa_X^{-1}(B_1)$ such that $\varpi_x(P) = \varpi_x(P')$, we introduce the winding number $\varepsilon_1(\gamma_{P',P})$, $(\gamma_{P',P} = \gamma_{P',\infty} - \gamma_{P,\infty})$, at b_1 in $\varpi_x \kappa_X \widetilde{X}$. Then it can be extended to $\varepsilon_1(\gamma_{P,\infty})$, and z can be expressed by $\zeta_3^{\varepsilon_1(\gamma_{(x,y),\infty})} \sqrt[3]{x - b_1}$.

Let us extend $\varpi_{\widehat{X}}$ to the projection $S^3\widehat{X} \to S^3 X$ naturally, and let it be denoted by the same symbol $\varpi_{\widehat{X}}$, i,e., $\varpi_{\widehat{X}}(P_1, \dots, P_3) = (\varpi_{\widehat{X}}(P_1), \dots, \varpi_{\widehat{X}}(P_3))$.

Henceforth we consider the $a = 3$ case without loss of generality.

Definition 2.214 With respect to the ramification point $(b_3, 0) \in X$, for $(P_1, P_2, P_3) \in S^g\widehat{X}$ and $P_i = (z_i, w_i)$, we define

$$\mathrm{al}_3^{(c)} := \zeta_3^c \frac{1}{z_1 z_2 z_3} \frac{\begin{vmatrix} 1 & z_1^3 & w_1 z_1 & z_1^6 \\ 1 & z_2^3 & w_2 z_2 & z_2^6 \\ 1 & z_3^3 & w_3 z_3 & z_3^6 \\ 1 & 0 & 0 \end{vmatrix}}{\begin{vmatrix} 1 & z_1^3 & w_1 z_1 \\ 1 & z_2^3 & w_2 z_2 \\ 1 & z_3^3 & w_3 z_3 \end{vmatrix}}, \qquad \mathrm{al}_3 := \mathrm{al}_3^{(0)}.$$

In Theorem 2.220, we show the properties of the al function.

Let us consider $\kappa_{\widehat{X}}^* \mathrm{al}_3$ as a function on $\mathbb{C}^3$ as the inverse of the Abelian integrals. As we will show in Proposition 2.217 and Theorem 2.220, it turns out that the periodicity of the al_3 function agrees with

$$\widehat{\Gamma}^{(3;0)} = 2\mathbb{Z}\omega_1' + 2\mathbb{Z}\omega_1'' + 2\mathbb{Z}\omega_2' + 2\mathbb{Z}\omega_2'' + 6\mathbb{Z}\omega_3' + 2\mathbb{Z}\omega_3''. \tag{2.141}$$

Let us show such properties of the al_3 function.

Consistent with the covering map, the actions on $\mathbf{H}_1(\widehat{X}, \mathbb{Z})$ are extended to $\mathbf{H}_1(\widehat{X}, \mathbb{Z}[\zeta_3])$. Thus we have a natural lift of the homology basis, $(\widehat{\alpha}_1^{(c)}, \widehat{\beta}_1^{(c)}, \widehat{\alpha}_2^{(c)}, \widehat{\beta}_2^{(c)}, \widehat{\alpha}_3, \widehat{\beta}_3) \in \mathbf{H}_1(\widehat{X}, \mathbb{Z}[\zeta_3])$, as shown in Fig. 2.13. Here we have used the fact that the actions of ι_{ζ_3} and $\widehat{\zeta}_3$ are exchanged at infinity due to the properties (2.139).

Up to homotopy, $\widehat{\alpha}_3$ equals $\varpi_{\widehat{X}}\widehat{\alpha}_3$; similarly $\varpi_{\widehat{X}}\widehat{\beta}_3 = \beta_3$. On the other hand, for $i = 1, 2$, Fig. 2.13 shows that

$$\begin{gathered} \varpi_{\widehat{X}}(\alpha_i^{(c)}) = \alpha_i, \quad \varpi_{\widehat{X}}(\beta_i^{(c)}) = \beta_i, \\ \varpi_{\widehat{X}}^{-1}(\alpha_i) = \widehat{\alpha}_i^{(0)} + \widehat{\alpha}_i^{(1)} + \widehat{\alpha}_i^{(2)}, \quad \varpi_{\widehat{X}}^{-1}(\beta_i) = (\widehat{\beta}_i^{(0)}, \widehat{\beta}_i^{(1)}, \widehat{\beta}_i^{(2)}). \end{gathered} \tag{2.142}$$

Let the Jacobian associated with $\widehat{X}$ be denoted by $J_{\widehat{X}}$ and a basis of holomorphic one-forms be given by[19]

$$\widehat{\nu}_1^{\mathrm{I}} = \frac{dz}{w^2}, \quad \widehat{\nu}_2^{\mathrm{I}} = \frac{zdz}{w^2}, \quad \widehat{\nu}_3^{\mathrm{I}} = \frac{z^2dz}{w^2}, \quad \widehat{\nu}_4^{\mathrm{I}} = \frac{dz}{w},$$

---

[19] For a curve of genus seven X_7 whose affine part is given by: $w^3 = \prod_{i=0}^8 (z - \widehat{c}_i)$, by letting

$$z' := \frac{1}{z - c_0}, \quad w' := \frac{1}{\sqrt[3]{\prod_{i=1}^8 (c_i - c_0)}} \frac{w}{(z - c_0)^3}, \quad c_i' := \frac{1}{c_i - c_0};$$

as another affine chart, cf. [Mum1984, Ch. IIIa], we have $w'^3 = \prod_{i=1}^8 (z' - c_i')$. The holomorphic one-forms are given by

$$\widehat{\nu}_1^{\mathrm{I}'} = \frac{dz'}{3w'^2}, \quad \widehat{\nu}_2^{\mathrm{I}'} = \frac{z'dz'}{3w'^2}, \quad \widehat{\nu}_3^{\mathrm{I}'} = \frac{z'^2dz'}{3w'^2}, \quad \widehat{\nu}_4^{\mathrm{I}'} = \frac{dz'}{3w'}, \quad \widehat{\nu}_5^{\mathrm{I}'} = \frac{z'^3dz'}{3w'^2}, \quad \widehat{\nu}_6^{\mathrm{I}'} = \frac{z'w'dz'}{3w'^2}, \quad \widehat{\nu}_7^{\mathrm{I}'} = \frac{z'^4dz'}{3w'^2}.$$

We state that when $c_0 = 0$, $\widehat{\nu}_i^{\mathrm{I}} = -3\widehat{\nu}_{8-i}^{\mathrm{I}'}$, writing $dz' = -dz/z^2$.

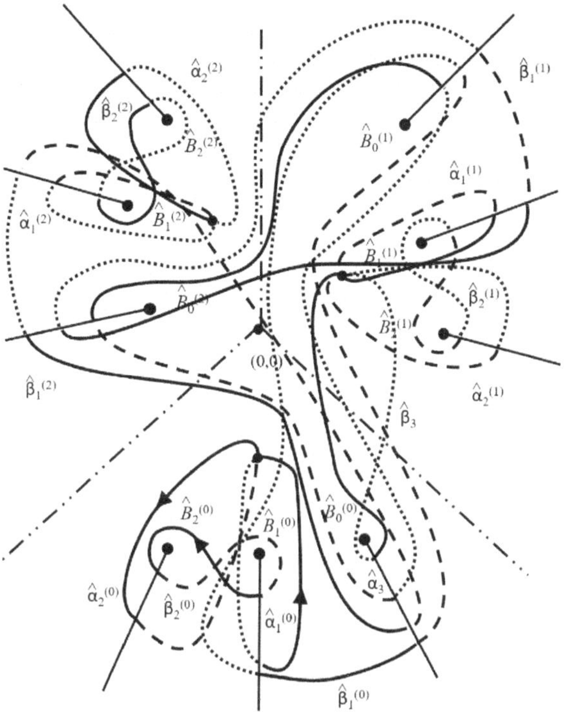

Fig. 2.13 Homology basis of $\widehat{X}$

$$\widehat{\nu}_5^{\mathrm{I}} = \frac{z^3 dz}{w^2}, \quad \widehat{\nu}_6^{\mathrm{I}} = \frac{zw dz}{w^2}, \quad \widehat{\nu}_7^{\mathrm{I}} = \frac{z^4 dz}{w^2}.$$

We have removed the factor 3 in the denominators for later convenience. The Abelian integral $\widehat{w} : \widetilde{X} = \Gamma_{\infty^{(0)}}\widehat{X} \to \mathbb{C}^7$ defined by these holomorphic one-forms is denoted by

$$\widehat{w}(\gamma_{P_1,\infty^{(0)}}, \ldots, \gamma_{P_n,\infty^{(0)}}) = \sum_{i=1}^{n} \widehat{w}(\gamma_{P_i,\infty^{(0)}}), \quad \widehat{w}(\gamma_{P_i,\infty^{(0)}}) = \int^{\gamma_{P_i,\infty^{(0)}}} \widehat{\nu}^{\mathrm{I}}.$$

Since $dx = 3z^2 dz$, we have the relations

$$\widehat{\nu}_1^{\mathrm{I}} = \varpi_{\widehat{X}}^* \nu_1^{\mathrm{I}}, \quad \widehat{\nu}_5^{\mathrm{I}} = \varpi_{\widehat{X}}^* \nu_2^{\mathrm{I}}, \quad \widehat{\nu}_6^{\mathrm{I}} = \varpi_{\widehat{X}}^* \nu_3^{\mathrm{I}}, \quad \varpi_{\widehat{X}}^*(\nu^{\mathrm{I}}) = \Pi_{\widehat{X}}(\nu^{\mathrm{I}}) = \begin{pmatrix} \widehat{\nu}_1^{\mathrm{I}} \\ \widehat{\nu}_5^{\mathrm{I}} \\ \widehat{\nu}_6^{\mathrm{I}} \end{pmatrix},$$

where $\Pi_{\widehat{X}} = \begin{pmatrix} 1\,0\,0\,0\,0\,0\,0 \\ 0\,0\,0\,0\,1\,0\,0 \\ 0\,0\,0\,0\,0\,1\,0 \end{pmatrix}$.

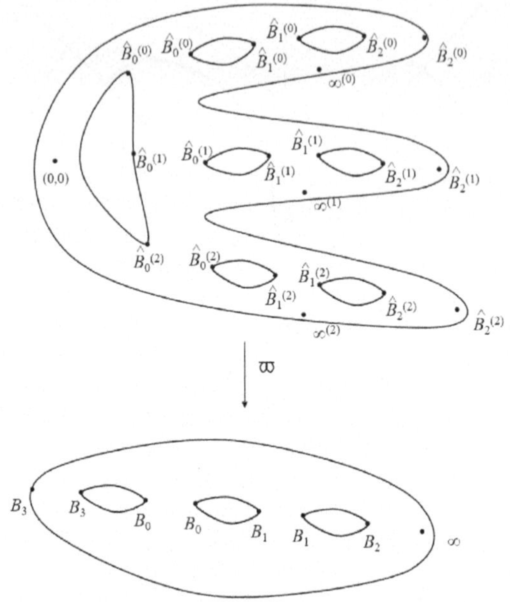

Fig. 2.14 $\varpi_{\widehat{X}} : \widehat{X} \to X$

However, we should note the $\widehat{\zeta_3}$-action according to the covering defined by z:

Lemma 2.215 *The maps* ι_{ζ_3} *and* $\widehat{\zeta_3}$ *induce*

$$\iota_{\zeta_3}\varpi_{\widehat{X}}^*(\nu^{\mathrm{I}}) = \varpi_{\widehat{X}}^*(\widehat{\zeta_3}\nu^{\mathrm{I}}) = \widehat{\zeta_3}\varpi_{\widehat{X}}^*(\nu^{\mathrm{I}}).$$

We consider the projections of the extended Abel–Jacobi maps,

$$\widehat{w}(\Gamma_{\widehat{P},\infty^{(0)}})|_{\varpi_{\widehat{X}}^*\nu^{\mathrm{I}}} = \int_{\Gamma_{\widehat{P},\infty^{(0)}}} \varpi_{\widehat{X}}^*\nu^{\mathrm{I}},$$

and we have the lemma:

Lemma 2.216

$$\int_\gamma \nu^{\mathrm{I}} = \frac{1}{3}\int_{\widehat{\gamma}} \varpi_{\widehat{X}}^*\nu^{\mathrm{I}}, \quad \widehat{\gamma} = \varpi_{\widehat{X}}^{-1}\gamma, \quad \gamma = \alpha_3, \alpha_i, \beta_i (i = 1, 2),$$

$$\int_\gamma \nu^{\mathrm{I}} = \int_{\widehat{\gamma}} \varpi_{\widehat{X}}^*\nu^{\mathrm{I}}, \quad \gamma = \varpi_{\widehat{X}}\widehat{\gamma}, \quad \widehat{\gamma} = \widehat{\beta}_3, \widehat{\alpha}_i^{(c)}, \widehat{\beta}_i^{(c)}, (i = 1, 2, c = 0, 1, 2).$$

Proof For a neighborhood U of a point in α_i, we have $U^{(c)}$ $(c = 0, 1, 2)$ such that $U = \varpi_{\widehat{X}}U^{(c)}$. Then $\varpi_{\widehat{X}}^*\nu^{\mathrm{I}}|_{U^{(c)}} = \iota_{\zeta_3}^c\varpi_{\widehat{X}}^*\nu^{\mathrm{I}}|_{U^{(0)}}$ and they cancel the phase difference between $\alpha_i^{(c)}$ and $\alpha_i^{(0)}$. Then we have

$$\int_{\widehat{\beta}_3} \varpi_{\widehat{X}}^* \nu^{\mathrm{I}} = \left(\int_{\widehat{B}_0^{(0)}}^{(0,0)} + \int_{(0,0)}^{\widehat{B}_0^{(1)}} + \int_{\widehat{B}_0^{(1)}}^{(0,0)} + \int_{(0,0)}^{\widehat{B}_0^{(2)}} + \int_{\widehat{B}_0^{(2)}}^{(0,0)} + \int_{(0,0)}^{\widehat{B}_0^{(0)}} \right) \varpi_{\widehat{X}}^* \nu^{\mathrm{I}}$$
$$= \left(\int_{\beta_3} + \widehat{\zeta_3^2} \int_{\widehat{\zeta}_3 \beta_1} + \widehat{\zeta}_3 \int_{\widehat{\zeta_3^2} \beta_1} \right) \nu^{\mathrm{I}} = 3 \int_{\beta_3} \nu^{\mathrm{I}}.$$

Here we used the fact that $\zeta_3 + \zeta_3^2 = -1$. Similarly,

$$\int_{\varpi_{\widehat{X}}^{-1} \alpha_3} \varpi_{\widehat{X}}^* \nu^{\mathrm{I}} = \int_{\alpha_3^{(0)} + \alpha_3^{(1)} + \alpha_3^{(2)}} \varpi_{\widehat{X}}^* \nu^{\mathrm{I}} = 3 \int_{\alpha_3} \nu^{\mathrm{I}}.$$

■

Then we have the period matrices for $\widehat{X}$,

$$(\widehat{\omega}') := \frac{1}{2} \Big(\int_{\widehat{\alpha}_1^{(0)}} \widehat{\nu}^{\mathrm{I}}, \int_{\widehat{\alpha}_1^{(1)}} \widehat{\nu}^{\mathrm{I}}, \int_{\widehat{\alpha}_1^{(2)}} \widehat{\nu}^{\mathrm{I}}, \int_{\widehat{\alpha}_2^{(0)}} \widehat{\nu}^{\mathrm{I}}, \int_{\widehat{\alpha}_2^{(1)}} \widehat{\nu}^{\mathrm{I}}, \int_{\widehat{\alpha}_2^{(2)}} \widehat{\nu}^{\mathrm{I}}, \int_{\widehat{\alpha}_3} \widehat{\nu}^{\mathrm{I}} \Big),$$

$$(\widehat{\omega}'') := \frac{1}{2} \Big(\int_{\widehat{\beta}_1^{(0)}} \widehat{\nu}^{\mathrm{I}}, \int_{\widehat{\beta}_1^{(1)}} \widehat{\nu}^{\mathrm{I}}, \int_{\widehat{\beta}_1^{(2)}} \widehat{\nu}^{\mathrm{I}}, \int_{\widehat{\beta}_2^{(0)}} \widehat{\nu}^{\mathrm{I}}, \int_{\widehat{\beta}_2^{(1)}} \widehat{\nu}^{\mathrm{I}}, \int_{\widehat{\beta}_2^{(2)}} \widehat{\nu}^{\mathrm{I}}, \int_{\widehat{\beta}_3} \widehat{\nu}^{\mathrm{I}} \Big).$$

Using these, we have the lattice $\widehat{\Gamma}$ in $\mathbb{C}^7$ and the Jacobian $J_{\widehat{X}}$ is: $J_{\widehat{X}} = \mathbb{C}^7 / \widehat{\Gamma}$.

Lemmas 2.215 and 2.216 give the following proposition:

Proposition 2.217 $(\widehat{\omega}', \widehat{\omega}'')|_{\varpi_{\widehat{X}}^* \nu^{\mathrm{I}}} = \Pi_{\widehat{X}}(\widehat{\omega}', \widehat{\omega}'')$, *and* $(\widehat{\omega}', \widehat{\omega}'')|_{\varpi_{\widehat{X}}^* \nu^{\mathrm{I}}}$ *is equal to*

$$(\omega_1', \widehat{\zeta}_3 \omega_1', \widehat{\zeta_3^2} \omega_1' \omega_2', \widehat{\zeta}_3 \omega_2', \widehat{\zeta_3^2} \omega_2', 3\omega_3', \omega_1'', \widehat{\zeta}_3 \omega_1'', \widehat{\zeta_3^2} \omega_1'', \omega_2'', \widehat{\zeta}_3 \omega_2'', \widehat{\zeta_3^2} \omega_2'', \omega_3'').$$

Let ${}^t\Pi_{\widehat{X}}^* := \begin{pmatrix} 1\,0\,0\,0\,0\,0\,0 \\ 0\,0\,0\,1\,0\,0\,0 \\ 0\,0\,0\,0\,0\,0\,1 \end{pmatrix}$ and then we define

$$(\widehat{\omega}'_{\mathrm{al}^{(3,0)}}, \widehat{\omega}'_{\mathrm{al}^{(3,0)}}) := \Pi_{\widehat{X}}(\widehat{\omega}', \widehat{\omega}'') \begin{pmatrix} \Pi_{\widehat{X}}^* & 0 \\ 0 & \Pi_{\widehat{X}}^* \end{pmatrix}.$$

In consequence, it is natural to introduce

$$\widehat{\Gamma}_{(3,0)} := 2 \sum_{i=1}^{3} \left(\mathbb{Z} \widehat{\omega}'_{\mathrm{al}^{(3,0)},i} + \mathbb{Z} \widehat{\omega}''_{\mathrm{al}^{(3,0)},i} \right),$$

which equals to (2.141).

On the other hand, we introduce a natural subvariety of $J_{\widehat{X}}$ as follows:

Definition 2.218 For the projection of the extended Abelian integral

$$\widehat{w}|_{\varpi_{\widehat{X}}^* \nu^{\mathrm{I}}} : S^3 \Gamma_{\infty^{(0)}} \widehat{X} = S^3 \widetilde{X} \to \mathbb{C}^3 \subset \mathbb{C}^7,$$

$\widehat{J}_X^{\mathrm{al};(3,0)}$ is defined by

$$\widehat{J}_X^{\mathrm{al};(3,0)} := \frac{\widehat{w}(S^3\widetilde{X})|_{\varpi_{\widehat{X}}^* \nu^{\mathrm{I}}}}{\widehat{\Gamma}^{(3,0)}}$$

for $c = 0, 1, 2$.

From the definition, we have the following proposition.

Proposition 2.219 *$\widehat{J}_X^{\mathrm{al};(3,0)}$ is a triple covering of J_X: $\varpi_{\widehat{J}} : \widehat{J}_X^{\mathrm{al};(3,0)} \to J_X$.*

The al_3 function is characterized by the following theorem:

Theorem 2.220 *1. $\mathrm{al}_3^3(P_1, P_2, P_3) = \mathrm{al}_3^{(0)3}(P_1, P_2, P_3) = \varpi_{\widehat{X}}^* \left(\frac{(\mu_3(B_3))^3}{F(B_3)} \right)$,*

2. $\mathrm{al}_3(\widehat{\zeta_{\mathrm{al}}}(P_1), P_2, P_3) = \zeta_3 \mathrm{al}_3(P_1, P_2, P_3)$,

3. for generic points (P_1, P_2, P_3), we find Q_1, Q_2, Q_3 such that $Q_1 + Q_2 + Q_3 - 3\infty \sim -(P_1 + P_2 + P_3 - 3\infty)$, $\deg_{B_1, Q_2, Q_3} \mathrm{al}_3^3(Q_1, Q_2, Q_3) = 1$, or it can be a local parameter at B_3,

4. al_3 can be regarded as a function of $S^3\widetilde{X}$, as $\kappa_{\widehat{X}}^* \mathrm{al}_3$, simply denoted by al_3 again, and it can be regarded as

$$\mathrm{al}_3(\gamma_1, \gamma_2, \gamma_3) = \zeta_3^{\varepsilon_1'(\gamma_1)+\varepsilon_1'(\gamma_2)+\varepsilon_1'(\gamma_3)} \frac{(\mu_3(B_3))}{\sqrt[3]{F(B_3)}},$$

where $(x_i, y_i) = \kappa_X(\gamma_i)$, and

5. by identifying $w(S^3\widetilde{X}) = \mathbb{C}^3$ with $\widehat{w}_{\varpi^* \nu^{\mathrm{I}}}(S^3\widetilde{X}) = \mathbb{C}^3$, the al_3 function is a function on $\widehat{J}_X^{\mathrm{al};(3,0)}$, i.e., for $\ell \in \widehat{\Gamma}^{(3,0)}$ and $u \in \mathbb{C}^3$, $\mathrm{al}_3(u) = \mathrm{al}_3(u + \ell)$.

Proof Let us consider $P_i = (z_i, w_i)$ $(\varpi_{\widehat{X}}(P_i) = (z_i^3, z_i w_i))$ and $P = (z, w)$ $(\varpi_{\widehat{X}}(P) = (z^3, zw))$ in the vicinity of $z_i = 0$ or (B_3),

$$\varpi_{\widehat{X}}^* \mu_3(P, P_1, P_2, P_3) = (z - z_1)(z - z_2)(z - z_3) w_0$$
$$\times \left(z^3 + e_1 z^2 + (e_1^2 - e_2) z + \left(e_1 \cdot e_2 - e_3 - \frac{e_2^2}{e_1} \right) \right) (1 + d_{>0}(z^3, z_1^3, z_2^3, z_3^3)),$$

where $w_0 = \sqrt[3]{a_1 a_2 a_3}$, and

$$e_1 = z_1 + z_2 + z_3, \quad e_2 = z_1 z_2 + z_2 z_3 + z_3 z_1, \quad e_3 = z_1 z_2 z_3.$$

Thus we let the solution of the trigonal equation,

$$\left(z^3 + e_1 z^2 + (e_1^2 - e_2) z + \left(e_1 \cdot e_2 - e_3 - \frac{e_2^2}{e_1} \right) \right) = 0,$$

be denoted by $Q_i := (z_i', w_i')$. For $u = w(P_1) + w(P_2) + w(P_2)$, $-u = w(Q_1) + w(Q_2) + w(Q_2)$. Then we have

$$z_1' z_2' z_3' = \left(e_1 \cdot e_2 - e_3 - \frac{e_2^2}{e_1} \right) (1 + d_{>0}(z^3, z_1^3, z_2^3, z_3^3)),$$

which vanishes at $(0, z_2', z_3')$. By putting $P = (0, w_0)$ in $\varpi_{\hat{X}}^* \mu_3(P, P_1, P_2, P_3)$ and evaluating $\mathrm{al}_3^{(c)}$ in Definition 2.214, we have

$$\begin{aligned} \mathrm{al}_3^{(c)}(u) &= \zeta_3^c \left(e_1 \cdot e_2 - e_3 - \frac{e_2^2}{e_1} \right) (1 + d_{>0}(z^3, z_1^3, z_2^3, z_3^3)), \\ \mathrm{al}_3^{(c)}(-u) &= \zeta_3^c z_1' z_2' z_3' (1 + d_{>0}(z^3, z_1^3, z_2^3, z_3^3)). \end{aligned} \tag{2.143}$$

Thus $\mathrm{al}_1^{(c)}(-u)$ agrees with local parameter z_i' at the ramification point B_3. ∎

Theorem 2.220 3 states that the al_r function associated with B_r is identical to the local parameter at the ramification point B_r in a certain sense.

Proposition 2.221 *By letting $f(x) = (x - b_0) \cdots (x - b_3)$, we have*

$$\frac{\mathrm{al}_0^3(u)}{f'(b_0)} + \frac{\mathrm{al}_1^3(u)}{f'(b_1)} + \frac{\mathrm{al}_2^3(u)}{f'(b_2)} + \frac{\mathrm{al}_3^3(u)}{f'(b_3)} = 1,$$

$$\frac{\prod_{r=0}^2 \mathrm{al}_0^{(r)}(u)}{f'(b_0)} + \frac{\prod_{r=0}^2 \mathrm{al}_1^{(r)}(u)}{f'(b_1)} + \frac{\prod_{r=0}^2 \mathrm{al}_2^{(r)}(u)}{f'(b_2)} + \frac{\prod_{r=0}^2 \mathrm{al}_3^{(r)}(u)}{f'(b_3)} = 1.$$

Proof We define a one-form

$$K((x, y); P_1, P_2.P_3) = \frac{\prod_{r=0}^2 \mu_3((x, \zeta_3^r y); P_1, P_2, P_3)}{3 f(x) F((x, y); P_1, P_2, P_3)} dx.$$

Its integral over ∂D_X is

$$\oint_{\partial D_X} K = 0,$$

and we have the result. ∎

2.9 Degenerating Family of $X_s : y^3 = x(x-s)(x-b_1)(x-b_2), s \to 0$

In this section, we take up the topic of the degenerating curves since the W-curves with $H_X = \langle r_1, r_2, r_3, \ldots, r_{m_X} \rangle$ may be obtained by degenerating X' of

$H_{X'} = \langle r_1, r_2, r_3, \dots, r_{m_X - 1}\rangle$ due to $H_{X'} \subset H_X$. Further, since there exists, at least, a sequence of the numerical semigroups,

$$\langle r_i, r_j\rangle \subset \cdots \subset \langle r_1, r_2, r_3, \dots, r_{m_X-1}\rangle \subset \langle r_1, r_2, r_3, \dots, r_{m_X}\rangle, \tag{2.144}$$

we may find a sequence of space W-curves.

We consider the simplest example or the family of curves, $X_s : y^3 = x(x-s)(x-b_1)(x-b_2)$ for s in a small disk D_ε and a normalized curve $X_{\widehat{0}}$ of X_0 as we mentioned in Sect. 2.3.3.1 following [FeKMPA2022]; $H_{X_s} = \langle 3, 4\rangle \subset \langle 3, 4, 5\rangle = H_{X_{\widehat{0}}}$. Further, though this topic should be described in terms of the mixed Hodge theory [Ste1986], we employ a classic approach.

2.9.1 Preliminaries for a Degenerating Family of X_s

In order to describe the degenerating family of curves X_s parameterized by s, we introduce some objects in this subsection. For a real parameter $\varepsilon > 0$, $\min(|b_1|, |b_2|) > \varepsilon$, we define the ε (punctured) disk

$$D_\varepsilon = \{s \in \mathbb{C} \mid |s| < \varepsilon\}, \quad D_\varepsilon^* = D_\varepsilon \setminus \{0\},$$

and consider the degenerating family of curves X_s,

$$\mathfrak{X} := \{(x, y, s) \mid (x, y) \in X_s, s \in D_\varepsilon\}$$

with the projection $\pi_{\mathfrak{X}} : \mathfrak{X} \to D_\varepsilon$. We also consider the trivial bundle

$$\mathbb{P}_{D_\varepsilon} = \mathbb{P} \times D_\varepsilon$$

with $\pi_{\mathbb{P}} : \mathbb{P} \times D_\varepsilon \to D_\varepsilon$ so that we define the bundle map $\pi_1 : \mathfrak{X} \to \mathbb{P}_{D_\varepsilon}$ which is induced from $\pi_1 : X_s \to \mathbb{P}$ $(\pi_1(P) = x)$. Similarly, we define

$$\widetilde{\mathfrak{X}} := \{(x, y, s) \mid (x, y) \in \widetilde{X}_s, s \in D_\varepsilon\},$$

and "symmetric products,"

$$S^k\mathfrak{X} := \{(x, y, s) \mid (x, y) \in S^k X_s, s \in D_\varepsilon\}, \quad S^k\widetilde{\mathfrak{X}} := \{(x, y, s) \mid (x, y) \in S^k \widetilde{X}_s, s \in D_\varepsilon\}.$$

We also have the family of Jacobians,

$$\mathfrak{J} := \{(v, s) \mid v \in J_s, s \in D_\varepsilon^*\}.$$

Among them, we also have the bundle maps as morphisms induced from each fiber.

Further, we consider a smooth section $P : D_\varepsilon \to \mathfrak{X}$ $(P_s = (x_s, y_s)$ for a point $s \in D_\varepsilon)$ which satisfies the commutative diagram

$$\begin{array}{ccc} X_s & \xrightarrow{s\to 0} & X_0 \\ & \searrow^{\pi_1} & \downarrow{\scriptstyle \pi_1} \\ & & \mathbb{P}, \end{array}$$

i.e., $\pi_1(P_s) = \pi_1(P_{s'}) = x \in \mathbb{P}$ for $s, s' \in D_\varepsilon$. We call it an *x-constant section* of D_ε.

Remark 2.222 It is noted that x-constant section means the flow of the differential D_x satisfies $D_x(x) = 0$ in the moduli space $\mathfrak{X}$. It is remarked that the flow is unique in the sense of quasi-isomorphism [Man1964, EilGOY2017].

We recall a normalized curve $X_{\hat{0}}$ of X_0 as mentioned in Sect. 2.3.3.1. By using the x-constant section, we will compare the sigma functions over $X_{s\to 0}$ of $s \in D_\varepsilon^*$ and $X_{\hat{0}}$ in Sect. 3.7.

2.9.2 *Integrals for X_s for $s \in D_\varepsilon^*$*

In this subsection, we consider $X_s := \pi_{\mathfrak{X}}^{-1}(s)$ by fixing the parameter $s \in D_\varepsilon^*$.

Using the notations in Sects. 2.4.2.5 and 2.4.3.8, we evaluate the integrals,

$$\omega_{X_s ij} = \int_\infty^{B_i} \nu^{\mathrm{I}}_{X_s j}, \quad \text{and} \quad \eta_{X_s ij} := -\int_\infty^{B_i} \nu^{\mathrm{II}}_{X_s j}, \quad (i = 0, 1, 2, 3,\ j = 1, 2, 3).$$

Noting $|s| < |b_1|, |b_2|$, we introduce the regions

$$V_s := \{(x, y) \in X_s \mid |x| \le |s|\}, \quad V_b := \{(x, y) \in X_s \mid |x| \le \min\{|b_1|, |b_2|\}\},$$

and $V_s^c := X_s \setminus V_s$. In the following expansions, we use the cubic root since the ambiguity of the cubic root is naturally fixed.

For the region V_s^c, we have the expansion due to the absolute convergence,

$$\frac{1}{\sqrt[3]{x-s}} = \frac{1}{x^{1/3}}\left(\sum_{\ell=0} \frac{(3\ell+1)!!!}{\ell!}\left(\frac{1}{3}\frac{s}{x}\right)^\ell\right) =: \frac{1}{x^{1/3}}\sum_{\ell=0} c_\ell^{(1)}\left(\frac{s}{x}\right)^\ell,$$

$$\frac{1}{\sqrt[3]{(x-s)^2}} = \frac{1}{x^{2/3}}\left(\sum_{\ell=0} \frac{(3\ell+2)!!!}{\ell!}\left(\frac{2}{3}\frac{s}{x}\right)^\ell\right) =: \frac{1}{x^{2/3}}\sum_{\ell=0} c_\ell^{(2)}\left(\frac{s}{x}\right)^\ell,$$

where $n!!! = n(n-3)(n-6)\cdots(\ell+3)\ell$ for $\ell \in \{0, 1, 2\}$ such that $\ell \equiv n$ modulo 3.

Lemma 2.223 *In V_s^c, the differentials are expressed as follows:*

$$\nu^{\mathrm{I}}_{X_s 1} = \frac{dx}{3y^2} = \frac{dx}{3\sqrt[3]{x^4(x-b_1)^2(x-b_2)^2}}\left(1+\sum_{\ell=1} c_\ell^{(2)}\left(\frac{s}{x}\right)^\ell\right),$$

$$\nu^{\mathrm{I}}_{X_s 2} = \frac{xdx}{3y^2} = \frac{dx}{3\sqrt[3]{x(x-b_1)^2(x-b_2)^2}}\left(1+\sum_{\ell=1} c_\ell^{(2)}\left(\frac{s}{x}\right)^\ell\right),$$

$$\nu^{\mathrm{I}}_{X_s 3} = \frac{dx}{3y} = \frac{dx}{3\sqrt[3]{x^2(x-b_1)(x-b_2)}}\left(1+\sum_{\ell=1} c_\ell^{(1)}\left(\frac{s}{x}\right)^\ell\right),$$

$$\nu^{\mathrm{II}}_{X_s 1} = \frac{(5x^2+2\lambda_3 x+\lambda_2)dx}{3y} = \frac{(5x^2+2\lambda_3 x+\lambda_2)dx}{3\sqrt[3]{x^2(x-b_1)(x-b_2)}}\left(1+\sum_{\ell=1} c_\ell^{(1)}\left(\frac{s}{x}\right)^\ell\right),$$

$$\nu^{\mathrm{II}}_{X_s 2} = \frac{2xdx}{3y} = \frac{2xdx}{3\sqrt[3]{x^2(x-b_1)(x-b_2)}}\left(1+\sum_{\ell=1} c_\ell^{(1)}\left(\frac{s}{x}\right)^\ell\right),$$

$$\nu^{\mathrm{II}}_{X_s 3} = \frac{x^2dx}{3y^2} = \frac{xdx}{3\sqrt[3]{x(x-b_1)^2(x-b_2)^2}}\left(1+\sum_{\ell=1} c_\ell^{(2)}\left(\frac{s}{x}\right)^\ell\right).$$

Let us consider the region V_b which includes V_s because we have assumed that $\min\{|b_1|, |b_2|\} > \varepsilon > |s|$. In this region V_b, $h_a(x) := \dfrac{1}{\sqrt[3]{(x-b_1)^a(x-b_2)^a}}$ is expanded as

$$h_a(x) = \frac{\zeta_3^a}{\sqrt[3]{(b_1b_2)^a}}\left(\sum_{\ell=0} \beta_\ell^{(a)} x^\ell\right). \tag{2.145}$$

Further, by letting $s = s_r e^{\sqrt{-1}\varphi_s}$ ($s_r \in \mathbb{R}$), the crosscuts in Fig. 2.8 are illustrated in Fig. 2.15(a).

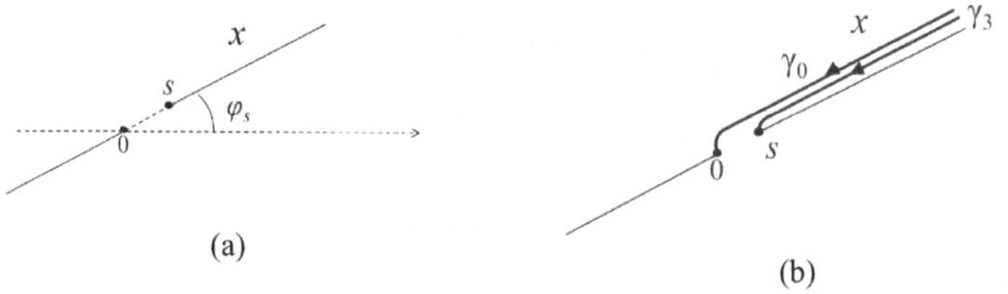

Fig. 2.15 Crosscuts and contours: (a) shows the crosscuts and (b) shows the contours of the integrals, $\gamma_a = \gamma_{\infty, B_a}$ ($a = 0, 3$)

We have the integrals along the contours from ∞ to B_a denoted by $\gamma_a := \gamma_{\infty, B_a}$; γ_0 and γ_3 are displayed in Fig. 2.15(b) as in Sects. 2.9.2.1 and 2.9.2.2.

The integrals $\omega_{X_s i, B_a} = \displaystyle\int_{\gamma_a} v^{\mathrm{I}}_{X_s i}$ $(i = 1, 2, 3, a = 0, 1, 2, 3)$ (2.118) are evaluated as follows.

Lemma 2.224 *The matrix of integrals has the expression,*

$$\begin{pmatrix} \omega_{X_s 1, B_0} & \omega_{X_s 1, B_1} & \omega_{X_s 1, B_2} & \omega_{X_s 1, B_3} \\ \omega_{X_s 2, B_0} & \omega_{X_s 2, B_1} & \omega_{X_s 2, B_2} & \omega_{X_s 2, B_3} \\ \omega_{X_s 3, B_0} & \omega_{X_s 3, B_1} & \omega_{X_s 3, B_2} & \omega_{X_s 3, B_3} \end{pmatrix} = \begin{pmatrix} A^{(0)}_{10} & A^{(0)}_{11} & A^{(0)}_{12} & s^{-1/3} A^{(0)}_{13} \\ A^{(0)}_{20} & A^{(0)}_{21} & A^{(0)}_{22} & A^{(0)}_{23} \\ A^{(0)}_{30} & A^{(0)}_{31} & A^{(0)}_{32} & A^{(0)}_{33} \end{pmatrix},$$

where $A^{(0)}_{ij} := A^{(0)}_{ij}(s^{1/3})$, and $A^{(0)}_{ij}(t)$ are holomorphic functions with respect to $t = s^{1/3}$ such that

$$A^{(0)}_{13}(s = 0) \neq 0.$$

Proof When we compute $\omega_{X_s a, B_b}$ $(a = 1, 2, 3, b = 1, 2)$, we apply Lemma 2.223 since we can regard γ_a $(a = 1, 2)$ as a subset of V^c_s. On the other hand, the cases of $a = 0$ and $a = 3$ are in Lemmas 2.229 and 2.230 in Sect. 2.9.2.1. ∎

2.9.2.1 Period Integrals

In this subsubsection, we estimate the dependence of the period integrals on s following [FeKMPA2022, Appendix A]. We assume that s is a positive sufficiently small real number satisfying $0 < s < \delta < \min\{|b_1|, |b_2|\}$. Further, without loss of generality, the imaginary parts of b_1 and b_2 are positive, i.e., $Im(b_1) > s$ and $Im(b_2) > s$. Let

$$h_2(x) := \frac{1}{\sqrt[3]{(x - b_1)^2 (x - b_2)^2}}$$

such that $h_2(0) = \dfrac{\zeta_3^2}{\sqrt[3]{(b_1)^2 (b_2)^2}}$ and we choose the branch so that $\sqrt[3]{x^2(x - s)^2}$ should be positive for $[x, \infty)$. Using $h_2(x)$, we consider two integrals for $[0, \infty)$ and (s, ∞),

$$I_1(s) := \int_0^\infty \frac{h_2(x)}{\sqrt[3]{x^2(x - s)^2}} dx, \quad I_2(s) := \int_s^\infty \frac{h_2(x)}{\sqrt[3]{x^2(x - s)^2}} dx.$$

More precisely, I_1 is defined as the upper half part of the contour integral with a positive $\varepsilon \to 0$,

$$I_1(s) = \lim_{\varepsilon \to 0} \frac{1}{\xi_3 - 1} \int_{\gamma_\varepsilon} \frac{h_2(x)}{\sqrt[3]{x^2(x - s)^2}} dx,$$

which is illustrated in Fig. 2.16.

Then we have the following proposition.

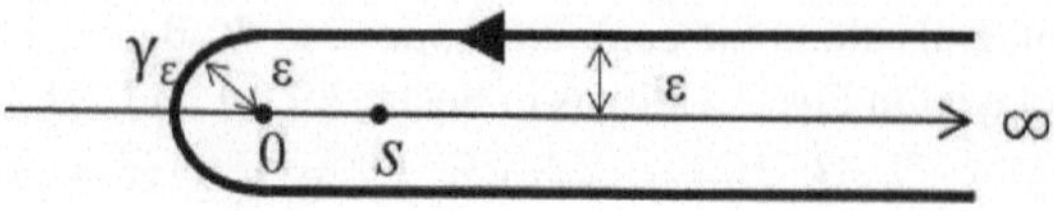

Fig. 2.16 The contour γ_ε of the integral

Proposition 2.225 *When* $\min\{|b_1|, |b_2|\} > s > 0$*,* $Im(b_1) > s$*, and* $Im(b_2) > s$*,*

$$I_1(s) = f(s), \quad I_2(s) = s^{-1/3} g(s^{1/3}),$$

where $f(t)$ *and* $g(t)$ *are regular functions with respect to* t *in the region*

$$U_\epsilon := \{t \in \mathbb{R} \,|0 \le t < \epsilon\},$$

for a certain $\epsilon > 0$ *and* $g(t)|_{t=0} \neq 0$*.*

We prove this proposition as follows. First, we note that the assumptions $Im(b_1) > 0$ and $Im(b_2) > 0$ mean that b_a $(a = 1, 2)$ does not belong to the x-axis.

It is obvious that $h_2(x)$ is holomorphic at $x = 0$ and, as in (2.145) its expansion given by

$$h_2(x) = \frac{\zeta_3^2}{\sqrt[3]{b_1^2 b_2^2}} \left(\sum_{i=0}^{\infty} \beta_i^{(2)} x^i \right), \quad \beta_0^{(2)} = 1,$$

converges in $V_b := \{x \in \mathbb{C} \mid |x| < \min\{|b_1|, |b_2|\}\}$. There is a certain neighborhood V of $\mathbb{R}_{\ge 0} := \{x \in \mathbb{R} \mid x \ge 0\}$ in the complex plane $\mathbb{C}$ such that

$$\Phi(x, s) := \frac{h_2(x)}{\sqrt[3]{x^2(x-s)^2}}$$

is a holomorphic function in $V \setminus \mathbb{R}_{\ge 0}$. Using a positive parameter ρ, $\big(\min(Im(b_1), Im(b_2)) > \rho > |s| > 0\big)$, we redefine the contour $\gamma \subset V$, $\gamma : (-\infty, \infty) \to \mathbb{C}$, given by

$$\gamma(t) = \begin{cases} -t - \dfrac{1}{2} + \rho\sqrt{-1} & \text{for } t \in \left(-\infty, -\dfrac{1}{2}\right], \\ \rho e^{\pi(t+1)\sqrt{-1}} & \text{for } t \in \left(-\dfrac{1}{2}, \dfrac{1}{2}\right], \\ t - \dfrac{1}{2} + \rho\sqrt{-1} & \text{for } t \in \left(\dfrac{1}{2}, \infty\right), \end{cases}$$

illustrated in Fig. 2.17. γ is disjoint to the points b_1 and b_2 in $\mathbb{C}$. Since γ is homotopic to γ_ε, we show the proposition for the integrals of γ.

In view of the multi-valued property of $\Phi(x, s)$, the Cauchy integral theorem gives

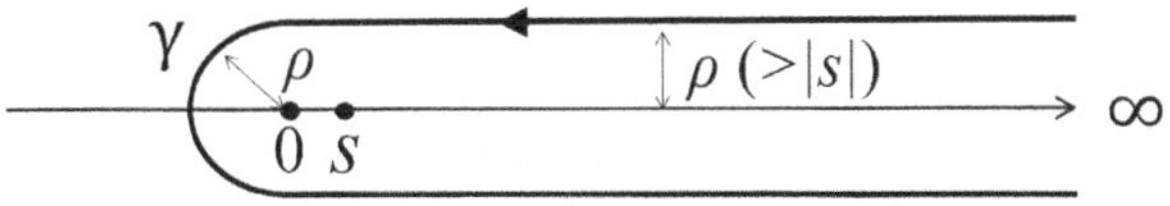

Fig. 2.17 The contour γ of the integral

$$\int_\gamma \Phi(x,s)dx = (\zeta_3 - 1)\int_s^\infty \frac{h_2(x)}{\sqrt[3]{x^2(x-s)^2}}dx + (-\zeta_3+1)\int_0^s \frac{h_2(x)}{|\sqrt[3]{x^2(x-s)^2}|}dx$$
$$=: (\zeta_3 - 1)I_\gamma(s). \tag{2.146}$$

Since γ is disjoint with $[0, s]$ and $\{b_1, b_2\}$ for $|s| < \rho$, $\Phi(x, s)$ is a regular function with respect to $(x, s) \in \gamma \times [0, \delta)$.

For the region, we have the expansion due to the absolute convergence,

$$\frac{1}{\sqrt[3]{(x-s)^2}} = \frac{1}{x^{2/3}}\left(\sum_{\ell=0}\frac{(3\ell+2)!!!}{\ell!}\left(\frac{2}{3}\frac{s}{x}\right)^\ell\right) =: \frac{1}{x^{2/3}}\sum_{\ell=0} c_\ell^{(2)}\left(\frac{s}{x}\right)^\ell,$$

where $(3\ell+2)!!! = (3\ell+2)(3\ell-1)\cdots 5\cdot 2$.

$$\Phi(x,s) = \Phi(x,0)\left(1 + \sum_{\ell=1} c_\ell^{(2)}\left(\frac{s}{x}\right)^\ell\right),$$

and

$$\Phi(x,0) = \frac{1}{\sqrt[3]{x^4(x-b_1)^2(x-b_2)^2}}.$$

Lemma 2.226 $\int_\gamma \Phi(x,s)dx$ *is a regular function with respect to s, i.e., the power expansion,*

$$I_\gamma(s) = \int_\gamma \Phi(x,s)dx = \sum_{i=0} v_i s^i,$$

has a finite radius of convergence δ', where

$$v_0 := \int_\gamma \frac{1}{\sqrt[3]{x^4(x-b_1)^2(x-b_2)^2}}dx.$$

Since γ_ε is homotopic to γ, we conclude that $I_1(s) = I_\gamma(s)/(\xi_3 - 1)$.
On the other hand, from the definition, we obtain

$$I_1(s) = \int_0^s \frac{h_2(x)}{3\sqrt[3]{|x^2(x-s)^2|}}dx + I_2(s). \tag{2.147}$$

It means that $I_2(s)$ is represented as the difference between the integrals over the generalized Lefschetz cycle and the detoured cycle,

$$I_2(s) = A_2 - A_1,$$

$$A_1 = \int_0^s \frac{h_2(x)}{\sqrt[3]{|x^2(x-s)^2|}}dx, \quad A_2 = \frac{1}{\zeta_3 - 1}\int_\gamma \Phi(x,s)dx.$$

By exchanging $x = st$, we have

$$\begin{aligned} A_1 &= \int_0^s \frac{h_2(x)dx}{\sqrt[3]{|x^2(x-s)^2|}} = \frac{\zeta_3^2}{\sqrt[3]{(b_1b_2)^2}}\sum_{\ell=0}\beta_\ell^{(2)}\int_0^s \frac{x^\ell dx}{\sqrt[3]{|x^2(x-s)^2|}} \\ &= s^{-1/3}\frac{\zeta_3^2}{\sqrt[3]{(b_1b_2)^2}}\sum_{\ell=0}\beta_\ell^{(2)}\int_0^1 \frac{(st)^\ell dt}{\sqrt[3]{|t^2(t-1)^2|}} \\ &= s^{-1/3}\frac{\zeta_3^2}{\sqrt[3]{(b_1b_2)^2}}\sum_{\ell=0}\beta_\ell^{(2)}s^\ell\int_0^1 t^{\ell+\frac{1}{3}-1}(1-t)^{\frac{1}{3}-1}dt \\ &= s^{-1/3}\frac{\zeta_3^2}{\sqrt[3]{(b_1b_2)^2}}\sum_{\ell=0}\beta_\ell^{(2)}s^\ell\frac{\Gamma(\ell+\frac{1}{3})\Gamma(\frac{1}{3})}{\Gamma(\ell+\frac{2}{3})}. \end{aligned} \tag{2.148}$$

We note that the series converges absolutely because, there is $0 < c < 1$ such that the ratio

$$\left|\beta_{\ell+1}^{(2)}\frac{\Gamma(\ell+\frac{4}{3})\Gamma(\frac{1}{3})}{\Gamma(\ell+\frac{5}{3})}\frac{1}{\beta_\ell^{(2)}}\frac{\Gamma(\ell+\frac{2}{3})}{\Gamma(\ell+\frac{1}{3})\Gamma(\frac{1}{3})}\right| = \left|\frac{\beta_{\ell+1}^{(2)}}{\beta_\ell^{(2)}}\frac{\ell+\frac{4}{3}}{\ell+\frac{5}{3}}\right| \tag{2.149}$$

is smaller than c for a certain L and $\ell > L$. Further, it is obvious that the leading term is not equal to zero, i.e.,

$$\frac{\zeta_3^2}{\sqrt[3]{(b_1b_2)^2}}\beta_0^{(2)}\frac{\Gamma(\frac{1}{3})\Gamma(\frac{1}{3})}{\Gamma(\frac{2}{3})} \neq 0.$$

On the other hand, the second term A_2 is given by

$$\sum_{\nu=0}^{\infty} u_\nu s^\nu, \quad u_0 := \frac{1}{\zeta_3 - 1}\int_\gamma \frac{1}{\sqrt[3]{x^4(x-b_1)^2(x-b_2)^2}}dx.$$

Then we obtain

$$I_2(s) = \sum_{\nu=0}^{\infty} u_\nu s^\nu - s^{-1/3}\frac{\zeta_3^2}{\sqrt[3]{(b_1b_2)^2}}\sum_{\ell=0}\beta_\ell^{(2)}s^\ell\frac{\Gamma(\ell+\frac{1}{3})\Gamma(\frac{1}{3})}{\Gamma(\ell+\frac{2}{3})}.$$

It means the complete proof of the proposition.

Remark 2.227 The case when $s < 0$ can be treated similarly, by replacing I_1 and I_2 with the corresponding integrals from $-\infty$ to s.

2.9.2.2 Period Integrals

In this subsubsection, we apply the results in Sect. 2.9.2.1 to our integrals, $\omega_{X_s,a,b}$. We assume that s is a complex number such that $|s| \ll \min\{|b_1|, |b_2|\}$. We employ the contours in Fig. 2.15. However, the phase factor $\log(s/|s|)$ is not crucial since it corresponds to rescaling the variables in the curve equation. Noting (2.145), we evaluate the integrals associated with the period matrices.

Lemma 2.228 *There are regular functions $f_a(t)$ for $t \in V_\epsilon$ such that*

$$\int_0^s \nu^{\mathrm{I}}_{X_s,1} = s^{-1/3} f_1(s^{1/3}), \quad \int_0^s \nu^{\mathrm{I}}_{X_s,2} = f_2(s^{1/3}), \quad \int_0^s \nu^{\mathrm{I}}_{X_s,3} = f_3(s^{1/3}),$$

and $f_1(t)|_{t=0}$ does not vanish.

Proof From (2.148), we have the first relation. Similarly, we also have

$$\begin{aligned}
\int_0^s \nu^{\mathrm{I}}_{X_s,2} &= \int_0^s \frac{h_2(x)x\,dx}{\sqrt[3]{x^2(x-s)^2}} = \frac{\zeta_3^2}{\sqrt[3]{(b_1b_2)^2}} \sum_{\ell=0} \beta_\ell^{(2)} \int_0^s \frac{x^{\ell+1}dx}{\sqrt[3]{x^2(x-s)^2}} \\
&= s^{-1/3} \frac{\zeta_3^2}{\sqrt[3]{(b_1b_2)^2}} \sum_{\ell=0} \beta_\ell^{(2)} \int_0^1 \frac{(st)^{\ell+1}dt}{\sqrt[3]{t^2(t-1)^2}} \\
&= s^{2/3} \frac{1}{\sqrt[3]{(b_1b_2)^2}} \sum_{\ell=0} \beta_\ell^{(2)} s^\ell \int_0^1 t^{\ell+\frac{4}{3}-1}(1-t)^{\frac{1}{3}-1}dt \\
&= s^{2/3} \frac{1}{\sqrt[3]{(b_1b_2)^2}} \sum_{\ell=0} \beta_\ell^{(2)} s^\ell \frac{\Gamma(\ell+\frac{4}{3})\Gamma(\frac{1}{3})}{\Gamma(\ell+\frac{5}{3})},
\end{aligned}$$

$$\begin{aligned}
\int_0^s \nu^{\mathrm{I}}_{X_s,3} &= \int_0^s \frac{h_1(x)x\,dx}{\sqrt[3]{x(x-s)}} = \frac{\zeta_3}{\sqrt[3]{(b_1b_2)}} \sum_{\ell=0} \beta_\ell^{(1)} \int_0^s \frac{x^{\ell+1}dx}{\sqrt[3]{x^2(x-s)^2}} \\
&= s^{1/3} \frac{\zeta_3}{\sqrt[3]{(b_1b_2)}} \sum_{\ell=0} \beta_\ell^{(1)} \int_0^1 \frac{(st)^{\ell}dt}{\sqrt[3]{t(t-1)}} \\
&= s^{2/3} \frac{1}{\sqrt[3]{(b_1b_2)}} \sum_{\ell=0} \beta_\ell^{(1)} s^\ell \int_0^1 t^{\ell+\frac{2}{3}-1}(1-t)^{\frac{2}{3}-1}dt \\
&= s^{1/3} \frac{1}{\sqrt[3]{(b_1b_2)}} \sum_{\ell=0} \beta_\ell^{(1)} s^\ell \frac{\Gamma(\ell+\frac{2}{3})\Gamma(\frac{2}{3})}{\Gamma(\ell+\frac{4}{3})}.
\end{aligned}$$

Except for the prefactor s^ℓ, the series absolutely converge as in (2.149). ■

Thus, we consider the integrals along the contours in Fig. 2.15(b). We have the following lemmas:

Lemma 2.229 *There are regular functions $f_a'(t)$ for $t \in V_\epsilon$ such that*

$$\int_{\gamma_3} \nu^{\mathrm{I}}_{X_s,1} = s^{-1/3} f_1'(s^{1/3}), \quad \int_{\gamma_3} \nu^{\mathrm{I}}_{X_s,2} = f_2'(s^{1/3}), \quad \int_{\gamma_3} \nu^{\mathrm{I}}_{X_s,3} = f_3'(s^{1/3}),$$

and $f_1'(t)|_{t=0}$ does not vanish.

Lemma 2.230 *There are regular functions $g_a(t)$ for $t \in V_\epsilon$ such that*

$$\int_{\gamma_0} \nu^{\mathrm{I}}_{X_s,1} = g_1(s^{1/3}), \quad \int_{\gamma_0} \nu^{\mathrm{I}}_{X_s,2} = g_2(s^{1/3}), \quad \int_{\gamma_0} \nu^{\mathrm{I}}_{X_s,3} = g_3(s^{1/3}).$$

2.9.3 *Periods for Degeneration*

In this subsection, we handle the period matrices for degeneration. Lemmas 2.231 and 2.232 below are corollaries of Lemma 2.224 by assuming $s \in D_\varepsilon^*$.

Lemma 2.231 *The matrices of the half-periods have the expression,*

$$\omega'_{X_s} = \begin{pmatrix} A^{(1)}_{11} & A^{(1)}_{12} & s^{-1/3}A^{(1)}_{13} \\ A^{(1)}_{21} & A^{(1)}_{22} & A^{(1)}_{23} \\ A^{(1)}_{31} & A^{(1)}_{32} & A^{(1)}_{33} \end{pmatrix}, \quad \omega''_{X_s} = \begin{pmatrix} s^{-1/3}A^{(2)}_{11} & A^{(2)}_{12} & s^{-1/3}A^{(2)}_{13} \\ A^{(2)}_{21} & A^{(2)}_{22} & A^{(2)}_{23} \\ A^{(2)}_{31} & A^{(2)}_{32} & A^{(2)}_{33} \end{pmatrix},$$

where $A^{(a)}_{ij} := A^{(a)}_{ij}(s^{1/3})$, and $A^{(a)}_{ij}(t)$ are holomorphic functions with respect to $t = s^{1/3}$ such that $A^{(a)}_{13}(0)$ $(a = 1, 2)$ and $A^{(2)}_{11}(0)$ are not zero.

Proof In the computation, we note that the contours of γ_a $(a = 0, 3)$ shown in Fig. 2.8 are given as Fig. 2.15. Due to Lemmas 2.198, 2.223, 2.224, and the fact that γ_a $(a = 1, 2)$ is a subset of V_s^c, we have the above results. Especially from (2.148), we have

$$A^{(1)}_{13} = \frac{\zeta_3^2}{\sqrt[3]{(b_1 b_2)^2}} \sum_{\ell=0} \beta^{(2)}_\ell s^\ell \frac{\Gamma(\ell+\frac{1}{3})\Gamma(\frac{1}{3})}{\Gamma(\ell+\frac{2}{3})}$$

using the notations in (2.148). ■

Lemma 2.232 *The determinant $|\omega'_{X_s}| = s^{-1/3} A(s^{1/3})$,*

$$\omega_{X_s}'^{-1} = \begin{pmatrix} s^{1/3}A_{11}^{(3)} & A_{12}^{(3)} & A_{13}^{(3)} \\ s^{1/3}A_{21}^{(3)} & A_{22}^{(3)} & A_{23}^{(3)} \\ s^{1/3}A_{31}^{(3)} & s^{1/3}A_{32}^{(3)} & s^{1/3}A_{33}^{(3)} \end{pmatrix}, \quad \omega_{X_s}'^{-1}\omega_{X_s}'' = \begin{pmatrix} A_{11}^{(4)} & A_{12}^{(4)} & A_{13}^{(4)} \\ A_{21}^{(4)} & A_{22}^{(4)} & s^{1/3}A_{23}^{(4)} \\ A_{31}^{(4)} & s^{1/3}A_{32}^{(4)} & A_{33}^{(4)} \end{pmatrix},$$

where for $a = 3, 4$, $A_{ij}^{(a)} := A_{ij}^{(a)}(s^{1/3})$ and $A_{ij}^{(a)}(t)$ and $A(t)$ are holomorphic functions with respect to $t = s^{1/3}$ such that $A(0) \neq 0$.

Proof The first relation is directly derived from Lemma 2.231. Thus we obtain the formula for $\omega_{X_s}'^{-1}$. Using it, we have

$$\omega_{X_s}'^{-1}\omega_{X_s}'' = \begin{pmatrix} A_{11}^{(4)} & A_{12}^{(4)} & A_{13}^{(4)} \\ A_{21}^{(4)} & A_{22}^{(4)} & A_{23}^{(4)\prime} \\ A_{31}^{(4)} & s^{1/3}A_{32}^{(4)} & A_{33}^{(4)} \end{pmatrix},$$

and the fact that $\omega_{X_s}'^{-1}\omega_{X_s}''$ is a symmetric matrix leads to the third result. ∎

From Lemma 2.198, we have the following lemma:

Lemma 2.233

$$\eta_{X_s}' = \begin{pmatrix} s^{1/3}A_{11}^{(5)} & s^{1/3}A_{12}^{(5)} & s^{1/3}A_{13}^{(5)} \\ A_{21}^{(5)} & A_{22}^{(5)} & A_{23}^{(5)} \\ A_{31}^{(5)} & A_{32}^{(5)} & A_{33}^{(5)} \end{pmatrix}, \quad \eta_{X_s}'' = \begin{pmatrix} s^{1/3}A_{11}^{(6)} & s^{1/3}A_{12}^{(6)} & s^{1/3}A_{13}^{(6)} \\ A_{21}^{(6)} & A_{22}^{(6)} & A_{23}^{(6)} \\ A_{31}^{(6)} & A_{32}^{(6)} & A_{33}^{(6)} \end{pmatrix},$$

where for $a = 5, 6$, $A_{ij}^{(a)} := A_{ij}^{(a)}(s^{1/3})$, and $A_{ij}^{(a)}(t)$ and $A(t)$ are holomorphic functions with respect to $t = s^{1/3}$ such that $A(0) \neq 0$.

Proof The terms coupled with $\omega'_{X_s 13}$ in Lemma 2.198 should be finite due to the Legendre relation and they consist of $\eta'_{X_s 1j}$ for $j = 1, 2, 3$. Thus η'_{X_s} behaves as above and Lemma 2.231 determines the η''_{X_s}. ∎

The following show that under the limit of $s \to 0$, some quantities of $\pi_{\mathfrak{X}}^{-1} D_\varepsilon^* \subset \mathfrak{X}$ correspond to those of $X_{\widehat{0}}$ using the x-constant sections, whereas others don't correspond to anything.

We have the following lemmas.

Lemma 2.234 *For the x-constant section P_s whose $\pi_1(P_s) = x$ belongs to the region V_s^c, and $\pi_1(P_0) = \widehat{\pi}_1(\widehat{P})$ of $\widehat{P} \in X_{\widehat{0}}$, we have the relations under the limit $s \to 0$,*

$$\lim_{s\to 0} \nu^{\mathrm{I}}_{X_s,i+1}(P_s) = \nu^{\mathrm{I}}_{X_{\widehat{0}},i}(\widehat{P}), \quad \lim_{s\to 0} \nu^{\mathrm{I}}_{X_s,i+1}(P_s) = \nu^{\mathrm{II}}_{X_{\widehat{0}},i}(\widehat{P}), \quad (i = 1, 2).$$

Lemma 2.235 *The $A^{(1)}$'s in Lemma 2.231 and $A^{(3)}$'s in Lemma 2.232 satisfy the following relations under the max-norm of the matrices,*

$$\lim_{s\to 0}\begin{pmatrix} A^{(1)}_{21} & A^{(1)}_{22} \\ A^{(1)}_{31} & A^{(1)}_{32} \end{pmatrix} = \omega'_{X_{\hat{0}}}, \quad \lim_{s\to 0}\begin{pmatrix} A^{(1)}_{21} & A^{(1)}_{22} \\ A^{(1)}_{31} & A^{(1)}_{32} \end{pmatrix} = \omega''_{X_{\hat{0}}},$$

$$\lim_{s\to 0} s^{1/3}|\omega'_{X_s}| = A^{(1)}_{13}|\omega'_{X_{\hat{0}}}|, \quad \lim_{s\to 0}\begin{pmatrix} A^{(3)}_{12} & A^{(3)}_{13} \\ A^{(3)}_{22} & A^{(3)}_{23} \end{pmatrix} = \omega'^{-1}_{X_{\hat{0}}}.$$

Proof In the computation, we note that the contours in Figs. 2.8 and 2.15 are consistent with those of $X_{\hat{0}}$ in Fig. 2.9. ■

This means that

$$\lim_{s\to 0}\omega'_{X_s,i+1,j} = \omega'_{X_{\hat{0}},i,j}, \qquad \lim_{s\to 0}\omega''_{X_s,i+1,j} = \omega''_{X_{\hat{0}},i,j}.$$

Similarly, we have the relations for the integrals of the second kind.

Lemma 2.236 *1.* $\lim_{s\to 0}\eta'_{X_s,i+1,j} = \eta'_{X_{\hat{0}},i,j}$ *and* $\lim_{s\to 0}\eta''_{X_s,i+1,j} = \eta''_{X_{\hat{0}},i,j}$ *for* $i, j = 1, 2,$ *and*

2. $\lim_{s\to 0}\eta X_s, 1, j' = \lim_{s\to 0}\eta X_s, 1, j'' = 0$ *for* $j = 1, 2, 3.$

Thus, it is easy to obtain the limit of $\tau_{X_s} := {\omega'_{X_s}}^{-1}\omega''_{X_s}$ for $\tau_{X_{\hat{0}}} := {\omega'_{X_{\hat{0}}}}^{-1}\omega''_{X_{\hat{0}}}$.

Lemma 2.237

$$\lim_{s\to 0}\tau_{X_s i,j} = \tau_{X_{\hat{0}} i,j}, \quad (i, j = 1, 2).$$

We have the list of the behavior of the period matrices of X_s for the degeneration $s \to 0$ and the their relation to those of $X_{\hat{0}}$.

2.10 Appendix: Normalization in Commutative Algebra

In this appendix, we review a normal ring and normalization in a commutative ring [Mat1986]. We assume that every ring is a commutative ring with unit.

B is a ring, and A is a subring of B. B is said to be an *extension* of A. An element b in B is said to be *integral* over A if b satisfies a monic polynomial over A, i.e., there exist n and $\{a_i\}_{i=1,\dots,n} \in A$ such that $b^n + a_1 b^{n-1} + \cdots a_n = 0$.

We say that B is *integral* over A, or B is an *integral ring* over A, or B is an *integral extension* of A if every element b in B is integral over A.

An *integral closure* in B over A is defined by $\widetilde{A} := \{b \in B \mid b \text{ is integral over } A\}$. If $A = \widetilde{A}$, A is *integrally closed* in B.

Definition 2.238 A is a ring and $\mathscr{Q}(A)$ is a quotient ring of A. We assume that A is an integral domain. A is normal if A is integrally closed in $\mathscr{Q}(A)$, i.e., for $\widetilde{A} := \{q \in \mathscr{Q}(A) \mid$ there exist n and $a_i \in A$ such that $q^n + a_1 q^{n-1} + \cdots a_n = 0\}$, $A = \widetilde{A}$.

We define the minimum extension $\widehat{A}$ of A in $\mathscr{Q}(A)$ so that $\widehat{A}$ is integrally closed in $\mathscr{Q}(A)$. We say that $\widehat{A}$ is *normalization* of A or the normalized ring of A.

Through the correspondence between an algebraic variety and a commutative ring, we have the well-known normalization theorem [Gri1989, p. 5, p. 68]:

Theorem 2.239 *For any irreducible algebraic curve $X \subset \mathbb{P}^2$, there exists a compact Riemann surface $\widetilde{X}$ and a holomorphic mapping $s : \widetilde{X} \to \mathbb{P}^2$ such that $s(\widetilde{X}) = X$ and s is injective on the inverse image of the set of smooth points of X. Further, the Riemann surface is unique up to its isomorphism; if there are two Riemann surfaces $\widetilde{X}$ and $\widetilde{X}'$ given by normalizations of X, there is a biholomorphic from $\widetilde{X}$ to $\widetilde{X}'$.*

As examples of Theorem 2.239, we give three examples.

Example 2.240 ($x^3 - y^2$): $R^\circ := \mathbb{C}[X, Y]/(X^3 - Y^2)$ is not normal because $\frac{Y}{X} \in \widetilde{R}^\circ \setminus R^\circ \subset \mathscr{Q}(R^\circ)$ due to $\left(\frac{Y}{X}\right)^2 - X = 0$. Since $R^\circ \approx \mathbb{C}[t^2, t^3]$, the normalized ring is $\widehat{R}^\circ = \mathbb{C}[t]$.

Example 2.241 ($y^3 = x^5 - 1$ and $w^3 = z - z^6$): Following Theorem 2.239, we consider the covering of a curve of $f(x, y) := y^3 - x^5 + 1$ and its behavior at ∞ as mentioned in Sect. 2.8.2. Let us consider a homogeneous polynomial $F(X, Y, Z) := Y^3 Z^2 - X^5 + Z^5 \in \mathbb{C}[X, Y, Z]$. For a Zariski open set given by $Z \neq 0$, we have $F(X, Y, Z) = Z^5 \left(\frac{Y^3}{Z^3} - \frac{X^5}{Z^5} + 1\right)$ and thus by regarding that $x = X/Z$ and $y = Y/Z$, we have $F(X, Y, Z) = Z^5 f(X/Z, Y/Z)$. $R^\circ := \mathbb{C}[x, y]/(f(x, y))$ is a normal ring. On the other hand, at $Z = 0$ and $X \neq 0$, we have $F(X, Y, Z) = X^5 \left(\frac{Y^3 Z^2}{X^5} - 1 + \frac{Z^5}{X^5}\right)$, and then we obtain a polynomial, $g(w, z) = w^3 z^2 - 1 + z^5$ by letting $w = Y/X$ and $z = Z/X$. However, $R^\circ_\infty := \mathbb{C}[w, z]/(g(w, z))$ is not a normal ring. As a vector space, R°_∞ is $\mathbb{C}1 + \mathbb{C}z + \mathbb{C}z^2 + \cdots + \mathbb{C}w + \mathbb{C}wz + \mathbb{C}wz^2 + \cdots + \mathbb{C}w^2 + \mathbb{C}w^2 z + \mathbb{C}w^2 z^2 + \cdots + \mathbb{C}w^3 + \mathbb{C}w^3 z$. We show that $q \in \mathscr{Q}(R^\circ_\infty) \setminus R^\circ_\infty$ exists such that $q^n + a_1 q^{n-1} + \cdots a_n = 0$ for certain $a_i \in R^\circ_\infty$. Since $\dfrac{1}{1-z} g(w, z) = \dfrac{w^3 z^2}{1-z} + 1 + z + z^2 + z^3 + z^4 = 0 \in \mathscr{Q}(R^\circ_\infty)$, it is obvious that $q := \frac{w^3}{1-z} + \frac{1+z}{z^2} \in \mathscr{Q}(R^\circ_\infty) \setminus R^\circ_\infty$, is integral over R°_∞. We define $\widehat{w} := wz = y/x^2$. $R_\infty = \widehat{R}^\circ_\infty := \mathbb{C}[\widehat{w}, z]/(\widehat{g}(w, z))$ is a normal ring, where $\widehat{g}(\widehat{w}, z) := \widehat{w}^3 - z + z^6$. The minimal condition is apparent. $\widehat{g}(\widehat{w}, z)$ is not a Weierstrass canonical form and thus Spec R_∞ is not a W-curve but R_∞ is an affine system for the projective curve at the Zariski open, $X \neq 0$; $ZF(X, Y, Z) = X^6 \widehat{g}(YZ/X, Z/X)$. This procedure is a projectivization of the W-curve $y^3 = x^5 - 1$.

Example 2.242 (a space curve; $y^3 = x^2(x^2 - 1)$ and $w^3 = x(x^2 - 1)^2$): Let us consider a polynomial $f(x, y) = y^3 - x^2(x^2 - 1)$ and show that $R^\circ := \mathbb{C}[x, y]/$

$(f(x, y))$ is not a normal ring. As a vector space, R° is $\mathbb{C}1 + \mathbb{C}x + \mathbb{C}x^2 + \cdots + \mathbb{C}y + \mathbb{C}yx + \mathbb{C}yx^2 + \cdots \quad + \mathbb{C}y^2 + \mathbb{C}y^2x + \mathbb{C}y^2x^2 + \cdots$. We show that $w \in \mathcal{Q}(R^\circ) \setminus R^\circ$ exists such that $w^n + a_1 w^{n-1} + \cdots a_n = 0$ for certain a_i's of R°. In other words, since $y = \sqrt[3]{x^2(x^2-1)}$ and $y^2 = x\sqrt[3]{x(x^2-1)}$ up to the ambiguity of $\sqrt[3]{1}$, one of such w must be $w := \frac{y^2}{x}$ which is integral over R° because $w^3 = \frac{y^6}{x^3} = x(x^2-1)^2$ or $w^3 - x(x^2-1)^2 = 0 \in R^\circ$. Let $g(x, w) = w^3 - x(x^2-1)^2$. Noting the relations that $w = \frac{y^2}{x}$, $y = \frac{w^2}{x^2-1}$, and $wy = (x^2-1)x$, we have $R = \widehat{R}^\circ := \mathbb{C}[x, y, w]/(f_1(x, y, z), f_2(x, y, z), f_3(x, y, z))$, as a normalized ring of R°, where $f_1(x, y, w) = y^2 - wx$, $f_2(x, y, w) = wy - (x^2-1)x$, and $f_3(x, y, w) = w^2 - y(x^2-1)$. The minimal condition is also obvious. This example corresponds to the special case of the affine (3, 4, 5) space curve in Sect. 2.3.3. Due to Theorem 2.239, the corresponding Riemann surface uniquely exists up to an isomorphism.

Example 2.243 (a space curve; $y^3 = (x+1)^2(x^3-1)$ and its projectivization): Since for $f(x, y) = y^3 - (x+1)^2(x^3-1)$, $\mathbb{C}[x, y]/(f(x, y))$ is not a normal ring, by following the procedure in Example 2.242, we obtain its normalized ring, $R = \mathbb{C}[x, y, w]/(f_1(x, y, z), f_2(x, y, z), f_3(x, y, z))$, where $f_1(x, y, w) = y^2 - w(x+1)$, $f_2(x, y, w) = wy - (x+1)(x^3-1)$, and $f_3(x, y, w) = w^2 - y(x^3-1)$. Spec R is a (3, 5, 7)-type space curve. Thus we consider homogeneous polynomials $F_1(X, Y, W, Z) := Y^2Z^2 - W(X+Z)Z^2$, $F_2(X, Y, W, Z) := YWZ^2 - (X+Z)(X^3-Z^3)$, and $F_3(X, Y, W, Z) := W^2Z^2 - Y(X^3-Z^3)$ in $\mathbb{C}[X, Y, W, Z]$. For a Zariski open set given by $Z \neq 0$ of the homogeneous ideals of $\widetilde{R} = \mathbb{C}[X, Y, W, Z]/(F_1, F_2, F_3)$, we recover the ideals of the affine ring R. Thus following the procedure in Example 2.241, we consider the Zariski open set given by $X \neq 0$ which covers the point $Z = 0$. Hence we have $F_1(X, Y, W, Z) = X^4\left(\frac{(YZ)^2}{X^4} - \frac{WZ^2}{X^3}(1 + \frac{Z}{X})\right)$, $F_2(X, Y, W, Z) = X^4\left(\frac{YWZ^2}{X^4} - ((1+\frac{Z}{X})(1-(\frac{Z}{X})^3)\right)$, $F_3(X, Y, W, Z) = X^4\left(\frac{(WZ)^2}{X^4} - \frac{Y}{X}(1-(\frac{Z}{X})^3)\right)$. Thus by letting $\widetilde{y} = Y/X$, $\widetilde{w} = W/X$, and $\widetilde{z} = Z/X$, we have a ring, which turns out to be not a normal. Following the arguments in Example 2.241, we normalize it again. We consider $Z^2F_1(X, Y, W, Z) = Z^2X^4\left(\frac{(YZ)^2}{X^4} - \frac{WZ^2}{X^3}(\frac{Z}{X})^2(1+\frac{Z}{X})\right)$, $Z^2F_2(X, Y, W, Z) = ZX^5\left(\frac{(YZ)(WZ^2)}{X^5} - (\frac{Z}{X})\times(1+\frac{Z}{X})(1-(\frac{Z}{X})^3)\right)$, $Z^2F_3(X, Y, W, Z) = X^6\left(\frac{(WZ^2)^2}{X^6} - \frac{Z}{X}\frac{YZ}{X^2}(1-(\frac{Z}{X})^3)\right)$. By letting $\eta = YZ/X^2 = y/x^2$ and $\xi = WZ^2/X^3 = w/x^3$, $z = 1/x$, we obtain the affine ring R^∞ given by $\mathbb{C}[\eta, \xi, z]/(f_1^\infty, f_2^\infty, f_3^\infty)$ as an affine ring around $\infty \in X$, where $f_1^\infty(\eta, \xi, z) = \eta^2 - \xi(z+1)$, $f_2^\infty(\eta, \xi, z) = \eta\xi - z(z+1)(1-z^3)$, and $f_3^\infty(\eta, \xi, z) = \xi^2 - \eta z(1-z^3)$. This procedure is a projectivization of the space W-curve $y^3 = (x+1)^2(x^3-1)$.

2.11 Appendix: Trace and Complementary Modules

In this appendix, we review Dedekind's trace and its related topics based on Kunz's book [Ku2004], and Dedekind different based on Neukirch's book [Neu1999] whose origin appeared in the paper by Dedekind and Weber [DedW1882].

Let R_P be an algebra over $\mathbb{C}$ and R_Y an R_P algebra such that $R_Y = \bigoplus_{i=0}^{r-1} R_P \mathbf{y}_i$, $\mathbf{y}_i \in R_Y$.

The dual of R_Y is defined by

$$\omega_{R_Y/R_P} = \mathrm{Hom}_{R_P}(R_Y, R_P)$$

with the basis $\{\mathbf{y}_i^*\}_{i\in\mathbb{Z}_r} \subset \omega_{R_Y/R_P}$ satisfying $\mathbf{y}_i^*\mathbf{y}_j = \delta_{ij}$. We assume that for $x, y \in R_Y$ and $z \in \omega_{R_Y/R_P}$, $(x \circ z)(y) = z(xy)$, and we also regard ω_{R_Y/R_P} as an R_Y-module.

Let us introduce *standard trace*, $\tau_{R_Y/R_P} := \sum_{i=0}^{r-1} \mathbf{y}_i \circ \mathbf{y}_i^*$, and its complementary module $R_Y^{\mathfrak{c}}$ with respect to τ_{R_Y/R_P} given by

$$R_Y^{\mathfrak{c}} := \{z \in \mathcal{Q}(R_Y) \mid \tau_{R_Y/R_P}(zR_Y) \subset R_P\}.$$

We construct the complementary module $R_Y^{\mathfrak{c}}$ as follows because $R_Y^{\mathfrak{c}}$ is directly connected with the Kähler differentials.

Lemma 2.244 *There are structure coefficients a_{ijk} in R_P satisfying $\mathbf{y}_i\mathbf{y}_j = \sum_k a_{ijk}\mathbf{y}_k$ and $a_{ijk} = a_{jik}$, which determines the structure of the R_P-module R_Y. Then the standard trace shows $\tau_{R_Y/R_P}(\mathbf{y}_j) = \sum_i a_{iji}$.*

Proof The former statement is obvious and due to the definition of the standard trace, we have $\tau_{R_Y/R_P}(\mathbf{y}_j) = \sum_i \mathbf{y}_i \circ \mathbf{y}_i^*(\mathbf{y}_j) = \sum_i \mathbf{y}_i^*(\mathbf{y}_i\mathbf{y}_j) = \sum_i \mathbf{y}_i^*(\sum_k a_{ijk}\mathbf{y}_k) = \sum_i a_{iji}$. ■

The R_Y-action on the R_P-module R_Y, $x : R_Y \to R_Y$ for $x \in R_Y$, has the matrix expression M_x, $(M_x)_{ij} = \sum_k x_k a_{jki}$ since $(xz)_i = \mathbf{y}_i^*(\sum_{kj} x_k z_j \mathbf{y}_k \mathbf{y}_j)$ $= \mathbf{y}_i^*(\sum_{kj\ell} x_k z_j a_{jk\ell}\mathbf{y}_\ell) = \sum_j(\sum_k x_k a_{jki}) z_j$. Lemma 2.244 asserts that the standard trace $\tau_{R_Y/R_P}(x)$ agrees with trace of M_x, i.e, $\tau_{R_Y/R_P}(x) = \sum_i (M_x)_{ii}$.

Definition 2.245 If there is an element τ in ω_{R_Y/R_P} such that $\omega_{R_Y/R_P} = R_Y \circ \tau$ as an R_P-module, i.e., ω_{R_Y/R_P} is a free R_P-module with the basis $\{\tau\}$, we say that R_Y has the trace τ.

We note that if R_Y/R_P is separable, the standard trace τ_{R_Y/R_P} is a trace in this definition.

Lemma 2.246 *If R_Y has a trace τ, the following hold.*

1. *If an element a in R_Y satisfies $a \circ \tau = 0$, then $a = 0$. It means that $\omega_{R_Y/R_P} \cong R_Y \circ \tau$ as an R_Y-module.*
2. *For a basis $\{\mathbf{y}_i\}$ of R_Y as an R_P-module, there exists a subset $\{\widehat{\mathbf{y}}_i\} \subset R_Y$ satisfying $\tau(\widehat{\mathbf{y}}_i\mathbf{y}_j) = \delta_{ij}$. (We call $\{\widehat{\mathbf{y}}_i\}$ the dual basis with respect to the trace τ.) Then*
$$\tau_{R_Y/R_P} = \left(\sum_{i=0}^{r-1} \widehat{\mathbf{y}}_i\mathbf{y}_i\right) \circ \tau.$$

Proof *1.* For any $x \in R_Y$, $0 = a \circ \tau(x) = \tau(ax) = x \circ \tau(a)$. For every $b \in \omega_{R_Y/R_P}$, $b(a) = 0$ implies that $a = 0$. The R_P-morphism $\omega_{R_Y/R_P} \to R_Y \circ \tau$ is injective. 2. For the dual basis $\{\mathbf{y}_i^*\}$ of ω_{R_Y/R_P} such that $\mathbf{y}_i^*(\mathbf{y}_j) = \delta_{ij}$, we can find an element $\widehat{\mathbf{y}}_i \in R_Y$ satisfying $\mathbf{y}_i^* = \widehat{\mathbf{y}}_i \circ \tau$. Then $\tau(\widehat{\mathbf{y}}_i\mathbf{y}_j) = \mathbf{y}_i^*(\mathbf{y}_j) = \delta_{ij}$, and $\tau_{R_Y/R_P} = \sum_{i=0}^{r-1} \mathbf{y}_i \circ \mathbf{y}_i^*$
$$= \left(\sum_{i=0}^{r-1} \widehat{\mathbf{y}}_i\mathbf{y}_i\right) \circ \tau.$$
■

We construct τ and τ_{R_Y/R_P} in terms of the enveloping algebra $R_Y^e := R_Y \otimes_{R_P} R_Y$. For R_Y^e, the standard multiplication $\mu : R_Y^e \to R_Y$ is defined by $\mu(a \otimes b) = ab$.

Lemma 2.247 *The kernel of μ, $\mathrm{Ker}\mu$, is generated by $\{\mathbf{y}_i \otimes 1 - 1 \otimes \mathbf{y}_i\}_{i=0,\cdots,r-1}$.*

Proof Following [Ku2004, Theorem G.7], we show it. Let $I := \langle\{a \otimes 1 - 1 \otimes a\}_{a \in R_Y}\rangle_{R_P}$. Clearly $I \subset \mathrm{Ker}\mu$. There are surjective R_P-homomorphisms,

$$R_Y \otimes_{R_P} R_Y \xrightarrow{p_\mu} (R_Y \otimes_{R_P} R_Y)/I \xrightarrow{p'_\mu} (R_Y \otimes_{R_P} R_Y)/\mathrm{Ker}\mu \cong R_Y.$$

For $a, b \in R_Y$, we have $a \otimes b = (a \otimes 1)(1 \otimes b) = -(a \otimes 1)(b \otimes 1 - 1 \otimes b) + (ab \otimes 1)$. Thus there is an injection $R_Y \xrightarrow{\iota_\mu} (R_Y \otimes_{R_P} R_Y) \to (R_Y \otimes_{R_P} R_Y)/I$. Thus p'_μ is the identity map as a set, and is bijective. Further, since

$$(ab \otimes 1) - (1 \otimes ab) = (b \otimes 1)(a \otimes 1 - 1 \otimes a) + (1 \otimes a)(b \otimes 1 - 1 \otimes b),$$

every element in $\{a \otimes 1 - 1 \otimes a\}_{a \in R_Y}$ is generated by $\{\mathbf{y}_i \otimes 1 - 1 \otimes \mathbf{y}_i\}_{i=0,\cdots,r-1}$. ■

We consider the annihilator of the kernel of μ,

$$\mathrm{Ann}_{R_Y^e}(\mathrm{Ker}\mu) := \{z \in R_Y^e \mid z \cdot \mathrm{Ker}\mu = 0\}.$$

For an element $\sum a_i \otimes b_i \in \mathrm{Ann}_{R_Y^e}(\mathrm{Ker}\mu)$, $(c \otimes 1 - 1 \otimes c) \cdot \sum a_i \otimes b_i = 0$ or $\sum ca_i \otimes b_i = \sum a_i \otimes cb_i$.

Lemma 2.248 *There is a natural embedding $\varphi : \mathrm{Ann}_{R_Y^e}(\mathrm{Ker}\mu) \to \mathrm{Hom}_{R_Y}(\omega_{R_Y/R_P}, R_Y)$.*

Proof Since $\sum ca_i \otimes b_i = \sum a_i \otimes cb_i$ for an element $\sum a_i \otimes b_i \in \mathrm{Ann}_{R_Y^e}(\mathrm{Ker}\mu)$ and $c \in R_Y$, it should be defined $\varphi(\sum_i a_i \otimes b_i)(\rho) = \sum_i \rho(a_i)b_i \in R_Y$ for $\rho \in \omega_{R_Y/R_P}$; for every $s \in R_Y$, we have $s\rho(a)b = \rho(a)sb = \rho(sa)b$. ■

Proposition 2.249 [Ku2004, Corollary H.20] *Suppose R_Y/R_P has a trace. Then φ induces a bijection between the set of all traces of R_Y/R_P and the set of all generators of the R_Y-module $\mathrm{Ann}_{R_Y^e}(\mathrm{Ker}\mu)$: Each trace $\tau \in \omega_{R_Y/R_P}$ is mapped to the unique element $\Delta_\tau := \sum_{i=0}^{r-1} \widehat{\mathbf{y}}_i \otimes \mathbf{y}_i \in \mathrm{Ann}_{R_Y^e}(\mathrm{Ker}\mu)$ associated with τ such that $\sum_{i=0}^{r-1} \tau(\widehat{\mathbf{y}}_i)\mathbf{y}_i = 1$. Furthermore:*

1. *Δ_τ generates the R_Y-module $\mathrm{Ann}_{R_Y^e}(\mathrm{Ker}\mu)$, and $\{\widehat{\mathbf{y}}_1, \ldots, \widehat{\mathbf{y}}_r\}$ is the dual basis of the basis $\{\mathbf{y}_1, \ldots, \mathbf{y}_r\}$ of R_X with respect to τ; i.e.,*
$$\tau(\widehat{\mathbf{y}}_i \mathbf{y}_j) = \delta_{i,j}, \quad (i, j = 1, \ldots, r).$$
2. *If $\sum_{i=0}^{r-1} \widehat{\mathbf{y}}'_i \otimes \mathbf{y}_i$ generates the R_Y-module $\mathrm{Ann}_{R_Y^e}(\mathrm{Ker}\mu)$, and if $\tau' \in \omega_{R_Y/R_P}$ is a linear form with $\sum_{i=0}^{r-1} \tau'(\widehat{\mathbf{y}}'_i)\mathbf{y}_i = 1$, then τ' is a trace of R_Y/R_P and $\Delta_{\tau'} = \sum_{i=0}^{r-1} \widehat{\mathbf{y}}'_i \otimes \mathbf{y}_i$ is associated with the trace τ'; hence $\{\widehat{\mathbf{y}}'_1, \ldots, \widehat{\mathbf{y}}'_r\}$ is the dual basis of the basis $\{\mathbf{y}_1, \ldots, \mathbf{y}_r\}$ of R_X with respect to τ'.*
3. *For each trace τ of R_Y/R_P,*
$$\tau_{R_Y/R_P} = \mu(\Delta_\tau) \circ \tau.$$

Proof Let us consider $1 \in \mathrm{Hom}_{R_Y}(\omega_{R_Y/R_P}, R_Y)$, and consider the inverse image of $\{1\}$, $\varphi^{-1}(1) \in \mathrm{Ann}_{R_Y^e}(\mathrm{Ker}\mu)$, which is denoted by Δ_τ. By using the basis $\{\mathbf{y}_i\} \subset R_Y$, we can express it as $\Delta_\tau = \sum_i \widehat{\mathbf{y}}_i \otimes \mathbf{y}_i$ for a certain $\{\widehat{\mathbf{y}}_i\}$. Thus $\varphi(\sum_i \widehat{\mathbf{y}}_i \otimes \mathbf{y}_i)(\tau) = \sum_i \tau(\widehat{\mathbf{y}}_i)\mathbf{y}_i = 1$. The fact that $\sum_i \tau(\widehat{\mathbf{y}}_i)\mathbf{y}_i = 1$ implies that $\mathbf{y}_j = \mathbf{y}_j \varphi(\sum_i \widehat{\mathbf{y}}_i \otimes \mathbf{y}_i)(\tau) = \varphi(\sum_i \widehat{\mathbf{y}}_i \otimes \mathbf{y}_i)(\mathbf{y}_j \circ \tau) = \sum_i \tau(\widehat{\mathbf{y}}_i \mathbf{y}_j)\mathbf{y}_i$. Thus $\tau(\widehat{\mathbf{y}}_i \mathbf{y}_j) = \delta_{ij}$. Hence the set $\{\widehat{\mathbf{y}}_i\}$ is a dual basis of R_Y with respect to τ.

Let us consider the case $\mathrm{Ann}_{R_Y^e}(\mathrm{Ker}\mu) = R_Y(\sum_i \widehat{\mathbf{y}}'_i \otimes \mathbf{y}_i)$ with $\sum_{i=0}^{r-1} \tau'(\widehat{\mathbf{y}}'_i)\mathbf{y}_i = 1$ for $\tau' \in \omega_{R_Y/R_P}$. Then by the above arguments, obviously $\varphi(\sum_i \widehat{\mathbf{y}}'_i \otimes \mathbf{y}_i) = 1$. we obtain the statements in 2.

Lemma 2.246 shows that $\tau_{R_Y/R_P} = \sum_i \mathbf{y}_i \circ \mathbf{y}_i^* = \sum_i \widehat{\mathbf{y}}_i \mathbf{y}_i \circ \tau = \mu(\Delta_\tau) \circ \tau$. ■

Finally, we sketch the proof of the Dedekind different, Proposition 2.40, following Neukirch's arguments [Neu1999]. Since we are dealing only with algebraic curves, we assume that R_Y and R_P are the Dedekind domains over $\mathbb{C}$, i.e., any nonzero prime ideal is a maximal ideal, and both Krull dimensions are one. Let $Y := \mathrm{Spec}\, R_Y$ and $P := \mathrm{Spec}\, R_P$.

We first introduce the ramification algebraically. The maximal ideal $\mathfrak{p} \in R_P$ is said to split completely in $\mathcal{Q}(R_Y)$ if $\mathfrak{p}$ is decomposed as $\mathfrak{p}R_Y = \mathfrak{P}_1^{e_1} \cdots \mathfrak{P}_\ell^{e_\ell}$ so that $r = \ell$ and $e_i = 1$ for $r = [\mathcal{Q}(R_Y) : \mathcal{Q}(R_P)]$. On the decomposition $\mathfrak{p}R_Y = \mathfrak{P}_1^{e_1} \cdots \mathfrak{P}_\ell^{e_\ell}$, the prime ideal $\mathfrak{P}_i \subset R_Y$ is unramified over $\mathcal{Q}(R_P)$ if $e_i = 1$ and is ramified if not. Then e_i is called the ramification index of $\mathfrak{P}_i$ over $\mathcal{Q}(R_P)$.

Second, we mention the Kähler differentials. Let $\Omega^1_{R_Y/R_P}$ be the module of differentials, the Kähler differentials, of an algebra R_Y/R_P, which is defined by $\Omega^1_{R_Y/R_P} := \mathrm{Ker}\mu/(\mathrm{Ker}\mu)^2$ for $\mu : R_Y \to R_Y^e$. We note that the R_P homomorphism, a derivation, $d : R_Y \to \Omega^1_{R_Y/R_P}$ defined by $dy = y \otimes 1 - 1 \otimes y$ satisfies $d(yz) = ydz + ydz$ and $da = 0$ for $a \in R_P$. Thus $\Omega^1_{R_Y/R_P}$ is a R_P-module and a R_Y-module. $\Omega^1_{R_Y/R_P}$ is generated by $\sum_i z_i dy_i$ for $z_i, y_i \in R_Y$.

Furthermore, we define the different $\mathfrak{D}_{R_Y/R_P}$ as the inverse of the complementary module $\mathfrak{D}_{R_Y/R_P} := R_Y^{\mathfrak{c}-1}$, i.e.,

$$\mathfrak{D}_{R_Y/R_P} = \{z \in \mathscr{Q}(R_Y) \mid zR_Y^{\mathfrak{c}} \subset R_Y\}.$$

For prime ideals $\mathfrak{p} \subset R_P$ and $\mathfrak{P} \subset R_Y$ such that $\mathfrak{P}$ divides $\mathfrak{p}R_y$, we consider the local rings $R_{Y,\mathfrak{P}}$ and $R_{P,\mathfrak{p}}$.

Due to [Neu1999, Chap. III, Proposition (2.2), Corollary (2.2)], the property of the different is known as follows: $\mathfrak{D}_{R_Y/R_P} R_{Y,\mathfrak{P}} = \mathfrak{D}_{R_{Y,\mathfrak{P}}/R_{P,\mathfrak{p}}}$, where $R_{Y,\mathfrak{P}}$ and $R_{P,\mathfrak{p}}$ are localized at $\mathfrak{P}$ and $\mathfrak{p}$ respectively and thus $\mathfrak{D}_{R_Y/R_P} = \bigsqcup_{\mathfrak{P}} \mathfrak{D}_{R_{Y,\mathfrak{P}}/R_{P,\mathfrak{p}}}$, since the complementary module is also locally determined.

Here we connect the different and the Kähler differentials on R_Y/R_P.

Proposition 2.250 $\mathfrak{D}_{R_Y/R_P} = \{x \in R_Y \mid xdy = 0 \textit{ for every } y \in R_Y\}$.

Proof For a commutative R_P algebra A_P, let A_Y be $A_Y := R_Y \otimes_{R_P} A_P$. Then $\Omega^1_{A_Y/A_P}$ is given by $\Omega^1_{A_Y/A_P} = \Omega^1_{R_Y/R_P} \otimes_{R_P} A_P$. The module of differentials is preserved under localization and completion. Thus we may consider its local version. Due to the assumptions for R_Y and R_P, $R_{Y,\mathfrak{P}}$ and $R_{P,\mathfrak{p}}$ are complete discrete valuation rings. Thus for a certain $\alpha \in R_{Y,\mathfrak{P}}$, $R_{Y,\mathfrak{P}} = R_{P,\mathfrak{p}}[\alpha]$ with the minimal polynomial $h(X) \in R_{P,\mathfrak{p}}[X]$ of α, i.e., $h(\alpha) = 0$ in $R_{Y,\mathfrak{P}}$. Then $\Omega^1_{R_{Y,\mathfrak{P}}/R_{P,\mathfrak{p}}}$ is generated by $d\alpha$. The annihilator of $d\alpha = \alpha \otimes 1 - 1 \otimes \alpha$ is given by $h'(\alpha)$, since $h'(\alpha)d\alpha = dh(\alpha) = 0$; it is consistent with the proof in Proposition 2.42, since $h'(\alpha) = \frac{h(\alpha)\otimes 1 - 1 \otimes h(\alpha)}{\alpha \otimes 1 - 1 \otimes \alpha}$ modulo $(\mathrm{Ker}\mu)^2$. Thus the annihilator of $\Omega^1_{R_{Y,\mathfrak{P}}/R_{P,\mathfrak{p}}}$ is generated by $(h'(\alpha))$. By interpreting the results in Proposition 2.42 for this system, we find $R^{\mathfrak{c}}_{Y,\mathfrak{P}} = \dfrac{1}{h'(t_{\mathfrak{P}})} R_{Y,\mathfrak{P}}$, and thus $\mathfrak{D}_{R_{Y,\mathfrak{P}}/R_{P,\mathfrak{p}}} = h'(t_{\mathfrak{P}})R_{Y,\mathfrak{P}}$. Hence we have the equality. ■

Now we state Dedekind's different theorem on $\mathbb{C}$, Proposition 2.40 noting $\mathrm{diff}(R_Y/R_P) = \mathrm{div}(\mathfrak{D}_{R_Y/R_P})$ [Neu1999, Chap. III, Theorem (2.6)].

Theorem 2.251 *A prime ideal $\mathfrak{P}$ of R_Y is ramified over $\mathscr{Q}(R_P)$ if and only if $\mathfrak{P}$ divides $\mathfrak{D}_{R_{Y,\mathfrak{P}}/R_{P,\mathfrak{p}}}$. Let $\mathfrak{P}^s$ be the maximal power of $\mathfrak{P}$ dividing $\mathfrak{D}_{R_{Y,\mathfrak{P}}/R_{P,\mathfrak{p}}}$, and e be the ramification index of $\mathfrak{P}$ over $\mathscr{Q}(R_P)$. Then $s = e - 1$ if $\mathfrak{P}$ is ramified.*

Proof Due to Proposition 2.250, we give a sketch of this theorem. We consider the local rings associated with $\mathfrak{P} \subset R_Y$. Assume that $\mathfrak{p}R_Y = \mathfrak{P}^e \mathfrak{P}_2^{e_2} \cdots \mathfrak{P}_\ell^{e_\ell}$. As mentioned in the proof of Proposition 2.250, we can find $\alpha \in R_{Y,\mathfrak{P}}$, such that $R_{Y,\mathfrak{P}} = R_{P,\mathfrak{p}}[\alpha]$ and its minimal polynomial $h(X) \in R_{P,\mathfrak{p}}[X]$ of α. Then due to the assumption in Theorem 2.251, we can find $t_{\mathfrak{p}} \in \mathfrak{p}/\mathfrak{p}^2$ and $t_{\mathfrak{P}} \in \mathfrak{P}/\mathfrak{P}^2$ so that $t_{\mathfrak{p}} \neq 0$, $t_{\mathfrak{P}} \neq 0$ and $c\, t_{\mathfrak{p}} = t_{\mathfrak{P}}^e$ in $R_{Y,\mathfrak{P}}$ where c is the unit of $R_{Y,\mathfrak{P}}$ i.e., $h = t_{\mathfrak{P}}^e - c\, t_{\mathfrak{p}}$. Then $R_{Y,\mathfrak{P}}$ turns out to be $R_{Y,\mathfrak{P}} = R_{P,\mathfrak{p}}[t_{\mathfrak{P}}]/(h(t_{\mathfrak{P}}))$. On the other hand, from the proof in Proposition 2.250, $\mathfrak{D}_{R_{Y,\mathfrak{P}}/R_{P,\mathfrak{p}}} = h'(t_{\mathfrak{P}}) R_{Y,\mathfrak{P}}$. Accordingly $s = e - 1$. ∎

Hence we have the proof of Theorem 2.251 and Proposition 2.40.

Since we can interpret this problem by the Kähler differentials for Y and P due to Proposition 2.250, this proof is basically the same as the proof in Proposition 2.27, which is geometrically investigated.

Further, we recall that for the sequence of the fields, $k \to K \to L$, we have the exact sequence $\Omega^1_{K/k} \otimes_K L \to \Omega^1_{L/k} \to \Omega^1_{L/K} \to 0$ [Mat1986, Chap. 9, §26, Lemma 1], and then we can interpret Lemma 2.30 and Corollary 2.35 in terms of the above arguments naturally. We also remark that Hurwitz studied the behavior of the ramification points by using the local parameters, e.g., to obtain Corollary 2.35 [Hur1888, Hur1891].

2.12 Appendix: Deformation of Curves

In this appendix, we review the deformation theory of curves with the action $\mathbb{G}_m$ by Schlessinger [Sch1968], and Pinkham [Pin1974] based on a report by Komeda [Kom1993]. In other words, this appendix gives a brief overview of a construction in Sect. 2.2.3. Komeda investigated the case generated by two elements, or plane W-curves [Kom1993]. Komeda's results also play an essential role in the heat equation treatment of sigma functions in Appendix 3.11.

2.12.1 Preliminaries

We show the basic theory of the deformation theory given by Schlessinger [Sch1968]. Let X be an affine scheme over a field k with only isolated singularities. Let $\mathscr{C}$ be the category of Artin local k-algebras A with $A/\mathfrak{m}_A = k$. where $\mathfrak{m}_A$ is the maximal ideal in A.

Consider a pro-representable covariant functor $\mathscr{F}$ from $\mathscr{C}$ to the category of set $\mathscr{S}ets$, i.e., $\mathscr{F}$ satisfies $\mathscr{F}(A) \cong \mathrm{Hom}_{\text{local } A\text{-alg.}}(R, A)$, where R is a complete local A-algebra such that $R/\mathfrak{m}^n$ is in $\mathscr{C}$ for all n, and $\mathfrak{m}$ is the maximal ideal in R. In general a covariant functor $\mathscr{F} : \mathscr{C} \to \mathscr{S}ets$ is not pro-representable, but one may find an R and a morphism $\mathrm{Hom}_{\mathscr{C}}(R, -) \to \mathscr{F}$ of functors such that $\mathrm{Hom}_{\mathscr{C}}(R, A) \to \mathscr{F}(A)$ is surjective for all A in $\mathscr{C}$. If R is chosen suitably "minimal" then R is called a *hull* of $\mathscr{F}$; R is then unique up to noncanonical isomorphism.

An infinitesimal deformation of X/k to A is defined by a fiber product diagram,

$$\begin{array}{ccc} X & \longrightarrow & Y \\ \downarrow & & \downarrow \\ \operatorname{Spec} k & \longrightarrow & \operatorname{Spec} A \end{array}$$

where Y is flat over Spec A, i.e., $X \xrightarrow{\sim} Y \times_{\operatorname{Spec} A} \operatorname{Spec} k$. We refer to this diagram as $Y \to \operatorname{Spec} A$ for short.

Let $\mathscr{D}_X(A)$ be the deformation functor from $\mathscr{C}$ to $\mathscr{S}ets$ by setting

$$\mathscr{D}_X(A) = \text{ the set of isomorphism classes of deformations of } X/k \text{ to } A.$$

We note that X has only an isolated singularity. Due to [Sch1968, Proposition 3.10], $\mathscr{D}_X$ has a hull; there exist a complete noetherian local k-algebra R for which $R/\mathfrak{m}_R^n \in \mathscr{C}$, all $n \geq 1$, and $\xi \in \widehat{\mathscr{D}}_X(R) := \text{proj. Lim } \mathscr{D}_X(R/\mathfrak{m}_R^n)$ such that:

1. for every surjective morphism $p : A \to B$ in $\mathscr{C}$, the map

$$h_R(A) \to h_R(B) \times \mathscr{D}_X(A), \quad u \mapsto (p \circ u, \varphi(A)(u)),$$

   is surjective, where $h_R(C) := \operatorname{Hom}_{\mathscr{C}}(R, C)$, φ is the morphism of functors induced by ξ, and
2. the induced map $\varphi(k[\epsilon]) : h_R(k[\epsilon]) \to \mathscr{D}_X(k[\epsilon])$ is bijection.

In this case, R or Spec R is called the formal moduli space of X or the parameter space of X.

2.12.2 *Deformations from the Monomial Curves*

Schlessinger [Sch1968], and Pinkham [Pin1974] investigated the deformations of the monomial curves associated with numerical semigroups H. We set $k = \mathbb{C}$.

H is a numerical semigroup $H = \langle r_1, \ldots, r_m \rangle$ and $g = \#(\mathbb{N}_0 \setminus H)$. Let $k[Z] := k[Z_{r_1}, \ldots, Z_{r_m}]$ and $R_H := k[Z]/\ker(\widetilde{\varphi}_H)$, where

$$\widetilde{\varphi}_H : k[Z_{r_1}, \ldots, Z_{r_m}] \to k[H] := k[t^a]_{a \in \{r_1, \ldots, r_m\}}.$$

Assume that $I = \ker(\widetilde{\varphi}_H)$ is generated by the minimal set $\{f_{c_1}^H, \ldots, f_{c_k}^H\}$, and Z_a has the weight a. For $Z := {}^t(Z_1, \ldots, Z_m)$, there is a set $\{h_j \in R_H\}$ such that $h_j \cdot Z = 0$ on R_H for every j where $h_j = (h_{j1}, \ldots, h_{jm})$ is the basis of the relation whose weight of h_{ji} is $(c_j - r_i)$, where the weight of $f_{c_i}^H = h_i \cdot Z$ is c_i. $\mathbb{G}_m$ acts on R_H such that for $g_m \in \mathbb{G}_m$ $Z_a \mapsto g_m^a Z_a$ and the weight of Z_a is a.

Let $X_H := \operatorname{Spec} R_H$. Let us consider $\operatorname{Hom}_{R_H}(I/I^2, R_H)$ and its R_H-submodule D_{R_H} generated by $d_{(1)}, \ldots, d_{(m)}$ where $d_{(\ell)}$ is defined by sending $h + I^2$ to $\partial h/\partial X_\ell + I$ for $\ell = 1, \ldots, m$, i.e.,

$$D_{R_H} := \left\langle \frac{\partial f^H_{c_1}}{\partial Z_{r_1}}, \ldots, \frac{\partial f^H_{c_1}}{\partial Z_{r_m}}, \ldots, \frac{\partial f^H_{c_k}}{\partial Z_{r_1}}, \ldots, \frac{\partial f^H_{c_k}}{\partial Z_{r_m}} \right\rangle_{R_H} \subset \mathrm{Hom}_{R_H}(I/I^2, R_H).$$

Let us consider the exact sequence,

$$0 \to D_{R_H} \to \mathrm{Hom}_{R_H}(I/I^2, R_H) \to \mathscr{D}_{X_H}(k[\epsilon]) \to 0,$$

where

$$\Phi : \mathrm{Hom}_{R_H}(I/I^2, R_H) \to \mathscr{D}_{X_H}(k[\epsilon]).$$

We construct Φ as follows: Let $\underline{f}_0^{(H)} := {}^t(f^H_{c_1}, \ldots, f^H_{c_k})$. Since $I = \sum_i k[Z_{r_1}, \ldots, Z_{r_m}] f^H_{c_i}$, $\mathrm{Hom}_{R_H}(I/I^2, R_H)$ is generated by $\vartheta_\nu : f^H_{c_i} + I^2 \to g_{c_i}(\nu) + I$, where $g_{c_i}(\nu)$ has the weight $c_i - \nu$ for a certain $\nu \in \mathbb{Z}$.

We consider the general first-order lifting of $\underline{f}_0^{(H)}$, which is given by $\underline{f}_1^{(H)} = \underline{f}_0^{(H)} + \sum_{\nu \in \mathbb{Z}} t_{-\nu} \underline{f}_X^{H(-\nu)}$, where $\underline{f}_{1,c_i}^{H(-\nu)} = g_{c_i}(\nu)$ modulo $(f^H_{c_i}, \partial f^H_{c_i}/\partial Z_{r_j})_{i,j}$. $\mathscr{D}_X(k[\epsilon])$ is graded and we refer to the ν-th graded piece as $T^1_{X_H}(\nu)$, and we have the decomposition $\mathscr{D}_X(k[\epsilon]) = \bigoplus_{\nu \in \mathbb{Z}} T^1_{X_H}(\nu)$.

Since X_H has a hull, $\mathscr{D}_{X_H}(k[\epsilon])$ is a finite k-vector space [Sch1968, Theorem 2.11]. Let $\mathfrak{n} = \dim_k \mathscr{D}_{X_H}(k[\epsilon])$. Choose a homogeneous k-basis $\underline{f}_1^{H(-\nu_j)} \in T^1_{X_H}(\nu_j)$ $1 \leq j \leq r$. We let the maximal ideal of $k[[t_\nu]] := k[[t_{\nu_1}, \ldots, t_{\nu_\mathfrak{n}}]]$ be $\mathfrak{m}_{k[[t_\nu]]}$. Here let the weight of t_j be $-\nu_j$. Let $\xi_2 \in \mathscr{D}_{X_H}(k[[t_\nu]]/\mathfrak{m}_{k[[t_\nu]]}{}^2)$ inducing a bijection of $\mathrm{Hom}_k(k[[t_\nu]]/\mathfrak{m}_{k[[t_\nu]]}{}^2, k[\epsilon])$ with $T^1_{X_H}$.

Let us consider the formal moduli space R of X_H. Schlessinger constructed it by induction on q and $J_q \subset k[[t_\nu]]$, $J_q = \mathfrak{m}_{k[[t_\nu]]}{}^q$ and a deformation $\xi_a \in \mathscr{D}_{X_H}(k[[t_\nu]]/J_q)$ such that J_q is the minimal ideal J' satisfying the following two conditions [Sch1968, pp. 213–214], [Pin1974, pp. 21–22];

1. $\mathfrak{m}_{k[[t_\nu]]} J_{q-1} \subset J' \subset J_q$, and
2. there exists a deformation $\xi \in \mathscr{D}_{X_H}(k[[t_\nu]]/J')$ inducing ξ_{q-1}.

Then by letting $J := \bigcap_{q=2}^{\infty} J_q$, we have $R = k[[t_\nu]]/J$. Let R' be the quotient of R obtained by making every t_j with $\nu_j > 0$ vanish. Then $\mathbb{G}_m$ acts on Spec R' naturally.

Pinkham showed that *for an algebraically closed field k of characteristic* 0, *there exists a $\mathbb{G}_m$-invariant open subset U of* Spec R' *such that C_H is isomorphic to $U/\mathbb{G}_m$* [Pin1974, Theorems 1.15, 13.9].

2.12.3 Plane W-Curves

Komeda investigated the case generated by two elements or plane W-curves with an associated numerical semigroup $H = \langle r, s\rangle$ [Kom1993]. Due to Komeda [Kom1993], we have the proposition:

Proposition 2.252 *For an element f of $k[Z_r, Z_s]$, let $R_H := k[Z_r, Z_s]/(f)$ and $X :=$ Spec R_H. If $k[Z_r, Z_s]/(f, \partial f/\partial Z_r, \partial f/\partial Z_s)$ is an Artinian ring, then the parameter space of X_H is $k[[\lambda_1, \dots, \lambda_{\mathfrak{n}_f}]]$, where*

$$\mathfrak{n}_f = \dim_k \mathbf{D}_X(k[\epsilon]) = \dim_k k[Z_r, Z_s]/(f, \partial f/\partial Z_r, \partial f/\partial Z_s).$$

We consider the case $k[Z_r, Z_s]/(Z_r^s - Z_s^r)$ with $f = Z_r^s - Z_s^r$ and thus $\mathfrak{n}_{\langle r,s\rangle} := \mathfrak{n}_{Z_r^s - Z_s^r}$.

Definition 2.253 Let $R'_H := \mathbb{C}[Z_r, Z_s]/(Z_r^s - Z_s^r, sZ_r^{s-1}, rZ_s^{r-1})$, $S'_\phi := \{x^i, y^j \mid Z_r^i Z_s^j \in R'_H\}$. $\mathfrak{n}_{\langle r,s\rangle} = \# S'_\phi$, and $\mathfrak{v} := \{\mathfrak{v}(i) = -\mathrm{wt}(\phi_i) \mid i = 1, 2, \dots, \mathfrak{n}_{\langle r,s\rangle}\}$. In the Weierstrass canonical form (2.8),

$$f_X(x, y) := y^r + A_1(x)y^{r-1} + A_2(x)y^{r-2} + \cdots + A_{r-1}(x)y + A_r(x),$$

with $A_0 = 1$ and $A_i = \sum_{j=0}^{\lfloor is/r\rfloor} \lambda_{i,j}x^j$ (2.9), we employ the coefficient $\lambda_{s-i,j}$ and define $\widehat{\lambda}_{sr-jr-is} := -\lambda_{s-i,j}$, $\mathfrak{q}(i, j) := sr - jr - is$ if $x^j y^i \in S'_\phi$ for (j, i). We define the Weierstrass-Schlessinger canonical form,

$$\widehat{f_{\lambda}}(x, y) := y^r - \sum_{i=1}^{r-2} \sum_{j=0}^{\lfloor is/r\rfloor} \widehat{\lambda}_{\mathfrak{q}(i,j)} x^j y^i - x^r - \sum_{j=0}^{r-2} \widehat{\lambda}_{\mathfrak{q}(0,j)} x^j. \tag{2.150}$$

Further, we define $S_{\widehat{\lambda}} := \{\widehat{\lambda}_\ell\}$. The number of $\{\widehat{\lambda}\}$ is $\mathfrak{n}_{\langle r,s\rangle}$.

We also define $\mathfrak{q}(i)$ ($i = 1, 2, \dots, \mathfrak{n}_{\langle r,s\rangle}$), such that $S_{\widehat{\lambda}} = \{\widehat{\lambda}_{\mathfrak{q}(i)} \mid i = 1, 2, \dots, \mathfrak{n}_{\langle r,s\rangle}\}$.

Example 2.254 Examples of the Weierstrass–Schlessinger canonical form are given as follows:

1. $\langle 2, 2g + 1\rangle$ [Sch1971]:

$$\widehat{f_{\lambda}}(x, y) = y^2 - (x^{2g+1} + \widehat{\lambda}_2 x^{2g-1} + \widehat{\lambda}_4 x^{2g-2} + \cdots + \widehat{\lambda}_{4g} x + \widehat{\lambda}_{4g+2}).$$

2. $\langle 3, 4\rangle$:

$$\widehat{f_{\lambda}}(x, y) = y^3 - (\widehat{\lambda}_2 x^2 + \widehat{\lambda}_5 x + \widehat{\lambda}_8)y - (x^4 + \widehat{\lambda}_6 x^2 + \widehat{\lambda}_9 x + \widehat{\lambda}_{12}).$$

We remark that these examples can be simply regarded as standard forms by Cardan's transformation [Cox2012], and Weierstrass gave his original versions of the Weierstrass canonical form as the above forms like his standard form of $g = 1$ [Wei1867].

Komeda computed the dimension of the formal moduli space associated with the numerical semigroup generated by two elements as follows [Kom1993]:

Theorem 2.255 *For $H = \langle r, s\rangle$, the dimension $\mathfrak{d}_{\langle r,s\rangle}$ of the moduli space is*

$$\mathfrak{d}_{\langle r,s\rangle} = g(H) + r + s - \left\lfloor \frac{s}{r} \right\rfloor - 4 = \frac{(r+1)(s+1)}{2} - \left\lfloor \frac{s}{r} \right\rfloor - 4$$

for the number $\mathfrak{n}_{\langle r,s\rangle}$ of the parameters $\mathfrak{d}_{\langle r,s\rangle} = \mathfrak{n}_{\langle r,s\rangle} - 1$.

Proof The number $\mathfrak{n}_{\langle r,s\rangle}$ of the parameters is given by

$$\begin{aligned}
&= \sum_{\nu_1=0}^{r-2} \#\{\nu_0 \in \mathbb{N}_0 \mid 0 \le \nu_0 \le s-2,\, \nu_0 < s - \frac{\nu_1 s}{r}\} \\
&= -2 + rs - \left\lfloor \frac{s}{r} \right\rfloor - \sum_{\nu_1=1}^{r-1} \left\lfloor \frac{\nu_1 s}{r} \right\rfloor \\
&= -2 + rs - \left\lfloor \frac{s}{r} \right\rfloor - \frac{(r-1)(s-1)}{2},
\end{aligned}$$

and thus $\mathfrak{d}_{\langle r,s\rangle} = \mathfrak{n}_{\langle r,s\rangle} - 1$. ■

Chapter 3
Theory of Sigma Function

In this chapter, we extend the Weierstrass sigma function of the elliptic curves to general W-curves (Weierstrass curves) given in the previous chapter and show their properties.

The Weierstrass sigma function of genus one is given by [WhW1927, 21·43][Wei1882, HurC1893] (cf. (1.28) in Sect. 1.3.2)

$$\begin{aligned}\sigma(u) &= \frac{2\omega'}{\theta_1'} \exp\left\{\frac{\eta'}{2\omega'}u^2\right\} \theta_1\left(\frac{u}{2\omega'}\Big|\frac{\omega''}{\omega'}\right) \\ &= \left(\frac{2\pi}{\omega'}\right)^{1/2} \Delta^{-1/8} \exp\left\{\frac{\eta'}{2\omega'}u^2\right\} \theta_1\left(\frac{u}{2\omega'}\Big|\frac{\omega''}{\omega'}\right),\end{aligned} \tag{3.1}$$

where $\Delta = -16(4\lambda_1^3 + 27\lambda_0^2)$ is the discriminant of the elliptic curve $y^2 = x^3 + \lambda_1 x + \lambda_0$), $\theta_1(z, \tau)$ is the theta function with characteristics $\begin{bmatrix}1/2\\1/2\end{bmatrix}$,

$$\theta_1(z, \tau) = -\sum_{n=-\infty}^{\infty} \exp\left(\pi\sqrt{-1}\tau\left(n + \frac{1}{2}\right)^2 + 2\pi\sqrt{-1}\left(n + \frac{1}{2}\right)\left(z + \frac{1}{2}\right)\right), \tag{3.2}$$

$\theta_1' := \frac{d}{dz}\theta_1'(z|\frac{\omega''}{\omega'})\Big|_{z=0}$, and $\omega', \omega'', \eta', and \eta''$ are the complete integrals of first and second kind, which satisfy the Legendre relation: $\eta'\omega'' - \eta''\omega' = \pi\sqrt{-1}/2$.

To generalize Weierstrass' sigma to the plane W-curves with higher genus, the authors in the papers [BucEL1999, BucLE1999, EilEL2000] use the W-normalized fundamental differential of the second kind[1] [BucEL1999, 2.2.3], to write a suitable

[1] They call it "fundamental second-order 2-differential".

S. Matsutani, *The Weierstrass Sigma Function in Higher Genus and Applications to Integrable Equations*, Springer Monographs in Mathematics,
https://doi.org/10.1007/978-981-97-8163-8_3

basis of $\mathbf{H}^1(X, \mathbb{C})$ which satisfies the generalized Legendre relations. Following them, the authors in the papers [KomMP2013, KomMP2019, KomMP2022a] applied the approach to several space W-curves. These views originated from Weierstrass [Wei1875a] and were developed by Baker [Ba1897]. Since in the previous chapter, we obtain explicit expressions of the W-normalized fundamental differential of the second kind, Ω in Definition 2.141 for the W-curves X, we define the sigma function for X and study their properties.

To investigate the sigma function for W-curves, we first define the sigma function of W-curves. Since Korotkin and Shramchenko [KorS2012], and Nakayashiki [Nak2016] trailblazed the way in defining the sigma function for every compact Riemann surface, we make use of it.

As we let the vector space $\mathbb{C}^g$ associated with the lattice $\widetilde{\Gamma}$ denoted by $(\mathbb{C}^g, \widetilde{\Gamma})$ in Definition 3.155, we handle both the spaces $(\mathbb{C}^g, \Gamma_X^\circ)$ and $(\mathbb{C}^g, \Gamma_X)$ of a W-curve X. In order to define the sigma function, we first define the Riemann theta function associated with $(\mathbb{C}^g, \Gamma_X^\circ)$ and show its properties in Sect. 3.1.

Since the Riemann constant ξ_X is not a half of the lattice Γ_X° if X is non-symmetric, we shift it to a half of the lattice Γ_X° and employ the shifted Abelian integrals (2.153) for the case following the results in [KomMP2016] in Sect. 3.1.2. (The shifted Riemann constant is naturally interpreted by the spin structure of the compact Riemann surface [At1971, KorS2012].) Using them, we show the properties of the Riemann theta function associated with $(\mathbb{C}^g, \Gamma_X^\circ)$, (1) its modular dependence in Sect. 3.1.3, (2) its translational formula in Sect. 3.1.4, (3) its Jacobi–Riemann fundamental relation in Sect. 3.1.5, (4) its relation to the Schur functions on W-curves in Sect. 3.1.6 (5) its Riemann–Kempf theorem on strata of the theta divisors in Sect. 3.1.7, and (6) normalized prime form in terms of the theta function in Sect. 3.1.8.

In Sect. 3.2, we define the sigma function based on the properties of the Riemann theta function by following Korotkin and Shramchenko [KorS2012], and Nakayashiki [Nak2016]. In Sect. 3.2.2, we also show an alternative definition of the sigma function following Buchstaber, Enolskii and Leykin [BucEL2012, BucL2008] and Eilbeck, Gibbons, Ônishi and Yasuda [EilGOY2017]. We comment on the consistency of both definitions in Sect. 3.2.2.2. We also provide the basic properties of the sigma function, (1) its modular invariance in Sect. 3.2.3, (2) its translational relation in Sect. 3.2.4, (3) its divisor in Sect. 3.2.5, (4) its Jacobi–Riemann fundamental relation in Sect. 3.2.6, (5) its parity in Sect. 3.2.8, (6) Galois action on sigma function in Sect. 3.2.9, and (7) the Riemann–Kempf theorem in Sect. 3.2.10.

Based on these basic properties, we show the Jacobi inversion formula in Sect. 3.3, and the addition theorems for the sigma function in Sect. 3.4 using the addition theorems given in Sect. 2.6.

In terms of the sigma function, we express the prime form more directly in Sect. 3.5, and in Sect. 3.6, we describe the al function introduced in Sect. 2.8. The description of the al function in terms of the sigma function connects the properties of the sigma function at the branch point. The connection enables us to investigate the behavior of the sigma function for the degenerating family of curves mentioned in Sect. 2.9. In Sect. 3.7, we demonstrate the behavior of the sigma function for a degenerating family of $(3, 4)$-curves.

In the Appendices, we give short overviews of topics related to the sigma function, using notations and conventions slightly different from those used in the main content. Appendix 3.8 is on the quadratic forms, and the Riemann theta and sigma functions. Appendix 3.9 briefly reviews the theory of the KP hierarchy, tau function, and the universal Grassmannian manifolds. Appendix 3.10 is devoted to the Schur polynomial and its derivative. Appendix 3.11 is on the heat equation for the sigma function.

3.1 Theta Functions for W-Curves and Their Properties

In order to define the sigma function, we, first, consider the Riemann theta function associated with the Weierstrass curves (W-curves) X as studied in the previous chapter. This section introduces the Riemann theta functions and investigates their properties.

3.1.1 Riemann Theta Function

This subsection shows the definition of the Riemann theta functions.

Let us fix a W-curve X, and its homology basis $\{\alpha_i, \beta_i\}$ in $\mathbf{H}_1(X, \mathbb{Z})$ as in (2.58) and consider the $g \times g$ matrix τ given by the Eq. (2.74). It is noted that τ depends on the basis. Due to Proposition 2.95, it belongs to the Siegel upper half space $\mathfrak{H}_g^\circ := \{Z \in \mathrm{GL}(g, \mathbb{C}) \mid {}^tZ = Z,\ \mathrm{Im}(Z) > 0\}$ in (3.87) (see Appendix 3.8). The Riemann theta function, analytic in both variables $z \in \mathbb{C}^g$ and $\tau \in \mathfrak{H}_g^\circ$, is defined by

$$\theta(z, \tau) = \sum_{n \in \mathbb{Z}^g} \exp\left(2\pi\sqrt{-1}({}^tnz + \frac{1}{2}{}^tn\tau n)\right). \tag{3.3}$$

(For a given W-curve X, we simply write it as $\theta(z) = \theta(z, \tau)$ by assuming that the homology basis is implicitly fixed.) Recalling Γ_X° and J_X° in (2.78), we let the zero-divisor of θ modulo Γ_X° denoted by $\Theta_X^\circ := \kappa_J\ \mathrm{div}(\theta) \subset J_X^\circ$.

The θ function with characteristic $\delta', \delta'' \in \mathbb{R}^g$ is defined as:[2]

[2] There are different conventions; we summarize them: A vector in $\mathbb{C}^g$ is $\circ + \bullet\tau$ and $\begin{bmatrix}\star\\ \diamond\end{bmatrix}$:

	$\circ$	$\bullet$	$\star$	$\diamond$
This book	δ'	δ''	δ''	δ'
[Mum1981, Mum1984]	δ''	δ'	δ''	δ'
[FarK1991]	$2\varepsilon'$	2ε	2ε	$2\varepsilon'$
[Ig1972]			${}^tn'$	${}^tn''$

$$\theta\begin{bmatrix}\delta''\\ \delta'\end{bmatrix}(z,\tau)=\sum_{n\in\mathbb{Z}^g}\exp\left[\pi\sqrt{-1}\{{}^t(n+\delta'')\tau(n+\delta'')+2\,{}^t(n+\delta'')(z+\delta')\}\right]. \tag{3.4}$$

We also basically write it as $\theta[\delta](z)=\theta[\delta](z,\tau)$. The relation between (3.3) and (3.4) is given in Sect. 3.8.3 in Appendix 3.8.

If $\delta=\begin{bmatrix}\delta''\\ \delta'\end{bmatrix}\in\left\{0,\dfrac{1}{2}\right\}^{2g}$, then $\theta\,[\delta]\,(z,\tau):=\theta\begin{bmatrix}\delta''\\ \delta'\end{bmatrix}(z,\tau)$ has definite parity in z, $\theta\,[\delta]\,(-z,\tau)=e(\delta)\theta\,[\delta]\,(z,\tau)$, where $e(\delta):=e^{4\pi\iota\delta'^t\delta''}$. There are 2^{2g} different characteristics of definite parity.

For the W-curve X of genus g, we regard the g-dimensional complex vector space $\mathbb{C}^g$ as the image of the Abelian integral $\widetilde{w}^\circ$ of $S^g\widetilde{X}$ (2.76) and (2.77) due to the Abel–Jacobi Theorem 2.98.

3.1.2 *Riemann Constant and Shifted Riemann Constant on W-Curves*

For the W-curve X, we consider the zero-divisor of the theta function, which is expressed by the Riemann constant ξ_X in this subsection [Rie1866, p. 162].

As mentioned in Theorem 2.175, the symmetry in the Jacobian J_X appears by means of the shifted Abel–Jacobi map $w^\circ_{\mathfrak{s}}$ in Definition 2.153 rather than the Abel–Jacobi map w° (2.76). We show that the characteristics in (3.4) are expressed by a half period in Γ°_X for every W-curve X, including a non-symmetric W-curve by shifting the Abel–Jacobi map and the Riemann constant as in Proposition 3.3.

We recall the basic results by Lewittes [Le1964] on the Riemann constant ξ_X (simply ξ) in the complex vector space $\mathbb{C}^g$ without proof [Le1964, Mum1981, On2013, Rie1866]:

Proposition 3.1 (Lewittes [Le1964])

1. *Using the Riemann constant $\xi_X\in\mathbb{C}^g$, we have the relation between the θ divisor Θ°_X and the standard θ divisor $w^\circ(S^{g-1}X)$,*

$$\Theta^\circ_X=w^\circ(S^{g-1}X)+\xi_X \quad \textit{modulo} \quad \Gamma^\circ_X,$$

i.e., for $P_i\in X$, $\theta(\widetilde{w}^\circ\circ\iota_X(P_1,\dots,P_{g-1})+\xi_X)=0$.

2. *The canonical divisor K_X of X in Proposition 2.117 and the Riemann constant ξ_X hold the relation*

$$w^\circ(K_X)+2\xi_X=0 \ \textit{modulo}\ \Gamma^\circ_X.$$

Atiyah showed that there are $2^{g-1}(2^g+1)$ spin-structures on a compact Riemann surface of genus g [At1971]. Since the spin structures are closely related to the *half-canonical divisor* or *Riemann divisor*.[3] Earle and Kra investigated the relation between the half-canonical divisor and the theta divisor [EaK1986]. We recall the divisor $\mathfrak{K}_\mathfrak{s}$ in Definition 2.116 and Proposition 2.117. We consider the spin line bundle $\mathscr{L}_X^{g-1}$ corresponding to the Riemann constant ξ_X:

Corollary 3.2 *There is an effective divisor $\mathfrak{K}_\xi$ of degree $g-1$ such that $2\mathfrak{K}_\xi = K_X$ of the canonical divisor K_X and $w^\circ(\mathfrak{K}_\xi) = -\xi_X$ modulo Γ_X°, which is called the* half-canonical divisor. *Then $\mathfrak{K}_\xi - \mathfrak{K}_\mathfrak{s} > 0$. Further, let $\mathfrak{K}_\xi^{(0)} := \mathfrak{K}_\xi - (g-1)\infty$, and then $w^\circ(\mathfrak{K}_\xi^{(0)}) = -\xi_X$ modulo Γ_X°.*

Proposition 3.1 with the shifted Abelian integral $\widetilde{w}_\mathfrak{s}^\circ$ and the Abel–Jacobi map $w_\mathfrak{s}^\circ$ in Definition 2.153 lead to the shifted Riemann constant [KomMP2016]:

Proposition 3.3 *1. If $\mathfrak{k}_X$ in Definition 2.81 is not zero or X is not symmetric (Lemma 2.110), the Riemann constant ξ_X is not a half period of Γ_X°.*

2. The shifted Riemann constant $\xi_{X,\mathfrak{s}} := \xi_X - \widetilde{w}_\mathfrak{s}^\circ(\iota_X\,\mathfrak{K}_\mathfrak{s})$ for every W-curve X is the half period of Γ_X°.

3. Using the shifted Abel–Jacobi map, we have

$$\Theta_X^\circ = w_\mathfrak{s}^\circ(S^{g-1}X) + \xi_{X,\mathfrak{s}} = w^\circ(S^{g-1}X) + \xi_X \quad \textit{modulo} \quad \Gamma_X^\circ,$$

i.e., for $P_i \in X$, $\theta\left(\widetilde{w}_\mathfrak{s}^\circ \circ \iota_X(P_1, \ldots, P_{g-1}) + \xi_{X,\mathfrak{s}}\right) = 0$ for every W-curve X.

4. The following holds:

$$2(w_\mathfrak{s}^\circ(\mathfrak{K}_\xi) + \xi_{X,\mathfrak{s}}) = 2\xi_{X,\mathfrak{s}} = 0 \quad \textit{modulo} \quad \Gamma_X^\circ.$$

5. There is a θ-characteristic δ_X of a half period which represents the shifted Riemann constant $\xi_{X,\mathfrak{s}}$, i.e., $\theta[\delta_X]\left(\frac{1}{2}\omega'^{-1}\widetilde{w}_\mathfrak{s} \circ \iota_X(P_1, \ldots, P_{g-1})\right) = 0$, i.e., $\delta_X = \begin{bmatrix} \delta_X'' \\ \delta_X' \end{bmatrix}$, $\xi_{X,\mathfrak{s}} \equiv \delta_X' + \tau_X \delta_X''$ modulo Γ_X°.

Proof For the symmetric W-curve, $\mathfrak{K}_\mathfrak{s} = 0$, $\widetilde{w}_\mathfrak{s}(\iota_X\mathfrak{K}_\mathfrak{s}) \in \Gamma_X^\circ$. When $\widetilde{w}_\mathfrak{s}(\iota_X\mathfrak{K}_\mathfrak{s}) \neq 0$, the direct computations give the results. ■

We have the following theorem as a corollary.

Theorem 3.4 *For every Weierstrass curve X, there is a theta characteristic δ_X,*

$$\delta_X := \begin{bmatrix} \delta'' \\ \delta' \end{bmatrix} \in \left(\tfrac{1}{2}\mathbb{Z}\right)^{2g}, \tag{3.5}$$

which is equal to the shifted Riemann constant $\xi_{X,\mathfrak{s}}$ module Γ_X° namely, for every $(P_1, P_2, \ldots, P_{g-1}) \in S^{g-1}\widetilde{X}$,

[3] The Riemann divisor is denoted by Δ in [Fay1973, p.10].

$$\theta(\widetilde{w}^\circ_\mathfrak{s} \circ \iota_X(P_1, \dots, P_{g-1}) + \xi_{X,\mathfrak{s}}, \tau) = \theta[\delta_X](\widetilde{w}^\circ_\mathfrak{s} \circ \iota_X(P_1, \dots, P_{g-1}), \tau) = 0.$$

We simply refer to δ_X as δ in this book.

This theorem is a key to our Abelian function theory since δ_X makes the sigma function have parity so that it is related to the spin structure in J_X.

Remark 3.5 Korotkin and Shramchenko defined the odd and even σ functions for a general compact Riemann surface based on Klein's investigation on his sigma function [KorS2012]. They assume that the θ characteristic δ as a half period of Jacobian is given by the Riemann constant ξ_X with $\widetilde{w}^\circ(\iota_X D_\mathfrak{s})$ for the divisor $D_\mathfrak{s}$ associated with the spin structure of the compact Riemann surface X. The spin structure $D_\mathfrak{s}$ corresponds to $\mathfrak{K}_\mathfrak{s}$ due to [At1971, Theorem 1], which is known as the half-canonical divisor [EaK1986]. In the paper [EaK1986], Proposition 3.3 was basically obtained, though they are described in the framework of the compact Riemann surfaces rather than W-curves.

Further, in the paper [KomMP2013], we introduced the shifted Abel integral and the shifted Riemann constant but called $\xi_\mathfrak{s}$ "Riemann constant" simply.

Proposition 3.6 (Lewittes [Le1964, p. 43], [Rie1866, p. 162]) *The Riemann constant ξ_X of X is given by*

$$\begin{aligned}\xi_{X,j} &:= \frac{1}{2}\tau_{Xjj} + \sum_{i=1}^{g} \int_{\alpha_i} \widetilde{w}^\circ_{Xi}(\iota_X P)\nu^{\mathrm{I}\circ}_j(P) + \widetilde{w}^\circ_{Xj}(\iota_X \beta_{j,0}) \\ &= \frac{1}{2}\tau_{Xjj} - \sum_{i=1}^{g} \int_{\alpha_i} \widetilde{w}^\circ_{Xj}(\iota_X P)\nu^{\mathrm{I}\circ}_i(P) + \widetilde{w}^\circ_{Xj}(\iota_X \beta_{j,0}) \quad \textit{modulo } \Gamma^\circ_X, \qquad (3.6)\end{aligned}$$

where $\beta_{j,0}$ is the beginning point of β_j as mentioned in Sect. 2.2.7.

We note that (3.6) is obtained by considering the contour integrals of $\nu^{\mathrm{I}\circ}$ in Fig. 2.3 as in Lemma 2.93 with Fig. 2.5. There is an ambiguity regarding where the beginning point or the terminal point we take is.

Even for non-symmetric W-curves, the theta characteristics δ_X belong to the half-period $(\mathbb{Z}/2)^{2g}$, and thus for a given W-curve X, we use the theta function θ in (3.4) with theta characteristics δ_X (simply denoted by δ) in this book.

3.1.2.1 Riemann Constant on Hyperelliptic Curves

As an example, we demonstrate the Riemann constant $\xi_X = \xi_{X,\mathfrak{s}}$ of the hyperelliptic curves X (2.65) [Mum1984, 3.75–85]:

Proposition 3.7 *The Riemann constant ξ_X is given by*

$$\xi_X = \sum_{j=1}^{g} \int_\infty^{\alpha_{j,0}} \nu^{\mathrm{I}\circ}(P) \quad \textit{module } \Gamma^\circ_X.$$

Here we take $\alpha_{j,0} = B_{2j}$ as in Fig. 2.7.

Proof Since (3.6) is obtained by handling ∂D_X in Fig. 2.3 [Le1964, p. 43], we will consider the contours in ∂D_X in this proof. After obtaining the integral formula, we realize the integrals by means of the contours in Fig. 2.7 by using homotopy equivalence.

By noting $\int_\infty^P \nu_i^{\mathrm{I}\circ}(P) = w_i^\circ(P)$, (3.6) is reduced to

$$\xi_{X,j} = \frac{1}{2}\tau_{jj} + w_j^\circ(\beta_{j,0}) - \sum_{i=1}^g \int_{\alpha_i} w_j^\circ(P)\nu_i^{\mathrm{I}\circ}(P) \quad \text{module } \Gamma_X^\circ.$$

1. $j = i$ case:

$$\int_{\alpha_j} w_j^\circ \nu_j^{\mathrm{I}\circ} = \frac{1}{2}\int_{\alpha_j} d(w_j^\circ(P))^2 = \frac{1}{2}(w_j^\circ(\alpha_{j,0})^2 - w_j^\circ(\alpha_{j,1})^2),$$

where $\alpha_{j,0}$ and $\alpha_{j,1}$ denote the beginning and ending points of α_j, respectively. Since the beginning and ending of α_j on X agree and $\int_{\alpha_j} \nu_j^{\mathrm{I}\circ} = 1$, we have

$$\begin{aligned}\int_{\alpha_j} w_j^\circ \nu_j^{\mathrm{I}\circ} =& \frac{1}{2}\left((w_j^\circ(\alpha_{j,0})^2 - (w_j^\circ(\alpha_{j,0}) - 1)^2\right) \\ =& w_j^\circ(\alpha_{j,0}) - \frac{1}{2} \quad \text{modulo } \Gamma_X^\circ \\ =& w_j^\circ(\beta_{j,0}) + \frac{1}{2} \quad \text{modulo } \Gamma_X^\circ.\end{aligned}$$

Here ∂D_X in Fig. 2.3 in Sect. 2.2.7 shows that the difference between $\alpha_{j,0}$ and $\beta_{j,0}$ is α_j. It means $\frac{1}{2}\tau_{jj} + w_j^\circ(\beta_{j,0}) - \int_{\alpha_j} w_j^\circ \nu_j^{\mathrm{I}\circ} = \frac{1}{2}(\tau_{jj} - 1)$ modulo Γ_X°. Figure 2.7 shows $\frac{1}{2}(\tau_{jj} - 1) = w_j^\circ(\alpha_{j,0})$ modulo Γ_X°.
2. $j \neq i$ case: Recalling the hyperelliptic involution $\widehat{\zeta}_H : (x, y) \mapsto (x, -y)$ in X and its action $\widehat{\zeta}_H \nu_j^{\mathrm{I}\circ}(P) = \nu_j^{\mathrm{I}\circ}(\widehat{\zeta}_H P) = -\nu_j^{\mathrm{I}\circ}(P)$, we have

$$\begin{aligned}\int_{\alpha_i} w_j^\circ(P)\nu_i^{\mathrm{I}\circ}(P) &= \int_{\alpha_{i,0}}^{\alpha_{i,1}} \left(w_j^\circ(P)\nu_i^{\mathrm{I}\circ}(P) - w_j^\circ(\widehat{\zeta}_H P)\nu_i^{\mathrm{I}\circ}(\widehat{\zeta}_H P)\right) \\ &= \int_{\alpha_{i,0}}^{\alpha_{i,1}} \left(w_j^\circ(P) + w_j^\circ(\widehat{\zeta}_H P)\right)\nu_i^{\mathrm{I}\circ}(P).\end{aligned}$$

The fact that $d(w_j^\circ(P) + w_j^\circ(\widehat{\zeta}_H P)) = \nu_j^{\mathrm{I}\circ}(P) + \nu_j^{\mathrm{I}\circ}(\widehat{\zeta}_H P) = 0$ meas that $(w_j^\circ(P) + w_j^\circ(\widehat{\zeta}_H P))$ is a constant ($= 2w_j^\circ(\alpha_{i,0})$). Since $\int_{\alpha_{i,0}}^{\alpha_{i,1}} \nu_i^{\mathrm{I}\circ} = \frac{1}{2}\int_{\alpha_i} \nu_i^{\mathrm{I}\circ} = \frac{1}{2}$, we have

$$\int_{\alpha_i} w_j^\circ(P)\nu_i^{\mathrm{I}\circ}(P) = w_j^\circ(\alpha_{i,0}) = \int_\infty^{\alpha_{i,0}} \nu_j^{\mathrm{I}\circ}.$$

The results in both cases show

$$\xi_{X,j} = \frac{1}{2}\tau_{jj} + w_j^\circ(\beta_{j,0}) - \sum_{i=1}^{g}\int_{\alpha_i} w_j^\circ(P)\nu_i^{\mathrm{I}\circ}(P) = \sum_{i=1}^{g}\int_\infty^{\alpha_{i,0}} \nu_j^{\mathrm{I}\circ}(P) \quad \text{modulo } \Gamma_X^\circ.$$

∎

For the homology basis in Fig. 2.7, we have explicit expressions of the Riemann constant [FarK1991, pp. 322–3][EnR2008, BucEL2012, Mum1984, On2013].

Theorem 3.8 *$\xi_X = \delta$ modulo Γ_X° for the theta characteristics $\begin{pmatrix}\delta''\\ \delta'\end{pmatrix}$, where $\delta' := {}^t\left[\frac{g}{2}\ \frac{g-1}{2}\ \cdots\ \frac{1}{2}\right]$, and $\delta'' := {}^t\left[\frac{1}{2}\ \cdots\ \frac{1}{2}\right]$.*

Proof As mentioned in Sect. 2.7.3, we can express the integrals in Proposition 3.7 in terms of τ. We multiply the relations in (2.121) by $(2\omega')^{-1}$ from the left-hand side and then we obtain

$$\int_\infty^{\alpha_{j,0}} \nu^{\mathrm{I}\circ} = \int_\infty^{B_{2j}} \nu^{\mathrm{I}\circ} = -\frac{1}{2}\sum_{i=1}^{j}\oint_{\alpha_i}\nu^{\mathrm{I}\circ} + \frac{1}{2}\oint_{\beta_j}\nu^{\mathrm{I}\circ}$$

$$= -\tfrac{1}{2}\,{}^t(1 \cdots \overset{j}{1}\ 0 \cdots 0) + \tfrac{1}{2}\tau\,{}^t(0 \cdots 0\ \overset{j}{1}\ 0 \cdots 0).$$

Proposition 3.7 proves the relation. ∎

3.1.2.2 Riemann Constant for $X_s : y^3 = x(x-s)(x-b_1)(x-b_2)$ and $X_{\widehat{0}}$

Let us consider the Riemann constant for the degenerating family of the curves X_s and the normalized curve $X_{\widehat{0}}$ of the singular fiber X_0 as mentioned in Sect. 2.9. We show that the shifted Riemann constant $\xi_{X_{\widehat{0}},\mathfrak{s}}$ for the non-symmetric curve $X_{\widehat{0}}$ appears as the degenerating limit of the Riemann constant ξ_{X_s} for the symmetric curve X_s. It means that the shifted Riemann constant is a natural object.

Due to (3.6) in Proposition 3.6, the Riemann constant ξ_X in X_s is defined by

$$\xi_{X_s j} := \frac{1}{2}\tau_{X_s,jj} + \sum_{i}^{3}\int_{\alpha_i} \widetilde{w}_{X_s i}^\circ(\iota_X P)\nu_{X_s,j}^{\mathrm{I}\circ}(P) + \widetilde{w}_{X_s j}^\circ(\iota_X\beta_{j,0}) \tag{3.7}$$

corresponds to a θ characteristic

$$\delta_{X_s} := \begin{bmatrix} \delta''_{X_s} \\ \delta'_{X_s} \end{bmatrix}, \quad \delta'_{X_s} \in (\mathbb{Z}/2)^3, \quad \delta''_{X_s} \in (\mathbb{Z}/2)^3,$$

for the curve X_s [Le1964, EilEMOP2007]. The vector ξ_{X_s} is unique modulo $\Gamma_s = \Gamma_{X_s}$.

On the other hand, as in (3.7), the Riemann constant $\xi_{X_{\hat{0}}}$ of $X_{\hat{0}}$ is evaluated by (3.6):

$$\xi_{X_{\hat{0}}j} := \frac{1}{2}\tau_{X_{\hat{0}}jj} + \sum_{i}^{2} \int_{\alpha_i} \widetilde{w}^{\circ}_{X_{\hat{0}}i}(\iota_X P)\nu^{I\circ}_{\hat{0},j}(P) + \widetilde{w}^{\circ}_{X_{\hat{0}}i}(\iota_X \widehat{\beta}_{j,0}), \tag{3.8}$$

where $\widehat{\beta}_{j,0}$ is the beginning point of β_j in $X_{\hat{0}}$.

From Proposition 3.3, we have the following facts:

1. Since $w^{\circ}_{X_{\hat{0}}}(K_{X_{\hat{0}}}) + 2\xi_{X_{\hat{0}}} = 0$ modulo $\Gamma_{\hat{0}} := \Gamma_{X_{\hat{0}}}$ and $w^{\circ}_{X_{\hat{0}}}(K_{X_{\hat{0}}}) = -w^{\circ}_{X_{\hat{0}}}(2B_0)$, we have $-2w^{\circ}_{X_{\hat{0}}}(B_0) + 2\xi_{X_{\hat{0}}} = 0$ modulo $\Gamma_{\hat{0}}$. Thus the Riemann constant $\xi_{X_{\hat{0}}}$ is not a half period; however,
$$\xi_{X_{\hat{0}}\mathfrak{s}} := \xi_{X_{\hat{0}}} - \widetilde{w}^{\circ}_{X_{\hat{0}}}(\iota_X B_0) \tag{3.9}$$
becomes a half period.
2. The relation between the theta divisor Θ°_X :=div(θ) and the standard theta divisor $w_{X_{\hat{0}}}(X_{\hat{0}})$ is alternatively expressed by
$$\Theta^{\circ}_X = w_{X_{\hat{0}}}(X_{\hat{0}}) + \xi_{X_{\hat{0}}} = w_{X_{\hat{0}},\mathfrak{s}}(X_{\hat{0}}) + \xi_{X_{\hat{0}}\mathfrak{s}} \quad \text{modulo} \quad \Gamma_{\hat{0}}.$$
3. The θ-characteristics of the half period $\begin{bmatrix} \delta''_{X_{\hat{0}}} \\ \delta'_{X_{\hat{0}}} \end{bmatrix} \in (\mathbb{Z}/2)^2$ corresponding to the vector $\xi_{X_{\hat{0}}\mathfrak{s}}$ represent the shifted Riemann constant $\xi_{X_{\hat{0}}\mathfrak{s}}$ [KomMP2016].

We show that the limit of the Riemann constant ξ_{X_s} in (3.7) turns out to be the shifted Riemann constant $\xi_{X_{\hat{0}}\mathfrak{s}}$ in (3.9), cf. Proposition 3.3; we analytically compute that the limit of the theta characteristics δ_{X_s} is $\delta_{X_{\hat{0}}}$; since the divisors at infinity correspond, we demonstrate naturalness of the shifted Riemann constant.

Proposition 3.9 [FeKMPA2022]

$$\lim_{s\to 0} \xi_{X_s,j} = \xi_{X_{\hat{0}},j} - \widetilde{w}^{\circ}_{X_{\hat{0}},j}(\iota_X B_0) = \xi_{X_{\hat{0}}\mathfrak{s},j} \quad (j = 1, 2).$$

$$\delta'_{X_s,j} \longrightarrow \delta'_{X_{\hat{0}},j}, \quad \delta''_{X_s,j} \longrightarrow \delta''_{X_{\hat{0}},j} \quad (j = 1, 2).$$

Proof For brevity, we omit ι_X in $\widetilde{w}^{\circ}_{X_s,3}$ here. Let us consider the limit of $s \to 0$ of (3.7),

$$\xi_{X_s j} = \frac{1}{2}\tau_{X_s jj} + \sum_{i=1}^{2}\int_{\alpha_i} \widetilde{w}^{\circ}_{X_s,i}(P)\nu_j^{\mathrm{I}\circ}(P) + \int_{\alpha_3} \widetilde{w}^{\circ}_{X_s,3}(P)\nu_j^{\mathrm{I}\circ}(P) + \widetilde{w}^{\circ}_{X_s,j}(\beta_{j,0}), \tag{3.10}$$

noting (3.8). Let the third term in the right-hand side be denoted by I. For sufficiently small $s \in \Delta_\varepsilon^*$, let $\widetilde{w}^{\circ}_{X_s,i}(P) = \widetilde{w}^{\circ}_{X_s,i}(B_0) + w_i(z)$ in terms of the local parameter z of $x = z^3$ at B_0. Though $w_3(z)$ is a linear combination of $(\widetilde{w}_{X_s,i}(P) - \widetilde{w}_{X_s,i}(B_0))$'s, the dominant term of $w_3(z)$ at B_0 is $\int_0^P \nu^{\mathrm{I}}_{X_s,1}$. Since $\nu^{\mathrm{I}}_{X_s,1} = \frac{dz}{z^2}(1 + d_{\geq 1}(z))$ and other entries are holomorphic at B_0, the partial integration at $s = 0$ gives

$$I = \oint_{\alpha_3} (w_3(z)dw_j(z)) = -\oint_{\alpha_3} (dw_3(z)(w_j(z) + \widetilde{w}^{\circ}_{X_s,j}(B_0))) = -\widetilde{w}^{\circ}_{X_s,j}(B_0)).$$

Since $\oint_{\alpha_3} dw_3(z) = 1$, the third equality implies $I' := \oint_{\alpha_3} dw_3(z)w_j(z) \to 0$ for $s \to 0$, which we now show.

Let $w_j(z) = \sum_{i=1}^{\infty} a_i^{(j)} z^i$ noting $w_j(0) = 0$; it has a non-vanishing radius of convergence. On the other hand, at $s = 0$, $dw_3(z) = \frac{s^{1/3}}{A_{13}^{(1)}}\nu^{\mathrm{I}}_{X_s,1}(1 + d_{\geq 1}(s^{1/3}))$. As in Sect. 2.9.2, we assume s to be positive without loss of generality. Using the notations in (2.148), we evaluate I',

$$\begin{aligned} I' &= \frac{(\zeta_3^2 - \zeta_3)s^{1/3}}{A_{13}^{(1)}} \int_0^s \frac{h_2(x)w_j(x^{1/3})}{\sqrt[3]{|x^2(x-s)^2|}}dx + d_{\geq 2}(s^{1/3}) \\ &= \frac{(\zeta_3^2 - \zeta_3)\zeta_3^2}{A_{13}^{(1)}\sqrt[3]{(b_1b_2)^2}} \sum_{\ell=0}\sum_{n=1} \beta_\ell^{(2)} a_n^{(j)} s^{\ell+n/3} \int_0^1 t^{\ell+\frac{n+1}{3}-1}(1-t)^{\frac{1}{3}-1}dt + d_{\geq 2}(s^{1/3}). \end{aligned}$$

Thus it is obvious that $\lim_{s\to 0} I' = 0$. The other terms in (3.10) become $\xi_{X_{\hat{0}},j}$ and thus we show the proposition. ■

Remark 3.10 We could conjecture that the shifted Riemann constant for every non-symmetric curve may appear due to the inductive computations of the degenerating limit of the Riemann constant from a symmetric curve since the W-curve X may have a degenerating sequence as in (2.144).

3.1.3 Modular Dependence

Here we describe the modular dependence of the theta function for X. We consider the symplectic group $\mathrm{Sp}(2g, \mathbb{Z})$ defined in (2.59).

There is an ambiguity regarding which loops are employed as the canonical basis $\{\alpha_i, \beta_i\}_{i=1,\dots,g}$ in $\mathbf{H}_1(X, \mathbb{Z})$. The symplectic group $\mathrm{Sp}(2g, \mathbb{Z})$ acts on the choice of the generators, and by the Abel–Jacobi map, the action induces the group action on Γ_X° and the Jacobian J_X° as mentioned in Lemma 2.87.

By letting $\eta^{\circ\prime} := 4\,{}^t\omega'\eta'$ and $\eta^{\circ\prime\prime} := 4\,{}^t\omega'\eta''$, we also consider the set $\widehat{\mathfrak{W}}_g$ of the complete Abelian integrals of X, which is defined by

$$\left\{M = \begin{pmatrix} 1_g & Z \\ \eta^{\circ\prime} & \eta^{\circ\prime\prime} \end{pmatrix} \in \mathrm{Mat}_{\mathbb{C}}(2g) \;\middle|\; M \begin{pmatrix} 0 & 1_g \\ -1_g & 0 \end{pmatrix} {}^tM = \frac{2\pi}{\sqrt{-1}} \begin{pmatrix} 0 & 1_g \\ -1_g & 0 \end{pmatrix},\ Z \in \mathfrak{H}_g^\circ \right\}.$$

The conditions in $\widehat{\mathfrak{W}}_g$ mean the Legendre relation in Theorem 2.148,

$$Z = {}^tZ, \qquad \eta^{\circ\prime}\,{}^t\eta^{\circ\prime\prime} = \eta^{\circ\prime\prime}\,{}^t\eta^{\circ\prime}, \qquad {}^tZ\eta^{\circ\prime} - \eta^{\circ\prime\prime} = 2\pi\sqrt{-1}1_g.$$

For $\gamma \in \mathrm{Sp}(2g, \mathbb{Z})$ and $Z \in \mathfrak{H}_g^\circ$, we have the action

$$\widehat{\varphi}_\gamma^\circ Z = ({}^td + Z\,{}^tc)^{-1}({}^tb + Z\,{}^ta) = (aZ + b)(cZ + d)^{-1}.$$

The theta functions depend on the action, which is known as modular dependence. From Proposition 3.179, the modular dependence of the theta functions is given by the following proposition.

Proposition 3.11 *For* $\gamma := \begin{pmatrix} a & b \\ c & d \end{pmatrix} \in \mathrm{Sp}(2g, \mathbb{Z})$ *and* $\tau \in \mathfrak{H}_g^\circ$*, the following hold:*

1. *For* $\widehat{\varphi}_\gamma^\circ\theta(z, \tau) := \theta(\widehat{\varphi}_\gamma^\circ z; \widehat{\varphi}_{\gamma_2}^\circ \tau) := \theta((\tau\,{}^tc + {}^td)^{-1}z; (a\tau + b)(c\tau + d)^{-1})$,

$$\widehat{\varphi}_\gamma^\circ\theta(z, \tau) = \mathrm{e}^{\rho_\gamma}\sqrt{\det[(c\tau + d)]}\exp\left[\frac{\pi}{2}\sqrt{-1}\,{}^tc\,{}^tz(c\tau + d)^{-1}cz\right]\theta(z + s_\gamma, \tau),$$

where the phase $\rho_\gamma = i_\gamma\pi\sqrt{-1}/2$ $(i_\gamma \in \mathbb{Z})$ *and* $s_\gamma \in \mathbb{C}^g$ *depend only on* γ.

2. *For* $\widehat{\varphi}_\gamma^\circ\theta[\delta](z, \tau) := \theta[\widehat{\varphi}_\gamma^\circ\delta](\widehat{\varphi}_\gamma^\circ z, \widehat{\varphi}_\gamma^\circ\tau)$,

$$\widehat{\varphi}_\gamma^\circ\theta[\delta](z, \tau) = \mathrm{e}^{\rho'_\gamma}\sqrt{\det[(c\tau + d)]}\exp\left[\frac{\pi}{2}\sqrt{-1}\,{}^tc\,{}^tz(c\tau + d)^{-1}cz\right]\theta[\delta](z, \tau), \tag{3.11}$$

where $\widehat{\varphi}_\gamma^\circ\begin{pmatrix} \delta'' \\ \delta' \end{pmatrix} = \begin{pmatrix} d & -c \\ -b & a \end{pmatrix}\begin{pmatrix} \delta'' \\ \delta' \end{pmatrix} + \frac{1}{2}\begin{pmatrix} (c\,{}^td)_D \\ (a\,{}^tb)_D \end{pmatrix}$, *and the phase* $\rho'_\gamma = \dfrac{i'_\gamma\pi\sqrt{-1}}{2}$ $(i'_\gamma \in \mathbb{Z})$ *depends only on* γ *and* δ*. For a* $g \times g$ *matrix* A, A_D *means the element in* $\mathbb{C}^g$ *whose component is the diagonal part of* A.

Proof See Proposition 3.179 in Appendix 3.8. ∎

3.1.4 Translational Formula

In this subsection, we show the translational property of the theta function for the lattice Γ_X°.

The theta function is associated with the line bundle $\mathscr{L}_J$ on J_X, and it satisfies the following translational formula that determines the structure of the line bundle.

Proposition 3.12 *For $\ell', \ell'' \in \mathbb{Z}^g$ and $\ell := \ell' + \tau\ell''$, the following translation formulae of the theta function hold:*

1. $\theta(z+\ell) = \exp\left(2\pi\sqrt{-1}\,{}^t\ell''\left(z - \frac{1}{2}\tau\ell''\right)\right)\theta(z)$, *and*
2. $\theta[\delta](z+\ell) = \exp\left(2\pi\sqrt{-1}\left({}^t\delta''\ell' - {}^t\ell''\left(z + \delta + \frac{1}{2}\tau\ell''\right)\right)\right)\theta[\delta](z)$.

Proof See Proposition 3.175 in Appendix 3.8. ■

3.1.5 Jacobi–Riemann Fundamental Relation

We mention here a fundamental relation between the theta functions and the differentials of the third kind following [Ba1897, Chap. X 188 p. 275]. We have called it the Jacobi–Riemann fundamental relation [KomM2020, KomMP2019] because Jacobi derived his theta function using the elliptic integrals of the third kind (1.21) [Jac1829], and following Jacobi, Riemann investigated the quotients of the theta function in [Rie1866], though Baker called it "Abel's theorem" [Ba1928, p. 588]. Weierstrass mentioned it in his lectures, at least of 1875–76, on the Riemann theta function (e.g., [Wei1875a, p. 518, p. 597, p. 605]), and Klein gave it explicitly in [Kl1886, p. 446], [Kl1888, p. 379]. It is also given in the book by [Fay1973, p. 23].

We display the relation, which connects the meromorphic functions on X and the entire function on $\mathbb{C}^g$ associated with J_X°:

Proposition 3.13 *For $2g+2$ points, $P_1, \ldots, P_g$, $Q_1, \ldots, Q_g$, P, and $Q \in X$, we let $z = \widetilde{w}_{\mathfrak{s}}^\circ(\iota_X(P_1, \ldots, P_g))$, $z' = \widetilde{w}_{\mathfrak{s}}^\circ(\iota_X(Q_1, \ldots, Q_g))$. The following holds:*

$$\exp\left(\sum_{j=1}^g \int_{\gamma_{P,Q}} \nu_{P_j,Q_j}^{\mathrm{III}\circ}\right) = \frac{\theta\left(\widetilde{w}^\circ(\iota_X(P)) - z + \xi_{X,\mathfrak{s}}\right)\theta\left(\widetilde{w}^\circ(\iota_X(Q)) - z' + \xi_{X,\mathfrak{s}}\right)}{\theta\left(\widetilde{w}^\circ(\iota_X(P)) - z' + \xi_{X,\mathfrak{s}}\right)\theta\left(\widetilde{w}^\circ(\iota_X(Q)) - z + \xi_{X,\mathfrak{s}}\right)}$$
$$= \frac{\theta[\delta](\widetilde{w}^\circ(\iota_X(P)) - z)\theta[\delta](\widetilde{w}^\circ(\iota_X(Q)) - z')}{\theta[\delta](\widetilde{w}^\circ(\iota_X(P)) - z')\theta[\delta](\widetilde{w}^\circ(\iota_X(Q)) - z)}.$$

Proof We note that the relation is on $\widetilde{X}$ and $\widetilde{w}^\circ(\widetilde{X}) = \mathbb{C}^g$ and depends on the choice of ι_X in general, though the dependence is not essential.

Here, we assume that u and v don't belong to Θ_X°, and we take P_i and Q_i for such u and v. We check the behavior of each side of the relation as a function of $P \in X$. We consider the divisor of the formula of each side. Due to Proposition 3.1, the divisor of the right-hand side is given by $\sum_{j=1}^{g} P_j - \sum_{j=1}^{g} Q_j$. Since for a meromorphic function f on X, the order of f at P' is given by $\mathrm{ord}_{P'} f := \oint_{\gamma_P} d \log f$, the definition of $\nu^{\mathrm{III}\circ}_{P_j,Q_j}$ leads to the order of the left-hand side is given by

$$\mathrm{ord}_{P_j} \exp\left(\sum_{j=1}^{g} \int_{\gamma_{P,Q}} \nu^{\mathrm{III}\circ}_{P_j,Q_j}\right) = \mathrm{res}_{P_j} \nu^{\mathrm{III}\circ}_{P_j,Q_j} = 1,$$

and similarly is -1 at each Q_j. Hence both sides have the same divisor.

Thus we investigate the behavior of the formula on each side. We take the path $\gamma_{P,Q} \in \widetilde{X}$ by moving P along α_ℓ and to the point P again. It means that we replace $\gamma_{P,Q}$ with $\gamma_{P,Q} + \alpha_\ell$. Recalling the definition of $\nu^{\mathrm{III}\circ}$, we have the relation $\int_{\gamma_{P,Q}+\alpha_\ell} \nu^{\mathrm{III}\circ}_{P_j,Q_j} = \int_{\gamma_{P,Q}} \nu^{\mathrm{III}\circ}_{P_j,Q_j}$, which does not generate any prefactor in the left-hand side. Since the translational formula of the theta function for the replacement is trivial, the right-hand side does not generate any prefactor either.

Similarly, we replace $\gamma_{P,Q}$ with $\gamma_{P,Q} + \beta_j$. Lemma 2.100 3 shows

$$\int_{\gamma_{P,Q}+\beta_j} \nu^{\mathrm{III}\circ}_{P_j,Q_j} = \int_{\gamma_{P,Q}} \nu^{\mathrm{III}\circ}_{P_j,Q_j} + \int_{\beta_\ell} \nu^{\mathrm{III}\circ}_{P_j,Q_j} = \int_{\gamma_{P,Q}} \nu^{\mathrm{III}\circ}_{P_j,Q_j} + \int_{Q_j}^{P_j} \nu^{\mathrm{I}\circ}_\ell.$$

The left-hand side generates the prefactor,

$$\exp\left(2\pi\sqrt{-1}\left[\sum_{j=1}^{g} \int_{Q_j}^{P_j} \nu^{\mathrm{I}\circ}_\ell\right]\right) = \exp\left(2\pi\sqrt{-1}(z_\ell - z'_\ell)\right).$$

The translational formula of the theta function for the replacement generates the prefactor in the right-hand side as

$$\exp\left(2\pi\sqrt{-1}\left[\left(\widetilde{w}^\circ_\ell(\iota_X(P)) + z_\ell + \xi_{X,\ell} - \frac{1}{2}\tau_{\ell\ell}\right) - \left(\widetilde{w}^\circ_\ell(\iota_X(P)) + z'_\ell + \xi_{X,\ell} - \frac{1}{2}\tau_{\ell\ell}\right)\right]\right).$$

Accordingly, we find that the relation turns out to be that on X rather than $\widetilde{X}$. The ratio of both sides of the relation is a meromorphic function on X. Since their divisors agree, the ratio does not have any poles. The ratio is a constant function on X. On the other hand, When $P = Q$, they are 1, and thus, the ratio is 1. It means that we prove equality on both sides. This relation clearly holds for any ι_X.

After this equality is proved, it is also shown that the case where P_i is $u \in W_X^{g-1}$ can also be determined by limit operations, thereby making use of l'Hôpital's theorem. For the operation, the Riemann–Kempf theorem is needed. The key to this is that differentiation of the theta function also yields that for W^k it is not identically zero, satisfying the same translation formula as mentioned in Proposition 3.24. The same is true for Q_j, and the proof is valid without any assumptions. ■

3.1.6 Theta and Schur Functions on W-Curves

It is known that the theta and the sigma functions are related to the so-called Schur functions. Following Nakayashiki [Nak2016], we show the relation between the theta and the Schur functions in this subsection.

The Schur functions or the Schur polynomials were studied by Schur 1901 [Sch1901, (44)] in his thesis on the representation theory of the symmetric and the general linear groups. The polynomials were known in those days; for example, Jacobi investigated it and found so-called the Jacobi–Trudi identity [Jac1841], and Baker used the Schur function for the studies on the hyperelliptic sigma functions [Ba1903] 1903 to find the KdV hierarchy as mentioned in Sect. 4.1. Hence it might be better to call it another name, but we employ the traditional name.

Let us consider $\mathbb{C}[\mathrm{t}_1, \dots, \mathrm{t}_g]$ and the symmetric polynomials, e.g., the power symmetric polynomials, $T_k^{(\ell)} := \dfrac{1}{k}\sum_{i=1}^{\ell} \mathrm{t}_i^k$ for $\mathrm{t} = {}^t(\mathrm{t}_1, \dots, \mathrm{t}_g)$, and $T_k = T_k^{(g)}$.

For a Young diagram $\Lambda = (\Lambda_1, \Lambda_2, \dots, \Lambda_n)$, the Schur function s_Λ is defined by the ratio of determinants of $n \times n$ matrices [Mac1985],

$$s_\Lambda(\mathrm{t}) = \frac{|\mathrm{t}_i^{\Lambda_j+n-j}|}{|\mathrm{t}_i^{j-1}|}.$$

When Λ is associated with the semigroup H as in Sect. 2.2.1, it is also regarded as a function of $T = {}^t(T_{\Lambda_1+n-1}, \dots, T_{\Lambda_n})$ due to Proposition 3.220, and thus, we express it by

$$S_\Lambda(T) = s_\Lambda(\mathrm{t}).$$

We mainly consider the Young diagram Λ_X associated with the numerical semigroup H_X for the W-curve X.

Segal and Wilson showed the expression of the theta function in terms of the Schur functions in the study of the tau functions and the loop group [SeW1985] as mentioned in Appendix 3.9. The Schur functions are associated with the universal Grassmannian manifolds (UGM), and meromorphic functions of general compact Riemann surfaces are embedded in the UGM.

On the other hand, Buchstaber, Leykin and Enolskii showed that the sigma functions of monomial curves of the numerical semigroups generated by two elements, e.g., $\langle r, s\rangle$, are related to the Schur functions [BucLE1999]. Subsequently, Eilbeck, Enolskii, and Gibbons showed his explicit relation [EilEG2010]. Nakayashiki, independently, showed that the sigma functions of plane W-curves are expanded in terms of Schur functions [Nak2010a, Nak2010b].

Based on these facts, Nakayashiki showed the relation among theta functions and the Schur functions on the compact Riemann surfaces whose Weierstrass non-gap sequences are given by the numerical semigroups H_X [Nak2016] (cf. Theorem 3.214 in Appendix 3.9); the surfaces completely correspond to our W-curves X with H_X. Nakayashiki explicitly represented the identifications for such surfaces in order to define the sigma functions explicitly [Nak2016, Definition 9]. We follow his results.

For W-curves X and $(\gamma_1, \ldots, \gamma_g) \in S^g \widetilde{X}$, we have the W-normalized Abelian integrals $u = \widetilde{w}(\gamma_1, \ldots, \gamma_g)$, and the normalized Abelian integrals $z = \widetilde{w}^\circ(\gamma_1, \ldots, \gamma_g)$; both are related as $z = \dfrac{1}{2}{\omega'}^{-1} u$.

We show the relation following [Nak2016, Theorem 10]. Since its proof is beyond the scope of this book, we provide only the statement (cf. Theorem 3.214).

Proposition 3.14 *With the notation of Proposition 3.1, for a general non-singular W-curve X associated with the numerical semigroup H_X and the Young diagram Λ_X, we have the following results: For a zero e of the theta function, i.e., $e \in (\kappa_J^\circ)^{-1}\Theta_X^\circ$, and for $u \in \mathbb{C}^g$,*

$$C\theta\left((2\omega')^{-1}u + e, \tau\right) = S_{\Lambda_X}(T)|_{T_{\Lambda_i+g-i}=-u_i} \\ + \textit{ higher-weight terms with respect to } \mathrm{wt}, \quad (3.12)$$

where C is a suitable constant.[4]

In the derivation of Proposition 3.14, Nakayashiki employed the transcendental approach and thus did not consider the W-curves explicitly. He derived this relation based on Fay's results and the Sato theory on the universal Grassmannian manifold for the general Riemann surfaces X with H_X using the relation between the prime form and theta function on the Riemann surface X [Fay1973]. In the Sato theory, it turns out that the tau function is given by the infinite linear combination of the Schur functions whose coefficients are provided by the Plücker coordinate [SeW1985]. We review these studies in Appendix 3.9.

3.1.7 Riemann–Kempf Theorem; Strata in the Theta Divisor

The precise expression of the Riemann–Kempf theory was carried out for the case of hyperelliptic curves by Ônishi in [On2002, On2005] and Enolskii and Gibbons [EnG2003]. Following their studies, Nakayashiki and Yori [NakY2012], the author of this book with Previato [MatP2014], and Nakayashiki [Nak2016] generalized them to more general curves. This subsection shows the Riemann–Kempf theorem and the strata in the theta divisor Θ_X°.

In [Nak2016], Nakayashiki showed the Riemann–Kempf properties of the Riemann theta function to define the sigma functions of the compact Riemann surface

[4] C^{-1} is given by $(-1)^{g+r_{\Lambda_X}} \partial_{\natural X,0}\theta(e, \tau)$ defined in Definition 3.18 and Theorem 3.21.

using Proposition 3.14. Here we mention the precise expression of the Riemann–Kempf theorem on the theta functions of W-curves following [MatP2014].

In Appendix 3.10, we show the Riemann–Kempf theorem on the theta functions of the W-curves based on [MatP2014] and Proposition 3.14. We use the results and state the Riemann–Kempf theorem.

We first state Riemann's singularity theorem (cf. [ArCGH1985, VI.1][Mum1999, p. 272]):

Theorem 3.15 *If D_k belongs to $S^k(X\backslash\infty) \setminus (S_1^k(X) \cap S^k(X\backslash\infty))$, and we let $u := \widetilde{w}(\iota_X D_k)$,*

$$n_k := \dim \mathbf{H}^0(X, D_k + (g-k-1)\infty) = \#\{\ell \mid 0 \le \ell, N(\ell) \le g-k-1\},$$

then:

1. *For every multi-index $(\alpha_1, \dots, \alpha_m) \in \mathrm{Index}(g, m) := \{1, \dots, g\}^m$ and $m < n_k$,*

$$\frac{\partial^m}{\partial u_{\alpha_1} \dots \partial u_{\alpha_m}} \theta\left((2\omega')^{-1}u + \xi_X, \tau\right) = 0.$$

2. *There exists a multi-index $I_{X,k} := (\beta_1, \dots, \beta_{n_k}) \in \mathrm{Index}(g, n_k)$, which in general depends on D_k, such that*

$$\frac{\partial^{n_k}}{\partial u_{\beta_1} \dots \partial u_{\beta_{n_k}}} \theta\left((2\omega')^{-1}u + \xi_X, \tau\right) \neq 0. \tag{3.13}$$

For D_k and n_k as in Theorem 3.15, Fay [Fay1984, Theorem 1.2] proved the following (cf. [SeW1985, Proposition 8.6]):

Theorem 3.16 *Let ν_i^+ $(0 \le \nu_1^+ < \nu_2^+ < \dots < \nu_{n_k}^+ \le g-1)$ be such that*

$$h^0(X, D_k + (g-k-\ell-1)\infty) = n_k - i + 1 \text{ for } \ell = \nu_i^+;$$

$$h^0(X, D_k + (g-k-\ell-1)\infty) \le n_k - i \qquad \text{for } \ell > \nu_i^+;$$

with ν_i^- defined for $[-1]D_k$ as ν_i^+ is for D_k and

$$N_k := n_k + \sum_{i=1}^{n_k} (\nu_i^+ + \nu_i^-).$$

For P and Q in X and $e - \omega_{\mathrm{R}} \in \Theta^k \setminus \left(\Theta_1^k \cup \Theta^{k-1}\right)$ and for every $t \in \mathbb{C}$,

$$\theta\left(t \int_Q^P \nu^{\mathrm{I}\mathrm{o}} + e\right) = t^{n_k} \prod_{i=1}^{n_k} \left(\prod_{k=1}^{\nu_i^+} (t-k) \prod_{\ell=1}^{\nu_i^-} (t-\ell) \right) E(P, Q)^{N_k} \Phi(P, Q, t), \tag{3.14}$$

where $\theta(z)$ is the Riemann θ function, $E(P, Q)$ is the prime form and $\Phi(P, Q, t)$ is an entire function of t for all $P \in X$ near Q, with

$$\Phi(P, P, t) = \frac{1}{N_k!} \sum_{i_1,\ldots,i_{N_k}=1}^{g} \frac{\partial^{N_k}\theta(e)}{\partial z_{i_1} \ldots \partial z_{i_{N_k}}} dz_{i_1}(P) \ldots dz_{i_{N_k}}(P) \neq 0,$$

where $e = (z_1, \ldots, z_g)$.

Recalling Definition 2.154, Remark 2.176, and Proposition 3.3 4, we define $\Theta_X^{\circ k}$ and $\Theta_X^{\circ k,1}$ by

$$\begin{aligned} \Theta_X^{\circ k} &:= (W_X^{\circ k}) \cup ([-1]W_{X,\mathfrak{s}}^{\circ k} - \widetilde{w}^{\circ}(\iota_X \mathfrak{K}_{\mathfrak{s}})), \\ \Theta_X^{\circ k,1} &:= (W_X^{\circ k,1}) \cup ([-1]W_{X,\mathfrak{s}}^{\circ k,1} - \widetilde{w}^{\circ}(\iota_X \mathfrak{K}_{\mathfrak{s}})). \end{aligned} \tag{3.15}$$

Here we note $\Theta_X^{\circ g} = J_X$ and $\Theta_X^{\circ g-1} = \Theta_X^{\circ}$ in Proposition 3.1.

In order to express the precise expression of the Riemann–Kempf theorem, we show the sequence of the indices for the Young diagram Λ_X associated with the numerical semigroup H_X introduced in Sect. 2.2.1. We recall the truncated Young diagrams $\Lambda_X^{(k)}$ and $\Lambda_X^{[k]}$ for the Young diagram Λ_X associated with the W-curve X in Definition 2.2. We show the result in Appendix 3.10, i.e., Propositions 3.225:

Proposition 3.17 *Let $r_k := r_{\Lambda_X^{[k]}}$, the rank of the Young diagram $\Lambda_X^{[k]}$. For the Young diagram Λ_X associated with the numerical semigroup H_X of genus g, an integer k $(0 \le k < g)$, and the characteristics of the partition of $\Lambda_X^{[k]} = (\Lambda_{k+1}, \Lambda_{k+2}, \ldots, \Lambda_g) = (a_{r_k}, \ldots, a_2, a_1; b_{r_k}, \ldots, b_2, b_1)$, the following holds:*

1. *$N^{\mathfrak{c}}(g - k - i) - N(i - 1)$ $(i = 1, 2, \ldots, r_k)$ is an element of the gap sequence $H_X^{\mathfrak{c}}$, and thus let $N^{\mathfrak{c}}(L_i^{[k]}) := N^{\mathfrak{c}}(g - k - i) - N(i - 1)$ and then we have*

$$\Lambda_{g-L_i^{[k]}} + g - L_i^{[k]} = a_i + b_i + 1$$

for every $i = 1, \ldots, r_k$, and

2. *$L_1^{[k]} = k + 1$.*

Definition 3.18 In terms of Lemma 3.224 2 we define the sequence $\natural_{X,k}$ (simply $\natural_k$) given by

$$\natural_k = \natural_{X,k} := \{L_1^{[k]}, L_2^{[k]}, \ldots, L_{r_k}^{[k]}\}, \qquad \natural_k^{(i)} = \natural_{X,k}^{(i)} := (\natural_{X,k} \setminus \{k+1\}) \bigcup \{i\}.$$

For $I \in \mathrm{Index}(g, r_k)$, let $|I| = \sum_{i \in I} |\Lambda_{X,i} + g - i|$.

Example 3.19 When the curve X is hyperelliptic of genus g, i.e., $(r, s) = (2, 2g + 1)$, the set of indices $\natural_{X,k}$, which was obtained by Ônishi [On2002, On2005], is equal to

$$\natural_k = \natural_{X,k} = \begin{cases} \{g, g-2, \ldots, k+3, k+1\} & \text{if } g-k \text{ is odd,} \\ \{g-1, g-3, \ldots, k+3, k+1\} & \text{otherwise.} \end{cases}$$

We define $\mathrm{u}^{[k]} = (\mathrm{u}_1^{[k]}, \ldots, \mathrm{u}_g^{[k]})$, where $\mathrm{u}_i^{[k]} := T^{(k)}_{\Lambda_i+g-i}$ and let

$$\mathbf{s}_{\Lambda_X^{(k)}}(\mathrm{u}^{[k]}) := S_{\Lambda_X^{(k)}}(T^{(k)})|_{T^{(k)}_{\Lambda_i+g-i}=\mathrm{u}_i^{[k]}}.$$

We also write the decomposition, $\mathrm{u}^{[g]} = \mathrm{u}^{[k]} + \mathrm{u}^{[g;k]} \in \mathbb{C}^g$.

Proposition 3.20 *For $\natural_{X,k}$ in Definition 3.18, the following holds:*

$$\mathbf{s}_{\Lambda_X^{(k)}}(\mathrm{u}^{[k]}) = \widehat{\varepsilon}_{\Lambda_X,\natural_{X,k}} \left(\prod_{i\in\natural_{X,k}} \frac{\partial}{\partial \mathrm{u}_i^{[g]}} \right) \mathbf{s}_{\Lambda_X}(\mathrm{u}^{[g]}) \Big|_{\mathrm{u}^{[g]}=\mathrm{u}^{[k]}}, \tag{3.16}$$

where $\widehat{\varepsilon}_{\Lambda_X,\natural_{X,k}} = (-1)^{g-k-r_k}$.

Proof See Lemma 3.224. ■

We note that r_k is equal to n_k as in Theorem 3.15 because of Propositions 3.14 and 3.20. We display the examples of Tables 3.1, 3.2, 3.3 and 3.4; Tables 3.1 and 3.2 are for the case $(r, s) = (5, 7)$, and Tables 3.3 and 3.4 are for the case $(r, s) = (7, 9)$.

Since the theta function is expressed by the Schur function due to Proposition 3.14, (3.16) and (3.127) give the theorem:

Table 3.1 Example; (5,7)-curve

i	0	1	2	3	4	5	6	7	8	9	10	11	12
$\phi(i)$	1	x	y	x^2	xy	y^2	x^3	x^2y	xy^2	x^4	y^3	x^3y	x^2y^2
$N(i)$	0	5	7	10	12	14	15	17	19	20	21	22	24
Λ_i	–	12	8	7	5	4	3	3	2	1	1	1	1

Table 3.2 Example; (5,7)-curve

k	$(a_1, \ldots, a_{n_k}; b_0, \ldots, b_{n_k})$	$(a_i + b_i + 1)_{1\le i\le n_k}$	$\sum(a_i + b_i + 1)$	$\natural_k$
0	(1, 4, 6, 11; 1, 4, 6, 11)	(3, 9, 13, 23)	48	(10, 6, 4, 1)
1	(0, 3, 5, 10; 0, 2, 5, 7)	(1, 6, 11, 18)	36	(12, 8, 5, 2)
2	(2, 4, 9; 1, 3, 6)	(4, 8, 16)	28	(9, 7, 3)
3	(1, 3, 8; 0, 2, 4)	(2, 6, 13)	21	(11, 8, 4)
4	(0, 2, 7; 0, 1, 3)	(1, 4, 11)	16	(12, 9, 5)
5	(1, 6; 1, 2)	(3, 9)	12	(10, 6)
6	(0, 5; 0, 2)	(1, 8)	9	(12, 7)
7	(4; 1)	(6)	6	(8)
8	(3; 0)	(4)	4	(9)
9	(2; 0)	(3)	3	(10)
10	(1; 0)	(2)	2	(11)
11	(0; 0)	(1)	1	(12)

Table 3.3 Example; (7,9)-curve

i	0	1	2	3	4	5	6	7	8	9	10	11	12
$\phi(i)$	1	x	y	x^2	xy	y^2	x^3	x^2y	xy^2	y^3	x^4	x^3y	x^2y^2
$N(i)$	0	7	9	14	16	18	21	23	25	27	28	30	32
Λ_i	–	24	18	17	13	12	11	9	8	7	6	6	5
i	13	14	15	16	17	18	19	20	21	22	23	24	
$\phi(i)$	xy^3	x^5	y^4	x^4y	x^3y^2	x^2y^3	x^6	xy^4	x^5y	y^5	x^4y^2	x^2y^4	
$N(i)$	34	35	36	37	39	41	42	43	44	45	46	48	
Λ_i	4	3	3	3	3	2	1	1	1	1	1	1	

Theorem 3.21 *For $u \in \mathbb{C}^g$ and a multi-index $J \in \mathrm{Index}(g, \ell)$, we define*

$$\partial_J := \prod_{j \in J} \frac{\partial}{\partial u_j}.$$

Let Λ_X be the Young diagram of the W-curve X and for given $k \in \{0, 1, \dots, g\}$, let $\Lambda_X = \Lambda_X^{[k]} \cup \Lambda_X^{(k)}$. We obtain that $n_k = r_k = r_{\Lambda_X^{[k]}}$ and $N_k = |\Lambda^{[k]}|$.

For every multi-index $I = \{\alpha_1, \dots, \alpha_m\} \in \mathrm{Index}(g, m)$, $m < n_k$, N_k, and $u \in \Theta_X^{\circ k}$, the following hold:

1.
$$\partial_I \theta\left((2\omega')^{-1}u + \xi_X, \tau\right) = 0,$$

whereas

$$\partial_{\natural_{X,k}} \theta\left((2\omega')^{-1}u + \xi_X, \tau\right) \neq 0;$$

2. *for $\ell < N_k$*

$$\partial_{u_g}^{\ell} \theta\left((2\omega')^{-1}u + \xi_X, \tau\right) = 0,$$

whereas

$$\partial_{u_g}^{N_k} \theta\left((2\omega')^{-1}u + \xi_X, \tau\right) \neq 0.$$

Proof From Proposition 3.14, Theorems 3.15 and 3.16 and Lemma 3.218, they are obtained. ■

Remark 3.22 As mentioned above, the latter one in Theorem 3.21 was obtained by Fay [Fay1984, Theorem 1.2], Segal and Wilson [SeW1985], and Birkenhake and Vanhaecke [BiV2003]. We demonstrate the non-vanishing property by investigating the leading term of the theta functions given by the Schur function. Our proof takes the position that the vanishing properties are guaranteed by Theorems 3.15 and 3.16 in this book. However, the vanishing properties can also proved by the fact that the theta series or the tau functions are expanded by the Schur functions as mentioned in Appendix 3.9.5. Segal and Wilson and Nakayashiki employed the alternative way

Table 3.4 Example; (7,9)-curve

k	$(a_1, ..., a_{n_k}; b_0, ..., b_{n_k})$	$(a_i + b_i + 1)_{1\le i\le n_k}$	$\natural_k$
0	(0, 2, 5, 7, 9, 14, 16, 23; 0, 2, 5, 7, 9, 14, 16, 23)	(1, 5, 11, 15, 19, 29, 33, 47)	(24, 20, 16, 12, 11, 6, 4, 1)
1	(1, 4, 6, 8, 13, 15, 22; 1, 3, 6, 8, 10, 15, 17)	(3, 8, 13, 17, 24, 31, 40)	(22, 18, 14, 12, 8, 5, 2)
2	(0, 3, 5, 7, 12, 14, 21; 0, 2, 4, 7, 9, 11, 16)	(1, 6, 10, 15, 22, 26, 38)	(24, 19, 17, 13, 9, 7, 3)
3	(2, 4, 6, 11, 13, 20; 1, 3, 5, 8, 10, 12)	(4, 8, 12, 20, 24, 33)	(21, 18, 15, 10, 8, 4)
4	(1, 3, 5, 10, 12, 19; 0, 2, 4, 6, 9, 11)	(2, 6, 10, 17, 22, 31)	(23, 19, 17, 12, 9, 5)
5	(0, 2, 4, 9, 11, 18; 0, 1, 3, 5, 7, 10)	(1, 4, 8, 15, 19, 29)	(24, 21, 18, 13, 11, 6)
6	(1, 3, 8, 10, 17; 1, 2, 4, 6, 8)	(2, 6, 14, 17, 26)	(22, 19, 14, 12, 7)
7	(0, 2, 7, 9, 16; 0, 2, 3, 5, 7)	(1, 5, 11, 15, 24)	(24, 20, 16, 15, 8)
8	(1, 6, 8, 15; 1, 3, 4, 6)	(3, 10, 13, 22)	(22, 17, 14, 9)
9	(0, 5, 7, 14; 0, 2, 4, 5)	(1, 8, 12, 20)	(24, 18, 15, 10)
10	(4, 6, 13; 1, 3, 5)	(6, 10, 19)	(19, 17, 11)
11	(3, 5, 12; 0, 2, 4)	(17, 8, 4)	(21, 18, 12)
12	(2, 4, 11; 0, 1, 3)	(3, 6, 15)	(22, 19, 13)
13	(1, 3, 10; 0, 1, 2)	(2, 5, 13)	(23, 20, 14)
14	(0, 2, 9; 0, 1, 2)	(1, 4, 12)	(24, 21, 15)
15	(1, 8; 1, 2)	(3, 11)	(22, 16)
16	(0, 7; 0, 2)	(1, 10)	(24, 17)
17	(6; 1)	(8)	(18)
18	(5; 0)	(6)	(19)
19	(4; 0)	(5)	(20)
20	(3; 0)	(4)	(21)
21	(2; 0)	(3)	(22)
22	(1; 0)	(2)	(23)
23	(0; 0)	(1)	(24)

based on the Sato theory. As choosing to do so, we thought, would spoil the traditional approach of the theta function theory for readers, we follow the above style.

Remark 3.23 Here, we note the symbol of $\partial_I h(\widetilde{u})$ for a function h on $\mathbb{C}^g$ and $\widetilde{u} \in \Theta_X^k$. It means that for $u \in \mathbb{C}^g$,

$$\partial_I h(\widetilde{u}) := \left[\prod_{j\in J} \frac{\partial}{\partial u_j} h(u)\right]\Big|_{u=\widetilde{u}}.$$

It should not be confused with $\displaystyle\prod_{j\in J} \frac{\partial}{\partial \widetilde{u}_j} h(\widetilde{u})$: They provide different values. Further, we use the latter one in Sect. 4.2.3.

The translation formulae of the theta functions of $\Theta_X^{\circ k}$ are preserved as the following proposition:

Proposition 3.24 *Let $N_k = |\Lambda^{[k]}|$, and $u \in 2\omega'\Theta_X^{\circ k}$. For $\ell', \ell'' \in \mathbb{Z}^g$ and $\ell := \ell' + \tau\ell''$, the following translation formulae of the theta function hold:*

1. $\partial_{\natural_{X,k}} \theta\left((2\omega')^{-1}u + \xi_X + \ell, \tau\right)$
$= \exp\left(2\pi\sqrt{-1}\,{}^t\ell''\left((2\omega')^{-1}u + \xi_X - \frac{1}{2}\tau\ell''\right)\right)\partial_{\natural_{X,k}}\theta\left((2\omega')^{-1}u + \xi_X, \tau\right),$
2. $\partial_{u_g}^{N_k}\theta\left((2\omega')^{-1}u + \xi_X + \ell, \tau\right)$
$= \exp\left(2\pi\sqrt{-1}\,{}^t\ell''\left((2\omega')^{-1}u + \xi_X - \frac{1}{2}\tau\ell''\right)\right)\partial_{u_g}^{N_k}\theta\left((2\omega')^{-1}u + \xi_X, \tau\right).$

Proof Propositions 3.12 and 3.13 show them. ■

3.1.8 Normalized Prime Form and Theta Function

As we mentioned the normalized prime forms in Sect. 2.4.1.4, we construct them in terms of the theta function in this subsection.

For the universal covering $\kappa_X : \widetilde{X} \to X$, the prime form is given as follows.

Proposition 3.25 *The normalized prime form $E^\circ(P, Q)$ for a point $P, Q \in \widetilde{X}$ is given by*

$$E^\circ(P, Q) = \frac{\theta\begin{bmatrix}\varepsilon''\\ \varepsilon'\end{bmatrix}\left(\int_Q^P \nu^{\mathrm{I}\circ}\right)}{\sqrt{\zeta^\circ(P)}\sqrt{\zeta^\circ(Q)}}, \tag{3.17}$$

where ε' and ε'' are a fixed nonsingular odd theta characteristic in $\mathbb{R}^g$ and $\zeta^\circ(P)$ is

$$\zeta^\circ(P) := \sum_{i=1}^{g}\left(\frac{\partial}{\partial z_i}\theta\begin{bmatrix}\varepsilon''\\ \varepsilon'\end{bmatrix}(0)\right)\nu_i^{\mathrm{I}\circ}(P)$$

for $z = \widetilde{w}^\circ(P) \in \mathbb{C}^g$.

We note that $\sqrt{\zeta^\circ(P)}$ is a holomorphic section of $\mathscr{L}_X(e) \otimes_{\mathbb{C}} \mathscr{L}_X(\xi_X)$. Further, we define

$$E^\circ(P, \infty) := c_0\frac{\theta\begin{bmatrix}\varepsilon''\\ \varepsilon'\end{bmatrix}\left(\int_\infty^P \nu^{\mathrm{I}\circ}\right)}{\sqrt{\zeta^\circ(P)}},$$

where c_0 is a certain constant factor.

We also refer to Fay's result without proof:

Theorem 3.26 [Fay1973, (43)(44)] *For $e \in \mathbb{C}^g$ such that $\theta(e) \neq 0$, and $P_1, \ldots, P_n$, $Q_1, \ldots, Q_n \in X$,*

$$\theta\left(\sum_{i=1}^n(\widetilde{w}^\circ(\iota_X P_i) - \widetilde{w}^\circ(\iota_X Q_i) - e\right)\theta(e)^{n-1}\frac{\prod_{i<j} E^\circ(P_i, P_j)E^\circ(Q_i, Q_j)}{\prod_{i,j} E^\circ(P_i, Q_j)}$$
$$= \det\left[\frac{\theta(\widetilde{w}^\circ(\iota_X P_i) - \widetilde{w}^\circ(\iota_X Q_i) - e)}{E^\circ(P_i, Q_j)}\right]. \tag{3.18}$$

In particular, if $e \in \mathbb{C}^g$ is a non-singular point of (θ) and $Q_1 = \cdots = Q_n = \infty$,

$$\theta\left(\sum_{i=1}^n \widetilde{w}^\circ(\iota_X P_i) - e\right)\frac{\zeta^\circ(P)^{n-1}\prod_{i<j} E^\circ(P_i, P_j)}{\prod_{i=1}^n \theta(\widetilde{w}^\circ(P_i) - e)E^\circ(P_i, \infty)^{n-1}} = \frac{(-1)^{\frac{(n-1)(n-2)}{2}}}{\prod_{k=1}^{n-1} k!}\det\left[\widetilde{\phi}_i(P_j)\right], \tag{3.19}$$

where $\widetilde{\phi}_1(P) = 1$, and for the arithmetic parameter t_Q of Q at ∞ and $z_k := \int^z \nu_k^{\mathrm{I}\circ}$,

$$\widetilde{\phi}_i(P) = \frac{\partial^{i-1}}{\partial t_Q^{i-1}}\left[\sum_{k=1}^g \frac{\partial \log\theta}{\partial z_k}(\widetilde{w}^\circ(\iota_X P) - \widetilde{w}^\circ(\iota_X Q) - e)\nu_k^{\mathrm{I}\circ}\right]_{Q=\infty}.$$

Remark 3.27 Baker mentioned this formula in [Ba1897, p. 432] and its history. Klein introduced the prime forms in [Kl1888, p. 351], [Kl1890, p. 40][5] by following Schtokky's proposal of a function [Sch1887, p. 242][Kl1888]. Thus, it is sometimes called the Schtokky–Klein prime form.

3.2 Sigma Function for W-Curves

In this section, we define the sigma function for W-curves X with higher genera and investigate their basic properties based on [KomMP2022b].

3.2.1 Definition of the Sigma Function for W-Curves

We are now ready to define the sigma function as a natural extension of the sigma function of elliptic curves, precisely speaking the first expression in (3.1). This subsection gives the definition of the sigma function for general W-curves X following [KomMP2022b].

[5] In defining the sigma function, he would have preferred to use the data of only a compact Riemann surface rather than an algebraic curve.

We basically mingle the definition of the sigma function by Korotkin and Shramchenko [KorS2012] and the definition by Nakayashiki [Nak2016, Definition 9] together. For the case of the symmetric W-curves, we employ the definition of the sigma function by Nakayashiki [Nak2016, Definition 9]. However, for the non-symmetric W-curves, Nakayashiki used the non-half integral theta characteristics to define his sigma function so that its expansions at the origin in $\mathbb{C}^g$ is naturally expressed by the Schur functions. For the connection with algebraic properties of X, we use the shifted Riemann constant and shifted Abelian integrals so that our sigma function is an even or odd function which is one of the sigma functions defined by Korotkin and Shramchenko [KorS2012] and is consistent with the spin structure of X. Even though we employ the definition that our sigma function is also normalized by its expansion at the Abel–Jacobi image of $(\infty)^g \in S^g X$ following Nakayashiki's definition, the point is apparently shifted from the origin in $\mathbb{C}^g$ with the shifted Riemann constant. Then our sigma function is connected with the algebraic properties mentioned in Chap. 2 as we will describe below [KomMP2022b]. As we show in Sect. 3.7.1, it is expected that our sigma function is connected for the degenerating family of curves, which is, at least, true for the case of (3,4,5)-type curves.

We consider a Weierstrass curve X, with either symmetric or non-symmetric Weierstrass semigroup H_X at its single ∞. As mentioned in Chap. 2, we have ω', ω'', η' and η'' for the curve X and the basis of $\mathbf{H}_1(X, \mathbb{Z})$.

Let the W-normalized Siegel upper half-space be $\mathfrak{H}_g$

$$\mathfrak{H}_g := \{(2\omega', 2\omega'') \in \mathrm{GL}(g, \mathbb{C}) \times \mathrm{GL}(g, \mathbb{C}) \mid \omega'^{-1}\omega'' \in \mathfrak{H}_g^\circ\},$$

and the symplectic space

$$\mathfrak{W}_g := \Bigg\{ M := \begin{pmatrix} 2\omega' & 2\omega'' \\ 2\eta' & 2\eta'' \end{pmatrix} \in \mathrm{Mat}_{\mathbb{C}}(2g) \;\Bigg|\; M \begin{pmatrix} 0 & 1_g \\ -1_g & 0 \end{pmatrix} {}^t M = \frac{2\pi}{\sqrt{-1}} \begin{pmatrix} 0 & 1_g \\ -1_g & 0 \end{pmatrix}, (2\omega', 2\omega'') \in \mathfrak{H}_g \Bigg\}.$$

For a given W-curve X, these ω', ω'', η' and η'' in Theorem 2.148 determines the points in $\mathfrak{H}_g$ and $\mathfrak{W}_g$; the condition in $\mathfrak{W}_g$, $M \begin{pmatrix} 0 & 1_g \\ -1_g & 0 \end{pmatrix} {}^t M = 2\pi\sqrt{-1} \begin{pmatrix} 0 & 1_g \\ -1_g & 0 \end{pmatrix}$, corresponds to the Legendre relations,

$${}^t\eta'\eta'' = {}^t\eta''\eta', \quad {}^t\omega'\omega'' = {}^t\omega''\omega', \quad {}^t\omega''\eta' - {}^t\omega'\eta'' = \frac{\pi\sqrt{-1}}{2} 1_g.$$

We fix a W-curve X for a while. Then we also find the theta characteristics δ associated with the shifted Riemann constant $\xi_{X,\mathfrak{s}}$ as in Theorem 3.4.

We generalize the first expression of the sigma function of genus one in (3.1) to a general W-curve X;[6] We show the definition of the sigma function [Nak2016].

Definition 3.28 [KomMP2022b] We define σ as an entire function of (a column-vector) $u = {}^t(u_1, u_2, \dots, u_g) \in \mathbb{C}^g$,

$$\sigma(u) = \sigma(u; M) = \frac{\varepsilon_{\Lambda,\natural_{X,0}} \exp(\frac{1}{2}\, {}^t u \eta' \omega'^{-1} u) \vartheta \begin{bmatrix} \delta'' \\ \delta' \end{bmatrix} (\frac{1}{2}\omega'^{-1} u;\ \omega'^{-1}\omega'')}{\partial_{\natural_{X,0}} \vartheta \begin{bmatrix} \delta'' \\ \delta' \end{bmatrix} (\frac{1}{2}\omega'^{-1} u;\ \omega'^{-1}\omega'')\Big|_{u=\widetilde{w}(\iota_X \mathfrak{K}_\mathfrak{s})}}$$

$$= \frac{\varepsilon_{\Lambda,\natural_{X,0}} \exp(\frac{1}{2}\, {}^t u \eta' \omega'^{-1} u)}{\partial_{\natural_{X,0}} \vartheta \begin{bmatrix} \delta'' \\ \delta' \end{bmatrix} (\frac{1}{2}\omega'^{-1} u;\ \omega'^{-1}\omega'')|_{u=\widetilde{w}(\iota_X \mathfrak{K}_\mathfrak{s})}}$$

$$\times \sum_{n \in \mathbb{Z}^g} \exp\left[\pi\sqrt{-1}\{\, {}^t(n+\delta'')\omega'^{-1}\omega''(n+\delta'') + {}^t(n+\delta'')(\omega'^{-1}u + 2\delta')\}\right], \tag{3.20}$$

where $\varepsilon_{\Lambda,\natural_{X,0}} := (-1)^g$ defined in (3.16), and $\natural_{X,0}$ is defined in Definition 3.18.

This sigma function has the required modular invariance and is homogeneous with respect to the extended weight of wt_λ.

Remark 3.29 1. Though we formally employ the definition of the sigma function following Nakayashiki, there are slight differences between them for the non-symmetric W-curves as mentioned in the introduction to this subsection. Even for the non-symmetric W-curves, we employ the shifted Riemann constant and the associated theta characteristics $\delta \in (\mathbb{Z}/2)^{2g}$, whereas Nakayashiki did not consider the shifted Riemann constant, and thus his δ does not belong to $(\mathbb{Z}/2)^{2g}$ in general. As Korotkin and Shramchenko defined their sigma function so that the sigma function is an even or odd function, our sigma function is also an even or odd function as mentioned in Sect. 3.2.8. The shifted Riemann constant is closely related to the half-canonical divisor and spin structure of the curve X as in Corollary 3.2.

Further, we note the fixed-point in the denominator differs from the origin of u, $u = 0$ since the point must be the same as the expansion of the sigma function as will be mentioned in Sect. 3.2.7.

As our sigma function (3, 4)-curves and (3, 4, 5)-curves are naturally connected by the degenerating family of curves as mentioned in Sect. 3.7 [FeKMPA2022], it is expected that our sigma function of W-curves whose Weierstrass semigroups

[6] In the definition of σ, the minus sign of the quadratic form appears in the exponential factor when we employ the different definitions of the Abelian differentials of the first or the second kind as mentioned in footnotes 10 and 12 of Chap. 2. In other words, the sign of the quadratic form differs from the convention in [KodMP2013, KomMP2013, Mat2002b, Mat2002c, Mat2008a, MatP2008, MatP2014] but is the same as the convention by Weierstrass (3.1), Baker [Ba1898a, p. 336], [Ba1907, p. 25], Buchstaber, Enolskii and Leykin [BucEL1997b, BucEL2012], and Nakayashiki [Nak2016].

Further, we note the convention of u in (2.81) which differs from Nakayaski's [Nak2016].

are generated by m_X generators might be connected by the degeneration limit of σ as mentioned in (2.144)

2. When applying the above definition to genus one, we notice the difference between the differentials of $\theta_1(z, \delta, Z)$ with respect to z and u.

3.2.2 Alternative Definition of Sigma Function

In this subsection, we also generalize the second definition of the sigma function of genus one in (3.1) to a certain W-curve X.

Buchstaber, Enolskii, and Leykin have studied this generalization to construct the general sigma function for three decades [BucEL1997a, BucEL1997b, BucEL1999, BucLE1999, BucEL2012, BucEL2020]. They define the sigma function for plane W-curves by [BucEL2012, Definition 1.4]

$$\widetilde{\sigma}(u) = \frac{1}{\sqrt[4]{D(X)}}\sqrt{\frac{\pi^g}{|\omega'|}}\exp(\tfrac{1}{2}{}^t u\eta'\omega'^{-1}u)\theta\begin{bmatrix}\delta''\\ \delta'\end{bmatrix}(\frac{1}{2}\omega'^{-1}u;\ \omega'^{-1}\omega''), \qquad (3.21)$$

where $D(X)$ is the discriminant of the defining equation $f_X(x, y)$.

Following these studies, Eilbeck, Gibbons, Ônishi, and Yasuda provided more precise descriptions of the sigma function [EilGOY2017].

Remark 3.30 We assume that X is a hyperelliptic curve of genus g or a (3, 4)-curve of genus $g = 3$. The sigma function of X is alternatively defined by

$$\widetilde{\sigma}(u) := c_0 e^{\frac{1}{2}u^t\omega'^{-1}\,{}^t\eta' u}\theta_X\begin{bmatrix}\delta''_X\\ \delta'_X\end{bmatrix}\left(\frac{1}{2}\omega'^{-1}u;\omega'^{-1}\omega''\right),\quad u\in\mathbb{C}^g,$$

where c_0 is the constant factor

$$c_0 = \left(\frac{(2\pi)^g}{|\omega'|}\right)^{1/2}\Delta^{-1/8}, \qquad (3.22)$$

and Δ is the discriminant of X [EilGOY2017].

Eilbeck, Gibbons, Ônishi, and Yasuda gave the computations of the discriminant in terms of the Sylvester resultant as a conjecture [EilGOY2017, Conjecture 1.14]. They define the sigma function using the heat equation based on the series of studies by Buchstaber and Leykin [BucL2002, BucL2004, BucL2005, BucL2008] (see Appendix 3.11).

Since the modular invariance of both definitions is proved, and the leading terms of the expansions in terms of the Schur polynomials of both sigma functions agree, both sigma functions agree as mentioned in Theorem 3.31. However, since the discriminant in the alternative definition cannot be clearly defined for every compact Riemann surface, we employ Definition 3.28 in this book.

We should note that both definitions of sigma functions are connected with the Thomae type formula [BucEL2012, FarZ2011, Ber2009].

3.2.2.1 Sigma Function for Cyclic Trigonal Curves of Genus Three X_s

Following the definition in [EilGOY2017, EilEMOP2007], we show the alternative definition of the sigma function for the cyclic trigonal curve of genus three X_s given by $y^3 = x(x-s)(x-b_1)(x-b_2)$. By using the integrals in Sect. 2.9, the sigma function of X_s is equal to [FeKMPA2022]

$$\widetilde{\sigma}(u) = c_s \mathrm{e}^{\frac{1}{2} u^t {\omega'_{X_s}}^{-1} {}^t\eta'_{X_s} u} \theta_{X_s} \begin{bmatrix} \delta''_{X_s} \\ \delta'_{X_s} \end{bmatrix} \left(\frac{1}{2} {\omega'_{X_s}}^{-1} u ; {\omega'_{X_s}}^{-1} \omega''_{X_s} \right), \quad u \in \mathbb{C}^3.$$

The ingredients of the formula are as follows: c_s is the constant factor

$$c_s := \left(\frac{(2\pi)^3}{|\omega'_{X_s}|} \right)^{1/2} \Delta_s^{-1/8}; \tag{3.23}$$

for the discriminant Δ_s which is equal to [EilGOY2017]

$$-729 s^4 b_2^4 b_1^4 \left((s+b_2)^3 b_1^3 + 3 s b_2 (s + \frac{1}{4} b_2)(s + 4 b_2) b_1^2 + 3 s^2 b_2^2 (s+b_2) b_1 + s^3 b_2^3 \right)^2. \tag{3.24}$$

3.2.2.2 On Both Sigma Functions

Theorem 3.31 *For hyperelliptic curves and* (3, 4)*-curves,* σ *is identical to* $\widetilde{\sigma}$. *From here,* $\widetilde{\sigma}$ *is also denoted by* σ.

Proof Since $\widetilde{\sigma}$ studied by [BucB2020, BucL2004, BucL2008, EilGOY2017] has the same properties as σ mentioned as follows, it is clear. In particular, both leading terms are given by the same Schur function. ■

3.2.3 *Modular Invariance*

We state, here, the modular invariance of the sigma function mentioned in [BucEL1997a, BucEL2012, Nak2010a, Nak2010b, Nak2016] (see Appendix 3.8). We recall the W-normalized Siegel upper half space $\mathfrak{H}_g$, and the related space $\mathfrak{W}_g$ in Sect. 3.2.1. We consider the period matrices $(2\omega', 2\omega'')$ and $(2\eta', 2\eta'')$ of the W-curve X as an element in $\mathfrak{H}_g$ and $\mathfrak{W}_g$. As mentioned in Sect. 3.1.3, there is an ambiguity regarding which loops are generators $\{\alpha_i, \beta_i\}_{i=1,\dots,g}$ in $\mathbf{H}_0(X, \mathbb{Z})$ in Lemma 2.87; There is a group element in $\mathrm{Sp}(2g, \mathbb{Z})$ in (2.59) acting on the generators and thus on Γ_X. The Jacobian J_X is invariant for the action.

For $\gamma = \begin{pmatrix} a & b \\ c & d \end{pmatrix} \in \mathrm{Sp}(2g, \mathbb{Z})$ and $M \in \mathfrak{W}_g$, $\mathrm{Sp}(2g, \mathbb{Z})$ acts on $\mathfrak{W}_g$ by φ_γ:

$$\varphi_\gamma M = M \begin{pmatrix} 0 & 1_g \\ -1_g & 0 \end{pmatrix} {}^t\gamma \begin{pmatrix} 0 & 1_g \\ -1_g & 0 \end{pmatrix} = 2 \begin{pmatrix} \omega' \, {}^t d + \omega'' \, {}^t c & \omega' \, {}^t b + \omega'' \, {}^t a \\ \eta' \, {}^t d + \eta'' \, {}^t c & \eta' \, {}^t b + \eta'' \, {}^t a \end{pmatrix}.$$

This corresponds to the exchange of the basis $\{\alpha_i, \beta_i\}_{i=1,\dots,g}$ of $\mathbf{H}_1(X, \mathbb{Z})$. The action is naturally extended to the action on the line bundle on the Jacobian J_X. Then the theta characteristics corresponding to the half period of Γ_X are transformed by

$$\varphi_\gamma \begin{pmatrix} \delta'' \\ \delta' \end{pmatrix} := \begin{pmatrix} d & -c \\ -b & a \end{pmatrix} \begin{pmatrix} \delta'' \\ \delta' \end{pmatrix} - \begin{pmatrix} \frac{1}{2}({}^t ac)_D \\ \frac{1}{2}({}^t db)_D \end{pmatrix},$$

where $B_D := {}^t(B_{11}, \dots, B_{gg}) \in \mathbb{Z}^g$ for $B \in \mathrm{Mat}_{\mathbb{Z}}(g)$.

From Proposition 3.194 in Appendix 3.8, we also have the following theorem:

Theorem 3.32 *For* $\gamma := \begin{pmatrix} a & b \\ c & d \end{pmatrix} \in \mathrm{Sp}(2g, \mathbb{Z})$, *we define the action of* $\mathrm{Sp}(2g, \mathbb{Z})$ *on the sigma function by*

$$\widehat{\varphi}_\gamma \sigma(u, \delta, M) = \sigma(u, \varphi_\gamma^\circ \delta, \widehat{\varphi}_\gamma M).$$

Then we have the modular invariance of the sigma function, i.e.,

$$\widehat{\varphi}_\gamma \sigma(u, \delta, M) = \sigma(u, \delta, M).$$

3.2.4 Translational Relation

As mentioned on the theta function in Sect. 3.1.4, we have the translational formula of the sigma function based on [EilEMOP2007, On2013].

Theorem 3.33 *For* $u, v \in \mathbb{C}^g$, *and* ℓ $(= 2\omega'\ell' + 2\omega''\ell'' \in \Gamma_X)$, *if we define* $L(u, v) := 2\, {}^t u(\eta' v' + \eta'' v'')$, $\chi(\ell) := \exp[\pi\sqrt{-1}\big(2({}^t\ell'\delta'' - {}^t\ell''\delta') + {}^t\ell'\ell''\big)]$, *the following holds:*

$$\sigma(u + \ell) = \sigma(u) \exp(L(u + \frac{1}{2}\ell, \ell))\chi(\ell). \tag{3.25}$$

Proof From Proposition 3.190, it is obtained. ■

We let $\varphi_\sigma(u, \ell) := \sigma(u + \ell)/\sigma(u)$ for later convenience.

In general, when the holomorphic theta function $\vartheta(z)$ associated with $\mathbb{C}^g$ and the lattice Γ (cf. Sect. 3.8.2) obeys the translational property,

$$\vartheta(z+\widetilde{\ell}) = \exp\left(2\pi\sqrt{-1}(\mathbf{L}(z,\widetilde{\ell})+\mathbf{J}(\widetilde{\ell}))\right)\vartheta(z)$$

for all $z \in \mathbb{C}^g$ and $\widetilde{\ell} \in \Gamma$, we call it the *holomorphic theta function* of $(\mathbf{L}, \mathbf{J})$ type as mentioned in Definition 3.159.

Lemma 3.34 *Let* $\widehat{K}_L(u) := \widehat{K}_L(u, \delta', \delta'') := ({}^t u'\delta'' - {}^t u''\delta') + \frac{1}{2}{}^t u' u''$, $\widehat{L}(u_1, u_2) := \dfrac{1}{2\pi\sqrt{-1}} L(u_1, u_2)$, *and* $\widehat{J}_L(u) := \widehat{J}_L(u, \delta', \delta'') := \widehat{K}_L(u, \delta', \delta'') + \widehat{L}(\frac{u}{2}, u)$. *Then the* σ *function is the theta function of* $(\widehat{L}, \widehat{J}_L)$ *type.*

Proof See Proposition 3.190. ■

Theorem 3.35 *The holomorphic theta function of type* $(\widehat{L}, \widehat{J}_L)$ *forms a one-dimensional* $\mathbb{C}$*-vector space.*

Proof Due to Theorem 3.166 and Lemma 3.183 3. ■

We let the normalized theta associated with $\sigma(u)$ be denoted by $\vartheta_\sigma(u)$ due to Proposition 3.164.

The translational formula characterizes the line bundle $\mathscr{L}_{J_X}$ of the Jacobian or $\mathscr{L}_{S^g X}$ of $S^g X$. Theorem 3.166 guarantees the uniqueness of the sigma function of J_X up to a constant factor.

3.2.5 Sigma Divisor

This subsection mentions the divisor of the sigma function, which is determined as follows:

Theorem 3.36 *The sigma function vanishes at the shifted standard theta divisor* $\kappa_J^{-1}\Theta_{X,\mathfrak{s}} \subset \mathbb{C}^g$, *where* $\Theta_{X,\mathfrak{s}} := w_{\mathfrak{s}}(S^{g-1}X) \subset J_X$.

Proof Proposition 3.3, Theorem 3.4, and Definition 3.28 lead us to the result. ■

We refer to $\Theta_{X,\mathfrak{s}}$ as the *sigma divisor*. This theorem and Theorem 3.162 show the uniqueness of the sigma function up to the trivial theta functions. Further, the agreement between the sigma divisor and the shifted standard theta divisor guarantees the direct correspondence between $S^g X$ and J_X. Corollary 3.2 and Proposition 3.3 show that the sigma divisor is connected to the half-canonical or Riemann divisor.

It is noted that for the symmetric W-curves X, $\Theta_{X,\mathfrak{s}}$ agrees with the standard theta divisor $w(S^{g-1}X)$.

3.2.6 Jacobi–Riemann Fundamental Relation

This subsection shows the Jacobi–Riemann fundamental relation of Proposition 3.13 in terms of the sigma functions.

Let $\widetilde{w} \circ \iota_X$ sometimes just be denoted by $\widetilde{w}$. Using Theorems 2.177 and 3.36, we show the version of sigma of the *Jacobi–Riemann fundamental relation* as a generalization of (1.42) [Rie1857], [Ba1897, §188, §195], though Baker called it "Abel's theorem" in [Ba1928, p. 587]:

Theorem 3.37 *For* $(P, Q, P_i, P_i') \in X^2 \times (S^g(X) \setminus S_1^g(X)) \times (S^g(X) \setminus S_1^g(X))$,

$$u := \widetilde{w}_{\mathfrak{s}}(\iota_X(P_1, \ldots, P_g)), \quad v := \widetilde{w}_{\mathfrak{s}}(\iota_X(P_1', \ldots, P_g')),$$

$$\begin{aligned}\exp\left(\sum_{i,j=1}^{g} \Pi_{P_i,P_j'}^{P,Q}\right) &= \frac{\sigma(\widetilde{w}(\iota_X(P)) - u)\sigma(\widetilde{w}(\iota_X(Q))) - v)}{\sigma(\widetilde{w}(\iota_X(Q)) - u)\sigma(\widetilde{w}(\iota_X(P)) - v)}\\ &= \frac{\sigma(\widetilde{w}(P) - \widetilde{w}_{\mathfrak{s}}(P_1, \cdots, P_g))\sigma(\widetilde{w}(Q) - \widetilde{w}_{\mathfrak{s}}(P_1', \cdots, P_g'))}{\sigma(\widetilde{w}(Q) - \widetilde{w}_{\mathfrak{s}}(P_1, \cdots, P_g))\sigma(\widetilde{w}(P) - \widetilde{w}_{\mathfrak{s}}(P_1', \cdots, P_g'))}.\end{aligned}$$

Proof The right-hand side can be expressed as

$$\exp(\text{bilinear term in } w\text{'s})\frac{\theta(\omega'^{-1}(\widetilde{w}(P) - u) + \xi_{X,\mathfrak{s}})\theta(\omega'^{-1}(\widetilde{w}(Q) - v) + \xi_{X,\mathfrak{s}})}{\theta(\omega'^{-1}(\widetilde{w}(Q) - u) + \xi_{X,\mathfrak{s}})\theta(\omega'^{-1}(\widetilde{w}(P) - v)) + \xi_{X,\mathfrak{s}})}.$$

By Riemann's theorem for the theta function in Proposition 3.13 the above becomes

$$\exp(\text{bilinear term in } w\text{'s}) \exp\left(\sum_{j=1}^{g} \int_Q^P \nu_{P_j,P_j'}^{\mathrm{III}\circ}\right).$$

The exponential part of the bilinear term turns out to be

$$-(u - v)^t \gamma(\widetilde{w}(\iota_X(P)) - \widetilde{w}(\iota_X(Q))) = -\sum_{i,j,k} \gamma_{ij} \int_{P_k'}^{P_k} \nu_i^{\mathrm{I}} \int_Q^P \nu_j^{\mathrm{I}},$$

which equals (Lemma (2.146))

$$\sum_{i=1}^{g} \Pi_{P_i,P_i'}^{P,Q} - \sum_{i=1}^{g} \int_Q^P \nu_{P_i,P_i'}^{\mathrm{III}\circ}.$$

The above integrals depend upon the paths we choose, but Lemma 2.146 shows that such dependence cancels. Thus the right-hand side coincides with the left-hand side. ■

We remark that this formula is the most crucial in the sigma function theory, which connects the algebraic and transcendental objects; it provides all of the data of the connection. We recall Remark 2.144 and point out that this relation directly leads to the Jacobi inversion formula in Sect. 3.3.

3.2.7 *Sigma Function and Schur Function*

We provide the relations between the sigma functions and the Schur functions in this subsection.

Buchstaber, Leykin and Enolskii [BucEL1999] showed that the σ function over the monomial curve given by $Z_s^r = Z_r^s$ is identified with a Schur function, cf. Proposition 3.14. Since it is known that the relations among the tau function, the Riemann theta function and the Schur function are studied well as mentioned in Appendix 3.9, Eilbeck, Enolskii and Gibbons studied the relation among the tau and the sigma functions in [EilEG2010] and Nakayashiki independently showed that the sigma function of plane W-curve curves are expanded in terms of the Schur functions in [Nak2010a] and showed the correspondence among the sigma function, the tau function, and the Schur polynomial in [Nak2010b, Nak2016].

In this book and this subsection, we provide the relation between the sigma function and the Schur function on W-curves X associated with the numerical semigroups H_X based on the final results by Nakayashiki [Nak2016] (cf. Theorem 3.214 in Appendix 3.9). The Young diagram Λ_X is associated with H_X.

As we mentioned in Sect. 3.1.6, the Riemann theta function is expressed by the Schur function in Proposition 3.14.

We note that $\xi_{X,\mathfrak{s}}$ corresponds to $\delta_X = \begin{bmatrix} \delta' \\ \delta'' \end{bmatrix}$ in Theorem 3.4 and ξ_X to e in Proposition 3.14, i.e., $2\omega' e \in \kappa_J^{-1}\Theta_X^{g-1}$. Then we rewrite the formula in Proposition 3.14 as

$$\frac{(-1)^{g-r_\Lambda}\theta\begin{bmatrix} \delta' \\ \delta'' \end{bmatrix}\left((2\omega')^{-1}u + e - \xi_{X,\mathfrak{s}}, \tau\right)}{e^{\pi\sqrt{-1}({}^t\delta''\tau\delta'' + 2\,{}^t\delta''(2\omega')^{-1}u + e - \xi_{X,\mathfrak{s}} + \delta')}\partial_{\natural_X,0}\theta(e,\tau)}$$
$$= S_{\Lambda_X}(T)|_{T_{\Lambda_i+g-i}=-u_i} + \text{higher-weight terms with respect to wt.}$$

Noting $\xi_X - \xi_{X,\mathfrak{s}} = \widetilde{w}(\iota_X(\mathfrak{K}_\mathfrak{s}))$, we have the expansion of the sigma function:[7]

[7] The expansion of the sigma function in [MatK2013, p. 87] and [KomMP2019, Prop. 5.2] should read with a shifted point as in Theorem 3.38.

Theorem 3.38 *The leading term in the expansion of the σ function associated with X is expressed by the Schur function*

$$\sigma(u+\widetilde{w}(\iota_X\mathfrak{K}_\mathfrak{s})) = (-1)^{r_{\Lambda_X}} S_{\Lambda_X}(T)|_{T_{\Lambda_i+g-i}=-u_i} + \sum_{|\mathrm{w}_g(\alpha)|>|\Lambda_X|} a_\alpha u^\alpha,$$

where $a_\alpha \in \mathbb{Q}[b_1,\cdots,b_{s+r}]$, $\alpha := (\alpha_1, ..., \alpha_g)$, $u^\alpha = u_1^{\alpha_1}\cdots u_g^{\alpha_g}$, and $\mathrm{w}_g(\alpha) := \sum \alpha_i \mathrm{wt}(u_i)$. Here $S_{\Lambda_X}(T)$ is the lowest-order term in the w-degree of the u_i.

$\sigma(u)$ is homogeneous of degree $|\Lambda|$ with respect to the λ-degrees.

For the symmetric W-curves X, the above formula is simply written as

$$\sigma(u) = S_{\Lambda_X}(T)|_{T_{\Lambda_i+g-i}=u_i} + \sum_{|\mathrm{w}_g(\alpha)|>|\Lambda_X|} a_\alpha u^\alpha.$$

We note that S_Λ is a function of $\{T_{\Lambda_i+g-i}\}_{i=1,...,g}$, even though *a priori* it depends on $\{T_i\}_{i=1,...,2g-1}$.

Remark 3.39 The line bundle associated with the sigma function is characterized by the translational formula in Theorem 3.33 up to a constant factor, the expansion in Theorem 3.38 determines the constant factor. This property is a natural generalization of the elliptic sigma function mentioned in Sect. 1.3.2.

The symmetric functions have a degree coming from the natural order of the multi-variables $(t_1, \dots, t_g)$ which corresponds to the Sato–Weierstrass weight wt.

For the case of a plane W-curve ((r, s)-curve), Ônishi showed that the coefficients of the power series of the sigma function at the origin have Hurwitz integrability; more precisely, for prime 2, the expansion is not Hurwitz integral, but its square is [On2018]. Ayano also showed the similar results for the telescopic curves [Ay2022].

3.2.8 Parity of Sigma Function

This subsection mentions the parity of the sigma functions.

Theorem 3.40 *The sigma function is an odd or even function, i.e.,*

$$\sigma(-u) = \pm\sigma(u).$$

We prove it in terms of the properties of the divisor of σ and $[-1]\Theta_{X,\mathfrak{s}} = \Theta_{X,\mathfrak{s}}$ in Theorems 2.177 and 3.36, which is essentially the same as the proof of the Galois action in the following subsection. We use the parity of the Riemann theta function with the characteristics of δ.

Proof Theorem 2.177 shows that $[-1]\Gamma_X = \Gamma_X$, $[-1]J_X = J_X$, and $[-1]W^i_{X,\mathfrak{s}} = W^i_{X,\mathfrak{s}}$ for $i = g-1, g$. We recall Theorem 3.35, i.e., Lemma 3.34 and $\mathrm{Pf}(\widehat{E}) = 1$ for

$\widehat{E}(u_1, u_2) := \dfrac{1}{2\pi\sqrt{-1}}(L(u_1, u_2) - L(u_2, u_1))$ due to Lemma 3.183; where $\mathrm{Pf}(\widehat{E})$ is the pfaffian of $\widehat{E}$ (see Definition 3.165).

Since for every holomorphic theta function ϑ, we have the normalized theta function $\widetilde{\vartheta}$ from Proposition 3.164, we let the normalized theta function $\widetilde{\vartheta}_1(u)$ for $\sigma(-u)$ whereas we recall that $\vartheta_\sigma(u)$ is the normalized theta function of $\sigma(u)$. On the other hand, in general, the normalized theta function is defined uniquely up to the trivial theta function, and $\sigma(u)$ is defined by $\theta[\delta](\frac{1}{2}\omega'^{-1}u)$ with the trivial theta function $\exp\left(\frac{1}{2}{}^t u\omega'^{-1}\eta' u\right)$. Thus $\sigma(u)$ and $\theta[\delta](\frac{1}{2}\omega'^{-1}u)$ have the same normalized theta function $\vartheta_\sigma(u)$. However, since $\theta[\delta](-z) = \pm\theta[\delta](z)$, $\sigma(-u)$ and $\sigma(u)$ have the same normalized theta function, i.e., $\vartheta_\sigma(u) = \widetilde{\vartheta}_1(u)$ (up to constant factor precisely).

Hence the difference between $\sigma(-u)$ and $\sigma(u)$ is given by a trivial theta function, i.e.,

$$[-1]\sigma(u) = \sigma(-u) = \exp({}^t u A u + Bu + c)\sigma(u).$$

However, since $[-1]$ is an involution, i.e., $[-1]^2 = 1$, the factor $\exp({}^t u A u + Bu + c)$ should be a unit, and $\exp(c) = \pm 1$. ■

It is remarked that $\sigma(-u + \widetilde{w}(\iota_X \mathfrak{K}_\mathfrak{s}))$ cannot be, in general, expressed by $\sigma(u + \widetilde{w}(\iota_X \mathfrak{K}_\mathfrak{s}))$ with a factor. Nakayashiki shows that the Schur function $S_{\Lambda_X}(-T)$ for $-T$ is expressed by $S_{{}^t\Lambda_X}(T)$ [Nak2010b]. Hence, the parity properties generally do not appear unless we introduce the shifted Riemann constant. For a particular case, the (3, 4, 5)-curve, given in Theorem 3.148, the parity of its sigma function is determined by that of the (3, 4)-curve. Thus the parity might be determined by a relation associated with (2.144).

Corollary 3.41 *If the W-curve X is symmetric, i.e., H_X is symmetric, we have*

$$\sigma(-u) = (-1)^{|\Lambda_X|}\sigma(u).$$

Proof We note that Lemma 2.3 4.1, if $|\Lambda_X|$ is even, the r_Λ is even. Due to the property of the Schur polynomial [Mac1985, p46], $S_\Lambda(-T) = S_{{}^t\Lambda}(T)$. Theorems 3.38 and 3.40 show it.

Theorem 3.40 means that we use some parts of the definition of the sigma function by Korotkin and Shramchenko [KorS2012] who defined the several sigma functions based on Klein's transcendental approaches.

3.2.9 *Galois Action on Sigma Function*

This subsection shows the Galois action on the sigma functions. Assume that $\varpi_r : X \to \mathbb{P}$ is the Galois covering and the Galois group G_X acts on X as an automorphism.

It induces the automorphism on the Jacobian J_X [On2009]. For an element $\widehat{\zeta} \in G_X$, $\widehat{\zeta}$ acts on Γ_X and also on J_X so that they are invariant, i.e., $\widehat{\zeta}\Gamma_X = \Gamma_X$ and $\widehat{\zeta} J_X = J_X$. The action also induces the line bundle on J_X and thus

$$\widehat{\zeta}\sigma(u) := \sigma(\widehat{\zeta} u).$$

Proposition 3.42 *If $\widehat{\zeta} \in G_X$ satisfies $\widehat{\zeta}^m = \mathrm{id}$ and $\widehat{\zeta}\varphi_\sigma(u, \ell) = \varphi_\sigma(u, \ell)$ for $\ell \in \Gamma_X$ and $u \in \mathbb{C}^g$, the action gives the one-dimensional representation such that*

$$\widehat{\zeta}\sigma(u) = \rho_{\widehat{\zeta}}\sigma(u),$$

where $\rho_{\widehat{\zeta}}^m = 1$.

Proof Due to the assumption and Lemma 3.34. σ and $\widehat{\zeta}\sigma$ belong to the same class of the theta functions of the $(\widehat{L}, \widehat{J}_L)$-type. Further, Proposition 3.164 shows

$$\widehat{\zeta}\sigma(u) = \exp({}^t u A u + B u + c)\sigma(u).$$

However, since $[\exp({}^t u A u + B u + c)]^m = 1$ for every u, $\exp(cm) = 1$. ■

Here we comment on the assumption in Proposition 3.42. Since $\widehat{\zeta}$ acts on J_X and Γ_X effectively, we can define the quotient spaces by $\Gamma_X^{\widehat{\zeta}} := \Gamma_X/\widehat{\zeta}$, $\varpi_{\widehat{\zeta}} : \Gamma_X \to \Gamma_X^{\widehat{\zeta}}$. Thus a point $u \in \mathbb{C}^g$ is decomposed to the point $[u] := \varpi_{\widehat{\zeta}}(\kappa_J u) \in \Gamma_X^{\widehat{\zeta}}$, and the fiber positions, $\varpi_{\widehat{\zeta}}^{-1}[u]$ and $\kappa_J^{-1}([u])$, and similarly ℓ is decomposed to $[\ell]$ and $\varpi_{\widehat{\zeta}}^{-1}[\ell]$. These configurations determine the value of the bilinear form. Thus the assumption in Proposition 3.42 means that the value must be invariant for the action of $\widehat{\zeta}$, which may be natural if there exists the Galois action on X, though we don't argue it more in this book.

Further, suppose $\widehat{\zeta} w(\mathfrak{K}_\mathfrak{s}) = w(\mathfrak{K}_\mathfrak{s})$, and then we obtain $\widehat{\zeta}(\mathfrak{K}_\mathfrak{s}) \sim \mathfrak{K}_\mathfrak{s}$. For $u \in \widetilde{w}(\iota_X S^g X)$, $\widehat{\zeta}(u + \widetilde{w}(\iota_X(\mathfrak{K}_\mathfrak{s}))) = \widehat{\zeta}(u) + \widetilde{w}(\iota_X(\mathfrak{K}_\mathfrak{s}))$ modulo Γ_X. Due to Theorem 2.181, the divisor $\kappa_J^{-1} W_{X,\mathfrak{s}}^{g-1}$ of σ is invariant for the action of $\widehat{\zeta}$. Then σ and $\widehat{\zeta}\sigma$ have the same divisor $\kappa_K^{-1} W_{X,\mathfrak{s}}^{g-1}$. Hence the assumption in Proposition 3.42 requires the condition $\widehat{\zeta} w(\mathfrak{K}_\mathfrak{s}) = w(\mathfrak{K}_\mathfrak{s})$. Further, it is expected that both might always have the same normalized theta function ϑ_σ.

3.2.10 Riemann–Kempf Theorem on Sigma Function

We state here Riemann's singularity theorem (cf. [ArCGH1985, VI.1]) for the sigma function based on Theorem 3.21, which is the precise expression of the Riemann–Kempf theory started by Ônishi in [On2002, On2005] as mentioned in Sect. 3.1.7 [On2009, On2010, On2011, NakY2012, MatP2014, Nak2016].

In this subsection, we mention it based on [MatP2014] and [Nak2016] (see Appendix 3.10).

Noting (3.15), we define the subvarieties in the Jacobian J_X.

Definition 3.43 As we recall the subvariety $W^k_{X,\mathfrak{s}} := w_{\mathfrak{s}}(S^k X)$ in Definition 2.154, we define the subvarieties in $\Theta^k_{X,\mathfrak{s}} \subset J_X$ as modulo Γ_X.

$$\Theta^k_{X,\mathfrak{s}} := W^k_{X,\mathfrak{s}} \bigcup [-1] W^k_{X,\mathfrak{s}}, \quad \Theta^{k,1}_{X,\mathfrak{s}} := W^{k,1}_{X,\mathfrak{s}} \bigcup [-1] W^{k,1}_{X,\mathfrak{s}}. \tag{3.26}$$

Since $[-1]W^k_{X,\mathfrak{s}} \subset W^k_{X,\mathfrak{s}}$ for $k = g-1$ and $k = g$, we note that $\Theta^g_{X,\mathfrak{s}} = W^g_{X,\mathfrak{s}} = J_X$ and $\Theta^{g-1}_{X,\mathfrak{s}} = W^{g-1}_{X,\mathfrak{s}} = \Theta_{X,\mathfrak{s}}$ in Theorem 3.36.

When the W-curve X is symmetric or its numerical semigroup H_X is symmetric, $W^k_{X,\mathfrak{s}} = W^k_X$ and thus we define Θ^k_X and $\Theta^{k,1}_X$ by omitting the symbol $\mathfrak{s}$.

When X is a hyperelliptic curve, $[-1]W^k_X \subset W^k_X = W^k_{X,\mathfrak{s}}$ for every $k = 0, 1, \dots, g$ and thus $\Theta^k_X = W^k_X = W^k_{X,\mathfrak{s}}$.

We recall in n_k and $\mathrm{Index}(g, m)$ in Theorem 3.15, $r_k := r_{\Lambda^{[k]}_X}$ and $L^{[k]}_i$ in Proposition 3.17, and $\natural_k$ and $\natural^{(i)}_k$ in Definition 3.18. From Theorem 3.21, we also state the Riemann–Kempf theorem for the sigma function noting Remark 3.23 [MatP2014, Nak2016]:

Theorem 3.44 *We recall $n_k = r_{\Lambda^{[k]}}$, $N_k = |\Lambda^{[k]}|$, and $u \in \Theta^k_{X,\mathfrak{s}}$ in Proposition 3.17, Theorems 3.15 and 3.21.*

1. *Let $\mathscr{I}_{X,g} = \{\emptyset\}$. For each $k = 1, 2, \dots, g$, there exists a subfamily $\mathscr{I}_{X,k}$ of* $\mathrm{Index}(g, n_k)$*, whose element I_k is such that $|I_k| \ge N_k$, and as a function over $\kappa_J^{-1}(\Theta^k_{X,\mathfrak{s}} \setminus (\Theta^{k,1}_{X,\mathfrak{s}} \cup \Theta^{k-1}_{X,\mathfrak{s}}))$,*

$$\sigma_{J_k} = \begin{cases} \neq 0 & \text{for } J_k = I_k, \\ = 0 & \text{for } J_k \in \cup_{\ell=1}^{n_k-1} \mathrm{Index}(g, \ell), \ J_k \subsetneqq I_k. \end{cases} \tag{3.27}$$

   Moreover, $\{\natural_k, \natural^{(k)}_k, \natural^{(k-1)}_k, \dots, \natural^{(2)}_k, \natural^{(1)}_k\} \subset \mathscr{I}_{X,k}$.
2. *As a function over $\kappa_J^{-1}(\Theta^k_{X,\mathfrak{s}} \setminus (\Theta^{k,1}_{X,\mathfrak{s}} \cup \Theta^{k-1}_{X,\mathfrak{s}}))$,*

$$\partial^\ell_{u_g} \sigma = \begin{cases} \neq 0 & \text{for } \ell \ge N_k, \\ = 0 & \text{otherwise.} \end{cases}$$

Remark 3.45 The property $\sigma_{\natural_k} \neq 0$ over $\kappa_J^{-1}(\Theta^k_{X,\mathfrak{s}} \setminus (\Theta^{k,1}_{X,\mathfrak{s}} \cup \Theta^{k-1}_{X,\mathfrak{s}}))$ is the generalization of results in [MatP2014]. Further, we note that there exists $J_k \in \mathrm{Index}(g, k)$ such that $\# J_k = n_k$ but $\sigma_{J_k}(u) = 0$ for $u \in \kappa_J^{-1}(\Theta^k_{X,\mathfrak{s}} \setminus (\Theta^{k,1}_{X,\mathfrak{s}} \cup \Theta^{k-1}_{X,\mathfrak{s}}))$.

Furthermore, we consider $u \in \kappa_J^{-1}(\Theta^k_{X,\mathfrak{s}} \setminus (\Theta^{k,1}_{X,\mathfrak{s}} \cup \Theta^{k-1}_{X,\mathfrak{s}}))$ satisfying $u = u' + v$ for $u' \in \kappa_J^{-1}(\Theta^{k-1}_{X,\mathfrak{s}})$ and $\kappa_J(v) = w(P)$ for a certain $P \in X$ in the neighborhood of ∞. Then Theorem 3.44 means that $\sigma_{\natural_k}(u' + v)$ is expanded by v_g,

$$\sigma_{\natural_k}(u' + v) = (-1)^{1 - r_k + r_{k-1}} \sigma_{\natural_{k-1}}(u')(v_g^{N_{k-1} - N_k} + d_{>N_{k-1} - N_k}(v_g)), \tag{3.28}$$

since we have

$$\sigma_{\natural_k}(u) = (-1)^{g-k-r_k} S_{\Lambda_X^{(k)}}(T^{(k)})\Big|_{T^{(k)}_{\Lambda_i+g-i}=-u_i+\widetilde{w}(\iota_X \mathfrak{K}_{\mathfrak{s}})} (1 + d_{>0}(u)), \tag{3.29}$$

where $T^{(k)} = {}^t(T^{(k)}_{\Lambda_1+g-1}, \ldots, T^{(k)}_{\Lambda_g})$.

Remark 3.46 Since the σ function is either even or odd due to Theorem 3.40, we can replace the assumption for D_k in Theorem 3.15 with $D_k \in (S^k(X\backslash\infty) \setminus (S^k_1(X) \cap S^k(X\backslash\infty)) \cup [-1](S^k(X\backslash\infty) \setminus (S^k_1(X) \cap S^k(X\backslash\infty)))$, and we also extend via the operator $[-1]$ the defining set of u or, $u \in \kappa_J^{-1}(W^k_{X,\mathfrak{s}}) \subset \kappa_J^{-1}(\Theta^k_{X,\mathfrak{s}})$, with the appropriate (extended) excluded subsets.

3.2.10.1 Examples of Riemann–Kempf Indices

Table 3.5 Example: hyperelliptic curves

g	$\sigma_{\natural 1}$	$\sigma_{\natural 2}$	$\sigma_{\natural 3}$	$\sigma_{\natural 4}$	$\sigma_{\natural 5}$	$\sigma_{\natural 6}$	$\sigma_{\natural 7}$	$\sigma_{\natural 8}$	$\cdots$
1	σ	σ	σ	σ	σ	σ	σ	σ	$\cdots$
2	σ_2	σ	σ	σ	σ	σ	σ	σ	$\cdots$
3	σ_2	σ_3	σ	σ	σ	σ	σ	σ	$\cdots$
4	σ_{24}	σ_3	σ_4	σ	σ	σ	σ	σ	$\cdots$
5	σ_{24}	σ_{35}	σ_4	σ_5	σ	σ	σ	σ	$\cdots$
6	σ_{246}	σ_{35}	σ_{46}	σ_5	σ_6	σ	σ	σ	$\cdots$
7	σ_{246}	σ_{357}	σ_{46}	σ_{57}	σ_6	σ_7	σ	σ	$\cdots$
8	σ_{2468}	σ_{357}	σ_{468}	σ_{57}	σ_{68}	σ_7	σ_8	σ	$\cdots$

Using Theorem 3.33, we have the following theorem, which shows $\sigma_{J_{k,i}}$ is also a normalized theta function over $\kappa_J^{-1}(\Theta^k_{X,\mathfrak{s}} \setminus (\Theta^{k,1}_{X,\mathfrak{s}} \cup \Theta^{k-1}_{X,\mathfrak{s}}))$, as Theorem 3.44:

Theorem 3.47 *For* $u \in \kappa_J^{-1}(\Theta^k_{X,\mathfrak{s}} \setminus (\Theta^{k,1}_{X,\mathfrak{s}} \cup \Theta^{k-1}_{X,\mathfrak{s}}))$, $\ell\ (= 2\omega'\ell' + 2\omega''\ell'') \in \Gamma_X$, *and* $J_k \in \mathscr{I}_{X,k}$, *we have*

$$\sigma_{J_k}(u + \ell) = \sigma_{J_k}(u) \exp(L(u + \frac{1}{2}\ell, \ell))\chi(\ell). \tag{3.30}$$

Proof After we apply the differential operators $\displaystyle\prod_{i \in I_k} \frac{\partial}{\partial u_i}$ on both sides of the equality in Theorem 3.33, we restrict the domain to $\kappa_J^{-1}(\Theta^k_{X,\mathfrak{s}} \setminus (\Theta^{k,1}_{X,\mathfrak{s}} \cup \Theta^{k-1}_{X,\mathfrak{s}}))$. Then by Theorem 3.44, the terms containing the lower-order derivatives of σ vanish, and thus we have the above equality. ■

From the proof in Theorem 3.40, we have the theorem:

Theorem 3.48 *For the case that* $[-1]W^k_{X,\mathfrak{s}} \subset W^k_{X,\mathfrak{s}}$, *we have* $\sigma_{J_k}(-u) = \pm\sigma_{J_k}(u)$ *for* $u \in W^k_{X,\mathfrak{s}}$.

From (3.29), we have the following results, for example.

Lemma 3.49 *Let* $u^{[g-1]} \in \kappa_J^{-1}(W_{X,\mathfrak{s}}^{g-1})$ *be a non-singular point. For* $v := \widetilde{w}(P)$, $u := u^{[g-1]} + v \in \kappa_J^{-1}(W_{X,\mathfrak{s}}^{g})$, $\dfrac{\sigma(u)^2}{\sigma_g(u)^2}$ *is expanded at* $v = 0$ *by*

$$\frac{\sigma(u)}{\sigma_g(u)} = (-1)^{1+r_1}(v_g)(1 + d_{>0}((v_g)).$$

Proof Corollary 3.226, Proposition 2.1 7 ($\Lambda_g = 1$) and $v_g = -t$ for the arithmetic parameter show it. ■

For the hyperelliptic curves, $\natural_k$ is listed in Table 3.5 and (3.28) is explicitly obtained as follows [On2005].[8]

Corollary 3.50 *For the hyperelliptic curves X given by (2.65), let* $v \in \kappa_J^{-1}\Theta_X^1$.

1.

$$\sigma_{\natural_1}(v) = (-1)^{g(g-1)/2} v_g^g + d_{>g+1}(v_g),$$

2. *for* $u \in \kappa_J^{-1}\Theta_X^k$, $(k < g)$,

$$\sigma_{\natural_{k+1}}(u + v) = (-1)^{-(g-k-1)}\sigma_{\natural_k}(u) v_g^{g-k} + d_{>g-k}(v_g).$$

Proof We note that for the hyperelliptic curves, $(-1)^{r_k} = (-1)^{(g-k)(g-k+1)/2}$ and thus $(-1)^{g-k+r_k} = (-1)^{(g-k)(g-k-1)/2}$. Further, $(g-k)(g-k-1) - (g-k-1)(g-k-2) = 2(g-k-1)$. ■

Corollary 3.51 *For the hyperelliptic curves X given by (2.65), let* $u \in \kappa_J^{-1}\Theta_X^1$ *and* $v \in \kappa_J^{-1}\Theta_X^2$. *We have*

$$\sigma_{\natural_1}(-u) = (-1)^g \sigma_{\natural_1}(u), \qquad \sigma_{\natural_2}(-v) = (-1)\sigma_{\natural_2}(v).$$

3.3 Jacobi Inversion Formula

This section shows the Jacobi inversion formula associated with W-curves X.

The Jacobi inversion formula for the general hyperelliptic curves was solved by Weierstrass as in Chap. 1 [Wei1854], and was revised by Baker in [Ba1897]. Buchstaber, Enolskii, and Leykin extend the sigma functions to plane W-curves, so-called (n, s)-curves, based on Baker's construction which we call EEL construction due to work by Eilbeck, Enolskii, and Leykin [EilEL2000] [BucEL2020, and its references] to find the Jacobi inversion formula. For the plane W-curves with the cyclic

[8] The following slightly differs from the formulae in [On2005] due to the different normalization in its Proposition 5.1.

symmetry using the sigma function of plane W-curves, the direct relations between the affine rings and the sigma functions were obtained as Jacobi inversion formulae even for subvarieties in [MatP2008, MatP2014]. The author of this book with Previato and Komeda further generalized Jacobi inversion formulae for particular space curves [MatK2013, KomMP2013, KomMP2019]. Based on these results, in [KomMP2022a, KomMP2022b], by using the structure of $R_X^{\mathfrak{c}}$, we defined the fundamental differential of the second kind Ω following the EEL-construction to obtain the Jacobi inversion formula for every W-curve as mentioned in the previous chapter. We show the Jacobi inversion formula following [KomMP2022b] for every W-curve as follows.

We also note that very recently, Bernatska and Leykin investigated the problem for the plane W-curves [BerL2022].

Here we generalize the elliptic $\wp$ and ζ functions to ones of a W-curve X:

Definition 3.52 For the W-curve X, we define the $\wp$ functions and the ζ functions for $u \in \mathbb{C}^g$ by

$$\zeta_i(u) := \frac{\sigma_i(u)}{\sigma(u)} = \frac{\partial}{\partial u_i} \log \sigma(u),$$

$$\wp_{ij}(u) := \frac{\sigma_i(u)\sigma_j(u) - \sigma(u)\sigma_{ij}(u)}{\sigma(u)^2} = -\frac{\partial^2}{\partial u_i \partial u_j} \log \sigma(u),$$

where

$$\sigma_{i_1,i_2,\ldots,i_n}(u) := \prod_{j=1}^{n} \frac{\partial}{\partial u_{i_j}} \sigma(u).$$

It is worth noting that these $\wp$ and ζ functions are natural generalizations of the elliptic ones (1.29) in the Weierstrass elliptic function theory.

Lemma 3.53 *The $\wp_{ij}$ function is a function of the Jacobian J_X; for ℓ $(= 2\omega'\ell' + 2\omega''\ell'') \in \Gamma_X)$,*

$$\wp_{ij}(u + \ell) = \wp_{ij}(u),$$

whereas

$$\zeta_i(u + \ell) = \zeta_i(u) + 2\sum_{j=1}^{g} (\eta'_{ij}\ell'_j + \eta''_{ij}\ell''_j).$$

Proof We apply "$\dfrac{\partial^2}{\partial u_i \partial u_j}$ log" to both sides of (3.25). Let $L_i := \dfrac{\partial}{\partial u_i} L(u + \dfrac{1}{2}\ell, \ell)$. The numerator in the right-hand side is given by

$$(\sigma_i + \sigma L_i)(\sigma_j + \sigma L_j) - \sigma(\sigma_{ij} + \sigma_i L_j + \sigma_j L_i + \sigma L_i L_j),$$

which is equal to $\sigma_i\sigma_j - \sigma\sigma_{ij}$. It means that $\wp_{ij}(u + \ell) = \wp_{ij}(u)$. Similarly, we obtain the ζ case. ■

Hence we sometimes regard the domain of $\wp_{ij}$ as the Jacobian J_X.

3.3.1 Jacobi Inversion Formulae on J_X

In terms of the sigma functions, this subsection gives the Jacobi inversion formulae on J_X.

We use the shifted Abel–Jacobi maps $w_{\mathfrak{s}}$ in Sect. 2.4.4 and F_Ω in (2.96) and then we obtain the Jacobi inversion formula on J_X:

Theorem 3.54 *For $(P, P_1, \ldots, P_g) \in X \times S^g(X) \setminus S_1^g(X)$ and $u := w_{\mathfrak{s}}(P_1, \ldots, P_g)$, the equality*

$$\sum_{i,j=1}^{g} \wp_{i,j}\,(w(P) - u)\,\widehat{\phi}_{i-1}(P)\widehat{\phi}_{j-1}(P_a) = \frac{F_\Omega(P, P_a)}{(x - x_a)^2}$$

holds for every $a = 1, 2, \ldots, g$.

Proof We use the relation in Lemma 2.161,

$$\sum_{i,j=1}^{g} \widehat{\phi}_{i-1}(P_1)\widehat{\phi}_{j-1}(P_2)\frac{\partial^2}{\partial w_i(P_1)\partial w_j(P_2)} = h_X(x_1, y_{1,\bullet})h_X(x_2, y_{2,\bullet})\frac{\partial^2}{\partial x_1 \partial x_2}.$$

We note that the above $\wp$ functions are defined over the Jacobian J_X, namely $\wp_{ij}(u) = \wp_{ij}(u + \ell)$ for $\ell \in \Gamma_X$, because of the functional relation satisfied by σ. Using the property of the vector fields in Lemma 2.161 and taking the logarithm of both sides in the Jacobi–Riemann fundamental relation in Theorem 3.37 and differentiating along $P_1 = P$ and $P_2 = P_a$, we obtain the claim. ■

We remark that the identity in Theorem 3.54,

$$-\sum_{i,j=1}^{g} \frac{\partial^2}{\partial u_i \partial u_j} \log \sigma\Big(w(P) - u\Big)\nu_i^{\mathrm{I}}(P)\nu_j^{\mathrm{I}}(P_a) = \Omega(P, P_a), \tag{3.31}$$

corresponds to (29) of Corollary 2.6 in [Fay1973]. In our case, $\Omega(P, P_a)$ has an explicit expression in terms of the affine coordinates on the curve.

Theorem 3.55 *For $(P, P_1, \cdots, P_g) \in X \times S^g(X) \setminus S_1^g(X)$,*

1. $\widehat{\mu}_g(P; P_1, \ldots, P_g) = \widehat{\phi}_g(P) - \sum_{j=1}^{g} \wp_{gj}(w_{\mathfrak{s}}(P_1, \ldots, P_g))\widehat{\phi}_{j-1}(P)$, *and*
2. $\wp_{g,k+1}(w_{\mathfrak{s}}(P_1, \ldots, P_g)) = (-1)^{g-k-1}\widehat{\mu}_{g,k}(P_1, \ldots, P_g)$, $(k = 0, \ldots, g-1)$, *i.e.,*

$$\frac{\sigma_i(u)\sigma_g(u) - \sigma_{gi}(u)\sigma(u)}{\sigma^2(u)} = (-1)^{g-i+1}\widehat{\mu}_{g,i-1}(P_1, \ldots, P_g), \quad \textit{for } 0 < i \le g.$$

Proof In Theorem 3.54, we let $P \to \infty$ after dividing both sides by $\widehat{\phi}_{g-1}(P)$, and then noting Lemma 2.142, we obtain

$$\sum_{j=1}^{g} \wp_{g,j}(u)\widehat{\phi}_{j-1}(P_i) = \widehat{\phi}_g(P_i)$$

for every i. Let Q_i be such that $u = w_{\mathfrak{s}}(P_1, \ldots, P_g) = -w_{\mathfrak{s}}(Q_1, \ldots, Q_g)$. Note that $\wp_{ij}(u)$ is an even function, i.e., $\wp_{g,j}(-u) = \wp_{g,j}(u)$. Thus both $\widehat{\mu}_g(P; P_1, \ldots, P_g)$ and

$$\widehat{\phi}_g(P) - \sum_{j=1}^{g} \wp_{g,j}(u)\widehat{\phi}_{j-1}(P) \tag{3.32}$$

are elements of R_X vanishing only at P_i and Q_i, viewed as functions of P. As in Propositions 2.172 and 2.173, we have the identity due to the uniqueness of $\widehat{\mu}_g(P; P_1, \ldots, P_g)$. ■

Proposition 3.56 *For the hyperelliptic curves (2.65), Theorem 3.55 and the formula in (3.31) are reduced to the following.*

1. $F_g(P; P_1, \ldots, P_g) = x^g - \sum_{j=1}^{g} \wp_{gj}(w(P_1, \ldots, P_g))x^{j-1}$.
2. $\wp_{g,k}(w(P_1, \ldots, P_g))$ *is the elementary symmetric function of* $x_1, x_2, \ldots, x_g$ *for every* $(k = 1, \ldots, g)$*, i.e.,*

$$\wp_{g,g} = (x_1 + x_2 + \cdots + x_g), \quad \wp_{g,g-1} = -(x_1x_2 + x_1x_3 + \cdots + x_{g-1}x_g),$$

$$\cdots, \quad \wp_{g,1} = (-1)^{g-1}(x_1x_2 \cdots x_{g-1}x_g).$$

3. $2y_k = -\left.\dfrac{\partial F_g(x)}{\partial u_g}\right|_{x=x_k} = \displaystyle\sum_{j=1}^{g} \wp_{ggj}(w(P_1, \ldots, P_g))x_k^{j-1}$.
4. $$\sum_{i=1}^{g}\sum_{j=1}^{g} \wp_{ij}(w(P') - u){x'}^{i-1}{x_a}^{j-1} = \frac{2y_a y' + f_\Omega(x', x_a)}{(x' - x_a)^2}, \tag{3.33}$$

   where $P' = (x', y')$ *and* $u = w(P_1, \ldots, P_g)$ *for* $f_\Omega(x', x_a)$ *in (2.101) and (2.102).*

Proof *1*, *2* and *4* are obvious. *3* was obtained by Baker [Ba1903, p. 147] and Buchstaber, Enolskii and Leykin [BucEL1997b]. By applying $\partial/\partial u_g$ to the relation in *1* and letting $x = x_k$, we obtain it. ■

Lemma 3.57 *For a polynomial* $g(X) = g_nX^n + \cdots + g_0$*, we introduce the* D_j *operator* $D_j(g) = \sum_{i=j}^{n} g_iX^{i-j}$*, and then we have*

$$\wp_{ggk}(u) = \sum_{i=1}^{g} 2y_i \left(\left.\frac{D_k(F(x))}{F'(x)}\right|_{x=x_i}\right).$$

Proof By considering the inversion of Proposition 3.56 3, we obtain the relation [BucEL1997b][BucEL2012, 2.55]. ■

Remark 3.58 The Jacobi inversion formulae for $\wp_{ij}$, $(i, j = 1, \ldots, g)$, for hyperelliptic curves are given by Proposition 3.85 and Corollary 3.88.

3.3.2 Jacobi Inversion Formulae on $\Theta^k_{X,\mathfrak{s}}$

In terms of differentials of the sigma functions, this subsection shows the Jacobi inversion formulae on the strata in J_X.

As we recall $\Theta^k_{X,\mathfrak{s}}$ in (3.26), we show the Jacobi inversion formulae for $\Theta^k_{X,\mathfrak{s}} \subset J_X$, which was partially given in [Jo1992] and [Fay1973].

Theorem 3.59 *The following relations hold.*

1. $\Theta^{g-1}_{X,\mathfrak{s}}$ *case: for* $(P_1, \ldots, P_{g-1}) \in S^{g-1}(X) \setminus S^{g-1}_1(X)$ *and* $u = \pm w_{\mathfrak{s}}(P_1, \ldots, P_{g-1}) \in \Theta^{g-1}_{X,\mathfrak{s}}$,

$$\frac{\sigma_i(u)}{\sigma_g(u)} = \begin{cases} (-1)^{g-i}\widehat{\mu}_{g-1,i-1}(P_1, \ldots, P_{g-1}) \text{ for } 0 < i \le g, \\ 1 \qquad\qquad\qquad\qquad\qquad\quad \text{for } i = g. \end{cases}$$

2. $\Theta^k_{X,\mathfrak{s}}$ *case* $(k < g - 1)$*: for* $(P_1, \ldots, P_k) \in S^k(X) \setminus S^k_1(X)$ *and* $u = \pm w_{\mathfrak{s}}(P_1, \ldots, P_k) \in \Theta^k_{X,\mathfrak{s}}$,

$$\frac{\sigma_{\natural^{(i)}_{X,k}}(u)}{\sigma_{\natural_{X,k}}(u)} = \begin{cases} (-1)^{k-i+1}\widehat{\mu}_{k,i-1}(P_1, \ldots, P_k) \text{ for } 0 < i \le k, \\ 1 \qquad\qquad\qquad\qquad\qquad\qquad \text{for } i = k+1, \\ 0 \qquad\qquad\qquad\qquad\qquad\qquad \text{for } k+1 < i \le g. \end{cases}$$

Proof Let $u := w_{\mathfrak{s}}(P_1, \cdots, P_g)$ for $(P_1, \cdots, P_g) \in S^g(X)$. By multiplying both sides of the formula in Theorem 3.55 2 by $\sigma(u)^2/\sigma_g(u)^2$ and taking the limit $P_g \to \infty$, we have

$$\begin{aligned} \lim_{P_g \to \infty} \left(\frac{\sigma(u)}{\sigma_g(u)}\right)^2 \left(\frac{\sigma_g(u)\sigma_i(u) - \sigma(u)\sigma_{gi}(u)}{\sigma(u)^2}\right) &= \lim_{P_g \to \infty} \left(\frac{\sigma_i(u)}{\sigma_g(u)} - \frac{\sigma(u)\sigma_{gi}(u)}{\sigma_g(u)^2}\right) \\ &= \lim_{P_g \to \infty} \left(\frac{\sigma_i(u)}{\sigma_g(u)}\right) \end{aligned} \tag{3.34}$$

noting Lemma 3.49. The right-hand side is a function on $W^{g-1}_{X,\mathfrak{s}}$. The fact $\widehat{N}(g) - \widehat{N}(g-1) = 2$ due to Lemma 2.109 8 allows us to consider the limit of the left-hand side:

$$(-1)^{g-i} \lim_{P_g \to \infty} \left(\frac{\widehat{\phi}_{g-1}(P_g)}{\widehat{\phi}_g(P_g)}\right) \widehat{\mu}_{g,i-1}(P_1, \ldots, P_g).$$

Proposition 2.182 shows that this limit gives 1. Similarly due to Proposition 2.1, Lemma 3.61 and Theorem 3.44 show $k < g$ case. ■

Corollary 3.60 *For* $k < g$, $(\gamma_1, \ldots, \gamma_k) \in S^k(\widetilde{X}) \setminus S_1^k(\widetilde{X})$, $u = \pm\widetilde{w}_{\mathfrak{s}}(\gamma_1, \ldots, \gamma_k) \in \kappa_J^{-1}(\Theta_{X,\mathfrak{s}}^k)$ *and* $\ell \in \Gamma_X$,

$$\frac{\sigma_{\natural_{X,k}^{(i)}}(u+\ell)}{\sigma_{\natural_{X,k}}(u+\ell)} = \frac{\sigma_{\natural_{X,k}^{(i)}}(u)}{\sigma_{\natural_{X,k}}(u)}.$$

Proof It is obvious from Theorem 3.47. ■

Hence we also have considered the ratio of the sigma as a function on J_X.

The following lemma will provide an induction step to prove Theorem 3.59.

Lemma 3.61 *For* $k < g - 1$, *a non-singular point* $u^{[k-1]} \in W_{X,\mathfrak{s}}^{k-1}$, *and* $u := u^{[k-1]} + w(P_k)$ *whose preimage is given by* $(P_1, \cdots, P_k) \in S^{k-1}(X) \times X$, *we have the relations in the meaning of L'Hôpital's rule,*

$$\frac{\sigma_i(u)}{\sigma_{k+1}(u)} = \begin{cases} (-1)^{k-i+1}\widehat{\mu}_{k,i-1} & \text{for } 0 < i \le k, \\ 1 & \text{for } i = k+1, \\ 0 & \text{for } k+1 < i \le g. \end{cases}$$

Proof Since the first equation holds for $k = g$ due to Theorem 3.59 1, we assume the first equation and then

$$\frac{\sigma_{k+1}(u)}{\sigma_k(u)} = -\frac{1}{\widehat{\mu}_{k,k-1}} = -\frac{\widehat{\psi}_k^{(k)}(P_1, \cdots, P_k)}{\widehat{\psi}_k^{(k-1)}(P_1, \cdots, P_k)}.$$

Let us consider the behavior around $P_k = \infty$. By an argument similar to the one used in Proposition 2.182, the right-hand side behaves like

$$\begin{aligned}\frac{1}{\widehat{\mu}_{k,k-1}} &= \frac{-\widehat{\phi}_{k-1}(P_k)\psi_{k-1}^{(k-1)}(P_1, \cdots, P_{k-1}) + d_{\ge 1}((w(P_k)_g)^{-\widehat{N}(k)+1})}{\widehat{\phi}_k(P_k)\psi_{k-1}^{(k-1)}(P_1, \cdots, P_{k-1}) + d_{\ge 1}((w(P_k)_g)^{-\widehat{N}(k)+1})} \\ &= -\frac{\widehat{\phi}_{k-1}(P_k)}{\widehat{\phi}_k(P_k)} + d_{\ge 1}((w(P_k)_g)^{\widehat{N}(k)-\widehat{N}(k-1)+1}).\end{aligned}$$

On the other hand, from Corollary 3.50, we have

$$\frac{\sigma_{k+1}(u)}{\sigma_k(u)} = (w(P_k)_g)^{\widehat{N}(k)-\widehat{N}(k-1)} + d_{\ge 1}((w(P_k)_g)^{\widehat{N}(k)-\widehat{N}(k-1)+1})$$

and

$$\frac{\sigma_{k+1}(u)}{\sigma_k(u)} = \frac{\widehat{\phi}_{k-1}(P_k)}{\widehat{\phi}_k(P_k)} + d_{\ge 1}((w(P_k)_g)^{\widehat{N}(k)-\widehat{N}(k-1)+1})$$

due to Lemma 2.109 4. Proposition 2.182 shows that the limit $P_k \to \infty$ lead to the $k-1$ case. ■

Thus we have the following corollary from Theorem 3.59:

Corollary 3.62 *($k = 1$ case) For $P = (x, y_\bullet) \in X$, and $u \in \widetilde{w}_{\mathfrak{s}}(P) \in \mathbb{C}^g$,*

$$\frac{\sigma_{\natural^{(1)}_{X,1}}(u)}{\sigma_{\natural_{X,1}}(u)} = \widehat{\phi}_1(P),$$

especially for a symmetric W-curve case,

$$\frac{\sigma_{\natural^{(1)}_{X,1}}(u)}{\sigma_{\natural_{X,1}}(u)} = x.$$

The final formula means the solution of the Jacobi inversion formula.

Remark 3.63 Bolza studied the reduction of the hyperelliptic integrals to the elliptic integrals in [Bol1887]. Bolza's result was re-evaluated by Belokolos and Enolskii based on the recent development of the sigma function theory [BelE2001, BelE2002]. They investigated the particular hyperelliptic Jacobian J_X of genus two, which is reduced to two elliptic Jacobians based on Bolza's study and related ones. Recently Ayano and Buchstaber showed the explicit relations among the $\wp$ functions of the hyperelliptic Jacobian and those of elliptic Jacobians for the particular curves [AyB2022] based on the study by Belokolo and Enolskii [BelE2001, BelE2002] whose origin is of Bolza [Bol1887]; it means that even for the particular curves, they surely showed the relationships among the $\wp$ functions in the Jacobian J_X and its subvarieties.

3.3.3 *Meromorphic Functions on J_X*

In this section, we show the properties of the meromorphic functions on J_X.

Theorem 3.64 *Every meromorphic function on J_X is described by the ratio of the differentials of the sigma function.*

Proof It is reduced to Theorems 3.35 and 3.161. ∎

As mentioned in Remark 2.166, we have equivalence between the differential rings on $\mathscr{D}_X$ and $\mathscr{D}_J$. We consider the left $\mathscr{D}_X$-modules $\mathscr{Q}(S^g R_X)$ and the left $\mathscr{D}_J$-module $\mathscr{M}_J$.

Lemma 3.65 *$\mathscr{M}_J$ is a $\mathscr{D}_J$-module, $\mathrm{Hom}_{\mathscr{D}_J}(\mathscr{M}_J, \mathscr{M}_J) = \mathbb{C}$. By the construction in Lemmas 2.167 and 2.168, we have $\varphi_M : \mathscr{D}_J \to \mathscr{D}_J$ such that $\varphi_M(P) = P$ for $M \in \mathrm{GL}(g, \mathbb{C})$.*

From Remark 2.169, we consider the category of the left $\mathscr{D}_X$-module $\mathrm{Mod}(\mathscr{D}_X)$, we also define the category of the left $\mathscr{D}_J$-module $\mathrm{Mod}(\mathscr{D}_J)$ similarly.

Theorem 3.66 *Both categories* $\mathrm{Mod}(\mathscr{D}_X)$ *and* $\mathrm{Mod}(\mathscr{D}_J)$ *are equivalent.*

Proof We can define the functors $\mu_{JX} : \mathrm{Mod}(\mathscr{D}_X) \to \mathrm{Mod}(\mathscr{D}_J)$ and $\mu_{XJ} : \mathrm{Mod}(\mathscr{D}_J) \to \mathrm{Mod}(\mathscr{D}_X)$. The Abel–Jacobi theorem gives the equivalence between $\mathscr{Q}(S^g R_X)$ and $\mathscr{M}_J$. In fact, Proposition 2.173 means that there is a μ_g function which provides the element in $\mathscr{Q}(S^g R_X)$ which has the divisors in $S^g X$. The Jacobi inversion formula shows that the element $\mathscr{Q}(S^g R_X)$ corresponds to $\wp_{gi}$ in $\mathscr{M}_J$. The morphisms φ_M in both categories naturally have correspondence for every $M \in \mathrm{GL}(g, \mathbb{C})$, so that $\mu_{JX} \circ \varphi_M = \varphi_M$. The lemmas in Sect. 2.5.3 obviously show $\mu_{JX} \circ \mu_{XJ} = id_{\mathrm{Mod}(\mathscr{D}_J)}$ and $\mu_{XJ} \circ \mu_{JX} = id_{\mathrm{Mod}(\mathscr{D}_X)}$, where $id_{\mathrm{Mod}(\mathscr{D}_J)}$ and $id_{\mathrm{Mod}(\mathscr{D}_X)}$ are identity morphisms in $\mathrm{Mod}(\mathscr{D}_J)$ and $\mathrm{Mod}(\mathscr{D}_X)$ respectively. ■

3.3.4 Relation Between ζ Function and ν^{II}

Here we provide the relation between the ζ function and ν^{II}. Following the study by Buchstaber, Enolskii and Leykin [BucEL1997b, pp. 33–35] (and its revised version) and [BucEL2012, Theorem 2.5], we show the relation between the hyperelliptic ζ functions, ζ_i's, and the differentials of the second kind ν^{II} of the hyperelliptic curves defined in (2.65), which was originally mentioned by Baker [Ba1897, p. 323]. However, Baker's results are not correct for $i \neq g$ or $g-1$. The following proposition was given in an improved version of [BucEL1997b], which cited in [Mat2002b] and published in [BucEL2012]. We briefly review it here [BucEL2012, Sect. 2.7].

Proposition 3.67 [BucEL1997b, pp. 33–35][BucEL2012, Sect. 2.7] *For* $P_i = (x_i, y_i)$*, the hyperelliptic* $\zeta_j(u)$ *function is expressed by*

$$-\zeta_i(\widetilde{w}_Q(\iota_X(P_1, \cdots, P_g))) = \sum_{k=1}^{g} \int_Q^{x_i} \nu_i^{\mathrm{II}} - \frac{1}{2} \det A_{g-i},$$

where $\widetilde{w}_Q(P) = \widetilde{w}(P) - \widetilde{w}(Q)$, $\det A_{g-i} = -\dfrac{1}{2} \displaystyle\sum_{j,k=1}^{g} x_j^k \frac{\partial x_j}{\partial u_{i+1}}$,

$$A_{n+1} := \begin{pmatrix} e_1 & -1 & & & & \\ 2e_2 & -e_1 & 1 & & 0 & \\ \vdots & \vdots & \vdots & \ddots & & \\ (n-2)e_{n-2} & -e_{n-3} & e_{n-4} & \cdots & \pm 1 & \\ (n-1)e_{n-1} & -e_{n-2} & e_{n-3} & \cdots & \pm e_1 & \mp 1 \\ ne_n & -e_{n-1} & e_{n-2} & \cdots & \pm e_2 & \mp e_1 & \pm 1 \\ (n+1)d_{n+1} & -d_n & d_{n-1} & \cdots & \pm d_3 & \mp d_2 & \pm d_1 \end{pmatrix},$$

and $e_i := \wp_{g,g+1-i}$ *and* $d_i := \wp_{g,g,g+1-i} := \partial\wp_{g,g+1-i}/\partial u_g$. *Then we have*

$$|A_g(u)| = 0, \quad |A_{g-1}(u)| = \wp_{ggg}(u), \quad |A_{g-2}(u)| = \wp_{gg}(u)\wp_{ggg}(u), \quad \cdots .$$

$$- \zeta_j(\widetilde{w}_Q(\iota_X(P, P_1, \cdots, P_g)))$$
$$= \int_Q^P \nu_j^{\mathrm{II}} + \sum_{i=0}^{g} \int_Q^{P_j} \nu_j^{\mathrm{II}} - \frac{1}{2}\sum_{i=0}^{g} y_i \left(\left.\frac{D_j(R'(z)) - jD_{j+1}(R(z))}{R'(z)}\right|_{z=x_i} \right),$$

where $R(z) = (z - x_0)F(z)$ *by using* F *in (2.127) for* $P = (x_0, y_0)$ *and* $P_j = (x_j, y_j)$.

There Buchstaber, Enolskii, and Leykin started from a fundamental relation among the hyperelliptic sigma functions and incomplete integrals of the first, second, and third kinds.

Proof We recall the D_j operator in Lemma 3.57. First, we start the Jacobi–Riemann fundamental formula of the σ in Theorem 3.37

$$\log\left(\frac{\sigma\left(\widetilde{w}(P) + u\right)\sigma\left(\widetilde{w}(Q) + u'\right)}{\sigma\left(\widetilde{w}(P) + u'\right)\sigma\left(\widetilde{w}(Q) + u\right)}\right) = \sum_{j=1}^{g} \Pi_{P_j,Q_J}^{P,Q} = \sum_{j=1}^{g} \Pi_{P,Q}^{P_j,Q_J},$$

where $u = -\widetilde{w}(P_1, P_2, \ldots, P_g)$ and $u' = -\widetilde{w}(Q_1, Q_2, \ldots, Q_g)$.

Noting (2.98). we have

$$\Pi_{Q_1,Q_2}^{P_1,P_2} = \int_{P_2}^{P_1} \left(\frac{y_P - y_{Q_1}}{x_P - x_{Q_1}} \frac{dx_P}{2y_P} - \frac{y_P - y_{Q_2}}{x_P - x_{Q_2}} \frac{dx_P}{2y_P} \right) - \sum_{i=1}^{g} \int_{P_2}^{P_1} \nu_i^{\mathrm{I}}(P) \int_{Q_2}^{Q_1} \nu_i^{\mathrm{II}}(P).$$

After differentiating it in u_j, we let all Q_j be Q and then we obtain

$$\zeta_j(\widetilde{w}(P) + u) - \zeta_j(\widetilde{w}(Q) + u) + \int_Q^P \nu_j^{\mathrm{II}} - \frac{\partial}{\partial u_j} \sum_{i=1}^{g} \int_Q^{P_i} \left(\frac{y - y_P}{x - x_P} \frac{dx}{2y} - \frac{y - y_Q}{x - x_Q} \frac{dx}{2y} \right) = 0. \tag{3.35}$$

Introducing $R(z) = (z - x_0)F(z)$ using F in (2.127) for $P = (x_0, y_0)$ and $P_j = (x_j, y_j)$, we have the relation

$$\sum_{k=1}^{g} \frac{1}{2y_k} \frac{\partial x_k}{\partial u_j} \frac{y_k - y_0}{x_k - x_0} = \sum_{k=1}^{g} \sum_{i=1}^{g} \left(\left.\frac{D_k(F(x))}{F'(x)}\right|_{x=x_i} \right) \frac{y_k - y_0}{x_k - x_0}$$
$$= \sum_{i=0}^{g} y_i \left(\left.\frac{D_j(R'(z)) - jD_{j+1}(R(z))}{R'(z)}\right|_{z=x_i} \right)$$
$$- \sum_{i=1}^{g} y_i \left(\left.\frac{D_j(F'(z)) - jD_{j+1}(F(z))}{F'(z)}\right|_{z=x_i} \right).$$

Then (3.35) becomes

$$\zeta_j(\widetilde{w}(P)+u)+\int_Q^P \nu_j^{\mathrm{II}}+\sum_{i=0}^g\int_Q^{P_j}\nu_j^{\mathrm{II}}-\frac{1}{2}\sum_{i=0}^g y_i\left(\frac{D_j(R'(z))-jD_{j+1}(R(z))}{R'(z)}\Big|_{z=x_i}\right)$$
$$=\zeta_j(\widetilde{w}(Q)+u)+\sum_{i=1}^g\int_Q^{P_j}\nu_j^{\mathrm{II}}-\frac{1}{2}\sum_{k=1}^g\frac{1}{2y_k}\frac{\partial x_k}{\partial u_j}\frac{1}{2y_i}\frac{y_i-y_Q}{x_i-x_Q}$$
$$-\frac{1}{2}\sum_{i=1}^g y_i\left(\frac{D_j(F'(z))-jD_{j+1}(F(z))}{F'(z)}\Big|_{z=x_i}\right).$$

The left-hand side is symmetrical in $x_0, x_1, \cdots x_g$, whereas the right-hand side is symmetrical in $x_1, x_2 \cdots x_g$ but does not depend on x_0. Hence the right-hand side neither depends on x_i. Though both sides depend on Q. For $Q \to \infty$, $\int_Q^P \nu_j^{\mathrm{II}}$ must be canceled with the third term in the right-hand side, and both side must be zero. Lemma 3.57 shows that the last term gives

$$\sum_{i=1}^g y_i\left(\frac{D_j(F'(z))-jD_{j+1}(F(z))}{R'(z)}\Big|_{z=x_i}\right)=\det A_{g-i}.$$

■

Corollary 3.68 [BucEL1997b, p. 37][BucEL2012, Sect. 3.1] *For* $P_i=(x_i, y_i)$, $-d\zeta_i(\widetilde{w}(\iota_X(P_1,\cdots,P_g)))$ *is expressed by*

$$\sum_{k=1}^g \wp_{ik}(\widetilde{w}(\iota_X(P_1,\cdots,P_g)))du_k=\sum_{k=1}^g \nu_i^{\mathrm{II}}-\frac{1}{2}\sum_{k=1}^g \wp_{gg,i+1,k}du_k.$$

3.4 Addition Theorems for Sigma Function

In this section, we generalize the addition theorems of the sigma functions of genus one, $E : y^2 = x^3 + \lambda_2 x^2 + \lambda_1 x + \lambda_0$, i.e., for $u = w((x_1, y_1))$, $v = w((x_2, y_2))$

$$-\frac{\sigma(u+v)\sigma(u-v)}{\sigma(u)^2\sigma(v)^2}=\wp(u)-\wp(v)=x_1-x_2, \tag{3.36}$$

and for $u^{(i)} \in J_E$,

$$\begin{aligned}
&\frac{\sigma(u^{(1)}+u^{(2)}+\cdots+u^{(n)})\prod_{i<j}\sigma(u^{(i)}-u^{(j)})}{\sigma(u^{(1)})^n\sigma(u^{(2)})^n\cdots\sigma(u^{(n)})^n} \\
&= \frac{(-1)^{(n-1)(n-2)/2}}{1!2!\cdots(n-1)!}\begin{vmatrix} 1 & \wp(u^{(1)}) & \wp'(u^{(1)}) & \wp''(u^{(1)}) & \cdots & \wp^{(n-2)}(u^{(1)}) \\ 1 & \wp(u^{(2)}) & \wp'(u^{(2)}) & \wp''(u^{(2)}) & \cdots & \wp^{(n-2)}(u^{(2)}) \\ \vdots & \vdots & \vdots & \vdots & \ddots & \vdots \\ 1 & \wp(u^{(n)}) & \wp'(u^{(n)}) & \wp''(u^{(n)}) & \cdots & \wp^{(n-2)}(u^{(n)}) \end{vmatrix} \\
&= (-1)^n\psi_n((x,y)(u^{(1)}),\cdots,(x,y)(u^{(n)}))
\end{aligned} \tag{3.37}$$

as mentioned in (1.34) and (1.35).

By applying the operators $\frac{\partial}{\partial u}\log$ and $\frac{\partial}{\partial v}\log$ on (3.36), we obtain the following relations (see (2.102)):

$$\zeta(u+v)+\zeta(u-v)-2\zeta(u) = -\frac{2y_1}{x_2-x_1}, \tag{3.38}$$

$$\zeta(u+v)-\zeta(u)-\zeta(v) = \frac{y_2-y_1}{x_2-x_1}, \tag{3.39}$$

$$\wp(u+v)+\wp(u-v) = \frac{2f_\Omega(x_1,x_2)}{(x_2-x_1)^2}, \tag{3.40}$$

$$\wp(u-v) = \frac{f_\Omega(x_1,x_2)+2y_1y_2}{(x_2-x_1)^2}, \tag{3.41}$$

$$\wp(u+v)+\wp(u)+\wp(v) = \left[\frac{y_2-y_1}{x_2-x_1}\right]^2 - \lambda_2. \tag{3.42}$$

Baker showed the genus two version of the first expression in (3.36) [Ba1898a, BucEL1997b, BucEL2012]:

$$-\frac{\sigma(u+v)\sigma(u-v)}{\sigma(u)^2\sigma(v)^2} = \wp_{11}(u)-\wp_{11}(v)+\wp_{21}(u)\wp_{22}(v)-\wp_{22}(u)\wp_{21}(v), \tag{3.43}$$

whereas Buchstaber, Enolskii and Leykin generalized it to genus three and four hyperelliptic curves [BucEL1997b, BucEL2012] and in [EilEMOP2007] trigonal curves of genus three, stated as

$$-\frac{\sigma(u+v)\sigma(u-v)}{\sigma(u)^2\sigma(v)^2} = \text{polynomial of } \wp_{ij}(u) \text{ and } \wp_{ij}(v). \tag{3.44}$$

Since this view is connected with the groupoid [BucEL2012] induced from the path space structure of the Abelian covering $\widetilde{X}$ on X, (3.44) plays a crucial role in the Abelian function theory. However, in this book, we generalize the second expressions in (3.36) and (3.37) respectively.

3.4.1 Addition Theorem on Hyperelliptic Curves

This section shows the addition theorem on hyperelliptic curves. We assume that the curve X is the hyperelliptic curve given by (2.65) $f_X(x, y) = y^2 - f(x) = 0$. Frobenius and Stickelberger gave the addition formula (3.37) [FroS1877]. We show the generalized Frobenius–Stickelberger formula of the elliptic function, which was given by Ônishi [On2005, Theorem 7.2], though Eilbeck, Enolskii and Previato also gave another type of a generalized Frobenius–Stickelberger addition formula in [EilEP2003]. In this section, we simply refer to $\natural_{X,k}$ as $\natural_k$.

3.4.1.1 Generalized Frobenius–Stickelberger Formula

Theorem 3.69 [On2005] *For a positive integer $n > 1$, let $(x_1, y_1), \cdots, (x_n, y_n)$ in X, and $u^{(1)}, \cdots, u^{(n)}$ in $\kappa_J^{-1}(W_X^1)$ be points such that $\kappa_J(u^{(i)}) = w((x_i, y_i))$. Then the following relation holds:*

$$\frac{\sigma_{\natural_n}(\sum_{i=1}^n u^{(i)}) \prod_{i<j} \sigma_{\natural_2}(u^{(i)} - u^{(j)})}{\prod_{i=1}^n \sigma_{\natural_1}(u^{(i)})^n} = \varepsilon_n \psi_n((x_1, y_1), \cdots, (x_n, y_n)), \tag{3.45}$$

where $\varepsilon_n = (-1)^{-g^2n/2+(n-2)g/2+1}$ for $n \le g$ and $\varepsilon_n = (-1)^{-g^2n/2+(n-2)g/2}$ for $n \ge g+1$.

Proof Let $P_i := (x_i, y_i)$. Let $A^n X$ be an n-th anti-symmetric product of X associated with $A^n R_X$ defined in Sect. 2.5.1. Due to the properties of $\sigma_{\natural_n}$, both vanish at $P_i = P_j$. The right-hand side is obviously meromorphic at $A^n X$, whereas the translational formula (3.30) means that the left-hand side is also meromorphic at $A^n X$ via the Abel–Jacobi map w and $\widetilde{w} \circ \iota_X$ since $\sigma_{\natural_2}(-u) = -\sigma_{\natural_2}(u)$ due to Corollary 3.51. Following Corollary 3.50, for any $n \le g$, the left-hand side of (3.45) is expanded by $u^{(n)}$,

$$\left[\sigma_{\natural_{n-1}}(\sum_{i=1}^{n-1} u^{(i)})(-1)^{(g-n)+(g-n)+1}(-u_g^{(n)})^{g-n+1} + d_{>g-n+2}(u_g^{(n)}))\right]$$

$$\times \frac{\prod_{i<j<n} \sigma_{\natural_2}(u^{(i)} - u^{(j)}) \cdot \prod_{i=1}^{n-1} [(-1)^{(g-2)}\sigma_{\natural_1}(u^{(i)})(-u_g^{(n)})^{g-1} + d_{>g-1}(u_g^{(n)}))]}{\prod_{i=1}^{n-1} \sigma_{\natural_1}(u^{(i)})^n \cdot \left[(-1)^{g+g(g-1)/2}(-u_g^{(n)})^g + d_{>g+1}(u_g^{(n)})\right]^n}$$

$$= (-1)^{1+(g-2)(n-1)-g(g+1)n/2}$$

$$\times \frac{\sigma_{\natural_{n-1}}(\sum_{i=1}^{n-1} u^{(i)}) \prod_{i<j<n} \sigma_{\natural_2}(u^{(i)} - u^{(j)})}{\prod_{i=1}^{n-1} \sigma_{\natural_1}(u^{(i)})^{n-1}} \frac{1}{(-u_g^{(n)})^{2n}}(1 + d_{>0}(u_g^{(n)})).$$

The exponent of the sign in the prefactor is reduced to $-g^2n/2 + (n-2)g/2 + 1$ modulo 2. On the other hand, the expansion of the right-hand side of (3.45) is also given as

$$\varepsilon_n \psi_{n-1}((x_1, y_1), \cdots, (x_{n-1}, y_{n-1}))\frac{1}{(u_g^{(n)})^{2n}}(1 + d_{>0}(-u_g^{(n)})).$$

Thus we have $\varepsilon_n = (-1)^{g-n-(g-1)(g-4)(n-1)/2}$ and (3.45) for $n \le g$. Similarly for $n > g$, the left-hand side of (3.45) is given by

$$\left[\sigma_{\natural_{n-1}}(\sum_{i=1}^{n-1} u^{(i)}) + d_{>0}(u_g^{(n)}))\right]$$

$$\times \frac{\prod_{i<j<n} \sigma_{\natural_2}(u^{(i)} - u^{(j)}) \cdot \prod_{i=1}^{n-1} [(-)^{(g-2)}\sigma_{\natural_1}(u^{(i)})(-u_g^{(n)})^{g-1} + d_{>g-1}(u_g^{(n)}))]}{\prod_{i=1}^{n-1} \sigma_{\natural_1}(u^{(i)})^n \cdot \left[(-1)^{(g-1)(g-2)/2}(-u_g^{(n)})^g + d_{>g+1}(u_g^{(n)})\right]^n}$$

$$= (-1)^{(g-2)(n-1)-g(g+1)n/2}$$

$$\times \frac{\sigma_{\natural_{n-1}}(\sum_{i=1}^{n-1} u^{(i)}) \prod_{i<j<n} \sigma_{\natural_2}(u^{(i)} - u^{(j)})}{\prod_{i=1}^{n-1} \sigma_{\natural_1}(u^{(i)})^{n-1}} \frac{1}{(-u_g^{(n)})^{g+n-1}}(1+d_{>0}(u_g^{(n)})).$$

We note that $\phi_n(P) = \dfrac{(-1)^n}{w_g(P)^n}(1 + d_{>0}(w_g(P)))$ at $P = \infty$. Hence even for $n > g$, we obtain the equality in (3.45). ■

Lemma 3.70 *For* $u \in \kappa_J^{-1}\Theta_X^1$,

$$\frac{\sigma_{\natural_2}(2u)}{\sigma_{\natural_1}(u)^4} = 2y.$$

Proof $\dfrac{\sigma_{\natural_2}(2u)}{\sigma_{\natural_1}(u)^4} = \dfrac{\sigma_{\natural_2}(2(u+2\ell))}{\sigma_{\natural_1}(u+\ell)^4}$ and thus the left-hand side is a meromorphic function on X. On the other hand, the leading term of its expansion at $u = 0$ is $\dfrac{1}{u_{2g+2}}$

due to Corollary 3.50, which agrees with the behavior of $2y$ at ∞. Hence we obtain the equality. ■

3.4.1.2 Addition Formulae of Hyperelliptic Curves

This subsubsection is based on [EilEMOP2008] and [EnMO2008].

The addition theorem of the hyperelliptic curves is given in Baker's paper [Ba1898a]. In this subsubsection, we show the addition theorem using Theorem 3.69. A particular case of this result appeared in [EilEP2003, (32) and (34)] in the process of deriving Baker's addition formula by the technique given in [On2002] for genus 2, and another particular case is in [BelES2005, (3.21)].

Theorem 3.71 *Assume that (m, n) is a pair of positive integers. Let (x_i, y_i) $(i = 1, \cdots, m)$, (x'_j, y'_j) $(j = 1, \cdots, n)$ in X and $u \in \kappa_J^{-1}(W_X^m)$, $v \in \kappa_J^{-1}(W_X^n)$ be points such that $\kappa_J(u) = w((x_1, y_1), \cdots, (x_m, y_m))$ and $\kappa_J(v) = w((x'_1, y'_1), \cdots, (x'_n, y'_n))$. Then the following relation holds:*

$$\begin{aligned}
&\frac{\sigma_{\natural_{m+n}}(u+v)\sigma_{\natural_{m+n}}(u-v)}{\sigma_{\natural_m}(u)^2\sigma_{\natural_n}(v)^2}\\
&= \delta(g, m, n)\frac{\prod_{i=0}^{1}\psi_{m+n}((x_1, y_1), \cdots, (x_m, y_m), (x'_1, (-1)^i y'_1), \cdots, (x'_n, (-1)^i y'_n))}{(\psi_m((x_1, y_1)\cdots, (x_m, y_m))\psi_n((x'_1, y'_1)\cdots, (x'_n, y'_n)))^2}\\
&\times \prod_{i=1}^{m}\prod_{j=1}^{n}\frac{1}{\psi_2((x_i, y_i), (x'_j, y'_j))},
\end{aligned} \tag{3.46}$$

where $\delta(g, m, n) = (-1)^{gn+\frac{1}{2}n(n-1)+mn}$.

To show this, we introduce the following symbol:

Definition 3.72 Let k be a positive integer. In general, for any k points $u^{(j)}$ $(j = 1, \cdots, k)$ in $\kappa_J^{-1}(W_X^1)$, we define

$$q(u^{(1)}, \cdots, u^{(k)}) = \frac{\sigma_{\natural_k}(u^{(1)} + \cdots + u^{(k)})}{\prod_{j=1}^{k}\sigma_{\natural_1}(u^{(j)})}.$$

Remark 3.73 By the translational relation (3.30), we see that $q(u^{(1)}, \cdots, u^{(k)})^2$ is a function on $S^k X$. However, $q(u^{(1)}, \cdots, u^{(k)})$ is in general only a function on the fiber product taken over J_X with respect to the Abel–Jacobi map $w : S^k X \to J_X$ and the duplication map $[2] : \widetilde{J_X} \to J_X$, where $\widetilde{J_X}$ is an Abelian variety which has J_X as a 2:1 image.

Proof *(of Theorem 3.71).* In the course of this proof, we let $u^{(i)}$ and $v^{(j)}$ be points such that $\kappa_J(u^{(i)}) = w((x_i, y_i))$ and $\kappa_J(u^{(j)}) = w((x'_j, y'_j))$ in W_X^1, respectively. Then we have $\kappa_J(-u^{(i)}) = w((x_i, -y_i))$ and $\kappa_J(-v^{(j)}) = w((x'_j, -y'_j))$. We note the simplest relation for the sign of u and v,

$$q(-u, v) = \frac{\sigma_{\natural_2}(-u+v)}{\sigma_{\natural_1}(-u)\sigma_{\natural_1}(v)} = \frac{-\sigma_{\natural_2}(u-v)}{\sigma_{\natural_1}(u)\sigma_{\natural_1}(-v)} = -q(u, -v), \tag{3.47}$$

where u, v in $\kappa_J^{-1}(W_X^1)$. By (3.45) and (3.47), we have

$$\begin{aligned}
&q(u^{(1)}, \cdots, u^{(m)}, \pm v^{(1)}, \cdots, \pm v^{(n)}) \\
&\quad \cdot \prod_{i_1<i_2} q(u^{(i_1)}, \pm u^{(i_2)}) \prod_{i,j} q(u^{(i)}, \pm v^{(j)}) \prod_{j_1<j_2} q(\pm v^{(j_1)}, v^{(j_2)}) \\
&= \frac{\sigma_{\natural_{m+n}}(u^{(1)} + \cdots + u^{(m)} \pm v^{(1)} \pm \cdots \pm v^{(n)})}{\prod_{i=1}^m \sigma_{\natural_1}(u^{(i)}) \prod_{j=1}^n \sigma_{\natural_1}(\pm v^{(j)})} \\
&\quad \cdot \prod_{i_1<i_2} \frac{\sigma_{\natural_2}(u^{(i_1)} - u^{(i_2)})}{\sigma_{\natural_1}(u^{(i_1)})\sigma_{\natural_1}(-u^{(i_2)})} \cdot \prod_{i,j} \frac{\sigma_{\natural_2}(u^{(i)} \mp v^{(j)})}{\sigma_{\natural_1}(u^{(i)})\sigma_{\natural_1}(\mp v^{(j)})} \cdot \prod_{j_1<j_2} \frac{\sigma_{\natural_2}(\pm v^{(j_1)} \mp v^{(j_2)})}{\sigma_{\natural_1}(u^{(j_1)})\sigma_{\natural}(\mp v^{(j_2)})} \\
&= (-1)^{g(\frac{1}{2}m(m-1)+mn+\frac{1}{2}n(n-1))} \varepsilon_{m+n} \\
&\qquad \cdot \psi_{m+n}((x_1, y_1), \cdots, (x_m, y_m), (x'_1, \pm y'_1), \cdots, (x'_n, \pm y'_n)),
\end{aligned} \tag{3.48}$$

and

$$\begin{aligned}
q(u^{(1)}, \cdots, u^{(m)}) \prod_{i_1<i_2} q(u^{(i_1)}, -u^{(i_2)}) &= (-1)^{\frac{1}{2}gm(m-1)} \varepsilon_m \psi_m((x_1, y_1), \cdots, (x_m, y_m)), \\
q(v^{(1)}, \cdots, v^{(n)}) \prod_{j_1<j_2} q(v^{(j_1)}, -v^{(j_2)}) &= (-1)^{\frac{1}{2}gn(n-1)} \varepsilon_n \psi_n((x'_1, y'_1), \cdots, (x'_n, y'_n)).
\end{aligned} \tag{3.49}$$

We regard both sides of (3.46) as functions on $S^m X \times S^n X$. By (3.48) and (3.49), the left-hand side of (3.46) is equal to

$$(-1)^{gn} \frac{q(u^{(1)}, \cdots, u^{(m)}, v^{(1)}, \cdots, v^{(n)})\, q(u^{(1)}, \cdots, u^{(m)}, -v^{(1)}, \cdots, -v^{(n)})}{q(u^{(1)}, \cdots, u^{(m)})^2\, q(v^{(1)}, \cdots, v^{(n)})^2}. \tag{3.50}$$

Using (3.48) and (3.49), we see that (3.50) is equal to

$$(-1)^{gn} \frac{\prod_\pm \psi_{m+n}((x_1, y_1), \cdots, (x_m, y_m), (x'_1, \pm y'_1), \cdots, (x'_n, \pm y'_n))}{\psi_n((x_1, y_1), \cdots, (x_m, y_m))^2 \psi_m((x'_1, y'_1), \cdots, (x'_n, y'_n))^2} \times Q,$$

where

$$Q = \frac{\prod_{i<j}^m q(u^{(i)}, -u^{(j)})^2 \prod_{i<j}^n q(v^{(i)}, -v^{(j)})^2}{\prod_\pm \left[\prod_{i<j}^m q(u^{(i)}, -u^{(j)}) \prod_{i<j}^n q(\pm v^{(i)}, \mp v^{(j)}) \prod_{i=1}^m \prod_{j=1}^n q(u^{(i)}, \mp v^{(j)})\right]}.$$

Due to (3.47), we have

$$
\begin{aligned}
Q &= (-1)^{n(n-1)/2} \frac{1}{\prod_{\pm} \left[\prod_{i=1}^{m} \prod_{j=1}^{n} q(u^{(i)}, \mp v^{(j)}) \right]} \\
&= (-1)^{n(n-1)/2} \frac{1}{\prod_{i=1}^{m} \prod_{j=1}^{n} q(u^{(i)}, v^{(j)})\, q(u^{(i)}, -v^{(j)})}.
\end{aligned}
$$

Using (3.47) and (3.45) again, we see

$$
\begin{aligned}
q(u, v) q(u, -v) &= \frac{\sigma_{\natural_2}(u+v)\sigma_{\natural_2}(u-v)}{\sigma_{\natural_1}(u)^2 \sigma_{\natural_1}(v) \sigma_{\natural_1}(-v)} \\
&= (-1)^g \frac{\sigma_{\natural_2}(u+v)\sigma_{\natural_2}(u-v)}{\sigma_{\natural_1}(u)^2 \sigma_{\natural_1}(v)^2} \qquad (3.51)\\
&= -\psi_2((x, y), (x', y'))
\end{aligned}
$$

for u, v in $\kappa_J^{-1}(W_X^1)$ and $(x, y), (x', y')$ in X such that $\kappa_J(u) = w((x, y))$ and $\kappa_J(v) = w((x', y'))$. Hence we obtain

$$
Q = (-1)^{n(n-1)/2} (-1)^{mn} \frac{1}{\prod_{i=1}^{m} \prod_{j=1}^{n} \psi_2((x_i, y_i), (x'_j, y'_j))}
$$

and have proved the assertion. ■

Remark 3.74 We note that Definition 3.72 for the q's is behind the Frobenius–Stickelberger relations. This is a relation that governs the automorphy factor of the σ function. In turn, such an automorphy factor defines the line bundle corresponding to the point of the Jacobian, cf. [Fay1973, pp. 24–36]. The Jacobian appears in the long exact sequence of cohomology for

$$
0 \to \mathbb{Z} \to \mathscr{O}_X \to \mathscr{O}_X^* \to 1,
$$

so the behavior of the automorphy factor derives from the exact sequence,

$$
0 \to \mathbb{Z} \to \mathbb{C} \to \mathbb{C}^* \to 1.
$$

Indeed, if $f : \mathbb{C} \to \mathbb{C}^*$ is a group homomorphism,

$$
\frac{f(u^{(1)} + u^{(2)} + \cdots + u^{(n)})}{\prod_{i=1}^{n} f(u^{(i)})} = 1.
$$

Corollary 3.75 [EnMO2008] *The formula in Theorem 3.71 for $m + n \le g + 1$ is simply equal to*

$$
\frac{\sigma_{\natural_{m+n}}(u+v)\sigma_{\natural_{m+n}}(u-v)}{\sigma_{\natural_m}(u)^2 \sigma_{\natural_n}(v)^2} = \delta(g, m, n) \prod_{i=1}^{m} \prod_{j=1}^{n} (x'_j - x_i), \qquad (3.52)
$$

where $\delta(g, m, n) = (-1)^{gn + \frac{1}{2}n(n-1) + mn}$.

Proof Under the condition $m+n \le g+1$, all ψ_{m+n}, ψ_m, ψ_n, ψ_2 are reduced to the Vandermonde determinant and thus we obtain the result. ■

We recall $F_k(x) = (x-x_1)(x-x_2)\cdots(x-x_k)$ and $F(x) = F_g(x)$ in (2.127) and Corollary 2.157 for $(x_i, y_i) \in S^k X$.

Corollary 3.76 *The formula in Theorem 3.71 for $n = 1$, $x := x_1'$, and $m = k$*

$$\frac{\sigma_{\natural_{k+1}}(u+v)\sigma_{\natural_{k+1}}(u-v)}{\sigma_{\natural_k}(u)^2\sigma_{\natural_1}(v)^2} = (-1)^{g+k}F_k(x)$$
$$= (-1)^{g+k}\begin{cases} \left(x^g - \displaystyle\sum_{i=1}^{g} \wp_{ig}(u)x^{i-1}\right) & \text{for } k = g, \\ \left(x^k - \displaystyle\sum_{i=1}^{k} \frac{\sigma_{\natural_k^{(i)}}(u)}{\sigma_{\natural_k}(u)}x^{i-1}\right) & \text{for } k < g. \end{cases} \tag{3.53}$$

Proof Corollary 3.75 gives the first equality. Proposition 3.56 and Theorem 3.59 show the second equality. ■

Corollary 3.77 *Let (x, y) and (x', y') be points in X and $(u, v) \in \mathbb{C}^g \times \mathbb{C}^g$ such that $u = \widetilde{w}(x, y)$ and $v = \widetilde{w}(x', y')$.*

$$\frac{\sigma_{\natural_2}(u+v)\sigma_{\natural_2}(u-v)}{\sigma_{\natural_1}(u)^2\sigma_{\natural_1}(v)^2} = (-1)^{g+1}(x'-x).$$

Corollary 3.77 is a natural generalization of the addition formula in the $g = 1$ case.

For the case of the ramification point $(x', y') = (b_r, 0) = B_r$ and $\omega_{B_r} = w(B_r)$, we have the following relation:

Corollary 3.78 *For $(x_i, y_i) \in X$ $(i = 1, \dots, k)$ and $u \in \mathbb{C}^g$ such that $u = \sum_{i=1}^{k} \widetilde{w}(x_i, y_i)$,*

$$\frac{\sigma_{\natural_{k+1}}(u+\omega_{B_r})\sigma_{\natural_{k+1}}(u-\omega_{B_r})}{\sigma_{\natural_k}(u)^2\sigma_{\natural_1}(\omega_{B_r})^2} = (-1)^{g+k}F_k(b_r).$$

Lemma 3.79 *The formula in Theorem 3.71 for $m = g$ and $n = 2$ is equal to*

$$\frac{\sigma(u+v)\sigma(u-v)}{\sigma(u)^2\sigma_{\natural_2}(v)^2} = -\Xi(u, v), \quad \Xi(u, v) := F(x_1')F(x_2')(\Delta_{x_1'x_2'}^2 - \widetilde{\Delta}_{x_1'x_2'}^2), \tag{3.54}$$

where $\Delta_{x_i'} := \displaystyle\sum_{r=1}^{g} \frac{y_r}{(x_i' - x_r)F'(x_r)}$,

$$\Delta_{x_1', x_2'} := -\frac{\Delta_{x_1'} - \Delta_{x_2'}}{x_1' - x_2'} = \sum_{r=1}^{g} \frac{y_r}{(x_1'-x_r)(x_2'-x_r)F'(x_r)},$$

$$\widetilde{\Delta}_{x'_1,x'_2} := \sum_{i=1, j\neq i}^{2} \frac{(-1)^i y'_i}{(x'_i - x'_j)F(x'_i)}.$$

Proof $\dfrac{\psi_{g+2}((x_1,y_1),\cdots,(x_g,y_g),(x'_1,(-1)^i y'_1),(x'_2,(-1)^i y'_2))}{\psi_g((x_1,y_1),\cdots,(x_g,y_g))\psi_2((x'_1,y'_1),(x'_2,y'_2))}$ is equal to

$$\frac{F(x'_1)F(x'_2)}{\prod_{k>\ell}(x_k - x_\ell)(x'_1 - x'_2)}\left(\sum_{j=1}^{g} \frac{(-1)^{j-1} y_j \prod_{k>\ell,\neq j}(x_k - x_\ell)}{(x'_1 - x_j)(x'_2 - x_j)} + (-1)^i \sum_{j=1}^{2} \frac{(-1)^{j+g-1} y'_j \prod_{k>\ell}(x_k - x_\ell)}{F(x'_j)}\right).$$

Thus we obtain the relation. ■

Lemma 3.79 leads to the following corollary.

Corollary 3.80 *The formula in Lemma 3.79 for the distinct branch points* $(x'_1, y'_1) = (b_a, 0) = B_a$ *and* $(x'_2, y'_2) = (b_b, 0) = B_b$ *is reduced to*

$$\frac{\sigma(u + \omega_{B_a} + \omega_{B_b})\sigma(u - \omega_{B_a} - \omega_{B_b})}{\sigma(u)^2 \sigma_{\natural_2}(\omega_{B_a} + \omega_{B_b})^2} = F(b_1)F(b_2)(\Delta^2_{b_1 b_2}), \tag{3.55}$$

$$\frac{\sigma(u + \omega_{B_a} + \omega_{B_b})\sigma(u - \omega_{B_a} - \omega_{B_b})\sigma(u)^2}{\sigma(u + \omega_{B_a})\sigma(u - \omega_{B_a})\sigma(u + \omega_{B_b})\sigma(u - \omega_{B_b})} \frac{\sigma_{\natural_2}(\omega_{B_a} - \omega_{B_b})}{\sigma_{\natural_2}(\omega_{B_a} + \omega_{B_b})} = (-1)^{g+1}(b_b - b_a)\Delta^2_{b_1 b_2}. \tag{3.56}$$

We note that this relation (3.56) was essentially obtained by Bolza [Bol1895] up to the factor $\sigma_{\natural_2}(\omega_{B_a} - \omega_{B_b})/\sigma_{\natural_2}(\omega_{B_a} + \omega_{B_b})$ and was studied by Baker [Ba1898a] precisely, whose summary is described in [BucEL2012, pp. 29–30]. The factor corresponds to the sign.

We generalize (3.38)–(3.42) noting Remark 3.23.

Lemma 3.81 *For the variables in Corollary 3.76 with* $k = g$*, the following hold:*

1. $\displaystyle\sum_{i=1}^{g} e^{i-1}\left(\zeta_i(u+v) + \zeta_i(u-v) - 2\zeta_i(u)\right) = -\sum_{\ell=1}^{g} \frac{2y_\ell F(e)}{(e - x_\ell)(x - x_\ell)F'(x_\ell)}$,
2. $\displaystyle\zeta_i(u+v) + \zeta_i(u-v) - 2\zeta_i(u) = -\sum_{\ell=1}^{g} \frac{2y_\ell \chi_{\ell,i-1}}{(x - x_\ell)F'(x_\ell)}$,
3. $\displaystyle\sum_{i=1}^{g} x_r^{i-1}\left(\zeta_i(u+v) + \zeta_i(u-v) - 2\zeta_i(u)\right) = -\frac{2y_r}{(x - x_r)}$,
4. $\displaystyle\sum_{i=1}^{g} x^{i-1}\left(\zeta_i(u+v) - \zeta_i(u-v) - 2\zeta_i^{\natural_1}(v)\right) = \frac{2yF'(x)}{F(x)} = \sum_{\ell=1}^{g} \frac{2y}{(x - x_\ell)}$,
5. $\displaystyle\sum_{i,j}^{g} x^{i-1} x_r^{i-1}\left(\wp_{ij}(u+v) - \wp_{ij}(u-v)\right) = -\frac{4y_r y}{(x - x_r)^2}$.

Here we let $\zeta_i^{\natural_j}(v) := \left[\dfrac{\partial}{\partial u_i}\log\sigma_{\natural_j}(u)\right]_{u=v}$*, and e a generic parameter, and* $\chi_{\ell,j}$ *is defined in Definition 4.7.*

Proof Lemma 4.11 shows *1* and *2*. We note that $\partial/\partial x = \sum_i (2y/x^{i-1})\partial/\partial v_i$, and then we obtain *3*, *4*, and *5*. ∎

Remark 3.82 As mentioned in Remark 3.23, we note the symbol of the differential operator here. If we confused $\zeta_i^{\natural_j}(v)$ with $\dfrac{\partial}{\partial v_i}\log\sigma_{\natural_j}(v)$, we would face a contradiction due to Corollary 3.50 and Lemma 4.91: Checking the degree of the pole at ∞ on both sides shows it.

Lemma 3.83 *For the variables in Corollary 3.76 with* $k = g$*, and* $x_1' = x$*, we introduce generic variables* x_2' *and* x_3'

$$\sum_{i=1}^{g} x_2'^{\,i-1}\left(\zeta_i(u+v) + \zeta_i(u-v) - 2\zeta_i(u)\right) = -F(x_2')\Delta_{x_1',x_2'}.$$

$$\sum_{i=1}^{g}\sum_{j=1}^{g} x_2'^{\,i-1}x_3'^{\,j-1}\left(\wp_{ij}(u+v) + \wp_{ij}(u-v) - 2\wp_{ij}(u)\right) = F(x_2')F(x_3')\Delta_{x_3',x_2'}\Delta_{x_1',x_2'}$$
$$- F(x_2')\left(\frac{(x_1'-x_3')\Delta^2_{x_1',x_3'} - (x_2'-x_3')\Delta^2_{x_2',x_3'}}{x_1'-x_2'}\right.$$
$$\left. + \frac{\varphi_{x_1'}}{(x_1'-x_2')(x_1'-x_3')} + \frac{\varphi_{x_2'}}{(x_2'-x_1')(x_2'-x_3')} + \frac{\varphi_{x_3'}}{(x_3'-x_1')(x_3'-x_2')}\right),$$

where $F(x) := F_g(x)$, $f_X(x, y) = y^2 - f(x) = 0$, *and*

$$\varphi_x := \frac{y_x^2}{F(x)} = \frac{f(x)}{F(x)}.$$

Proof The first one is the same as Lemma 3.81 1. This proof will be done in Proposition 4.10 and Lemma 4.11. ∎

3.4.2 *Fay's Identity on Hyperelliptic Curves*

We show the so-called Fay's identity on hyperelliptic curves based on the investigations by Baker. In Theorem 3.71 for $m = g$ and $n = 2$, i.e., Lemma 3.79, we have the following relation of the Ξ in (3.54):

Lemma 3.84 [Ba1903]

$$\Xi(u,v)|_{x_i=x'_j}=0.$$

Proof Following Baker [Ba1903], we consider the limit $x'_1\to x_1$:

$$F^2(x'_1)\left(\sum_{i=1}^{g}\frac{y_i}{(x_i-x'_1)(x_i-x'_2)F'(x_i)}\right)^2|_{x_1=x'_1}=\frac{y_1^2}{(x_1-x'_2)^2},$$

$$F^2(x'_1)\left(\sum_{i=1}^{2}\frac{(-1)^i y'_i}{(x'_1-x'_2)F(x'_i)}\right)^2|_{x_1=x'_1}=\frac{y_1^2}{(x_1-x'_2)^2}.$$

Thus $\Xi(u,v)$ vanishes. ∎

Baker [Ba1897, §11.217], [Ba1903, p. 138] proved the following:

Proposition 3.85 *Let $(x_i,y_i)\in X_g$ $(i=1,\dots,g)$ and $u,v\in\mathbb{C}^g$ such that $u=\widetilde{w}((x_1,y_1),\dots,(x_g,y_g))$ and $v=\widetilde{w}((x'_1,y'_1),(x'_2,y'_2))$.*

The following relation holds for generic x'_i $(i=1,2)$,

$$\begin{aligned}\sum_{i=1}^{g}\sum_{j=1}^{g}\wp_{ij}(u){x'_1}^{i-1}{x'_2}^{j-1}&=F(x'_1)F(x'_2)\left(\sum_{i=1}^{g}\frac{y_i}{(x'_1-x_i)(x'_2-x_i)F'(x_i)}\right)^2\\&\quad-\frac{f(x'_1)F(x'_2)}{(x'_1-x'_2)^2F(x'_1)}-\frac{f(x'_2)F(x'_1)}{(x'_1-x'_2)^2F(x'_2)}+\frac{f_\Omega(x'_1,x'_2)}{(x'_1-x'_2)^2}\\&=\Xi(u,v)+\frac{f_\Omega(x'_1,x'_2)-2y'_1y'_2}{(x'_1-x'_2)^2},\end{aligned}\tag{3.57}$$

where $f_\Omega(x_1,x_2)$ is given by (2.102).

Proof Since the proof of Baker in [Ba1897, §11.217 Ex. xi] is simple though it can be obtained from Fay's formula [Fay1973, (39)], we show Baker's proof as follows. Let $G(\xi):=(\xi-x'_1)(\xi-x'_2)R(\xi)$, and $R(\xi):=(\xi-x_0)F_g(\xi)$. Then for $u:=w((x_0,y_0),(x_1,y_1),\dots,(x_g,y_g))$, he obtained

$$\sum_{i=1}^{g}\sum_{j=1}^{g}\wp_{ij}(u){x'_1}^{i-1}{x'_2}^{j-1}=R(x'_1)R(x'_2)\sum_{i=0}^{g}\sum_{j=0}^{g}\frac{2y_iy_j-f_\Omega(x_i,x_j)}{G'(x_i)G'(x_j)}\tag{3.58}$$

as the generating relation of the Jacobi inversion formula (3.33). We show (3.58).

In the derivation, the following Lagrange interpolation formula is used. From (3.33) and the facts that $-\widetilde{w}(x,y)=\widetilde{w}(x,-y)$ and $\wp_{ij}(-u)=\wp_{ij}(u)$, we have

$$\sum_{i=1}^{g}\sum_{j=1}^{g}\wp_{ij}(\widetilde{w}(P_0)+u){x_0}^{i-1}{x_a}^{j-1}(x_0-x_a)^2=-2y_ay_0+f_\Omega(x_0,x_a)$$

for every $a = 0, \ldots, g$, by noting that both sides vanish for the $a = 0$ case. We interpolate it by $\mathscr{P}(x_1') = x_1'^{g+1} + \gamma_g x_1'^g + \cdots + \gamma_0$ such that

$$\mathscr{P}(x_a) = \sum_{i=1}^{g}\sum_{j=1}^{g} \wp_{ij}\left(\sum_{i=0}^{g}\widetilde{w}(P_i)\right) x_0{}^{i-1}x_a{}^{j-1}(x_0 - x_a)^2 = f_\Omega(x_0, x_a) - 2y_a y_0.$$

$\mathscr{P}(x_1')$ is given by

$$\sum_{a=0}^{g}\sum_{i=1}^{g}\sum_{j=1}^{g} \wp_{ij}\left(\sum_{i=0}^{g}\widetilde{w}(P_i)\right) \frac{x_0{}^{i-1}x_a{}^{j-1}(x_0 - x_a)^2 R(x_1')}{(x_1' - x_a)R'(x_a)} = \sum_{a=0}^{g} \frac{(f_\Omega(x_0, x_a) - 2y_0 y_a)R(x_1')}{(x_1' - x_a)R'(x_a')}.$$

We have

$$\sum_{a=0}^{g}\sum_{i=1}^{g}\sum_{j=1}^{g} \wp_{ij}\left(\sum_{i=0}^{g}\widetilde{w}(P_i)\right) \frac{x_b{}^{i-1}x_a{}^{j-1}(x_b - x_a)^2 R(x_1')}{(x_1' - x_a)R'(x_a)} = \sum_{a=0,\neq b}^{g} \frac{(f_\Omega(x_b, x_a) - 2y_b y_a)R(x_1')}{(x_1' - x_a)R'(x_a)} \tag{3.59}$$

for every $b = 0, 1, \ldots, g$. We interpolate it by $\mathscr{Q}(x_2') = x_2'^{g+1} + \gamma_g' x_2'^g + \cdots + \gamma_0'$ such that $\mathscr{Q}(x_b)$ equals (3.59). $\mathscr{Q}(x_2')$ is given as

$$\sum_{a,b=0}^{g}\sum_{i=1}^{g}\sum_{j=1}^{g} \wp_{ij}\left(\sum_{i=0}^{g}\widetilde{w}(P_i)\right) \frac{x_b{}^{i-1}x_a{}^{j-1}(x_b - x_a)^2 R(x_1')R(x_2')}{(x_1' - x_a)(x_2' - x_b)R'(x_a)R'(x_b)}$$

$$= \sum_{a,b=0}^{g} \frac{(f_\Omega(x_b, x_a) - 2y_b y_a)R(x_1')R(x_2')}{(x_1' - x_a)(x_2' - x_b)R'(x_a)R'(x_b)}.$$

We note the fact that $\wp_{ij}\left(\sum_{i=0}^{g}\widetilde{w}(P_i)\right) = \wp_{ij}(u)$ is factored out from the summation concerning x_a and x_b, and the relation

$$\sum_{a=0}^{g} x_a{}^{i-1}(x_b - x_a)^2 \frac{R(x_1')}{(x_1' - x_a)R'(x_a)} = x_1'^{i-1}(x_b - x_1')^2$$

from the Lagrange interpolation formula for the $(g+1)$-degree polynomial of x_a. We have

$$\sum_{i=1}^{g}\sum_{j=1}^{g} \wp_{ij}(u) x_1'^{i-1} x_2'^{j-1}(x_2' - x_1')^2 = \sum_{a,b=0}^{g} \frac{(f_\Omega(x_b, x_a) - 2y_b y_a)R(x_1')R(x_2')}{(x_1' - x_a)(x_2' - x_b)R'(x_a)R'(x_b)},$$

which is reduced to (3.58). The right-hand side in (3.58) is equal to

$$R(x_1')R(x_2')\left(\sum_{i=1}^{g} \frac{y_i}{(x_1' - x_i)(x_2' - x_i)R'(x_i)}\right)^2 - \frac{f(x_1')R(x_2')}{(x_1' - x_2')^2 R(x_1')} - \frac{f(x_2')R(x_1')}{(x_1' - x_2')^2 R(x_2')} + \frac{f_\Omega(x_1', x_2')}{(x_1' - x_2')^2}.$$

By taking the limit of $P_0 \to \infty$ for the equality, we obtain the formula (3.57). ■

Lemma 3.86 (Lagrange interpolation formula 1795) *For a polynomial* $h(x) = a_1x^{n-1} + a_2x^{n-2} + \cdots + a_{n-1}x + a_n$ *and a set* $\{(x_i, h_i := h(x_i))\}_{i=1,\cdots,n}$, *let* $F_n(x) = (x - x_1)(x - x_2)\cdots(x - x_n)$, $g_i(x) = \dfrac{F_n(x)}{(x - x_i)}$, *and* $z_i(x) = \dfrac{g_i(x)}{g_i(x_i)}$. *Then we have the identities*

$$h(x) = h_1z_1(x) + h_2z_2(x) + \cdots + h_nz_n(x) = \sum_{i=1}^{n} \frac{h_iF_n(x)}{(x - x_i)F_n'(x_i)}.$$

Proof It is proved from the facts

$$z_i(x) = \begin{cases} 1 & x = x_i, \\ 0 & x = x_j (j \neq i), \end{cases}$$

and $g_i(x_i) = F_n'(x_i)$. ■

Lemma 3.79 and Proposition 3.85 yield the key proposition.

Proposition 3.87 *For the variables in Proposition 3.85, the following relation holds:*

$$\frac{\sigma(u+v)\sigma(u-v)}{\sigma(u)^2\sigma_{\natural_2}(v)^2} = \frac{f_\Omega(x_1', x_2') - 2y_1'y_2'}{(x_1' - x_2')^2} - \sum_{i=1}^{g}\sum_{j=1}^{g} \wp_{ij}(u){x_1'}^{i-1}{x_2'}^{j-1}. \tag{3.60}$$

This corresponds to Fay's formula [Fay1973, (39)], which is the basis of "Fay's trisecant identity".

By expansion of Proposition 3.85 in x_a', $(a = 1, 2)$, we have the identities including $\wp_{ij}$ $(i, j = 1, 2, \ldots, g)$ with respect to x_a', $(a = 1, 2)$.

Corollary 3.88 *Proposition 3.85 leads to the expressions of* $\wp_{ij}$ $(i, j = 1, 2, \ldots, g)$ *in terms of* $\mathscr{Q}(S^gR_X)$, *e.g., for* $g = 2$,

$$\wp_{11}(u) = \frac{f_\Omega(x_1, x_2) - 2y_1y_2}{(x_1 - x_2)^2}.$$

Proof The former statement is obvious. For the $g = 2$ case, we replace $((x_i, y_i))_{i=1,2}$ with $((x_i', y_i'))_{i=1,2}$ by considering the symmetry between u and v, and then we obtain the expression of $\wp_{11}(u)$ which does not depend on $((x_i', y_i'))_{i=1,2}$. ■

Corollary 3.89 *We have the addition formula (3.43) of genus two.*

Proof Equation (3.60) provides it from $\wp_{22}(u) = x_1 + x_2$, $\wp_{21}(u) = -x_1x_2$ and Corollary 3.88. ■

3.4.3 *Generalized Frobenius–Stickelberger Formulae on Plane W-Curves*

In this subsection, we generalize Theorem 3.69, the generalized Frobenius–Stickelberger formula for the hyperelliptic curves, to plane W-curves, (r, s)-curves, X by two types, given by Nakayashiki and Yori [NakY2012] and Ônishi [On2010].

3.4.3.1 Generalized Frobenius–Stickelberger Formula, the First Type

In [NakY2012], Nakayashiki and Yori showed a generalization of the Frobenius–Stickelberger formula based on the investigation of the Schur functions for every plane W-curve; Ayano and Nakayashiki further generalized it to telescopic curves [AyN2013]. We state their results without proof because the proof is beyond the scope of this book.

Nakayashiki and Yori showed the following proposition by investigating by the Schur function precisely [NakY2012]:

Proposition 3.90 *Let X be a plane W-curve of (r, s)-type. For u, v in $\kappa_J^{-1}(W_X^1)$, $\partial_{u_g}^{|\Lambda_X^{[1]}|-g-1}\sigma(u-v) \neq 0$, and for a non-negative integer $n(< |\Lambda_X^{[1]}| - g - 1)$, $\partial_{u_g}^n \sigma(u-v) = 0$.*

Using this proposition, they also proved the following generalized Frobenius–Stickelberger formula [NakY2012, Corollary 10]:

Theorem 3.91 [NakY2012] *Let X be a plane W-curve of (r, s)-type. For a positive integer $n > 1$, let $(x_1, y_1), \cdots, (x_n, y_n)$ in X, and $u^{(1)}, \cdots, u^{(n)}$ in $\kappa_J^{-1}(W_X^1)$ be points such that $\kappa_J(u^{(i)}) = w((x_i, y_i))$. Then the following relation holds:*

$$\frac{\sigma_{\natural_n}(\sum_{i=1}^n u^{(i)}) \prod_{i<j} \partial_{u_g}^{|\Lambda_X^{[1]}|-g-1}\sigma(u^{(i)} - u^{(j)})}{\prod_{i=1}^n \sigma_{\natural_1}(u^{(i)})^n} = \varepsilon\psi_n((x_1, y_1), \cdots, (x_n, y_n)), \tag{3.61}$$

where ε is a certain constant depending on the type of curve and n.

This formula could be regarded as the precise version of Fay's addition formula (44) in his book [Fay1973, p. 33].

3.4.3.2 Generalized Frobenius–Stickelberger Formula, the Second Type

We have another generalization, which is associated with the group action $G_{X,x}$ for $\varpi_r : X \to \mathbb{P}$ and $x \in \mathbb{P}$ on the plane W-curves given by Ônishi. Here we handle only the case that there is a Galois action G_X on X.

Theorem 3.92 [On2010] *We consider a plane W-curve X of (r, s)-type $(r, s) = (2, *), (3, 4), (3, 5)$. For a positive integer $n > 1$, let $(x_1, y_1), \cdots, (x_n, y_n)$ in X, and $u^{(1)}, \cdots, u^{(n)}$ in $\kappa_J^{-1}(W_X^1)$ be points such that $\kappa_J(u^{(i)}) = w((x_i, y_i))$. Then the following relation holds:*

$$\frac{\sigma_{\natural_n}(\sum_{i=1}^n u^{(i)}) \prod_{i<j} \prod_{\widehat{\zeta} \in G_X, \widehat{\zeta} \neq \mathrm{id}} \sigma_{\natural_2}(u^{(i)} + \widehat{\zeta} u^{(j)})}{\prod_{i=1}^n \left(\sigma_{\natural_1}(u^{(i)})^{(r-1)(2-i)+1} \prod_{\widehat{\zeta} \in G_X, \widehat{\zeta} \neq \mathrm{id}} \sigma_{\natural_1}(u^{(i)})^{i-1} \right)} \tag{3.62}$$
$$= \varepsilon \psi_n((x_1, y_1), \cdots, (x_n, y_n)) \cdot |(x^{j-1}(u^{(i)}))_{1 \le i, j \le n}|^{r-2},$$

where ε is a certain sign.

Proof The proof of the (3, 4) and (3, 5) cases is given in [On2011]. ∎

The cases $(r, s) = (4, 5)$ and $(5, 6)$ are also asserted by Ônishi due to [On2010]. It is expected that the formula (3.62) could be proved for every general plane W-curve though no one did.

This addition structure is connected with the al functions and the degenerating family of the plane W-curves as we show them for the cyclic (3, 4)-curve case in Sects. 3.6.3 and 3.7.

3.4.4 Addition Theorem on Cyclic Trigonal Curves of Genus Three

We handle the addition theorem on cyclic trigonal curves of genus three in this subsection. We consider Theorem 3.92 for the case of cyclic (3,4)-curve X given by (2.67), $y^3 = x^4 + \lambda_3 x^3 + \lambda_2 x^2 + \lambda_1 x + \lambda_0$, based on [On2009].

3.4.4.1 Generalized Frobenius–Stickelberger Formula on Cyclic Trigonal Curves of Genus Three

The following generalized Frobenius–Stickelberger formula was given by Ônishi [On2009, Theorem 4.3],

Definition 3.93 For a positive integer $n \geq 1$ and a point $(x_1, y_1) \cdots, (x_n, y_n)$ in X, we define

$$\Delta_n((x_1, y_1), \cdots, (x_n, y_n)) := \psi_n((x_1, y_1), \cdots, (x_n, y_n)) \varphi_n((x_1, y_1), \cdots, (x_n, y_n)),$$

$$\varphi_n((x_1, y_1), \cdots, (x_n, y_n)) := \begin{vmatrix} 1 & x_1 & x_1^2 & \cdots & x_1^{n-2} & x_1^{n-1} \\ 1 & x_2 & x_2^2 & \cdots & x_2^{n-2} & x_2^{n-1} \\ \vdots & \vdots & \vdots & \ddots & \vdots & \vdots \\ 1 & x_{n-1} & x_{n-1}^2 & \cdots & x_{n-1}^{n-2} & x_{n-1}^{n-1} \\ 1 & x_n & x_n^2 & \cdots & x_n^{n-2} & x_n^{n-1} \end{vmatrix}.$$

Theorem 3.94 *For a positive integer $n > 1$, let $(x_1, y_1), \cdots, (x_n, y_n)$ in X and $u^{(1)}, \cdots, u^{(n)}$ in $\kappa_J^{-1}(W_X^1)$ be points such that $\kappa_J(u^{(i)}) = w((x_i, y_i))$. Then the following relation holds:*

$$\frac{\sigma_{\natural_n}(\sum_{i=1}^n u^{(i)}) \prod_{i<j} \sigma_{\natural_2}(u^{(i)} + \widehat{\zeta_3} u^{(j)}) \sigma_{\natural_2}(u^{(i)} + \widehat{\zeta_3^2} u^{(j)})}{\prod_{i=1}^n \sigma_{\natural_1}(u^{(i)})^{2n-1}} = \Delta_n((x_1, y_1), \cdots, (x_n, y_n)).$$

3.4.4.2 Addition Formulae of Cyclic Trigonal Curves

For our trigonal curve X, the following generalized addition formulae hold over $W_X^m \times W_X^n$.

Theorem 3.95 [EilEMOP2008] *Assume that (m, n) is a pair of positive integers $(n, m \geq 1)$. Let (x_i, y_i) $(i = 1, \cdots, m)$, (x'_j, y'_j) $(j = 1, \cdots, n)$ be in X and $u \in \kappa_J^{-1}(W_X^m)$, $v \in \kappa_J^{-1}(W_X^n)$ be points such that $\kappa_J(u) = w((x_1, y_1), \cdots, (x_m, y_m))$, $\kappa_J(v) = w((x'_1, y'_1), \cdots, (x'_n, y'_n))$. Then the following relation holds:*

$$\begin{aligned} &\frac{\sigma_{\natural_{m+n}}(u + v)\, \sigma_{\natural_{m+n}}(u + \widehat{\zeta_3} v)\, \sigma_{\natural_{m+n}}(u + \widehat{\zeta_3^2} v)}{\sigma_{\natural_m}(u)^3\, \sigma_{\natural_n}(v)^3} \\ &\quad = \frac{\prod_{i=0}^2 \Delta_{m+n}((x_1, y_1), \cdots, (x_m, y_m), (x'_1, \zeta_3^i y'_1), \cdots, (x'_n, \zeta_3^i y'_n))}{\left(\Delta_m((x_1, y_1) \cdots, (x_m, y_m)) \Delta_n((x'_1, y'_1) \cdots, (x'_n, y'_n))\right)^3} \\ &\quad \times \prod_{i=1}^m \prod_{j=1}^n \frac{1}{\Delta_2((x_i, y_i), (x'_j, y'_j))^2}. \end{aligned} \tag{3.63}$$

Proof This is proved in the same way as Theorem 3.71 [EilEMOP2008]. ■

Proposition 3.96 *Let (x_i, y_i) $(i = 1, \cdots, 3)$, (x'_j, y'_j) $(j = 1, \cdots, n)$ be in X and $u \in \kappa_J^{-1}(W_X^3)$, $v \in \kappa_J^{-1}(W_X^1)$ be points such that $\kappa_J(u) = w((x_1, y_1), \cdots, (x_3, y_3))$, $\kappa_J(v) = w((x'_1, y'_1))$. Then the following relation holds:*

$$\frac{\sigma(u + v)\, \sigma(u + \widehat{\zeta_3} v)\, \sigma(u + \widehat{\zeta_3^2} v)}{\sigma(u)^3\, \sigma_{\natural_1}(v)^3} = \frac{\prod_{i=0}^2 \mu_3((x'_1, \zeta_3^i y'_1); (x_1, y_1), \cdots, (x_3, y_3))}{F(x'_1; x_1, x_2, x_3)}, \tag{3.64}$$

where $F(x'_1; x_1, x_2, x_3) = \prod_{j=1}^3 (x'_1 - x_j)$.

Corollary 3.97 [On2009, Theorem 5.3, Lemma 4.3] *For a cyclic $(3, 4)$-curve, in particular, let (x, y) and $(x_i, y_i)_{i=1,2}$ be points in $X \times S^2 X$ and (u, v_1, v_2) be a point in $\kappa^{-1} W_X^1 \times \kappa^{-1} W_X^1 \times \kappa^{-1} W_X^1$ such that $u = w(x, y)$ and $v_i = w(x_i, y_i)$. Then the following three relations hold:*

$$\begin{aligned} &\frac{\sigma_{\natural_3}(u + v_1 + v_2) \prod_{a=1}^2 \left(\sigma_{\natural_2}(u + \widehat{\zeta_3^a} v_1) \sigma_{\natural_2}(u + \widehat{\zeta_3^a} v_2) \sigma_{\natural_2}(v_1 + \widehat{\zeta_3^a} v_2)\right)}{\sigma_{\natural_1}(u)^5 \sigma_{\natural_1}(v_1)^5 \sigma_{\natural_1}(v_2)^5} \\ &= -(x - x_1)(x - x_2)(x_1 - x_2) \left(y(x_1 - x_2) - y_1(x - x_2) + y_2(x - x_1)\right), \end{aligned}$$

$$\frac{\sigma_{\natural_2}(v_1+v_2)\sigma_{\natural_2}(v_1+\widehat{\zeta_3}v_2)\sigma_{\natural_2}(v_1+\widehat{\zeta_3^2}v_2)}{\sigma_{\natural_1}(v_1)^3\sigma_{\natural_1}(v_2)^3}=\begin{vmatrix}1 & x_1\\ 1 & x_2\end{vmatrix}^2,$$

$$\frac{\sigma_{\natural_2}(2u)}{\sigma_{\natural_1}(u)^4}=3y^2.$$

3.4.5 Kiepert Relation

This subsection describes the Kiepert relation in terms of the sigma functions. Using the Frobenius–Stickelberger formula, we can define the division polynomial $\widetilde{\psi}$ which is studied in Sect. 2.6.2,

$$\widetilde{\psi}_n(u):=\frac{\sigma_{\natural_n}(nu)}{\sigma_{\natural_1}(u)^{n^2}}, \tag{3.65}$$

for the $(2, 2g+1)$- and $(3, 4)$-curves.

Ônishi found the generalized Frobenius–Stickelberger formula mentioned in Sect. 3.4.1.1 and investigated the generalization of the Kiepert formula [On2002, On2005]. We recall a variant of the Kiepert formula, i.e., the Brioschi–Hermite formula as mentioned in Sects. 2.6.2 and 2.6.3.

We have the expressions of the generalized Kiepert formula:

Lemma 3.98 [On2005, Appendix A] *Assume X is the hyperelliptic curve of genus g. For $P \in X$ and $u := \widetilde{w}(\iota_X(P))$, we have the expressions of $\widetilde{\psi}_n(u)$,*

$$\widetilde{\psi}_n(u)=(-1)^{(2n-g)(g-1)/2}\frac{x^{(g-1)n(n-1)/2}}{(1!2!\cdots(n-1)!)}\mathfrak{w}_\infty^{n-1}(P)$$

in Definition 2.183 and

$$\widetilde{\psi}_n=\varepsilon_n(2y)^{n(n-1)/2}\times\begin{cases}\begin{vmatrix}y^{[(n+g+1)/2]} & y^{[(n+g-1)/2]} & \dots & y^{[g+2]}\\ y^{[(n+g+3)/2]} & y^{[(n+g+1)/2]} & \dots & y^{[g+3]}\\ \vdots & \vdots & \ddots & \vdots\\ y^{[n-1]} & y^{[n]} & \dots & y^{[(n+g+1)/2]}\end{vmatrix} & \text{for } n+g \text{ is odd},\\[2ex] \begin{vmatrix}y^{[(n+g)/2]} & y^{[(n+g-2)/2]} & \dots & y^{[g+1]}\\ y^{[(n+g+2)/2]} & y^{[(n+g)/2]} & \dots & y^{[g+2]}\\ \vdots & \vdots & \ddots & \vdots\\ y^{[n-1]} & y^{[n]} & \dots & y^{[(n+g)/2]}\end{vmatrix} & \text{for } n+g \text{ is even},\end{cases}$$

where $y^{[n]} = \dfrac{1}{n!}\dfrac{d^n}{dx^n} y$ *and* ε_n *is a sign* ± 1 *given by the following table:*

g *modulo* 4	$\varepsilon_{n=0 \text{ modulo } 4}$	$\varepsilon_{n=1 \text{ modulo } 4}$	$\varepsilon_{n=2 \text{ modulo } 4}$	$\varepsilon_{n=3 \text{ modulo } 4}$
0	−1	−1	1	−1
1	−1	1	1	1
2	1	−1	−1	−1
3	1	1	−1	1

Proof This can be obtained by the limit computations for (3.4.1.1). The latter follows the proof in Corollary 2.193. ε_n is due to [On2005, Appendix A]. ■

Due to the arguments in Sects. 2.6.2 and 2.6.3, X can be more general W-curves. From Theorem 2.184, the following is obvious:

Theorem 3.99 *Let* $n \geq g - 1$ *be a positive integer. For a plane-model curve* X, *the vanishing of* $\widetilde{\psi}_n(u)$ *at* $u \in \kappa_J^{-1} w(P)$ *for* $P \in X$ *is the necessary and sufficient condition for* $w((n+1)\cdot P)$ *to belong to* W_X^{g-1}.

The assumption $n \geq g - 1$ *can be weakened to* $n > 0$ *for the sufficiency in the following sense: if* P *is a zero of* $\mathfrak{w}_\infty^{(n)}$ *then* $\omega((n+1)\cdot P)$ *belongs to* W_X^m, *where* $m = N(N(n) - n - 1) - (N(n) - n - 1) < g$.

Lemma 3.100 *Assume that* X *is cyclic plane W-curve. For* $P \in X$ *and* $u := \widetilde{w}(\iota_X(P))$, *we have*

$$\widetilde{\psi}_{n+1}(u) = c' \psi_\infty^n(P)$$

in Definition 2.189, where c' *is a constant.*

Theorem 3.101 *For a positive integer* $n \geq g$, $\widetilde{\psi}_{n+1}(u)$ *is an element in* R_X *whose vanishing at* $u \in \kappa_J^{-1} w(P)$ *for* P *on the curve* X *is the necessary and sufficient condition for the normalized image of* $(n+1)\cdot P$ *in the Picard group to belong to* W_X^{g-1}.

We have the recursion formula of $\widetilde{\psi}$'s of genera one and two. The case of genus $g = 1$ ($r = 2, s = 3$), it is given in [Web1908, pp. 196–200]. Cantor generalized it to the $g = 2$ ($r = 2, s = 5$) case in terms of Padé expansions [Can1994].

Proposition 3.102 $g = 1$ *case:*

$$\widetilde{\psi}_{m+n}\widetilde{\psi}_{m-n} = \begin{vmatrix} \widetilde{\psi}_{m-1}\widetilde{\psi}_n & \widetilde{\psi}_m\widetilde{\psi}_{n+1} \\ \widetilde{\psi}_m\widetilde{\psi}_{n-1} & \widetilde{\psi}_{m+1}\widetilde{\psi}_n \end{vmatrix},$$

and $g = 2$ *case* [Can1994, Mat2003a]:

$$\widetilde{\psi}_2^2\widetilde{\psi}_m\widetilde{\psi}_n\widetilde{\psi}_{m+n}\widetilde{\psi}_{m-n} = \begin{vmatrix} \widetilde{\psi}_{m-2}\widetilde{\psi}_n & \widetilde{\psi}_{m-1}\widetilde{\psi}_{n+1} & \widetilde{\psi}_m\widetilde{\psi}_{n+2} \\ \widetilde{\psi}_{m-1}\widetilde{\psi}_{n-1} & \widetilde{\psi}_m\widetilde{\psi}_n & \widetilde{\psi}_{m+1}\widetilde{\psi}_{n+1} \\ \widetilde{\psi}_m\widetilde{\psi}_{n-2} & \widetilde{\psi}_{m+1}\widetilde{\psi}_{n-1} & \widetilde{\psi}_{m+2}\widetilde{\psi}_n \end{vmatrix}.$$

Proof The addition formula (3.36) becomes

$$\wp(nu) - \wp(u) = -\frac{\widetilde{\psi}_{n+1}(u)\widetilde{\psi}_{n-1}(u)}{\widetilde{\psi}_n(u)^2}, \quad \wp(mu) - \wp(nu) = -\frac{\widetilde{\psi}_{n+m}(u)\widetilde{\psi}_{m-n}(u)}{\widetilde{\psi}_m(u)^2\widetilde{\psi}_n(u)^2}.$$

The identity

$$[-\wp(mu) + \wp(nu)] + [\wp(mu) - \wp(u)] - [\wp(nu) - \wp(u)] = 0$$

is reduced to the relation of $g = 1$.

Similarly the $g = 2$ case can be obtained based on (3.43) in [Mat2003a], though the derivation is very complicated. Thus we demonstrate the proof in Sect. 4.4.1. ■

Remark 3.103 Proposition 3.102 suggests the possibility of recursion relations and difference equations, along the lines of [BraEH2005, Mat2002a], cf. Sect. 4.4.

Further, we note that there is a variant of the division polynomial $\sigma(nu)/\sigma(u)^{n^2}$ for $u \in \mathbb{C}^g$ proposed and investigated by Uchida [U2011].

3.5 W-Normalized Prime Form in Terms of the Sigma Function

In this section, we construct the prime form in terms of the sigma function [GiMO2013]. Following Nakayashiki's definition of the prime form [Nak2010a, Definition 3], we introduce $\mathscr{E}$:

Definition 3.104 We define another prime form for $\widetilde{X} \times \widetilde{X}$:

$$\mathscr{E}(P, Q) := \mathrm{e}^{\frac{1}{2}{}^t(\widetilde{w}(P)-\widetilde{w}(Q))\gamma(\widetilde{w}(P)-\widetilde{w}(Q))} E(P, Q), \tag{3.66}$$

where $\gamma := \omega'^{-1}\eta'$.

Proposition 3.105 *1. $\mathscr{E}(P_1, P_2)$ vanishes if and only if $\kappa_X(P_1) = \kappa_X(P_2)$.*
2. $\mathscr{E}(P_1, P_2)$ has a first-order zero along the diagonal $\kappa_X^{-1}\Delta$ of $\Delta \subset X \times X$.
3. $\mathscr{E}(P_1, P_2) = -\mathscr{E}(P_2, P_1)$.
4. Let t_i be a local parameter at P_i of X such that $\zeta(P_i) = dt_i$, $(i = 1, 2)$, and

$$\mathscr{E}(P_1, P_2) = \frac{t_1 - t_2}{\sqrt{dt_1}\sqrt{dt_2}}(1 + d_{>0}(t_1 - t_2)\,.$$

5. $\mathscr{E}(P, Q)$ has the properties for the action γ_ℓ on $\widetilde{X}$,

$$\mathscr{E}(P + \gamma_\ell, Q) := \chi(\ell)\exp\left(L\left(w(P) - w(Q) + \frac{1}{2}\ell, \ell\right)\right)\mathscr{E}(P, Q),$$

where $P + \gamma_\ell = \widetilde{w}(P) + \ell$ for $\ell \in \Gamma_X$.

Proof *1–4* are obvious. The translational formula is the same as the difference between the sigma and the theta functions; similarly, we obtain it. ■

Proposition 3.106 *If a $(-1/2)\otimes(-1/2)$-form on $\widetilde{X}\times\widetilde{X}$ has the properties 1, 2, 3, 4, and 5 in Proposition 3.105, it has the form of (3.66).*

Proof If there is another $(-1/2)\otimes(-1/2)$-form $\mathscr{E}'(P,Q)$ on $\widetilde{X}\times\widetilde{X}$ with these properties, the quasi-periodic property *5* shows that $\mathscr{E}'(P,Q)/\mathscr{E}(P,Q)$ must be a meromorphic function over $\mathbb{C}^g$. Further, the vanishing property *1* determines that the meromorphic function has no zero and thus must be a constant function. By the property *4*, this constant is equal to 1. ■

As we show in the following subsections, for any hyperelliptic $(2,2g+1)$-curves, we have

$$\mathscr{E}(P,Q)=\frac{\sigma_{\natural_2}(\widetilde{w}(P)-\widetilde{w}(Q))}{\sqrt{\nu^{\mathrm{I}}_1(P)}\sqrt{\nu^{\mathrm{I}}_1(Q)}},$$

and for a cyclic (3, 4)-curve, we have

$$\mathscr{E}(P,Q)=\frac{\sigma_{\natural_3}(\widetilde{w}(P)-\widetilde{w}(Q))}{\sqrt{\nu^{\mathrm{I}}_1(P)}\sqrt{\nu^{\mathrm{I}}_1(Q)}}=\frac{\sigma(\widetilde{w}(P)-\widetilde{w}(Q))}{\sqrt{\nu^{\mathrm{I}}_1(P)}\sqrt{\nu^{\mathrm{I}}_1(Q)}}.$$

Following Nakayashiki [Nak2010a, Definition 3], we introduce the prime function on $\widetilde{X}\times\widetilde{X}$,

$$\widetilde{\mathscr{E}}(P,Q):=\mathscr{E}(P,Q)\sqrt{\nu^{\mathrm{I}}_1(P)}\sqrt{\nu^{\mathrm{I}}_1(Q)},\quad \widetilde{\mathscr{E}}(\infty,Q):=\mathrm{e}^{\frac{1}{2}{}^t(\widetilde{w}(Q))\gamma(\widetilde{w}(Q)))}E(\infty,Q), \tag{3.67}$$

where $E(\infty,Q)$ is introduced in Sect. 2.4.1.4.

3.5.1 W-Normalized Prime Form: Hyperelliptic Curves

In terms of the sigma function, we show the W-normalized prime form on hyperelliptic curves in this subsection. Let us consider the hyperelliptic curves X. Let (x,y) and (x',y') be points in X and $(u,v)\in\mathbb{C}^g\times\mathbb{C}^g$ such that $u=\widetilde{w}(x,y)$ and $v=\widetilde{w}(x',y')$.

From [On2005, Lemma 9.1], we have the relation:

$$\lim_{u\to v}\frac{\sigma_{\natural_2}(u-v)}{u_1-v_1}=1.$$

This can be proved by Corollary 3.75,

$$\frac{1}{u_1-v_1}\frac{\sigma_{\natural_2}(u+v)\sigma_{\natural_2}(u-v)}{\sigma_{\natural_1}(u)^2\sigma_{\natural_1}(v)^2}=\frac{x_1-x_2}{u_1-v_1},$$

$$\frac{\sigma_{\natural_2}(u-v)}{u_1-v_1}=\frac{\sigma_{\natural_1}(u)^2\sigma_{\natural_1}(v)^2}{\sigma_{\natural_2}(u+v)}\frac{x_1-x_2}{u_1-v_1},$$

$\lim_{u\to v}\dfrac{x_1-x_2}{u_1-v_1}=\dfrac{dx}{du_1}(v)=2y(v)$. On the other hand, $\dfrac{\sigma_{\natural_2}(2u)}{\sigma_{\natural_1}(u)^4}=2y$ due to Lemma 3.70, and thus we have the relation.

This implies the following relation:

Lemma 3.107 *Let (x, y) and (x', y') be points in X and $(u, v) \in \mathbb{C}^g \times \mathbb{C}^g$ such that $u = \widetilde{w}(x, y)$ and $v = \widetilde{w}(x', y')$. Further, let $du_1 := \nu_1^{\mathrm{I}}(x, y)$, and $dv_1 := \nu_1^{\mathrm{I}}(x', y')$.*

$$\frac{\sigma_{\natural_2}(u-v)}{\sqrt{du_1}\sqrt{dv_1}}=\frac{u_1-v_1}{\sqrt{du_1}\sqrt{dv_1}}\left(1+d_>((u_1-v_1)^0)\right).$$

Remark 3.108 This establishes conditions *2* and *3* of 3.106. Further, Lemma 4.3 in [On2005] shows that the sigma function is an odd or even function with respect to $u \in \mathbb{C}^g$. Hence this also establishes *3* of 3.106.

Secondly, from [On2005, Proposition 7.5], we have the relation:

Lemma 3.109 [On2005, Proposition 7.5] *Let (x, y) and (x', y') be points in X and $(u, v) \in \mathbb{C}^g \times \mathbb{C}^g$ such that $u = \widetilde{w}(x, y)$, and $v = \widetilde{w}(x', y')$. Further, let $du_1 := \nu_1^{\mathrm{I}}(x, y)$, $dv_1 := \nu_1^{\mathrm{I}}(x', y')$. and $\ell\ (= 2\omega'\ell' + 2\omega''\ell'') \in \Pi$. Then the following quasi-periodicity relation holds:*

$$\frac{\sigma_{\natural_2}(u+\ell-v)}{\sqrt{du_1}\sqrt{dv_1}}=\exp(L(u+\frac{1}{2}\ell,\ell))\chi(\ell)\frac{\sigma_{\natural_2}(u+\ell-v)}{\sqrt{du_1}\sqrt{dv_1}}.$$

Remark 3.110 This establishes condition *(v)* of 3.106.

Lemma 3.111 *Let (x, y) and (x', y') be points in X and $(u, v) \in \kappa_J^{-1}\left(\Theta^1 \setminus \Theta^0\right) \times\mathbb{C}^g$ such that $u = \widetilde{w}(x, y)$ and $v = \widetilde{w}(x', y')$. Further, let $du_1 := \nu_1^{\mathrm{I}}(x, y)$, and $dv_1 := \nu_1^{\mathrm{I}}(x', y')$. Then the following relation holds:*

$$\lim_{v\to 0}\frac{\sigma_{\natural_2}(u-v)}{\sqrt{du_1}\sqrt{dv_1}}\neq 0.$$

Proof From Theorem 3.44, we have

$$\sigma_{\natural_2}(u-v)=c_1\sigma_{g^{N_2}}(u-v)=c_2\sigma_{g^{N_1}}(u)v_g^{g-1}+d_>(v_g^g),$$

where N_1, N_2. c_1 and c_2 are certain constants. Since the weight of v_1 is $2g-1$, we have the relation. ■

Remark 3.112 Thus $\dfrac{\sigma_{\natural_2}(u-v)}{\sqrt{du_1}\sqrt{dv_1}}$ has zeros only where $(x, y) = (x', y')$, and by Lemma 3.107, these is a simple giving condition *1* of Proposition 3.106.

By assembling these results, $\mathscr{E}(P, Q)$ is equal to $\dfrac{\sigma_{\natural_2}(u-v)}{\sqrt{du_1}\sqrt{dv_1}}$ due to Proposition 3.106. Thus we have proved the following:

Theorem 3.113 *Let (x, y) and (x', y') be points in X and $(u, v) \in \mathbb{C}^g \times \mathbb{C}^g$ such that $u = \widetilde{w}(x, y)$ and $v = \widetilde{w}(x', y')$. Then the following relation holds:*

$$\mathscr{E}(P, Q) = \frac{\sigma_{\natural_2}(u-v)}{\sqrt{du_1}\sqrt{dv_1}}.$$

Here $du_1 := \nu_1^{\mathrm{I}}(x, y)$ and $dv_1 := \nu_1^{\mathrm{I}}(x', y')$.

3.5.2 W-Normalized Prime Form: Cyclic (3, 4)-Curves

In this subsection, we show the W-normalized prime form on cyclic $(3, 4)$-curves X. For $u, v \in \widetilde{w}(\widetilde{X})$, the sigma function with the argument "$u - v$" should be treated following Nakayashiki's result in Proposition 3.90. However, for the cyclic (3,4)-curve case, we can also treat it using Theorem 3.94 and thus we show it. In order to state Theorem 3.120, we show some preliminary results.

First, we consider the action of the Galois group on the sigma function and its derivatives due to [On2009, Lemma 4.1, 4.2] and [EilEMOP2008, Lemma 4.2]:

Lemma 3.114 *For $u \in \mathbb{C}^3$ and action $\widehat{\zeta_3}$ on u via its Abel–Jacobi preimage $(x_i, y_i) \to (x_i, \zeta_3 y_i)$, we have*

$$\sigma(\widehat{\zeta_3} u) = \zeta_3 \sigma(u) = \zeta_3 \sigma_{\natural_3}(u), \quad \sigma_{\natural_2}(\widehat{\zeta_3} u) = \zeta_3^2 \sigma_{\natural_2}(u), \quad \sigma_{\natural_1}(\widehat{\zeta_3} u) = \sigma_{\natural_1}(u), \tag{3.68}$$

and $\sigma(-u) = -\sigma(u)$.

Proof From the equations of the Galois action in Sect. 3.2.9, (3.68) is obtained. ■

Secondly, from [On2009, Lemma 6.1], we have:

Lemma 3.115 *Let (x, y) and (x', y') be points in X and $(u, v) \in \mathbb{C}^3 \times \mathbb{C}^3$ such that $u = \widetilde{w}(x, y)$ and $v = \widetilde{w}(x', y')$. Then the following relations hold:*

$$\lim_{u \to v} \frac{\sigma_{\natural_1}(u)(x - x')}{\sigma_{\natural_2}(u + \widehat{\zeta_3} v)} = (\zeta_3 - 1)y, \quad \lim_{u \to v} \frac{\sigma_{\natural_1}(u)(x - x')}{\sigma_{\natural_2}(u + \widehat{\zeta_3^2} v)} = (\zeta_3^2 - 1)y.$$

Proof Noting $\sigma_{\natural_2} = \sigma_3$ and $\sigma_{\natural_2^{(2)}} = \sigma_2 = \sigma_{\natural_1}$, from Theorem 5.1 in [MatP2008] and Theorem 5.23 in [MatP2014], we have the relation

$$\frac{\sigma_{\natural_1}(u+v)}{\sigma_{\natural_2}(u+v)} = (-1)^{3-2+1} \frac{y - y'}{x - x'}.$$

Thus

$$\lim_{u\to v}\frac{x-x'}{\sigma_{\natural_2}(u+\widehat{\zeta_3 v})}=\lim_{u\to v}\frac{y-\zeta_3 y'}{\sigma_{\natural_1}(u+\widehat{\zeta_3 v})}$$
$$=\frac{(1-\zeta_3)y'}{\sigma_{\natural_1}(\widehat{(1+\zeta_3)v})}=\frac{(1-\zeta_3)y'}{\sigma_{\natural_1}(\widehat{-\zeta_3^2 v})}=-\frac{(1-\zeta_3)y'}{\sigma_{\natural_1}(v)}.$$

The stated relations then follow. ∎

From the Frobenius–Stickelberger relation in Theorem 3.94, we have the following proposition:

Proposition 3.116 *Let (x, y) and (x', y') be points in X and $(u, v) \in \mathbb{C}^3 \times \mathbb{C}^3$ such that $u = \widetilde{w}(x, y)$ and $v = \widetilde{w}(x', y')$. Then the following relation holds:*

$$\frac{\sigma(u-v)\sigma_{\natural_2}(u+v)}{\sigma_{\natural_1}(u)^2\sigma_{\natural_1}(v)^2}=\frac{\sigma_{\natural_3}(u+\widehat{\zeta_3 v}+\widehat{\zeta_3^2 v})\sigma_{\natural_2}(u+v)}{\sigma_{\natural_1}(u)^2\sigma_{\natural_1}(v)^2}=(x-x').$$

Proof By letting $v^{(a)} = \widetilde{w}(x_a, y_a)$ $(a = 1, 2)$, the left-hand side is equal to

$$\lim_{v^{(a)}\to\zeta_3^a v}\varepsilon\psi_3((x,y),(x_1,y_1),(x_2,y_2))\varphi_3((x,y),(x_1,y_1),(x_2,y_2))$$
$$\times\frac{\sigma_{\natural_1}(u)^3\sigma_{\natural_1}(v)^3}{\left(\prod_{a=0}^{2}\sigma_{\natural_2}(u+\widehat{\zeta_3^a v})\right)}\frac{\sigma_{\natural_1}(v^{(1)})^2\sigma_{\natural_1}(v^{(2)})^2}{\sigma_{\natural_2}(v^{(1)}+\widehat{\zeta_3^2 v^{(2)}})}\frac{\sigma_{\natural_1}(v^{(1)})}{\sigma_{\natural_2}(\widehat{\zeta_3 v^{(1)}}+\widehat{\zeta_3^2 v^{(2)}})}$$
$$=(x-x')^3(\zeta_3-\zeta_3^2)y'\lim_{v^{(a)}\to\zeta_3^a v}(x_1-x_2)\cdot\frac{1}{(x-x')^2}\frac{1}{3(y')^2}\frac{\sigma_{\natural_1}(v^{(1)})}{\zeta_3^2\sigma_{\natural_2}(v^{(1)}+\widehat{\zeta_3 v^{(2)}})}$$
$$=(\zeta_3-\zeta_3^2)(x-x')\frac{\zeta_3-1}{3\zeta_3^2}=(x-x').$$
∎

Remark 3.117 Proposition 3.116 is the trigonal analogue of Corollary 3.77, and similarly generalizes the addition formula in the $g = 1$ case.

We now look at the order of vanishing of $\sigma_{\natural_3} = \sigma$, and $\sqrt{du_1}$.

Lemma 3.118 *Let (x, y) and (x', y') be points in X and $(u, v) \in \mathbb{C}^3 \times \mathbb{C}^3$ such that $u = \widetilde{w}(x, y)$ and $v = \widetilde{w}(x', y')$. Then the following relation holds:*

$$\lim_{u\to v}\frac{\sigma_{\natural_3}(u+\widehat{\zeta_3 v}+\widehat{\zeta_3^2 v})}{u_1-v_1}=\lim_{u\to v}\frac{\sigma_{\natural_3}(u-v)}{u_1-v_1}=1.$$

Proof Noting $\lim_{u\to v}\dfrac{(x-x')}{u_1-v_1}=\dfrac{d}{dv_1}x(v)=3y(v)^2$, we have

$$\lim_{u\to v}\frac{\sigma_{\natural_3}(u+\widehat{\zeta_3 v}+\widehat{\zeta_3^2 v})}{u_1-v_1}=\lim_{u\to v}\frac{\sigma_{\natural_1}(u)^2\sigma_{\natural_1}(v)^2}{\sigma_{\natural_2}(2v)}\frac{(x-x')}{u_1-v_1}=1.$$
∎

Lemma 3.119 *Let (x, y) and (x', y') be points in X and $(u, v) \in \kappa_J^{-1}\left(\Theta^1 \setminus \Theta^0\right) \times \mathbb{C}^3$ such that $u = \widetilde{w}(x, y)$, and $v = \widetilde{w}(x', y')$. Further, let $du_1 := \nu_1^{\mathrm{I}}(x, y)$, and $dv_1 := \nu_1^{\mathrm{I}}(x', y')$. Then the following relation holds:*

$$\lim_{v \to 0} \frac{\sigma_{\natural_3}(u - v)}{\sqrt{du_1}\sqrt{dv_1}} \neq 0.$$

Proof Around $v = 0$, we have $\sigma(u - v) = \frac{1}{20}\sigma_{33}(u)v_3^2 + \cdots$, whereas $dv_1 = v_3^4 dv_3 + \cdots$. ■

Thus, again using Proposition 3.106 we have proved the following:

Theorem 3.120 *For the cyclic trigonal* (3, 4)*-curve,*

$$\mathscr{E}(P, Q) = \frac{\sigma(u + \widehat{\zeta}_3 v + \widehat{\zeta}_3^2 v)}{\sqrt{-3}\sqrt{du_1}\sqrt{dv_1}} = \frac{\sigma(u - v)}{\sqrt{du_1}\sqrt{dv_1}} = \frac{\sigma_{\natural_3}(u - v)}{\sqrt{du_1}\sqrt{dv_1}}.$$

Remark 3.121 As mentioned in the beginning of this subsection, the sigma function with the argument "$u - v$" should be treated following Nakayashiki and Yori's results in Proposition 3.90 for $u, v \in \widetilde{w}(\widetilde{X})$ [NakY2012]. In fact, our approach encounters the difficulty for higher genus g and larger r of (r, s)-curves ($g > 5$ for $r > 2$, and $r > 3$). However, the Galois action on the Jacobi variety is concerning because it is directly connected with the cyclic Weierstrass canonical form (2.8). Theorem 3.120 is also associated with the Galois action, whereas the properties in Proposition 3.90 and Theorem 3.91 basically come from the properties of general Jacobians. Thus we pick up the result in this book, though Nakayashiki and Yori's results are applicable for every plane W-curve, and probably for every W-curve.

As mentioned in Remarks 2.103 and 2.144, the W-normalized differentials of the third kind and thus, the sigma functions, as well as the prime forms are connected with the Schiffer and the Bergman kernels. Some of them or their differentials are also described by the structure ring R_X directly. Thus these results might bring precise properties of these kernels. Recently D. Korotkin investigated "Bergman tau functions" associated with kernel functions [Kor2020].

3.5.2.1 Sigma Function in Terms of W-Normalized Prime Forms

In terms of the expression of Theorem 3.26, Nakayashiki showed the beautiful relations between the sigma function and the W-normalized prime forms for the plane W-curves. We also refer to two of Nakayashiki's results [Nak2010a, Theorems 1 and 2] without proofs:

Theorem 3.122 *We assume that X is a plane W-curve with $H_X = \langle r, s\rangle$ and $n \geq g$. For $\widetilde{P}_1, \ldots, \widetilde{P}_n \in \widetilde{X}$ and $P_i = \kappa_X \widetilde{P}_i$,*

$$\sigma\left(\sum_{i=1}^{n}\widetilde{w}(\widetilde{P}_i)\right)=\frac{\prod_{i=1}^{n}\mathscr{E}(\widetilde{P}_i,\infty)^n}{\prod_{i<j}\mathscr{E}(\widetilde{P}_i,\widetilde{P}_j)}\psi_n(P_1,\cdots,P_n). \tag{3.69}$$

Proposition 3.123 *For* $P_1=(x_{P,1},y_{P,1}),\dots,P_n=(x_{P,n},y_{P,n})$, $Q_1=(x_{Q,1},y_{Q,1}),\dots,Q_n=(x_{Q,n},y_{Q,n})\in X$,

$$\sigma\left(\sum_{i=1}^{n}(\widetilde{w}(\iota_X P_i)-\widetilde{w}(\iota_X Q_i))\right)=C_n\frac{\prod_{i,j}^{n}\mathscr{E}(\iota_X P_i,\iota_X Q_j)}{\prod_{i<j}\mathscr{E}(\iota_X P_i,\iota_X P_j)\mathscr{E}(\iota_X Q_i,\iota_X Q_j)\prod_{i,j}^{n}(x_{P,i}-x_{Q,j})}$$
$$\times\frac{\psi_{rn}(P_1,\dots,P_n,Q_1^{(1)},Q_1^{(2)},\dots,Q_1^{(r)},Q_2^{(1)},\dots,Q_n^{(r)})}{\prod_{i<j}(x_{Q,i}-x_{Q,j})^{n-2}\prod_{k=1}^{n}\prod_{i\le i<j\le r-1}(y_{Q,k}^{(i)}-y_{Q,k}^{(j)})}, \tag{3.70}$$

where $Q_i^{(j)};=\widehat{\zeta}^j Q_i$, *and for a group action* $\widehat{\zeta}\in G_{X,\varpi_r(Q_i)}$, $\zeta\neq id$, $\zeta_r=\mathrm{e}^{\pi\sqrt{-1}/r}$, C_n *is given by*

$$(-1)^{rn(n-1)/2}\frac{\prod_{1\le i<j\le r-1}(\zeta_r^{si}-\zeta_r^{sj})}{\prod_{1\le i<j\le r-1}(\zeta_r^{i}-\zeta_r^{j})}\zeta_r^{n(n-1)/2-n(n-1)(r-1)(r-2)/4+nr(r-1)/2-gnr(r-3)/2}.$$

3.6 al Function in Terms of Sigma Function for Finite Covering of X

In this section, we review the hyperelliptic al function mainly following [MatP2015], which was written based on [Ba1897, Ba1898a]. The al function is a generalization of Jacobi sn, cn, dn functions of genus one, which are defined by the ratio of the sigma functions.

3.6.1 al Function on Hyperelliptic Curves

This subsection provides the al functions in terms of the sigma functions. We consider a hyperelliptic curve X of genus g ($g>0$) be defined by (2.65) with the parameterization (2.65). The periodic matrices of the hyperelliptic curve is given by Lemma 2.195. We note $(b_j, 0) = B_j \in X$.

Definition 3.124 We define the double coverings of J_X by

$$J_X^{(a)} = \mathbb{C}^g/\Gamma^{(a)},$$

where $\Gamma^{(0)} := \bigcap_{a=1}^{2g} \Gamma^{(a)}$,

$$\Gamma^{(0)} := \sum_{b=1} (2\mathbb{Z}\omega'_b + 4\mathbb{Z}\omega''_b),$$

$$\Gamma^{(a)} := \sum_{b=1,\neq i} 2\mathbb{Z}\omega'_b + 4\mathbb{Z}\omega'_i + \sum_{b=1}^{i} 4\mathbb{Z}\omega''_b + \sum_{b=i+1}^{g} 2\mathbb{Z}\omega''_b \quad \text{for } a = 2i, i = 1, 2, \ldots, g,$$

$$\Gamma^{(a)} := \sum_{b=1,\neq i} 2\mathbb{Z}\omega'_b + 4\mathbb{Z}\omega'_i + \sum_{b=1}^{i-1} 4\mathbb{Z}\omega''_b + \sum_{b=i}^{g} 2\mathbb{Z}\omega''_b \quad \text{for } a = 2i-1, i = 1, \ldots, g.$$

Lemma 3.125 *$J_X^{(1)}$ agrees with $\widehat{J}_X^{\mathrm{al},(1)}$ in Definition 2.204.*

Proof The periods of both agree, and thus it is proved. ■

Following [Ba1898a, 340], we express the al function in terms of the sigma function:

Proposition 3.126 *For the branch-point integral $\omega_{B_r} := \int_\infty^{B_r} \nu^{\mathrm{I}}$ in Sect. 2.7.1, we have the relation*

$$\mathrm{al}_r(u) = \gamma''_r \frac{\exp(-{}^t u \eta_{B_r})\sigma(u + \omega_{B_r})}{\sigma(u)}, \quad r = 1, 2, \ldots, 2g, \tag{3.71}$$

where γ''_r is a certain constant, $\eta_{B_r} = -\displaystyle\int_{B_r,\infty} \nu^{\mathrm{II}}$, and

$$\eta_{B_r} = \begin{cases} \sum_{b=1}^{g} \eta'_b & r = 0, \\ -\sum_{b=1}^{a} \eta'_b + \eta''_a & r = 2a, \qquad a = 1, 2, \ldots, g, \\ -\sum_{b=1}^{a-1} \eta'_b + \eta''_a & r = 2a-1, \ a = 1, 2, \ldots, g. \end{cases}$$

Proof Due to the proof of Proposition 3.127, the right-hand side is a function on $J_X^{(a)}$. By comparing zeros and poles of both sides, we have the result. We derive the relations on η_{B_r} similar to Lemma 2.195. ■

Proposition 3.127 *For a lattice point ℓ in $\Gamma^{(r)}$*

$$\mathrm{al}_r(u) = \mathrm{al}_r(u + \ell).$$

Proof We have

$$\frac{\sigma(u+\omega_{B_r}+\ell)}{\sigma(u+\ell)}=\frac{\sigma(u+\omega_{B_r})}{\sigma(u)}\exp(L(\omega_{B_r},\ell)).$$

Let $2\omega'\ell'+2\omega''\ell''\in\Gamma^{(r)}$, i.e., some of ℓ'_i, ℓ''_j belong to $2\mathbb{Z}$. For the $r=2a$ case, $\omega_{B_{2a}}=-\sum_{b=1}^{a}\omega'_b+\omega''_b$, and

$$\begin{aligned}L(\omega_{B_{2a}},\ell)&=-2\sum_{b=1}^{a}{}^t\omega'_b(\eta'\ell'+\eta''\ell'')+2\,{}^t\omega''_a(\eta'\ell'+\eta''\ell'')\\&=-2\sum_{b=1}^{a}[{}^t\omega'_b\eta'\ell'+2\,{}^t\omega''_b\eta'\ell''-\frac{\pi}{2}\sqrt{-1}\ell''_b]\\&\quad+2({}^t\omega'_a\eta''\ell'+{}^t\omega''_a\eta''\ell''+\frac{\pi}{2}\sqrt{-1}\ell'_a]),\end{aligned}\tag{3.72}$$

whereas

$$\begin{aligned}&2\,{}^t(\omega'\ell'+\omega''\ell'')(-\sum_{b=1}^{a}\eta'_b+\eta''_a)\\&=-2\sum_{b=1}^{a}[{}^t\omega'_b\eta'\ell'+2\,{}^t\omega''_b\eta'\ell'']+2({}^t\omega'_a\eta''\ell'+{}^t\omega''_a\eta''\ell'').\end{aligned}\tag{3.73}$$

Hence we have the equality. Other cases are similarly obtained. ■

Using the al function, we can find the behavior of the sigma function at the Abelian integral image of the ramification points in X.

Theorem 3.128

$$\sigma(u+\omega_{B_r})=z_1\cdots z_g(u)\cdot\gamma''_r\exp({}^tu\eta_{B_r})\sigma(u),\quad r=1,2,\ldots,2g,\tag{3.74}$$

where z_i is a local parameter at the ramification point B_r.

As a generalization of the relation $\mathrm{sn}^2u+\mathrm{cn}^2u=1$, we have the following relation in Proposition 2.208 for $\{b_r\mid b_r$ is zero in $f_{\mathrm{II}}(x)$ in (2.66)$\}$, which is known as the Frobenius relation [Mum1984].

Theorem 3.129 *For $\{b_r\mid b_r$ is zero in $f_{\mathrm{II}}(x)$ in (2.66)$\}$*

$$\sum_{b_r\in\mathrm{div}(f_{\mathrm{II}})}\frac{(\gamma''_r)^2\mathrm{e}^{-2\,{}^tu\eta_{B_r}}}{f'_{\mathrm{II}}(b_r)}\sigma(u+\omega_{B_r})^2=\sigma(u)^2_;$$

which shows the embedding of J_X in $\mathbb{C}^{g+1}$ by the homogeneous equation

$$Z_1^2 + Z_2^2 + \cdots + Z_{g+1}^2 = Z_0^2.$$

There are combinations in which we select $(g + 1)$ *elements as the zeros of* $f_{\rm II}$ *of (2.66) in* $\{b_1, b_2, \ldots, b_{2g+1}\}$ *and thus we have* $(2g + 1)!/(g + 1)!g!$ *relations.*

3.6.2 al_{ab} Function

In this subsection, we deal with al_{ab} function. Let us focus on Corollary 3.80 to define the other al functions which are investigated by Baker and Bolza [Ba1898a, Bol1895]:

Lemma 3.130 $\omega_{B_a} + \omega_{B_b}$ *is in the half lattice* $\frac{1}{2}\Gamma_X$.

Proof The divisor of

$$\mu_3(P, B_a, B_b) = \frac{\begin{vmatrix} 1 & b_a & b_a^2 \\ 1 & b_b & b_b^2 \\ 1 & x & x^2 \end{vmatrix}}{\begin{vmatrix} 1 & b_a \\ 1 & b_b \end{vmatrix}} = (x - b_a)(x - b_b)$$

is given by $2(B_a + B_b) - 4\infty$, which is divided by two. The twice of $(B_a + B_b) - 2\infty$ is linearly equivalent to zero or at the lattice point. ■

We, thus, look at the function $\dfrac{\sigma(u + \omega_{B_a} + \omega_{B_b})}{\sigma(u)\sigma_{\natural 2}(\omega_{B_a} + \omega_{B_b})}$ [Ba1898a, Bol1895][Ba1897, p. 318].

Proposition 3.131 *There are a vector* $\varphi_{ab} \in \mathbb{C}^g$ *and constant factor* $\gamma''_{ab} \in \mathbb{C}^\times$ *satisfying*

$$\gamma''_{ab}\frac{\exp(-{}^t u\varphi_{ab})\sigma(u + \omega_{B_a} + \omega_{B_b})}{\sigma(u)} = \sqrt{F(b_a)F(b_b)}\sum_{r=1}^{g}\frac{y_r}{(b_a - x_r)(b_b - x_r)F'(x_r)}.$$

Proof Lemma 3.79 shows it. ■

These φ_{ab} and γ''_{ab} are determined in Lemma 4.48. We thus introduce the other al function, $\mathrm{al}_{ab}(u) := \gamma''_{ab}\dfrac{\exp(-{}^t u\varphi_{ab})\sigma(u + \omega_a + \omega_b)}{\sigma(u)}$. Though we don't argue the properties of this al_{ab} here, it is easy to deal with it by extending the arguments of the al functions in the previous subsection.

3.6.3 Trigonal al Function of Genus One

In this subsection, we represent the al function associated with the elliptic curve E_s : $y(y-s)=x^3$ in terms of the sigma function as mentioned in Sect. 2.8.4 following [FeKMPA2022] based on [EilMO2011].

$$\frac{\sigma(u-v)\sigma(u-\zeta_3 v)\sigma(u-\zeta_3^2 v)}{\sigma(u)^3\sigma(v)^3}=(y(u)-y(v)). \tag{3.75}$$

The translation formula of the sigma function is given by [WhW1927]

$$\sigma(u+2\omega' n+2\omega'' m)=(-1)^{n+m+nm}\mathrm{e}^{(2n\eta'+2m\eta'')(u+n\omega'+m\omega'')}\sigma(u). \tag{3.76}$$

We recall ω_{B_s} defined in (2.133) and Lemma 2.209. and thus we have

$$\begin{aligned}\sigma(u-\zeta_3\omega_{B_s})&=\sigma(u-\omega_{B_s}+2\omega')=(-1)\mathrm{e}^{2\eta'(u+\omega_{B_s}+\omega')}\sigma(u-\omega_{B_s}),\\ \sigma(u-\zeta_3^2\omega_{B_s})&=\sigma(u-\omega_{B_s}+2(\omega'+\omega''))\\ &=(-1)\mathrm{e}^{2(\eta'+\eta'')(u+\omega_{B_s}+\omega'+\omega'')}\sigma(u-\omega_{B_s}).\end{aligned} \tag{3.77}$$

3.6.3.1 al Function of E_s

It is well-known that the Jacobian J_{E_s}, the fundamental domain of the $\wp$ function, is given by $J_{E_s}=\mathbb{C}/\Gamma_s$ but the fundamental domains of Jacobi's sn, cn, dn functions differ from J_{E_s}. In this appendix, we introduce a meromorphic function al, which is the elliptic function version of the al function in Sect. 3.6.4. Its domain also differs from J_{E_s}. By identifying its domain, we give a crucial relation in Proposition 3.136.

Equations (3.75) and (3.77) give the following lemma:

Lemma 3.132

$$\begin{aligned}\frac{\sigma(u-\omega_{B_s})\sigma(u-\zeta_3\omega_{B_s})\sigma(u-\zeta_3^2\omega_{B_s})}{\sigma(u)^3\sigma(\omega_{B_s})^3}&=y(u)-s,\\ -\frac{\sigma(u+\omega_{B_s})\sigma(u+\zeta_3\omega_{B_s})\sigma(u+\zeta_3^2\omega_{B_s})}{\sigma(u)^3\sigma(\omega_{B_s})^3}&=y(u),\\ \frac{\mathrm{e}^{2((2+\zeta_3)\eta')(u+\omega_{B_s})+\pi\sqrt{3}}\sigma(u-\omega_{B_s})^3}{\sigma(u)^3\sigma(\omega_{B_s})^3}&=y(u)-s.\end{aligned}$$

Proof The first and second equalities are directly obtained from (3.75), and we have the third one by the computation

$$\begin{aligned}\sigma(u-\omega_{B_s})\sigma(u-\zeta_3\omega_{B_s})\sigma(u-\zeta_3^2\omega_{B_s})&=\mathrm{e}^{2[2\eta'+\eta'')(u+\omega_{B_s})+2\eta'\omega'+\eta''\omega'']}\sigma(u+\omega_{B_s})^3\\ &=\mathrm{e}^{2(2+\zeta_3^2)\eta')(u+\omega_{B_s})+6\eta'\omega'}\sigma(u-\omega_{B_s})^3.\end{aligned}$$

■

Noting the relation $\sigma(u+\zeta_3^\ell\omega_{B_s}) = \sigma(\zeta_3^\ell(\zeta_3^{-\ell}u+\omega_{B_s})) = \zeta_3^\ell\sigma(\zeta_3^{-\ell}u+\omega_{B_s})$, we define the al function:

Definition 3.133

$$\mathrm{al}_s^{(r)}(u) := \frac{\mathrm{e}^{-\varphi_r u}\sigma(u-\zeta_3^{-r}\omega_{B_s})}{\sigma(u)\sigma(\omega_{B_s})} = \frac{\mathrm{e}^{-\varphi_r u}\zeta_3^{-r}\sigma(\zeta_3^r u-\omega_{B_s})}{\sigma(u)\sigma(\omega_{B_s})},$$

where

$$\varphi_0 := -\frac{2}{3}(2\eta'+\eta''), \quad \varphi_1 := \frac{2}{3}(\eta'+2\eta''), \quad \varphi_2 := \frac{2}{3}(\eta'-\eta'').$$

Then $\mathrm{al}_s^{(r)}(u)$ is a meromorphic function of $\mathbb{C}$ with double periods and has the properties:

Proposition 3.134 *1.* $\prod_{r=0}^{2}\mathrm{al}_s^{(r)}(u) = y - s,$

2. for every $\ell, k \in \mathbb{Z}$,

$$\begin{aligned}
&\mathrm{al}_s^{(0)}(u+2\ell(2\omega'+\omega'')+6k\omega'') = \mathrm{al}_s^{(0)}(u),\\
&\mathrm{al}_s^{(1)}(u+2\ell(\omega'+2\omega'')+6k\omega') = \mathrm{al}_s^{(1)}(u),\\
&\mathrm{al}_s^{(2)}(u+2\ell(\omega'-\omega'')-6k(\omega'+\omega'')) = \mathrm{al}_s^{(2)}(u),
\end{aligned}$$

3. and

$$\varphi_0 = \frac{2}{3}(2+\zeta_3^2)\eta', \quad \varphi_1 = -\frac{2}{3}(1+2\zeta_3^2)\eta', \quad \varphi_2 = -\frac{2}{3}(1-\zeta_3^2)\eta'.$$

It follows that the fundamental domain $J_{E_s,r}$ of $\mathrm{al}_s^{(r)}(u)$ is as given in Fig. 3.1.

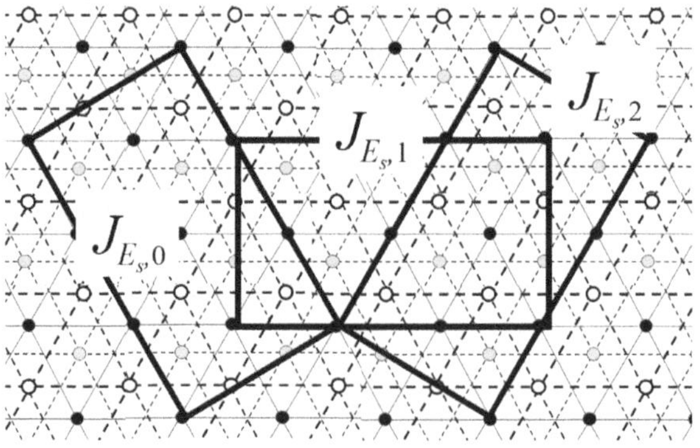

Fig. 3.1 The fundamental domain $J_{E_s,r}$ of $\mathrm{al}_s^{(r)}(u)$: $J_{E_s,r}$ shows the fundamental domain of $\mathrm{al}_s^{(r)}(u)$ $(r = 0, 1, 2)$

Proof *1* and *3* are obvious from Lemma 3.132 and properties of η's in Sect. 2.8.4 respectively. Thus we show *2*. Noting $\omega_{B_s} = \frac{1}{3}(2+\zeta_3)\omega'$ and $(\zeta_3\eta' - \eta'')\omega' = \frac{\pi}{2}\sqrt{-1}$ and letting $\ell_{m,n} := m\omega' + n\omega''$, we have

$$\frac{\sigma(u - \omega_{B_s} + \ell_{2m,2n})}{\sigma(u + \ell_{2m,2n})} = \frac{\mathrm{e}^{-2(m\eta'+n\eta'')\omega_{B_s}}\sigma(u - \omega_{B_s})}{\sigma(u)}.$$

The factor on the right-hand side is determined by

$$\begin{aligned}2(m\eta' + n\eta'')\omega_{B_s} &= \frac{1}{3}(2+\zeta_3)\eta'(2m\omega') + \frac{1}{3}(2+\zeta_3)\eta''(2n\omega')\\ &= \frac{1}{3}(4m\eta'\omega' + m\pi\sqrt{-1} + 2m\eta''\omega') + \frac{1}{3}(-2n\pi\sqrt{-1} + 4n\eta'\omega'' + 2n\zeta_3\eta''\omega').\end{aligned}$$

Then we obtain

$$\frac{\sigma(u - \omega_{B_s} + \ell_{4\ell,2\ell+6k})}{\sigma(u + \ell_{4\ell,2\ell+6k})} = \frac{\mathrm{e}^{\frac{4}{3}\eta'(2\ell\omega'+(\ell+3k)\omega'')+\frac{2}{3}\eta''(2\ell\omega'+(\ell+3k)\omega'')}\sigma(u - \omega_{B_s})}{\sigma(u)},$$

which corresponds to the first relation in *2*. Similarly, since

$$\begin{aligned}2(m\eta' + n\eta'')\zeta_3^2\omega_{B_s} &= -\frac{1}{3}(1+2\zeta_3)\eta'(2m\omega') - \frac{1}{3}(1+2\zeta_3)\eta''(2n\omega')\\ &= -\frac{1}{3}(2m\eta'\omega' + 4m\eta''\omega' + 2m\pi\sqrt{-1}) - \frac{1}{3}(2n\zeta_3\omega'\eta' + 4n\eta''\zeta_3\omega' - n\pi\sqrt{-1}),\end{aligned}$$

we have the second relation in *2* due to

$$\frac{\sigma(u - \zeta_3^2\omega_{B_s} + \ell_{6k+2\ell,4\ell})}{\sigma(u + \ell_{6k+2\ell,4\ell})} = \frac{\mathrm{e}^{-\frac{1}{3}\eta'((6k+2\ell)\omega'+4\ell\omega'')-\frac{2}{3}\eta''((6k+2\ell)\omega'+4\ell\omega'')}\sigma(u - \zeta_3^2\omega_{B_s})}{\sigma(u)}.$$

The third relation in *2* is also obtained because the relation

$$\begin{aligned}2(m\eta' + n\eta'')\zeta_3\omega_{B_s} &= \frac{1}{3}(\zeta_3 - 1)\eta'(2m\omega') + \frac{1}{3}(\zeta_3 - 1)\eta''(2n\omega')\\ &= \frac{1}{3}(2m\eta''\omega' - 2m\eta'\omega' + m\pi\sqrt{-1}) + \frac{1}{3}(2n\omega''\eta'' - 2n\eta'\zeta_3\omega' + n\pi\sqrt{-1})\end{aligned}$$

shows the identity

$$\begin{aligned}&\frac{\sigma(u - \zeta_3\omega_{B_s} + \ell_{3k+\ell,3k-\ell})}{\sigma(u + \ell_{6k+2\ell,6k-2\ell})}\\ &= \frac{\mathrm{e}^{-\frac{1}{3}\eta'((6k+2\ell)\omega'+(6k-2\ell)\omega'')+\frac{1}{3}\eta''((6k+2\ell\omega'+(6k-2\ell)\omega'')}\sigma(u - \zeta_3\omega_{B_s})}{\sigma(u)}.\end{aligned}$$

■

Remark 3.135 We can employ the alternative definition of the "al" function

$$\hat{\mathrm{al}}_s^{(r)}(u) := \frac{\mathrm{e}^{\varphi_r u}\sigma(u+\zeta_3^{-r}\omega_{B_s})}{\sigma(u)\sigma(\omega_{B_s})} = \frac{\mathrm{e}^{-\varphi_r u}\zeta_3^{-r}\sigma(\zeta_3^r u+\omega_{B_s})}{\sigma(u)\sigma(\omega_{B_s})}.$$

Instead of u, we use $-u$ for the $\mathrm{al}_s^{(r)}$ function and have the same relations of $\hat{\mathrm{al}}_s^{(r)}(u)$ as $\mathrm{al}_s^{(r)}$'s. Then we have $\hat{\mathrm{al}}_s^{(r)}(u) = \mathrm{al}_0(u)$ in Proposition 2.212.

Proposition 3.136 $\sigma(u-\zeta_3^{-r}\omega_{B_s}) = \mathrm{e}^{\varphi_r u}\mathrm{al}_s^{(r)}\sigma(u)\sigma(\omega_{B_s})$ *shows the behavior of the sigma function at* $\zeta_3^{-r}\omega_{B_s}$ *and its leading term is the local parameter* z *of the ramification point* $(0, s)$.

Proposition 2.212 and Lemma 3.132, $\mathrm{al}_s^3/s - \mathrm{al}_0^3/s = 1$, show the following theorem as a generalization of the Frobenius θ relation:

Theorem 3.137

$$\frac{\prod_{r=0}^2\sigma(u-\zeta_3^r\omega_{B_s})}{s\sigma(\omega_{B_s})^3} + \frac{\prod_{r=0}^2\sigma(u+\zeta_3^r\omega_{B_s})}{s\sigma(\omega_{B_s})^3} = \sigma(u)^3.$$

$$\frac{(\mathrm{e}^{-\varphi_r u}\sigma(u-\zeta_3^{-r}\omega_{B_s}))^3}{s\sigma(\omega_{B_s})^3} + \frac{(\mathrm{e}^{-\varphi_r u}\sigma(u+\zeta_3^{-r}\omega_{B_s}))^3}{s\sigma(\omega_{B_s})^3} = \sigma(u)^3.$$

This gives the embedding of the Jacobian J_{E_s} *in* $\mathbb{C}^3$ *as a Fermat curve,*

$$Z_1^3 + Z_2^3 = Z_0^3.$$

3.6.4 Non-hyperelliptic al Function of Genus Three

In this subsection, we handle the al function for non-hyperelliptic curves of genus three in terms of the sigma functions. We introduce a meromorphic function on an unramified covering of the Jacobian J_{X_s}, which is a generalization of Jacobi's sn, cn, dn functions, i.e., the al function [MatP2015]; it is also regarded as a generalization of the hyperelliptic al function [Ba1898a, Wei1854]. (Since the simplest non-rational curve with an order-three automorphism is realized by an elliptic curve (E: $y(y-s) = x^3$), we compute the al function of E in Sect. 3.6.3 above, and note similar properties to the ones derived in [MatP2015].) In the same way as the fundamental domains of sn, cn, dn functions are double covering spaces of the Jacobian of genus one, or the fundamental domain of the $\wp$ function, so is the fundamental domain of the al_a function also given as a certain triple covering of J_{X_s}, denoted by $J_{X_s}^{(a,c)}$; $\pi_{\mathrm{al}}^{(a,c)} : J_{X_s}^{(a,c)} \to J_{X_s}$ for a branch point B_a ($a = 0, 1, 2, 3$) and $c = 0, 1, 2$. For a fixed B_a, there is an order-three cyclic action on $J_{X_s}^{(a,c)}$ with respect to c so that the origin of $J_{X_s}^{(a,c)}$ is translated by $\widehat{\zeta_3'}$,

$$\begin{array}{ccccc} & & \overset{\widehat{\zeta_3'}}{\curvearrowleft} & & \\ J_{X_s}^{(a,0)} & \xrightarrow{\widehat{\zeta_3'}} & J_{X_s}^{(a,1)} & \xrightarrow{\widehat{\zeta_3'}} & J_{X_s}^{(a,2)} \\ & \searrow & \downarrow & \swarrow & \\ & & J_{X_s} & & \end{array} \quad . \tag{3.78}$$

Corresponding to $J_{X_s}^{(a,c)}$, we have the triple covering $\widehat{X}_s^{(a)}$ of X_s with respect to the branch point B_a $(a = 0, 1, 2, 3)$; $\varpi_{\widehat{X}} : \widehat{X}_s^{(a)} \to X_s$. There is a natural projection $\widetilde{X}_s \to \widehat{X}_s^{(a)}$ and there exist three different ∞'s on the three sheets in $\widehat{X}_s^{(a)}$. The above action $\widehat{\zeta_3'}$ on $J_{X_s}^{(a,c)}$ corresponds to the choice for which ∞ is assigned to the fixed point in $\widetilde{X}_s$ modulo the triple covering. Further, for each $a = 0, 1, 2, 3$ and $c = 0, 1, 2$, we introduce a certain vector $\varphi_{a;c} \in \mathbb{C}^3$ defined in [MatP2015, Definition 8.2] which is related to the periods of the Jacobian $J_{X_s}^{(a,c)}$:

$$\varphi_{a;c} := \frac{2}{3} \sum_{b=1}^{3} \left(h_{a,b}^{(c)\prime} \eta'_{X_s b} + h_{a,b}^{(c)\prime\prime} \eta''_{X_s b} \right),$$

where h's are defined in Lemma 2.213; this formula corresponds to φ_r in Definition 3.133 of the elliptic curve E case. Proposition 3.96 provides the following theorem:

Theorem 3.138 *For $\gamma_{\infty, P_i} \in \widetilde{X}_s$ of $P_i = (x_i, y_i)$ $(i = 1, 2, 3)$ of X_s, and*

$$u = \widetilde{w}_{\mathfrak{S}}(\gamma_{\infty,P_1}, \gamma_{\infty,P_2}, \gamma_{\infty,P_3}),$$

the al *function defined in Definition 2.214 has the equality*

$$\mathrm{al}_a^{(c)}(u) = \frac{\mathrm{e}^{-{}^t u \varphi_{a;c}} \sigma(u + \widehat{\zeta_3^c} \omega_{B_a})}{\sigma(u) \sigma_{33}(\widehat{\zeta_3^c} \omega_{B_a})}, \quad (a = 0, 1, 2, 3, c = 0, 1, 2)$$

as a function of $u \in J_{X_s}^{(a,c)}$, where $\sigma_{33}(u) := \dfrac{\partial^2}{\partial u_3^2} \sigma(u)$,

$$\sigma_{33}(\omega_{B_a}) = \frac{\sqrt{2}}{\sqrt[3]{f'(b_a)}}.$$

Remark 3.139 We note that the al_a function is a function on a covering space $S^3 \widehat{X}_s^{(a)}$ of $S^3 X_s$ and thus could be viewed as a function on $S^3 \widetilde{X}_s$ because there is a natural projection $\widetilde{X}_s \to \widehat{X}_s^{(a)}$; a point γ of $\widetilde{X}_s$ corresponds to a path in X_s from $\infty \in X_s$ to a point $P \in X_s$ and thus we refer to it as $\gamma = \gamma_{\infty,P}$. If we fix the points γ_2 and γ_3 for $(\gamma_1, \gamma_2, \gamma_3) \in S^3\widetilde{X}_s$ and regard the al_a function as a function of $\gamma_1 = \gamma_{\infty,P}$, the al_a function is a function on $\widehat{X}_s^{(a)}$ and locally agrees with the local parameter at the branch point $\pi_{al;a}^{-1} B_a$ up to ζ_3^c. The al_a function could be characterized as being a meromorphic function over $S^3\widetilde{X}_s$ so that it agrees with the local parameter at B_c if and only if $c = a$.

Theorem 2.220 3 shows the following theorem:

Theorem 3.140

$$\sigma(-u + \widehat{\zeta_3^c}\omega_{B_a}) = \mathrm{e}^{{}^t u \varphi_{a:c}} \mathrm{al}_a^{(c)}(u)\sigma(u)\sigma_{33}(\widehat{\zeta_3^c}\omega_{B_a}). \tag{3.79}$$

Proposition 2.221 shows the generalization of the Frobenius theta relation:

Proposition 3.141

$$\sum_{a=1}^{4} \frac{\mathrm{e}^{-3\,{}^t u \varphi_{a;c}}(\sigma(u + \widehat{\zeta_3^c}\omega_{B_a}))^3}{\sigma_{33}(\widehat{\zeta_3^c}\omega_{B_a})^3} = \sigma(u)^3,$$

or

$$\sum_{a=1}^{4} \frac{\prod_{c=0}^{3} \sigma(u + \widehat{\zeta_3^c}\omega_{B_a})}{\sigma_{33}(\widehat{\zeta_3^c}\omega_{B_a})^3} = \sigma(u)^3.$$

This relation shows the embedding of J_X into $\mathbb{C}^5$, as

$$Z_1^3 + Z_2^3 + Z_3^3 + Z_4^3 = Z_0^3.$$

3.7 Sigma Function for a Degenerating Family of Curves

There are several papers on the degeneration of the sigma functions, e.g., by J. Bernatska and D. Leykin [BerL2019] and J. Bernatska, V. Enolskii, and A. Nakayashiki [BerLN2020]. In this section, we consider the sigma function of the degenerating family of curves of genus three to genus two following [FeKMPA2022]. We deal with the degenerating family of $X_s : y^3 = x(x-s)(x-b_1)(x-b_2)$ for s in the punctured disk D_ε^* as mentioned in Sect. 2.9. We also recall that $X_{\widehat{0}}$ is the normalized curve of X_0 as in Sect. 2.3.3.1.

3.7.1 Sigma Function at a Ramification Point in Terms of al Function

This subsection shows the expression of the sigma function at a ramification point in terms of the al function. From (3.23) and (3.24), the discriminant Δ_s vanishes at $s \to 0$, whereas c_s diverges.

Lemma 3.142 *At $s = 0$, we have*

$$\Delta_s = (-729 b_1^{10} b_2^{10}) s^4 (1 + d_{\geq 1}(s)), \quad c_s = \sqrt{\frac{(2\pi)^3}{A(0)} \frac{s^{-1/3}}{(-729 b_1^{10} b_2^{10})^{1/8}}} (1 + d_{\geq 1}(s)),$$

where $d_{\geq n}(s)$ means formal series of s whose degree is greater than or equal to n in $\mathbb{C}[b_1, b_2][[s]]$.

We recall Theorem 3.38, and then we have the expansions of σ_{X_s} and $\sigma_{X_{\hat{0}}}$ [MatK2013, FeKMPA2022]:[9]

Proposition 3.143 *For every $s \in D_s$, the sigma function of X_s in Sect. 2.9 is expanded as*

$$\sigma_{X_s}(u) = u_1 - u_2^2 u_3 + \frac{1}{20} u_3^5 + \textit{higher-order terms with respect to the weight of } u$$

at the origin of $\mathbb{C}^3$, whereas the sigma function of $X_{\hat{0}}$ in Sect. 2.3.3.1 is expanded as

$$\sigma_{X_{\hat{0}}}(v + \omega_{B_0}) = v_1 + v_2^2/2 + \textit{higher-order terms with respect to the weight of } v$$

at the origin of $\mathbb{C}^2$, where ω_{B_0} is the branch-point integral of $X_{\hat{0}}$ (2.118) in Fig. 2.9.

Remark 3.144 Theorem 3.38 entails that the sigma function can be defined even for X_0. One of the reasons is that the Schur polynomial can be regarded as a sigma function of the monomial curve $Y^3 = X^4$ [BucL2008, EilGOY2017]. Theorem 3.38 and Lemma 3.142 imply that for the limit of $s \to 0$, the c_s diverges, whereas the theta function also vanishes to the order of $s^{1/3}$ but the sigma function is defined even in the limit. Thus this property of the sigma function in Theorem 3.38 is very crucial. However, the proposition only shows the local behavior of σ_{X_s} at a point, or in the image of $\widetilde{w}_{X_s}$ for $\kappa_X^{-1}(\infty, \infty, \infty)$ in $S^3 \widetilde{X}_s$; this does not capture the degeneration phenomenon, and moreover, the argument of $\sigma_{X_{\hat{0}}}$ is translated. We are concerned with the behavior of σ_{X_s} at ω_{B_s}. We investigate instead the relation between $\sigma_{X_{\hat{0}}}$ and σ_{X_s} directly using the result in Theorem 3.38 and the $\mathrm{al}_a^{(c)}$ function in Theorem 3.138.

We now evaluate the sigma function, σ_{X_s} at the Abel–Jacobi map image of the ramification point $\omega_{B_s} = \omega_{B_3} = \displaystyle\int_{\gamma_3} \nu^{\mathrm{I}}$ for γ_3 in Fig. 2.15(b) using the $\mathrm{al}_3^{(0)}$ function.

9 The expansion of the sigma function in [MatK2013] and [KomMP2019] should be read in Proposition 3.143.

Lemma 3.145 *For* $P_i = (x_i, y_i) \in X_s$ $(i = 1, 2, 3)$,

$$x_i = \frac{1}{t_i^3}, \quad y_i = \frac{1}{t_i^4}(1 + d_{\geq 1}(t_i)),$$

and $u^{(i)} = \widetilde{w}_{X_s}(P_i)$ $(i = 1, 2)$, *we have the following relation:*

$$\sigma_{X_s}(u^{(1)} + u^{(2)} + \omega_{B_s}) = -\frac{\sqrt{2}}{\sqrt[3]{b_1 b_2}} s^{-1/3} t_1 t_2 (1 + d_{\geq 1}(s, t_1, t_2, t_3)).$$

Proof For $P_i \in X_s$ and $u^{(i)} = \widetilde{w}_{X_s}(P_i)$, let us consider

$$u = u^{(1)} + u^{(2)} + u^{(3)} + \omega_{B_s},$$

and the limit $u^{(3)} \to 0$ or $P_3 \to \infty$. For the limit,

$$\sigma_{X_s}(u + \omega_{B_s}) = \frac{\sqrt{2} \mathrm{e}^{{}^t u \varphi_{s;0}}}{\sqrt[3]{s(s - b_1)(s - b_2)}} \mathrm{al}_3^{(0)}(u) \sigma_{X_s}(u) \tag{3.80}$$

is reduced to the above relation because a direct computation shows

$$\mathrm{al}_3^{(0)}(u) = \frac{1}{t_1^2 + t_1 t_2 + t_2^2} \frac{1}{t_3}(1 + d_{\geq 1}(s, t_1, t_2, t_3))$$

and due to Proposition 3.143,

$$\sigma_{X_s}(u) = -t_1 t_2 (t_1^2 + t_1 t_2 + t_2^2) t_3 (1 + d_{\geq 1}(s, t_1, t_2, t_3)).$$

Note that the $\varphi_{s;0}$ of $\mathrm{e}^{{}^t u \varphi_{s;0}}$ does not have any effect on this evaluation. We can arrange for the ambiguous third root of unity to 1 by considering proper contours in the integral. ■

By using the arithmetic local parameter t_i at $\infty \in X_{\widehat{0}}$ $(i = 1, 2)$, we note that $v = {}^t(v_1, v_2) \in \mathbb{C}^2$ behaves like $v_1 = -(t_1^2/2 + t_2^2/2)(1 + d_{\geq 1}(t_1, t_2))$ and $v_2 = -(t_1 + t_2)(1 + d_{\geq 1}(t_1, t_2))$, and $\sigma_{X_{\widehat{0}}}$ is expanded by $\sigma_{X_{\widehat{0}}}(v) = t_1 t_2 (1 + d_{\geq 1}(t_1, t_2))$; $v_1 + v_2^2/2 = t_1 t_2 (1 + d_{\geq 1}(t_1, t_2))$. Now let us consider the limit of $s \to 0$ of J_{X_s}. Now let us consider the limit of $s \to 0$ of J_{X_s}.

Remark 3.146 For $s \to 0$, the rank of the matrix $\omega_{X_3}^{\prime -1}$ is reduced to two. This implies that for $s \to 0$, we have a 2-dimensional space. Hence every vector is a combination of two vectors and

$$\mathbb{C}^3 \ni \omega_{X_s}^{\prime -1} \begin{pmatrix} u_1 \\ u_2 \\ u_3 \end{pmatrix} \longrightarrow \begin{pmatrix} z_0 + s^{1/3} A_{11}^{(3)} u_1 \\ z_1 + s^{1/3} A_{21}^{(3)} u_1 \\ s^{1/3} z_2 + s^{1/3} A_{31}^{(3)} u_1 \end{pmatrix} \in \mathbb{C}^3,$$

for certain z's. However, Proposition 3.143 and Remark 3.144 mean that by rescaling u_1 in each component of the image of $\omega'^{-1}_{X_s}$, u_1 survives in the expansion of the sigma function and contributes to the expression in Proposition 3.143.

For $s \in D^*_\varepsilon$, let us consider the image of

$$\widetilde{w}^{(s)}_{X_s} : S^2\widetilde{X}_s \to \mathbb{C}^3, \quad \widetilde{w}^{(s)}_{X_s}(\gamma_1, \gamma_2) := \widetilde{w}_{X_s}(\gamma_1, \gamma_2, \iota_X(B_s)),$$

and consider the projections $p_{2,3} : \mathbb{C}^3 \to \mathbb{C}^2$ and $p^\perp_{2,3} : \mathbb{C}^3 \to \mathbb{C}$ such that $p_{2,3}(u_1, u_2, u_3) = (u_2, u_3)$ and $p^\perp_{2,3}(u_1, u_2, u_3) = (u_1)$ in the meaning of Lemma 2.234.

Proposition 3.147 *The map* $(p_{2,3}\widetilde{w}^{(s)}_{X_s})$ *is well-defined even for the limit* $s \to 0$. *The image of* $\lim_{s\to 0}(p_{2,3}\widetilde{w}^{(s)}_{X_s})$ *is* $\mathbb{C}^2$.

Proof It is obvious that $p_{2,3}\widetilde{w}^{(s)}_{X_s}$ agrees with $\widetilde{w}_{\hat{0},\mathfrak{s}}$ for the limit $s \to 0$ due to Lemmas 2.234 and 2.235. The Abel–Jacobi theorem for $X_{\hat{0}}$ shows that the image of $\widetilde{w}_{\hat{0},\mathfrak{s}}$ is $\mathbb{C}^2$. ■

We state our main theorem in this section.

Theorem 3.148 *For every* $v = {}^t(v_1, v_2) \in \mathbb{C}^2$, *there are* x*-constant sections* γ_1 *and* γ_2 *of* $\mathbf{H}^0(D^*_\varepsilon, S^2\widetilde{\mathfrak{X}})$, *such that*

$$v + \omega_{B_0} = \lim_{s\to 0} p_{2,3}\widetilde{w}^{(s)}_{X_s}(\gamma_1, \gamma_2),$$

and $u = \widetilde{w}^{(s)}_{X_s}(\gamma_1, \gamma_2)(= \widetilde{w}_{X_s}(\gamma_1, \gamma_2) + \omega_{B_s})$, *we have the relation*

$$\sigma_{X_{\hat{0}}}(v + \omega_{B_0}) = \lim_{s\to 0}\left(-\frac{\sqrt[3]{sb_1b_2}}{\sqrt{2}}\sigma_{X_s}(u)\right). \tag{3.81}$$

Proof The limit $\lim_{s\to 0} p_{2,3}(\omega_{B_s})$ in X_s is equal to ω_{B_0} in $X_{\hat{0}}$. The first equation is obtained by the Abel theorem for $\widetilde{X}_{\hat{0}}$ and Proposition 3.147. The sigma function of $X_{\hat{0}}$ is expanded as

$$\sigma_{X_{\hat{0}}}(v + \omega_{B_0}) = v_1 + v_2^2/2 + \cdots = t_1t_2 + \cdots,$$

where t_1 and t_2 are the local parameters at ∞. By noting $u = \widetilde{w}_{X_s}(\gamma_1, \gamma_2) + \omega_{B_s}$, Lemma 3.145, the expansions of both sides at the origin in $\mathbb{C}^2$ agree. On the other hand, the translational formula of σ_{X_s} of $X_s|_{s=0}$ on $(u_1, u_2 + \ell_2, u_3 + \ell_3)$ for ${}^t(\ell_2, \ell_3) \in p_{2,3}\Gamma_s|_{s=0}$ agrees with that of $\sigma_{\hat{0}}$ on $(v_1 + \ell_2, v_2 + \ell_3)$ for ${}^t(\ell_2, \ell_3) \in \Gamma_{\hat{0}} = p_{2,3}\Gamma_s|_{s=0}$ because of Lemmas 2.235, 2.236 and 2.237 and Proposition 3.3. ■

Remark 3.149 The factor $s^{1/3}$ of the right-hand side in (3.81) means that the function is well defined on a triple cover $\widetilde{D^*_\varepsilon}$ of D^*_ε. This corresponds to the group action $\widehat{\zeta_3^*}$ in Lemma 2.90. Corresponding to the group action $\widehat{\zeta_3^*}$, i.e.,$X_{\widehat{0}}^* = \widehat{\zeta_3^*} X_{\widehat{0}}$ and $X_{\widehat{0}}^{**} = \widehat{\zeta_3^{*2}} X_{\widehat{0}}$, the left-hand side of $\sigma_{\widehat{0}}$ of $X_{\widehat{0}}$ in (3.81) is replaced with the sigma functions $\sigma_{X_{\widehat{0}}^*}$ and $\sigma_{X_{\widehat{0}}^{**}}$ of $X_{\widehat{0}}^*$ and $X_{\widehat{0}}^{**}$ rather than $X_{\widehat{0}}$.

On the other hand, since the factor $s^{1/3}$ in the right-hand side in (3.81) comes from $f'(b_s)$ in Theorem 3.138 and the proof of Lemma 3.145, this action $\widehat{\zeta_3^*}$ can be regarded as the action $\widehat{\zeta_3'}$ on the triple covering of the domain of the al function (3.78) in the limit of the degenerating family of curves because z in $X_{\widehat{0}}$ and $\mathrm{al}_3^{(0)}$ in X_0 are the local parameters of the branch points $(B_{X_{\widehat{0}},0})$ and $\pi_{al,3}^{-1}(B_s)$. The symmetries $\widehat{\zeta_3^*}$ and $\widehat{\zeta_3'}$ act on these local parameters. After B_s and B_0 collapse, Theorem 3.148 shows the identification of $\widehat{\zeta_3'}$ with $\widehat{\zeta_3^*}$.

Accordingly, by introducing $\widetilde{w}_{X_s}^{(s,c)}(\gamma_1, \gamma_2) := \widetilde{w}_{X_s}(\gamma_1, \gamma_2) + \widehat{\zeta_3^c}\omega_{B_s}$ for $c = 0, 1, 2$ and using the expansions of $\mathrm{al}_3^{(c)}$, we have other versions of the sigma function as the left-hand side in (3.81) under the action of $\widehat{\zeta_3'}$. We encounter three sigma functions

$$\sigma_{X_{\widehat{\zeta}}}(v), \quad \sigma_{X_{\widehat{\zeta}}^*}(v), \quad \sigma_{X_{\widehat{\zeta}}^{**}}(v) \tag{3.82}$$

of three curves of genus two.

Remark 3.150 In Appendix C of [FeKMPA2022], we investigate the behavior of the Weierstrass sigma function of the degenerating family of the elliptic curves $y(y - s) = x^3$ for $s \to 0$, which behaves similarly to the sigma function for X_s as in Sect. 3.7.1.1. We make some remarks on this comparison.

We note that ${}^t u\varphi_{s;0}$ in the proof of Lemma 3.145 is decomposed to ${}^t u\varphi_{s;0} = p_{2,3}^{\perp}({}^t u\varphi_{s;0}) + p_{2,3}({}^t u\varphi_{s;0})$ by letting $p_{2,3}^{\perp}({}^t u\varphi_{s;0}) := u_1\varphi_{s;0,1}$ and $p_{2,3}({}^t u\varphi_{s;0}) := u_2\varphi_{s;0,2} + u_2\varphi_{s;0,2}$ and in the genus one case, $\mathrm{e}^{\varphi_0 u}$ behaves like the sigma function of $\mathbb{P}$, cf. Appendix C in [FeKMPA2022]. Further, the quasi-periodicity of the sigma function $\sigma(u + \omega_{B_s})$ in the left-hand side of (3.80) for the image of $p_{2,3}$ is determined by $\mathrm{e}^{p_{2,3}({}^t u\varphi_{s;0})}$, al function and $\sigma(u)$. Though due to Lemma 2.236, η'_{1j} and η''_{1j} $(j = 1, 2, 3)$ in $\varphi_{s;0,2}$ become zero for the limit $s \to 0$, we can rewrite the relation (3.81) in Theorem 3.148 at $s = 0$ as

$$\sigma_{X_s}(u) = \frac{\sqrt{2}\mathrm{e}^{p_{2,3}^{\perp}({}^t u\varphi_{s;0})}}{\sqrt[3]{sb_1b_2}} \sigma_{X_{\widehat{0}}}(p_{2,3}(u))(1 + d_{>0}(s^{1/3})) \tag{3.83}$$

for $u = \widetilde{w}_{X_s}(\gamma_1, \gamma_2) + \omega_{B_s}$ and x-constant sections $\gamma_1, \gamma_2 \in \mathbf{H}^0(D_\varepsilon^*, S^2\widetilde{\mathfrak{X}})$, where $d_{>0}(s^{1/3})$ means a formal power series in $s^{1/3}$ with certain coefficients. For the parameter $t = \omega_I' u$ $(u = \omega' t)$, $\mathrm{e}^{p_{2,3}^{\perp}({}^t t\, {}^t\omega'\varphi_{s;0})}$ is the sigma function of $\mathbb{P}$ for the limit $s \to 0$. The expression (3.83) is consistent with the generalized theta function [CleG1866, Ig1956].

3.7.1.1 The Degenerating Family of Curves E_s

Similarly, we consider the degenerating family of the curve $E_s : y(y-s) = x^3$ for $s \to 0$, which is associated with Kodaira's classification of the singular fibers of elliptic surfaces, the IV-type [Kod1963]. We consider the fibers $\pi_{\mathfrak{E}} : \mathfrak{E} \to D_\varepsilon^*$, and the behavior of σ on $\pi_{\mathfrak{E}}^{-1} D_\varepsilon^*$ and its limit $s \to 0$ as in Sect. 2.9.1. More precisely, we also consider the x-constant section over D_ε.

The section of the line bundle on the Jacobian of each $s \in D_\varepsilon^*$ at the branch point ω_s can be evaluated by the relation due to Remark 3.135,

$$\sigma(u + \omega_s) = \mathrm{e}^{\frac{2}{3}((2+\zeta_3^2)\eta')u} \mathrm{al}_0(-u)\sigma(\omega_s)\sigma(-u).$$

In order to evaluate it, we compute $\sigma(\omega_s)$.

Lemma 3.151

$$\sigma(\omega_s) = \frac{\mathrm{e}^{2\sqrt{3}\pi/9}}{\sqrt[9]{-12\sqrt[3]{s}}}.$$

Proof We note $\sigma'(0) = 1$ and due to the translational formula, $\sigma(4\omega' + 2\omega'') = 0$ and $\sigma'(4\omega' + 2\omega'') = -\mathrm{e}^{(4\eta'+2\eta'')(4\omega'+2\omega'')}\sigma(0)$. Using Kiepert's relation, we have the identity [On1998],

$$\frac{\sigma(3u)}{\sigma(u)^9} = 3\wp(u)(\wp(u)^3 - 12s^2).$$

When $u = \omega_s$, both sides vanish. Thus we consider

$$\frac{\sigma(3u)}{3\wp(u)} = (\wp(u)^3 - 12s^2)\sigma(u)^9$$

and its limit $u \to \omega_s$ and then we have

$$\frac{\sigma'(3\omega_s)}{\wp'(\omega_s)} = \frac{-\mathrm{e}^{(4\eta'+2\eta'')(4\omega'+2\omega'')}}{s} = \sigma(\omega_s)^9(-12s^2).$$

Here we use $\sigma'(4\omega' + 2\omega'') = -\mathrm{e}^{(4\eta'+2\eta'')(4\omega'+2\omega'')}\sigma'(0)$ and $\sigma'(0) = 1$, and thus we have

$$\frac{4(2\eta' + \eta'')(2\omega' + \omega'')}{9} = \frac{4(2+\zeta_3^2)(2+\zeta_3)\eta'\omega'}{9} = \frac{2\sqrt{3}\pi}{9}.$$

■

Since we have the expansions of σ and y for E_s,

$$\sigma(u) = u - \frac{1}{120}s^2u^7 + d_{\ge 11}(u), \qquad y(u) = -\frac{1}{u^3} + \frac{1}{2}s^2u^3 + d_{\ge 9}(u), \tag{3.84}$$

the al function is given by

$$\mathrm{al}_0(u) = \frac{\zeta_3}{u}(1 + \frac{1}{3}su^3 - \frac{5}{18}s^2u^6 + d_{\geq 9}(u)),$$

and from Definition 3.133, we have the relation

$$\sigma(u + \omega_s) = -\sigma(-u - \omega_s) = \sigma(-u)\sigma(\omega_s)\mathrm{e}^{\varphi_r u}\mathrm{al}_r(-u).$$

Using these relations, we evaluate its behavior at the branch point for $s \to 0$.

Proposition 3.152 *The sigma function at the branch point is given for* $s \to 0$.

$$\sigma(u + \omega_s) = -\frac{\mathrm{e}^{2\sqrt{3}\pi/9}}{\sqrt[9]{-12\sqrt[3]{s}}}\mathrm{e}^{\frac{2}{3}(2+\zeta_3^2)\eta_0 s^{1/3} u}(1 - \frac{1}{3}su^3 - \frac{103}{360}s^2u^6 + d_{\geq 9}(u)),$$

where $\eta_0 := \dfrac{\pi\sqrt{-1}}{3\sqrt{3}}\dfrac{\Gamma\left(\frac{2}{3}\right)}{\Gamma\left(\frac{1}{3}\right)^2}$, $(\eta' = \eta_0 s^{1/3})$.

We note that this proposition corresponds to Theorem 3.148.

Remark 3.153 First, we note that the Weierstrass sigma function could be defined even for $s = 0$ if we regard it as an expansion of u at the point in Γ_s $(u = 0)$ as in (3.84). Second, we note the behavior at $s = 0$ and $u = \omega_s$. The normalized curve of the singular fiber associated with $\mathbb{C}[x, y]/(x^3 - y^2)$ is the rational curve $\mathbb{P}^1$ whose affine ring is $\mathbb{C}[z]$. In this parametrization z, we have three choices $z = \zeta_3^a x/y$ of $a = 0, 1, 2$, and three biholomorphic normalized rational curves

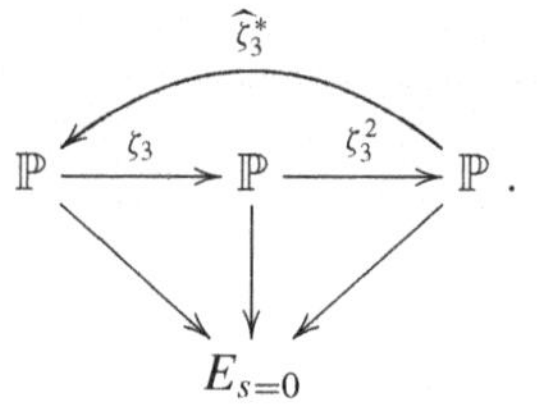

This corresponds to Kodaira's result in [Kod1963].

3.8 Appendix: Quadratic Forms, Riemann Theta and Sigma Functions

This appendix reviews the symplectic group, the quadratic forms, theta series, and sigma function following [Si1972, Mum1981, BucEL1997b, Ig1972, Lan1982, On2013, Ros1986].

3.8.1 Symplectic Group

This subsection mentions the symplectic group and its properties. We recall the symplectic group [GuiS1984, Mum1981]. Let 1_g be the identity matrix of $\mathrm{GL}(g, \mathbb{C})$ and

$$J := \begin{pmatrix} 0 & 1_g \\ -1_g & 0 \end{pmatrix}. \tag{3.85}$$

In this appendix, we use the symbols J and the Jacobian J_X of the curve X. We recall the symplectic group defined in (2.59),

$$\mathrm{Sp}(2g, \mathbb{Z}) = \left\{ \gamma := \begin{pmatrix} a & b \\ c & d \end{pmatrix} \,\middle|\, {}^t\gamma J \gamma = J, a, b, c, d \in \mathrm{Mat}_{\mathbb{Z}}(g) \right\}. \tag{3.86}$$

The condition ${}^t\gamma J \gamma = J$ means, in the notation, ${}^tca = {}^tac$, ${}^tdb = {}^tbd$, and ${}^tad - {}^tcb = 1_g$. Due to $J^{-1} = {}^tJ = -J$, ${}^t\gamma$ belongs to $\mathrm{Sp}(2g, \mathbb{Z})$ because of $\gamma^{-1} = -J\,{}^t\gamma J$. Further, $J\gamma J$ belongs to $\mathrm{Sp}(2g, \mathbb{Z})$ and $\gamma J\,{}^t\gamma = J$. Thus, we also have the relations that $b\,{}^ta = a\,{}^tb$, $d\,{}^tc = c\,{}^td$, and $a\,{}^td - b\,{}^tc = 1_g$. We modify the results in [Ig1972, §10 Lemma 15], [HuaR1949], and [Mum1981, Appendix] to consider the generators in $\mathrm{Sp}(2g, \mathbb{Z})$:

Lemma 3.154 *The symplectic group* $\mathrm{Sp}(2g, \mathbb{Z})$ *is generated by*

$$G_1 := \left\{ \begin{pmatrix} 0 & b \\ -{}^tb^{-1} & 0 \end{pmatrix} \,\middle|\, b \in \mathrm{GL}(g, \mathbb{Z}) \right\}, \quad \textit{and}$$

$$G_2 := \left\{ \begin{pmatrix} a & b \\ 0 & {}^ta^{-1} \end{pmatrix} \,\middle|\, a \in \mathrm{GL}(g, \mathbb{Z}),\ b \in \mathrm{Mat}_{\mathbb{Z}}(g),\ {}^tb = b \right\}.$$

Proof It is known that for $A \in \mathrm{Sp}(2g, \mathbb{Z})$, AJ and JA belong to $\mathrm{Sp}(2g, \mathbb{Z})$ because of ${}^tJ = -J$. Due to Mumford [Mum1981, Appendix], and Hua and Reiner [HuaR1949], $\mathrm{Sp}(2g, \mathbb{Z})$ is generated by the sets of types,

$$\begin{pmatrix} 1_g & b \\ 0 & 1_g \end{pmatrix}, \quad \begin{pmatrix} u & 0 \\ 0 & {}^tu^{-1} \end{pmatrix}, \quad \begin{pmatrix} I_\varepsilon & 1_g - I_\varepsilon \\ I_\varepsilon - 1_g & I_\varepsilon \end{pmatrix},$$

where u is the unimodular, I_ε is a diagonal matrix whose elements are given by 0 or 1 and thus $(1_g - I_\varepsilon)^2 = 1_g - I_\varepsilon$. G_2 contains the first type and the second type. (For $A \in G_1$, AJ is the second type.) For certain $A_1, A_2 \in G_1$, we can construct $A_1 J A_2$ which is the third type. ■

The Siegel upper half space $\mathfrak{H}_g^\circ$ is defined by [Ig1972, Lan1982]

$$\mathfrak{H}_g^\circ := \{ Z \in \mathrm{GL}(g, \mathbb{C}) \mid {}^tZ = Z,\ \mathrm{Im}(Z) > 0 \}. \tag{3.87}$$

The condition means the Riemann bilinear relation (2.75). Here $\mathrm{Im}(Z) > 0$ means that it is positively defined; due to ${}^tZ = Z$, there is a matrix U in $\mathrm{GL}(g, \mathbb{C})$ such that $U^{-1} = {}^tU$ and $Z = {}^tU Z_d U$, where Z_d is a diagonal matrix, $Z_d = \mathrm{diag}(z_1, \cdots, z_g)$, and $\mathrm{Im}(z_i) > 0$. We also introduce the W-normalized Siegel upper half space $\mathfrak{H}_g$,

$$\mathfrak{H}_g := \{(2\omega', 2\omega'') \in \mathrm{GL}(g, \mathbb{C}) \times \mathrm{GL}(g, \mathbb{C}) \mid \omega'^{-1}\omega'' \in \mathfrak{H}_g^\circ\}. \tag{3.88}$$

Here there are a natural projection $\pi_{\mathfrak{H}} : \mathfrak{H}_g \to \mathfrak{H}_g^\circ$, i.e., $\pi_{\mathfrak{H}}(\omega', \omega'') = \omega'^{-1}\omega''$, and an embedding $\iota_{\mathfrak{H}} : \mathfrak{H}_g^\circ \to \mathfrak{H}_g$, i.e., $\iota_{\mathfrak{H}}(1, Z) = (1, Z)$. The condition ${}^tZ = Z$ corresponds to

$$\omega J\, {}^t\omega = 0, \quad \omega''\, {}^t\omega' - \omega'\, {}^t\omega'' = 0$$

for $\omega \in \mathfrak{H}_g$, whereas the condition $\mathrm{Im}(Z) > 0$ means

$$\omega J\, {}^t\overline{\omega} > 0, \quad \omega''\, {}^t\overline{\omega}' - \omega'\, {}^t\overline{\omega}'' > 0.$$

Both are associated with Lemma 2.113 for the W-curves X. Let us mainly consider

$$\mathfrak{W}_g := \left\{ M := \begin{pmatrix} 2\omega' & 2\omega'' \\ 2\eta' & 2\eta'' \end{pmatrix} \in \mathrm{Mat}_{\mathbb{C}}(2g) \;\middle|\; M J\, {}^tM = \frac{2\pi}{\sqrt{-1}} J, \ (2\omega', 2\omega'') \in \mathfrak{H}_g \right\}.$$

Here $M J\, {}^tM = -2\pi\sqrt{-1} J$ corresponds to the Legendre relations,

$$\eta'\, {}^t\eta'' = \eta''\, {}^t\eta', \quad \omega'\, {}^t\omega'' = \omega''\, {}^t\omega', \quad \omega''\, {}^t\eta' - \omega'\, {}^t\eta'' = \frac{\pi\sqrt{-1}}{2} 1_g.$$

It is manifest that because of $J^{-1} = -J$, we have ${}^tM J M = -2\pi\sqrt{-1} J$, i.e.,

$${}^t\omega'\eta' = {}^t\eta'\omega', \quad {}^t\omega''\eta'' = {}^t\eta''\omega'', \quad {}^t\omega''\eta' - {}^t\omega'\eta'' = \frac{\pi\sqrt{-1}}{2} 1_g.$$

For $\gamma = \begin{pmatrix} a & b \\ c & d \end{pmatrix} \in \mathrm{Sp}(2g, \mathbb{Z})$ and $M \in \mathfrak{W}_g$, $\mathrm{Sp}(2g, \mathbb{Z})$ acts on $\mathfrak{W}_g$ by φ_γ,

$$\varphi_\gamma M = M J\, {}^t\gamma J = 2 \begin{pmatrix} \omega'\, {}^td + \omega''\, {}^tc & \omega'\, {}^tb + \omega''\, {}^ta \\ \eta'\, {}^td + \eta''\, {}^tc & \eta'\, {}^tb + \eta''\, {}^ta \end{pmatrix}.$$

This action is related to the exchange of the basis $\{\alpha_i, \beta_i\}_{i=1,\ldots,g}$ of the homology $H_1(X, \mathbb{Z})$ of the W-curves X in Sect. 2.2.7. The action φ_γ manifestly preserves the Legendre relation, $(\varphi_\gamma M) J\, {}^t(\varphi_\gamma M) = 2\pi J/\sqrt{-1}$, especially

$$\omega(J\, {}^t\gamma J) J\, {}^t(\omega(J\, {}^t\gamma J)) = \omega J\, {}^t\omega = 0.$$

By noting $\overline{\gamma} = \gamma$, we also have

$$\omega(J\,{}^t\gamma J)J\,{}^t\overline{(\omega(J\,{}^t\gamma J))} = \omega J\,{}^t\overline{\omega} > 0.$$

By letting $\eta^{\circ\prime} := 4\,{}^t\omega'\eta'$ and $\eta^{\circ\prime\prime} := 4\,{}^t\omega'\eta''$, we also consider

$$\mathfrak{W}_g^\circ := \left\{ M := \begin{pmatrix} 1_g & Z \\ \eta^{\circ\prime} & \eta^{\circ\prime\prime} \end{pmatrix} \in \mathrm{Mat}_{\mathbb{C}}(2g) \;\middle|\; MJ\,{}^tM = \frac{2\pi}{\sqrt{-1}}J,\ (2\omega', 2\omega'') \in \mathfrak{H}_g \right\},$$

which is related to the Legendre relation,

$$Z = {}^tZ, \qquad \eta^{\circ\prime}\,{}^t\eta^{\circ\prime\prime} = \eta^{\circ\prime\prime}\,{}^t\eta^{\circ\prime}, \qquad {}^tZ\eta^{\circ\prime} - \eta^{\circ\prime\prime} = 2\pi\sqrt{-1}1_g.$$

This action φ_γ on $\mathfrak{W}_g$ induces the action of $\mathrm{Sp}(2g, \mathbb{Z})$ on $\mathfrak{H}_g^\circ$, for $\gamma \in \mathrm{Sp}(2g, \mathbb{Z})$ and $(1, Z) \in \mathfrak{H}_g^\circ$,

$$\widehat{\varphi}_\gamma^\circ Z = ({}^td + Z\,{}^tc)^{-1}({}^tb + Z\,{}^ta) = (aZ + b)(cZ + d)^{-1}.$$

3.8.2 Bilinear Forms and Normalized Theta Function

This subsection describes the bilinear forms and theta functions for an Abelian variety following [Lan1982, Ros1986]. We recall that the Jacobi variety J_X of W-curve X is an Abelian variety, $\kappa_J^\circ : \mathbb{C}^g \to J_X = \mathbb{C}^g/\Gamma_X^\circ$ in (2.78).

Let consider a $2g$ free $\mathbb{Z}$-module $\widetilde{\Gamma}$ which is realized as $\widetilde{\Gamma} : \mathbb{Z}^{2g} \to \mathbb{C}^g$ so that $\kappa^\circ : \mathbb{C}^g \to \mathbb{C}^g/\widetilde{\Gamma} =: T$ is an Abelian variety.

Definition 3.155 We let the vector space $\mathbb{C}^g$ associated with the lattice $\widetilde{\Gamma}$ be denoted by $(\mathbb{C}^g, \widetilde{\Gamma})$, and simply $\mathbb{C}^g$ again.

Definition 3.156 [Lan1982, p. 90] We call the bilinear form $\mathbf{E} : \mathbb{C}^g \times \mathbb{C}^g \to \mathbb{R}$ with respect to $\widetilde{\Gamma}$ a *Riemann form* if it satisfies the following conditions:

1. The form $\mathbf{E}$ is alternating, i.e., $\mathbf{E}(z_1, z_2) = -\mathbf{E}(z_2, z_1)$,
2. It takes integral values on $\widetilde{\Gamma} \times \widetilde{\Gamma}$, i.e., $\mathbf{E}|_{\widetilde{\Gamma}\times\widetilde{\Gamma}} : \widetilde{\Gamma} \times \widetilde{\Gamma} \to \mathbb{Z}$, and
3. The form $(z_1, z_2) \mapsto \mathbf{E}(\sqrt{-1}z_1, z_2)$ is symmetric positive.

Definition 3.157 [Ros1986, p. 85] We call the two-form $\mathbf{H} : \mathbb{C}^g \times \mathbb{C}^g \to \mathbb{C}$ with respect to $\widetilde{\Gamma}$ a *Hermitian Riemann form* associated with the Riemann form $\mathbf{E}$ if it satisfies the following conditions:

1. for fixed $z_2 \in \mathbb{C}^g$, $z_1 \mapsto \mathbf{H}(z_1, z_2)$ is a linear map $\mathbb{C}^g \to \mathbb{C}$,
2. for fixed $z_1 \in \mathbb{C}^g$, $z_2 \mapsto \mathbf{H}(z_1, z_2)$ is a antilinear map $\mathbb{C}^g \to \mathbb{C}$,
3. $\mathbf{H}(z_1, z_2) = \overline{\mathbf{H}(z_2, z_1)}$ for all z_1, z_2 in $\mathbb{C}^g$, and
4. the imaginary part of $\mathbf{H}$ is the Riemann form $\mathbf{E}$, i.e., $\mathbf{E} = \mathrm{Im}(\mathbf{H})$.

Further, if $\mathbf{H}(z, z) \geq 0$ for all $z \in \mathbb{C}^g$, we say $\mathbf{H}$ is a positive Hermitian Riemann form, and if $\mathbf{H}(z, z) > 0$ for all $z \in \mathbb{C}^g$, $z \neq 0$, we say $\mathbf{H}$ is a non-degenerate Hermitian Riemann form.

Theorem 3.158 *A complex torus T is the manifold of complex points on an Abelian variety if and only if T possesses a non-degenerate Hermitian Riemann form.*

It is noted that divisors on $\mathbb{C}^g$ are easier to analyze than those on T. Rosen provided two examples [Ros1986, p. 87]:

1. Suppose that g is a meromorphic function over $\mathbb{C}^g$ with a trivial divisor. Then g is a nowhere vanishing holomorphic function and we can write

$$g(z) = \exp(2\pi\sqrt{-1}h(z)),$$

   where $h(z)$ is a holomorphic function on $\mathbb{C}^g$.
2. (Cousin's Theorem) Every divisor on $\mathbb{C}^g$ is principal, i.e. for a family $\{(U_a, f_a)\}$ where the U_a form an open covering of $\mathbb{C}^g$, f_a is meromorphic on U_a not identically zero, we can set all $f_a = f|_{U_a}$, a single function meromorphic on $\mathbb{C}^g$. In cohomological terms, this says $\mathbf{H}^1(\mathbb{C}^g, \mathscr{O}^*_{\mathbb{C}^g}) = (0)$, where $\mathscr{O}^*_{\mathbb{C}^g}$ is the sheaf of nowhere vanishing holomorphic functions on $\mathbb{C}^g$.

Suppose D' is a divisor on T and $\kappa^{\circ *}(D') = D$ is the corresponding periodic divisor on $\mathbb{C}^g$. By Cousin's theorem, $D = \mathrm{div}(f)$. Since $D_{\widetilde{\ell}} = D$ for all $\widetilde{\ell} \in \widetilde{\Gamma}$, we see $f(z + \widetilde{\ell}) = F_{\widetilde{\ell}}(z) f(z)$ for all $\widetilde{\ell} \in \widetilde{\Gamma}$ where $F_{\widetilde{\ell}}(z)$ is a nowhere vanishing holomorphic function. Thus, by above *1*, $F_{\widetilde{\ell}}(z) \exp(2\pi\sqrt{-1}h_{\widetilde{\ell}}(z))$ where $h_{\widetilde{\ell}}(z)$ is holomorphic. Setting $\widetilde{\ell} = \widetilde{\ell}_1 + \widetilde{\ell}_2$, we find the following consistency condition:

$$h_{\widetilde{\ell}_1+\widetilde{\ell}_2}(z) = h_{\widetilde{\ell}_1}(z + \widetilde{\ell}_2) + h_{\widetilde{\ell}_2}(z) \quad \text{modulo } \mathbb{Z}.$$

By choosing $h_{\widetilde{\ell}}(z)$ to be as simple as possible, conventionally we describe it by

$$h_{\widetilde{\ell}}(z) = \mathbf{L}(z, \widetilde{\ell}) + \mathbf{J}(\widetilde{\ell}),$$

where $\mathbf{L}(z, \widetilde{\ell})$ is linear in z, and $\mathbf{J}(\widetilde{\ell})$ is a constant.

Definition 3.159 Let $\mathbf{L} : \mathbb{C}^g \times \widetilde{\Gamma} \to \mathbb{C}$ and $\mathbf{J} : \widetilde{\Gamma} \to \mathbb{C}$ be maps with $\mathbf{L}(z, \widetilde{\ell})$ linear in z for all $\widetilde{\ell} \in \widetilde{\Gamma}$. A *holomorphic theta function* for $\widetilde{\Gamma}$ of type $(\mathbf{L}, \mathbf{J})$ is a holomorphic function ϑ on $\mathbb{C}^g$ such that

$$\vartheta(z + \widetilde{\ell}) = \exp\left(2\pi\sqrt{-1}(\mathbf{L}(z, \widetilde{\ell}) + \mathbf{J}(\widetilde{\ell}))\right)\vartheta(z)$$

for all $z \in \mathbb{C}^g$ and $\widetilde{\ell} \in \widetilde{\Gamma}$.

Further, if $\vartheta(z)$ is given by

$$\vartheta(z) = \exp\left(2\pi\sqrt{-1}(\mathbf{q}(z) + \mathbf{l}(z) + c)\right),$$

where $\mathbf{q}(z)$ is a quadratic form, $\mathbf{l}(z)$ is linear and c is a constant, we call it a *trivial theta function*.

Lemma 3.160 *The trivial theta function has a trivial Riemann form. For* $\vartheta(z) = \exp\left(2\pi\sqrt{-1}(\mathbf{q}(z) + \mathbf{l}(z) + c)\right)$*, let* $\mathbf{B}(z_1, z_2) := \mathbf{q}(z_1 + z_2) - \mathbf{q}(z_1) - \mathbf{q}(z_2)$*. Then* $\mathbf{B}(z_1, z_2)$ *is* $\mathbb{C}$*-bilinear and symmetric, and we have*

$$\vartheta(z + \widetilde{\ell}) = \exp\left(2\pi\sqrt{-1}(\mathbf{B}(z, \widetilde{\ell}) + \mathbf{q}(\widetilde{\ell}) + \mathbf{l}(z))\right)\vartheta(z).$$

Further, $\mathbf{E}(z_1, z_2) = \mathbf{B}(z_1, z_2) - \mathbf{B}(z_2, z_1) = 0$.

A function f on $\mathbb{C}^g$ is periodic with respect to $\widetilde{\Gamma}$ if $f(z + \widetilde{\ell}) = f(z)$ for all $z \in \mathbb{C}^g$, $\ell \in \widetilde{\Gamma}$. Such a function gives rise to a function on $\mathbb{C}^g/\widetilde{\Gamma}$ and conversely, if g is a function on $\mathbb{C}^g/\widetilde{\Gamma}$ then $f = g \circ \kappa^\circ$ is a periodic function on $\mathbb{C}^g$. A Cartier divisor D on $\mathbb{C}^g$ is given by a family $\{(U_a, f_a)\}$ where the U_a form an open covering of $\mathbb{C}^g$, f_a is meromorphic on U_a not identically zero, and f_a/f_b is holomorphic on $U_a \cap U_b$ for all a and b. The divisor is called positive if the function f_a is holomorphic. If $\ell \in \mathbb{C}^g$ the translation of the divisor D by ℓ, D_ℓ, is given by $\{(U_a + \ell, f_a(z - \ell)\}$. If $D_{\widetilde{\ell}} = D$ for all $\widetilde{\ell} \in \widetilde{\Gamma}$, we say D is a periodic divisor. It is obvious that the divisor of a periodic meromorphic function is a periodic divisor. The group of divisors on $\mathbb{C}^g$ (T) are denoted by $\mathscr{D}(\mathbb{C}^g)$ $(\mathscr{D}(T))$. Then κ° induces a homomorphism,

$$\kappa^{\circ *} : \mathscr{D}(T) \to \mathscr{D}(\mathbb{C}^g).$$

The image of κ° consists of the periodic divisors.

Theorem 3.161 [Ros1986, Theorem B] (Poincaré) *For every divisor* D' *on* T*, the periodic divisor* $\kappa^{\circ *}(D') = D$ *is the divisor of a meromorphic theta function,* $\kappa^{\circ *}(D') = (\vartheta)$*. If* D' *is a positive (holomorphic) divisor, then* ϑ *is a holomorphic theta function.*

Theorem 3.162 *For every divisor* D' *on* T*, the periodic divisor* $\kappa^{\circ *}(D') = D$ *is the divisor of a holomorphic theta function,* $\kappa^{\circ *}(D') = (\vartheta)$*. If* D' *is a positive divisor, then* ϑ *is a holomorphic theta function.*

We know that

$$\mathscr{D}(T) \approx \frac{\text{group of theta functions}}{\text{trivial theta functions}}.$$

The holomorphic theta function ϑ of $(\mathbf{L}, \mathbf{J})$ type is related to the Hermitian Riemann form $\mathbf{H}$ as follows:

The consistency conditions for $\widetilde{\ell}_1, \widetilde{\ell}_2 \in \widetilde{\Gamma}$ means

1. $\mathbf{J}(\widetilde{\ell}_1 + \widetilde{\ell}_2) - \mathbf{J}(\widetilde{\ell}_1) - \mathbf{J}(\widetilde{\ell}_2) = \mathbf{L}(\widetilde{\ell}_1, \widetilde{\ell}_2)$ modulo $\mathbb{Z}$,
2. $\mathbf{L}(\widetilde{\ell}_1, \widetilde{\ell}_2) = \mathbf{L}(\widetilde{\ell}_2, \widetilde{\ell}_1)$ modulo $\mathbb{Z}$, and
3. $\mathbf{L}(z, \widetilde{\ell}_1 + \widetilde{\ell}_2) = \mathbf{L}(z, \widetilde{\ell}_1) + \mathbf{L}(z, \widetilde{\ell}_2)$.

Lemma 3.163 *For a given holomorphic theta function ϑ of* $(\mathbf{L}, \mathbf{J})$ *type,* $\widetilde{\mathbf{E}}(z_1, z_2) := \mathbf{L}(z_1, z_2) - \mathbf{L}(z_2, z_1)$ *is a Riemann form, and* $\widetilde{\mathbf{H}}(z_1, z_2) := \widetilde{\mathbf{E}}(\sqrt{-1}z_1, z_2) + \sqrt{-1}\widetilde{\mathbf{E}}(z_1, z_2)$ *is Hermitian Riemann form.*

Thus we call them Riemann forms associated with ϑ.

We show the translational formula of the normalized theta function.

Proposition 3.164 *Let ϑ be a theta function and $\widetilde{\mathbf{H}}$ the associated Hermitian Riemann form. Then there is a theta function $\widetilde{\vartheta}$, unique up to a multiplication constant, such that $\widetilde{\vartheta}/\vartheta$ is a trivial theta function, and*

$$\widetilde{\vartheta}(z + \widetilde{\ell}) = \exp\left(\pi\sqrt{-1}(\widetilde{\mathbf{H}}(z + \frac{\widetilde{\ell}}{2}, \widetilde{\ell}) + 2\pi\sqrt{-1}\widetilde{\mathbf{K}}(\widetilde{\ell}))\right)\widetilde{\vartheta}(z),$$

where $\widetilde{\mathbf{K}}(\widetilde{\ell})$ is real valued and

$$\widetilde{\mathbf{K}}(\widetilde{\ell}_1 + \widetilde{\ell}_2) - \widetilde{\mathbf{K}}(\widetilde{\ell}_1) - \widetilde{\mathbf{K}}(\widetilde{\ell}_2) = \frac{1}{2}\widetilde{\mathbf{E}}(\widetilde{\ell}_1, \widetilde{\ell}_2) \text{ modulo } \mathbb{Z}.$$

Here $\widetilde{\vartheta}$ is called the normalized theta function *associated with ϑ.*

Let $\mathrm{Th}(\mathbf{L}, \mathbf{J})$ be the vector space of theta functions of type $(\mathbf{L}, \mathbf{J})$ and $\mathrm{Th}^\circ(\widetilde{\mathbf{H}}, \widetilde{\mathbf{K}})$ the space of normalized theta functions with the associated Hermitian Riemann form $\widetilde{\mathbf{H}}$ and $\widetilde{\mathbf{K}}$. Both are isomorphic in the meaning of the multiplication of trivial theta functions.

Definition 3.165 For the bases $\{\gamma_i\}_{i=1,2,\ldots,2g}$ of $\widetilde{\Gamma}$, we define $\mathrm{Pf}(\widetilde{\mathbf{E}}) := \mathrm{Pf}(\widetilde{\mathbf{E}}(\gamma_i, \gamma_j))$, where $\mathrm{Pf}(G)$ is the pfaffian for an antisymmetric matrix ${}^tG = -G$ such that $(\mathrm{Pf}(G))^2 = |G|$.

Theorem 3.166 *The vector space of holomorphic theta functions of type* $(\mathbf{L}, \mathbf{J})$ *over $\mathbb{C}$ has dimension* $\mathrm{Pf}(\widetilde{\mathbf{E}})$.

3.8.3 Riemann Theta Function and Related Bilinear Forms

This subsection provides the relations Riemann theta function and related bilinear forms. We apply the previous facts to the Jacobi variety associated with a W-curve X. Assume that $z \in \mathbb{C}^g$, Z (${}^tZ = Z$) belongs to the Siegel upper half space, $Z \in \mathfrak{H}_g^\circ$, and $\Gamma^\circ : \mathbb{Z}^{2g} \to \mathbb{C}^g$ is given by

$$\Gamma^\circ(\ell', \ell'') := \ell' + Z\ell'' = \ell_x + \sqrt{-1}\ell_y,$$

where $\Gamma_x^\circ : \mathbb{Z}^{2g} \to \mathbb{R}^g$ and $\Gamma_y^\circ : \mathbb{Z}^{2g} \to \mathbb{R}^g$. We identify image $\Gamma^\circ(\mathbb{Z}^{2g})$ with Γ° by fixing Z. We let $\ell^\circ = \ell' + Z\ell'' = \ell_x + \sqrt{-1}\ell_y \in \Gamma^\circ$.

Definition 3.167 For $z \in (\mathbb{C}^g, \Gamma^\circ)$, we use the convention,

$$z = z' + Zz'' = x + \sqrt{-1}y \in \mathbb{C}^g,$$

where z', z'', x and y belongs to $\mathbb{R}^g$. We also define $X := (Z + \overline{Z})/2 \in \mathrm{Mat}_{\mathbb{R}}(g)$ and $Y := (Z - \overline{Z})/2\sqrt{-1} \in \mathrm{Mat}_{\mathbb{R}}(g)$, i.e., $Z = X + \sqrt{-1}Y$.

Then we have the following lemma:

Lemma 3.168 *1.* $Y \in \mathrm{GL}(g, \mathbb{R})$, $Y > 0$.
2. ${}^tX = X$, ${}^tY = Y$.
3. $\begin{pmatrix} x \\ y \end{pmatrix} = \begin{pmatrix} 1_g & X \\ 0 & Y \end{pmatrix} \begin{pmatrix} z' \\ z'' \end{pmatrix}$.
4. $\begin{pmatrix} z' \\ z'' \end{pmatrix} = \begin{pmatrix} 1_g & -XY^{-1} \\ 0 & Y^{-1} \end{pmatrix} \begin{pmatrix} x \\ y \end{pmatrix}$.
5. *The map* $z \mapsto \sqrt{-1}z$ *has the matrix representation for* ${}^t(z', z'')$ *as* $\widehat{\sqrt{-1}} \begin{pmatrix} z' \\ z'' \end{pmatrix} = \begin{pmatrix} -XY^{-1} & -XY^{-1}X - Y \\ Y^{-1} & Y^{-1}X \end{pmatrix} \begin{pmatrix} z' \\ z'' \end{pmatrix}$.
6. ${}^tz_1'z_2'' = {}^tx_1 Y^{-1} y_2 + {}^ty_1 Y^{-1} X Y^{-1} y_2$.
7. ${}^tz_1'z_2'' - {}^tz_2'z_1'' = {}^tx_1 Y^{-1} y_2 - {}^tx_2 Y^{-1} y_1$.

Proof Since $Y = \mathrm{Im}Z$ is positive definite, *1* is obtained. *2* and *3* are obvious. We note $\begin{pmatrix} 1_g & -XY^{-1} \\ 0 & Y^{-1} \end{pmatrix} = \begin{pmatrix} 1_g & X \\ 0 & Y \end{pmatrix}^{-1}$. We obtain *5* because of ${}^t\begin{pmatrix} 1_g & -XY^{-1} \\ 0 & Y^{-1} \end{pmatrix} \begin{pmatrix} 0 & 1_g \\ -1_g & 0 \end{pmatrix}$ $\times \begin{pmatrix} 1_g & X \\ 0 & Y \end{pmatrix} = \begin{pmatrix} -XY^{-1} & -XY^{-1}X - Y \\ Y^{-1} & Y^{-1}X \end{pmatrix}$. Since we have

$${}^t\begin{pmatrix} 1_g & -XY^{-1} \\ 0 & Y^{-1} \end{pmatrix} \begin{pmatrix} 0 & 1_g \\ -1_g & 0 \end{pmatrix} \begin{pmatrix} 1_g & -XY^{-1} \\ 0 & Y^{-1} \end{pmatrix} = \begin{pmatrix} 0 & Y^{-1} \\ -Y^{-1} & 0 \end{pmatrix},$$

we obtain *6* and *7*. The others are shown by direct computations. ■

We introduce some quantities.

Definition 3.169 1. For $z_1, z_2 \in (\mathbb{C}^g, \Gamma^\circ)$, the symplectic bilinear form is defined by

$$\langle z_1, z_2 \rangle^\circ := ({}^tz_1', {}^tz_1'') J \begin{pmatrix} z_2' \\ z_2'' \end{pmatrix} = {}^tz_1'z_2'' - {}^tz_1''z_2'.$$

2. For $z_1, z_2 \in (\mathbb{C}^g, \Gamma^\circ)$, we define the Hermitian form, $\widehat{H}^\circ(z_1, z_2) := {}^tz_1 Y^{-1} \overline{z_2}$.
3. For $z_1, z_2 \in (\mathbb{C}^g, \Gamma^\circ)$, we define the skew-symmetric $\mathbb{R}$-bilinear form,

$$\widehat{E}^\circ(z_1, z_2) := \widehat{L}^\circ(z_1, z_2) - \widehat{L}^\circ(z_2, z_1) = \mathrm{Im}(\widehat{H}^\circ(z_1, z_2)),$$

where $\widehat{L}^\circ(z_1, z_2) := \dfrac{1}{2}\widehat{H}^\circ(z_1, z_2) = \dfrac{1}{2}{}^tz_1 Y^{-1} \overline{z_2}$.

The properties of these quantities are displayed in the following two lemmas:

Lemma 3.170 *For $z_1, z_2 \in (\mathbb{C}^g, \Gamma^\circ)$, the following hold:*

1. *$\widehat{E}^\circ$ is a Riemann form on the complex vector space $\mathbb{C}^g$ with respect to Γ°, i.e.,*
   a. $\widehat{E}^\circ(z_1, z_2) = ({}^t y_1 Y^{-1} x_2 - {}^t x_1 Y^{-1} y_2) = -({}^t z_1' z_2'' - {}^t z_2' z_1'') = -\langle z_1, z_2\rangle^\circ$,
   b. $\widehat{E}^\circ(z_1, z_2) = -\widehat{E}^\circ(z_2, z_1) \in \mathbb{R}$, *and*
   c. *for* $\ell_1^\circ, \ell_2^\circ \in \Gamma^\circ$, $\widehat{E}^\circ(\ell_1^\circ, \ell_2^\circ) \in \mathbb{Z}$.
2. *$\widehat{H}^\circ$ is a nondegenerate Hermitian Riemann form associated with $\widehat{E}^\circ$, i.e.,*
   a. $\widehat{H}^\circ(\sqrt{-1}z_1, z_2) = \sqrt{-1}\widehat{H}^\circ(z_1, z_2)$,
   b. $\widehat{H}^\circ(z_1, \sqrt{-1}z_2) = -\sqrt{-1}\widehat{H}^\circ(z_1, z_2)$, *and*
   c. $\overline{\widehat{H}^\circ(z_1, z_2)} = \widehat{H}^\circ(z_2, z_1)$, *and* $\widehat{H}^\circ(z, z) \geq 0$ *with equality only if* $z = 0$.
3. $\mathrm{Pf}(\widehat{E}^\circ) = 1$.
4. $\widehat{H}^\circ(z + \ell^\circ, Z) - \widehat{H}^\circ(z, Z) = 2\sqrt{-1}\,{}^t(z + \frac{1}{2}\ell^\circ)Y^{-1}\overline{\ell^\circ} - 2\sqrt{-1}\langle z, \ell^\circ\rangle^\circ$.

Proof *1* is proved as follows: $\widehat{E}^\circ(z_1, z_2) = ({}^t y_1 Y^{-1} x_2 - {}^t x_1 Y^{-1} y_2) = -({}^t z_1' z_2'' - {}^t z_2' z_1'') = -\langle z_1, z_2\rangle^\circ$, and thus $\widehat{E}^\circ(z_1, z_2) = -\widehat{E}^\circ(z_2, z_1)$. For $\ell_a^\circ = \ell_a' + Z\ell_a'' \in \Gamma^\circ$, $(a = 1, 2)$, $\widehat{E}^\circ(\ell_1^\circ, \ell_2^\circ)$ belongs to $\mathbb{Z}$.

Let us consider *2*. Further, $\widehat{E}^\circ(\sqrt{-1}z_1, \sqrt{-1}z_2) = \widehat{E}^\circ(z_1, z_2)$ (compatibility of the complex structure) and $\widehat{E}^\circ(\sqrt{-1}z_1, z_2) = \mathrm{Re}(\widehat{H}^\circ(z_1, z_2))$; it means that $\widehat{H}^\circ(z_1, z_2) = \widehat{E}^\circ(\sqrt{-1}z_1, z_2) + \sqrt{-1}\widehat{E}^\circ(z_1, z_2)$.

We note that $Y > 0$, ${}^t Y = Y$. Since $\widehat{H}^\circ(z_1, z_2) = {}^t(z_1' + Xz_1'')Y^{-1}(z_2' + Xz_2'') - {}^t z_1'' Y z_2'' + \sqrt{-1}({}^t(z_1' + Xz_1'')z_2'' - {}^t z_1''(z_2' + Xz_2''))$. and $Y \in GL(n, \mathbb{R})$, it is obvious that $\widehat{H}^\circ(\sqrt{-1}z_1, z_2) = \sqrt{-1}\widehat{H}^\circ(z_1, z_2)$, $\widehat{H}^\circ(z_1, \sqrt{-1}z_2) = -\sqrt{-1}\widehat{H}^\circ(z_1, z_2)$, $\overline{\widehat{H}^\circ(z_1, z_2)} = \widehat{H}^\circ(z_2, z_1)$, and $\widehat{H}^\circ(z, z) \geq 0$ with equality only if $z = 0$.

Moreover, from the definition, we obtain that $\mathrm{Pf}(\widehat{E}^\circ) = 1$ in *3*. The relation *4* is directly obtained from the definition of $\widehat{H}^\circ$. ■

In order to introduce the Riemann theta functions, we define the bilinear form on $(\mathbb{C}^g, \Gamma^\circ)$.

Definition 3.171 For $z, w \in (\mathbb{C}^g, \Gamma^\circ)$,

$$D(z, w) := D(z, w', w'') := D(z, w', w''; Z) := \frac{1}{2}{}^t w'' Z w'' + {}^t w''(z + w').$$

Lemma 3.172 *For $z, w_1, w_2 \in \mathbb{C}^g$, the following hold:*

1. $D(z, w) = {}^t w''(z + w' + \frac{1}{2}Zw'') = {}^t w''(z + \frac{1}{2}w) + \frac{1}{2}{}^t w'' w'$.
2. $D(z, w_1 + w_2) = D(z + w_2, w_1) + D(z + w_1', w_2)$.
3. $D(z + w_1, w_2) = D(z, w_2) + {}^t w_2'' w_1$.

Proof The relations *1*, *2*, and *3* are obtained from the computations

$$
\begin{aligned}
D(z, w_1 + w_2) &= \frac{1}{2}{}^t(w_1'' + w_2'')Z(w_1'' + w_2'') + {}^t(w_1'' + w_2'')(z + w_1' + w_2') \\
&= \frac{1}{2}{}^t w_1'' z w_1'' + {}^t w_1''(z + w_2' + w_1') \\
&\quad + \frac{1}{2}{}^t w_2'' Z w_2'' + {}^t w_2''(z + w_1' + w_2') + \frac{1}{2}({}^t w_1'' Z w_2'' + {}^t w_2'' Z w_1'') \\
&= \frac{1}{2}{}^t w_1'' Z w_1'' + {}^t w_1''(z + w_2' + w_1' + Z w_2'') + \frac{1}{2}{}^t w_2'' Z w_2'' + {}^t w_2''(z + w_1' + w_2') \\
&= D(z + w_2, w_1) + D(z + w_1', w_2).
\end{aligned}
$$
■

Definition 3.173 The Riemann theta function and its shifted version over $\mathbb{C}^g \times \mathfrak{H}_g^\circ$ are defined by

$$
\theta(z) := \theta(z; Z) := \sum_{n'' \in \mathbb{Z}^g} \exp\left(2\pi\sqrt{-1}D(z, 0, n''; Z)\right),
$$

$$
\theta(z, \delta) := \theta(z, \delta; Z) := \theta\begin{bmatrix}\delta'' \\ \delta'\end{bmatrix}(z; Z) := \sum_{n'' \in \mathbb{Z}^g} \exp\left(2\pi\sqrt{-1}D(z, \delta', n'' + \delta'', Z)\right).
$$

We may regard the latter as

$$
\theta(z, \delta) = \sum_{n'' \in \mathbb{Z}^g} \exp\left(2\pi\sqrt{-1}D(z, Zn'' + \delta)\right) = \sum_{n'' \in \mathbb{Z}^g} \exp\left(2\pi\sqrt{-1}D(z, n + \delta)\right)
$$

using $\delta = \delta' + Z\delta''$ and $n = 0_g + Zn''$. From Lemma 3.172 3, by letting $(w_1', w_1'', w_2', w_2'') = (0_g, n, \delta', \delta'')$, we have $D(z, n + \delta) = D(z + \delta, n) + D(z + 0_g, \delta)$, and thus obtain the relation between both theta functions,

$$
\begin{aligned}
\theta(z, \delta) &= \exp\left(2\pi\sqrt{-1}(D(z, \delta)\right) \sum_{n'' \in \mathbb{Z}^g} \exp\left(2\pi\sqrt{-1}D(z + \delta, Zn'')\right) \\
&= \exp\left(2\pi\sqrt{-1}D(z, \delta)\right)\theta(z + \delta).
\end{aligned} \tag{3.89}
$$

Proposition 3.174 *Let* $\frac{1}{2}\mathbb{Z}_2 := \{0, \frac{1}{2}\}$.

1. *For* $\delta = (0, 0)$, $\theta(z, \delta) = \theta(z)$ *and* $\theta(-z, \delta) = \theta(z, \delta)$.
2. *For* $\delta', \delta'' \in (\frac{1}{2}\mathbb{Z}_2)^g$, $\theta(-z, \delta) = (-1)^{4{}^t\delta''\delta'}\theta(z, \delta)$.
3.
$$
2^{(g-1)}(2^g - 1) = \#\left\{\begin{pmatrix}\delta'' \\ \delta'\end{pmatrix} \in (\frac{1}{2}\mathbb{Z}_2)^{2g} \mid 4\,{}^t\delta''\delta' = 1 \text{ modulo } 2\right\},
$$

$$
2^{(g+1)}(2^g + 1) = \#\left\{\begin{pmatrix}\delta'' \\ \delta'\end{pmatrix} \in (\frac{1}{2}\mathbb{Z}_2)^{2g} \mid 4\,{}^t\delta''\delta' = 0 \text{ modulo } 2\right\}.
$$

Proof *1* is proved by $D(-z, 0, n''; Z) = \frac{1}{2}{}^t(n'')Z(n'') - {}^t n'' z = D(z, 0, -n''; Z)$. *2* is given by

$$\begin{aligned}
&D(-z, \delta', n'' + \delta'', Z)\\
&= \frac{1}{2}{}^t(n'' + \delta'')Z(n'' + \delta'') + {}^t(n'' + \delta'')(-z + \delta')\\
&= \frac{1}{2}{}^t(n'' + \delta'')Z(n'' + \delta'') - {}^t(n'' + \delta'')(z - \delta')\\
&= \frac{1}{2}{}^t(n'' + \delta'')Z(n'' + \delta'') - {}^t(n'' + \delta'')(z + \delta') + 2\,{}^t n''\delta' + 2\,{}^t\delta''\delta'\\
&= \frac{1}{2}{}^t(m'' + \delta'')Z(m'' + \delta'') + {}^t(m'' + \delta'')(z + \delta') + 2\,{}^t\delta''\delta' \text{ modulo } \mathbb{Z},
\end{aligned}$$

where $m'' \in \mathbb{Z}^g$ is an element such that $m'' + \delta'' = -n'' - \delta''$. On *3*, see [FarK1991, p. 303] for example. ■

Proposition 3.175 *We have the following translation formulae for* $\ell^\circ \in \Gamma^\circ$:

1. *The phase factor* $\varphi(z, \ell^\circ)$ *given by* $\theta(z + \ell^\circ) = \exp(\varphi(z, \ell^\circ))\theta(z)$ *is expressed by*

$$\varphi(z, \ell^\circ) = 2\pi\sqrt{-1}\,{}^t\ell''\left(z - \frac{1}{2}Z\ell''\right).$$

2. *The phase factor* $\varphi(z, \ell^\circ, \delta)$ *given by* $\theta(z + \ell^\circ, \delta) = \exp(\varphi(z, \ell^\circ, \delta))\theta(z, \delta)$ *is expressed by*

$$\begin{aligned}
\varphi(z, \ell^\circ, \delta) &= 2\pi\sqrt{-1}\left({}^t\delta''\ell' - {}^t\ell''\left(z + \delta + \frac{1}{2}Z\ell''\right)\right)\\
&= 2\pi\sqrt{-1}\left({}^t(\ell'' + \delta'')\ell' - D(z + \delta', \ell^\circ)\right)\\
&= 2\pi\sqrt{-1}\left({}^t(\ell'' + \delta'')\ell' - D(z, \ell^\circ) - {}^t\ell''\delta'\right).
\end{aligned}$$

Proof Let $n = 0_g + Zn''$ and $\widetilde{n}'' := \ell'' + n''$. Lemma 3.172 3 (substituting $w_1 = n + \delta$, $w_2 = \ell^\circ$) means that $D(z + \ell^\circ, n + \delta)$ is equal to

$$\begin{aligned}
&D(z, n + \delta + \ell^\circ) - D(z + 0_g + \delta', \ell^\circ)\\
&= D(z, Z(n'' + \ell'') + \delta) + {}^t(n'' + \ell'' + \delta'')\ell' - {}^t\ell''\left(z + \delta' + \ell' + \frac{1}{2}Z\ell''\right)\\
&= D(z, Z\widetilde{n}'' + \delta) + {}^t\delta''\ell' - {}^t\ell''\left(z + \delta' + \frac{1}{2}Z\ell''\right) \quad \text{module } \mathbb{Z}.
\end{aligned}$$

Thus $\varphi(z, \ell^\circ, \delta) = 2\pi\sqrt{-1}\left({}^t\delta''\ell' - {}^t\ell''\left(z + \delta' + \frac{1}{2}Z\ell''\right)\right)$. The second term is equal to

$$ {}^t\ell''\left(z + \delta' + \frac{1}{2}Z\ell''\right) = {}^t\ell''\left(z + \delta' + \ell' + \frac{1}{2}Z\ell''\right) - {}^t\ell''\ell' = D(z + \delta', \ell^\circ) - {}^t\ell''\ell' $$

from Lemma 3.172 3. By letting $\delta = 0$ formally, we have $\varphi(z, \ell^\circ) = 2\pi\sqrt{-1}\,{}^t\ell''\left(z - \frac{1}{2}Z\ell''\right)$. ■

3.8.4 Symplectic Action on Theta Function

This subsection shows the symplectic action on the Riemann theta functions. Thus, we express the theta functions in terms of $G(A, c)$ in Lemma 3.197.

Lemma 3.176 *For $A \in \mathrm{GL}(g, \mathbb{C})$ with $\mathrm{Im} A > 0$, ${}^tA = A$ and $c \in \mathbb{C}^g$,*

$$ G(A, z) = \sum_{m \in \mathbb{Z}^g} \exp(2\pi\sqrt{-1}D(z, 0, m; A)) = \theta(z, A). $$

$$ G(A, z) = \sqrt{\frac{\sqrt{-1}^g}{\det[A]}} \exp\left[-\pi\sqrt{-1}\,{}^tzA^{-1}z\right] G\left(-A^{-1}, A^{-1}z\right). $$

Lemma 3.177 *Let $\gamma_1 := \begin{pmatrix} a & b \\ 0 & {}^ta^{-1} \end{pmatrix} \in \mathrm{Sp}(2g, \mathbb{Z})$, $({}^ta = a)$ and $\widehat{\varphi}^\circ_{\gamma_1} Z = (aZ + b)\,{}^ta$, (because of ${}^t(\widehat{\varphi}^\circ_{\gamma_1} Z) = \widehat{\varphi}^\circ_{\gamma_1} Z$).*

1. *For the action,*

$$ \widehat{\varphi}^\circ_{\gamma_1}\theta(z; Z) := \theta(\widehat{\varphi}^\circ_{\gamma_1} z; \widehat{\varphi}^\circ_{\gamma_1} Z) := \theta(az; (aZ + b)\,{}^ta), $$

we have

$$ \widehat{\varphi}^\circ_{\gamma_1}\theta(z; Z) = \theta(z + \frac{1}{2}a^{-1}(a\,{}^tb)_D; Z), $$

where $B_D := {}^t(B_{11}, \ldots, B_{gg}) \in \mathbb{Z}^g$ for $B \in \mathrm{Mat}_{\mathbb{Z}}(g)$.

2. *For $\widehat{\varphi}^\circ_{\gamma_1}\begin{pmatrix} \delta'' \\ \delta' \end{pmatrix} := \begin{pmatrix} {}^ta^{-1} & 0 \\ -b & a \end{pmatrix}\begin{pmatrix} \delta'' \\ \delta' \end{pmatrix} + \begin{pmatrix} 0 \\ \frac{1}{2}(a\,{}^tb)_D \end{pmatrix}$, which is consistent with $\widehat{\varphi}^\circ_{\gamma_1}(\delta) = a\delta = \widehat{\varphi}^\circ_{\gamma_1}(\delta') + \widehat{\varphi}^\circ_{\gamma_1}(Z)\widehat{\varphi}^\circ_{\gamma_1}(\delta'')$, and the action,*

$$ \widehat{\varphi}^\circ_{\gamma_1}\theta(z, \delta, Z) := \theta(\widehat{\varphi}^\circ_{\gamma_1} z, \widehat{\varphi}^\circ_{\gamma_1}\delta; \widehat{\varphi}^\circ_{\gamma_1} Z) := \theta(az, \widehat{\varphi}^\circ_{\gamma_1}\delta; (aZ + b)\,{}^ta), $$

we have

$$\widehat{\varphi}^{\circ}_{\gamma_1}\theta(z,\delta,Z) = (-1)^{{}^t\delta'' a^{-1}((a\,{}^tb)_D - b\delta'')}\theta(z,\delta,Z).$$

The result is consistent with genus one case, e.g., [Ra1973, (80.2)].

Proof Since a is regular, the map $a : \mathbb{Z}^g \to \mathbb{Z}^g$ is bijective, and thus we let ${}^tan =: \widetilde{n}$. Further, we have $(b\,{}^ta) = a\,{}^tb = {}^t(b\,{}^ta)$ from the definition. Noting the proof of Lemma 3.197 1, for $n \in \mathbb{Z}^g$, we use the relation, $(-1)^{{}^tnb\,{}^tan} = (-1)^{{}^tn(a\,{}^tb)_D} = (-1)^{{}^t\widetilde{n}a^{-1}(a\,{}^tb)_D}$. Then we have

$$\begin{aligned}
D(az,0,n;\widehat{\varphi}^{\circ}_{\gamma_1}Z) &= \frac{1}{2}{}^tnaZ\,{}^tan + \frac{1}{2}{}^tnb\,{}^tan + {}^tnaz\\
&= \frac{1}{2}{}^t\widetilde{n}Z\widetilde{n} + {}^t\widetilde{n}(z + \frac{1}{2}a^{-1}(a\,{}^tb)_D) \quad \text{modulo } \mathbb{Z}\\
&= D(z + \frac{1}{2}a^{-1}(a\,{}^tb)_D, 0, \widetilde{n}; Z),
\end{aligned}$$

$$\begin{aligned}
&D(az, -b\delta'' + a\delta', n + {}^ta^{-1}\delta'' + \frac{1}{2}(a\,{}^tb)_D; (aZ+b)\,{}^ta)\\
&= \frac{1}{2}{}^t(n + {}^ta^{-1}\delta'')(aZ+b)\,{}^ta(n + {}^ta^{-1}\delta'')\\
&\quad + {}^t(n + {}^ta^{-1}\delta'')(az - b\delta'' + a\delta' + \frac{1}{2}(a\,{}^tb)_D)\\
&= \frac{1}{2}{}^t(\widetilde{n}+\delta'')Z(\widetilde{n}+\delta'') + \frac{1}{2}\left({}^tn(b\,{}^ta)n + {}^tnb\delta'' + {}^t\delta''a^{-1}b\,{}^tan + {}^t\delta''a^{-1}b\delta''\right)\\
&\quad + {}^t(\widetilde{n}+\delta'')(z+\delta') - {}^tnb\delta'' - {}^t\delta''a^{-1}b\delta'' + \frac{1}{2}\left({}^tn(a\,{}^tb)_D + {}^t\delta''a^{-1}(a\,{}^tb)_D\right)\\
&= \frac{1}{2}{}^t(\widetilde{n}+\delta'')Z(\widetilde{n}+\delta'') + {}^t(\widetilde{n}+\delta'')(z+\delta') + \frac{1}{2}{}^t\delta''a^{-1}((a\,{}^tb)_D - b\delta'') \quad \text{module } \mathbb{Z}\\
&= D(z,\delta',\widetilde{n}+\delta''; Z) + \frac{1}{2}{}^t\delta''a^{-1}((a\,{}^tb)_D - b\delta'').
\end{aligned}$$

Here we use the facts that 1) ${}^t\delta''a^{-1}(b\,{}^ta)n = {}^t\delta''\,{}^tbn = {}^tnb\delta''$, and 2) ${}^tn(a\,{}^tb)_D = 0$ module $\mathbb{Z}$, and the trick in the proof of *1*. ■

Lemma 3.178 *Let* $\gamma_2 := \begin{pmatrix} 0 & b \\ -{}^tb^{-1} & 0 \end{pmatrix} \in \mathrm{Sp}(2g,\mathbb{Z})$, *(* ${}^tb = b$ *) and* $\widehat{\varphi}^{\circ}_{\gamma_2}Z = -bZ^{-1}\,{}^tb$, *(because of* ${}^t(\widehat{\varphi}^{\circ}_{\gamma_2}Z) = \widehat{\varphi}^{\circ}_{\gamma_2}Z$*).*

1. For the action,

$$\widehat{\varphi}^{\circ}_{\gamma_2}\theta(z,Z) := \theta(\widehat{\varphi}^{\circ}_{\gamma_2}z; \widehat{\varphi}^{\circ}_{\gamma_2}Z) = \theta(bZ^{-1}z; -bZ^{-1}\,{}^tb),$$

we have

$$\widehat{\varphi}^{\circ}_{\gamma_2}\theta(z,Z) = \mathrm{e}^{\rho_c}\sqrt{\det[Z]}\exp\left[\pi\sqrt{-1}\,{}^tzZ^{-1}z\right]\theta(z,Z),$$

where the phase $\rho_c = i_c\pi\sqrt{-1}/2$ *is indeterminate in this stage.*

2. *For* $\widehat{\varphi}^\circ_{\gamma_2}\begin{pmatrix}\delta''\\ \delta'\end{pmatrix} := \begin{pmatrix}0 & -{}^t b^{-1}\\ {}^t b & 0\end{pmatrix}\begin{pmatrix}\delta''\\ \delta'\end{pmatrix}$, *which is consistent with* $\widehat{\varphi}^\circ_{\gamma_2}(\delta)=bZ^{-1}\delta = \widehat{\varphi}^\circ_{\gamma_2}(\delta') + \widehat{\varphi}^\circ_{\gamma_2}(Z)\widehat{\varphi}^\circ_{\gamma_2}(\delta'')$, *and the action,*

$$\widehat{\varphi}^\circ_{\gamma_2}\theta(z,\delta,Z) := \theta(\widehat{\varphi}^\circ_{\gamma_2}z, \widehat{\varphi}^\circ_{\gamma_2}\delta; \widehat{\varphi}^\circ_{\gamma_2}Z) = \theta(bZ^{-1}z, \widehat{\varphi}^\circ_{\gamma_2}\delta; -bZ^{-1}b),$$

we have

$$\widehat{\varphi}^\circ_{\gamma_2}\theta(z,\delta,Z) = \mathrm{e}^{\rho_c}\sqrt{\det[Z]}\exp\left(2\pi\sqrt{-1}\left[\frac{1}{2}{}^t zZ^{-1}z - {}^t\delta''\delta'\right]\right)\theta(z,\delta,Z),$$

and by employing $\rho_c = 2\pi\sqrt{-1}\delta'\delta''$ *noting* $\delta'\delta'' \in \frac{1}{4}\mathbb{Z}$,

$$\widehat{\varphi}^\circ_{\gamma_2}\theta(z,\delta,Z) = \sqrt{\det[Z]}\exp\left(2\pi\sqrt{-1}\left[\frac{1}{2}{}^t zZ^{-1}z\right]\right)\theta(z,\delta,Z). \tag{3.90}$$

Proof *1*: Here for ${}^t bn =: \widetilde{n}$,

$$\begin{aligned} D(bZ^{-1}z, 0, n; -bZ^{-1}\,{}^t b) &= -\frac{1}{2}{}^t nbZ^{-1}\,{}^t bn + {}^t nbZ^{-1}z \\ &= -\frac{1}{2}{}^t\widetilde{n}Z^{-1}\widetilde{n} + +\widetilde{n}(Z^{-1}z) \\ &= D(Z^{-1}z, 0, \widetilde{n}; -Z^{-1}). \end{aligned}$$

Since b is also regular, the map $b : \mathbb{Z}^g \to \mathbb{Z}^g$ is bijective. Using the connection between G and the theta function in Lemma 3.176, we have

$$\begin{aligned} &\sum_{n\subset\mathbb{Z}^g}\exp\left(2\pi\sqrt{-1}D(\widehat{\varphi}^\circ_{\gamma_2}Z; bZ^{-1}z, 0, n)\right) = \sum_{\widetilde{n}\in\mathbb{Z}^g}\exp\left(2\pi\sqrt{-1}D(Z^{-1}z, 0, \widetilde{n}; -Z^{-1})\right) \\ &= \sum_{\widetilde{n}\in\mathbb{Z}^g}\exp\left(2\pi\sqrt{-1}(-{}^t\widetilde{n}Z^{-1}\widetilde{n} + {}^t\widetilde{n}Z^{-1}z)\right) = G(-Z^{-1}, Z^{-1}z) \\ &= \sqrt{(-\sqrt{-1})^g\det[Z]}\exp\left[\pi\sqrt{-1}\,{}^t zZ^{-1}z\right]G(Z,z) \\ &= \sqrt{(-\sqrt{-1})^g\det[Z]}\exp\left[\pi\sqrt{-1}\,{}^t zZ^{-1}z\right]\theta(z;Z). \end{aligned}$$

2: We have

$$\begin{aligned} &D(bZ^{-1}z, b\delta'', n - b^{-1}\delta'; -bZ^{-1}\,{}^t b) \\ &= -\frac{1}{2}{}^t(n - b^{-1}\delta')bZ^{-1}\,{}^t b(n - b^{-1}\delta') + {}^t(n - b^{-1}\delta')(bZ^{-1}z + b\delta'') \end{aligned}$$

$$
\begin{aligned}
&= -\frac{1}{2}{}^t(\widetilde{n}-\delta')Z^{-1}(\widetilde{n}-\delta') + {}^t(\widetilde{n}-\delta')(Z^{-1}z+\delta'') \\
&= D(Z^{-1}z, \delta'', \widetilde{n}-\delta'; -Z^{-1}) \\
&= D(Z^{-1}z, \delta'', -\delta'; -Z^{-1}) + D(Z^{-1}z + \delta'' + Z^{-1}\delta', 0_g, \widetilde{n}; -Z^{-1}).
\end{aligned}
$$

Hence $\sum_{n\in\mathbb{Z}^g} \exp\left(2\pi\sqrt{-1}D(bZ^{-1}z, b\delta'', n - b^{-1}\delta'; -bZ^{-1}\,{}^tb)\right)$ is equal to

$$
\begin{aligned}
&\sum_{\widetilde{n}\in\mathbb{Z}^g} \exp\left(2\pi\sqrt{-1}D(Z^{-1}z, \delta'', \widetilde{n}-\delta'; -Z^{-1})\right) \\
&= \exp\left(\pi\sqrt{-1}\left[-{}^t\delta' Z^{-1}\delta' - 2\delta'(Z^{-1}z+\delta'')\right]\right) \\
&\quad \times \sum_{\widetilde{n}\in\mathbb{Z}^g} \exp\left(2\pi\sqrt{-1}D(Z^{-1}(z+\delta'+Z\delta''), 0, \widetilde{n}; -Z^{-1})\right) \\
&= \exp\left(\pi\sqrt{-1}\left[-{}^t\delta' Z^{-1}\delta' - 2\,{}^t\delta'(Z^{-1}z+\delta'')\right]\right) G(-Z^{-1}, Z^{-1}(z+\delta'+Z\delta'')) \\
&= \sqrt{(-\sqrt{-1})^g \det[Z]} \exp\left(\pi\sqrt{-1}\left[{}^t(z+\delta'+Z\delta'')Z^{-1}(z+\delta'+Z\delta'')\right]\right) \\
&\qquad \exp\left(-\pi\sqrt{-1}\left[-{}^t\delta' Z^{-1}\delta' - 2\,{}^t\delta'(Z^{-1}z+\delta'')\right]\right)\theta\left(z+\delta'+Z\delta'', Z\right) \\
&= \sqrt{(-\sqrt{-1})^g \det[Z]} \exp\left(\pi\sqrt{-1}\left[{}^tzZ^{-1}z + 2\,{}^t\delta' Z^{-1}(z+Z\delta'') + {}^t\delta' Z^{-1}\delta' + 2\,{}^t\delta'' z + {}^t\delta'' Z\delta''\right]\right) \\
&\qquad \exp\left(-\pi\sqrt{-1}\left[-{}^t\delta' Z^{-1}\delta' - 2\,{}^t\delta'(Z^{-1}z+\delta'')\right]\right)\theta\left(z+\delta'+Z\delta'', Z\right) \\
&= \sqrt{(-\sqrt{-1})^g \det[Z]} \exp\left(\pi\sqrt{-1}\left[{}^tzZ^{-1}z + 2\,{}^t\delta'' z + {}^t\delta'' Z\delta''\right]\right)\theta\left(z+\delta'+Z\delta'', Z\right) \\
&= \sqrt{(-\sqrt{-1})^g \det[Z]} \exp\left(\pi\sqrt{-1}\left[{}^tzZ^{-1}z + 2D(z, \delta', \delta''; Z) - 2\,{}^t\delta''\delta'\right]\right)\theta\left(z+\delta'+Z\delta'', Z\right). \quad\blacksquare
\end{aligned}
$$

From Lemmas 3.154, 3.177 and 3.178, we obtain the following result [Ig1972]:

Proposition 3.179 *Let* $\gamma := \begin{pmatrix} a & b \\ c & d \end{pmatrix} \in \mathrm{Sp}(2g, \mathbb{Z})$. *We have the following relations:*

1. *For the action,*

$$
\widehat{\varphi}^\circ_\gamma\theta(z, Z) := \theta(\widehat{\varphi}^\circ_\gamma z; \widehat{\varphi}^\circ_\gamma Z) := \theta((Z\,{}^tc + {}^td)^{-1}z; (aZ+b)(cZ+d)^{-1}),
$$

we have

$$
\widehat{\varphi}^\circ_\gamma\theta(z, Z) = \mathrm{e}^{\rho_\gamma}\sqrt{\det[(cZ+d)]}\exp\left[\frac{\pi}{2}\sqrt{-1}\,{}^tc\,{}^tz(cZ+d)^{-1}cz\right]\theta(z+s_\gamma, Z),
$$

where the phase $\rho_\gamma = i_\gamma\pi\sqrt{-1}/2$ $(i_\gamma \in \mathbb{Z})$ *and* $s_\gamma \in \Gamma^\circ/2$ *depend only on* γ.

2. *For the action,*

$$
\widehat{\varphi}^\circ_\gamma\theta(z, \delta, Z) := \theta(\widehat{\varphi}^\circ_\gamma z, \widehat{\varphi}^\circ_\gamma\delta; \widehat{\varphi}^\circ_\gamma Z),
$$

with $\widehat{\varphi}^\circ_\gamma \begin{pmatrix} \delta'' \\ \delta' \end{pmatrix} = \begin{pmatrix} d & -c \\ -b & a \end{pmatrix} \begin{pmatrix} \delta'' \\ \delta' \end{pmatrix} + \frac{1}{2} \begin{pmatrix} (c\,{}^t d)_D \\ (a\,{}^t b)_D \end{pmatrix}$, *we have*

$$\widehat{\varphi}^\circ_\gamma \theta(z, \delta, Z) = \mathrm{e}^{\rho'_\gamma} \sqrt{\det[(cZ + d)]} \exp\left[\frac{\pi}{2}\sqrt{-1}\,{}^t c\,{}^t z (cZ + d)^{-1} cz\right] \theta(z, \delta, Z), \tag{3.91}$$

where the phase $\rho'_\gamma = i'_\gamma \pi \sqrt{-1}/2$ ($i'_\gamma \in \mathbb{Z}$) *depends only on* γ *and* δ.

3.8.5 W-Normalized Bilinear Forms and Sigma Function

In this subsection, we handle the W-normalized bilinear forms to investigate the properties of the sigma functions. Let us fix $M \in \mathfrak{W}_g$. We consider the lattice $\Gamma = \langle 2\omega', 2\omega'' \rangle_{\mathbb{Z}}$ and its relation to $\Gamma^\circ = \dfrac{1}{2}\omega'^{-1}\Gamma$ in $\mathbb{C}^g$ or $Z = \omega'^{-1}\omega''$.

Definition 3.180 1. We also let the $\mathbb{C}$ vector spaces $\mathbb{C}^g$ associated with Γ be denoted by $(\mathbb{C}^g, \Gamma)$ and its correspondence to $(\mathbb{C}^g, \Gamma^\circ)$ by

$$u_z : (\mathbb{C}^g, \Gamma^\circ) \to (\mathbb{C}^g, \Gamma), \quad z \to u_z(z) = 2\omega' z, \quad \omega' Z = \omega''.$$

2. For $u \in (\mathbb{C}^g, \Gamma)$, we let $u = 2\omega' u' + 2\omega'' u''$ for $u', u'' \in \mathbb{R}^g$.
3. For $u_1, u_2 \in (\mathbb{C}^g, \Gamma)$, we define the symplectic product,

$$\langle , \rangle : (\mathbb{C}^g, \Gamma) \times (\mathbb{C}^g, \Gamma) \to \mathbb{R},$$

by $\langle u_1, u_2 \rangle = {}^t u'_1 u''_2 - {}^t u''_1 u'_2$.

Lemma 3.181 *1. For* $z \in (\mathbb{C}^g, \Gamma^\circ)$ *such that* $z = z' + Zz''$ *we let*

$$u_z(z) := 2\omega' z = 2\omega' u_z(z)' + 2\omega'' u_z(z)'',$$

and then $z' = u_z(z)'$ *and* $z'' = u_z(z)''$.
2. *For* $z_1, z_2 \in (\mathbb{C}^g, \Gamma^\circ)$,

$$\langle u_z(z_1), u_z(z_2) \rangle = \langle z_1, z_2 \rangle^\circ.$$

Definition 3.182 We define the following bilinear forms on $(\mathbb{C}^g, \Gamma)$ with M:

$$L(u_1, u_2) := {}^t u_1 (2\eta' u' + 2\eta'' u'') = 2\,{}^t u_1 (\eta' u' + \eta'' u''),$$

$$L_z(z_1, z_2) := 2L(2\omega' z_1, 2\omega' z_2),$$

$$E(u_1, u_2) := (L(u_1, u_2) - L(u_2, u_1)),$$

$$\widehat{L}(u_1,u_2) := \frac{1}{2\pi\sqrt{-1}} L(u_1,u_2), \quad \widehat{L}_z(z_1,z_2) := \frac{1}{2\pi\sqrt{-1}} L_z(z_1,z_2),$$

$$\widehat{E}(u_1,u_2) := \frac{1}{2\pi\sqrt{-1}} E(u_1,u_2).$$

Lemma 3.183 *1.* $L_z(z_1,z_2) = {}^t z_1(\eta^{\circ\prime} z_2' + \eta^{\circ\prime\prime} z_2'')$ *by noting* $\eta^{\circ\prime} = 8\,{}^t\omega'\eta'$, *and* $\eta^{\circ\prime\prime} = 8\,{}^t\omega'\eta''$.

2. $\widehat{E}(u_1,u_2)$ *is a Riemann form, i.e.,*

   a. $E(u_1,u_2) = 2\pi\sqrt{-1}\langle u_1,u_2\rangle = 2\pi\sqrt{-1}\langle z_1,z_2\rangle$, $\widehat{E}(u_1,u_2) = \widehat{E}^\circ(z_1,z_2) \in \mathbb{R}$ *for* $u_a = 2\omega' z_a$,
   b. $E(u_1,u_2) = -E(u_2,u_1)$,
   c. *for* $\ell_1,\ell_2 \in \Gamma$, $\widehat{E}(\ell_1,\ell_2) \in \mathbb{Z}$, *and*
   d. $E(\sqrt{-1}u_1,u_2) = E(\sqrt{-1}u_2,u_1)$.

3. $\mathrm{Pf}(\widehat{E}) = 1$.
4. $\widehat{H}_E(u_1,u_2) := \widehat{E}(\sqrt{-1}u_1,u_2) - \sqrt{-1}\widehat{E}(u_1,u_2)$ *is the Hermitian Riemann form associated with* $\widehat{E}(u_1,u_2)$, *i.e.,*

   a. $\widehat{H}_E(u_1,u_2) = \frac{1}{4}({}^t u_1'\omega'^{-1}, {}^t u_1''\omega'^{-1}) \begin{pmatrix} Y^{-1} & Y^{-1}X + \sqrt{-1}1_g \\ XY^{-1} - \sqrt{-1}1_g & XY^{-1}X + Y \end{pmatrix} \begin{pmatrix} \omega'^{-1}u_2' \\ \omega'^{-1}u_2'' \end{pmatrix}$,
   b. $\widehat{H}_E(u_z(z_1), u_z(z_2)) = \widehat{H}^\circ(z_1,z_2)$,
   c. $\widehat{L}(\ell_1,\ell_2) = \widehat{L}(\ell_2,\ell_1)$ *modulo* $\mathbb{Z}$, *and*
   d. $\widehat{L}(u,\ell_1+\ell_2) = \widehat{L}(u,\ell_1) + \widehat{L}(u,\ell_2)$.

Proof $L_z(z_1,z_2) = 8\,{}^t(\omega' z_1' + \omega' Z z_1'')(\eta' z_2' + \eta'' z_2'')$ because of ${}^t z_1'\,{}^t\omega'\eta' z_2 = {}^t(\omega' z_1')\eta' z_2$.

$$\begin{aligned} L(u_1,u_2) &= 4\,{}^t(\omega' u_1' + \omega u_1'')(\eta' u_2' + \eta'' u_2'') \\ &= 4\,{}^t(\eta' u_2' + \eta'' u_2'')(\omega' u_1' + \omega u_1'') \\ &= 4({}^t u_2'\,{}^t\eta'\omega' u_1' + {}^t u_2'\,{}^t\eta'\omega'' u_1'' + {}^t u_2''\,{}^t\eta''\omega' u_1' + {}^t u_2''\,{}^t\eta''\omega'' u_1'') \\ &= 4\,{}^t u_2'\,{}^t\eta'\omega' u_1' + {}^t u_2'(4\,{}^t\omega'\eta'' - 2\pi\sqrt{-1}1_g)u_1'' \\ &\quad + {}^t u_2''(4\,{}^t\omega''\eta' + 2\pi\sqrt{-1}1_g)u_1' + 4\,{}^t u_2''\,{}^t\eta''\omega'' u_1'' \\ &= 4({}^t u_2'\,{}^t\omega'\eta' u_1' + {}^t u_2'\,{}^t\omega'\eta'' u_1'' + {}^t u_2''\,{}^t\omega''\eta' u_1' + {}^t u_2''\,{}^t\omega''\eta'' u_1'') \\ &\quad - 2\pi\sqrt{-1}({}^t u_2' u_1'' - {}^t u_2'' u_1') \\ &= L(u_2,u_1) + 2\pi\sqrt{-1}({}^t u_1' u_2'' - {}^t u_1'' u_2'). \end{aligned}$$

Hence $L_z(z_1,z_2) - L_z(z_2,z_1) = 4\pi\sqrt{-1}({}^t z_1' z_2'' - {}^t z_1'' z_2') = 4\pi\sqrt{-1}\langle z_1,z_2\rangle^\circ$ and $E(u_1,u_2) = 2\pi\sqrt{-1}\langle u_1,u_2\rangle$ or $\widehat{E}(u_1,u_2) = \widehat{E}^\circ(z_1,z_2)$. Thus we show *1*, *2* and *4*. From Lemma 3.170 3, $\mathrm{Pf}(\widehat{E}) = 1$. ■

Remark 3.184 In $(\mathbb{C}^g, \Gamma^\circ)$, we find the symplectic structure $\langle z_1,z_2\rangle^\circ$ $(\widehat{E}^\circ)$ as the imaginary part of the nondegenerate Hermitian Riemann form $\widehat{H}^\circ$ in Definition 3.169.

These quantities are represented well by the so-called Hodge theory on the compact Riemann surfaces [Gri1989]. However, due to Lemma 3.168 7, Definition 3.169, Lemma 3.170 1a, and 2c, $\widehat{E}$ (L) and $\widehat{H}_E$ defined in Definition 3.182 and Lemma 3.183 in terms of $(\omega', \omega'', \eta', \eta'')$ recover those structures as in Lemma 3.183 2a, and 4b. This fact supports the vision in Remark 2.136.

Definition 3.185 We define the parameterized maps $\mathbb{C}^g \to \mathbb{C}$,

$$Q(u, M) := \frac{1}{2} {}^t u({}^t\omega')^{-1}\, {}^t\eta' u = \frac{1}{2} {}^t u \eta' {\omega'}^{-1} u, \quad \widehat{Q}(u, M) := \frac{1}{2\pi\sqrt{-1}} Q(u, M),$$

$$\widehat{K}_L(u) := \widehat{K}_L(u, \delta', \delta'') := ({}^t u'\delta'' - {}^t u''\delta') + \frac{1}{2} {}^t u' u'',$$

$$\widehat{J}_L(u) := \widehat{J}_L(u, \delta', \delta'') := \widehat{K}_L(u, \delta', \delta'') + \widehat{L}(\frac{u}{2}, u),$$

where δ' and δ'' are elements of $(\frac{1}{2}\mathbb{Z})^g$.

Lemma 3.186 *For $\ell_1, \ell_2 \in \Gamma$, we have the following facts:*

1. $\widehat{K}_L(\ell_1 + \ell_2) - \widehat{K}_L(\ell_1) - \widehat{K}_L(\ell_2) = \widehat{E}(\ell_1, \ell_2)$ *modulo* $\mathbb{Z}$,
2. $\widehat{J}_L(\ell_1 + \ell_2) - \widehat{J}_L(\ell_1) - \widehat{J}_L(\ell_2) = \widehat{L}(\ell_1, \ell_2)$ *modulo* $\mathbb{Z}$.

Proof ${}^t(\ell_1' + \ell_2')(\ell_1'' + \ell_2'') = {}^t\ell_1'\ell_1'' + {}^t\ell_2'\ell_2'' + {}^t\ell_1'\ell_2'' + {}^t\ell_2'\ell_1'' = {}^t\ell_1'\ell_1'' + {}^t\ell_2'\ell_2'' - {}^t\ell_1'\ell_2'' + {}^t\ell_2'\ell_1''$ modulo $2\mathbb{Z}$. ${}^t(\ell_1' + \ell_2')(\ell_1'' + \ell_2'') = {}^t\ell_1'\ell_1'' + {}^t\ell_2'\ell_2'' + {}^t\ell_1'\ell_2'' + {}^t\ell_2'\ell_1'' = {}^t\ell_1'\ell_1'' + {}^t\ell_2'\ell_2'' - {}^t\ell_1'\ell_2'' + {}^t\ell_2'\ell_1''$ modulo $2\mathbb{Z}$. ∎

Lemma 3.187 *For $u = 2\omega u' + 2\omega'' u''$ and $w = 2\omega w' + 2\omega'' w''$, the following hold:*

1. $Q(u + w, M) - Q(u, M) = {}^t(u + \frac{1}{2}w)\eta' {\omega'}^{-1} w = {}^t(u + \frac{1}{2}w)\, {}^t{\omega'}^{-1}\, {}^t\eta' w$, *and*
2. $L(u + \frac{1}{2}w, w) = -2\pi\sqrt{-1} D(\frac{1}{2}{\omega'}^{-1} u, \frac{1}{2}{\omega'}^{-1} w)$
$+Q(u + w, M) - Q(u, M) + \pi\sqrt{-1}\, {}^t w' w''.$

Proof We note $\eta' {\omega'}^{-1} = {}^t(\eta' {\omega'}^{-1}) = {}^t{\omega'}^{-1}\, {}^t\eta$. Then the difference of Q's in *1* is equal to

$$\begin{aligned} Q(u + w, M) - Q(u, M) &= \frac{1}{2} {}^t(u + w)({}^t\omega')^{-1}\, {}^t\eta'(u + w) - \frac{1}{2} {}^t u({}^t\omega')^{-1}\, {}^t\eta' u \\ &= {}^t(u + \frac{1}{2}w)({}^t\omega')^{-1}\, {}^t\eta' w. \end{aligned}$$

The formula in *2* is equal to

$$
\begin{aligned}
L(u+\frac{1}{2}w, w) &= 2\,{}^t(u+\frac{1}{2}w)(\eta' w' + \eta'' w'')\\
&= 2\,{}^t(u+\frac{1}{2}w)({}^t\omega')^{-1}({}^t\omega'\eta' w' + {}^t\omega'\eta'' w'')\\
&= {}^t(u+\frac{1}{2}w)({}^t\omega')^{-1}(2\,{}^t\omega'\eta' w' + 2\,{}^t\omega''\eta' w'' - \pi\sqrt{-1}w'')\\
&= {}^t(u+\frac{1}{2}w)({}^t\omega')^{-1}\,{}^t\eta' w - \pi\sqrt{-1}\,{}^t(\omega'^{-1}(u+\frac{1}{2}w))w''.
\end{aligned}
$$

$$
{}^t(\omega'^{-1}(u+\frac{1}{2}w))w'' = 2\,{}^t(\frac{1}{2}\omega'^{-1}(u+\frac{1}{2}w))w'' = 2D(\frac{1}{2}\omega'^{-1}u, \frac{1}{2}\omega'^{-1}w) - w'w''. \blacksquare
$$

We fix the W-curve X with associated with the numerical semigroup H_X and the Young diagram Λ_X. We let $k_\mathfrak{s}$ be the minimal positive integer such that there exists an effective divisor $D \in S^{k_\mathfrak{s}}X$ satisfying $D \sim -\mathfrak{K}_\mathfrak{s}$, i.e., $w_\mathfrak{s}(D) = 0$ for non-vanishing $\mathfrak{K}_\mathfrak{s}$, whereas for the case $\mathfrak{K}_\mathfrak{s} = 0$, we let $k_\mathfrak{s} = 0$.

Proposition 3.188 *From Theorem 3.21, we have*

$$
\widehat{\varphi}^{\circ}_{\gamma_1}\partial_{\natural_{k_\mathfrak{s}}}\theta(\frac{1}{2}\omega'^{-1}\widetilde{w}_\mathfrak{s}(0), \delta) = (-1)^{{}^t\delta'' a^{-1}((a\,{}^tb)_D - b\delta'')}\partial_{\natural_{k_\mathfrak{s}}}\theta(\frac{1}{2}\omega'^{-1}\widetilde{w}_\mathfrak{s}(0), \delta), \quad \textit{and}
$$

$$
\widehat{\varphi}^{\circ}_{\gamma_2}\partial_{\natural_{k_\mathfrak{s}}}\theta(\frac{1}{2}\omega'^{-1}\widetilde{w}_\mathfrak{s}(0), \delta) = \mathrm{e}^{\rho_c}\sqrt{\det[Z]}\partial_{\natural_{k_\mathfrak{s}}}\theta(\frac{1}{2}\omega'^{-1}\widetilde{w}_\mathfrak{s}(0), \delta),
$$

where

$$
\partial_{\natural_{k_\mathfrak{s}}} := \prod_{i\in\natural_{k_\mathfrak{s}}}\frac{\partial}{\partial u_i}
$$

for the Riemann–Kempf coefficient I_0 for the associated Young diagram given in Theorem 3.44.

For a W-curve X, we define $\delta \in (\mathbb{Z}/2)^{2g}$ connected with the shifted Riemann constant $\xi_{X,\mathfrak{s}}$ as in Proposition 3.3, which is one of δ in the sigma functions defined by Korotkin and Shramchenko [KorS2012].

Modifying Nakayashiki's definition [Nak2016, Definition 9], we define our sigma function:

Definition 3.189 For $\delta \in (\mathbb{Z}/2)^{2g}$ of X, we define

$$
\sigma(u) := (-1)^g \exp(Q(u, M))\frac{\theta(\frac{1}{2}\omega'^{-1}u, \delta)}{[\partial_{\natural_{k_\mathfrak{s}}}\theta(\frac{1}{2}\omega'^{-1}u, \delta)]_{u=\widetilde{w}(\iota_X\mathfrak{K}_\mathfrak{s})}}.
$$

Here we note Remark 3.29 on this definition of the sigma function.

We show the translational formula of the sigma function.

Proposition 3.190 *1. For* $\ell = 2\omega'\ell' + 2\omega''\ell''(= 2\omega'\ell) \in \Gamma$ $(\ell \in \Gamma^\circ)$,

$$\begin{aligned}\sigma(u+\ell) &= \exp\Big(L(u+\frac{\ell}{2},\ell) + 2\pi\sqrt{-1}\widehat{K}_L(\ell,\delta',\delta'')\Big)\sigma(u)\\ &= \exp\Big(L(u+\frac{\ell}{2},\ell)\Big)\chi(\ell,\delta',\delta'')\sigma(u)\\ &= \exp\Big(2\pi\sqrt{-1}(\widehat{L}(u,\ell) + \widehat{J}_L(\ell,\delta',\delta''))\Big)\sigma(u),\end{aligned}$$

where

$$\chi(\ell) := \chi(\ell,\delta',\delta'') := \exp(2\pi\sqrt{-1}\widehat{K}_L(\ell,\delta',\delta'')).$$

Then the σ *function is the theta function of* $(\widehat{L}, \widehat{J}_L)$ *type.*

2. *For* $\ell_1, \ell_2 \in \Gamma$,

$$\chi(\ell_1,\delta',\delta'')\chi(\ell_2,\delta',\delta'')\chi(\ell_1+\ell_2,\delta',\delta'')^{-1} = \exp(\pi\sqrt{-1}\widehat{E}(\ell_1,\ell_2)).$$

Proof We note

$$\varphi(\frac{1}{2}\omega'^{-1}u,\ell,\delta) = 2\pi\sqrt{-1}\left({}^t(\ell''+\delta'')\ell' - D(\frac{1}{2}\omega'^{-1}u,\ell) - {}^t\ell''\delta'\right),$$

$$\begin{aligned}Q(u+\ell,M) - Q(u,M) = {}& L(u+\frac{1}{2}\ell,\ell)\\ &+ 2\pi\sqrt{-1}D(\frac{1}{2}\omega'^{-1}u,\frac{1}{2}\omega'^{-1}\ell) - \pi\sqrt{-1}\,{}^t\ell'\ell''.\end{aligned}$$

Hence, the phase difference is given by

$${}^t(\ell''+\delta'')\ell' - {}^t\ell''\delta' - \frac{1}{2}{}^t\ell'\ell'' = \frac{1}{2}{}^t\ell'\ell'' + \delta''\ell' - {}^t\ell''\delta' = -K_L(\ell,\delta',\delta''),$$

and thus *2* is obtained by the relation

$$\begin{aligned}&K_L(\ell_1+\ell_2,\delta',\delta'') - K_L(\ell_1,\delta',\delta'') - K_L(\ell_2,\delta',\delta'')\\ &= -\frac{1}{2}({}^t(\ell_1'+\ell_2')(\ell_1''+\ell_2'') - {}^t\ell_1'\ell_1'' - {}^t\ell_2'\ell_2'')\\ &= -\frac{1}{2}({}^t\ell_1'\ell_2'' + {}^t\ell_1''\ell_2') = \frac{1}{2}({}^t\ell_1'\ell_2'' - {}^t\ell_1''\ell_2') \quad \text{modulo } \mathbb{Z}.\end{aligned}$$

■

Proposition 3.191 σ *belongs to* $\mathrm{Th}(\widehat{L}, \widehat{J}_L)$ *whose dimension is 1.*

Proof Theorem 3.166 and Lemma 3.183 show it. ■

3.8.6 Symplectic Action on Sigma Function

This subsection mentions the symplectic action on the sigma functions.

Lemma 3.192 *Let* $\gamma_1 := \begin{pmatrix} a & b \\ 0 & a^{-1} \end{pmatrix} \in \mathrm{Sp}(2g, \mathbb{Z})$, *(* ${}^t a = a$*). We have* $\widehat{\varphi}^{\circ}_{\gamma_1} Z = (aZ + b)a$,

$$\widehat{\varphi}_{\gamma_1}\omega' = \omega'\,{}^t a^{-1}, \quad \widehat{\varphi}_{\gamma_1}{\omega'}^{-1} = {}^t a{\omega'}^{-1}, \quad \widehat{\varphi}_{\gamma_1}\eta' = \eta'\,{}^t a^{-1},$$

$$\widehat{\varphi}_{\gamma_1} M = 2\begin{pmatrix} \omega'\,{}^t a^{-1} & \omega'\,{}^t b + \omega''\,{}^t a \\ \eta'\,{}^t a^{-1} & \eta'\,{}^t b + \eta''\,{}^t a \end{pmatrix},$$

and $\widehat{\varphi}^{\circ}_{\gamma_1}\begin{pmatrix}\delta'' \\ \delta'\end{pmatrix} := \begin{pmatrix} a^{-1} & 0 \\ -b & a \end{pmatrix}\begin{pmatrix}\delta'' \\ \delta'\end{pmatrix} - \begin{pmatrix} 0 \\ \frac{1}{2}(a^{-1}b)_D \end{pmatrix}$, *which is consistent with* $\widehat{\varphi}^{\circ}_{\gamma_1}(\delta) = a\delta = \widehat{\varphi}^{\circ}_{\gamma_1}(\delta') + \widehat{\varphi}^{\circ}_{\gamma_1}(Z)\widehat{\varphi}^{\circ}_{\gamma_1}(\delta'')$. *For the action,*

$$\widehat{\varphi}_{\gamma_1}\sigma(u, \delta, M) := \sigma(u, \widehat{\varphi}^{\circ}_{\gamma_1}\delta, \widehat{\varphi}_{\gamma_1} M),$$

we have

$$\widehat{\varphi}_{\gamma_1}\sigma(u, \delta, M) = \sigma(u, \delta, M).$$

Proof Since the action $\widehat{\varphi}_{\gamma_1}({\omega'}^{-1}u) = a{\omega'}^{-1}u$ corresponds to the action $\widehat{\varphi}^{\circ}_{\gamma_1} z = az$ in Lemma 3.177. Further, it is obvious that $\widehat{\varphi}_{\gamma_1} Q(u; M) = Q(u; M)$. Thus we show the formula. ■

Lemma 3.193 *Let* $\gamma_2 := \begin{pmatrix} 0 & b \\ -b^{-1} & 0 \end{pmatrix} \in \mathrm{Sp}(2g, \mathbb{Z})$, *(* ${}^t b = b$*). We have* $\widehat{\varphi}^{\circ}_{\gamma_2} Z = -bZ^{-1}b$,

$$\widehat{\varphi}_{\gamma_2}\omega' = -\omega''\,{}^t b^{-1}, \quad \widehat{\varphi}_{\gamma_2}{\omega'}^{-1} = -\,{}^t b{\omega''}^{-1}, \quad \widehat{\varphi}_{\gamma_2}\eta' = -\eta''\,{}^t b^{-1},$$

$$\widehat{\varphi}_{\gamma_2} M = 2\begin{pmatrix} -\omega''\,{}^t b^{-1} & \omega'\,{}^t b \\ -\eta''\,{}^t b^{-1} & \eta'\,{}^t b \end{pmatrix},$$

and $\widehat{\varphi}^{\circ}_{\gamma_2}\begin{pmatrix}\delta'' \\ \delta'\end{pmatrix} := \begin{pmatrix} 0 & -b^{-1} \\ b & 0 \end{pmatrix}\begin{pmatrix}\delta'' \\ \delta'\end{pmatrix}$, *which is consistent with* $\widehat{\varphi}^{\circ}_{\gamma_2}(\delta) = bZ^{-1}\delta = \widehat{\varphi}^{\circ}_{\gamma_2}(\delta') + \widehat{\varphi}^{\circ}_{\gamma_2}(Z)\widehat{\varphi}^{\circ}_{\gamma_2}(\delta'')$. *For the action*

$$\widehat{\varphi}_{\gamma_2}\sigma(u, \delta, M) = \sigma(u, \widehat{\varphi}^{\circ}_{\gamma_2}\delta, \widehat{\varphi}_{\gamma_2} M),$$

we have

$$\widehat{\varphi}_{\gamma_2}\sigma(u, \delta, M) = \sigma(u, \delta, M).$$

Proof $\widehat{\varphi}_{\gamma_2} Q(u; M) = \frac{1}{2} {}^t u \eta'' \omega''^{-1} u$ is equal to

$$
\begin{aligned}
&= \frac{1}{2} {}^t u \eta'' \, {}^t\omega' \, {}^t\omega'^{-1} \omega''^{-1} u \\
&= \frac{1}{2} {}^t u (\eta' \, {}^t\omega'' - \frac{1}{2} \pi \sqrt{-1}) \omega'^{-1} \omega''^{-1} u \\
&= \frac{1}{2} {}^t u \eta' \, {}^t\omega'' \, {}^t\omega'^{-1} \omega''^{-1} u - \frac{1}{4} \pi \sqrt{-1} \, {}^t u \, {}^t\omega'^{-1} \omega''^{-1} u \\
&= \frac{1}{2} {}^t u \eta' \omega'' \omega'^{-1} \omega''^{-1} u - \pi \sqrt{-1} \, {}^t u \frac{1}{2} \omega'^{-1} \omega''^{-1} \omega' \frac{1}{2} \omega'^{-1} u \\
&= \frac{1}{2} {}^t u \eta' \omega'^{-1} u - \pi \sqrt{-1} \, {}^t z \omega''^{-1} \omega' z,
\end{aligned}
$$

$$
\widehat{\varphi}^{\circ}_{\gamma_2} \theta(0, \delta) = \sqrt{|Z|} \theta(0, \delta).
$$

Using the relation in (3.90), we show the invariance. ∎

From Lemmas 3.154, 3.192 and 3.193, we have the following result:

Proposition 3.194 *For* $\gamma := \begin{pmatrix} a & b \\ c & d \end{pmatrix} \in \mathrm{Sp}(2g, \mathbb{Z})$*, and its action,*

$$
\widehat{\varphi}_{\gamma} \sigma(u, \delta, M) = \sigma(u, \widehat{\varphi}^{\circ}_{\gamma} \delta, \widehat{\varphi}_{\gamma} M),
$$

we have

$$
\widehat{\varphi}_{\gamma} \sigma(u, \delta, M) = \sigma(u, \delta, M).
$$

Since $\widehat{\varphi}_{\gamma}(u) = u$, the sigma function is a function of $u \in \mathbb{C}^g$ which does not depend on the choice of the basis of $\mathbf{H}_1(X, \mathbb{Z})$. It means that the sigma function $\sigma \circ \widetilde{w}$ is a function of $S^g \widetilde{X}$.

3.8.7 Analytic Preliminaries

In this subsection, we review the results in the theta series based on some formal computations [MatO2003b, HanB1980], though we could justify these results mathematically rigorously [Ig1972, Chap. 1].

Let us consider the set of complex-valued smooth functions $\mathscr{C}^\infty(\mathbb{R}, \mathbb{C})$ over $\mathbb{R}$, and its linear map $p_x : \mathscr{C}^\infty(\mathbb{R}, \mathbb{C}) \to \mathbb{C}$ for $x \in \mathbb{R}$ by $p_x(f) = f(x)$. The Dirac distribution, $\delta(x)\, dx$ is defined by $\int_{\mathbb{R}} f(x)\delta(x)dx := p_0(f)$ for a function f belonging to a certain subclass of $\mathscr{C}^\infty(\mathbb{R}, \mathbb{C})$ [Yo1970]. The so-called Dirac comb $\delta_c(x)$ defined in the following lemma is studied by Córboa [Cor1989], which is related to the Poisson summation formula. We show it without proof.

Lemma 3.195 *For*

$$\delta_c(x)dx := \sum_{m=-\infty}^{\infty} \delta(x-m)\,dx,$$

we have

$$\delta_c(x) = \sum_{m=-\infty}^{\infty} \mathrm{e}^{2\pi\sqrt{-1}xm}.$$

We show some results formally, which can be justified by the limit of function spaces.

Lemma 3.196 *For $a, c \in \mathbb{C}$ with $\mathrm{Im}(a) > 0$, let*

$$G(a,c) := \sum_{m=-\infty}^{\infty} \exp(\pi\sqrt{-1}(am^2 + 2cm)).$$

Then the following relations hold:

1. *For $\ell \in \mathbb{Z}$, $G(a+\ell, c) = G(a, c + \frac{1}{2}\ell)$.*
2. $G(a,c) = \sqrt{\dfrac{\sqrt{-1}}{a}} \exp\left[-\dfrac{\pi\sqrt{-1}c^2}{a}\right] G\left(-\dfrac{1}{a}, -\dfrac{c}{a}\right)$.

Proof We have the following computations:

$$\begin{aligned} G(a+\ell,c) &= \sum_{m=-\infty}^{\infty} \exp(\pi\sqrt{-1}(am^2+2cm))(-1)^{\ell m^2} \\ &= \sum_{m=-\infty}^{\infty} \exp(\pi\sqrt{-1}(am^2+2cm))(-1)^{\ell m} \\ &= \sum_{m=-\infty}^{\infty} \exp(\pi\sqrt{-1}(am^2+(2c+\ell)m)). \end{aligned}$$

$$\begin{aligned} G(a,c) &= \int_{\infty}^{\infty} dx \delta_c(x) \mathrm{e}^{\pi\sqrt{-1}(ax^2+2cx)} \\ &= \int_{\infty}^{\infty} dx \sum_{m=-\infty}^{\infty} \mathrm{e}^{\pi\sqrt{-1}(ax^2+2(m+c)x)} \\ &= \sum_{m=-\infty}^{\infty} \int_{\infty}^{\infty} dx \exp\left(\pi\sqrt{-1}a\left(x - \frac{(m+c)}{a}\right)^2 - \pi\sqrt{-1}\frac{(m+c)^2}{a}\right) \end{aligned}$$

$$= \sum_{m=-\infty}^{\infty} \sqrt{\frac{1}{-a\sqrt{-1}}} \exp\left(-\pi\sqrt{-1}\frac{(m+c)^2}{a}\right)$$

$$= \sum_{m=-\infty}^{\infty} \sqrt{\frac{1}{-a\sqrt{-1}}} \exp\left(-\pi\sqrt{-1}\left(\frac{m^2}{a} + 2m\frac{c}{a} + \frac{c^2}{a}\right)\right).$$

■

Lemma 3.197 *For* $A \in \mathrm{GL}(g, \mathbb{C})$ *with* $\mathrm{Im} A > 0$, ${}^t A = A$ *and* $c \in \mathbb{C}^g$, *we let*

$$G(A, c) := \sum_{m \in \mathbb{Z}^g} \exp(\pi\sqrt{-1}({}^t m A m + 2\,{}^t c m)).$$

Then we have the following:

1. *For* $B \in \mathrm{Mat}_{\mathbb{Z}}(g)$ $({}^t B = B)$,

$$G(A + B, c) = G(A, c + \frac{1}{2}B_D),$$

where $B_D := {}^t(B_{11}, B_{22}, \ldots, B_{gg}) \in \mathbb{Z}^g$.

2. $$G(A, c) = \sqrt{\frac{\sqrt{-1}^g}{\det[A]}} \exp\left[-\pi\sqrt{-1}\,{}^t c A^{-1} c\right] G\left(-A^{-1}, -A^{-1} c\right),$$

where the phase with respect to $\mathrm{e}^{2\pi\sqrt{-1}/4}$ *is undetermined in this stage.*

Proof *1* is proved by

$$(-1)^{{}^t m B m} = (-1)^{\sum_{i=1}^{g} m_i B_{i,i} m_i + \sum_{i<j} 2 m_i B_{i,j} m_j}$$
$$= (-1)^{\sum_{i=1}^{g} m_i B_{i,i}}.$$

For a unitary matrix U, we consider the diagonal matrix $D = {}^*U A U$, and then the right-hand side in *2* is reduced to Lemma 3.196. ■

3.9 Appendix: KP Hierarchy, Tau Function and Universal Grassmannian Manifolds

In this appendix, we review the theory of KP (Kadomtsev–Petviashvili) hierarchy, tau functions, and Sato's theory of the universal Grassmannian manifolds based on Shiota's paper [Shi1986], Nakayashiki [Nak2010b, Nak2016] and Segal and Wilson [SeW1985]. Sato and Sato discovered the relation between the KP hierarchy and the universal Grassmannian manifolds [SatM1980, SatS1981, SatS1982, SatN1984]. On the basis of their results, it turns out that the theta series, the Schur polynomials, and the integrable equations are closely connected in the following way.

3.9.1 Kadomtsev–Petviashvili Hierarchy

In this subsection, we review the modern theory of KP hierarchy. Let us define the formal ring of the pseudo-differential operators over the formal power series $\mathbb{C}[[z]]$ over $\mathbb{C}$. Let $\check{R} := \mathbb{C}[[z]]$, $\partial_z := \dfrac{d}{dz}$ and

$$\mathscr{E}_{\check{R}} := \check{R}\left((\partial_z^{-1})\right),$$

whose element is given by

$$P = \sum_{j=-\infty}^{m} a_j(z)\left(\frac{d}{dz}\right)^j, \quad m \in \mathbb{Z}.$$

Using the Leibniz rule, we have

$$a(z)\partial_z^j = \sum_k (-1)^k \binom{j}{k} \partial^{j-k} a^{(k)}(z), \quad a^{(k)}(z) = \frac{d^k\, a(z)}{dz^k}.$$

For $\ell \in \mathbb{Z}$, let $\mathscr{E}_{\check{R}}(\ell) := \{P \in \mathscr{E}_{\check{R}} \mid \mathrm{ord} P \leq \ell\}$ and $\mathscr{E}_{\check{R}}^- := \mathscr{E}_{\check{R}}(-1)$. Further, we consider the set of the ordinary differential operators, $D := \mathbb{C}[[z]][d/dz]$. We have the decomposition $\mathscr{E}_{\check{R}} = D \oplus \mathscr{E}_{\check{R}}^-$, and for $P \in \mathscr{E}_{\check{R}}$, also have $P = P^+ + P^-$ ($P^+ \in D$ and $P^- \in \mathscr{E}_{\check{R}}^-$).

The multiplicative group $1 + \mathscr{E}_{\check{R}}^-$ acts on $\mathscr{E}_{\check{R}}$ by $P \mapsto SPS^{-1}$ for $S \in 1 + \mathscr{E}_{\check{R}}^-$. For the $m(\neq 0)$-order $P \in \mathscr{E}_{\check{R}}$, i.e., $P = (d/dt)^m + \cdots$, the orbit of the action is given as $P + \mathscr{E}_{\check{R}}(m-2)$, and the stability subgroup of P is $(1 + \mathscr{E}_{\check{R}}^-) \cap Z_P = 1 + \mathbb{C}[[P^{-1/m}]]P^{-1/m}$, where $Z_P := \ker(\mathrm{ad}\, P : \mathscr{E}_{\check{R}} \to \mathscr{E}_{\check{R}}) = \{Q \in \mathscr{E}_{\check{R}} \mid PQ = QP\} = \mathbb{C}((P^{-1/m}))$, the commutative subring of P.

Further, we introduce $\mathfrak{u}_j$ such that only finite many $\mathfrak{u}_j$ are in $\mathbb{C} \setminus \{0\}$ and $\mathfrak{u} \in \mathbb{C}^\infty = \varinjlim \mathbb{C}^n$.

Let us consider deformation of the differential operator L in $\mathscr{E}_{\check{R}}$; we consider the differential operator $L = L(\mathfrak{u}) \in d/dz + \mathscr{E}_{\check{R}}^- \otimes \mathbb{C}[[\mathfrak{u}]]$ which obeys the KP hierarchy,

$$\partial_{\mathfrak{u}_n} L = [B_n, L], \tag{3.92}$$

where z is identified with $\mathfrak{u}_1$, $\partial_{\mathfrak{u}_n} := \dfrac{\partial}{\partial \mathfrak{u}_n}$, and $B_n := (L^n)_+$. By letting $B_n^{\mathrm{c}} := -(L^n)^- \in \mathscr{E}_{\check{R}}^-$, we have $L^n = B_n - B_n^{\mathrm{c}}$ and $[B_n, L] = [B_n^{\mathrm{c}}, L] \in \mathscr{E}_{\check{R}}^-$.

Using the correspondence $\varphi : T_0\mathbb{C}^\infty \to D$ by $\varphi(\sum c_n \partial_{\mathfrak{u}_n}) = \sum c_n B_n$, we set $dL : T_0\mathbb{C}^\infty \to \mathscr{E}_{\check{R}}^-$ by

$$dL\left(\sum c_n \frac{\partial}{\partial \mathfrak{u}_n}\right) = [\sum c_n B_n, L],$$

$K_L := \ker dL$ and $R_L := \mathbb{C} \oplus \varphi(K_L)$.

The solution of the KP hierarchy is given by the "wave function" satisfying

$$L\psi = \mathfrak{z}\psi, \quad \partial_{\mathfrak{u}_n}\psi = B_n\psi, \tag{3.93}$$

which is related to a scalar-valued function $\tau(\mathfrak{u})$ over $\mathbb{C}^\infty$ by

$$\psi(\mathfrak{u}, \mathfrak{z}) = \frac{\tau(\mathfrak{u}_1 - \mathfrak{z}, \mathfrak{u}_2 - \mathfrak{z}^2/2, \mathfrak{u}_3 - \mathfrak{z}^3/3, \ldots)}{\tau(\mathfrak{u})} \exp\left(\sum_{i=1}^{\infty} \mathfrak{u}_i \mathfrak{z}^{-i}\right). \tag{3.94}$$

Here $\mathfrak{z}$ is a formal scalar parameter. The $\tau(\mathfrak{u})$ satisfies the infinite many Baker–Hirota bilinear equations of the KP-hierarchy (see Definition 4.22)

$$\sum_{j=0} \mathfrak{p}_j(-2y)\mathfrak{p}_{j+1}(\mathring{\mathfrak{D}}) \exp\left(\sum_{i=1}^{\infty} y_i \mathfrak{D}_i\right) \tau \cdot \tau = 0, \tag{3.95}$$

where $\mathring{\mathfrak{D}} := (\mathfrak{D}_1, 2^{-1}\mathfrak{D}_2, 3^{-1}\mathfrak{D}_3, \cdots)$ with the Baker–Hirota bilinear operator $\mathfrak{D}_i$ defined by

$$\mathfrak{D}_i f(\mathfrak{u}) \cdot g(\mathfrak{u}) = \left[\frac{\partial}{\partial \mathfrak{v}_i} f(\mathfrak{u} + \mathfrak{v}) \cdot g(\mathfrak{u} - \mathfrak{v})\right]_{\mathfrak{v}=0}, \tag{3.96}$$

and $\mathfrak{p}_j$ is the elementary Schur polynomial (which is equal to the complete homogeneous symmetric polynomial $\mathfrak{p}_i = h_i$ (3.119) if $\mathfrak{u}_i = \frac{1}{i}\sum_{j=1}^{\infty} \mathfrak{t}_j^i$,[10] defined by

$$\exp\left(\sum_{i=1}^{\infty} \mathfrak{u}_i \mathfrak{z}^i\right) = \sum_{i=0}^{\infty} \mathfrak{p}_i(\mathfrak{u})\mathfrak{z}^i. \tag{3.97}$$

The coefficient of y_3 provides the KP-equation in terms of the bilinear form,

$$(\mathfrak{D}_1^4 + 3\mathfrak{D}_2^2 - 4\mathfrak{D}_1\mathfrak{D}_3)\tau(\mathfrak{u})\tau(\mathfrak{u}) = 0.$$

The series of the equations which we obtain from (3.95) by setting $\mathfrak{D}_{jp} = 0$ for every $j \geq 1$ is called the p-reduced KP-hierarchy.

Let $G_1 := \left\{\exp\left(c_0 + \sum_{j=1}^{\infty} c_j \mathfrak{u}_j\right) \,\middle|\, c_j \in \mathbb{C}\right\}$. When $L = L(\mathfrak{u})$ is a solution of the KP hierarchy, we regard $K_{L(\mathfrak{u})}$ and $R_{L(\mathfrak{u})}$ as functions of $\mathfrak{u}$. Then by the uniqueness

---

[10] A formal computation $\exp\left(\sum_{n=1}^{\infty}\sum_{j}^{N} \frac{1}{n}\mathfrak{t}_j^n \mathfrak{z}^n\right) = \exp\left(\sum_{j}^{N}\sum_{n=1}^{\infty} \frac{1}{n}\mathfrak{t}_j^n \mathfrak{z}^n\right) = \prod_j^N \frac{1}{1 - \mathfrak{t}_j \mathfrak{z}}$ and (3.119) show $\mathfrak{p}_i = h_i$ for $N \to \infty$ [Kod2017, Remark 1.5].

of the solution to the initial value problem and the commuting flow property, it is shown that $K_{L(\mathfrak{u})}$ is independent of $\mathfrak{u}$, and thus we go on to write it by K_L.

Then the following facts are known:

1. For $Z_L := \ker(\mathrm{ad}L : \mathscr{E}_{\check{R}} \to \mathscr{E}_{\check{R}})$, $R_L = Z_L \cap D$.
2. R_L is a maximal commutative subring of D if $R_L \neq \mathbb{C}$.

R_L has a filter structure $R_L(n) \subset R_L(n+1)$ induced from $\mathscr{E}_{\check{R}}(n)$.

Definition 3.198 $L \in \partial_z + \mathscr{E}_{\check{R}}^-$ is called g-dimensional if it satisfies the following equivalent conditions:

1. dL is of rank g.
2. $\dim_{\mathbb{C}}(\mathbb{C}^\infty/K_L) = g$.
3. There are $w_1, \ldots, w_g \in \mathbb{N}$ $(w_i < w_{i+1})$ such that
$$\dim_{\mathbb{C}} R_L(n)/R_L(n-1) = \begin{cases} 0, \text{ if } n = w_j \text{ for some } j, \\ 1, \text{ otherwise.} \end{cases}$$
4. $\dim_{\mathbb{C}} R_L(2g+1) = g+2$.
5. $\dim_{\mathbb{C}}(K_L \cap \mathbb{C}^{2g+1}) = g+1$.

Since $K_{L(\mathfrak{u})}$ does not depend on $\mathfrak{u}$, we regard $\mathbb{C}^\infty/K_L \cong \mathbb{C}^g$ as $L : \mathbb{C}^\infty \to \mathbb{C}^\infty/K_L \xrightarrow{\overline{L}} (d/dz) + \mathscr{E}_{\check{R}}^-$ with $d\overline{L}$ injective. Let the discrete subgroup $\Gamma^\circ := \{\mathfrak{v} \in \mathbb{C}^\infty/K_L \mid \overline{L}(\mathfrak{u}+\mathfrak{v}) = \overline{L}(\mathfrak{u})\}$, and $T_L := (\mathbb{C}^\infty/K_L)/\Gamma^\circ$.

Let $A \in \mathrm{Mat}_{\mathbb{C}}(g, \infty)$ such that $A : \mathbb{C}^\infty \to \mathbb{C}^g$ and $A\Gamma^\circ \in \mathbb{Z}^g + \widehat{\tau}\mathbb{Z}^g$, where $\widehat{\tau} \in \mathrm{Mat}_{\mathbb{C}}(g, g)$, and then $\tau(\mathfrak{u}+\mathfrak{v}) = \rho_{\mathfrak{v}}\tau(\mathfrak{u})$ for $\rho_{\mathfrak{v}} \in G_1$.

The following is due to Burchnall and Chaundy [BurC1922, BurC1928], [BurC1931], Baker [Ba1928], Novikov [Nov1974], Dubrovin [Dub1975], Krichever [Kr1997], Mumford [Mum1977], Mulase [Mul1984], and Shiota [Shi1986, p. 348]:

Theorem 3.199 *There exists a natural bijection between the following two sets of data:*

Data A:

(a) *X a complete reduced and irreducible curve over $\mathbb{C}$,*
(b) *$p \in X$ a smooth point, and $[z] \in (\mathfrak{m}_p - \mathfrak{m}_p^2)/\mathfrak{m}_p^3$ an equivalence class of local coordinate at p,*
(c) *$\mathscr{F}$ a torsion free rank 1 sheaf on X such that $\mathbf{H}^0(X, \mathscr{F}) = 0$ and $\mathbf{H}^1(X, \mathscr{F}) = 0$; thus $\mathscr{H}om(\mathscr{F}, \mathscr{F}) \cong \mathscr{O}_X$.*

Data B: *A commutative subring $R \in D$ with $\mathbb{C} \subset R$ such that for any $n \gg 0$, there exists $P \in R$ of the form*

$$P = (d/dz)^n + c(d/dz)^{n-1} + (\textit{lower order terms}), \quad c \in \mathbb{C}.$$

We note that T_L for a compact L has a natural principal polarization as the Abelian variety.

Due to Krichever, we have the following lemma [Kr1997], [Shi1986, p. 354]:

Lemma 3.200 *If X is a complete smooth curve over $\mathbb{C}$, and $p \in X$, then there exists a matrix $A \in \mathrm{Mat}_{\mathbb{C}}(g, \infty)$ with rank $A = g$ and a quadratic form $\widetilde{q}(\mathrm{u}) = \sum \widetilde{q}_{i,j}\mathrm{u}_i\mathrm{u}_j$ such that for any $\xi \in \mathbb{C}^g$,*

$$\tau(\mathrm{u}) = \mathrm{e}^{\widetilde{q}(\mathrm{u})}\theta(A\mathrm{u} + \xi) \tag{3.98}$$

gives a tau function for the KP hierarchy associated with a solution belonging to the pair (X, p) and any solution belonging to (X, p) is given, up to equivalence, by the equation for some ξ. Hence θ is the Riemann theta function of X defined in (3.3), and $\widetilde{q}_{i,j}$ is identified with $q^{\circ}_{ij}/2$ of the normalized differential of the second kind in Lemma 2.102.

We let $q^{\circ}(\mathrm{u}) := \sum \frac{1}{2}q^{\circ}_{i,j}\mathrm{u}_i\mathrm{u}_j$ for later convenience. Then, the following is due to Mulase [Mul1984] and Shiota [Shi1986, p. 351]:

Theorem 3.201 *For a principally polarized Abelian variety $\mathbb{A} = \mathbb{C}^g/(\mathbb{Z}^g + \Gamma^{\circ}\mathbb{Z}^g)$, $\Gamma^{\circ} \in \mathfrak{H}_g$, the following conditions are equivalent:*

1. *$\mathbb{A} \cong J_X$ ($\Gamma^{\circ} = \Gamma^{\circ}_X$) for a complete smooth curve X over $\mathbb{C}$.*
2. *$\mathbb{A} \cong T_L$ for a compact solution $L = L(\mathrm{u})$ to the KP-hierarchy.*
3. *There exist a matrix $A \in \mathrm{Mat}_{\mathbb{C}}(g, \infty)$ with rank $A = g$ and a quadratic form $\widetilde{q}(\mathrm{u}) = \sum \widetilde{q}_{i,j}\mathrm{u}_i\mathrm{u}_j$ such that $\tau(\mathrm{u}) = \mathrm{e}^{\widetilde{q}(\mathrm{u})}\theta(A\mathrm{u} + \xi, \Gamma^{\circ})$ is a tau function for the KP hierarchy for some $\xi \in \mathbb{C}^g$.*

Using them, Shiota finally proved the Novikov conjecture, which means that (A) implies (B) in the following theorem:

Theorem 3.202 *The following two conditions (A) and (B) for a principally polarized Abelian variety $\mathbb{A} = \mathbb{C}^g/(\mathbb{Z}^g + \Gamma^{\circ}\mathbb{Z}^g)$, $\Gamma^{\circ} \in \mathfrak{H}_g$ are equivalent:*

A a. *There exist vectors $a_1, a_2, a_3 \in \mathbb{C}^g$, $a_1 \neq 0$, and a quadratic form $\widetilde{q}(\mathrm{u}) = \sum_{i,j=1}^{3} \widetilde{q}_{i,j}\mathrm{u}_i\mathrm{u}_j \in \mathbb{C}$ such that for any $\xi \in \mathbb{C}^g$,*

$$\tau(\mathrm{u}) = \exp(\widetilde{q}(\mathrm{u}))\theta(\mathrm{u}_1 a_1 + \mathrm{u}_2 a_2 + \mathrm{u}_3 a_3 + \xi)$$

gives a solution to the KP equation in Hirota's form

$$(\mathfrak{D}_1^4 + 3\mathfrak{D}_2^2 - 4\mathfrak{D}_1\mathfrak{D}_3)\tau \cdot \tau = 0;$$

b. *The theta divisor Θ of $\mathbb{A}$ is irreducible.*

B *$\mathbb{A}$ is isomorphic to the Jacobian variety J_X ($\Gamma^{\circ} = \Gamma^{\circ}_X$) of a completely smooth curve X of genus g over $\mathbb{C}$.*

Remark 3.203 The correspondence in Theorem 3.202 reminds us of the problem posed by Baker at the end of the 1903 paper [Ba1903] (see Sect. 1.6.4). The problem was also proved by the theorem with Nakayashiki's results on the relation among the tau, sigma, and theta functions [Nak2016] (see the following subsections).

In the KP hierarchy, the subspace of $\mathfrak{u} \in \mathbb{C}^\infty$ corresponds to $\mathbb{C}^g$ and its tau function characterizes the Jacobian $J_X^\circ = \mathbb{C}^g/\Gamma_X^\circ$, whereas $\mathfrak{z}$ corresponds to the local parameter at a point p in a curve X; $\mathfrak{z}$ is connected with $\mathfrak{u}$ via (3.93) and the differentials (and the Abelian integrals) of the second kind (cf. Sect. 2.4.3.6).

3.9.2 Universal Grassmannian Manifold

In this subsection, let us follow [Nak2010b, Nak2016] to review the universal Grassmannian manifold (UGM). Here we remark that the non-commutative relation

$$[\partial_z, z] = \frac{d}{dz}z - z\frac{d}{dz} = 1 \tag{3.99}$$

causes the following properties. Let $V := \mathscr{E}_{\check{R}}/\mathscr{E}_{\check{R}}z \cong \mathbb{C}((\partial_z^{-1}))$ be the left $\mathscr{E}_{\check{R}}$ module. We define the element $\varepsilon_i \in V$ by $\varepsilon_i = \partial^{-i-1}$ on which the action of $\mathscr{E}_{\check{R}}$ is

$$\partial_z \varepsilon_i = \varepsilon_{i-1}, \quad z\varepsilon_i = (i+1)\varepsilon_i.$$

By introducing $V^\phi := \bigoplus_{i<0} \mathbb{C}\varepsilon_i$ and $V^{(0)} := \prod_{i\geq 0} \mathbb{C}\varepsilon_i$, we have the decomposition $V = V^\phi \oplus V^{(0)}$. For a sub-vector space U of V, we define the linear map,

$$\pi_U : U \to V/V^{(0)} \cong V^\phi,$$

induced from the inclusion $U \hookrightarrow V$ and the projection $V \to V/V^{(0)}$.

Definition 3.204 The universal Grassmannian manifold, UGM, is the set of subspaces U of V whose $\mathrm{Ker}(\pi_U)$ and $\mathrm{Coker}(\pi_U)$ are finite-dimensional and $\dim_{\mathbb{C}} \mathrm{Ker}(\pi_U) = \dim_{\mathbb{C}} \mathrm{Coker}(\pi_U)$.

For a point U of UGM, a frame $\boldsymbol{\xi}$ of U is a basis of U:

$$\boldsymbol{\xi} = \{\xi_j\}_{j<0} = (\dots, \xi_{-3}, \xi_{-2}, \xi_{-1}), \quad \xi_j = \sum_{i\in\mathbb{Z}} \xi_{ij}\varepsilon_i,$$

such that for $j \ll 0$,

$$\xi_{ij} = \begin{cases} 0, & (i < j), \\ 1, & (i = j). \end{cases}$$

For a point U of UGM, there exists a unique strictly increasing sequence $\boldsymbol{\rho}(U) = \{\rho_i \in \mathbb{Z} \mid i \in -\mathbb{N}\}$ and a unique frame $\boldsymbol{\xi}$ of U such that

$$\rho_{-1} > \rho_{-2} > \rho_{-3} > \cdots, \quad \rho_{-i} = -i \text{ for } -i \ll 0, \tag{3.100}$$

$$\xi_{ij} = \begin{cases} 0, \ i < \rho_j \text{ or } i = \rho_{j'} \text{ for some } j' > j, \\ 1, \ i = \rho_j. \end{cases} \tag{3.101}$$

We say that the frame is normalized if ξ satisfies the condition (3.101).

The relation $\dim_{\mathbb{C}} \mathrm{Ker}\pi_U = \dim_{\mathbb{C}} \mathrm{Coker}\pi_U$ is reduced to

$$\#(\boldsymbol{\rho}(U) \cap \mathbb{N}_0) = \#((-\mathbb{N}) \setminus \boldsymbol{\rho}(U)). \tag{3.102}$$

For a point $U \in \mathrm{UGM}$, the sequence $\boldsymbol{\rho}$ satisfying (3.100) corresponds to a partition $\Lambda(\boldsymbol{\rho}) = (\Lambda_1, \Lambda_2, \cdots,)$ by

$$\Lambda_i = \rho_{-i} + i.$$

We identify the partition $\Lambda = (\Lambda_1, \cdots, \Lambda_\ell)$ with $(\Lambda_1, \cdots, \Lambda_\ell, 0^n)$ for any n. For a partition Λ, let

$$\mathrm{UGM}^\Lambda := \{U \in \mathrm{UGM} \mid \Lambda(\boldsymbol{\rho}(U)) = \Lambda\}.$$

Then the UGM is disjoint union of UGM^Λ's, $\mathrm{UGM} = \bigsqcup_\Lambda \mathrm{UGM}^\Lambda$. Given a point $U \in \mathrm{UGM}$ and a sequence $\boldsymbol{\rho}$ satisfying (3.100), the Plücker coordinate $\xi_{\Lambda(\boldsymbol{\rho})}$ of the normalized frame $\boldsymbol{\xi}$ is given by the determinant of the $(-\mathbb{N}) \times (-\mathbb{N})$-matrix,

$$\xi_{\Lambda(\boldsymbol{\rho})} = \det(\xi_{\rho_i, j}). \tag{3.103}$$

For two partitions $\Lambda = (\Lambda_1, \Lambda_2, \cdots,)$, and $\Lambda' = (\Lambda'_1, \Lambda'_2, \cdots,)$, we define $\Lambda \geq \Lambda'$ if $\Lambda_i \geq \Lambda'_i$ for any i.

It is known that for $U \in \mathrm{UGM}^\Lambda$, the Plücker coordinates satisfy

$$\xi_{\Lambda'} = \begin{cases} 0, \ \text{unless } \Lambda' \geq \Lambda, \\ 1, \ \Lambda' = \Lambda. \end{cases}$$

Following [Nak2016, Definition 7], we define the tau function based on the frame $\boldsymbol{\xi}$ of a point $U \in \mathrm{UGM}$:

Definition 3.205 For $U \in \mathrm{UGM}$ and the partition $\Lambda(\boldsymbol{\rho}(U))$, we define

$$\begin{aligned} \tau(\mathrm{u}; \boldsymbol{\xi}(U)) &:= \sum_{\Lambda \geq \Lambda(\boldsymbol{\rho}(U))} \xi_\Lambda \mathbf{s}_\Lambda(\mathrm{u}) \\ &= \mathbf{s}_{\Lambda(\boldsymbol{\rho}(U))}(\mathrm{u}) + \sum_{\Lambda > \Lambda(\boldsymbol{\rho}(U))} \xi_\Lambda \mathbf{s}_\Lambda(\mathrm{u}), \end{aligned}$$

where $\boldsymbol{\xi}(U)$ is the normalized frame $\boldsymbol{\xi}$ of U, and $\mathbf{s}_\Lambda(\mathrm{u})$ is the Schur polynomial or Schur function defined by

$$\mathrm{s}_\Lambda(\mathrm{u}) := |\mathrm{p}_{\Lambda_i - i + j}(\mathrm{u})|_{1 \le i, j, \le \ell}. \tag{3.104}$$

We have the two tau functions, $\tau(\mathrm{u}; \boldsymbol{\xi}(U))$ in Definition 3.205 and $\tau(\mathrm{u})$ obeying (3.95) in Theorem 3.201, which are essentially the same up to a constant factor [Nak2016, Theorem 9] ([SeW1985, Proposition 8.3]):

Theorem 3.206 [Nak2016, Theorem 9] *For every* $U \in$ UGM, $\tau(\mathrm{u}; \boldsymbol{\xi}(U))$ *satisfies the KP hierarchy (3.92), whereas for every solution* $\tau(\mathrm{u}) \in \mathbb{C}[[\mathrm{u}]]$ *of (3.95) or (3.98), there exist a point* $U \in$ UGM *and* $C \in \mathbb{C}^\times$ *such that* $\tau(\mathrm{u}) = C\tau(\mathrm{u}; \boldsymbol{\xi}(U))$.

It means that each point $U \in$ UGM corresponds to the theta functions of a compact Riemann surface X, such that [Nak2016, (31)]

$$\tau(\mathrm{u} : U) = \mathrm{e}^{q^\circ(\mathrm{u})}\theta(A\mathrm{u} + \xi, \Gamma_X^\circ),$$

where q_{ij}° and $A = (a_{ij}) \in \mathrm{Mat}_{\mathbb{C}}(g, \infty)$ are given by [Nak2016, (30)],

$$\Omega^\circ(P, Q) = \left(\frac{1}{(t_P - t_Q)^2} + \sum_{i,j=1}^{\infty} q_{ij}^\circ t_P^{i-1} t_Q^{j-1}\right) dt_P dt_Q, \tag{3.105}$$

$$\nu_i^{\mathrm{I}\circ}(P) = \left(\sum_{j=1}^{\infty} a_{ij} t_P^{j-1}\right) dt_P \tag{3.106}$$

for the arithmetic parameters t_P and t_Q of $\widetilde{X}$ for P and Q at ∞ in X, and the normalized holomorphic one-forms $\nu_i^{\mathrm{I}\circ}$ in Sect. 2.4.2. Here we note Lemma 2.102.

Let $\check{K} = \mathbb{C}((s))$ and $\mathscr{E}_{\check{K}} := \check{K}((\partial_z^{-1}))$. We introduce $\mathscr{W}$ following [Nak2016]:

Definition 3.207 Let $\mathscr{W}$ be the subset of $\mathscr{E}_{\check{K}}$ whose element W satisfies the following:

1. $W := \sum_{i \le 0} w_i \partial_z^i$, $w_0 = 1$, and
2. there exist non-negative integers ℓ and m such that $z^\ell W \in \mathscr{E}_{\check{K}}^\circ$ and $W^{-1} z^m \in \mathscr{E}_{\check{K}}^\circ$.

Proposition 3.208 *There is a bijection* $\gamma : \mathscr{W} \to$ UGM *by* $\gamma(W) = W^{-1} z^m V^\phi$, *where* m *satisfies* $W^{-1} z^m \in \mathscr{E}_{\check{K}}^\circ$.

Lemma 3.209 *For any partition* $\Lambda = (\Lambda_1, \Lambda_2, \ldots)$ *whose maximal nonvanishing index is* g, *i.e.,* $\Lambda_g \ne 0$ *and* $\Lambda_i = 0$ $i > g$, *we have*

$$\mathrm{s}_\Lambda(\mathrm{u}_1, 0, 0, \ldots) = d_\Lambda \mathrm{u}_1^{|\Lambda|},$$

where $d_\Lambda = \dfrac{\prod_{i<j}((\Lambda_i + g - i) - (\Lambda_j + g - j))}{\prod_{i=1}^{g}(\Lambda_i + g - i)!}$.

3.9.3 Numerical Semigroups and UGM

In this subsection, following [SeW1985, p. 46], we consider the relation between the numerical semigroup and UGM. (In [SeW1985], S corresponds to $\boldsymbol{\rho}$ such that $S(i) = -\rho_{-i}$ for $i \in \mathbb{N}_0$.) We fix a numerical semigroup H. Recalling that $N^{\mathfrak{c}}(i)$ shows the gap-sequence $H^{\mathfrak{c}}$ in Sect. 2.2.1, we have the strictly decreasing ordered set in Definition 2.4,

$$\overline{H}^{\mathfrak{c}} = H^{\mathfrak{c}} \cup (-\mathbb{N}) = \{\ell_i = N^{\mathfrak{c}}(g-i) \mid i = 1, 2, \ldots, g\} \cup \{\ell_{i+g} = -i \mid i \in \mathbb{N}\}.$$

From Proposition 2.1 4, we obviously have the following lemma:

Lemma 3.210 *The strictly increasing ordered set,*

$$\boldsymbol{\rho}^H := \overline{H}^{\mathfrak{c}} - g = \{\rho_{-i}^H = \ell_i - g \mid \ell_i \in \overline{H}^{\mathfrak{c}}\},$$

satisfies (3.100) and (3.101), i.e., $\boldsymbol{\rho}_{-i}^H = -i$ *for* $i \gg 0$ *and* $\#(\boldsymbol{\rho}^H \cap \mathbb{N}_0) = \#((-\mathbb{N}) \setminus \boldsymbol{\rho}^H)$.

Example 3.211 For $H = \langle 3, 4, 5\rangle$, we have $\overline{H}^{\mathfrak{c}} = \{\cdots, -3, -2, -1, 1, 2\}$, $S^H = \{\rho_{-1} = 0, \rho_{-2} = -1, \rho_{-3} = -3, \ldots\}$.

Thus it determines $\mathrm{UGM}^{A_H} \subset \mathrm{UGM}$. As mentioned in Lemma 2.5 and [SeW1985, p. 46], the numerical semigroup H is characterized a maximal set whose element r satisfies $\boldsymbol{\rho}^H - r \subset \boldsymbol{\rho}^H$ (or $\overline{H}^{\mathfrak{c}} - r \subset \overline{H}^{\mathfrak{c}}$). The affine algebra A_H is the algebra spanned by $\{z^r\}_{r \in H}$ is regarded as the numerical semigroup ring R_H in Sect. 2.2.3.

Since there is a Hurwitz problem on the numerical semigroup (cf. Sect. 2.2.1), we handle only the numerical semigroup H_X corresponding to the non-singular Weierstrass curve X in this book. We consider the points in UGM^{A_H} which correspond to the numerical semigroup ring R_H and R_X, respectively.

3.9.4 Monomial Curves X_H and Schur Function

In this subsection, we consider again the numerical semigroup ring R_H as mentioned in Sect. 2.2.3.

By fixing the arithmetic local parameter t at ∞ in X following Ônishi [On2018] and by using the ring-homomorphism φ_∞, we may consider the correspondence between R_X and R_H. However, since every numerical semigroup H satisfies Lemma 3.210, we consider a general monomial curve in this subsection.

We use the correspondence $z \to \frac{1}{t}$ in this appendix. and the numerical semigroup ring R_H as a subring of $\mathbb{C}[t^{-1}]$ is expressed as

$$R_H = \bigoplus_{i \in H} \mathbb{C}t^{-i} = \{\sum_i a_i t^{-i} \mid a_i \in \mathbb{C},\ a_j = 0 \text{ for } j \in H^{\mathrm{c}}\}.$$

We also call it the "monomial curve" $X_H = \mathrm{Spec}\, R_H$ and its symmetric product $S^g X_H$ as mentioned in Sect. 2.2.3 We introduce the differentials formally,

$$\widehat{\nu}_i^{\mathrm{I}} = d\widehat{u}_i = -t^{N^{\mathrm{c}}(g-i)-1}dt, \quad \widehat{\nu}_i^{\mathrm{II}} = -\frac{1}{t^{N^{\mathrm{c}}(g-i)}}dt, \quad (i = 1, 2, \dots, g).$$

Lemma 3.212 *For the symplectic product,*

$$\langle \widehat{\nu}, \widehat{\nu}' \rangle := \frac{1}{2\pi\sqrt{-1}} \oint \left(\int_0^t \widehat{\nu} \right) \widehat{\nu}'(t),$$

we have

$$\langle \widehat{\nu}_i^{\mathrm{I}}, \widehat{\nu}_j^{\mathrm{I}} \rangle = 0, \quad \langle \widehat{\nu}_i^{\mathrm{II}}, \widehat{\nu}_j^{\mathrm{II}} \rangle = 0, \quad \langle \widehat{\nu}_i^{\mathrm{I}}, \widehat{\nu}_j^{\mathrm{II}} \rangle = \delta_{i,j}.$$

We introduce the set of differentials of the monomial curve X_H,

$$\mathscr{A}_H := \sum_{i \in \overline{H}^{\mathrm{c}}} \mathbb{C}t^{i-1}dt, \tag{3.107}$$

as a $\mathbb{C}$ vector space and a dual R_H^* of R_H,

$$R_H^* := \int_0^t \mathscr{A}_H = \bigoplus_{n \in \overline{H}^{\mathrm{c}}} \mathbb{C} \int_0^t t^{n-1}dt = \bigoplus_{n \in \overline{H}^{\mathrm{c}}} \mathbb{C}\frac{1}{n}t^n.$$

For the monomial curve X_H, we find the parameters d_h in Lemma 2.45 with an ambiguity. We fix it. Then we define $\mathfrak{k}_H := d_h - 2g - r + 1$ which is modeled after $\mathfrak{k}_X$ in Lemma 2.82 and introduce

$$\widehat{R}_H = R_H^* t^{-2g+1-\mathfrak{k}_H}.$$

Since every element ϕ in $\widehat{R}_H$ is given by

$$\phi = \sum_{\ell \in \overline{H}^{\mathrm{c}} - 2g + 1 - \mathfrak{k}_H} a_\ell t^\ell,$$

we can find the embedding $\iota_H : \widehat{R}_H \to V$ by

$$\iota_H\left(\sum_{\ell\in\overline{H}^{\mathrm{c}}-2g+1-\mathfrak{k}_H} a_\ell t^\ell\right)=\sum a_\ell \varepsilon_{\ell+g-1+\mathfrak{k}_H},\tag{3.108}$$

from Lemma 3.210. Since the generators of $\widehat{R}_H=\oplus\mathbb{C}Z_i$ as the $\mathbb{C}$-vector space are given by $\{Z_i=t^{-\widehat{N}(i)-\mathfrak{k}_H}=t^{N^{\mathrm{c}}(i)-2g+1-\mathfrak{k}_H}\}_{i\in\mathbb{N}_0}$ from Lemma 2.109 7, we consider the transformation matrix Ξ_H,

$$\Xi_H:=(\delta_{i,N^{\mathrm{c}}(i)-g}).\tag{3.109}$$

This embedding is the same as the one in Lemma 3.210.

When H is symmetric, $\overline{H}^{\mathrm{c}}-2g+1=-H$, $\widehat{R}_H=R_H$ and $\mathfrak{k}_H=0$ and thus the embedding is simply written

$$\iota_H\left(\sum_{\ell\in H} a_\ell t^\ell\right)=\sum_{\ell\in H} a_\ell \varepsilon_{\ell+g-1}.\tag{3.110}$$

We implicitly consider the vector space $\bigoplus_{\ell\in\overline{H}^{\mathrm{c}}-g+1}\mathbb{C}z^\ell$. The index $\ell-g+1$ can be regarded as the Euler characteristic of degree ℓ [KawNTY1988]. Following the convention in this study, we let $w_i:=N^{\mathrm{c}}(i-1)$, $(i=1,\dots,g)$ noting $w_1=1$.

We note that ν_j^{I} and ν_j^{II} belong to $\mathscr{A}_H$. We define the "Abelian integral",

$$\widehat{w}:X_H\to\mathbb{C}^g,\quad \widehat{w}(t_i)=\begin{pmatrix}\widehat{u}_{w_1}\\ \widehat{u}_{w_2}\\ \vdots\\ \widehat{u}_{w_g}\end{pmatrix}=-\begin{pmatrix}t_i\\ \frac{1}{w_2}t_i^{w_2}\\ \vdots\\ \frac{1}{w_g}t_i^{w_g}\end{pmatrix},$$

and for $(t_1,t_2,\dots,t_k)\in S^kX_H$, $\widehat{w}:S^kX_H\to\mathbb{C}^g$, $\widehat{w}(t_1,\dots,t_k):=\sum_{i=1}^k\widehat{w}(t_i)$. In particular,

$$-\widehat{w}(t_1,t_2,\dots,t_g)=\begin{pmatrix}t_1+\cdots+t_g\\ \frac{1}{w_2}(t_1^{w_2}+\cdots+t_g^{w_2})\\ \vdots\\ \frac{1}{w_g}(t_1^{w_g}+\cdots+t_g^{w_g})\end{pmatrix}=:\begin{pmatrix}\widehat{\mathfrak{u}}_{w_1}\\ \widehat{\mathfrak{u}}_{w_2}\\ \vdots\\ \widehat{\mathfrak{u}}_{w_g}\end{pmatrix}=:\widehat{\mathfrak{u}}.$$

For consistency with our convention, we also introduce the exchanging matrix, $\widehat{j}$ such that $\widehat{j}\,{}^t(u_1,\dots,u_g)=(u_g,\dots,u_1)$.

Let $\widehat{\mathfrak{u}}(t)=\widehat{\mathfrak{u}}_{w_1}t+\mathfrak{u}_{w_2}t^{w_2}+\cdots+\widehat{\mathfrak{u}}_{w_{g-1}}t^{w_{g-1}}+\widehat{\mathfrak{u}}_{w_g}t^{w_g}$, and $\exp(\widehat{\mathfrak{u}}(t))=\sum_{i=0}\widehat{\mathfrak{p}}_i(\widehat{\mathfrak{u}})t^i$. Using the above embedding, we have the normalized frame $\boldsymbol{\xi}$ of UGM, and thus the tau function for the monomial curve X_H by the Schur function.

Proposition 3.213 *For the monomial curve, we have the tau function,*

$$\tau(\mathrm{u};\xi) = \mathbf{s}_\Lambda(\mathrm{u}). \tag{3.111}$$

Proof From Definition 3.205, it is obvious. (See also Proposition 8.2 in [SW p. 49].) It is also obtained by the determinant of the multiplication of the matrices (3.109) and (3.117) due to the Binet–Cauchy lemma [Ga1984], [Kod2017, Chap. 1]. ■

Bukhshtaber, Enolskii, and Leykin studied the sigma functions of the monomial curves generated by two elements in [BucEL1999]. We emphasize that the study was the beginning of the discovery of a beautiful correspondence between the sigma functions and the Schur polynomials, though Baker implicitly used the correspondence (cf. Sect. 4.1.3), and it could be interpreted as a corollary in the above lemma after Nakayashiki's final results [Nak2016].

3.9.5 Weierstrass Curves X and Schur Function

This subsection shows the relation between Weierstrass curves X and Schur functions. Let us consider W-curves discussed in Chap. 2. A W-curve X with ∞ can be regarded as a compact Riemann surface with the Weierstrass semigroup H_X at ∞. Thus we regard it as a compact Riemann surface with H_X with which Nakayashiki dealt in the paper [Nak2016]. By following [Nak2016], we consider the theta functions and Schur functions for the W-curve X.

We recall that the normalized differentials of the second kind in Sect. 2.4.1.2 satisfy $\nu^{\mathrm{II}\circ}_{\infty,t,i+1} = d\left(\dfrac{1}{t^i} + d_{>0}(t)\right)$ at ∞ and $\lim\limits_{P\to\infty}\left(\displaystyle\int_\infty^P \nu^{\mathrm{II}\circ}_{\infty,t,i+1} - t_\infty^{-i}\right) = 0$.

For a point $u = {}^t(u_1, u_2, \dots, u_g) \in \mathbb{C}^g$, we implicitly refer to $\widehat{u} = {}^t(\widehat{u}_{w_1}, \dots, \widehat{u}_{w_g}) = -\widehat{j}\,{}^t(u_1, \dots, u_g) \in \widehat{j}\,\mathbb{C}^g$ rather than $u \in \mathbb{C}^g$ itself.

Noting (3.106), and following [Nak2016, (32)][KawNTY1988, (6.3) Corollary], we rewrite the wave function (3.94) for $\mathrm{u} \in \mathbb{C}^\infty$ and and the local parameter $t = t_P$ at P in X,

$$\psi(\mathrm{u}, t)\frac{\sqrt{dt}}{t} = \frac{1}{E(\infty, P)}\frac{\theta(\widetilde{w}^\circ(P) + A\mathrm{u} + e)}{\theta(\widetilde{w}^\circ(P) + e)} \exp\left(-\sum_{i\in H_X^c}^{\infty} \mathrm{u}_i \int_\infty^P \nu^{\mathrm{II}\circ}_{\infty,t,i+1}\right), \tag{3.112}$$

where e is a zero $e \in \Theta$ or $e = w^\circ(\mathfrak{e})$, $\mathfrak{e} := (P_1 + \cdots + P_{g-1} - \mathfrak{K}_\xi)$ due to Corollary 3.2 for certain $P_1, \dots, P_{g-1} \in X$.[11]

[11] An example of the wave function is given by (4.42), which can be generalized due to Theorem 3.91.

Then $\theta(\widetilde{w}^\circ(P) + A\mathfrak{u} + e)\exp\left(\sum_{i\in H_X^c}^{g} \mathfrak{u}_i \int_\infty^P \nu^{\mathrm{II}\circ}_{\infty,t,i+1}\right)$ is a section of the holomorphic line bundle $w^{\circ *}(\Theta) \otimes \mathscr{L}_X(-e)$. We conclude that $\psi(\mathfrak{u}, t)\frac{\sqrt{dt}}{t}$ is the section of the line bundle $\mathbf{H}^0(X, (\mathscr{L}_X^{g-1}(\mathfrak{K}_\xi) \otimes \mathscr{L}_X^0(-\mathfrak{e}))(*\infty))$. Nakayashiki embedded it into UGM, i.e., $\iota_{\mathrm{spin}} : \mathbf{H}^0(X, (\mathscr{L}_X^{g-1}(\mathfrak{K}_\xi) \otimes \mathscr{L}_X^0(-\mathfrak{e}))(*\infty)) \to V$, as in Theorem 8 in [Nak2016]. It corresponds to the spin structure as mentioned in Corollary 3.2 and the Fermionic expressions in [KawNTY1988].

Segal and Wilson showed that the tau function is expressed by the expansion of the Schur polynomials and the theta function [SeW1985, Propositions 8.3 and 9.10]. Nakayashiki explicitly showed the relation between the theta functions and the Schur polynomials [Nak2016, Theorem 10] for the compact Riemann surface X associated with the numerical semigroup H_X.

For the arithmetic local parameter t and $\mathfrak{u}_i$, we let

$$\mathfrak{u}(t) = \mathfrak{u}_{w_1} t + \mathfrak{u}_{w_2} t^{w_2} + \cdots + \mathfrak{u}_{w_{g-1}} t^{w_{g-1}} + \mathfrak{u}_{w_g} t^{w_g}.$$

By using this, the elementary Schur polynomial $\mathfrak{p}(\mathfrak{u})$

$$\exp(\mathfrak{u}(t)) = \sum_{i=0} \mathfrak{p}_i(\mathfrak{u}) t^i \tag{3.113}$$

as in (3.97), and $s_\Lambda(\mathfrak{u}) = |\mathfrak{p}_{\Lambda_i+j-i}(\mathfrak{u})|$.

We assume the expansion of the basis of $\mathbf{H}^0(X, \mathscr{A}_X)$ given by

$$\nu_i^{\mathrm{I}} = \sum_{j=0}^{\infty} b_{w_{g-i+1},j} t^{j-1} dt,$$

and $\widetilde{B} = (b_{w_{g-i+1},j}) \in \mathrm{Mat}_{\mathbb{C}}(g, \infty)$ defined by

$$\widetilde{B} = \begin{bmatrix} b_{w_1,1} & b_{w_1,2} & b_{w_1,3} & b_{w_1,4} & \cdots \\ b_{w_2,1} & b_{w_2,2} & b_{w_2,3} & b_{w_2,4} & \cdots \\ \vdots & \vdots & \vdots & \vdots & \ddots \\ b_{w_{g-2},1} & b_{w_{g-2},2} & b_{w_{g-2},3} & b_{w_{g-2},4} & \cdots \\ b_{w_{g-1},1} & b_{w_{g-1},2} & b_{w_{g-1},3} & b_{w_{g-1},4} & \cdots \\ b_{w_g,1} & b_{w_g,2} & b_{w_g,3} & b_{w_g,4} & \cdots \end{bmatrix}.$$

From (3.98) and (3.105), we obviously have

$$\tau(\mathfrak{u}; U) = e^{q^\circ(\mathfrak{u})}\theta((2\omega')^{-1}\widehat{j}\,\widetilde{B}\mathfrak{u} + \xi).$$

By letting $\widetilde{B}_0 := (b_{w_{g-i+1}, w_{g-j+1}}) \in \mathrm{Mat}_{\mathbb{C}}(g)$, which is invertible due to Lemma 2.109, we introduce the truncated vector $\widehat{\mathrm{u}} := {}^t(\mathrm{u}_{w_1}, \mathrm{u}_{w_2}, \ldots, \mathrm{u}_{w_g})$ and the transformation between $\widehat{\mathrm{u}}$ and $u \in \mathbb{C}^g$

$$u = u(\widehat{\mathrm{u}}) = \widehat{j}\,\widetilde{B}_0 \widehat{\mathrm{u}} \in \mathbb{C}^g, \quad \widehat{\mathrm{u}} = \widehat{\mathrm{u}}(u) := \widetilde{B}_0^{-1} \widehat{j}\, u.$$

Nakayashiki showed that [Nak2016, (51)][12]

$$\theta((2\omega')^{-1} u(\widehat{\mathrm{u}}) + \xi) = e^{-q^\circ(\widehat{\mathrm{u}})} \tau(\widehat{\mathrm{u}}; U)$$

for a certain U. From Definition 3.205, we obtain $\tau(\mathrm{u}; U) = \mathbf{s}_{\Lambda_X}(\mathrm{u}) + \sum_{\Lambda_X < \Lambda'} \xi_{\Lambda'} \mathbf{s}_{\Lambda'}(\mathrm{u})$ for certain ξ_Λ. We obviously have the same equation if we restrict $\mathrm{u} \in \mathbb{C}^\infty$ to the truncated vector space $\widehat{\mathrm{u}} \in \mathbb{C}^g$, (if necessary, we could introduce the projection $\varpi_{g,\infty} : \mathbb{C}^\infty \to \mathbb{C}^g$), i.e.,

$$\tau(\widehat{\mathrm{u}}; U) = \mathbf{s}_{\Lambda_X}(\widehat{\mathrm{u}}) + \sum_{\Lambda_X < \Lambda'} \xi_{\Lambda'} \mathbf{s}_{\Lambda'}(\widehat{\mathrm{u}}). \tag{3.114}$$

Due to Theorem 3.206 ([Nak2016, Theorem 9]), the $\tau(\mathrm{u}, U)$ is expressed by the theta function via (3.98). Finally, we obtain the expansion of the theta function via Schur functions in the following theorem from [Nak2016, Theorem 10]:

Theorem 3.214 [Nak2016, Theorem 10]

$$(-1)^{|\Lambda_X|} C \theta((2\omega')^{-1} u - e | \tau) = s_{\Lambda_X}(\mathrm{u})|_{\mathrm{u}_{w_i} = -u_{g-i+1}} + \textit{higher weight terms}.$$

We mention the embedding of the affine algebra of X into UGM further. In terms of the arithmetic local parameter t at ∞ in X, every element in $\widehat{R}_X$ defined in Definition 2.108 2 is expressed by t, e.g.,

$$\widehat{\phi}_i = \frac{1}{t^{N^c(i) + \mathfrak{k}_X}} (1 + d_{>0}(t)),$$

due to Definition 2.108 3, whereas $\overline{H}_X^c = \{\mathrm{wt}(\varphi) + 2g - 1 + \mathfrak{k}_X \mid \varphi \in \widehat{R}_X\}$, and $\widehat{R}_X \cong \mathscr{A}_X(*\infty)$. As in the case of the monomial curve X_H, we may embed $\widehat{R}_X$ into UGM.

Here we note $-\mathfrak{e} + \mathfrak{K}_\xi = -(P_1 + \cdots + P_{g-1}) + 2\mathfrak{K}_\xi = -(P_1 + \cdots + P_{g-1}) + K_X$, and $\mathscr{L}_X(-\xi_X)^{g-1} \otimes \mathscr{L}_X(-e)^0 \cong \mathscr{L}_X^{g-1}(K_X - (P_1 + \cdots + P_{g-1}))$. Then we have $\mathscr{L}_X^{g-1}(K_X - (P_1 + \cdots + P_{g-1}))(*\infty) \cong \mathscr{L}_X(K_X - (P_1 + \cdots + P_{g-1}) - (g-1)\infty)(*\infty)$. From Proposition 2.117, $K_X = -\mathfrak{K}_X + (2g - 2 + \mathfrak{k}_X)\infty$, we regard $\widehat{R}_X = \mathbf{H}^0(X, \mathscr{O}_X(-\mathfrak{K}_X)(*\infty))$ in Lemma 2.109 5. By using (3.108), the embedding ι_{spin} might be replaced with

---

12 We note that ω'_N, $\widetilde{B}_N$, and u_N in Nakayashiki's convention correspond to $\omega' = -\omega'_N$, $\widetilde{B} = -\widetilde{B}_N$, and $u = -u_N$.

$$\widetilde{\iota}_X : \widehat{R}_X \to V, \quad \left(\widetilde{\iota}_X\left(\sum_{\ell=-N}^{\infty} a_\ell t^\ell\right) = \sum_\ell a_\ell \varepsilon_{\ell+g-1+\mathfrak{k}_X}\right).$$

It is related to the Euler characteristic $\chi(\ell') = \ell' + 1 - g$ of the degree ℓ', and $\overline{H}_X^{\mathfrak{c}} - g$ in Lemma 3.210 because $\ell + g - 1 + \mathfrak{k}_X \in \overline{H}_X^{\mathfrak{c}} - g$ [KawNTY1988].

When H_X is symmetric, $R_X = \widehat{R}_X$, $\mathfrak{k}_X = 0$ and thus the embedding is interpreted as $\iota_X : R_X \to V$, $\left(\widetilde{\iota}_X\left(\sum_{\ell=-N}^{\infty} a_\ell t^\ell\right) = \sum_\ell a_\ell \varepsilon_{g+\ell-1}\right)$ [Nak2010b, p. 383]. Let $U_X := \widetilde{\iota}_X(R_X)$. Nakayashiki showed U_X belongs to the UGM [Nak2010b, Lemma 2] and let $\boldsymbol{\xi}_X$ be the normalized frame of U_X in the UGM.

$$\tau(\widehat{\mathrm{u}}; \boldsymbol{\xi}_X) = \mathbf{s}_{\Lambda_X}(\widehat{\mathrm{u}}) + \sum_{\Lambda > \Lambda_X} \xi_\Lambda \mathbf{s}_\Lambda(\widehat{\mathrm{u}}), \tag{3.115}$$

for any $\widehat{\mathrm{u}} \in \widetilde{B}_0^{-1} \widetilde{j}\, \mathbb{C}^g$. Here we note that ξ_Λ is defined by (3.103).

3.9.6 Plane Weierstrass Curves X and Schur Function

In this subsection, we specify the W-curve by the (r, s)-curve or the Weierstrass plane curve and show the explicit expressions of the σ function in terms of the Schur functions following [On2013, On2018] which is written based on [Nak2010a, Nak2010b]. The expression is extended to the telescopic curves by Ayano [Ay2022].

We recall the normalized fundamental differential of the second kind given in Proposition 2.101, its expansion in (3.105), and the Baker function (3.112), which is expressed by [Nak2010b, Lemma 3],

$$\psi(\widehat{\mathrm{u}}, t) = \frac{t^g}{\widetilde{\mathscr{E}}(\infty, P)} \frac{\sigma(\widetilde{w}(P) + \widehat{j}\widetilde{B}_0\widehat{\mathrm{u}})}{\sigma(\widehat{j}\widetilde{B}_0\widehat{\mathrm{u}})} \exp\left(-\sum_{i=1}^{g} \mathrm{u}_{w_{g-i+1}} \int_\infty^P \widehat{\nu}_i^{\mathrm{II}}\right), \tag{3.116}$$

in terms of the prime function $\widetilde{\mathscr{E}}(\infty, P)$ in (3.67), the arithmetic parameter t for P, and

$$\widehat{\nu}_i^{\mathrm{II}} := \nu_{\infty,t,w_{g-i+2}}^{\mathrm{II}\circ} + \sum_{k,\ell} \widetilde{B}_{0 w_{g-i+1},k}(\eta'\omega'^{-1})_{k,\ell}\nu_\ell^{\mathrm{I}}.$$

For hyperelliptic curves, it is explicitly expressed in (4.42).

Noting that ν_1^{I} is the derivative of the local parameter $X \setminus \{\infty\}$ and thus the prefactor of the sigma function turns out to be the prime form with $\sqrt{\nu_1^{\mathrm{I}}}$, we consider the expansion

$$\log\left(t^{-c_X/2+1}\sqrt{\frac{\nu_1^{\mathrm{I}}}{dt}}\right)=\sum_j \frac{c_j}{j}t^j.$$

Using them, we define

$$\widehat{q}(\widehat{\mathrm{u}}):=\sum_{i=1}^{g} c_{w_i}\mathrm{u}_{w_i}+q(\widehat{\mathrm{u}}),\quad q(\widehat{\mathrm{u}}):=\frac{1}{2}\sum_{i=1}^{g}\sum_{j=1}^{g} q_{w_i,w_j}\mathrm{u}_{w_i}\mathrm{u}_{w_j},$$

where from Lemma 2.102 and Corollary 2.147,

$$\sum_{i,j} q_{w_i,w_j}\mathrm{u}_{w_i}\mathrm{u}_{w_j}:=\sum_{i,j} q^{\circ}_{w_i,w_j}\mathrm{u}_{w_i}\mathrm{u}_{w_j}-{}^t u(\widehat{\mathrm{u}})\gamma u(\widehat{\mathrm{u}}).$$

Then Nakayashiki showed the following theorem [Nak2010b, Theorem 8]:

Theorem 3.215 *The tau and sigma functions are connected by*

$$\tau(\widehat{\mathrm{u}}(u):\xi_X)=\exp(-\widehat{q}(\widehat{\mathrm{u}}(u)))\sigma(u).$$

Here we find that the term of c_j's in q comes from the factor $t^g/\widetilde{\mathscr{E}}(\infty,P)$ in (3.112) when we compare (3.94) and (3.116).

We consider the power expansions of the basis $\{\widehat{\phi}_i\}=\widehat{S}_X$ of the R_X-module $\widehat{R}_X$ in Definition 2.108 in terms of the arithmetic local parameter t, $\widehat{\phi}_i(t)=\sum_j \widehat{\phi}_i^{(j)}t^j$. Using the data of the expansions, we introduce the $\infty\times 2\infty$ matrix Ξ,

$$\Xi:=\left[\begin{array}{cccccccc|cccc}
\ddots & \vdots & \vdots & \vdots & \vdots & \vdots & \vdots & \vdots & \\
\cdots & \widehat{\phi}_5^{(-g-3)} & \widehat{\phi}_5^{(-g-2)} & \widehat{\phi}_5^{(-g-1)} & \widehat{\phi}_5^{(-g)} & \widehat{\phi}_5^{(-g+1)} & \widehat{\phi}_5^{(-g+2)} & \widehat{\phi}_5^{(-g+3)} & \cdots \\
\cdots & \widehat{\phi}_4^{(-g-3)} & \widehat{\phi}_4^{(-g-2)} & \widehat{\phi}_4^{(-g-1)} & \widehat{\phi}_4^{(-g)} & \widehat{\phi}_4^{(-g+1)} & \widehat{\phi}_4^{(-g+2)} & \widehat{\phi}_4^{(-g+3)} & \cdots \\
\cdots & \widehat{\phi}_3^{(-g-3)} & \widehat{\phi}_3^{(-g-2)} & \widehat{\phi}_3^{(-g-1)} & \widehat{\phi}_3^{(-g)} & \widehat{\phi}_3^{(-g+1)} & \widehat{\phi}_3^{(-g+2)} & \widehat{\phi}_3^{(-g+3)} & \cdots \\
\cdots & \widehat{\phi}_2^{(-g-3)} & \widehat{\phi}_2^{(-g-2)} & \widehat{\phi}_2^{(-g-1)} & \widehat{\phi}_2^{(-g)} & \widehat{\phi}_2^{(-g+1)} & \widehat{\phi}_2^{(-g+2)} & \widehat{\phi}_2^{(-g+3)} & \cdots \\
\cdots & \widehat{\phi}_1^{(-g-3)} & \widehat{\phi}_1^{(-g-2)} & \widehat{\phi}_1^{(-g-1)} & \widehat{\phi}_1^{(-g)} & \widehat{\phi}_1^{(-g+1)} & \widehat{\phi}_1^{(-g+2)} & \widehat{\phi}_1^{(-g+3)} & \cdots \\
\cdots & \widehat{\phi}_0^{(-g-3)} & \widehat{\phi}_0^{(-g-2)} & \widehat{\phi}_0^{(-g-1)} & \widehat{\phi}_0^{(-g)} & \widehat{\phi}_0^{(-g+1)} & \widehat{\phi}_0^{(-g+2)} & \widehat{\phi}_0^{(-g+3)} & \cdots
\end{array}\right],$$

and the $2\infty\times\infty$ matrix S,

$$S(\mathrm{u}) = \left[\begin{array}{cccccccc}
\ddots & \vdots & \vdots & \vdots & \vdots & \vdots & \vdots & \vdots \\
\cdots & 1 & & & & & & \\
\cdots & \mathfrak{p}_1 & 1 & & & & & \\
\cdots & \mathfrak{p}_2 & \mathfrak{p}_1 & 1 & & & & \\
\cdots & \mathfrak{p}_3 & \mathfrak{p}_2 & \mathfrak{p}_1 & 1 & & & \\
\cdots & \mathfrak{p}_4 & \mathfrak{p}_3 & \mathfrak{p}_2 & \mathfrak{p}_1 & 1 & & \\
\cdots & \mathfrak{p}_5 & \mathfrak{p}_4 & \mathfrak{p}_3 & \mathfrak{p}_2 & \mathfrak{p}_1 & 1 & \\
\cdots & \mathfrak{p}_6 & \mathfrak{p}_5 & \mathfrak{p}_4 & \mathfrak{p}_3 & \mathfrak{p}_2 & \mathfrak{p}_1 & 1 \\
\hline
\cdots & \mathfrak{p}_7 & \mathfrak{p}_6 & \mathfrak{p}_5 & \mathfrak{p}_4 & \mathfrak{p}_3 & \mathfrak{p}_2 & \mathfrak{p}_1 \\
\cdots & \mathfrak{p}_8 & \mathfrak{p}_7 & \mathfrak{p}_6 & \mathfrak{p}_5 & \mathfrak{p}_4 & \mathfrak{p}_3 & \mathfrak{p}_2 \\
\cdots & \mathfrak{p}_9 & \mathfrak{p}_8 & \mathfrak{p}_7 & \mathfrak{p}_6 & \mathfrak{p}_5 & \mathfrak{p}_4 & \mathfrak{p}_3 \\
\cdots & \mathfrak{p}_{10} & \mathfrak{p}_9 & \mathfrak{p}_8 & \mathfrak{p}_7 & \mathfrak{p}_6 & \mathfrak{p}_5 & \mathfrak{p}_4 \\
\cdots & \mathfrak{p}_{11} & \mathfrak{p}_{10} & \mathfrak{p}_9 & \mathfrak{p}_8 & \mathfrak{p}_7 & \mathfrak{p}_6 & \mathfrak{p}_5 \\
\ddots & \vdots & \vdots & \vdots & \vdots & \vdots & \vdots & \vdots
\end{array}\right], \tag{3.117}$$

due to (3.113). By using the Binet–Cauchy lemma [Ga1984], [Kod2017, Chap. 1], we obtain the expansion formulae of the sigma functions.

Using (3.115) and Theorem 3.215, Nakayashiki showed that the sigma function for the (r, s)-curves can be expressed the following relation from Theorem 3.206 [Nak2010a, Nak2010b], [On2018, Theorem 8.1]:

Theorem 3.216

$$\sigma(u) = \det(\Xi\ S(\widehat{\mathrm{u}}(u))) \cdot \exp(\widehat{q}(\widehat{\mathrm{u}}(u))),$$

where the column and the row lines should be matched in the product.

This provides the power expansions of the sigma functions.

For the case of a (3, 5)-curve for $g = 4$ as an example, $\Lambda_{\langle 3,4\rangle} = (4, 2, 1, 1)$, we have

$$\Xi = \left[\begin{array}{ccccccc|cccccc}
\vdots & \vdots & \vdots & \vdots & \vdots & \vdots & \vdots & \vdots & \vdots & \vdots & \vdots & \vdots & \vdots \\
\cdots & 1 & \widehat{\phi}_5^{(-8)} & \widehat{\phi}_5^{(-7)} & \widehat{\phi}_5^{(-6)} & \widehat{\phi}_5^{(-5)} & \widehat{\phi}_5^{(-4)} & \widehat{\phi}_5^{(-3)} & \widehat{\phi}_5^{(-2)} & \widehat{\phi}_5^{(-1)} & \widehat{\phi}_5^{(-0)} & \widehat{\phi}_5^{(1)} & \cdots \\
\cdots & & 1 & \widehat{\phi}_4^{(-7)} & \widehat{\phi}_4^{(6)} & \widehat{\phi}_4^{(-5)} & \widehat{\phi}_4^{(-4)} & \widehat{\phi}_4^{(-3)} & \widehat{\phi}_4^{(-2)} & \widehat{\phi}_4^{(-1)} & \widehat{\phi}_4^{(0)} & \widehat{\phi}_4^{(1)} & \cdots \\
\cdots & & & & 1 & \widehat{\phi}_3^{(-5)} & \widehat{\phi}_3^{(-4)} & \widehat{\phi}_3^{(-3)} & \widehat{\phi}_3^{(-2)} & \widehat{\phi}_3^{(-1)} & \widehat{\phi}_3^{(0)} & \widehat{\phi}_3^{(1)} & \cdots \\
\cdots & & & & & 1 & \widehat{\phi}_2^{(-4)} & \widehat{\phi}_2^{(-3)} & \widehat{\phi}_2^{(-2)} & \widehat{\phi}_2^{(-1)} & \widehat{\phi}_2^{(0)} & \widehat{\phi}_2^{(1)} & \cdots \\
\cdots & & & & & & & 1 & \widehat{\phi}_1^{(-2)} & \widehat{\phi}_1^{(-1)} & \widehat{\phi}_1^{(0)} & \widehat{\phi}_1^{(1)} & \cdots \\
\cdots & & & & & & & & & & 1 & &
\end{array}\right],$$

and the same S as (3.117); the Binet–Cauchy lemma shows

$$\det(\Xi S) = \det \begin{pmatrix} \mathfrak{p}_1 & 1 & & \\ \mathfrak{p}_2 & \mathfrak{p}_1 & 1 & \\ \mathfrak{p}_4 & \mathfrak{p}_3 & \mathfrak{p}_2 & \mathfrak{p}_1 \\ \mathfrak{p}_7 & \mathfrak{p}_6 & \mathfrak{p}_5 & \mathfrak{p}_4 \end{pmatrix} + \sum_{\Lambda > \Lambda_{\langle 3,4\rangle}} \Xi_\Lambda s_\Lambda.$$

3.10 Appendix: Schur Polynomials and Their Derivative

Following [MatP2014] we consider the derivative properties of the Schur function in this appendix. The study of the derivation of the Schur functions was started by Ônishi for the precise expression of the Riemann–Kempf theory in [On2002, On2005]. Ônishi [On2002, On2005, On2009, On2010, On2011], and Nakayashiki and Yori [NakY2012] studied the same topics. Our approach in [MatP2014] is basically the same as Ônishi's peal method [On2010], though ours is based on the Frobenius characteristics for the Young diagram associated with a numerical semigroup. As we treated the Young diagram associated with the numerical semigroup generated by two elements in [MatP2014], our approach can be easily applied to general Schur functions associated with the numerical semigroups; we basically follow the approaches in [MatP2014] in this appendix. Further, we note that though the approach by Nakayashiki and Yori [NakY2012] is different from ours, they obtained the same results.

Let us consider the Young diagram Λ which is associated with a numerical semigroup H. For the Young diagram $\Lambda = (\Lambda_1, \cdots, \Lambda_g)$, we recall the Schur polynomial $\mathbf{s}_\Lambda(\mathrm{u})$ (3.104) but redefine it as a finite polynomial of $t := (t_1, \ldots, t_g)$ in $R_t := \mathbb{Q}[t_1, \ldots, t_g]$ by

$$\mathbf{s}_\Lambda(t) := \frac{|t_j^{\Lambda_i+g-i}|_{1\le i,j\le g}}{|t_j^{i-1}|_{1\le i,j\le g}}. \tag{3.118}$$

The relation between the definitions (3.104) and (3.118) are obtained by Lemma 3.221 and Propositions 3.217 and 3.220.

Further, we introduce the complete homogeneous symmetric function $h_n^{\langle \ell_1,\ell_2\rangle} = h_n(t_{\ell_1}, \ldots, t_{\ell_2})$ for positive integers ℓ_1 and ℓ_2 ($\ell_1 < \ell_2$) is given by

$$\prod_{i=\ell_1}^{\ell_2} \frac{1}{(1-zt_i)} = \sum_{n\ge 0} h_n^{\langle \ell_1,\ell_2\rangle} z^n, \quad h_n^{\langle \ell_1,\ell_2\rangle} = 0 \text{ for } n < 0. \tag{3.119}$$

Proposition 3.217 *Using the complete homogeneous symmetric functions $h_n := h_n^{\langle 1,g\rangle}$, we can express $\mathbf{s}_\Lambda$ by a $(g \times g)$ Jacobi–Trudi Determinant, $|a_{ij}|_{1\le i,j\le g}$ with $a_{ij} = h_{\Lambda_i+j-i}$:*

$$\mathbf{s}_\Lambda(t) := |h_{\Lambda_i+j-i}|, \quad h_n = \frac{1}{n!}\begin{vmatrix} T_1 & -1 & 0 & \cdots & \\ 2T_2 & T_1 & -2 & \cdots & \\ \vdots & \vdots & \vdots & \ddots & \vdots \\ (n-1)T_{n-1} & (n-2)T_{n-2} & (n-3)T_{n-3} & \cdots & 1-n \\ nT_n & (n-1)T_{n-1} & (n-2)T_{n-2} & \cdots & T_1 \end{vmatrix},$$

where $h_0 = 1$, $h_{i<0} = 0$ *and* $T_k := T_k^{\langle 1,g\rangle}$,

$$T_k^{\langle \ell_1,\ell_2\rangle} := \frac{1}{k}\sum_{j=\ell_1}^{\ell_2} t_j^k.$$

Proof This follows from Theorem 4.5.1 in [Sag2001, p. 158] and Lemma 3.221. ∎

We note that $h_n = T_n + \cdots$, and $h_n^{\langle k,k\rangle} = \mathrm{t}_k^n$.

We recall results in Lemma 2.3. Then we have the following lemma:

Lemma 3.218 *Let the Young diagrams* $\Lambda = (\Lambda_1, \ldots, \Lambda_g)$ *associated with the numerical semigroup* H *be expressed by the Frobenius representation* $(a_{r_\Lambda}, \ldots, a_1; b_{r_\Lambda}, \ldots, b_1)$ *with rank* r_Λ.

1. *In the pattern in the matrix* $((h_{\Lambda_i+j-i})_{ij})$, h_0 *in the* i*-th row appears if* $i > r_\Lambda$ *and its position is* $(i, i - \Lambda_i)$.
2. *By the Laplace expansion of the determinant of* (h_{Λ_i+j-i}), *we have*

$$\begin{aligned}\mathbf{s}_{\Lambda[\ell]}(t) &= (-1)^{g_{\Lambda[\ell]}+1} h_{a_\ell+b_\ell+1}\mathbf{s}_{\Lambda[\ell+1]}(t) + \text{ terms of lower power}\\ &= (-1)^{g_{\Lambda[\ell]}+1} T_{a_\ell+b_\ell+1}\mathbf{s}_{\Lambda[\ell+1]}(t) + \text{ terms of lower power},\end{aligned}$$

where we recall $\Lambda[\ell] = (a_{r_\Lambda}, \ldots, a_\ell; b_{r_\Lambda}, \ldots, b_\ell)$ *in Lemma 2.3.*

3. *We have*

$$\left(\prod_{i=1}^{r_\Lambda} \frac{\partial}{\partial T_{a_i+b_i+1}}\right)\mathbf{s}_\Lambda(t_1, \ldots, t_g) = \varepsilon_\Lambda,$$

where $\varepsilon_\Lambda = (-1)^{g_\Lambda - r_\Lambda}$, *and for every proper subsequence* J *of* $\{1, 2, \ldots, r_\Lambda\}$, *we have*

$$\left(\prod_{i\in J} \frac{\partial}{\partial T_{a_i+b_i+1}}\right)\mathbf{s}_\Lambda(t_1, \ldots, t_g)\Big|_{t_j=0: j=1,\ldots,g} = 0.$$

Proof We consider the pattern on the right-hand side of $\mathbf{s}_\Lambda(t) := |h_{\Lambda_i+j-i}|$:

$$s_\Lambda(t) = \begin{vmatrix} h_{\Lambda_1} & h_{\Lambda_1+1} & \cdots & h_{\Lambda_1+g-2} & h_{\Lambda_1+g-1}\\ h_{\Lambda_2-1} & h_{\Lambda_2} & \cdots & h_{\Lambda_2+g-3} & h_{\Lambda_2+g-2}\\ \vdots & \vdots & \ddots & \vdots & \vdots\\ h_{\Lambda_{g-1}-g+2} & h_{\Lambda_{g-1}-g+3} & \cdots & h_{\Lambda_{g-1}} & h_{\Lambda_{g-1}+1}\\ h_{\Lambda_g-g+1} & h_{\Lambda_g-g+2} & \cdots & h_{\Lambda_g-1} & h_{\Lambda_g}\end{vmatrix}$$

$$
= \left|\begin{array}{cccccccc}
\cdots & \cdots & & & & & & h_{\Lambda_1+g-1} \\
\cdots & \cdots & & h_{\Lambda_2+g-\ell'} & & & & \\
\cdots & \cdots & & & & & \cdots & \\
\cdots h_0 & & & & & & & \\
\vdots & \vdots & \ddots & \vdots & \vdots & \vdots & & \\
& \cdots h_0 & & h_1 & & & & \\
& \cdots & & & h_0 & & & \\
& \cdots & & & & \ddots & & \\
& \cdots & & & & & h_0 & h_1
\end{array}\right| .
$$

Since the elements in the lower-left part vanish, we can reduce the determinant to a combination of ones of smaller size. The pattern of h_0's has the property that there is one situated Λ_i entries to the left of the diagonal elements in each i-th row for $i > r_\Lambda$ because of the configuration of the lower boundary of the diagram Λ. In the determinant calculation, a series of h_0 in the diagonal direction occurs in a unique position and contributes to the determinant a factor of 1. In order to omit that factor, we observe the following properties about the pattern of the matrix.

We introduce a hierarchy of submatrices starting from the above matrix whose determinant is $\mathbf{s}_\Lambda(t)$. We define , $i_e[\ell] := \ell + g_{\Lambda[\ell]}$, $j_e[\ell] := g_{\Lambda[\ell]}$, and

$$H_\ell := (h_{\Lambda_i+j-i})_{\ell \le i \le i_e[\ell], 1 \le j \le j_e[\ell]} \in \mathrm{Mat}_{R_t}(g_{\Lambda[\ell]}).$$

Then we obtain $H_1 = (h_{\Lambda_i+j-i})$, and $\mathbf{s}_\Lambda(t) = |h_{\Lambda_i+j-i}|$. Obviously $H_{\ell+1}$ is the submatrix of H_ℓ. Further,

$$H_\ell = (h_{\Lambda[\ell]_i+j-i})$$

and thus $\mathbf{s}_{\Lambda[\ell]}(t) = |H_\ell|$. The above sequence of matrices stops when the number of entries becomes negative, i.e., $\ell = r_\Lambda$, i.e., $H_{r_\Lambda+1} = \emptyset$, where r_Λ is the rank of Λ. $H_1 = (h_{\Lambda_i+j-i})$ for $\Lambda = \Lambda[1]$,

For the example of the $H = \langle 5, 7, 13, 16 \rangle$ of $g_H = 8$, we have

$$
H_1 = \begin{pmatrix}
h_4 & h_5 & h_6 & h_7 & h_8 & h_9 & h_{10} & h_{11} \\
h_2 & h_3 & h_4 & h_5 & h_6 & h_7 & h_8 & h_9 \\
h_1 & h_2 & h_3 & h_4 & h_5 & h_6 & h_7 & h_8 \\
 & h_0 & h_1 & h_2 & h_3 & h_4 & h_5 & h_6 \\
 & & & h_0 & h_1 & h_2 & h_3 & h_4 \\
 & & & & h_0 & h_1 & h_2 & h_3 \\
 & & & & & h_0 & h_1 & h_2 \\
 & & & & & & h_0 & h_1
\end{pmatrix},
$$

$$H_2 = \begin{pmatrix} h_2 & h_3 & h_4 \\ h_1 & h_2 & h_3 \\ & h_0 & h_1 \end{pmatrix}, \quad H_3 = \begin{pmatrix} h_1 \end{pmatrix}, \quad H_4 = \emptyset.$$

For the example of $H = \langle 7, 9, 15, 18, 19, 20 \rangle$ of $g_H = 11$, we have

$$H_1 = \begin{pmatrix} h_3 & h_4 & h_5 & h_6 & h_7 & h_8 & h_9 & h_{10} & h_{11} & h_{12} & h_{13} \\ h_2 & h_3 & h_4 & h_5 & h_6 & h_7 & h_8 & h_9 & h_{10} & h_{11} & h_{12} \\ h_1 & h_2 & h_3 & h_4 & h_5 & h_6 & h_7 & h_8 & h_9 & h_{10} & h_{11} \\ h_0 & h_1 & h_2 & h_3 & h_4 & h_5 & h_6 & h_7 & h_8 & h_9 & h_{10} \\ & & h_0 & h_1 & h_2 & h_3 & h_4 & h_5 & h_6 & h_7 & h_8 \\ & & & & h_0 & h_1 & h_2 & h_3 & h_4 & h_5 & h_6 \\ & & & & & h_0 & h_1 & h_2 & h_3 & h_4 & h_5 \\ & & & & & & h_0 & h_1 & h_2 & h_3 & h_4 \\ & & & & & & & h_0 & h_1 & h_2 & h_3 \\ & & & & & & & & h_0 & h_1 & h_2 \\ & & & & & & & & & h_0 & h_1 \end{pmatrix},$$

$$H_2 = \begin{pmatrix} h_2 & h_3 & h_4 & h_5 \\ h_1 & h_2 & h_3 & h_4 \\ h_0 & h_1 & h_2 & h_3 \\ & & h_0 & h_1 \end{pmatrix}, \quad H_3 = \begin{pmatrix} h_1 & h_2 \\ h_0 & h_1 \end{pmatrix}, \quad H_4 = \emptyset.$$

For example, the subscript of the upper-right corner in H_ℓ is given by $a_\ell + b_\ell + 1$. Every $(c+1, c)$ entry of H_ℓ is equal to $h_0 = 1$ if it is not in $H_{\ell+1}$ and the number of h_0's in H_ℓ is $g_{\Lambda[\ell]} - g_{\Lambda[\ell+1]} - 1$. These facts imply that the term containing $(h_0)^{g-r_\Lambda}$ in the expansion of the determinant given by products of all the entries with indices permuted, is given by the determinant of an $(r_\Lambda \times r_\Lambda)$ matrix, as shown:

$$|H_\ell| = \varepsilon''(h_0)^{g_{\Lambda[\ell]}-g_{\Lambda[\ell+1]}-1} h_{a_{\ell 1}+b_\ell+1} |H_{\ell+1}| + \text{lower power terms},$$

where ε'' is

$$\begin{aligned} \varepsilon'' &= (-1)^{2g_{\Lambda[\ell]}-1}(-1)^{2g_{\Lambda[\ell]}-3} \cdots (-1)^{2g_{\Lambda[\ell+1]}+3} \\ &= (-1)^{\sum_{i=1}^{g_{\Lambda[\ell]}-g_{\Lambda[\ell+1]}-1}(2g_{\Lambda[\ell]}+1-2i)} = (-1)^{g_{\Lambda[\ell]}-g_{\Lambda[\ell+1]}-1} \end{aligned}$$

due to the Laplace expansion. The subscripts of the elements on the straight line from the upper-right corner to the lower-left corner, $\Lambda_1 + g - 1, \ldots, \Lambda_{r_\Lambda} - r_\Lambda + 1$, are given by $a_1 + b_1 + 1, a_2 + b_2 + 1, \ldots, a_{r_\Lambda} + b_{r_\Lambda} + 1$. The determinant is given by

$$\varepsilon' \prod_{i=1}^{r_\Lambda} h_{a_i+b_i+1} + \ldots, \tag{3.120}$$

where ε' is

$$\varepsilon' = \prod_{\ell=1}^{r_\Lambda}(-1)^{g_{\Lambda[\ell]}-g_{\Lambda[\ell+1]}-1} = (-1)^{g_\Lambda - r_\Lambda}.$$

We note that each term in the determinant has the same multi-degree in the t variables, with each t_i weighting one, $\sum_{i=1}^{r_\Lambda}(a_i + b_i + 1) = |\Lambda^{[k]}|$. Moreover, $h_{a_i+b_i+1} = (-1)^{a_i+b_i}T_{a_i+b_i+1} + \cdots$. The subscript of the upper-right corner of H_1 is characterized by $\Lambda_k + g - k = a_{r_\Lambda} + b_{r_\Lambda} + 1$ and is the largest among the subscripts of the elements in H_1. In other words the first term in (3.120) is the only term that contains T_{Λ_k+g-k} and cannot be canceled. Hence we have

$$H_1 = (-1)^{g_\Lambda - r_\Lambda}\prod_{i=1}^{r_\Lambda} T_{a_i+b_i+1} + \ldots = (-1)^{g_\Lambda - r_\Lambda}\prod_{i=1}^{r_\Lambda} T_{a_i+b_i+1} + \ldots,$$

$$\left(\prod_{i=1}^{r_\Lambda}\frac{\partial}{\partial T_{a_i+b_i+1}}\right)\mathbf{s}_\Lambda(t_1,\ldots,t_g) = \varepsilon_\Lambda, \quad \varepsilon_\Lambda = (-1)^{g_\Lambda - r_\Lambda}.$$

Moreover, for every proper subsequence J of $\{1, 2, \ldots, r_\Lambda\}$, we have

$$\left(\prod_{i\in J}\frac{\partial}{\partial T_{a_i+b_i+1}}\right)\mathbf{s}_\Lambda(t_1,\ldots,t_g)\Big|_{t_j=0:j=1,\ldots,g} = 0.$$

■

The following is known as the Frobenius formula:

Lemma 3.219

$$\mathbf{s}_\Lambda = \sum_{\rho=\{1^{r_1},2^{r_2},3^{r_3},\cdots\}} \chi_\rho^\Lambda \prod_k \frac{1}{r_k!}T_k^{r_k},$$

where χ_ρ^Λ is the character of the representation, $\rho = \{1^{r_1}, 2^{r_2}, 3^{r_3}, \cdots\}$ means that the partition ρ has r_k parts of length k.

Proposition 3.220 *For the Young diagram Λ_X given by the numerical semigroup H_X, $\mathbf{s}_\Lambda$ is a function of $T_{\Lambda_1+g-1}, T_{\Lambda_2+g-2}, \cdots, T_{\Lambda_{g-1}+1}, T_{\Lambda_g}$.*

Proof First, we note $\{\Lambda_i + g - i\}_{i=1,\ldots,g} = H_X^c$. From Lemma 3.224, we inductively show it. Let us consider the decomposition Λ into $\Lambda^{(g-j)}$ and $\Lambda^{[g-j]}$. If Λ is the Young diagram associated with a numerical semigroup, $\Lambda^{[g-j]}$ is also the Young diagram associated with a numerical semigroup. Further, when we express $\Lambda^{[g-j]} = (a_{r_j}, \ldots, a_1; b_{r_j}, \ldots, b_1)$, $a_i + b_i + 1$ is given by $\Lambda_{j_i} + g - j_i$ for a certain j_i. For $\Lambda^{[g-1]}$, it means that $\mathbf{s}_{\Lambda^{[g-1]}}$ is a function of $T_{\Lambda_g} = T_1$ due to Lemma 3.224. We assume that $\mathbf{s}_{\Lambda^{[g-i]}}$ is a function only of T_{Λ_j+g-j} $(j = g, g-1, \ldots, i)$. Then $\mathbf{s}_{\Lambda^{[g-i-1]}}$ depends on $T_{\Lambda_{g-i-1}+i+1}$ and thus $\mathbf{s}_{\Lambda[g-i-1]}$ is a function only of T_{Λ_j+g-j} $(j = g, g-1, \ldots, i-1)$. ■

When regarded as a function of T, we rename $\mathbf{s}$, $S_\Lambda(T) := \mathbf{s}_\Lambda(t)$.

We now give an earlier version of the Jacobi–Trudi formula of Proposition 3.217; by connecting the two, we provide proof of Proposition 3.217, as well as a modified formula which will be used in Sects. 3.1.7 and 3.2.10.

Lemma 3.221

$$\mathbf{s}_\Lambda(t) = |h^{\langle j,g\rangle}_{\Lambda_i+j-i}|_{1\le i,j\le g}.$$

Proof By using a Vandermonde determinant, (3.118) is reduced to

$$\frac{\begin{vmatrix} t_1^{\Lambda_1+g-1} & t_2^{\Lambda_1+g-1} & \cdots & t_{g-1}^{\Lambda_1+g-1} & t_g^{\Lambda_1+g-1} \\ \vdots & \vdots & \ddots & \vdots & \vdots \\ t_1^{\Lambda_{g-1}+1} & t_2^{\Lambda_{g-1}+1} & \cdots & t_{g-1}^{\Lambda_{g-1}+1} & t_g^{\Lambda_{g-1}+1} \\ t_1^{\Lambda_g} & t_2^{\Lambda_g} & \cdots & t_{g-1}^{\Lambda_g} & t_g^{\Lambda_g} \end{vmatrix}}{\prod_{i<j}(t_i - t_j)}.$$

This equals

$$\frac{\begin{vmatrix} t_1^{\Lambda_1+g-1} - t_g^{\Lambda_1+g-1} & t_2^{\Lambda_1+g-1} - t_g^{\Lambda_1+g-1} & \cdots & t_{g-1}^{\Lambda_1+g-1} - t_g^{\Lambda_1+g-1} & t_g^{\Lambda_1+g-1} \\ \vdots & \vdots & \ddots & \vdots & \vdots \\ t_1^{\Lambda_{g-1}+1} - t_g^{\Lambda_{g-1}+1} & t_2^{\Lambda_{g-1}+1} - t_g^{\Lambda_{g-1}+1} & \cdots & t_{g-1}^{\Lambda_{g-1}+1} - t_g^{\Lambda_{g-1}+1} & t_g^{\Lambda_{g-1}+1} \\ t_1^{\Lambda_g} - t_g^{\Lambda_g} & t_2^{\Lambda_g} - t_g^{\Lambda_g} & \cdots & t_{g-1}^{\Lambda_g} - t_g^{\Lambda_g} & t_g^{\Lambda_g} \end{vmatrix}}{\prod_{i<j}(t_i - t_j)}.$$

By noting $t^\ell - t'^\ell = (t - t')h_\ell(t, t')$, this becomes

$$\frac{\begin{vmatrix} h_{\Lambda_1+g-2}(t_1, t_g) & h_{\Lambda_1+g-2}(t_2, t_g) & \cdots & h_{\Lambda_1+g-2}(t_{g-1}, t_g) & t_g^{\Lambda_1+g-1} \\ \vdots & \vdots & \ddots & \vdots & \vdots \\ h_{\Lambda_{g-1}}(t_1, t_g) & h_{\Lambda_{g-1}}(t_2, t_g) & \cdots & h_{\Lambda_{g-1}}(t_{g-1}, t_g) & t_g^{\Lambda_{g-1}+1} \\ h_{\Lambda_g-1}(t_1, t_g) & h_{\Lambda_g-1}(t_2, t_g) & \cdots & h_{\Lambda_g-1}(t_{g-1}, t_g) & t_g^{\Lambda_g} \end{vmatrix}}{\prod_{i<j<g}(t_i - t_j)}.$$

We go on similarly, to co-factorize by $\prod_{i}^{g-2}(t_{g-1} - t_i)$,

$$\frac{\begin{vmatrix} h_{A_1+g-3}(t_1,t_{g-1},t_g) & h_{A_1+g-3}(t_2,t_{g-1},t_g) & \cdots & h_{A_1+g-3}(t_{g-2},t_{g-1},t_g) & h_{A_1+g-2}(t_{g-1},t_g) & t_g^{A_1+g-1} \\ \vdots & \vdots & \ddots & \vdots & \vdots & \vdots \\ h_{A_{g-1}-1}(t_1,t_{g-1},t_g) & h_{A_{g-1}-1}(t_2,t_{g-1},t_g) & \cdots & h_{A_{g-1}-1}(t_{g-2},t_{g-1},t_g) & h_{A_{g-1}}(t_{g-1},t_g) & t_g^{A_{g-1}+1} \\ h_{A_g-2}(t_1,t_{g-1},t_g) & h_{A_g-2}(t_2,t_{g-1},t_g) & \cdots & h_{A_g-2}(t_{g-2},t_{g-1},t_g) & h_{A_g-1}(t_{g-1},t_g) & t_g^{A_g} \end{vmatrix}}{\prod_{i<j<g-1}(t_i-t_j)}$$

and derive the statement. ■

Lemma 3.221 will yield the Jacobi–Trudi determinant formula in Proposition 3.217. In order to relate the two expressions, we prove:

Lemma 3.222

1. $h_n^{\langle \ell_1,\ell_2\rangle} = h_n^{\langle \ell_1+1,\ell_2\rangle} + h_{n-1}^{\langle \ell_1,\ell_2\rangle} t_{\ell_1} = h_n^{\langle \ell_1,\ell_2-1\rangle} + h_{n-1}^{\langle \ell_1,\ell_2\rangle} t_{\ell_2}$.

2. For positive integers ℓ_1, ℓ_2, n *and* m *satisfying* $\ell_2 > \ell_1$ *and* $m < \ell_2 - \ell_1$, $h_n^{\langle \ell_1,\ell_2\rangle} = h_n^{\langle \ell_1+m,\ell_2\rangle} + h_{n-1}^{\langle \ell_1+m-1,\ell_2\rangle} h_1^{\langle \ell_1,\ell_1+m-1\rangle} + h_{n-2}^{\langle \ell_1+m-2,\ell_2\rangle} h_2^{\langle \ell_1,\ell_1+m-2\rangle} + \cdots + h_{n-m}^{\langle \ell_1,\ell_2\rangle} h_m^{\langle \ell_1,\ell_1\rangle}$.

3. For positive integers ℓ_1, ℓ_2 *and* n *satisfying* $\ell_2 > \ell_1$, $h_n^{\langle \ell_1,\ell_2\rangle} = h_n^{\langle \ell_2,\ell_2\rangle} + h_{n-1}^{\langle \ell_2-1,\ell_2\rangle} h_1^{\langle \ell_1,\ell_2-1\rangle} + h_{n-2}^{\langle \ell_2-2,\ell_2\rangle} h_2^{\langle \ell_1,\ell_2-2\rangle} + \cdots + h_{n-\ell_2+\ell_1}^{\langle \ell_1,\ell_2\rangle} h_{\ell_2-\ell_1}^{\langle \ell_1,\ell_1\rangle}$.

Proof The product $\prod_{i=\ell_1}^{\ell_2} \frac{1}{(1-zt_i)} = \frac{1}{(1-zt_{\ell_1})} \prod_{i=\ell_1+1}^{\ell_2} \frac{1}{(1-zt_i)}$ is expanded by using

$$(1-t_{\ell_1}z)(1+h_1^{\langle \ell_1,\ell_2\rangle}z + h_2^{\langle \ell_1,\ell_2\rangle}z^2 + \cdots) = (1 + h_1^{\langle \ell_1+1,\ell_2\rangle}z + h_2^{\langle \ell_1+1,\ell_2\rangle}z^2 + \cdots),$$

which shows the first equality in the statement of *1*; the second equality is similarly obtained. 2 for the $m = 1$ case is reduced to the first statement for every n. We give a proof by induction: assume that 2 holds for $m \geq m_0$,

$$h_n^{\langle \ell_1,\ell_2\rangle} = h_n^{\langle \ell_1+m,\ell_2\rangle} + h_{n-1}^{\langle \ell_1+m-1,\ell_2\rangle} h_1^{\langle \ell_1+1,\ell_1+m-1\rangle} + h_{n-2}^{\langle \ell_1+m-2,\ell_2\rangle} h_2^{\langle \ell_1,\ell_1+m-2\rangle}$$
$$+ \cdots + h_{n-m}^{\langle \ell_1,\ell_2\rangle} h_m^{\langle \ell_1,\ell_1\rangle}.$$

We consider the $m+1$ case: *1* implies that the individual terms in the right-hand side satisfy

$$\begin{aligned} h_n^{\langle \ell_1,\ell_2\rangle} &= (h_n^{\langle \ell_1+m+1,\ell_2\rangle} + h_{n-1}^{\langle \ell_1+m,\ell_2\rangle} t_{\ell_1+m}) \\ &+ (h_{n-1}^{\langle \ell_1+m,\ell_2\rangle} + h_{n-2}^{\langle \ell_1+m-1,\ell_2\rangle} t_{\ell_2-m-1}) h_1^{\langle \ell_1+1,\ell_1+m+1\rangle} \\ &+ (h_{n-2}^{\langle \ell_1+m-1,\ell_2\rangle} + h_{n-3}^{\langle \ell_1+m-2,\ell_2\rangle} t_{\ell_2-m-2}) h_2^{\langle \ell_1+2,\ell_1+m-1\rangle} \\ &+ \cdots + (h_{n-m}^{\langle \ell_1+1,\ell_2\rangle} + h_{n-m-1}^{\langle \ell_1,\ell_2\rangle} t_{\ell_1}) h_m^{\langle \ell_1,\ell_1\rangle} \\ &= h_n^{\langle \ell_1+m+1,\ell_2\rangle} + h_{n-1}^{\langle \ell_1+m,\ell_2\rangle} (t_{\ell_1+m} + h_1^{\langle \ell_1+1,\ell_1+m+1\rangle}) \\ &+ h_{n-2}^{\langle \ell_1+m-1,\ell_2\rangle} (t_{\ell_1-m-1} h_1^{\langle \ell_1+1,\ell_1+m+1\rangle} + h_2^{\langle \ell_1+2,\ell_1+m-1\rangle}) \\ &+ \cdots + h_{n-m-1}^{\langle \ell_1,\ell_2\rangle} t_{\ell_1} h_m^{\langle \ell_1,\ell_1\rangle}, \end{aligned}$$

which gives the $m+1$ case. *3* is obtained by setting $m = \ell_2 - \ell_1 - 1$. ■

A modified version of the Jacobi–Trudi relation follows:

Lemma 3.223 *For any fixed k, $1 \le k \le g$, juxtaposing two matrices to obtain a $g \times g$ matrix,*

$$\mathbf{s}_\Lambda(t_1, \ldots, t_g) = |H_{g,k}(t_1, \ldots, t_g), H_{g,g-k}(t_{k+1}, \ldots, t_g)|,$$

where $H_{a,b}(t_\ell, \ldots, t_g)$'s are $a \times b$ matrices defined by

$$H_{g,k}(t_1, \ldots, t_g) := (h_{\Lambda_i+j-i}(t_1, \ldots, t_g))_{1\le i\le g, 1\le j\le k},$$
$$H_{g,g-k}(t_{k+1}, \ldots, t_g) := (h_{\Lambda_i+j-i}(t_{k+1}, \ldots, t_g))_{1\le i\le g, k+1\le j\le g}.$$

We recall $\Lambda^{(k)}$ and $\Lambda^{[k]}$ in Definition 2.2 to extract the Schur polynomial $\mathbf{s}_{\Lambda^{(k)}}$ from $\mathbf{s}_\Lambda$ by differentials $\mathbf{s}_{\Lambda^{[k]}}$ using Lemma 3.218. We note the $\Lambda^{[k]}$ is associated with a numerical semigroup if Λ is also.

Using the Schur polynomials,

$$\mathbf{s}_{\Lambda^{(k)}}(t) := \frac{|t_j^{\Lambda_i+k-i}|_{1\le i,j\le k}}{|t_j^{i-1}|_{1\le i,j\le k}}, \quad \tilde{\mathbf{s}}_{\Lambda^{[k]}}(t) := \frac{|t_{k+j}^{\Lambda_{k+i}+g-k-i}|_{1\le i,j\le g-k}}{|t_{k+j}^{i-1}|_{1\le i,j\le g-k}},$$

and letting $T_j^{(k)} := \frac{1}{j}(t_1^j + \cdots + t_k^j) = T_j^{\langle 1,k\rangle}$, and $T_j^{(g;k)} := \frac{1}{j}(t_{k+1}^j + \cdots + t_g^j) = T_j^{\langle k+1,g\rangle}$, we define the following quantities:

$$S_{\Lambda^{(k)}}(T^{(k)}) := \mathbf{s}_{\Lambda^{(k)}}(t), \quad \tilde{S}_{\Lambda^{[k]}}(T^{(g;k)}) := \tilde{\mathbf{s}}_{\Lambda^{[k]}}(t).$$

The notation above for the left-hand sides $S_{\Lambda^{(k)}}(T^{(k)})$ and $\tilde{S}_{\Lambda^{[k]}}(T^{(k)})$ is consistent with Proposition 3.217, because they are functions of the $T^{(k)}$. For a given $(t_\ell)_{\ell=1,2,\ldots,k} \in \mathbb{C}^k$, letting $\mathrm{u}^{[k]} := (\mathrm{u}_i^{[k]})_{i=1,\ldots,g} \in \mathbb{C}^g$ be defined by $\mathrm{u}_i^{[k]} := T_{\Lambda_i+g-i}^{(k)}$, we can also define consistently, in view of Lemma 3.224 proven below: Due to Proposition 3.220, we can regard $\mathbf{s}_\Lambda(t)$ as a function of $\mathrm{u}^{[k]}$ and recall (3.104) noting footnote 10 in this chapter. We use the same symbol, i.e., $\mathbf{s}_\Lambda(\mathrm{u}^{[k]}) = \mathbf{s}_\Lambda(\mathrm{u}^{[k]}(t))$, since there can be no confusion: We refer to it by

$$\mathbf{s}_{\Lambda^{(k)}}(\mathrm{u}^{[k]}) = S_{\Lambda^{(k)}}(T^{(k)})|_{T_{\Lambda_i+g-i}^{(k)}=\mathrm{u}_i^{[k]}}. \tag{3.121}$$

We also introduce the complete symmetric polynomial $h_n = h^{\langle k+1,g\rangle}$ such that $h_{n\ge 1} = (-1)^{n-1}(n-1)!T_n^{(g;k)} + \cdots, h_0 = 1$ and $h_{n<0} = 0$ (in particular, h_n depends only on $t_{k+1}, \ldots, t_g$).

Lemma 3.224 *For a given $(t_\ell)_{\ell=1,2,\ldots,k} \in \mathbb{C}^k$, we let $\mathrm{u}^{[k]} := (\mathrm{u}_i^{[k]})_{i=1,\ldots,g} \in \mathbb{C}^g$ be defined by $\mathrm{u}_i^{[k]} := T^{(k)}_{\Lambda_i+g-i}$, and $\mathrm{u}^{[g;k]} := (\mathrm{u}_i^{[g;k]})_{i=1,\ldots,g} \in \mathbb{C}^g$ be defined by $\mathrm{u}_i^{[g;k]} := T^{(g;k)}_{\Lambda_i+g-i}$. For brevity, we let I denote a sequence of indices (among $\{k+1, \ldots, g\}$) which may be repeated, and the notation $i \in I$ means that i runs through the sequence with given repetitions, if any; the order of the indices in the sequence is irrelevant. One such sequence for which I has the smallest number of elements $r_k := r_{\Lambda^{[k]}}$ and the sum of the degrees in t (each variable t_ℓ having degree 1), $\sum_{i\in I} deg(\mathrm{u}_i) = \sum_{i\in I}(\Lambda_i + g - i) = |\Lambda^{[k]}|$ is also minimum, is given in the proof. We use the notation: $\widehat{\varepsilon}_{\Lambda,I} := (-1)^{g-k-r_k}$.*

1. *There exists some (possibly non-unique) finite sequence I (whose length may not be unique but is at least $r_k = r_{\Lambda^{[k]}}$) such that by using the decomposition $\mathrm{u}^{[g]} = \mathrm{u}^{[k]} + \mathrm{u}^{[g;k]} \in \mathbb{C}^g$,*

$$\mathbf{s}_{\Lambda^{(k)}}(\mathrm{u}^{[k]}) = \widehat{\varepsilon}_{\Lambda,I} \left(\prod_{i\in I} \frac{\partial}{\partial \mathrm{u}_i^{[g;k]}} \right) \mathbf{s}_{\Lambda^{(g)}}(\mathrm{u}^{[g]}) \Big|_{\mathrm{u}^{[g]}=\mathrm{u}^{[k]}}.$$

In performing this partial derivative, we restrict the independent variables to a subspace $\mathbb{C}^{g-k} \subset \mathbb{C}^g$ spanned by $\left\{ \frac{\partial}{\partial \mathrm{u}_i^{[g;k]}} \right\}_{i=k+1,\ldots,g}$.

2. *There exists some (possibly non-unique) finite sequence I (whose length may not be unique but is at least $r_k = r_{\Lambda^{[k]}}$) such that*

$$\mathbf{s}_{\Lambda^{(k)}}(\mathrm{u}^{[k]}) = \widehat{\varepsilon}_{\Lambda,I} \left(\prod_{i\in I} \frac{\partial}{\partial \mathrm{u}_i^{[g]}} \right) \mathbf{s}_{\Lambda^{(g)}}(\mathrm{u}^{[g]}) \Big|_{\mathrm{u}^{[g]}=\mathrm{u}^{[k]}}.$$

Proof We treat both statements *1* and *2* together.

Using the modified Jacobi–Trudi determinant formula in Lemma 3.223, we see

$$\mathbf{s}_\Lambda(t_1, \ldots, t_g) = \begin{vmatrix} H_{k,k} & H_{k,g-k} \\ H_{g-k,k} & H_{g-k,g-k} \end{vmatrix},$$

where

$$\begin{aligned} H_{k,k}(t_1, \ldots, t_g) &:= (h_{\Lambda_i+j-i}(t_1, \ldots, t_g))_{1\le i,j\le k}, \\ H_{k,g-k}(t_1, \ldots, t_g) &:= (h_{\Lambda_i+j-i}(t_1, \ldots, t_g))_{1\le i\le k, k+1\le j\le g}, \\ H_{g-k,k}(t_{k+1}, \ldots, t_g) &:= (h_{\Lambda_i+j-i}(t_{k+1}, \ldots, t_g))_{k+1\le i\le g, 1\le j\le k}, \\ H_{g-k,g-k}(t_{k+1}, \ldots, t_g) &:= (h_{\Lambda_i+j-i}(t_{k+1}, \ldots, t_g))_{k+1\le i\le g, k+1\le j\le g}, \end{aligned}$$

the right-hand side of I without the constant $\widehat{\varepsilon}_{\Lambda,I}$, is given by

$$\left(\prod_{i\in I}\frac{\partial}{\partial u_i^{[g;k]}}\right)\left|\begin{matrix} H_{k,k} & H_{k,g-k} \\ H_{g-k,k} & H_{g-k,g-k}\end{matrix}\right|\Bigg|_{t_j=0:j=k+1,\dots,g}. \tag{3.122}$$

Since we have $\mathbf{s}_{\Lambda^{(k)}}(u^{[k]}) = |H_{k,k}(t_1,\dots,t_k)|$, we are to consider the derivative of

$$\tilde{\mathbf{s}}_{\Lambda^{[k]}}(t) = |H_{g-k,g-k}(t_{k+1},\dots,t_g)|;$$

we will settle one case, in which I has the smallest number of elements and least degree in t as given in the statement.

However, it is not difficult because $\Lambda^{[k]}$ is a Young diagram associated with a numerical semigroup due to Lemma 2.3 and thus

$$\tilde{\mathbf{s}}_{\Lambda^{[k]}}(t_{k+1},\dots,t_g) = \mathbf{s}_{\Lambda^{[k]}}(t_{k+1},\dots,t_g).$$

Lemma 3.218 shows

$$\left(\prod_{i=1}^{r_{\Lambda^{[k]}}}\frac{\partial}{\partial T_{a_i+b_i+1}^{[g;k]}}\right)\mathbf{s}_{\Lambda^{[k]}}(t_{k+1},\dots,t_g) = (-1)^{g_{\Lambda^{[k]}}-r_{\Lambda^{[k]}}},$$

and for every proper subsequence J of $\{1,2,\dots,r_\Lambda\}$, we have

$$\left(\prod_{i\in J}\frac{\partial}{\partial T_{a_i+b_i+1}^{[g;k]}}\right)\mathbf{s}_{\Lambda^{[k]}}(t_{k+1},\dots,t_g)\Big|_{t_j=0:j=k,\dots,g} = 0.$$

Indeed, by producing a sequence I such that I satisfies

$$\left(\prod_{i\in I}\frac{\partial}{\partial u_i^{[g;k]}}\right)\left|H_{g-k,g-k}\right|\Big|_{t_j=0:j=k+1,\dots,g} = \varepsilon, \tag{3.123}$$

and for any proper subsequence $J\subset I$, we have

$$\left(\prod_{i\in J}\frac{\partial}{\partial u_i^{[g;k]}}\right)\left|H_{g-k,g-k}\right|\Big|_{t_j=0:j=k+1,\dots,g} = 0, \tag{3.124}$$

then

$$\left(\prod_{i\in I}\frac{\partial}{\partial u_i^{[g;k]}}\right)\mathbf{s}_\Lambda(t_1,\dots,t_g)\Big|_{t_j=0:j=k+1,\dots,g} = (-1)^{g_{\Lambda^{[k]}}-r_{\Lambda^{[k]}}}\mathbf{s}_\Lambda^{(k)}(u^{[k]}) + \text{lower-order}. \tag{3.125}$$

The lower-order (in t) term vanishes for the following reasons: firstly, $H_{g-k,k}(t_{k+1},\dots,t_g)$ becomes the zero matrix and $H_{g-k,g-k}$ becomes a matrix whose every entry is zero except $h_0(=1)$ when $t_j=0: j=k+1,\dots,g$. The derivative lowers the order, so we have (3.123) and (3.124). The vanishing order of the entries in each column in $H_{g-k,k}$ is larger than that of the entries in $H_{g-k,g-k}$, thus the lowest-degree property of I means that any term in the lower-order part of (3.125) which comes from the derivative of an entry in $H_{g-k,k}$ in (3.125) vanishes. Secondly, due to (3.124), any term which comes from the derivative of an entry in $H_{k,k}$ or $H_{g-k,k}$ gives no contribution to the right-hand side in (3.125).

Since the element $h^{(g;k)}_{\Lambda_i+j-i}$ $(k+1\le i\le g,\ 1\le j\le k)$ in $H_{k,g-k}(t_{k+1},\dots,t_g)$ has sufficiently large degree in t, the contribution from $\left(\prod_{i=1}^{r_k}\frac{\partial}{\partial T^{(g;k)}_{a_i+b_i+1}}\right)H_{k,g-k}(t_{k+1},\dots,t_g)$ vanishes when we compute (3.122). In other words, noting $r_k=r_{\Lambda^{[k]}}$, we have

$$\left(\prod_{i=1}^{r_k}\frac{\partial}{\partial T^{(g;k)}_{a_i+b_i+1}}\right)\mathbf{s}_{\Lambda^{(g)}}(t_1,\dots,t_g)\bigg|_{t_j=0:j=k+1,\dots,g}=(-1)^{g-k-r_k}\mathbf{s}_{\Lambda^{(k)}}(t_1,\dots,t_k).$$

We recall Lemma 2.3 2 and 3. Since $\Lambda^{[k]}$ is associated with a numerical semigroup, there exists an integer L_i such that $\Lambda_{L_i}+g-L_i=a_i+b_i+1$ for every i due to Lemma 2.3. By naming I the sequence $(L_1,L_2,\dots,L_{r_k})$, we obtain this way the smallest degree in t and the least number of derivatives, because, given the configuration of the H_i's, in the determinant (3.120) we have the largest number of h_0's (which equal 1 hence have degree zero). Since $\Lambda_{k+1}+g-k-1=a_1+b_1+1$ and $a_i+b_i+1\le a_1+b_1+1$, each element $j\in I$ belongs to $\{k+1,\dots,g\}$.

For this I, we obtain

$$\left(\prod_{i\in I}\frac{\partial}{\partial u_i^{[g;k]}}\right)\mathbf{s}_{\Lambda^{(g)}}(t_1,\dots,t_g)\bigg|_{t_j=0:j=k+1,\dots,g}=(-1)^{g-k-r_k}\mathbf{s}_{\Lambda^{(k)}}(t_1,\dots,t_k).$$

Similarly, for every proper subsequence J of I,

$$\left(\prod_{i\in J}\frac{\partial}{\partial u_i^{[g;k]}}\right)\mathbf{s}_{\Lambda^{(g)}}(t_1,\dots,t_g)\bigg|_{t_j=0:j=k+1\dots g}=0.$$

The proof of *1* is complete.

For statement 2, we use the same sequence I. Noting that $j\in I$ belongs to $\{k+1,\dots,g\}$, we have

$$\begin{pmatrix}du_1^{[g]}\\ du_2^{[g]}\\ \vdots\\ du_g^{[g]}\end{pmatrix}=M_T^{(g)}\begin{pmatrix}dt_1\\ dt_2\\ \vdots\\ dt_g\end{pmatrix},\qquad \begin{pmatrix}du_{k+1}^{[g;k]}\\ du_{k+2}^{[g;k]}\\ \vdots\\ du_g^{[g;k]}\end{pmatrix}=M_T^{(g;k)}\begin{pmatrix}dt_{k+1}\\ dt_{k+2}\\ \vdots\\ dt_g\end{pmatrix},$$

where

$$M_T^{(g)} := \begin{pmatrix} t_1^{\Lambda_1+g-2} & t_2^{\Lambda_1+g-2} & \cdots & t_g^{\Lambda_1+g-2} \\ t_1^{\Lambda_2+g-3} & t_2^{\Lambda_2+g-3} & \cdots & t_g^{\Lambda_2+g-3} \\ \vdots & \vdots & \ddots & \vdots \\ t_1^{\Lambda_{g-1}} & t_2^{\Lambda_{g-1}} & \cdots & t_g^{\Lambda_{g-1}} \\ t_1^{\Lambda_g-1} & t_2^{\Lambda_g-1} & \cdots & t_g^{\Lambda_g-1} \end{pmatrix},$$

$$M_T^{(g;k)} := \begin{pmatrix} t_{k+1}^{\Lambda_{k+1}+g-2} & t_{k+2}^{\Lambda_{k+1}+g-2} & \cdots & t_g^{\Lambda_{k+1}+g-2} \\ t_{k+1}^{\Lambda_{k+1}+g-3} & t_{k+2}^{\Lambda_{k+1}+g-3} & \cdots & t_g^{\Lambda_{k+1}+g-3} \\ \vdots & \vdots & \ddots & \vdots \\ t_{k+1}^{\Lambda_{g-1}} & t_{k+2}^{\Lambda_{g-1}} & \cdots & t_g^{\Lambda_{g-1}} \\ t_{k+1}^{\Lambda_g-1} & t_{k+2}^{\Lambda_g-1} & \cdots & t_g^{\Lambda_g-1} \end{pmatrix}.$$

Since the one-forms are given by

$$dt_j = \sum_{i=1}^{g} \left[\frac{\partial t_j}{\partial \mathrm{u}_i^{[g]}} \right] d\mathrm{u}_i^{[g]} = \sum_{i=1}^{g} ((M_T^{(g)})^{-1})_{ji} d\mathrm{u}_i^{[g]}, \quad dt_j = \sum_{i=k+1}^{g} ((M_T^{(g;k)})^{-1})_{ji} d\mathrm{u}_i^{[g]},$$

we have

$$\frac{\partial}{\partial \mathrm{u}_i^{[g]}} = \sum_{j=1}^{g} \left[\frac{\partial t_j}{\partial \mathrm{u}_i^{[g]}} \right] \frac{\partial}{\partial t_j} = \sum_{j=1}^{g} ((M_T^{(g)})^{-1})_{ji} \frac{\partial}{\partial t_j}, \quad \frac{\partial}{\partial \mathrm{u}_i^{[g;k]}} = \sum_{j=k+1}^{g} ((M_T^{(g;k)})^{-1})_{ji} \frac{\partial}{\partial t_j}.$$

We note that the $\frac{\partial}{\partial \mathrm{u}_i^{[g;k]}}$ span a $\mathbb{C}^{g-k}$. $((M_T^{(g)})^{-1})_{ji}$ is given by

$$((M_T^{(g)})^{-1})_{ji} = (-1)^{i+j} \frac{\begin{vmatrix} t_1^{\Lambda_1+g-2} & \cdots & \check{t}_j^{\Lambda_1+g-2} & \cdots & t_g^{\Lambda_1+g-2} \\ \vdots & \ddots & \vdots & \ddots & \vdots \\ t_1^{\Lambda_{i-1}+g-i} & \cdots & \check{t}_j^{\Lambda_{i-1}+g-i} & \cdots & t_g^{\Lambda_{i-1}+g-i} \\ t_1^{\Lambda_{i+1}+g-i-2} & \cdots & \check{t}_j^{\Lambda_{i+1}+g-i-2} & \cdots & t_g^{\Lambda_{i+1}+g-i-2} \\ \vdots & \ddots & \vdots & \ddots & \vdots \\ t_1^{\Lambda_g-1} & \cdots & \check{t}_j^{\Lambda_g-1} & \cdots & t_g^{\Lambda_g-1} \end{vmatrix}}{\begin{vmatrix} t_1^{\Lambda_1+g-2} & t_2^{\Lambda_1+g-2} & \cdots & t_g^{\Lambda_1+g-2} \\ t_1^{\Lambda_2+g-3} & t_2^{\Lambda_2+g-3} & \cdots & t_g^{\Lambda_2+g-3} \\ \vdots & \vdots & \ddots & \vdots \\ t_1^{\Lambda_{g-1}} & t_2^{\Lambda_{g-1}} & \cdots & t_g^{\Lambda_{g-1}} \\ t_1^{\Lambda_g-1} & t_2^{\Lambda_g-1} & \cdots & t_g^{\Lambda_g-1} \end{vmatrix}}.$$

We claim that for a symmetric function $h(\mathrm{u}^{[g]})$, and subsequence $J \subset I$, we have

$$\left(\prod_{i\in J}\frac{\partial}{\partial \mathrm{u}_i^{[g]}}\right) h(\mathrm{u}^{[g]})\Big|_{t_{k+1}=0,\dots,t_g=0} = \left(\prod_{i\in J}\frac{\partial}{\partial \mathrm{u}_i^{[g;k]}}\right) h(\mathrm{u}^{[g]})\Big|_{t_{k+1}=0,\dots,t_g=0}. \tag{3.126}$$

We prove (3.126) directly as follows: Let $t_{k+1}, \dots, t_g$ have the same order ε, written as $\varepsilon_i = t_i$, $i = k+1, \dots, g$.

We use the property of the chosen sequence I, namely that every $i \in I$ satisfies $k < i \le g$. Then for $1 \le j \le k$, let $\Xi_{k,g,i} := \sum_{\ell=k,\ell\neq i}^{g} (\Lambda_\ell + g - \ell - 1)$ and $\Xi_{k+1,g} := \sum_{\ell=k+1}^{g} (\Lambda_\ell + g - \ell - 1)$. For $1 \le j \le k$, noting that $\Xi_{k,g,i} > \Xi_{k+1,g}$, we have that the element of $((M_T^{(g)})^{-1})_{ji}$ is equal to

$$(-1)^{i+j}\frac{\begin{vmatrix} t_1^{\Lambda_1+g-2} & \cdots & \check{t}_j^{\Lambda_1+g-2} & \cdots & t_k^{\Lambda_1+g-2} \\ \vdots & \ddots & \vdots & \ddots & \vdots \\ t_1^{\Lambda_{k-1}+g-k} & \cdots & \check{t}_j^{\Lambda_{k-1}+g-k} & \cdots & t_g^{\Lambda_{k-1}+g-k} \end{vmatrix} d_{\ge}(\varepsilon^{\Xi_{k,i,g}}) + d_>(\varepsilon^{\Xi_{k,i,g}})}{\begin{vmatrix} t_1^{\Lambda_1+g-2} & \cdots & t_k^{\Lambda_1+g-2} \\ \vdots & \ddots & \vdots \\ t_1^{\Lambda_k+g-k-1} & \cdots & t_k^{\Lambda_k+g-k-1} \end{vmatrix} d_{\ge}(\varepsilon^{\Xi_{k+1,g}}) + d_>(\varepsilon^{\Xi_{k+1,g}})},$$

which vanishes when ε vanishes. Here $d_>(z^\ell) \in \{\sum_{|\alpha|>\ell} a_\alpha z^\alpha\}$, $d_\ge(z^\ell) \in \{\sum_{|\alpha|\ge\ell} a_\alpha z^\alpha\}$ for $\alpha = (\alpha_1, \dots, \alpha_m)$, $z^\alpha = z_1^{\alpha_1} \cdots z_g^{\alpha_m}$, and $|\alpha| = \alpha_1 + \cdots + \alpha_m$. Then for $k+1 \le j \le g$, $\Xi_{k+1,g,i} = \sum_{\ell=k+1,\ell\neq i}^{g} (\Lambda_\ell + g - \ell - 1)$. For $k+1 \le j \le g$, noting $\Xi_{k+1,g,i} < \Xi_{k+1,g}$, we have

$$((M_T^{(g)})^{-1})_{ji} = (-1)^{i+j}\frac{\begin{vmatrix} t_1^{\Lambda_1+g-2} & \cdots & t_k^{\Lambda_1+g-2} \\ \vdots & \ddots & \vdots \\ t_1^{\Lambda_k+g-k-1} & \cdots & t_k^{\Lambda_k+g-k-1} \end{vmatrix} d_{\ge}(\varepsilon^{\Xi_{k+1,i,g}}) + d_>(\varepsilon^{\Xi_{k+1,i,g}})}{\begin{vmatrix} t_1^{\Lambda_1+g-2} & \cdots & t_k^{\Lambda_1+g-2} \\ \vdots & \ddots & \vdots \\ t_1^{\Lambda_k+g-k-1} & \cdots & t_k^{\Lambda_k+g-k-1} \end{vmatrix} d_{\ge}(\varepsilon^{\Xi_{k+1,g}}) + d_>(\varepsilon^{\Xi_{k+1,g}})}$$

which is singular for small ε. Its leading term becomes $((M_T^{(g;k)})^{-1})_{ji}$. This means that we have (3.126) and we proved the second statement.

In our last example, to give a visual description of the general pattern, rather than the (5,7) or (7,9) cases given above which would occupy several pages, we treat the trigonal (3,7) case for $k = 2$:

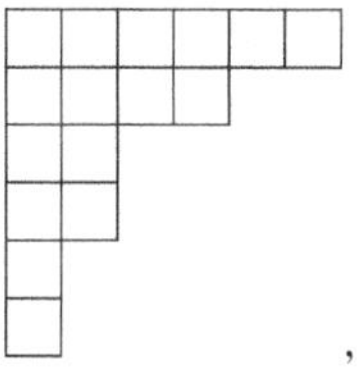

$$, \quad H_1 = \begin{pmatrix} h_2^{(6,2)} & h_3^{(6,2)} & h_4^{(6,2)} & h_5^{(6,2)} \\ h_1^{(6,2)} & h_2^{(6,2)} & h_3^{(6,2)} & h_4^{(6,2)} \\ & h_0^{(6,2)} & h_1^{(6,2)} & h_2^{(6,2)} \\ & & h_0^{(6,2)} & h_1^{(6,2)} \end{pmatrix}, \quad H_2 = \left(h_1^{(6,2)}\right), \quad H_3 = \emptyset.$$

Then the transition matrices expand as follows:

$$M_T^{(6)} = \begin{pmatrix} t_1^{10} & t_2^{10} & \varepsilon_3^{10} & \varepsilon_4^{10} & \varepsilon_5^{10} & \varepsilon_6^{10} \\ t_1^7 & t_2^7 & \varepsilon_3^7 & \varepsilon_4^7 & \varepsilon_5^7 & \varepsilon_6^7 \\ t_1^4 & t_2^4 & \varepsilon_3^4 & \varepsilon_4^4 & \varepsilon_5^4 & \varepsilon_6^4 \\ t_1^3 & t_2^3 & \varepsilon_3^3 & \varepsilon_4^3 & \varepsilon_5^3 & \varepsilon_6^3 \\ t_1 & t_2 & \varepsilon_3 & \varepsilon_4 & \varepsilon_5 & \varepsilon_6 \\ 1 & 1 & 1 & 1 & 1 & 1 \end{pmatrix},$$

$$[M_T^{(6)-1}]_{1,6} = -\frac{\left|t_2^{10}\right| d_{\geq}(\varepsilon^{7+4+3+1}) + d_{>}(\varepsilon^{15})}{\begin{vmatrix} t_1^{10} & t_2^{10} \\ t_1^7 & t_2^7 \end{vmatrix} d_{\geq}(\varepsilon^{4+3+1+0}) + d_{>}(\varepsilon^8)}, \quad [M_T^{(6)-1}]_{2,6} = \frac{\left|t_1^{10}\right| d_{\geq}(\varepsilon^{7+4+3+1}) + d_{>}(\varepsilon^{15})}{\begin{vmatrix} t_1^{10} & t_2^{10} \\ t_1^7 & t_2^7 \end{vmatrix} d_{\geq}(\varepsilon^{4+3+1+0}) + d_{>}(\varepsilon^8)},$$

$$[M_T^{(6)-1}]_{3,6} = -\frac{\begin{vmatrix} t_1^{10} & t_2^{10} \\ t_1^7 & t_2^7 \end{vmatrix} \begin{vmatrix} \varepsilon_4^4 & \varepsilon_5^4 & \varepsilon_6^4 \\ \varepsilon_4^3 & \varepsilon_5^3 & \varepsilon_6^3 \\ \varepsilon_4^1 & \varepsilon_5^1 & \varepsilon_6^1 \end{vmatrix} + d_{>}(\varepsilon^8)}{\begin{vmatrix} t_1^{10} & t_2^{10} \\ t_1^7 & t_2^7 \end{vmatrix} \begin{vmatrix} \varepsilon_3^4 & \varepsilon_4^4 & \varepsilon_5^4 & \varepsilon_6^4 \\ \varepsilon_3^3 & \varepsilon_4^3 & \varepsilon_5^3 & \varepsilon_6^3 \\ \varepsilon_3^1 & \varepsilon_4^1 & \varepsilon_5^1 & \varepsilon_6^1 \\ 1 & 1 & 1 & 1 \end{vmatrix} + d_{>}(\varepsilon^8)}, \quad [M_T^{(6)-1}]_{4,6} = \frac{\begin{vmatrix} t_1^{10} & t_2^{10} \\ t_1^7 & t_2^7 \end{vmatrix} \begin{vmatrix} \varepsilon_3^4 & \varepsilon_5^4 & \varepsilon_6^4 \\ \varepsilon_3^3 & \varepsilon_5^3 & \varepsilon_6^3 \\ \varepsilon_3^1 & \varepsilon_5^1 & \varepsilon_6^1 \end{vmatrix} + d_{>}(\varepsilon^8)}{\begin{vmatrix} t_1^{10} & t_2^{10} \\ t_1^7 & t_2^7 \end{vmatrix} \begin{vmatrix} \varepsilon_3^4 & \varepsilon_4^4 & \varepsilon_5^4 & \varepsilon_6^4 \\ \varepsilon_3^3 & \varepsilon_4^3 & \varepsilon_5^3 & \varepsilon_6^3 \\ \varepsilon_3^1 & \varepsilon_4^1 & \varepsilon_5^1 & \varepsilon_6^1 \\ 1 & 1 & 1 & 1 \end{vmatrix} + d_{>}(\varepsilon^8)},$$

$$\cdots\cdots,$$

$$[M_T^{(6)-1}]_{1,3} = \frac{\left|t_2^{10}\right| d_{\geq}(\varepsilon^{7+3+1+0}) + d_{>}(\varepsilon^{11})}{\begin{vmatrix} t_1^{10} & t_2^{10} \\ t_1^7 & t_2^7 \end{vmatrix} d_{\geq}(\varepsilon^{4+3+1+0}) + d_{>}(\varepsilon^8)}, \quad [M_T^{(6)-1}]_{2,3} = -\frac{\left|t_1^{10}\right| d_{\geq}(\varepsilon^{7+3+1+0}) + d_{>}(\varepsilon^{11})}{\begin{vmatrix} t_1^{10} & t_2^{10} \\ t_1^7 & t_2^7 \end{vmatrix} d_{\geq}(\varepsilon^{4+3+1+0}) + d_{>}(\varepsilon^8)},$$

$$[M_T^{(6)-1}]_{3,3} = \frac{\begin{vmatrix} t_1^{10} & t_2^{10} \\ t_1^7 & t_2^7 \end{vmatrix} \begin{vmatrix} \varepsilon_4^3 & \varepsilon_5^3 & \varepsilon_6^3 \\ \varepsilon_4^1 & \varepsilon_5^1 & \varepsilon_6^1 \\ 1 & 1 & 1 \end{vmatrix} + d_>(\varepsilon^4)}{\begin{vmatrix} t_1^{10} & t_2^{10} \\ t_1^7 & t_2^7 \end{vmatrix} \begin{vmatrix} \varepsilon_3^4 & \varepsilon_4^4 & \varepsilon_5^4 & \varepsilon_6^4 \\ \varepsilon_3^3 & \varepsilon_4^3 & \varepsilon_5^3 & \varepsilon_6^3 \\ \varepsilon_3^1 & \varepsilon_4^1 & \varepsilon_5^1 & \varepsilon_6^1 \\ 1 & 1 & 1 & 1 \end{vmatrix} + d_>(\varepsilon^8)}, \quad [M_T^{(6)-1}]_{4,3} = -\frac{\begin{vmatrix} t_1^{10} & t_2^{10} \\ t_1^7 & t_2^7 \end{vmatrix} \begin{vmatrix} \varepsilon_3^3 & \varepsilon_5^3 & \varepsilon_6^3 \\ \varepsilon_3^1 & \varepsilon_5^1 & \varepsilon_6^1 \\ 1 & 1 & 1 \end{vmatrix} + d_>(\varepsilon^4)}{\begin{vmatrix} t_1^{10} & t_2^{10} \\ t_1^7 & t_2^7 \end{vmatrix} \begin{vmatrix} \varepsilon_3^4 & \varepsilon_4^4 & \varepsilon_5^4 & \varepsilon_6^4 \\ \varepsilon_3^3 & \varepsilon_4^3 & \varepsilon_5^3 & \varepsilon_6^3 \\ \varepsilon_3^1 & \varepsilon_4^1 & \varepsilon_5^1 & \varepsilon_6^1 \\ 1 & 1 & 1 & 1 \end{vmatrix} + d_>(\varepsilon^8)},$$

$$\cdots\cdots.$$

■

The above proof and Lemma 2.3 yields the following Proposition 3.225:

Proposition 3.225 *For the Young diagram Λ associated with the numerical semigroup H of genus g, an integer k $(0 \le k < g)$, and the characteristics of the partition of $\Lambda^{[k]} = (\Lambda_{k+1}, \Lambda_{k+2}, \dots, \Lambda_g)$ with rank $r_k = r_{\Lambda^{[k]}}$,*

$$(a_{r_k}, \dots, a_2, a_1; b_{r_k}, \dots, b_2, b_1),$$

the following holds:

1. *$N^c(g-k-i) - N(i-1)$ is the complementary element of the numerical semigroup and thus let $N^c(L_i^{[k]}) := N^c(g-k-i) - N(i-1)$ and then we have*

$$\Lambda_{g-L_i^{[k]}} + g - L_i^{[k]} = a_i + b_i + 1$$

for every $i = 1, \dots, r_k$, and

2. *$L_1^{[k]} = k+1$.*

Proof We recall Lemma 2.3. First, we note that since Λ is associated with the numerical semigroup H, $\Lambda^{[k]}$ is also associated with a certain numerical semigroup, which is denoted by $H^{[k]}$ due to Lemma 2.3 3. Then its complement $H^{[k]c}$ is a subset of H^c. Applying Lemma 2.3 1 and 2 to $\Lambda^{[k]}$, we obtain these results. ■

From (3.122) of $k = g-1$ case, and the fact $h_n^{\langle g,g\rangle}(t_g) = t_g^n$, we have the following corollary.

Corollary 3.226

$$\mathbf{s}_{\Lambda^{(k+1)}}(t_1, \dots, t_{k+1}) = \mathbf{s}_{\Lambda^{(k)}}(t_1, \dots, t_k) t_{k+1}^{\Lambda_{k+1}} + d_{>\Lambda_{k+1}}(t_{k+1}).$$

In the proof of Lemma 3.224, we note that $h_{a_i+b_i+1}^{(g;k)}$ is given by $\cdots + \dfrac{1}{(a_i+b_i+1)!}(T_1^{(g;k)})^{a_i+b_i+1} + \cdots$, which appears in (3.120) for the partition characteristics $(a_{r_k}, \dots, a_1; b_{r_k}, \dots, b_1)$ of $\Lambda^{[k]}$. We have the following lemma:

Lemma 3.227 *By using the notations $(\mathfrak{u}_i^{[k]} := T_{\Lambda_i+g-i}^{(k)}, \mathfrak{u}_i^{[g]} := T_{\Lambda_i+g-i}^{(g)})$ in Lemma 3.224, the characteristics of the partition $(a_{r_k}, \dots, a_1; b_{r_k}, \dots, b_1)$ of $\Lambda^{[k]}$, and a subsequence $J_\ell = \{i_1, i_2, \dots, i_{r_\Lambda}\} \subset \natural_k$ $(\ell \le r_k)$, we have*

$$\mathbf{s}_{\Lambda^{(k)}}(\mathfrak{u}^{[k]}) = \widehat{\varepsilon}'_{\Lambda, J_\ell, \natural_k} \left(\frac{\partial}{\partial \mathfrak{u}_g^{[g]}}\right)^{\mathrm{wt}(\natural_k \setminus J_\ell)} \left(\prod_{i \in J_\ell} \frac{\partial}{\partial \mathfrak{u}_i^{[g]}}\right) \mathbf{s}_{\Lambda^{(g)}}(\mathfrak{u}^{[g]})\Big|_{\mathfrak{u}^{[g]}=\mathfrak{u}^{[k]}}, \tag{3.127}$$

where $\widehat{\varepsilon}'_{\Lambda, J_\ell, \natural_k}$ *is a certain non-vanishing rational number. Particularly,*

$$\mathbf{s}_{\Lambda^{(k)}}(\mathfrak{u}^{[k]}) = (-1)^{g-k-r_k} \left(\prod_{j \in \natural_k} j!\right) \left(\frac{\partial}{\partial \mathfrak{u}_g^{[g]}}\right)^{|\Lambda^{[k]}|} \mathbf{s}_{\Lambda^{(g)}}(\mathfrak{u}^{[g]})\Big|_{\mathfrak{u}^{[g]}=\mathfrak{u}^{[k]}}. \tag{3.128}$$

Proof With notation as in the proof of Lemma 3.224, using the characteristics of the partition, we introduce the Young diagrams $\Lambda^{[k]}[i]$, $(i = 1, \dots, r_k)$,given by $(a_{r_k}, \dots, a_i; b_{r_k}, \dots, b_i)$ such that $\Lambda^{[k]}[1] = \Lambda^{[k]}$. Then by letting $t^{(g;k)} := (t_{k+1}, \dots, t_g)$, the determinant of the matrix in the proof of Lemma 3.224, H_j, equals the Schur function of $\Lambda^{[k]}[i]$,

$$s_{\Lambda^{[k]}[i]}(t^{(g;k)}) = |H_i|.$$

Since we are concerned only with the term of $(\frac{1}{n!}T_1^{(g;k)})^n$ in $h_n^{(g;k)} = \cdots + \frac{1}{n!}(T_1^{(g;k)})^n + \cdots$, we analyze the behavior of T_1.

From the Jacobi–Trudi determinant expression in Proposition 3.217, for a Young diagram Λ, s_Λ contains the term

$$S_{\Lambda,T_1} := \left|\frac{1}{(\Lambda_i + j - i)!} T_1^{\Lambda_i+j-i} \delta_{(\Lambda_i+j-i)}\right|,$$

where $\delta_i = 1$ for $i \geq 0$ and , $\delta_i = 0$ for $i < 0$. Then the term S_{Λ,T_1}, i.e., the determinant of the matrix $\left(\frac{1}{(\Lambda_i + j - i)!} T_1^{\Lambda_i+j-i}\right)$, does not vanish as a polynomial of T_1 because the vectors $U_i := \left(\frac{1}{(\Lambda_i + j - i)!} T_1^{\Lambda_i+j-i}\right)$, viewed as columns, are independent as their coordinates show. By considering their weight, S_{Λ,T_1} is a monomial $T_1^{|\Lambda|}$ with a non-vanishing rational factor.

On the other hand, (3.120) shows that $\tilde{\mathbf{s}}_{\Lambda^{[k]}}(t^{(g;k)})$ has a term $\left(\mathfrak{u}_g^{[g;k]}\right)^{\mathrm{wt}(\natural_k \setminus J_\ell)} \prod_{i \in J_\ell} \mathfrak{u}_i^{[g;k]}$ up to a non-vanishing rational factor, where $\mathfrak{u}^{[g;k]} := \mathfrak{u}^{[g]} - \mathfrak{u}^{[k]}$. The claim follows. ∎

3.11 Appendix: Heat Equation of Sigma Function

In this appendix, we show the recent studies on the heat equations of the sigma functions [EilO2020, EilGOY2017, BucL2002, BucL2004] based on [EilGOY2017].

It is well-known that the Jacobi theta function (3.2) satisfies the heat equation,

$$(L - H)\theta_1(z, \tau) = 0,$$

where $L := 4\pi\sqrt{-1}\frac{\partial}{\partial\tau}$ and $H := \frac{\partial^2}{\partial z^2}$. On the other hand, as mentioned in Sect. 1.3.2, Weierstrass also found the heat Eq. (1.32) to show the expansion of the sigma function (1.33). Buchstaber and Leykin generalized the heat equation of the sigma function to plane W-curves from Lie-algebraic viewpoints [BucL2002, BucL2004], and Eilbeck, Gibbons, Ônishi, and Yasuda studied it rigorously [EilO2020], [EilGOY2017].

We consider only plane W-curves X associated with the numerical semigroup $H = \langle r, s\rangle$, $(r < s)$.

We bear in mind Torelli's theorem [FarK1991, p. 163 III.12.3]:

Theorem 3.228 *For a compact Riemann surface X of genus $g \ge 2$ with an Abel–Jacobi map $w : X \to J_X$, $W^1_X := w(X)$ is determined up to translation and reflection by the canonical polarizations of J_X; that is, by the class of translates W^{g-1}_X.*

It means that X can be characterized by the Jacobi variety J_X in the Abelian variety.

Let the associated numerical semigroup H be generated by (r, s). As Pinkham and Komeda investigated the deformation of the curves from the monomial curves $X_H = \operatorname{Spec}\mathbb{C}[H]$ as in Appendix 2.12, we investigate the same mathematical phenomena from analytic viewpoints using the so-called the heat equations connected with the deformation of the Jacobi varieties in this appendix.

Due to Theorem 2.255 and Definition 2.253, we consider the parameter space $\widehat{\lambda} := (\widehat{\lambda}_{\mathfrak{q}(1)}, \widehat{\lambda}_{\mathfrak{q}(2)}, \cdots, \widehat{\lambda}_{\mathfrak{q}(\mathfrak{n})})$, $(\mathfrak{n} := \mathfrak{n}_{\langle r,s\rangle})$ and the Weierstrass-Schlessinger canonical form (2.150),

$$\widehat{f_{\widehat{\lambda}}}(x, y) := y^r - \sum_{i=1}^{r-2}\sum_{j=0}^{\lfloor is/r\rfloor} \widehat{\lambda}_{\mathfrak{q}(i,j)}x^j - x^r - \sum_{j=0}^{r-2}\widehat{\lambda}_{\mathfrak{q}(0,j)}x^j.$$

Here we note that $\mathfrak{n}(= \mathfrak{n}_H)$ is the dimension of the parameter space.

In order to describe the family of curves given by $\widehat{f_{\widehat{\lambda}}}(x, y)$, we introduce some objects as follows. We consider the parameter space of $\widehat{\lambda}$, i.e., $\mathfrak{L} := \{\widehat{\lambda} \in \mathbb{C}^{\mathfrak{n}}\}$. We introduce the set $S_{\phi} \subset \{x^i y^j \mid i, j = 0, 1, 2, \ldots\}$ whose element entries in the Weierstrass-Schlessinger canonical form with coefficient $\widehat{\lambda}$. Let $\widehat{\mathbb{A}}_{\mathbb{C}} := \mathbb{C}[\widehat{\lambda}] = \mathbb{C}[\widehat{\lambda}_{\mathfrak{q}(1)}, \cdots, \widehat{\lambda}_{\mathfrak{q}(\mathfrak{n})}]$ and $\widehat{\mathbb{A}}_{\mathbb{Q}} := \mathbb{Q}[\widehat{\lambda}] = \mathbb{Q}[\widehat{\lambda}_{\mathfrak{q}(1)}, \cdots, \widehat{\lambda}_{\mathfrak{q}(\mathfrak{n})}]$. For every $\widehat{\lambda} \in \mathfrak{L}$, we define $R_{\widehat{\lambda}} := \widehat{\mathbb{A}}_{\mathbb{Q}}[x, y]/(\widehat{f_{\widehat{\lambda}}}(x, y))$ and then we can regard $X_{\widehat{\lambda}} = \operatorname{Spec} R_{\widehat{\lambda}}$. Then we consider the degenerating family of curves $X_{\widehat{\lambda}}$,

$$\mathfrak{X} := \{(x, y, \widehat{\lambda}) \mid (x, y) \in X_{\widehat{\lambda}} = \operatorname{Spec} R_{\widehat{\lambda}}, \widehat{\lambda} \in \mathfrak{L}\}.$$

We also have the family of the lattice $\Gamma_{\widehat{\lambda}} := \Gamma_{X_{\widehat{\lambda}}}$ and the Jacobians $J_{\widehat{\lambda}} := J_{X_{\widehat{\lambda}}} = \mathbb{C}^g/\Gamma_{\widehat{\lambda}}$,

$$\mathfrak{J} := \{(v, \widehat{\lambda}) \mid v \in J_{\widehat{\lambda}}, \widehat{\lambda} \in \mathfrak{L}\},$$

with the projection $\pi_{\mathfrak{J}} : \mathfrak{J} \to \mathfrak{L}$. Since each Jacobian $J_{\widehat{\lambda}}$ and its tangent bundle $T J_{\widehat{\lambda}}$ are embedded into $\mathbb{C}^g$ and $\mathbb{C}^{2g}$ respectively, we also consider the trivial bundles,

$$\mathbb{C}^g_{\mathfrak{L}} = \mathbb{C}^g \times \mathfrak{L}, \quad \mathbb{C}^{2g}_{\mathfrak{L}} = \mathbb{C}^{2g} \times \mathfrak{L},$$

with $\pi_{\mathbb{C}^\ell} : \mathbb{C}^\ell \times \mathfrak{L} \to \mathfrak{L}$.

We don't explicitly use these vector bundles $\mathbb{C}^g_{\mathfrak{L}}$ or $\mathbb{C}^{2g}_{\mathfrak{L}}$ in the following arguments, but they are, implicitly, crucial in the formulations as mentioned in Remark 3.236.

By introducing $f_{\widehat{\lambda},x} := \dfrac{\partial f_{\widehat{\lambda}}}{\partial x}$ and $f_{\widehat{\lambda},y} := \dfrac{\partial f_{\widehat{\lambda}}}{\partial y}$, we define the cohomology,

$$\mathbf{H}^1(X_{\widehat{\lambda}}, \widehat{\mathbb{A}}_{\mathbb{Q}}) := \left\{ \frac{h(x, y)dx}{f_{\widehat{\lambda},y}(x, y)} \;\middle|\; h(x, y) \in R_{\widehat{\lambda}} \right\}.$$

For the deformation of the Jacobi varieties, we introduce $\left\{ \partial_{\widehat{\lambda}_i} := \dfrac{\partial}{\partial \widehat{\lambda}_i} \right\}$. We let $\mathscr{K}_{\mathfrak{X}}$ denote the function field $\mathbb{Q}(\widehat{\lambda}, x, y)$ on $\mathfrak{X}$ so that each (x, y) satisfies $f_{\widehat{\lambda}}(x, y) = 0$.

Let ξ be an element in $\mathscr{K}_{\mathfrak{X}}$. For a derivation D on $\mathscr{K}_{\mathfrak{X}}$ such that $D\xi = 0$, and $\nu \in \mathscr{K}_{\mathfrak{X}} d\xi$, we define

$$D(\nu) = D\left(\frac{\nu}{d\xi}\right) d\xi.$$

We recall Chevalley's lemma [Ch1951, Man1964]:

Lemma 3.229 *Any derivation is independent of the choice of the element ξ in the following sense. Let ξ and ξ' be two elements in $\mathscr{K}_{\mathfrak{X}}$ transcendental over $\mathbb{Q}(\widehat{\lambda})$. Let D and D' be derivations on $\mathscr{K}_{\mathfrak{X}}$ such that $Ds = D's$ for $s \in \mathbb{Q}(\widehat{\lambda})$, and hence in particular $D\xi = D'\xi' = 0$. Then we have for any $w \in \mathscr{K}_{\mathfrak{X}}$,*

$$D(wd\xi) - D'(wd\xi) = d(-wD'\xi).$$

Proof The operator $D'' := D - D' + (D'\xi)\frac{d}{d\xi}$ is a derivation on $\mathscr{K}_{\mathfrak{X}}$ which vanishes on $\mathbb{Q}(\widehat{\lambda})$ and also $D''\xi = 0$. Hence $D''w = 0$. Moreover, $(D - D')(wd\xi) = ((D - D')w)d\xi + w(D - D')d\xi$. Since ξ' is transcendental, we have $[\frac{d}{d\xi'}, D'] = 0$, $Dd\xi = D(\frac{d\xi}{d\xi})d\xi = 0$, and $D'd\xi = D'(\frac{d\xi}{d\xi'})d\xi' = \frac{d}{d\xi'}(D'\xi)d\xi' = d(D'\xi)$. Thus $(D - D')(wd\xi) = -(D'\xi)\frac{dw}{d\xi}d\xi - wd(D'\xi) = -d(wD'\xi)$. ■

Let L be a first-order operator as an element in $\oplus_j \mathbb{Q}[\widehat{\lambda}]\frac{\partial}{\partial \widehat{\lambda}_{\mathfrak{q}(j)}}$. For such a L, we define $\Gamma^L \in \mathrm{Mat}_{\widehat{\mathbb{A}}_{\mathbb{Q}}}(2g)$ by

$$L\nu = \Gamma^L \nu \quad \text{for} \quad \nu \in \mathbf{H}^1(X_{\widehat{\lambda}}, \widehat{\mathbb{A}}_{\mathbb{Q}}).$$

Here Buchstaber and Leykin called ${}^t(\Gamma^L)$ *the Gauss–Manin connection* [EilGOY2017, BucL2002, BucL2004], which satisfies

$$L(\Omega) = \Gamma^L \Omega, \quad \Omega := \begin{pmatrix} \omega' & \omega'' \\ \eta' & \eta'' \end{pmatrix},$$

where Ω is the period matrix of $X_{\widehat{\lambda}}$; ${}^t\Omega J \Omega = 2\pi\sqrt{-1}J$, where $J = \begin{pmatrix} & 1_g \\ -1_g & \end{pmatrix}$. Here we note that the components in the period matrix differ from our conventions by a factor of two.

For $\Gamma = \begin{pmatrix} -\beta & \alpha \\ -\gamma & {}^t\beta' \end{pmatrix} \in \mathfrak{sp}(2g, \widehat{\mathbb{A}}_{\mathbb{Q}})$, the symplectic Lie algebra, we have

$$L = \sum_{i,j=1}^{g} \left[(\Gamma\Omega)_{ij} \frac{\partial}{\partial \omega'_{i,j}} + (\Gamma\Omega)_{ij+g} \frac{\partial}{\partial \omega''_{i,j}} + (\Gamma\Omega)_{g+ij} \frac{\partial}{\partial \eta'_{i,j}} + (\Gamma\Omega)_{g+ij+g} \frac{\partial}{\partial \eta''_{i,j}} \right].$$

The deformation of the curves has difficulty because we have to consider it as a fibered category. However, we investigate it via their Jacobi varieties J_X embedded in $\mathbb{C}^g$ and its tangent space $T\mathbb{C}^g = \mathbb{C}^{2g}$ since it is a $\mathbb{C}$ vector bundle over $\mathfrak{L}$ whose rank is $2g$. Instead of (3.99), we consider a point $u = (u_1, \ldots, u_g)$ in $\mathbb{C}^g$ with the relations

$$[u_i, u_j] = [\partial_{u_i}, \partial_{u_j}] = 0, \quad [\partial_{u_i}, u_j] = \delta_{i,j}. \tag{3.129}$$

This relations correspond to the intersection relations of the homology basis (2.58).

For a symmetric matrix $\Gamma J = \begin{pmatrix} \alpha & \beta \\ {}^t\beta & \gamma \end{pmatrix}$, we also introduce the second-order differential operator,

$$\begin{aligned} H^L &= \frac{1}{2}[{}^t\partial_u, {}^tu] \begin{pmatrix} \alpha & \beta \\ {}^t\beta & \gamma \end{pmatrix} \begin{pmatrix} \partial_u \\ u \end{pmatrix} \\ &= \sum_{i,j=1}^{g} \left(\frac{1}{2}\alpha_{ij} \frac{\partial^2}{\partial u_i, \partial u_j} + \beta_{ij} u_i \frac{\partial}{\partial u_i} + \frac{1}{2}\gamma_{ij} u_i u_j \right) + \frac{1}{2}\mathrm{tr}\beta. \end{aligned}$$

Lemma 3.230 *Let* $G_0(u, \Omega) = \left(\frac{(2\pi)^g}{\det \omega'} \right)^{1/2} \exp\left(\frac{1}{2} {}^tu\eta'\omega'^{-1}u \right)$, *and* $F(z, p, q)$ *that satisfies*

$$F^{-1}(z, p, q)(L - H^L)F(z, p, q) = L - H^L.$$

$F(z, p, q)$ depends on a constant vector $b := (b', b'')$. We define

$$
\begin{aligned}
G(b, u, \Omega) &= F(z(b), p(b), q(b)) G_0(u, \Omega) \\
&= \left(\frac{(2\pi)^g}{\det \omega'}\right)^{1/2} \exp\left(\frac{1}{2} {}^t u \eta' \omega'^{-1} u\right) \\
&\quad \times \exp\left(2\pi\sqrt{-1}\left(\frac{1}{2} {}^t b'' \omega'^{-1} \omega'' b'' + {}^t b'' (\omega'^{-1} u + b')\right)\right),
\end{aligned}
$$

and then

$$(L - H^L) G(b, u, \Omega) = 0.$$

Then the following is obvious:

Lemma 3.231

$$(L - H^L) \sum_{n \in \mathbb{Z}^g} G(\delta' n + \delta'', u, \Omega) = 0.$$

Let $\boldsymbol{\phi}(P) = {}^t(\boldsymbol{\phi}_1(P), \boldsymbol{\phi}_2(P), \ldots, \boldsymbol{\phi}_{\mathfrak{n}}(P))$ for $\boldsymbol{\phi}_i \in S_{\boldsymbol{\phi}}$ and $P \in X_{\widehat{\lambda}}$. For $P = (x, y)$ and $Q = (z, w)$ in $X_{\widehat{\lambda}}$, we define

$$
h(P, Q) := \frac{1}{2} \begin{vmatrix} \dfrac{\widehat{f_{\lambda,x}}(x, y) - \widehat{f_{\lambda,x}}(z, w)}{x - z} & \dfrac{\widehat{f_{\lambda,y}}(x, y) - \widehat{f_{\lambda,y}}(z, w)}{x - z} \\ \dfrac{\widehat{f_{\lambda,x}}(x, y) - \widehat{f_{\lambda,x}}(z, w)}{y - w} & \dfrac{\widehat{f_{\lambda,y}}(x, y) - \widehat{f_{\lambda,y}}(z, w)}{y - w} \end{vmatrix},
$$

and the matrices T, H, V of $\mathrm{Mat}_{\widehat{\mathbb{A}_{\mathbb{Q}}}}(\mathfrak{n})$ satisfying

$$-rs \widehat{f_{\lambda}} \boldsymbol{\phi}_i = \sum_j T_{ij} \boldsymbol{\phi}_{\mathfrak{n}+1-j} \quad \text{modulo} \quad (\widehat{f_{\lambda,x}}, \widehat{f_{\lambda,y}}),$$

$$h(P, Q) = {}^t\boldsymbol{\phi}(P) H \boldsymbol{\phi}(Q), \quad \widehat{f_{\lambda}}(P) h(P, Q) = {}^t\boldsymbol{\phi}(P) V \boldsymbol{\phi}(Q),$$

in the commutative ring $\widehat{\mathbb{A}_{\mathbb{Q}}}[x, y, x, w]/(\widehat{f_{\lambda,x}}(x, y), \widehat{f_{\lambda,x}}(z, w), \widehat{f_{\lambda,y}}(x, y),$ $\widehat{f_{\lambda,y}}(z, w))$. Then the direct computations shows $H = \begin{pmatrix} * & \cdots & * & -rs \\ * & \cdots & -rs & \\ * & \iddots & & \\ -rs & & & \end{pmatrix}$.

We assume that $\mathfrak{n} = \mathfrak{n}_H = 2g$ as follows: The cases $\mathfrak{n} = 2g$ correspond to the hyperelliptic curves $H = \langle 2, 2g+1 \rangle$ and $H = \langle 3, 4 \rangle$.

For $\mathfrak{v}(i) \in \mathfrak{v}$ in Definition 2.253 and the matrix V defined above, we introduce differential operator

$$L_{\mathfrak{v}(i)} = \sum_{j=1}^{\mathfrak{n}} V_{ij} \frac{\partial}{\partial \widehat{\lambda}_{rs - \mathfrak{v}(j)}}, \qquad (i = 1, 2, \ldots, \mathfrak{n}).$$

The discriminant Δ of $\widehat{f_\lambda}$ is characterized by the following lemma:

Lemma 3.232 *The determinant* $\det(V) = \det(T)$ *is a constant multiple of the discriminant* Δ *of* $\widehat{f_\lambda}$.

The hessian of g is denoted by Hess g, i.e, Hess $g := \begin{vmatrix} \frac{\partial^2 g}{\partial x^2} & \frac{\partial^2 g}{\partial x \partial y} \\ \frac{\partial^2 g}{\partial x \partial y} & \frac{\partial^2 g}{\partial y^2} \end{vmatrix}$.

Using this expression, Buchstaber and Leykin showed the relations $\{L_{\mathfrak{v}(i)}\}$ and the following discriminant Δ fact [BucL2002, BucL2004]:

Proposition 3.233 *Let* $\mathbf{L} := {}^t\boldsymbol{\phi}\,{}^t(L_{\mathfrak{v}(1)}, L_{\mathfrak{v}(2)}, \cdots, L_{\mathfrak{v}(\mathfrak{n})})$,

$$\mathbf{L}(\Delta) = \mathrm{Hess}\,\widehat{f_\lambda} \cdot \Delta.$$

This proposition may be related to the deformation theory of the monomial curves in Appendix 2.12.

We have the simplest version of the heat equation governing the sigma functions.

Lemma 3.234 *For the sigma function* σ *defined (3.21) in Sect. 3.2.2, we have the relation*

$$(L_0 - H^{L_0} + ML_0 \log \Delta)\sigma(u) = 0, \tag{3.130}$$

where

$$H^{L_0} = -\sum_{j=1}^{g} \mathrm{wt}(u_j) u_j \frac{\partial}{\partial u_j} - \frac{1}{2}\sum_{j=1}^{g} \mathrm{wt}(u_j).$$

Then we have the following theorem as the heat equations of the sigma functions for the hyperelliptic curves and (3, 4)-curves.

Theorem 3.235 *The sigma function (3.21) can be characterized by the solutions of the following equations:*

$$(L_{\mathfrak{v}(i)} - H_{\mathfrak{v}(i)})\sigma = 0 \quad (i = 1, \ldots, \mathfrak{n}), \tag{3.131}$$

where

$$H_{\mathfrak{v}(i)} := H^{L_{\mathfrak{v}(i)}} + \frac{1}{8} L_{\mathfrak{v}(i)} \log \Delta. \tag{3.132}$$

We call (3.131) with (3.132) *the heat equations* of the sigma functions of (r, s)-curves [BucL2002, BucL2004]. By obtaining the heat Eq. (3.131), the expansions of the sigma functions are obtained. The solutions of the heat Eq. (3.131) were studied by Bunkova, and Buchstaber for genus two [BunB2009] and its generalizations [BucB2020], and by Eilbeck, Gibbons, Ônishi and Yasuda for general hyperelliptic curves and (3,4)-curves [EilGOY2017].

Remark 3.236 The sigma function characterizes the Jacobian $J_{\widehat{\lambda}}$ at each point $\widehat{\lambda} \in \mathfrak{L}$. Recalling Theorem 3.228, it also characterizes the curves $X_{\widehat{\lambda}}$. Due to de Rham cohomology, we consider the relations (3.129) corresponding to the intersection relations of the homology basis (2.58). Thus we implicitly consider $\mathbb{C}^g_{\mathfrak{L}}$ and $\mathbb{C}^{2g}_{\mathfrak{L}}$ in this study of the heat equations. This should be compared with the deformation theory of curves mentioned in Sects. 2.9 and 3.1.2.2. Further we should recall Remark 2.136.

Chapter 4
Application of the Sigma Function Theory to Integrable Systems

This chapter shows some applications of the sigma function theory constructed in the previous chapter to integrable systems. We investigate the differential and difference relations among Abelian functions on J_X which are expressed by the sigma function. They correspond to some solutions of integrable equations in terms of the sigma function and their related functions.

As we mentioned in its history in Sect. 1.8, the differential relations known as the integrable nonlinear differential equations have been studied since the 1950s. The Korteweg–de Vries (KdV) equation was discovered by D. J. Korteweg and G. de Vries in 1895 [KorV1885], and was studied in the middle of the 20th century by various researchers. From the studies, it turned out that the KdV equation has a marvelous property, i.e., the complete integrability [AblS1981, Lam1980] and there is a class of differential equations that have the same property, such as the sine-Gordon equation, the Toda lattice equation, the modified Korteweg–de Vries (mKdV) equation. Further, there is a series of each differential equation; e.g., the KdV equation has a hierarchical structure called the KdV hierarchy. We give a short review in Appendix 3.9.2 on a theory related to the Kadomtsev–Petviashvili (KP) equation [KadP1970]; It is noted that the KP–hierarchy in the theory includes the KdV hierarchy and the KdV equation naturally. Through the studies of these equations, it is clarified that these differential equations are connected with the algebraic curves. These results in the integrable systems in Appendix 3.9.2 have a crucial effect on the modern theory of the sigma function, as mentioned in Chap. 3.

However, in this chapter, we provide the differential and difference relations among Abelian functions associated with W-curves in classical style because the KdV hierarchy (including KP–hierarchy partially) and several differential equations were discovered in the flow of the (incomplete) Weierstrass Abelian function theory long before the studies in the middle of the 20th century, though they are not familiar in the study of the integrable system. The prototype is Baker's 1898 paper [Ba1898b] on the KdV hierarchy as a generalization of genus one (1.31), whose origin was the

S. Matsutani, *The Weierstrass Sigma Function in Higher Genus and Applications to Integrable Equations*, Springer Monographs in Mathematics,
https://doi.org/10.1007/978-981-97-8163-8_4

static sine-Gordon equation (1.11) by Gauss and (1.24) by Weierstrass. Its derivation was published in the paper [Ba1903] in 1903. (Further, back in the past, we encounter the elastica problem as its origin.) Their importance has not been diminished in the slightest today. On the contrary, we believe that they are growing in importance. In this chapter, we review and reevaluate them from a contemporary perspective.

We note that there are several excellent books on the algebro-geometric solutions of the integral system, e.g., by D. Mumford [Mum1981, Mum1984], by Tanaka and Date [TanD1979], by E. D. Belokolos, A. I. Bobenko, V. Z. Enolskii, A. R. Its, and V. B. Matveev [BelBEIM1994], by F. Gesztesy, H. Holden, J. Michor and G. Teschl [GeH2003, GeHMT2008], and by V. M. Buchstaber, V. Z. Enolskii, and D. V. Leykin [BucEL2012]. Except for [BucEL2012] they showed the algebro-geometric solutions mainly using the Riemann theta function. In particular, the book [BucEL2012] is a brilliant book on the sigma function and its applications to integrable systems. In this book we basically demonstrate the algebro-geometric solutions following more classical ways and as straightforward extensions of the elliptic function solutions.

In Sect. 4.1, we give, first, a short review of Baker's work in 1903, which provided the KdV hierarchy and KP hierarchy [Mat2001a]. Based on the approach, we develop the approach to obtain the algebraic relations on W-curves, mainly hyperelliptic curves given by (2.65), which are considered as the algebraic solutions of the integrable differential equations in this chapter. Section 4.1 reviews the typical approach on the $\wp$ functions. Section 4.1.3 provides a connection between the KdV hierarchy and the sigma function, whereas Sect. 4.1.4 shows the hyperelliptic solutions of the KP hierarchy based on Baker's investigation [Ba1897, Ba1898b]. Section 4.1.5 shows the similar relations among $\wp$'s for (3,4) plane curve based on [EilEMOP2007]. Additionally, in Sect. 4.1.6, we review the so-called soliton solutions of the integrable equations in terms of the sigma function, mainly the KdV equation.

Following the approaches, we develop it to provide the quasi-periodic or the algebro-geometric solutions of the integrable systems, the mKdV equation in Sect. 4.2.1 [Mat2002c], the generalized Lamé equation in Sect. 4.2.2, the sine-Gordon equation in Sect. 4.2.3 [Mat2002d, Mat2005], the Neumann system in Sect. 4.2.4 [Mat2008a], the Toda equation in Sect. 4.3 [KodMP2013, Mat2002a, Mat2003b, Mat2020], difference relations from the addition theorem in Sect. 4.4, and dispersionless KP equation in Sect. 4.5 [MatP2009b].

Further, we show a generalization of the Euler–Bernoulli elastica problem in Sect. 4.6. As mentioned in Chap. 1, the origin of the Abelian theory is the elastica studied by Jacob and Daniell Bernoulli, Euler, Lagrange, and Gauss. It is also remarkable that (1) Weierstrass found the Al functions, the primitive version of the sigma functions, by investigating the static sine-Gordon equation which governs the elastica, and (2) Baker expressed his thanks to A. E. H. Love, in the preface of his book [Ba1897] in 1897 that Love read the original version for revision, whereas Love was an applied mathematician who wrote a book on elasticity, which is known as the standard textbook of Euler–Bernoulli elastica [Lo1892]. The elastica problem is still crucial in the Abelian function theory. In Sect. 4.6, we take up a modern problem of elastic rods, i.e., the shapes of the supercoiled DNA. Beyond the Euler–Bernoulli theory, we investigate the excited states of Euler–Bernoulli energy, which obey the mKdV hierarchy. Though we cannot overcome the difficulty of the real analytic property of the generalized elastica in this stage, we show its difficulty and the roles of our Abelian function theory in the generalization of the elastica problem. We illustrate some real curves associated with the generalized elastica, which resemble the shapes of the supercoiled DNA [Mat1998, Mat2010, MatO2003a, MatP2016, MatP2022].

Remark 4.1 Before we get into the following sections, we summarize the approach in the 19th century. As in Sects. 2.5.3 and 3.3.3, we know the "equivalence" between J_X and $S^g X$ in a certain sense. By using the data of R_X, we may find a differential relation and interpret it as a solution of a certain differential equation. The differential equations represents the algebraic or the geometric structure of the Jacobian J_X or the sigma function. Thus finding an interesting relation is crucial to obtain knowledge of the considered structure of the sigma function. The idea was in Weierstrass' 1854 paper [Wei1854] as shown in Sect. 4.6, and we can demonstrate the solutions explicitly.

To obtain the nonlinear differential equations such as the KdV hierarchy, Baker used the following flow, which is the same as Weierstrass' approach in the study on the Jacobi inversion problem [Wei1854]:

1. We consider the nonlinear differential equations, e.g., the KdV hierarchy, and the sine-Gordon equations, as relations in the differential ring $\mathscr{D}_J = \mathscr{M}_J\left[\dfrac{\partial}{\partial u_1}, \cdots, \dfrac{\partial}{\partial u_g}\right]$, where $\mathscr{M}_J$ is the quotient field of the meromorphic functions over J_X, which are generated by ratio of the differentials of σ function.
2. By using the relation $\dfrac{\partial}{\partial u_r} = \sum_{i=1}^{g} \dfrac{\partial x_i}{\partial u_r}\dfrac{\partial}{\partial x_i} =: \dfrac{\widetilde{\partial}}{\partial u_r, x}\left(\dfrac{\partial x_i}{\partial u_r} \in \mathscr{Q}(S^g R_X)\right)$, and the Jacobi inversion formula in Proposition 4.2, we recognize them as relations in the $\mathscr{D}_X$-module, where
$$\mathscr{D}_X = \mathscr{Q}(S^g R_X)\left[\frac{\widetilde{\partial}}{\partial u_1, x}, \cdots, \frac{\widetilde{\partial}}{\partial u_g, x}\right],$$
and $\mathscr{Q}(S^g R_X)$ is the quotient field of affine symmetric algebra $S^g R_X$.
3. We translate everything defined on J_X into one in $S^g X$, which means the transcendental relations on J_X are described by the algebraic relations on $S^g X$, and vice versa.

This picture is essential in this book and the theory of Abelian function in the 19th century. If we regard the sets of differential modules $Mod(\mathscr{D}_X)$ and $Mod(\mathscr{D}_J)$ as categories, the picture can be interpreted as the equivalence between the categories $Mod(\mathscr{D}_X)$ and $Mod(\mathscr{D}_J)$. When we solve some problems in physics, geometry, and number theory in terms of elliptic function, we use the typical properties in which every transcendental relation of an elliptic function is reduced to an algebraic one in a certain sense. This chapter demonstrates that we can straightforwardly extend the flow in the elliptic function theory to the Abelian function theory by using the equivalence between $\mathscr{D}_X$ and $\mathscr{D}_J$ in a certain sense.

In the following section, to obtain the relations in the ring $\mathscr{D}_J$, we start with some algebraic relations in the ring $\mathscr{D}_X$. By considering the relations in $\mathscr{D}_X$, we can also investigate relations in its subring,

$$\mathscr{D}_X^{[k]} := \mathscr{Q}(S^k R_X)\left[\frac{\widetilde{\partial}}{\partial u_{g-k+1}^{(k)}, x}, \dots, \frac{\widetilde{\partial}}{\partial u_g^{(k)}, x}\right], \quad \frac{\partial}{\partial u_r^{(k)}} = \sum_{i=1}^{k} \frac{\partial x_i}{\partial u_r^{(k)}} \frac{\partial}{\partial x_i} =: \frac{\widetilde{\partial}}{\partial u_r^{(k)}, x},$$

of the differential ring $\mathscr{D}_X^{[k]}$ ($k \le g$). We demonstrate the generalization of the integrable differential equations associated with the Jacobian J_X to ones in its subvariety Θ_X^k in Sect. 4.2.3 for the case of the sine-Gordon equation [Mat2005].

Recently, Ayano and Buchstaber studied the same generalization of the KdV hierarchy of the subvariety of the hyperelliptic Jacobian of genus three [AyB2019]. Further, Buchstaber and Mikhailov used the equivalence of these categories to investigate the infinite-dimensional Lie algebras related to them [BucM2017, BucM2021].

In the following sections, we encounter typical examples of such a view.

4.1 Differential Relations Among $\wp$ Functions

In this section, we consider the differential relations among $\wp$ functions, which are defined in Definition 3.52 for the hyperelliptic curves X given by $f_X(x, y) = y^2 - f(x) = 0$ in (2.65) and (2.66) of genus g. We show that $\wp$'s obey the Korteweg–de Vries (KdV) hierarchy in Theorem 4.25 and Kadomtsev–Petviashvili (KP) hierarchy in Theorem 4.29 as partially mentioned in [Mat2001a].

In [Ba1898b, Ba1903], Baker derived identities including $\wp_{\mu\nu\rho\lambda}$ and $\wp_{\mu\nu}$ as a generalization of

$$\frac{d^2\wp(u)}{du^2} = 6\wp(u)^2 + 4\lambda_2\wp + \frac{1}{2}\lambda_1$$

for the curve $y^2 = x^3 + \lambda_2 x^2 + \lambda_1 x + \lambda_0$. Since Weierstrass used this relation to evaluate the power expansion of the elliptic sigma function essentially, he expected that the identity determines the property of the hyperelliptic sigma function as mentioned in Sect. 1.6.4.

Baker showed the derivations of these equations in 1903 [Ba1903]. We follow the derivations. Taking his cue from Weierstrass' 1854 approaches [Wei1854], Baker used the equivalence between $\mathscr{D}_X$ and $\mathscr{D}_J$ as in Remark 4.1. We state that the approach is critical in the contemporary Weierstrass Abelian function theory, i.e., in the view illustrated in Fig. 1.9. Thus we show its sketch of the derivations.

4.1.1 KdV Hierarchy in Baker's 1898 and 1903 Papers

In this subsection, we handle the so-called KdV hierarchy based on Baker's investigations on the hyperelliptic sigma function. In [Ba1903], Baker started with the following proposition [Ba1897, §11.217], [Ba1903, p. 138], which is essentially the same as Fay's formula in Proposition 3.87 [Fay1973, (39)], as in mentioned in Sect. 3.4.2.

Let us consider an element $((x, y), (x_1, y_2), \dots, (x_g, y_g))$ in $X \times S^g X$ and its meromorphic function $F(x) = (x - x_1) \cdots (x - x_g)$ as in Corollary 2.157. We recall Proposition 3.85:

Proposition 4.2 *Let* $(x_i, y_i) \in X$ $(i = 1, \dots, g)$ *and* $u \in \mathbb{C}^g$ *such that* $u = \widetilde{w}((x_1, y_1), \dots, (x_g, y_g))$. *The following relation holds for generic* x_i' $(i = 1, 2)$,

$$\sum_{i=1}^{g}\sum_{j=1}^{g} \wp_{ij}(u) {e_1}^{i-1} {e_2}^{j-1} = F(e_1)F(e_2)\left(\sum_{i=1}^{g} \frac{y_i}{(e_1 - x_i)(e_2 - x_i)F'(x_i)}\right)^2 - \frac{f(e_1)F(e_2)}{(e_1 - e_2)^2 F(e_1)} - \frac{f(e_2)F(e_1)}{(e_1 - e_2)^2 F(e_2)} + \frac{f_\Omega(e_1, e_2)}{(e_1 - e_2)^2}, \tag{4.1}$$

where $f_\Omega(x_1, x_2) = \sum_{i=0}^{g} x_1^i x_2^i (\lambda_{2i+1}(x_1 + x_2) + 2\lambda_{2i})$.

Section I in [Ba1903] devotes the derivation of his fundamental formula as in Theorem 4.3, which corresponds to the generating relations of the KdV hierarchy of the hyperelliptic curves. Baker considered the relation in Proposition 4.2 as the identity of the meromorphic functions over $X^2 \times J_X \cong X^2 \times S^g X$ whose element $((e_1, y_{e_1}), (e_2, y_{e_2}), u)$ corresponding to $((e_1, y_{e_1}), (e_2, y_{e_2}), (x_1, y_1) \dots, (x_g, y_g))$.[1] He introduced the operator,

$$\delta_e = \frac{1}{F(e)} \sum_{j=1}^{g} e^{j-1} \frac{\partial}{\partial u_j} \tag{4.2}$$

for points $(e_3, y_{e_3}), (e_4, y_{e_4}) \in X$ and act δ_{e_3}, δ_{e_4} on both sides of the relation in Proposition 4.2 to find his fundamental formula in the following theorem:

Theorem 4.3

$$\frac{1}{2}\left(\prod_{i,j=1,i<j}^{4} (e_j - e_i)\right) \sum_{\lambda,\mu,\nu,\rho=1}^{g} \wp_{\lambda\mu\nu\rho} e_1^{\lambda-1} e_2^{\mu-1} e_3^{\nu-1} e_4^{\rho-1}$$

[1] His method can be interpreted from a modern point of view as an investigation by means of the Poincaré line bundle [ArCGH1985, IV].

$$
\begin{aligned}
&= (e_2 - e_3)(e_4 - e_1)\left[f_\Omega(e_2, e_3) - (e_2 - e_3)^2 \sum_{\lambda,\nu=1}^{g} \wp_{\lambda\nu} e_2^{\lambda-1} e_3^{\nu-1} \right] \\
&\quad \times \left[f_\Omega(e_4, e_1) - (e_4 - e_1)^2 \sum_{\lambda,\nu=1}^{g} \wp_{\lambda\nu} e_4^{\lambda-1} e_1^{\nu-1} \right] \\
&+ (e_3 - e_1)(e_4 - e_2)\left[f_\Omega(e_3, e_1) - (e_3 - e_1)^2 \sum_{\lambda,\nu=1}^{g} \wp_{\lambda\nu} e_3^{\lambda-1} e_1^{\nu-1} \right] \\
&\quad \times \left[f_\Omega(e_4, e_2) - (e_4 - e_2)^2 \sum_{\lambda,\nu=1}^{g} \wp_{\lambda\nu} e_4^{\lambda-1} e_2^{\nu-1} \right] \\
&+ (e_1 - e_2)(e_4 - e_3)\left[f_\Omega(e_1, e_2) - (e_1 - e_2)^2 \sum_{\lambda,\nu=1}^{g} \wp_{\lambda\nu} e_1^{\lambda-1} e_2^{\nu-1} \right] \\
&\quad \times \left[f_\Omega(e_4, e_3) - (e_4 - e_3)^2 \sum_{\lambda,\nu=1}^{g} \wp_{\lambda\nu} e_4^{\mu-1} e_3^{\nu-1} \right].
\end{aligned}
$$

As we show the derivation of the fundamental relation in Theorem 4.3 in the following subsections, the strategy is to interpret differential relations in the Jacobian in terms of the curve coordinate x_i and y_i. In other words, we deal with relations in $\mathscr{Q}(R_X \times S^g R_X)\left[\frac{\widetilde{\partial}}{\partial u_{1,x}}, \dots, \frac{\widetilde{\partial}}{\partial u_{g,x}} \right]$.

After he obtained the fundamental relation, Baker recomputed the results for the curves given by $y^2 = x^{2g+2} + \cdots$ (4.6) to obtain a variant of the fundamental relations for the curves. In Section II in [Ba1903], Baker considered the genus three case to extract part of the KP hierarchy (including the KdV hierarchy) explicitly. By comparing the coefficients of each $e_1^a e_2^b e_3^c e_4^d$, he discovered the differential equations in Theorems 4.25 and 4.29. In the comparison, Baker used the symmetric functions, Pfaffian and bilinear operators [Ba1898b]. We also show its idea in Sect. 4.1.3.

4.1.2 Fundamental Relation in Differential Ring on $S^g X$

This subsection aims to prove Theorem 4.3 from Proposition 4.2. Since Theorem 4.3 is naturally obtained by Proposition 4.6, we prove the latter. Further, we use a relation equivalent to that in Proposition 4.2 as in Lemma 4.5. To describe them, we prepare the quantities, though some of them are defined in Lemmas 3.79 and 3.83.

Definition 4.4 We define the following:

1. $\Delta_{e_i} := \sum_{r=1}^{g} \frac{y_r}{(e_i - x_r)F'(x_r)}$,

2. $\Delta_{e_ie_j} := -\dfrac{\Delta_{e_i} - \Delta_{e_j}}{e_i - e_j} = \sum_{r=1}^{g} \dfrac{y_r}{(e_i - x_r)(e_j - x_r)F'(x_r)}$,
3. $\widetilde{\Delta}_{e_1e_2} := \left(\sum_{i=1, j\neq i}^{2} \dfrac{(-1)^i y_i'}{(e_i - e_j)F(x_i')} \right)^2$,
4. $\varphi_{e_i} := \dfrac{y_{e_i}^2}{F(e_i)} = \dfrac{f(e_i)}{F(e_i)}$,
5. $\varpi_{e_1,e_2,e_3} := (e_2 - e_3)(e_3 - e_1)(e_1 - e_2)$, and
6. $\Omega_{e_ie_j} := (e_i - e_j)^2 \Delta_{e_ie_j}^2 - \varphi_{e_i} - \varphi_{e_j} = \left(\sum_{r=1}^{g} \dfrac{(e_i - e_j)y_r}{(e_i - x_r)(e_j - x_r)F'(x_r)} \right)^2 - \varphi_{e_i}$
$-\varphi_{e_j}$.

Using these definitions, we rewrite Proposition 4.2:

Lemma 4.5

$$\sum_{i=1}^{g}\sum_{j=1}^{g} \wp_{ij}(u) {e_1}^{i-1} {e_2}^{j-1} = \frac{f_\Omega(e_1, e_2)}{(e_1 - e_2)^2} + \frac{F(e_1)F(e_2)\Omega_{e_1,e_2}}{(e_1 - e_2)^2}$$
$$= \frac{f_\Omega(e_1, e_2)}{(e_1 - e_2)^2} + F(e_1)F(e_2)\Delta_{e_1,e_2}^2 - \frac{F(e_1)F(e_2)(\varphi_{e_1} + \varphi_{e_2})}{(e_1 - e_2)^2}.$$

Though tedious computations are required, Theorem 4.3 is obtained from the following proposition.

Proposition 4.6

$$\frac{1}{F(e_1)F(e_2)F(e_3)F(e_4)} \sum_{\lambda,\mu,\nu,\rho=1}^{g} \wp_{\lambda\mu\nu\rho} e_1^{\lambda-1} e_2^{\mu-1} e_3^{\nu-1} e_4^{\rho-1}$$

is equal to

$$\begin{aligned}
&-(\Delta_{e_1,e_4} + \Delta_{e_2,e_4} + \Delta_{e_3,e_4}) \left[\Delta_{e_2,e_3}\Delta_{e_3,e_1}\Delta_{e_1,e_2} + \frac{1}{\varpi_{e_1,e_2,e_3}} \sum_{e_1,e_2,e_3} \varphi_{e_1}(e_2 - e_3)\Delta_{e_2,e_3} \right] \\
&+ \frac{1}{\varpi_{e_1,e_2,e_3}} \sum_{e_1,e_2,e_3} \varphi_{e_1}(e_2 - e_3)\Delta_{e_2,e_3}\Delta_{e_1,e_4} \\
&+ \left[\Delta_{e_1,e_2}\Delta_{e_1,e_3} + \frac{1}{\varpi_{e_1,e_2,e_3}} \varphi_{e_1}(e_2 - e_3) \right] \\
&\quad \times \left[\frac{(e_2 - e_4)\Delta_{e_2,e_4}^2 - (e_3 - e_4)\Delta_{e_3,e_4}^2}{e_2 - e_3} - \frac{1}{\varpi_{e_2,e_3,e_4}} \sum_{e_2,e_3,e_4} (e_2 - e_3)\varphi_{e_4} \right] \\
&+ \left[\Delta_{e_2,e_3}\Delta_{e_2,e_1} + \frac{1}{\varpi_{e_1,e_2,e_3}} \varphi_{e_2}(e_3 - e_1) \right]
\end{aligned}$$

$$\times \left[\frac{(e_3 - e_4)\Delta^2_{e_3,e_4} - (e_1 - e_4)\Delta^2_{e_1,e_4}}{e_3 - e_1} - \frac{1}{\varpi_{e_3,e_1,e_4}} \sum_{e_3,e_1,e_4} (e_3 - e_1)\varphi_{e_4} \right]$$

$$+ \left[\Delta_{e_3,e_1}\Delta_{e_3,e_2} + \frac{1}{\varpi_{e_1,e_2,e_3}}\varphi_{e_3}(e_1 - e_2) \right]$$

$$\times \left[\frac{(e_1 - e_4)\Delta^2_{e_1,e_4} - (e_2 - e_4)\Delta^2_{e_2,e_4}}{e_1 - e_2} - \frac{1}{\varpi_{e_2,e_3,e_4}} \sum_{e_1,e_2,e_4} (e_1 - e_2)\varphi_{e_4} \right].$$

Let us prove this proposition; the proof is finished by Lemma 4.18. For the proof, we further prepare several tools and investigate their properties.

Definition 4.7 1. A polynomial associated with $F(x)$ is introduced by

$$\begin{aligned} \pi_i(x) &:= \frac{F(x)}{x - x_i} \\ &= \chi_{i,g-1}x^{g-1} + \chi_{i,g-2}x^{g-2} + \cdots + \chi_{i,1}x + \chi_{i,0}, \end{aligned} \tag{4.3}$$

where $\chi_{i,g-1} = 1$, $\chi_{i,g-2} = -(x^1 + \cdots + x_g) + x_i$, and so on.

2. We define $g \times g$-matrices,

$$\mathscr{X}_g := \begin{pmatrix} \chi_{1,0} & \chi_{1,1} & \cdots & \chi_{1,g-1} \\ \chi_{2,0} & \chi_{2,1} & \cdots & \chi_{2,g-1} \\ \vdots & \vdots & \ddots & \vdots \\ \chi_{g,0} & \chi_{g,1} & \cdots & \chi_{g,g-1} \end{pmatrix}, \quad \mathscr{Y}_g := \begin{pmatrix} y_1 & & & \\ & y_2 & & \\ & & \ddots & \\ & & & y_g \end{pmatrix} = \frac{1}{2}\mathscr{H}_g,$$

$$\mathscr{F}'_g := \begin{pmatrix} F'(x_1) & & & \\ & F'(x_2) & & \\ & & \ddots & \\ & & & F'(x_g) \end{pmatrix}, \quad \mathscr{V}_g := \begin{pmatrix} 1 & 1 & \cdots & 1 \\ x_1 & x_2 & \cdots & x_g \\ x_1^2 & x_2^2 & \cdots & x_g^2 \\ \cdot & \cdot & & \cdot \\ x_1^{g-1} & x_2^{g-1} & \cdots & x_g^{g-1} \end{pmatrix},$$

where $F'(x) := dF(x)/dx$.

3.

$$\mathscr{E}_g := \begin{pmatrix} 1 & & & & \\ \varepsilon_{g-1} & 1 & & 0 & \\ \varepsilon_{g-2} & \varepsilon_{g-1} & 1 & & \\ \vdots & \vdots & \vdots & \ddots & \\ \varepsilon_1 & \varepsilon_2 & \cdots & \varepsilon_{g-1} & 1 \end{pmatrix}, \quad \mathscr{K}_g := \begin{pmatrix} x_1^{g-1} & x_1^{g-2} & \cdots & 1 \\ x_2^{g-1} & x_2^{g-2} & \cdots & 1 \\ \vdots & \vdots & \ddots & \vdots \\ x_g^{g-1} & x_g^{g-2} & \cdots & 1 \end{pmatrix},$$

where $F(x) = x^g + \varepsilon_{g-1}x^{g-1} + \cdots \varepsilon_1 x + \varepsilon_0$, $\varepsilon_i = (-1)^{g-i}\wp_{g,i+1}$.

Following Lemma 2.158, we interpret the relations in $\mathscr{D}_J$ in terms of $\mathscr{D}_X$. Using the symmetric functions and the differential of the first kind ν^{I}_i, we construct the differential relations in $\mathscr{D}_X$ following Baker [Ba1903].

Lemma 4.8 *1. The inverse matrix of* $\mathscr{V}_g$ *is given by* $\mathscr{V}_g^{-1} = \mathscr{F}_g'^{-1}\mathscr{X}_g$, *i.e.,*

$$\sum_{k=1}^{g} \frac{1}{F'(x_i)} \chi_{i,k-1} x_j^{k-1} = \delta_{i,j}, \qquad \sum_{k=1}^{g} \frac{x_k^{i-1}}{F'(x_k)} \chi_{k,j-1} = \delta_{i,j}.$$

2. Let $\partial_{u_i} := \partial/\partial u_i$, $\partial_{x_i} := \partial/\partial x_i$ *and* $\partial_{u_i}^{(r)} := \partial/\partial u_i^{(r)}$,

$$\begin{pmatrix} du_1 \\ du_2 \\ \vdots \\ du_g \end{pmatrix} = \frac{1}{2}\mathscr{Y}_g^{-1}\, {}^t\mathscr{V}_g \begin{pmatrix} dx_1 \\ dx_2 \\ \vdots \\ dx_g \end{pmatrix}, \qquad \begin{pmatrix} \partial_{u_1} \\ \partial_{u_2} \\ \vdots \\ \partial_{u_g} \end{pmatrix} = 2\mathscr{Y}_g \mathscr{F}_g'^{-1} \mathscr{X}_g \begin{pmatrix} \partial_{x_1} \\ \partial_{x_2} \\ \vdots \\ \partial_{x_g} \end{pmatrix}.$$

3.

$$dx_i = \sum_{r=1}^{g} \left[\frac{2y_i}{F'(x_i)} \chi_{i,r-1} \right] du_r, \quad du_i = \sum_{j=1}^{g} \frac{x_j^{i-1}}{2y_j} dx_j,$$

$$\frac{\partial x_i}{\partial u_r} = \frac{2y_i}{F'(x_i)} \chi_{i,r-1}, \qquad \frac{\partial y_i}{\partial u_r} = \frac{f'(x_i)}{F'(x_i)} \chi_{i,r-1}. \tag{4.4}$$

Proof *1* can be seen using the properties of the Vandermonde matrix, i.e., $\left.\dfrac{F(x)}{x - x_i} F'(x_i)\right|_{x=x_j} = \delta_{ij}$. *2* is directly obtained from Lemma 2.160. From the relation $\left(\dfrac{F(x)}{x - x_i}\right) \cdot (x - x_i) = F(x)$, we have

$$\chi_{i,j} = \varepsilon_{j+1} + x_i \chi_{i,j+1}.$$

Then we obtain the relations (4.4) in *3*. ■

Remark 4.9 To prove Proposition 4.6, we use the flow in Remark 4.1 and operate δ_{e_i} of (4.2) on the relation in Lemma 4.5 for $i = 3, 4$.

Proposition 4.10 *1.* $\delta_{e_2}\Delta_{e_1}$ *is equal to*

$$(e_2 - e_1)\Delta_{e_2,e_1}^2 + \frac{1}{e_2 - e_1}\left(\varphi_{e_2} - \varphi_{e_1} + e_2 - e_1\right).$$

2. $\delta_{e_3}\Delta_{e_1 e_2}$ *is equal to*

$$\left[\frac{(e_1 - e_3)\Delta_{e_1,e_3}^2 - (e_2 - e_3)\Delta_{e_2,e_3}^2}{e_1 - e_2} - \frac{1}{\varpi_{e_1,e_2,e_3}} \sum_{i:cyclic} (e_i - e_{i+1})\varphi_{e_{i+2}} \right].$$

Proof These formulae are asserted by Lemmas 4.14 and 4.16. ■

To prove Lemmas 4.14 and 4.16, we show the following lemmas.

Lemma 4.11 *1.* $\delta_{e_2}F(e_1) = -2F(e_1)\Delta_{e_1e_2}$,
2. $\delta_{e_2}\varphi_{e_1} = 4\varphi_{e_1}\Delta_{e_1e_2}/F(e_2)$,
3. $\delta_{e_3}\Omega_{e_1e_2}$ *is equal to* $2(e_1-e_2)\Delta_{e_1e_2}\Bigg[(e_1-e_3)\Delta^2_{e_1e_3} - (e_2-e_3)\Delta^2_{e_2e_3}$

$$+(e_1-e_2)\sum_{1,2,3}\frac{\varphi_{e_1}}{(e_1-e_2)(e_1-e_3)}\Bigg] - 4\varphi_{e_1}\Delta_{e_1e_3} - 4\varphi_{e_2}\Delta_{e_2e_3},$$

4.
$$-\frac{\delta_{e_3}(\Omega_{e_1e_2}F(e_1)F(e_2))}{2(e_1-e_2)^2F(e_1)F(e_2)}$$
$$=\Delta_{e_1e_2}\Delta_{e_2e_3}\Delta_{e_3e_1}+\frac{(e_2-e_3)\varphi_{e_1}\Delta_{e_2e_3}+(e_3-e_1)\varphi_{e_2}\Delta_{e_3e_1}+(e_1-e_2)\varphi_{e_3}\Delta_{e_1e_2}}{\varpi_{e_1e_2e_3}}.$$

Proof *1.* $\delta_{e_3}F(e_1) = \dfrac{1}{F(e_3)}\displaystyle\sum_{j=1}^{g}e_3^{j-1}\sum_{i=1}^{g}\left[\prod_{k\neq i}(e_1-x_k)\right]\left[-\frac{2y_i}{F'(x_i)}\chi_{g-j}(x_i)\right]$. The definition of χ_{g-j} (4.3) shows the relation *1.*

2. $\delta_{e_3}\varphi_{e_1} = \displaystyle\sum_{j=1}^{g}\sum_{r=1}^{g}e_3^{r-1}\frac{\partial x_r}{\partial u_j}\frac{\partial}{\partial x_r}\left[\frac{f(e_1)}{[F(e_1)^2]}\right]$. The direct computation shows it.

3. Due the definition of $\Omega_{e_1e_2}$, $\delta_{e_3}\Omega_{e_1,e_2} = 2(e_1-e_2)^2\Delta_{e_1,e_2}\delta_{e_3}\Delta_{e_1,e_2} - \delta_{e_3}\varphi_{e_1} - \delta_{e_3}\varphi_{e_2}$. We use Proposition 4.10 and then we obtain it.

4. $\dfrac{1}{2F(e_1)F(e_2)}\delta_{e_3}[\Omega_{e_1e_2}F(e_1)F(e_2)] = (e_1-e_2)\Delta_{e_1e_2}\Bigg[(e_1-e_3)\Delta^2_{e_1e_3}$

$$-(e_2-e_3)\Delta^2_{e_2e_3} - (e_1-e_2)\sum_{1,2,3}\frac{\varphi_{e_1}}{(e_1-e_2)(e_1-e_3)}\Bigg]$$
$$-2\varphi_{e_1}\Delta_{e_1e_3} - 2\varphi_{e_2}\Delta_{e_2e_3} - (\Delta_{e_1e_3}+\Delta_{e_2e_3})[(e_1-e_2)^2\Delta^2_{e_1e_2} - \varphi_{e_1} - \varphi_{e_2}].$$

Here $(e_2-e_3)\Delta_{e_2e_3} + (e_3-e_1)\Delta_{e_3e_1} + (e_1-e_2)\Delta_{e_1e_2} = 0$. From the definition, $(e_1-e_2)\Delta_{e_1e_2} = -\Delta_{e_1} + \Delta_{e_2}$. ∎

We note the fact $\left.\dfrac{\partial}{\partial x}F'(x)\right|_{x=x_k} \neq \dfrac{\partial}{\partial x_k}F'(x_k)$. Then the following is obvious:

Lemma 4.12

$$\frac{1}{F'(x_k)}\frac{\partial}{\partial x_k}F'(x_k) = \frac{1}{2F'(x_k)}\left[\frac{\partial^2}{\partial x^2}F(x)\right]_{x=x_k} = \sum_{i=1,\neq k}^{g}\frac{1}{x_k-x_i} = \sum_{i=1}^{g}{}^{(k)}\frac{1}{x_k-x_i},$$

where $\displaystyle\sum_{i=1}^{g}{}^{(k)} := \sum_{i=1,i\neq k}$.

By using Lemma 4.12, the direct computation leads to the following lemma:

Lemma 4.13 *1.* $\delta_{e_1}\dfrac{y_k}{F'(x_k)}$ *is equal to*

$$\frac{f'(x_k)F'(x_k)-f(x_k)F''(x_k)}{(e_1-x_k)(F'(x_k))^3}+\frac{2y_k}{F'(x_k)}\sum_{i=1}^{g\ (k)}\frac{y_i}{(e_1-x_i)(x_k-x_i)F'(x_i)}.$$

2. $\delta_{e_2}\dfrac{y_k}{(e_1-x_k)F'(x_k)}$ *is equal to*

$$\frac{2f'(x_k)F'(e_2)}{(e_1-x_k)^2(e_2-x_k)(F'(x_k))^2}+\frac{1}{e_1-x_k}\delta_{e_1}\frac{y_k}{F'(x_k)}.$$

In order to prove Proposition 4.10, we evaluate the right-hand side of those formulae:

Lemma 4.14 *1.* $\delta_{e_2}\Delta_{e_1}$ *is equal to*

$$\left((e_2-e_1)\Delta^2_{e_1e_2}+\sum_{k=1}^{g}\frac{1}{F'(x_k)}\frac{\partial}{\partial x}\left[\frac{f(x)}{(e_2-x)(e_1-x)F'(x)}\right]\Big|_{x=x_k}\right).$$

2. $\delta_{e_3}\Delta_{e_1e_2}$ *is equal to*

$$=\frac{(e_1-e_3)\Delta^2_{e_1e_3}-(e_2-e_3)\Delta^2_{e_2e_3}}{e_1-e_2}+\sum_{k=1}^{g}\frac{1}{F'(x_k)}\frac{\partial}{\partial x_k}\left[\frac{f(x)}{(e_1-x)(e_2-x)(e_3-x)F'(x)}\right]\Big|_{x=x_k}.$$

Proof Due to Lemma 4.13, $\delta_{e_2}\Delta_{e_1}=\displaystyle\sum_{k=1}^{g}\delta_{e_2}\frac{y_k}{(e_1-x_k)F'(x_k)}$ is equal to

$$\sum_{k=1}^{g}\left\{\frac{f'(x_k)F'(x_k)-f(x_k)F''(x_k)}{(e_2-x_k)(e_1-x_k)(F'(x_k))^3}+\sum_{i=1}^{g^{(k)}}\frac{2f(x_k)}{(e_2-x_k)^2(e_1-x_i)[F'(x_k)]^2}\right\}$$

$$+2\sum_{k=1}^{g}\frac{y_k}{(e_1-x_k)F'(x_k)}\sum_{i=1}^{g^{(k)}}\frac{y_i}{(e_2-x_i)(x_k-x_i)F'(x_i)}.$$

The second term is expressed by

$$\sum_{k=1}^{g}\sum_{i=1}^{g^{(k)}}\frac{(e_2-e_1)y_iy_k}{(e_2-x_k)(e_1-x_k)F'(x_k)(e_2-x_i)(e_1-x_i)F'(x_i)},$$

because we have

$$\left(\frac{1}{(e_1-x_i)(e_2-x_k)}-\frac{1}{(e_2-x_i)(e_1-x_k)}\right)\frac{1}{x_i-x_k}=\frac{(e_2-e_1)}{(e_1-x_i)(e_2-x_k)(e_2-x_i)(e_1-x_k)}.$$

The term becomes

$$(e_2-e_1)\left(\left[\sum_{k=1}^{g}\frac{y_k}{(e_1-x_k)(e_2-x_k)F'(x_k)}\right]^2-\sum_{k=1}^{g}\frac{f(x_k)}{[(e_2-x_k)(e_1-x_k)F'(x_k)]^2}\right).$$

By using $\Delta_{e_1e_2}$ in Definition 4.4, we have the expression of $\delta_{e_2}\Delta_{e_1}$ as

$$(e_2-e_1)\Delta_{e_1e_2}^2+\sum_{k=1}^{g}\left[\frac{f'(x_k)F'(x_k)-f(x_k)F''(x_k)}{(e_2-x_k)(e_1-x_k)(F'(x_k))^3}+\frac{(e_1+e_2-2x_k)f(x_k)}{[(e_2-x_k)(e_1-x_k)F'(x_k)]^2}\right].$$

The second term turns out to be the second term we desire by straightforward computation in *1*. Similarly, we obtain the formula in *2* by noting

$$\frac{1}{(e_1-e_2)}\left(\frac{1}{(e_1-x_k)}-\frac{1}{(e_2-x_k)}\right)=\frac{1}{(e_1-x_k)(e_2-x_k)}.$$ ■

The following is obvious.

Lemma 4.15 *Let $h(x)$ be a meromorphic function over X whose divisor does not contain x_k. Then at $(x,y)=(x_k,y_k)$,*

$$\mathrm{res}_{x=x_k}\frac{h(x)}{F(x)^2}=\frac{1}{F'(x_k)}\frac{\partial}{\partial x}\left[\frac{h(x)}{F'(x)}\right]_{x=x_k}.$$

By using the expression, we have the following lemma:

Lemma 4.16 *1.* $\displaystyle\sum_{k=1}^{g}\frac{1}{F'(x_k)}\frac{\partial}{\partial x}\left[\frac{f(x)}{(e_2-x)(e_1-x)F'(x)}\right]\Bigg|_{x=x_k}=\frac{\wp_{e_2}-\wp_{e_1}}{e_2-e_1}+1.$

2. $\displaystyle\sum_{k=1}^{g}\frac{1}{F'(x_k)}\frac{\partial}{\partial x}\left[\frac{f(x)}{(e_1-x)(e_2-x)(e_3-x)F'(x)}\right]\Bigg|_{x=x_k}=\sum_{i=1,2,3}\frac{\wp_{e_i}}{(e_i-e_{i+1})(e_i-e_{i+2})}.$

Proof For the cases $h(x):=\dfrac{f(x)}{(e_1-x)(e_2-x)}$ and $h(x):=\dfrac{f(x)}{(e_1-x)(e_2-x)(e_3-x)}$, we consider the integral over ∂D_X in Fig. 2.3,

$$\int_{\partial D_X}\frac{h(x)}{F(x)^2}dx,$$

whose residues show the results. ■

We thus prove Proposition 4.10. Here we note that Lemma 4.16 is also proved by direct computations though the computation process is quite complicated.

We operate δ_{e_3} to both sides in Lemma 4.5 to obtain the following lemma:

Lemma 4.17

$$-\frac{1}{2F(e_1)F(e_2)F(e_3)}\sum_{\lambda,\mu,\nu=1}^{g}\wp_{\lambda\mu\nu}e_1^{\lambda-1}e_2^{\mu-1}e_3^{\nu-1}$$
$$=\Delta_{e_2,e_3}\Delta_{e_3,e_1}\Delta_{e_1,e_2}+\frac{1}{\varpi_{e_1,e_2,e_3}}\sum\varphi_{e_1}(e_2-e_3)\Delta_{e_2,e_3},$$

where $\sum$ means the summation over the symmetric action on $\{1,2,3\}$.

Proof We operate δ_{e_3} on both sides in Lemma 4.5 noting that the first term in the right-hand side behaves as a constant for the operator, i.e., $\delta_{e_3}f_\Omega(e_1,e_2)=0$. Lemma 4.11 4 directly shows the right-hand side in the formula. ■

We operate δ_{e_4} on both sides of the formula in Lemma 4.17 again to obtain the following lemma:

Lemma 4.18

$$\delta_{e_4}\left[F(e_1)F(e_2)F(e_3)\left(\Delta_{e_2,e_3}\Delta_{e_3,e_1}\Delta_{e_1,e_2}+\frac{1}{\varpi_{e_1,e_2,e_3}}\sum_{1,2,3}\varphi_{e_1}(e_2-e_3)\Delta_{e_2,e_3}\right)\right]$$

is equal to

$$\begin{aligned}
&F(e_1)F(e_2)F(e_3)\times\\
&\left\{-2(\Delta_{e_1e_4}+\Delta_{e_2e_4}+\Delta_{e_3e_4})\left[\Delta_{e_2,e_3}\Delta_{e_3,e_1}\Delta_{e_1,e_2}+\frac{1}{\varpi_{e_1,e_2,e_3}}\sum_{1,2,3}\varphi_{e_1}(e_2-e_3)\Delta_{e_2,2_3}\right]\right.\\
&+\frac{4}{\varpi_{e_1,e_2,e_3}}\sum_{1,2,3}\varphi_{e_1}(e_2-e_3)\Delta_{e_1e_4}\Delta_{e_2,2_3}\\
&+\left[\Delta_{e_1,e_2}\Delta_{e_3,e_1}+\frac{1}{\varpi_{e_1,e_2,e_3}}\varphi_{e_1}(e_2-e_3)\right]\\
&\times\left[\frac{(e_2-e_4)\Delta_{e_2,e_4}^2-(e_3-e_4)\Delta_{e_3,e_4}^2}{e_2-e_3}-\frac{1}{\varpi_{e_2,e_3,e_4}}\sum_{2,3,4}\varphi_{e_4}(e_2-e_3)\right]\\
&+\left[\Delta_{e_2,e_3}\Delta_{e_1,e_2}+\frac{1}{\varpi_{e_1,e_2,e_3}}\varphi_{e_2}(e_3-e_1)\right]\\
&\times\left[\frac{(e_3-e_4)\Delta_{e_3,e_4}^2-(e_1-e_4)\Delta_{e_1,e_4}^2}{e_3-e_1}-\frac{1}{\varpi_{e_3,e_1,e_4}}\sum_{4,3,1}\varphi_{e_4}(e_3-e_1)\right]\\
&+\left[\Delta_{e_2,e_3}\Delta_{e_3,e_1}+\frac{1}{\varpi_{e_1,e_2,e_3}}\varphi_{e_3}(e_1-e_2)\right]\\
&\left.\times\left[\frac{(e_1-e_4)\Delta_{e_1,e_4}^2-(e_2-e_4)\Delta_{e_2,e_4}^2}{e_1-e_2}-\frac{1}{\varpi_{e_1,e_3,e_4}}\sum_{4,1,2}\varphi_{e_4}(e_1-e_2)\right]\right\}.
\end{aligned}$$

Proof Direct computations show it. ■

We note the relations

$$(e_2 - e_3)\Delta_{e_2e_3} + (e_3 - e_1)\Delta_{e_3e_1} + (e_1 - e_2)\Delta_{e_1e_2} = 0,$$

and $\Delta_{e_1e_2} = \Delta_{e_2e_1}$, and then Lemma 4.18 means Proposition 4.6.

Further calculations lead us to the following lemma.

Lemma 4.19

$$-\frac{1}{2F(e_1)F(e_2)F(e_3)F(e_4)} \sum_{\lambda,\mu,\nu,\rho=1}^{g} \wp_{\lambda\mu\nu\rho} e_1^{\lambda-1} e_2^{\mu-1} e_3^{\nu-1} e_3^{\rho-1}$$
$$= \Delta_{e_1e_2}\Delta_{e_1e_3}\Delta_{e_4e_2}\Delta_{e_4e_3} + \Delta_{e_2e_3}\Delta_{e_2e_1}\Delta_{e_4e_3}\Delta_{e_4e_1} + \Delta_{e_3e_1}\Delta_{e_3e_2}\Delta_{e_4e_1}\Delta_{e_4e_2}$$
$$- \sum_{1,2,3,4} \varphi_{e_h} \left[\frac{\Delta_{e_ie_j}\Delta_{e_ie_k}}{((e_h - e_j)(e_h - e_k)} + \frac{\Delta_{e_je_k}\Delta_{e_je_i}}{((e_h - e_k)(e_h - e_i)} + \frac{\Delta_{e_ke_i}\Delta_{e_ke_j}}{((e_h - e_i)(e_h - e_j)} \right]$$
$$+ \sum_{1,2,3,4} \frac{\varphi_{e_h}\varphi_{e_i} + \varphi_{e_j}\varphi_{e_k}}{(e_i - e_j)(e_i - e_k)(e_h - e_j)(e_h - e_k)}.$$

Proof We evaluate the right-hand side in Lemma 4.18 by checking the coefficients of each degree of φ, and then we have the relation in the lemma. ■

Furthermore, we consider the multiplication of $\varpi_{e_1,e_2,e_3,e_4} := \prod_{i=1,i<j}^{4} (e_i - e_j)$ on the relation in Lemma 4.19. We let $\lambda_{e_ie_j} := (e_i - e_j)^2 \Delta_{e_i,e_j}^2$.

Then we have the lemma:

Lemma 4.20

$$\frac{\varpi_{e_1,e_2,e_3,e_4}}{2F(e_1)F(e_2)F(e_3)F(e_4)} \sum_{\lambda,\mu,\nu,\rho=1}^{g} \wp_{\lambda\mu\nu\rho} e_1^{\lambda-1} e_2^{\mu-1} e_3^{\nu-1} e_3^{\rho-1}$$
$$= (e_2 - e_3)(e_4 - e_1) \left[\lambda_{e_2e_3} - (\varphi_{e_2} + \varphi_{e_3})\right] \left[\lambda_{e_4e_1} - (\varphi_{e_4} + \varphi_{e_1})\right]$$
$$+ (e_3 - e_1)(e_4 - e_2) \left[\lambda_{e_3e_1} - (\varphi_{e_3} + \varphi_{e_1})\right] \left[\lambda_{e_4e_2} - (\varphi_{e_4} + \varphi_{e_2})\right]$$
$$+ (e_1 - e_2)(e_4 - e_3) \left[\lambda_{e_1e_2} - (\varphi_{e_1} + \varphi_{e_2})\right] \left[\lambda_{e_4e_3} - (\varphi_{e_4} + \varphi_{e_3})\right].$$

Proof We note the identity

$$\sum_{1,2,3,4} (e_2 - e_3)(e_4 - e_1)\Delta_{e_2e_3}\Delta_{e_4e_1} = 0.$$

The left-hand side in Lemma 4.19 with ϖ_{e_1,e_2,e_3,e_4} is given by the symmetric summation of

$$(e_2 - e_3)(e_4 - e_1)\Big[\lambda_{e_2e_3}\lambda_{e_4e_1} - \lambda_{e_2e_3}(\varphi_{e_1} + \varphi_{e_4}) - \lambda_{e_4e_1}(\varphi_{e_2} + \varphi_{e_3}) \\ - (\varphi_{e_2}\varphi_{e_3} + \varphi_{e_4}\varphi_{e_1})\Big]$$

$$= (e_2 - e_3)(e_4 - e_1)\left\{\left[\lambda_{e_2e_3} - (\varphi_{e_2} + \varphi_{e_3})\right]\left[\lambda_{e_4e_1} - (\varphi_{e_4} + \varphi_{e_1})\right] - \sum_{i,j}\varphi_{e_i}\varphi_{e_j}\right\}.$$

By using the identity

$$(e_2 - e_3)(e_4 - e_1) + (e_3 - e_1)(e_4 - e_2) + (e_1 - e_2)(e_4 - e_3) = 0,$$

we have the formula in the lemma. ■

Lemma 4.20 under the relation in Lemma 4.5 with $\Omega_{e_ie_j} = \lambda_{e_ie_j} - (\varphi_{e_i} + \varphi_{e_j})$ shows Theorem 4.3.

4.1.3 Bilinear Operator and KdV Hierarchy 1898

This subsection comments on Baker's study in terms of the symmetric functions and the bilinear operators. It is crucial that the relation in Theorem 4.3 is the identity for every e_i. The components are symmetric, and thus the identity can be decomposed to the sum of the Schur polynomials. Baker defined the Schur polynomials using the Jacobi–Trudi determinant, though Schur introduced the Schur polynomials in his thesis 1901 [Sch1901]. We can derive the relations as the identities in $\mathscr{Q}(S^g R_X)$ or $\mathscr{M}(J_X)$ by using mathematical software, e.g., Maple [Map2019] but we mention Baker's approach.

Let $e_{(a,b,c,d)}$ be a "minimal monic symmetric polynomial containing $e_1^a e_2^b e_3^c e_4^d$ in $e_1, \ldots, e_4$". Let the Schur polynomial be denoted by $\mathbf{s}_\Lambda = \mathbf{s}_{(\Lambda_1,\Lambda_2,\ldots,\Lambda_\ell)}$.

We have the following expansions:

Lemma 4.21 [Ba1903, p. 152]

$$\mathbf{s}_\emptyset = e_{(0,0,0,0)} = 1, \quad \mathbf{s}_{(1)} = e_{(1,0,0,0)}, \quad \mathbf{s}_{(1,1)} = e_{(1,1,0,0)}, \\ \mathbf{s}_{(1,1,1)} = e_{(1,1,1,0)}, \quad \mathbf{s}_{(1,1,1,1)} = e_{(1,1,1,1)},$$

$$\mathbf{s}_{(2)} = e_{(2,0,0,0)} + e_{(1,1,0,0)}, \quad \mathbf{s}_{(2,1)} = e_{(2,1,0,0)} + 2e_{(1,1,1,1)}, \\ \mathbf{s}_{(2,1,1)} = e_{(2,1,1,0)} + 3e_{(1,1,1,1)}, \quad \mathbf{s}_{(2,2)} = e_{(2,2,0,0)} + e_{(2,1,1,0)} + 2e_{(1,1,1,1)}, \\ \mathbf{s}_{(2,2,1)} = e_{(2,2,1,0)} + 2e_{(2,1,1,1)}, \quad \mathbf{s}_{(2,2,2)} = e_{(2,2,2,0)} + e_{(2,2,1,1)},$$

$$\mathbf{s}_{(2,1,1,1)} = e_{(2,1,1,1)}, \quad \mathbf{s}_{(2,2,1,1)} = e_{(2,2,1,1)}, \\ \mathbf{s}_{(2,2,2,1)} = e_{(2,2,2,1)}, \quad \mathbf{s}_{(2,2,2,2)} = e_{(2,2,2,2)}.$$

It can be extended for the cases containing the Young diagram $\emptyset, \ldots, (g, g, g, g)$.

We assume the genus $g = 3$. The formula in Theorem 4.3 is expanded by these symmetric polynomials. Since they are linear independent, we have fifteen identities among $\wp_{\mu\nu\rho\lambda}$ and $\wp_{\mu\rho}$'s.

However, there is a simple question; since $\wp_{\mu\nu}$ belongs to $\mathbf{H}^0(J_X, \mathscr{O}_X(2\Theta_X))$ whereas $\wp_{\mu\nu\rho\lambda}$ is an element of $\mathbf{H}^0(J_X, \mathscr{O}_X(4\Theta_X))$, the relations might connect elements in both function spaces and bring trivial relations.

Baker introduced the bilinear operator in the papers [Ba1898b, p. 519] [Ba1903, p. 151], which is known nowadays as "Hirota bilinear operator" based on Hirota's method [Hir1971, Hir1976]. We refer to it as the Baker–Hirota bilinear operator:

Definition 4.22 [Ba1898b, p. 519][Ba1903, p. 151] The differential operator $\mathfrak{D}_i := \dfrac{\partial}{\partial u_i} - \dfrac{\partial}{\partial u_i'}$ and let

$$\mathfrak{D}_{\lambda_1} \cdots \mathfrak{D}_{\lambda_\ell} \sigma(u)\sigma(u')$$

mean that u' is to be replaced by u after differentiation.

Proposition 4.23 [Ba1903, p. 151]

$$\wp_{\mu\nu} = -\frac{\mathfrak{D}_\mu \mathfrak{D}_\nu \sigma(u)\sigma(u')}{2\sigma(u)^2},$$

$$Q_{\mu\nu\rho\lambda} = -\frac{\mathfrak{D}_\mu \mathfrak{D}_\nu \mathfrak{D}_\lambda \mathfrak{D}_\rho \sigma(u)\sigma(u')}{2\sigma(u)^2},$$

where

$$Q_{\mu\nu\rho\lambda} := \wp_{\mu\nu\rho\lambda} - 2(\wp_{\mu\nu}\wp_{\rho\lambda} + \wp_{\mu\rho}\wp_{\mu\lambda} + \wp_{\mu\lambda}\wp_{\nu\rho}),$$

and thus $\wp_{\mu\nu}$ and $Q_{\mu\nu\rho\lambda}$ belong to $\mathbf{H}^0(J_X, \mathscr{O}_X(2\Theta_X))$.

Proposition 4.24 $\Delta = \wp_{32}\wp_{21} - \wp_{31}\wp_{22} + \wp_{31}^2 - \wp_{33}\wp_{11}$ *which is equal to*

$$\frac{\sigma_{11}\sigma_{33} - \sigma_{12}\sigma_{23} - \sigma_{13}^2 + \sigma_{13}\sigma_{22}}{\sigma^2}$$
$$+ \frac{(2\sigma_1\sigma_3 - \sigma_2^2)\sigma_{13} + (\sigma_2\sigma_{23} - \sigma_3\sigma_{22})\sigma_1 + \sigma_2\sigma_3\sigma_{12} - \sigma_{11}\sigma_3^2 - \sigma_1^2\sigma_{33}}{\sigma^3}$$

belongs to $\mathbf{H}^0(J_X, \mathscr{O}_X(2\Theta_X))$.

Proof It is obvious that it belongs to $\mathbf{H}^0(J_X, \mathscr{O}_X(3\Theta_X))$. However, due to the particular symmetry of the hyperelliptic curves, the numerator belongs to $\mathbf{H}^0(\mathbb{C}^g, \kappa_J^{-1}\Theta_X)$ due to Theorem 4.25. Indeed there appears Δ in Theorem 4.25. ■

Hence the relations in Theorem 4.3 are among the elements in $\mathbf{H}^0(J_X, \mathscr{O}_X(2\Theta_X))$, whose dimension is much less than the number of elements $\{\wp_{\mu\nu}, Q_{\mu\nu\rho\lambda} \mid \mu, \ldots, \lambda = 1, \ldots g\}$.

In other words, the following KdV hierarchy showed the linear dependence among the elements in $\mathscr{M}_J$ with the Sato–Weierstrass weight (2.22).

Baker investigated the coefficients a_Λ of $\sum_{\Lambda \le (g,g,g,g)} a_\Lambda \mathbf{s}_\Lambda$ using the pfaffian structure. We note that $a_{\mu\nu\rho\lambda}$ contains $Q_{\mu\nu\rho\lambda}$. Here the Baker–Hirota bilinear operator plays the essential roles. Baker derived the following of genus two in 1898 [Ba1898b] and of genus three in 1903 [Ba1903]:

Theorem 4.25

$$
\begin{aligned}
&(a.1)\quad \wp_{3333} - 6\wp_{33}^2 = 2\lambda_5\lambda_7 + 4\lambda_6\wp_{33} + 4\lambda_7\wp_{32},\\
&(a.2)\quad \wp_{3332} - 6\wp_{33}\wp_{32} = 4\lambda_6\wp_{32} + 2\lambda_7(3\wp_{31} - \wp_{22}),\\
&(a.3)\quad \wp_{3331} - 6\wp_{31}\wp_{33} = 4\lambda_6\wp_{31} - 2\lambda_7\wp_{21},\\
&(a.4)\quad \wp_{3322} - 4\wp_{32}^2 - 2\wp_{33}\wp_{22} = 2\lambda_5\wp_{32} + 4\lambda_6\wp_{31} - 2\lambda_7\wp_{21},\\
&(a.5)\quad \wp_{3321} - 2\wp_{33}\wp_{21} - 4\wp_{32}\wp_{31} = 2\lambda_5\wp_{31},\\
&(a.6)\quad \wp_{3311} - 4\wp_{31}^2 - 2\wp_{33}\wp_{11} = 2\Delta,\\
&(a.7)\quad \wp_{3222} - 6\wp_{32}\wp_{22} = -4\lambda_2\lambda_7 - 2\lambda_3\wp_{33} + 4\lambda_4\wp_{32} + 4\lambda_5\wp_{31} - 6\lambda_7\wp_{11},\\
&(a.8)\quad \wp_{3221} - 4\wp_{32}\wp_{21} - 2\wp_{31}\wp_{22} = -2\lambda_1\lambda_7 + 4\lambda_4\wp_{31} - 2\Delta,\\
&(a.9)\quad \wp_{3211} - 4\wp_{31}\wp_{21} - 2\wp_{32}\wp_{11} = -4\lambda_0\lambda_7 + 2\lambda_3\wp_{31},\\
&(a.10)\quad \wp_{3111} - 6\wp_{31}\wp_{11} = 4\lambda_0\wp_{33} - 2\lambda_1\wp_{32} + 4\lambda_2\wp_{31},\\
&(a.11)\quad \wp_{2222} - 6\wp_{22}^2 = -8\lambda_2\lambda_6 + 2\lambda_3\lambda_5\\
&\qquad - 6\lambda_1\lambda_7 - 12\lambda_2\wp_{33} + 4\lambda_3\wp_{32} + 4\lambda_4\wp_{22} + 4\lambda_5\wp_{21} - 12\lambda_6\wp_{11} + 12\Delta,\\
&(a.12)\quad \wp_{2221} - 6\wp_{22}\wp_{21} = -4\lambda_1\lambda_6 - 8\lambda_0\lambda_7 - 6\lambda_1\wp_{33} + 4\lambda_3\wp_{31}\\
&\qquad\qquad + 4\lambda_4\wp_{21} - 2\lambda_5\wp_{11},\\
&(a.13)\quad \wp_{2211} - 4\wp_{21}^2 - 2\wp_{22}\wp_{11} =\\
&\qquad\qquad - 8\lambda_0\lambda_6 - 8\lambda_0\wp_{33} - 2\lambda_1\wp_{32} + 4\lambda_2\wp_{31} + 2\lambda_3\wp_{21},\\
&(a.14)\quad \wp_{2111} - 6\wp_{21}\wp_{11} = -4\lambda_0\lambda_5 - 8\lambda_0\wp_{32} + 2\lambda_1(3\wp_{31} - \wp_{22}) + 4\lambda_2\wp_{21},\\
&(a.15)\quad \wp_{1111} - 6\wp_{11}^2 = - 8\lambda_0\lambda_4 + 2\lambda_1\lambda_3 + 4\lambda_0(4\wp_{31} - 3\wp_{22}) + 4\lambda_1\wp_{21} + 4\lambda_2\wp_{11},
\end{aligned}
$$

where $\lambda_7 = 1$, *and*

$$\Delta = \wp_{32}\wp_{21} - \wp_{31}\wp_{22} + \wp_{31}^2 - \wp_{33}\wp_{11}.$$

Proof We derived these relations as the identities in $\mathscr{Q}(S^g R_X)$ or $\mathscr{M}(J_X)$ by using mathematical software, Maple [Map2019]. We evaluated the coefficients of $e_1^{2+i} e_2^{1+j} e_3^k e_4^{\ell-1}$ for the direct expansion of Theorem 4.3 up to the symmetric properties in the indices, e.g., $\wp_{3,2,1,1} = \wp_{3,1,2,1}$, which are relevant to $\wp_{ijk\ell}$. (a.6) is obtained by $\wp_{3311} - 6\wp_{31}^2 + 2\wp_{33}\wp_{11} - 2\wp_{32}\wp_{21} = 0$ directly and is written including Δ. Similarly we obtain (a.8).

If one employs the method on the direct evaluations of Theorem 4.3 by using mathematical software like Maple, it is an easy task to generalize Theorem 4.25 to $g = 4$ and $g = 5$, like a quite nice exercise for undergraduate students.

Theorem 4.26 $v(\mathrm{t}_1, \mathrm{t}_3, \mathrm{t}_5) := -2\wp_{33}\begin{pmatrix} 16\mathrm{t}_5 \\ 4\mathrm{t}_3 + 16\lambda_6\mathrm{t}_5/3 \\ \mathrm{t}_1 + 4\lambda_6\mathrm{t}_3 \end{pmatrix} - \frac{2}{3}\lambda_6$ *obeys first and second KdV equations:*

$$\partial_{\mathrm{t}_3} v + 6v\partial_{\mathrm{t}_1} v + \partial_{\mathrm{t}_1}^3 v = 0,$$

$$\partial_{\mathrm{t}_5} v + 30v^2\partial_{\mathrm{t}_1} v + 20\partial_{\mathrm{t}_1} v\partial_{\mathrm{t}_1}^2 v + 10v\partial_{\mathrm{t}_1}^3 v + \partial_{\mathrm{t}_1}^5 v = 0.$$

Proof (a.1) is given as

$$\wp_{3333} = 6(\wp_{33} + \lambda_6/3)^2 + 2\lambda_5 - \frac{2}{3}\lambda_6^2 + 4\lambda_6\wp_{33} + 4\wp_{32},$$

$$\wp_{33333} = 12(\wp_{33} + \lambda_6/3)\wp_{333} + 4\lambda_6\wp_{333} + 4\wp_{332}.$$

By letting $v = -2(\wp_{33} + \lambda_6/3)$, we have the first equation.

∂_{u_2} (a.1) and $2\partial_{u_3}$ (a.2) are equal to

$$3\wp_{33332} - 12\wp_{333}\wp_{32} - 24\wp_{33}\wp_{332} = +12\lambda_6\wp_{332} + 12\wp_{331},$$

which is also expressed as

$$(\partial_3^3 - 4\wp_{333} - 8(\wp_{33} + \lambda_6/3)\partial_3)\wp_{32} = (4\lambda_6/3)\wp_{332} + 4\wp_{331}.$$

Thus we have

$$\begin{aligned}
&(\partial_3^3 - 4\wp_{333} - 8(\wp_{33} + \lambda_6/3)\partial_3) \\
&\quad \times (\wp_{3333} - 6(\wp_{33} + \lambda_6/3)^2 - 2\lambda_5 + \lambda_6^2 - 4\lambda_6(\wp_{33} + \lambda_6/3)) \\
&\quad = (16\lambda_6/3)\wp_{332} + 16\wp_{331}.
\end{aligned}$$

Let $\frac{\partial}{\partial \mathrm{t}_5} := 4\lambda_6\frac{\partial}{\partial u_2} + 16\frac{\partial}{\partial u_1}$. We consider the linear relation among ∂_{u_i} and ∂_{t_i} in terms of the matrix $M := (\partial_{\mathrm{t}_{2i+1}} u_{3-j})$,

$$\begin{pmatrix} \partial_{\mathrm{t}_1} \\ \partial_{\mathrm{t}_3} \\ \partial_{\mathrm{t}_5} \end{pmatrix} = \begin{pmatrix} \partial_{\mathrm{t}_1} u_3 & \partial_{\mathrm{t}_1} u_2 & \partial_{\mathrm{t}_1} u_1 \\ \partial_{\mathrm{t}_3} u_3 & \partial_{\mathrm{t}_3} u_2 & \partial_{\mathrm{t}_3} u_1 \\ \partial_{\mathrm{t}_5} u_3 & \partial_{\mathrm{t}_5} u_2 & \partial_{\mathrm{t}_5} u_1 \end{pmatrix} \begin{pmatrix} \partial_{u_3} \\ \partial_{u_2} \\ \partial_{u_1} \end{pmatrix} = \begin{pmatrix} 1 & 0 & 0 \\ 4\lambda_6 & 4 & 0 \\ 0 & 16\lambda_6/3 & 16 \end{pmatrix} \begin{pmatrix} \partial_{u_3} \\ \partial_{u_2} \\ \partial_{u_1} \end{pmatrix}.$$

By considering ${}^t M$ and its inverse, we have the linear relations among du_i and $d\mathfrak{t}_i$,

$$\begin{pmatrix} du_3 \\ du_2 \\ du_1 \end{pmatrix} = \begin{pmatrix} 1 & 4\lambda_6 & 0 \\ 0 & 4 & 16\lambda_6/3 \\ 0 & 0 & 16 \end{pmatrix} \begin{pmatrix} d\mathfrak{t}_1 \\ d\mathfrak{t}_3 \\ d\mathfrak{t}_5 \end{pmatrix}, \quad \begin{pmatrix} d\mathfrak{t}_1 \\ d\mathfrak{t}_3 \\ d\mathfrak{t}_5 \end{pmatrix} = \begin{pmatrix} 1 & -\lambda_6 & \lambda_6^2/3 \\ 0 & 1/4 & -\lambda_6/12 \\ 0 & 0 & 1/16 \end{pmatrix} \begin{pmatrix} du_3 \\ du_2 \\ du_1 \end{pmatrix}.$$

■

Remark 4.27 These higher order KdV equations mean the linear dependence of these $\wp_{\mu\nu}$ and $Q_{\mu\nu\rho\lambda}$ as the $\mathbb{C}$ vector space $\mathbf{H}^0(J, \mathscr{O}_J(2\Theta))$, which Grant pointed out [Gra1990]. However, we also note that due to the multiplicity of the singularity at ∞ for hyperelliptic curves of higher genus, there are quantities that belong to $\mathbf{H}^0(J_X, \mathscr{O}_X(2\Theta_X))$ such as Proposition 4.24. Thus Grant's approaches for hyperelliptic curves of genus two mentioned in Sect. 1.9 encounter difficulties at higher genus cases, $g \geq 3$.

Further, these higher order KdV equations in terms of the sigma function were investigated by Buchstaber, Enolskii, and Leykin [BucEL1997a, BucEL1997b], [BucEL2012, BucEL2020]; they derived these equations more directly based on the relation in Corollary 3.68. The more precise computations are obtained by Eilbeck, England and Ônishi [EilEO2011].

Moreover, we also regard the first relation in Theorem 4.25 as

$$((\partial_g^2 - 6\wp_{gg})\partial_g - 4\lambda_{2g-1}\partial_g - 4\partial_{g-1})\zeta_g = -2\lambda_{2g-1}. \tag{4.5}$$

4.1.4 KP Hierarchy 1903

This subsection shows the hyperelliptic solutions of the KP hierarchy by Baker. Instead of the curve of $(2g+1)$-degree, we investigate hyperelliptic curves $\widehat{X}$ of $(2g+2)$-degree in this subsection following Baker [Ba1903].

$$\begin{aligned} y^2 &= \widehat{f}(x) \\ &= \widehat{\lambda}_0 + \widehat{\lambda}_1 x + \bar{\lambda}_2 x^2 + \cdots + \bar{\lambda}_{2g+2} x^{2g+2} \\ &= (x-\alpha_1)\cdots(x-\alpha_g)(x-\alpha_{g+1})\cdots(x-\alpha_{2g})(x-\alpha_{2g+1})(x-\alpha_{2g+2}), \end{aligned} \tag{4.6}$$

where $\widehat{\lambda}_{2g+2} = 1$ and $\widehat{\lambda}_j$'s and α_j's are complex values. Athorne investigated the form of the ℘ functions for the action of PSL$(2, \mathbb{C})$ mentioned in Sect. 2.8 [Ath2008, AthEE2003, AthEE2004].

Remark 4.28 [Ba1897, p. 195], [Ba1903, pp. 144–145]

1. The transformation between $y^2 = f(x)$ and $\zeta^2 = \widehat{f}(\xi)$ is given in Sect. 2.8.2 with the W-normalized differentials of the first kind,

$$\nu_1^{\rm I} = \frac{dx}{2y}, \quad \nu_2^{\rm I} = \frac{xdx}{2y}, \quad \cdots, \quad \nu_g^{\rm I} = \frac{x^{g-1}dx}{2y}.$$

2. The definition of σ and $\wp$ functions are the same as those in Definition 3.52, where we regard that y obeys the equation $y^2 = \widehat{f}(x)$ instead of $y^2 = f(x)$.
3. Baker rewrote the fundamental relation in Theorem 4.3 in terms of (4.6) and derived the following equations.

Theorem 4.29 [Ba1903, pp. 145–146] *The hyperelliptic $\wp$ functions of a curve $y^2 = \widehat{f}(x)$ $(g = 3)$ obey the relations*

$$\begin{aligned}
&(b.1) \quad \wp_{3333} - 6\wp_{33}^2 = 2\widehat{\lambda}_5\widehat{\lambda}_7 + 4\widehat{\lambda}_6\wp_{33} + 4\widehat{\lambda}_7\wp_{32} - 8\widehat{\lambda}_4\widehat{\lambda}_8 + 4\widehat{\lambda}_8(4\wp_{31} - 3\wp_{22}),\\
&(b.2) \quad \wp_{3332} - 6\wp_{33}\wp_{32} = 4\widehat{\lambda}_6\wp_{32} + 2\widehat{\lambda}_7(3\wp_{31} - \wp_{22}) - 4\widehat{\lambda}_3\widehat{\lambda}_8 + 8\widehat{\lambda}_8\wp_{21},\\
&(b.3) \quad \wp_{3331} - 6\wp_{31}\wp_{33} = 4\widehat{\lambda}_6\wp_{31} - 2\widehat{\lambda}_7\wp_{21} + 4\widehat{\lambda}_8\wp_{11},\\
&(b.4) \quad \wp_{3322} - 4\wp_{32}^2 - 2\wp_{33}\wp_{22} = 2\widehat{\lambda}_5\wp_{32} + 4\widehat{\lambda}_6\wp_{31} - 2\widehat{\lambda}_7\wp_{21} - 8\widehat{\lambda}_2\widehat{\lambda}_8 - 8\widehat{\lambda}_8\wp_{11},\\
&(b.5) \quad \wp_{3321} - 2\wp_{33}\wp_{21} - 4\wp_{32}\wp_{31} = 2\widehat{\lambda}_5\wp_{31} - 4\widehat{\lambda}_1\widehat{\lambda}_8,\\
&(b.6) \quad \wp_{3311} - 4\wp_{31}^2 - 2\wp_{33}\wp_{11} = 2\Delta - 8\widehat{\lambda}_8\widehat{\lambda}_0,\\
&(b.7) \quad \wp_{3222} - 6\wp_{32}\wp_{22} =\\
&\qquad - 4\widehat{\lambda}_2\widehat{\lambda}_7 - 2\widehat{\lambda}_3\wp_{33} + 4\widehat{\lambda}_4\wp_{32} + 4\widehat{\lambda}_5\wp_{31} - 6\widehat{\lambda}_7\wp_{11} - 8\widehat{\lambda}_1\widehat{\lambda}_8,\\
&(b.8) \quad \wp_{3221} - 4\wp_{32}\wp_{21} - 2\wp_{31}\wp_{22} = -2\widehat{\lambda}_1\widehat{\lambda}_7 + 4\widehat{\lambda}_4\wp_{31} - 2\Delta - 8\widehat{\lambda}_0\widehat{\lambda}_8,\\
&(b.11) \quad \wp_{2222} - 6\wp_{22}^2 = -8\widehat{\lambda}_2\widehat{\lambda}_6 + 2\widehat{\lambda}_3\widehat{\lambda}_5 - 6\widehat{\lambda}_1\widehat{\lambda}_7 - 12\widehat{\lambda}_2\wp_{33} + 4\widehat{\lambda}_3\wp_{32}\\
&\qquad + 4\widehat{\lambda}_4\wp_{22} + 4\widehat{\lambda}_5\wp_{21} - 12\widehat{\lambda}_6\wp_{11} + 12\Delta - 16\widehat{\lambda}_0\widehat{\lambda}_8,
\end{aligned} \tag{4.7}$$

(a.9),(a.10), (a.12)–(a,15) and Δ, which have the same form as those in Theorem 4.25 by replacing λ's with $\widehat{\lambda}$'s in (4.6).

Proof They are obtained as in the proof of Theorem 4.3. ■

Proposition 4.30 $v(\mathrm{t}_1, \mathrm{t}_2, \mathrm{t}_3) = -2\wp_{33}\begin{pmatrix} -16\mathrm{t}_3 \\ 12\mathrm{t}_2 - 4\widehat{\lambda}_7\mathrm{t}_3 \\ \mathrm{t}_1 \end{pmatrix} - \frac{2}{3}\widehat{\lambda}_6$ *obeys the KP equation:*

$$\partial_{\mathrm{t}_1}(\partial_{\mathrm{t}_3}v + 6v\partial_{\mathrm{t}_1}v + \partial_{\mathrm{t}_1}^3 v) + \partial_{\mathrm{t}_2}^2 v = 0.$$

Proof We differentiate (b.1) in u_3 twice.

$$\partial_{u_3}(\wp_{33333} - 12\wp_{33}\wp_{333} - 4\widehat{\lambda}_6\wp_{333} - 4\widehat{\lambda}_7\wp_{332} - 16\wp_{331}) = -12\wp_{3322},$$

■

Remark 4.31 1. Theorem 4.29 means that we obtain an explicit function form of the hyperelliptic solution of the KP equation.

2. We note the dispersion relation. Since the curve $y^2 = \widehat{f}(x)$ is not ramified at infinity point. There t_j behaves like $(1/x)^{(g-j)}$ up to a constant factor. By comparing the degree at ∞ of x and y, we have the relations

$$\deg_{\infty,0}(\mathrm{t}_2) = 2\deg_{\infty,0}(\mathrm{t}_3), \quad \deg_{\infty,0}(\mathrm{t}_1) = 3\deg_{\infty,0}(\mathrm{t}_3).$$

These are the dispersion relations of the KP equation.

3. The KP hierarchy associated with the sigma function is investigated by Buchstaber, Enolskii, and Leykin [BucEL2012, BucEL2020].

4.1.5 Weierstrass (3,4)-Curve

In this subsection, we show the similar results for a (3,4)-curve [EilEMOP2007]. For a convention, a different parametrization is used than that of the plane (3,4)-curve defined in (2.8) as

$$\begin{aligned} f_X(x, y) = y^3 &+ (\mu_1 x + \mu_4)y^2 + (\mu_2 x^2 + \mu_5 x + \mu_8)y \\ &- (x^4 + \mu_3 x^3 + \mu_6 x^2 + \mu_9 x + \mu_{12}), \end{aligned} \tag{4.8}$$

where the μ_j's are constants.

As we remark in Remark 4.27 and Sect. 1.9.1, Grant showed that these relations mean the linear dependence of these $\wp_{\mu\nu}$ and $Q_{\mu\nu\rho\lambda}$ as the $\mathbb{C}$ vector space $\mathbf{H}^0(J, \mathscr{O}_J(2\Theta))$ in Proposition 4.23 for curves of genus two [Gra1990]. Even for a (3, 4)-curve, Grant's approach can be applied. We care about the determinant $\widetilde{\Delta}$ of 2×2 matrices of $\wp_{ij}$ belonging to $\mathbf{H}^0(J, \mathscr{O}_J(3\Theta))$. After expanding σ at the lattice point in Γ_X, we compare the coefficients of these $\wp_{ij}$ and $\wp_{ijkl}$'s to obtain the following formulae.

Theorem 4.32 [EilEMOP2007] *The* 4*-index functions* $\wp_{ijk\ell}$ *associated with* (4.8) *satisfy the following relations:*

$$\wp_{3333} = 6\wp_{33}{}^2 + \mu_1{}^2\wp_{33} - 3\wp_{22} + 2\mu_1\wp_{23} - 4\mu_2\wp_{33} - 2\mu_4,$$

$$\wp_{2333} = 6\wp_{23}\wp_{33} + \mu_1{}^2\wp_{23} + 3\mu_3\wp_{33} - \mu_2\wp_{23} - \mu_5 - \mu_1\wp_{22},$$

$$\wp_{2233} = 4\wp_{23}{}^2 + 2\wp_{33}\wp_{22} + \mu_1\mu_3\wp_{33} - \mu_2\wp_{22} + 2\mu_6 + 3\mu_3\wp_{23} + \mu_1\mu_2\wp_{23} + 4\wp_{13},$$

$$\begin{aligned}\wp_{2223} = {}& 6\wp_{22}\wp_{23} + 4\mu_1\wp_{13} + \mu_1\mu_3\wp_{23} + \mu_2\mu_3\wp_{33} + 2\mu_3\mu_4 + \mu_2{}^2\wp_{23} + 4\mu_4\wp_{23} + 3\mu_3\wp_{22} \\ & + 2\mu_1\mu_6 + \mu_2\mu_5 - 2\mu_5\wp_{33},\end{aligned}$$

$$\begin{aligned}\wp_{2222} = {}& 6\wp_{22}{}^2 - 2\mu_2\mu_3\wp_{23} + \mu_1\mu_2\mu_5 + 2\mu_1\mu_3\mu_4 + 24\wp_{13}\wp_{33} + 4\mu_1{}^2\wp_{13} - 4\mu_2\wp_{13} - 4\wp_{1333} \\ & + 4\mu_5\wp_{23} + 2\mu_1{}^2\mu_6 - 2\mu_2\mu_6 + \mu_3\mu_5 - 3\mu_3{}^2\wp_{33} + 12\mu_6\wp_{33} + 4\mu_4\wp_{22} \\ & + \mu_2{}^2\wp_{22} + 4\mu_1\mu_3\wp_{22},\end{aligned}$$

$$
\begin{aligned}
\wp_{1233} &= 4\wp_{13}\wp_{23} + 2\wp_{33}\wp_{12} - 2\mu_1\wp_{33}\wp_{13} - \tfrac{1}{3}\mu_1{}^3\wp_{13} + \tfrac{1}{3}\mu_1\wp_{1333} + \tfrac{1}{3}\mu_1{}^2\wp_{12} + 3\mu_3\wp_{13} \\
&\quad + \tfrac{1}{3}\mu_1\mu_8 + \tfrac{4}{3}\mu_1\mu_2\wp_{13} - \mu_2\wp_{12} + \mu_9, \\
\wp_{1223} &= 4\wp_{23}\wp_{12} + 2\wp_{13}\wp_{22} - 2\mu_2\wp_{33}\wp_{13} - 2\mu_8\wp_{33} - \tfrac{2}{3}\mu_8\mu_2 + \tfrac{1}{3}\mu_2\wp_{1333} + 3\mu_3\wp_{12} + 4\mu_4\wp_{13} \\
&\quad + \tfrac{4}{3}\mu_2{}^2\wp_{13} - 2\wp_{11} - \tfrac{1}{3}\mu_1{}^2\mu_2\wp_{13} + \tfrac{1}{3}\mu_1\mu_2\wp_{12} + \mu_1\mu_3\wp_{13}, \\
\wp_{1222} &= 6\wp_{22}\wp_{12} + 6\mu_9\wp_{33} - \mu_3\wp_{1333} + 4\mu_5\wp_{13} + \mu_2{}^2\wp_{12} - \mu_2\mu_9 + 4\mu_4\wp_{12} - 2\mu_1\wp_{11} \\
&\quad + 6\mu_3\wp_{33}\wp_{13} - 3\mu_2\mu_3\wp_{13} + \mu_1{}^2\mu_3\wp_{13} + 3\mu_1\mu_3\wp_{12} - \mu_1\mu_2\mu_8, \\
\wp_{1133} &= 4\wp_{13}{}^2 + 2\wp_{33}\wp_{11} - \mu_9\wp_{23} + 2\mu_6\wp_{13} + \mu_8\wp_{22} - \mu_5\wp_{12} + \tfrac{2}{3}\mu_4\wp_{1333} + \tfrac{2}{3}\mu_4\mu_8 \\
&\quad + 2\mu_2\mu_8\wp_{33} - 4\mu_4\wp_{13}\wp_{33} + \tfrac{2}{3}\mu_2\mu_4\wp_{13} + \mu_1\mu_9\wp_{33} - \mu_1\mu_8\wp_{23} + \mu_1\mu_5\wp_{13} \\
&\quad - \tfrac{2}{3}\mu_1{}^2\mu_4\wp_{13} + \tfrac{2}{3}\mu_1\mu_4\wp_{12}, \\
\wp_{1123} &= 4\wp_{12}\wp_{13} + 2\wp_{23}\wp_{11} + 2\mu_3\mu_4\wp_{13} - \mu_3\mu_8\wp_{33} - 2\mu_5\wp_{13}\wp_{33} + \mu_2\mu_8\wp_{23} + \tfrac{4}{3}\mu_2\mu_5\wp_{13} \\
&\quad - \mu_9\wp_{22} + 2\mu_6\wp_{12} + \tfrac{1}{3}\mu_5\wp_{1333} + \tfrac{1}{3}\mu_5\mu_8 + \mu_1\mu_9\wp_{23} - \tfrac{1}{3}\mu_1{}^2\mu_5\wp_{13} + \tfrac{1}{3}\mu_1\mu_5\wp_{12}, \\
\wp_{1122} &= 4\wp_{12}{}^2 + 2\wp_{11}\wp_{22} + \tfrac{2}{3}\mu_1{}^2\mu_6\wp_{13} + \tfrac{4}{3}\mu_1\mu_6\wp_{12} + \mu_3\mu_9\wp_{33} + \mu_2\mu_9\wp_{23} + 8\mu_{12}\wp_{33} \\
&\quad + 2\mu_3\mu_4\wp_{12} - \tfrac{2}{3}\mu_6\wp_{1333} + 4\mu_8\wp_{13} - \tfrac{2}{3}\mu_6\mu_8 + 4\mu_6\wp_{33}\wp_{13} - \mu_3\mu_8\wp_{23} + \mu_3\mu_5\wp_{13} \\
&\quad - \tfrac{8}{3}\mu_2\mu_6\wp_{13} + \mu_2\mu_8\wp_{22} + \mu_2\mu_5\wp_{12}, \\
\wp_{1113} &= 6\wp_{13}\wp_{11} + 6\mu_2\mu_8\wp_{13} - 2\mu_2\mu_{12}\wp_{33} - \mu_1{}^2\mu_8\wp_{13} + 4\mu_1\mu_{12}\wp_{23} + \mu_1\mu_8\wp_{12} + \mu_5\mu_9\wp_{33} \\
&\quad + \mu_5{}^2\wp_{13} - 2\mu_4\mu_9\wp_{23} + \mu_1\mu_9\wp_{13} - 6\mu_8\wp_{33}\wp_{13} - 2\mu_6\mu_8\wp_{33} + \mu_8\wp_{1333} - 4\mu_4\mu_{12} \\
&\quad + 3\mu_9\wp_{12} - 6\mu_{12}\wp_{22} - \mu_5\mu_8\wp_{23} + 4\mu_4\mu_6\wp_{13}, \\
\wp_{1112} &= 6\wp_{12}\wp_{11} + 6\mu_3\mu_{12}\wp_{33} + 3\mu_3\mu_8\wp_{13} - 2\mu_6\mu_8\wp_{23} - \mu_1\mu_8{}^2 + 5\mu_2\mu_8\wp_{12} + 4\mu_2\mu_{12}\wp_{23} \\
&\quad - 2\mu_1\mu_{12}\wp_{22} + 4\mu_4\mu_6\wp_{12} - \mu_5\mu_8\wp_{22} + \mu_5{}^2\wp_{12} + 4\mu_5\mu_{12} - \mu_9\wp_{1333} - 4\mu_1\mu_{12}\mu_4 \\
&\quad + \mu_1{}^2\mu_9\wp_{13} + 3\mu_1\mu_9\wp_{12} - 2\mu_4\mu_9\wp_{22} + \mu_5\mu_9\wp_{23} - 4\mu_2\mu_9\wp_{13} + 6\mu_9\wp_{13}\wp_{33} - 3\mu_8\mu_9, \\
\wp_{1111} &= 6\wp_{11}{}^2 + 4\mu_4\mu_9\wp_{12} - 8\mu_4{}^2\mu_{12} - 2\mu_2{}^2\mu_4\mu_{12} - 3\mu_8{}^2\wp_{22} - 2\mu_4\mu_8{}^2 + \mu_5{}^2\wp_{11} - 3\mu_9{}^2\wp_{33} \\
&\quad - 4\mu_{12}\wp_{1333} + 24\mu_{12}\wp_{33}\wp_{13} + 12\mu_5\mu_{12}\wp_{23} + \mu_2\mu_4\mu_5\mu_9 - 6\mu_1\mu_3\mu_4\mu_{12} \\
&\quad + \mu_1\mu_2\mu_5\mu_{12} + 2\mu_6{}^2\mu_8 + 2\mu_2{}^2\mu_8{}^2 - \mu_5\mu_6\mu_9 - 2\mu_5\mu_9\wp_{13} + 4\mu_4\mu_6\wp_{11} + 4\mu_6\mu_8\wp_{13} \\
&\quad + 8\mu_2\mu_8\wp_{11} - 6\mu_2\mu_6\mu_{12} - 12\mu_2\mu_{12}\wp_{13} + 4\mu_1{}^2\mu_{12}\wp_{13} + 2\mu_1{}^2\mu_6\mu_{12} + 2\mu_8\mu_5\wp_{12} \\
&\quad - 6\mu_8\mu_9\wp_{23} - 12\mu_4\mu_{12}\wp_{22} + \mu_2\mu_5{}^2\mu_8 + 2\mu_1\mu_4\mu_6\mu_9 + \mu_1\mu_5\mu_6\mu_8 + 12\mu_6\mu_{12}\wp_{33} \\
&\quad + 4\mu_1\mu_9\wp_{11} + 2\mu_3\mu_4{}^2\mu_9 + 9\mu_3\mu_5\mu_{12} - 2\mu_1\mu_3\mu_8{}^2 - 6\mu_3\mu_8\mu_9 + 2\mu_1\mu_2\mu_8\mu_9 \\
&\quad + \mu_3\mu_4\mu_5\mu_8 + 2\mu_2\mu_4\mu_6\mu_8 + 2\mu_2\mu_9{}^2.
\end{aligned}
$$

Similar computations for $(3, 5)$-curves are obtained by Baldwin, Eilbeck, Gibbons, and Ônishi [BalEGO2008], and for $(4, 5)$ by England and Eilbeck [EnE2009]. However, as we mentioned in Remark 4.27, for higher genus cases, even of trigonal $(3, s)$-curves, Grant's approach may encounter difficulty.

4.1.6 *Soliton Solutions of Korteweg–de Vries Equation*

Though it is not directly related to Baker's investigation, in this subsection, we consider a degenerate curve that is connected with the so-called soliton solution as in Its and Matveev [It1986, ItM1976]. We show the relation between the sigma function and Hirota's tau function in [Hir1971, Hir1976] based on an unpublished note [Mat2001b].

We consider a degenerate hyperelliptic curve X of genus g, $(g > 0)$, given by an affine equation,

$$y^2 = f_{\mathrm{I}}(x)^2 x = x(x^g + \lambda_{g-1}x^{g-1} + \lambda_{g-2}x^{g-2} + \cdots \lambda_1 x + \lambda_0)^2, \tag{4.9}$$

i.e., $f_{\mathrm{II}}(x) = f_{\mathrm{I}}(x)x$ in (2.65) and (2.66), or $c_0 = 0$, $c_1 = a_1$, $c_2 = a_2, \ldots, c_g = a_g$.

For later convenience, we introduce parameters $(k_i)_{i=1,\cdots,g}$, $k_i = \sqrt{a_i}$. Here we note that since the affine equation has symmetries, i.e., translation, dilatation, inversion, in a projective space, it should be regarded as a representative element in an equivalent relations. In this subsection, we use the representation for simplicity.

For a degenerate curve, the differentials in Sects. 2.4.2.5 and 2.4.3.8 become singular as we show later, but the sigma function can be defined as a limit of nonsingular curves even though it requires regularization. This subsection will show that the sigma function is associated with Hirota's original tau function [Hir1971, Hir1976].

Definition 4.33 1. For g-th polynomial $f_{\mathrm{I}}(x)$ in (2.66), we define

$$\begin{aligned} f_{\mathrm{I},i}(x) &:= \frac{f_{\mathrm{I}}(x)}{x - a_i} \\ &= \chi_{i,g-1}x^{g-1} + \chi_{i,g-2}x^{g-2} + \cdots + \chi_{i,1} + \chi_{i,0}. \end{aligned}$$

2. We will introduce $g \times g$-matrices

$$\mathscr{W} := \begin{pmatrix} \chi_{1,0} & \chi_{1,1} & \cdots & \chi_{1,g-1} \\ \chi_{2,0} & \chi_{2,1} & \cdots & \chi_{2,g-1} \\ \vdots & \vdots & \ddots & \vdots \\ \chi_{g,0} & \chi_{g,1} & \cdots & \chi_{g,g-1} \end{pmatrix}, \quad \mathscr{M} := \begin{pmatrix} 1 & & & & \\ \lambda_{g-1} & 1 & & 0 & \\ \lambda_{g-2} & \lambda_{g-1} & 1 & & \\ \vdots & \vdots & \vdots & \ddots & \\ \lambda_1 & \lambda_2 & \cdots & \lambda_{g-1} & 1 \end{pmatrix},$$

$$\mathscr{P} = \begin{pmatrix} f_{\mathrm{I}}'(a_1) & & & \\ & f_{\mathrm{I}}'(a_2) & & \\ & & \ddots & \\ & & & f_{\mathrm{I}}'(a_g) \end{pmatrix}, \quad \mathscr{K}(l) := \begin{pmatrix} k_1^{2g+l-2} & k_1^{2g+l-4} & \cdots & k_1^l \\ k_2^{2g+l-2} & k_2^{2g+l-4} & \cdots & k_2^l \\ \vdots & \vdots & \ddots & \vdots \\ k_g^{2g+l-2} & k_g^{2g+l-4} & \cdots & k_g^l \end{pmatrix},$$

where $f_{\mathrm{I}}'(x) := df_{\mathrm{I}}(x)/dx$ and λ's are defined in (2-2).

Lemma 4.34 *1.* $\mathscr{W}_{ig} = {}^t(1, 1, \cdots, 1)$ *and*

$$\mathscr{W} = \mathscr{K}(0)\mathscr{M}. \tag{4.10}$$

2. *The W-normalized differentials analogue of "the first and the second kind" in Sects. 2.4.2.5 and 2.4.3.8 become*

$$\nu_i^{\mathrm{I}} = \frac{s^{2i-2}ds}{f_{\mathrm{I}}(s^2)}, \quad \widetilde{\nu}_i^{\mathrm{II}} = \frac{s^{2g+2i-2}ds}{f_{\mathrm{I}}(s^2)},$$

where $s^2 := x$.

3. *The twisted W-normalization differential analogues of (4.51) defined by*

$$dv_i := \sum_{j=1}^{g} \mathscr{W}_{ij}\nu_j^{\mathrm{I}}, \quad d\widetilde{v}_i := \sum_{j=1}^{g} \mathscr{W}_{ij}\widetilde{\nu}_j^{\mathrm{II}}, \tag{4.11}$$

are expressed by

$$\begin{aligned}(dv_1, dv_2, \cdots, dv_g) &= \left(\frac{ds}{(s^2-a_1)}, \frac{ds}{(s^2-a_2)}, \cdots, \frac{ds}{(s^2-a_g)}\right),\\ (d\widetilde{v}_1, d\widetilde{v}_2, \cdots, d\widetilde{v}_g) &= \left(\frac{s^{2g}ds}{(s^2-a_1)}, \frac{s^{2g}ds}{(s^2-a_2)}, \cdots, \frac{s^{2g}ds}{(s^2-a_g)}\right).\end{aligned} \tag{4.12}$$

4. *The inverse matrix of* $\mathscr{W}$ *is given by* $\mathscr{W}^{-1} = \mathscr{P}^{-1}V$, *where* V *is Vandermonde matrix,*

$$\mathscr{V}_a = \begin{pmatrix} 1 & 1 & \cdots & 1 \\ a_1 & a_2 & \cdots & a_g \\ a_1^2 & a_2^2 & \cdots & a_g^2 \\ \vdots & \vdots & & \vdots \\ a_1^{g-1} & a_2^{g-1} & \cdots & a_g^{g-1} \end{pmatrix}.$$

5. *The W-normalized differentials of the second kind dr's are*

$$\nu_j^{\mathrm{II}} = \Lambda \begin{pmatrix} \mathscr{W}^{-1} & 0 \\ 0 & \mathscr{W}^{-1} \end{pmatrix} \begin{pmatrix} dv \\ d\widetilde{v} \end{pmatrix}.$$

Proof We consider *1*. Noting the relation $f_{\mathrm{I},i}(x)(x-a_i) = f_{\mathrm{I}}(x)$, we have

$$\chi_{i,k} = \lambda_{k+1} - a_i\chi_{i,k-1},$$

and the initial condition $\chi_{i,g-1} = 1$. Then we have *1*. By noting the properties of the inverse matrix of the Vandermonde matrix, others are obtained by direct computations. ∎

Remark 4.35 1. For the degenerate curve in (4.9), the ν_i^{I}'s are no longer holomorphic over X due to the behavior of dv_i.

2. $\dfrac{1}{k_i}dv_i$ is a differential of the third kind, i.e., a differential with two singular points whose residues are $+1$ and -1 respectively.
3. dv's are composed only of the local data of a single zero of $f_{\rm I}(x)$ and do not depend upon the y explicitly.

Lemma 4.36 *For the contour α's and β's as illustrated in Fig. 2.4, we have the following results.*

1.

$$\int_{\alpha_j} dv_i = \pi\sqrt{-1}\frac{1}{k_i}\delta_{j,i}, \quad \int_{\alpha_j} d\widetilde{v}_i = \pi\sqrt{-1}k_i^{2g-1}\delta_{j,i}.$$

2.

$$\begin{aligned} \int_{\beta_j} dv_i &= \frac{1}{k_i}\log\left|\frac{(k_i-k_j)}{(k_i+k_j)}\right|, \\ \int_{\beta_j} d\widetilde{v}_i &= \sum_{r=0}^{g-1}\frac{k_i^{2r}k_j^{2g-2r-1}}{g-r-1/2} + k_i^{2g-1}\log\left|\frac{(k_i-k_j)}{(k_i+k_j)}\right|. \end{aligned} \tag{4.13}$$

Proof We consider *1*. The contour α contains only one singularity,

$$\int_{\alpha_j} dv_i = \oint_{s=k_i}\frac{ds}{(s-k_i)(s+k_i)} = 2\pi\sqrt{-1}\frac{1}{2k_i}.$$

Similarly, we have integration of $d\widetilde{v}$. We have *1*. On 2, the contour β can be restricted over the real line if a_i are real. Real-valued integration (a_j, ∞) gives (4.13) and it is also true even if a_i is not real valued. The factor 2 in (4.13) comes from the returned path. ∎

Using Lemma 4.36, we have ω', ω'', τ, and η' by direct computations following these definitions in Chap. 2 and (4.11), as in the next lemma.

Lemma 4.37 *1.*

$$\omega' = \mathscr{W}^{-1}\left(\frac{\pi\sqrt{-1}}{k_i}\delta_{ij}\right), \quad \omega'' = \mathscr{W}^{-1}\left(\frac{1}{k_i}\log\left|\frac{(k_i-k_j)}{(k_i+k_j)}\right|\right). \tag{4.14}$$

2. *All components of the diagonal part of $\tau = \omega'^{-1}\omega''$, which is denoted by $\tau_{\rm diag}$, are diverge $\tau_{ii} = \sqrt{-1}\infty$ and off diagonal part $\tau_{\rm off} := \tau - \tau_{\rm diag}$ is given by*

$$\tau_{ij} = \frac{\sqrt{-1}}{\pi}\log\left|\frac{(k_i+k_j)}{(k_i-k_j)}\right|, \quad i \neq j. \tag{4.15}$$

3. *The components of* η', $\eta' = (\eta_i(k_j))$, *are given by*

$$\eta_i(k) = \frac{\pi\sqrt{-1}k^{2i-3}}{f_{\rm I}'(k^2)}\left[\frac{d}{dx}\left(\frac{f(x)}{x^{2i}}\right)_+\right]_{x=k^2}, \tag{4.16}$$

where $()_+$ *is the polynomial part of rational function of* x.

4. *The matrix* $\gamma = \eta'\omega'^{-1}$ *in Lemma 2.146 is equal to*

$$\gamma = \left(\frac{k_j}{\pi\sqrt{-1}}\eta_i(k_j)\right)\mathscr{W},$$

and especially $\gamma_{gg} = 1$.

Here we note that the origin of the truncated polynomial in (4.16) is the Λ matrix in Sect. 2.4.3.8 1, i.e.,

$$\nu_i^{\rm II} = \frac{x^i}{2y}dx\frac{d}{dx}\left[\left(\frac{f(x)}{x^{2i}}\right)_+\right],$$

where for $h \in \mathbb{C}(x)$, $(h)_+$ means the polynomial part of h.

Further, we remark that due to the properties of $d\nu$'s, the components of matrices $\mathscr{W}\omega''$, $\mathscr{W}\omega'$ and τ in (4.14), (4.15) and (4.16) are also composed only of the local data of one or two zeros of $f_{\rm I}(x)$ and do not depend upon other global structures. In other words, for different curves expressed by the form (4.9), the components have the same form if the corresponding zeros are the same.

Lemma 4.38 *By introducing a new coordinate in the Jacobian* J_X,

$$\mathrm{t} := \mathscr{M}^{-1}u, \tag{4.17}$$

where $u = {}^t(u_1, u_2, \cdots, u_g)$ *and* $\mathrm{t} := {}^t(\mathrm{t}_g, \mathrm{t}_{g-1}, \cdots, \mathrm{t}_1)$, *i.e.,* $u_g = \mathrm{t}_1$, $u_{g-1} = \mathrm{t}_2 + \lambda_{g-1}\mathrm{t}_1, \cdots$, *we have the relations:*

1.
$$\pi\sqrt{-1}\omega'^{-1}u = \mathscr{K}(1)\mathrm{t}. \tag{4.18}$$

2.
$$\begin{pmatrix}\partial/\partial u_1\\ \partial/\partial u_2\\ \vdots\\ \partial/\partial u_g\end{pmatrix} = {}^t\mathscr{M}\begin{pmatrix}\partial/\partial \mathrm{t}_g\\ \partial/\partial \mathrm{t}_{g-1}\\ \vdots\\ \partial/\partial \mathrm{t}_1.\end{pmatrix}.$$

3.
$$\widetilde{\wp}_{\mu\nu} := \frac{\partial^2}{\partial u_i\partial u_j}\log\theta(u) = \sum_{kl}\mathscr{M}_{ki}\mathscr{M}_{lj}\frac{\partial^2}{\partial \mathrm{t}_k\partial \mathrm{t}_l}\log\theta(\mathrm{t}),$$

especially,

$$\wp_{gg} = -\frac{\partial^2}{\partial u_g \partial u_g} \log \theta(u) - 1 = -\frac{\partial^2}{\partial \mathrm{t}_1 \partial \mathrm{t}_1} \log \theta(\mathrm{t}) - 1.$$

Proof *1* is obvious from (4.10) and (4.14). We obtain *2* by comparing

$$d\mathrm{t}_i = \sum_{j=1}^{g} \left(\frac{\partial \mathrm{t}_i}{\partial u_j}\right) du_j, \quad \left(\frac{\partial}{\partial u_i}\right) = \sum_{j=1}^{g} \left(\frac{\partial \mathrm{t}_j}{\partial u_i}\right) \frac{\partial}{\partial \mathrm{t}_i} \tag{4.19}$$

with the definition (4.17). *2* shows *3*. ■

Here we note that from (4.15) $\theta(u)$ itself vanishes but $\widetilde{\theta}(u) := \mathrm{e}^{(\pi\sqrt{-1}\mathrm{tr}\tau_{\mathrm{diag}})/2}\theta(u)$ is reduced to a finite summation,

$$\widetilde{\theta}(u) = \sum_{n \in [-1,0]^g} \exp\left[2\pi\sqrt{-1}\left\{\frac{1}{2}\ {}^t(n+\delta'')\tau_{\mathrm{off}}(n+\delta'') + {}^t(n+\delta'')(\omega'^{-1}u+\delta')\right\}\right]. \tag{4.20}$$

Recalling Theorem 3.8, we introduce the vector

$$\varepsilon := {}^t(\varepsilon_1, \varepsilon_2, \ldots, \varepsilon_g),$$

where $\varepsilon_i \in \{-1, 1\}$. Equation (4.20) turns out to be

$$\widetilde{\theta}(u) = \sum_{\epsilon \in \{-1,1\}^g} \exp\left[2\pi\sqrt{-1}\left\{\frac{1}{8}\ {}^t\varepsilon\, \tau_{\mathrm{off}}\, \varepsilon + {}^t\varepsilon(\omega'^{-1}u+\delta')\right\}\right].$$

We summarize the above results.

Theorem 4.39 *The $\widetilde{\wp}$ function for the degenerate curve, $y^2 = f_1(x)^2 x$, is given by*

$$\widetilde{\wp}_{\mu\nu}(\mathrm{t}) = -\frac{\partial^2}{\partial u_\mu \partial u_\nu} \log \widetilde{\theta}(u),$$

where

$$\widetilde{\theta}(u(\mathrm{t})) = \sum_{\varepsilon \in \{-1,1\}^g} \left[\prod_{i<j} \left(\frac{k_i - k_j}{k_i + k_j}\right)^{\varepsilon_i \varepsilon_j/2}\right] \exp\left(\sum_i \varepsilon_i \xi_i(\mathrm{t})\right), \tag{4.21}$$

and ξ_i is a component of a vector $\xi := {}^t(\xi_g, \xi_{g-1}, \cdots, \xi_1)$ given by

$$\xi = \mathscr{K}(1)\mathrm{t} + \pi\sqrt{-1}\delta'. \tag{4.22}$$

Proof We note that $(\varepsilon_i\varepsilon_j + \varepsilon_j\varepsilon_i)/4 = \varepsilon_j\varepsilon_i/2$. Since we have the relation

$$\frac{\partial^2}{\partial u_i \partial u_j} \log \theta(u(\mathrm{t})) = \frac{\partial^2}{\partial u_i \partial u_j} \log \widetilde{\theta}(u(\mathrm{t})),$$

the above theorem is obtained. ∎

Here we note that the coefficient of the linear term of t in the exponential function in (4.21) is replaced by a hyperelliptic integral of the second kind whose integrand behaves $\sqrt{x}^j$ of t_j around the infinite point. On the other hand, ours is roughly the inverse of the integral of the first kind due to (4.18) and (4.22). Both are connected by the Legendre relation in Theorem 2.148,

$${}^t\eta'' = 2\pi\sqrt{-1}\omega'^{-1} + {}^t\eta'\omega''\omega'^{-1}. \tag{4.23}$$

Example 4.40 We demonstrate the case of genus $g = 2$.

$$\widetilde{\theta}(\mathrm{t}) = -\Bigg[\left(\frac{k_1 - k_2}{k_1 + k_2}\right)^{1/2} \mathrm{e}^{-\xi_1 - \xi_2} + \left(\frac{k_1 + k_2}{k_1 - k_2}\right)^{1/2} \mathrm{e}^{\xi_1 - \xi_2}$$
$$+ \left(\frac{k_1 + k_2}{k_1 - k_2}\right)^{1/2} \mathrm{e}^{-\xi_1 + \xi_2} \left(\frac{k_1 - k_2}{k_1 + k_2}\right)^{1/2} \mathrm{e}^{\xi_1 + \xi_2}\Bigg].$$

Let us introduce

$$\theta_\tau(\mathrm{t}) := -\left(\frac{k_1 - k_2}{k_1 + k_2}\right)^{1/2} \mathrm{e}^{\xi_1 + \xi_2} \widetilde{\theta}(\mathrm{t}),$$

and

$$\xi_i' := \xi_i + \frac{1}{2} \log\left(\frac{k_1 - k_2}{k_1 + k_2}\right). \tag{4.24}$$

Here the transformation (4.24) corresponds to the shift of the origin in the Jacobi variety J_X, which appears in Theorem 3.148. Then the new theta function θ_τ is expressed by

$$\theta_\tau(\mathrm{t}) = 1 + \mathrm{e}^{2\xi_1'} + \mathrm{e}^{2\xi_2'} + \left(\frac{k_1 - k_2}{k_1 + k_2}\right)^2 \mathrm{e}^{2\xi_1' + 2\xi_2'}.$$

This agrees with the tau function of two soliton solutions [Hir1971, Hir1976], which generates the same $\widetilde{\wp}_{\mu\nu}$ in Theorem 4.39. Since $\wp_{\mu\nu} = \widetilde{\wp}_{\mu\nu} - \gamma_{\mu\nu}$ obeys the KdV equation in Theorem 4.26, it is obvious that $\theta_\tau(\mathrm{t})$ generates the solutions of the KdV equations.

Remark 4.41 1. Example 4.40 can be easily extended to an arbitrary genus g case.
2. It is easy to construct the $\widetilde{\theta}$ function over a curve $Y\colon y^2 = f_1(x)^2(x - a_{g+1})^2 x$ if one uses the coordinate t and knows the data of $\widetilde{\theta}$ over $X\colon y^2 = f_1(x)^2 x$ because all matrices for t appearing in $\widetilde{\theta}_{\mathrm{t}}(\mathrm{t}) := \widetilde{\theta}(\mathscr{M}\mathrm{t})$ in (4.20) for X are sub-matrices

of corresponding ones of Y. The origin of this fact comes from the existence of dv in (4.12) and the degenerate type of curves $y^2 = f_1(x)^2 x$. As long as we restrict ourselves to considering the such simple degenerate curves, we have the natural inclusion of matrices for curves of any genera g and g'. Then it might be easy to construct a filter space and its inverse limit of the filter space for set of the degenerate curves. Then one can regard this operation for X and Y as a transformation in the filter space.

3. Recently, Nakayashiki investigated again the degenerate hyperelliptic curves and soliton solutions from the viewpoint of the universal Grassmannian manifold (UGM) [Nak2019] after his a similar investigation on trigonal curves [Nak2018]. On the other hand, in [BerL2019], Bernatska and Leykin also independently studied the degenerate sigma functions by means of the heat equations given by Buchstaber and Leykin [BucL2002, BucL2004, BucL2005, BucL2008]. Subsequently, Bernatska, Enolskii and Nakayashiki generalized to more general degenerated curves in the framework of UGM theory [BerLN2020]. Further, Nakayashiki studied a related problem [Nak2020].
   On the other hand, Kodama has investigated the moduli space of the KP-soliton solutions associated with real two-dimensional shallow waves in nature [Kod2017]. There it is shown that the pattern of the two-dimensional shallow waves is combinatorially parameterized in the moduli space. In other words, he classified the moduli space of the soliton solutions of the KP equations.
   Further, Kodama and Xie found the soliton solutions of the KP equations for the space curves [KodX2021], which is very interesting from the viewpoint of the application of the numerical semigroup to the integrable system.

4.2 Differential Relations Among al Functions

In this section, we consider the differential relations associated with the al functions of the hyperelliptic curves X given by (2.65) and (2.66), $f_X(x, y) = y^2 - f(x) = 0$. Precisely, we investigate $\mathscr{D}_{\widehat{J}}$ and $\mathscr{D}_{\widehat{X}}$ instead of $\mathscr{D}_J$ and $\mathscr{D}_X$ for the covering spaces $\widehat{X}$ and $\widehat{J}_X (:= \widehat{J}_X^{\mathrm{al},(1)})$, i.e., $\varpi_{\widehat{X}} : \widehat{X} \to X$ and $\varpi_{\widehat{J}} : \widehat{J}_X \to J_X$ as mentioned in (2.130) and Proposition 2.204, However, these al functions basically appear as al^2 and thus we go on to consider the hyperelliptic curve given by $f_X(x, y) = 0$, and study relations in $\mathscr{D}_J$ and $\mathscr{D}_X$.

Euler implicitly studied the static modified KdV equation or the static sine-Gordon equation 1744 to solve the elastica problem, whereas Gauss derived the static sine-Gordon equation and investigated it to discover his elliptic theta functions as mentioned in Sect. 1.1.1. Taking Gauss' investigation for the model to obtain the theta function, Weierstrass defined the al functions and his sigma function, Al, to solve the Jacobi inversion formula for the hyperelliptic curves. In the studies by Weierstrass [Wei1854], the solutions of the sine-Gordon equation in terms of the al functions were implicitly obtained.

Section 4.2.1 is on the modified KdV equation, Sect. 4.2.2 on the generalized Lame equations, Sect. 4.2.3 on the sine-Gordon equations, and Sect. 4.2.4 on the Neumann system.

4.2.1 Modified KdV Equation

We take up the modified KdV (MKdV) equation in this subsection. The MKdV equation is known as the partial nonlinear differential equation

$$\partial_t q + 6q^2 \partial_s q + \partial_s^3 q = 0, \tag{4.25}$$

where $\partial_t := \partial/\partial t$ and $\partial_s := \partial/\partial s$.

We recall that X is given by $f_X(x, y) = y^2 - f(x)$, where $f(x) = \prod_{r=0}^{2g+1}(x - b_r)$, and $F(x) = (x - x_1)\cdots(x - x_g)$ for $(x_i, y_i)_{i=1,\dots,g} \in S^g X$.

Theorem 4.42 *Let $(b_r, 0) \in X$ and $F(b_r) = \mathrm{al}_r(u)^2$.*

1. *By letting*

$$\mu^{(r)} := \frac{1}{2}\frac{\partial}{\partial u_g}\phi^{(r)}, \quad \phi^{(r)}(u) := \frac{1}{\sqrt{-1}}\log F(b_r), \tag{4.26}$$

   $\mu^{(r)}$ obeys the modified KdV equation.[2]

$$-4\left(\frac{\partial}{\partial u_{g-1}} - \frac{1}{2}(\lambda_{2g} + 3b_r)\frac{\partial}{\partial u_g}\right)\mu^{(r)} + 6\mu^{(r)2}\frac{\partial}{\partial u_g}\mu^{(r)} + \frac{\partial^3}{\partial u_g^3}\mu^{(r)} = 0. \tag{4.27}$$

2. *We have the Miura relation,*

$$-(2\wp_{gg} + \lambda_{2g} + b_r) = \mu^{(r)2} - \sqrt{-1}\frac{\partial}{\partial u_g}\mu^{(r)}. \tag{4.28}$$

Now let us prove the theorem. From Lemma 4.11 and (4.3), we express u's by the affine coordinate x_i's,

$$\frac{\partial}{\partial u_g} = \sum_{i=1}^{g}\frac{2y_i}{F'(x_i)}\frac{\partial}{\partial x_i}, \quad \frac{\partial}{\partial u_{g-1}} = \sum_{i=1}^{g}\frac{2y_i\chi_{i,g-2}}{F'(x_i)}\frac{\partial}{\partial x_i}.$$

Hence we have

$$\mu^{(r)} = \frac{1}{2\sqrt{-1}}\frac{\partial}{\partial u_g}\log F(b_r) = \frac{1}{\sqrt{-1}}\sum_{i=1}^{g}\frac{y_i}{F'(x_i)(x_i - b_r)},$$

[2] In the paper [Mat2002c], $\left(\frac{\partial}{\partial u_{g-1}} - (\lambda_{2g} + b_r)\frac{\partial}{\partial u_g}\right)$ should be read as the first term in (4.27).

$$\frac{\partial}{\partial u_{g-1}} \log F(b_r) = \sum_{i=1}^{g} \frac{2y_i \chi_{i,g-2}}{F'(x_i)(x_i - b_r)}. \tag{4.29}$$

Let us evaluate the right-hand side of (4.29):

$$\frac{1}{2}\frac{\partial^2}{\partial u_g^2} \log F(b_r) = \sum_{j=1,i=1}^{g} \frac{y_j}{F'(x_j)} \frac{\partial}{\partial x_j} \frac{2y_i}{F'(x_i)(x_i - b_r)}, \tag{4.30}$$

and express it as a lower derivative.

Remark 4.43 In this section, we show the MKdV equation in terms of the al$_r$ functions for the generalized elastica problem in Sect. 4.6. However, we may have more general solutions of the MKdV equation for $(\widetilde{x}, \widetilde{y}) \in X$, $\mu = \frac{1}{\sqrt{-1}}$ $\left[\sum_{i=1}^{g} \frac{y_i}{F'(x_i)(x_i - \widetilde{x})} + \frac{\widetilde{y}}{F(\widetilde{x})}\right]$ by noting Sect. 4.2.2.3. Its derivation is essentially the same as the following.

In order to evaluate (4.30), we first show the following lemma.

Lemma 4.44 *1.*

$$\sum_{k=1}^{g} \frac{1}{F'(x_k)} \left[\frac{\partial}{\partial x}\left(\frac{f(x)}{(x - b_r)F'(x)}\right)\right]_{x=x_k} = \lambda_{2g} + b_r + 2\wp_{gg}.$$

2.

$$\left[\sum_{k} \frac{y_k}{(x - x_k)F'(x_k)}\right]^2 = \sum_{k,l,k \neq l} \frac{2y_k y_l}{(x - x_k)(x_k - x_l)F'(x_k)F'(x_l)} + \sum_{k} \frac{y_k^2}{(x - x_k)^2 F'(x_k)^2}.$$

Proof In order to prove *1*, let us consider an integral over a contour in X. The following four steps provide *1*.

a. Let ∂D_X be a boundary of a polygon representation D_X of the curve X as illustrated in Fig. 2.3,

$$\oint_{\partial D_X} \frac{f(x)}{(x - b_r)F(x)^2} dx = 0. \tag{4.31}$$

b. The divisor of the integrand of (4.31) is

$$\left(\frac{f(x)}{(x - b_r)F(x)^2} dx\right) = 3 \sum_{i=1, b_i \neq b_r}^{2g+1} (b_i, 0) + (b_r, 0) - 2\sum_{i=1}^{g}(x_i, y_i) - 2\sum_{i=1}^{g}(x_i, -y_i) - 3\infty. \tag{4.32}$$

c. Noting that the local parameter t at ∞ is $x = 1/t^2$, we have

$$\mathrm{res}_{\infty} \frac{f(x)}{(x-b_r)F(x)^2}dx = -2(\lambda_{2g} + b_r + 2\wp_{gg}).$$

d. Noting that the local parameter t at $(x_k, \pm y_k)$ is $t = x - x_k$, we obtain

$$\mathrm{res}_{(x_k,\pm y_k)} \frac{f(x)}{(x-b_r)F(x)^2}dx = \frac{1}{F'(x_k)}\left[\frac{\partial}{\partial x}\left(\frac{f(x)}{(x-b_r)F'(x)}\right)\right]_{x=x_k}.$$

By arranging them, we obtain the first equation in the lemma. On the other hand, 2 can be proved by using a trick: for $i \neq j$,

$$\frac{1}{(x_j - x_i)(x_i - b_r)} + \frac{1}{(x_i - x_j)(x_j - b_r)} = \frac{1}{(x_j - b_r)(x_i - b_r)}.$$

Then we obtain 2. ■

Let us compute the differential of the right-hand side of (4.30), which is $\sqrt{-1}\times \frac{\partial}{\partial u_g}\mu^{(r)}$, and then we find that $\frac{1}{2}\frac{\partial^2}{\partial u_g^2}\log F(b_r)$ is equal to

$$2\sum_{i=1}^{g}\frac{y_i}{F'(x_i)}\frac{\partial}{\partial x_i}\frac{y_i}{F'(x_i)(x_i-b_r)} + 2\sum_{i=1,j=1,j\neq i}^{g}\frac{y_j}{F'(x_j)}\frac{\partial}{\partial x_j}\frac{y_i}{F'(x_i)(x_i-b_r)}. \tag{4.33}$$

The first term in the right-hand side in (4.33) becomes

$$\sum_{i=1}^{g}\frac{y_i}{F'(x_i)}\left(\frac{1}{y_iF'(x_i)(x_i-b_r)}\frac{df(x_i)}{dx_i} - \frac{y_i}{F'(x_i)^2(x_i-b_r)}\left[\frac{\partial^2}{\partial x^2}F(x)\right]_{x=x_i} - \frac{2y_i}{F'(x_i)(x_i-b_r)^2}\right).$$

It is modified by Lemma 4.12 to

$$\sum_{i=1,j=1,j\neq i}^{g}\frac{1}{F'(x_k)}\left[\frac{\partial}{\partial x}\left(\frac{f(x)}{(x-b_r)F'(x)}\right)\right]_{x=x_i} - \sum_{i=1}^{g}\frac{f(x_i)}{F'(x_i)^2}\left(\frac{1}{(x_i-b_r)^2}\right). \tag{4.34}$$

Lemma 4.44 1 shows that the first term in (4.34) is expressed well. On the other hand, the second term in the right-hand side of (4.33) is reduced to

$$-2\sum_{i=1,j=1,j\neq i}^{g}\frac{y_iy_j}{F'(x_j)F'(x_i)}\frac{1}{(x_j-x_i)(x_i-b_r)}. \tag{4.35}$$

From Lemma 4.44 2, the second term in (4.34) and (4.35) becomes

$$-\frac{1}{4}\left(\sum_{i=1}^{g}\frac{2y_i}{F'(x_i)}\frac{1}{(x_i-b_r)}\right)^2=\mu^{(r)2}.$$

Hence we show Theorem 4.42 2 as follows: from Lemma 4.44, ($\varepsilon_{g-1}=-\wp_{gg}$ in Definition 4.7), we obtain

$$\frac{\partial}{\partial u_g}\mu^{(r)}=\frac{1}{\sqrt{-1}}\left(-2\varepsilon_{g-1}+\lambda_{2g}+b_r+\mu^{(r)2}\right). \tag{4.36}$$

We consider Theorem 4.42 1. We apply the differential in u_g to both sides of (4.36) again,

$$\begin{aligned}\frac{\partial^2}{\partial u_g^2}\mu^{(r)}&=\frac{1}{\sqrt{-1}}\frac{\partial}{\partial u_g}\left(-2\varepsilon_{g-1}+\lambda_{2g}+b_r+\mu^{(r)2}\right)\\&=\frac{1}{\sqrt{-1}}\left(2\sum_{i=1}^{g}\frac{2y_i}{F'(x_i)}+2\mu^{(r)}\frac{\partial}{\partial u_g}\mu^{(r)}\right)\\&=\frac{2}{\sqrt{-1}}\left(\sum_{i=1}^{g}\frac{2y_i}{F'(x_i)}+\frac{1}{\sqrt{-1}}\mu^{(r)}\left[-2\varepsilon_{g-1}+\lambda_{2g}+b_r+\mu^{(r)2}\right]\right).\end{aligned}$$

In the second step, we use (4.36) again and

$$\frac{\partial}{\partial u_g}\varepsilon_{g-1}=-\sum_{i=1}^{g}\frac{2y_i}{F'(x_i)}.$$

Due to the fact $\chi_{i,g-2}=\varepsilon_{g-1}+x_i$, the first and second terms in the parentheses are

$$\begin{aligned}\sum_{i=1}^{g}\frac{2y_i}{F'(x_i)}-\frac{2}{\sqrt{-1}}\varepsilon_{g-1}\mu^{(r)}&=\sum_{i=1}^{g}\frac{2y_i}{F'(x_i)}\frac{x_i-b_r}{x_i-b_r}+\varepsilon_{g-1}\sum_{i=1}^{g}\frac{2y_i}{F'(x_i)(x_i-b_r)}\\&=\sum_{i=1}^{g}\frac{2y_i}{F'(x_i)}\frac{\varepsilon_{g-1}+x_i}{x_i-b_r}-b_r\sum_{i=1}^{g}\frac{2y_i}{F'(x_i)(x_i-b_r)}\\&=\frac{\partial}{\partial u_{g-1}}\log F(b_r)+\frac{2}{\sqrt{-1}}b_r\mu^{(r)}.\end{aligned} \tag{4.37}$$

Thus we obtain Theorem 4.42 1,

$$\frac{1}{4}\frac{\partial^3}{\partial u_g^3}\phi^{(r)}=\left(\frac{\partial}{\partial u_{g-1}}-\frac{1}{2}(\lambda_{2g}+3b_r)\frac{\partial}{\partial u_g}\right)\phi^{(r)}-\frac{1}{8}\left(\frac{\partial}{\partial u_g}\phi^{(r)}\right)^3. \tag{4.38}$$

We differentiate (4.38) in u_g again and then obtain (4.27). Hence we completely prove our Theorem 4.42.

Remark 4.45 We note that $\mu^{(r)} = \frac{1}{\sqrt{-1}}(\zeta_g(u+\omega_{B_r}) - \zeta_g(u) - \eta_{B_r,g})$, and thus (4.5) directly shows

$$\begin{aligned}((\partial_g^2 - 6\wp_{gg}(u+\omega_{B_r}))\partial_g - 4\lambda_{2g-1}\partial_g - 4\partial_{g-1})\frac{1}{\sqrt{-1}}\zeta_g(u+\omega_{B_r})\\ - ((\partial_g^2 - 6\wp_{gg}(u))\partial_g - 4\lambda_{2g-1}\partial_g - 4\partial_{g-1})\frac{1}{\sqrt{-1}}\zeta_g(u) = 0.\end{aligned} \tag{4.39}$$

With (4.28), i.e., (4.36), $-\wp_{gg}(u+\omega_{B_r}) = \wp_{gg}(u) + \lambda_{2g} + b_r + \mu^{(r)2}$, (4.39) is also reduced to the MKdV equation (4.27).

4.2.2 *Generalized Lamé Equation of $n = 1$ and Other Relations*

We show the Lamé equation and related formulae in this subsection.

The Lamé equation (of genus one) is given by

$$\frac{d^2}{du^2}\psi(u) - n(n+1)\wp(u)\psi(u) = \lambda\psi(u).$$

It is known that for general λ and n, $\psi(u)$ is not periodic but quasi-periodic for the periodicity of $\wp(u)$. Further, when $n = 1$, ψ for a certain λ is given by the al function, i.e., $\psi = \mathrm{al}_a(u) = \sqrt{\wp(u) - e_a}$ function for $y^2 = (x-e_1)(x-e_2)(x-e_3)$. The periodicity of al_a is double the periodicity of $\wp$. Due to the property, it gives the general solution $\psi(u; v) = \dfrac{\mathrm{e}^{-\zeta(v)u}\sigma(u-v)}{\sigma(u)\sigma(v)}$ for $n = 1$ as in (4.40).

The Lamé equation of $n = 1$ in the sigma function theory for a general hyperelliptic curve of genus g is described well in [BelE2002, BucEL1997a, BucEL1997b] as the Baker function due to [Ba1897, p. 421] or Baker–Akhiezer function [BelE2002]. There $\psi(u, v) = \dfrac{\mathrm{e}^{{}^t\zeta(v)u}\sigma(u-v)}{\sigma(u)\sigma(v)}$ is considered, e.g., in [BucEL1997b, §7.1], where ${}^t\zeta(u) = (\zeta_1(u), \ldots, \zeta_g(u))$. Due to Lemma 3.81, we employ an alternative definition (4.42) following (3.116) by Nakayashiki [Nak2010b, Lemma 3] and Buchstaber, Enolskii, and Leykin [BucEL2012, Chap. 12], which is originally obtained by Krichever and Its [Kr1997] and Dubrovin [Dub1975, Dub1981] as follows.

We concentrate on $n = 1$ and, first, al function solutions in this book. Only in this subsection, we use the conventions $h' := \dfrac{\partial}{\partial u_g}h$, $h'' := \dfrac{\partial^2}{\partial u_g^2}h, \ldots h^{(n\prime)} := \dfrac{\partial^n}{\partial u_g^n}h$, and $\dot{h} := \dfrac{\partial}{\partial u_{g-1}}h$.

4.2.2.1 The Lamé Equation and al_a Functions

From the viewpoint of the Lamé equation, we rewrite Theorem 4.42 2 as follows. Let $\phi^{[a]} := \log \mathrm{al}_a = \frac{1}{2}\log F(b_a) = \frac{\sqrt{-1}}{2}\phi^{(a)}$ for the hyperelliptic curves X of genus g.

Theorem 4.46 *For* $\mathrm{al}_a(u) = \gamma_a'' \dfrac{\exp(-{}^t u\varphi_a)\sigma(u+\omega_a)}{\sigma(u)\sigma_{\natural_1}(\omega_a)}$ *in (3.71), we have the following relations:*

1. $\dfrac{\partial}{\partial u_g}\mathrm{al}_a = \left(\dfrac{\partial}{\partial u_g}\phi^{[a]}\right)\mathrm{al}_a$, *i.e.,* $\dfrac{\partial}{\partial u_g}\mathrm{al}_a = \left(\phi^{[a]\prime}\right)\mathrm{al}_a$,
2. $\dfrac{\partial^2}{\partial u_g^2}\phi^{[a]} + \left(\dfrac{\partial}{\partial u_g}\phi^{[a]}\right)^2 = 2(x_1 + x_2 + \cdots + x_g) + \lambda_{2g} + b_a = 2\wp_{gg} + \lambda_{2g} + b_a$, *and*
3. $\dfrac{\partial^2}{\partial u_g^2}\mathrm{al}_a = \left(\dfrac{\partial^2}{\partial u_g^2}\phi^{[a]} + \left(\dfrac{\partial}{\partial u_g}\phi^{[a]}\right)^2\right)\mathrm{al}_a$, *which means a solution of the generalized Lamé equations of* $n = 1$, $\dfrac{\partial^2}{\partial u_g^2}\mathrm{al}_a - 2\wp_{gg}(u)\mathrm{al}_a = (\lambda_{2g} + b_a)\mathrm{al}_a$.

We rewrite (4.37) and the above Lamé equation as a proposition:

Proposition 4.47

$$\frac{1}{2}\frac{\partial}{\partial u_g}\wp_{gg} - (\wp_{gg} - b_a)\frac{\partial}{\partial u_g}\phi^{[a]} = \frac{\partial}{\partial u_{g-1}}\phi^{[a]}, \quad 2\wp_{gg} + \lambda_{2g} + b_a = \frac{\mathrm{al}_a''}{\mathrm{al}_a}.$$

Using these relations, we reobtain the MKdV equation, and have the MKdV hierarchy by inserting the second relation into Theorem 4.25.

4.2.2.2 The Lamé Equation and al_{ab} Functions

Noting Corollary 3.80 and Proposition 3.131, we consider the differential properties of $\mathrm{al}_{ab}(u) = \gamma_{ab}'' \dfrac{\exp(-{}^t u\varphi_{ab})\sigma(u+\omega_a+\omega_b)}{\sigma(u)\sigma_{\natural_2}(\omega_a+\omega_b)}$:

Lemma 4.48 *For certain* φ_{ab} *and constant factor* γ_{ab}'', *we have*

$$\mathrm{al}_{ab} = \frac{1}{b_b - b_a}\left(\mathrm{al}_a \frac{\partial}{\partial u_g}\mathrm{al}^{[b]} - \mathrm{al}_b \frac{\partial}{\partial u_g}\mathrm{al}^{[a]}\right).$$

Proof Recalling (4.29), we have

$$\frac{1}{b_b - b_a}\left[\mathrm{al}_a \frac{\partial}{\partial u_g}\mathrm{al}_b - \mathrm{al}_b \frac{\partial}{\partial u_g}\mathrm{al}_a\right] = -\sqrt{F(b_a)F(b_b)}\sum_{i=1}^{g}\frac{y_i}{F'(x_i)(b_a - x_i)(b_b - x_i)}.$$

∎

Theorem 4.49 *Let*

$$\phi^{\circ}_{[ab]} := \log \mathrm{al}_{ab}, \quad \kappa^{\circ}_{ab} := \frac{\partial}{\partial u_g}\phi^{\circ}_{[ab]}, \quad \Delta^{\circ}_{ab} := \frac{(b_b - b_a)(\phi^{[b]\prime} + \phi^{[a]\prime})}{\phi^{[b]\prime} - \phi^{[a]\prime}}.$$

1. $\mathrm{al}_{ab} = \dfrac{1}{b_b - b_a}\left(\dfrac{\partial}{\partial u_g}\phi^{[b]} - \dfrac{\partial}{\partial u_g}\phi^{[a]}\right)\mathrm{al}_a\mathrm{al}_b$, *i.e.*, $\mathrm{al}_{ab} = \dfrac{\phi^{[b]\prime} - \phi^{[a]\prime}}{b_b - b_a}\mathrm{al}_a\mathrm{al}_b$,
2. $\kappa^{\circ}_{ab} = \dfrac{\partial}{\partial u_g}\phi^{\circ}_{[ab]} = \dfrac{\mathrm{al}_a\mathrm{al}_b}{\mathrm{al}_{ab}} = \dfrac{b_b - b_a}{\frac{\partial}{\partial u_g}\phi^{[b]} - \frac{\partial}{\partial u_g}\phi^{[a]}} = \dfrac{b_b - b_a}{\phi^{[b]\prime} - \phi^{[a]\prime}}$,
3. $\dfrac{\partial}{\partial u_g}\mathrm{al}_{ab} = \mathrm{al}_a\mathrm{al}_b$, *i.e.*, $\dfrac{\partial}{\partial u_g}\mathrm{al}_{ab} = \left(\dfrac{\partial}{\partial u_g}\phi^{\circ}_{[ab]}\right)\mathrm{al}_{ab}$, *i.e.*, $\dfrac{\partial}{\partial u_g}\mathrm{al}_{ab} = \phi^{\circ\prime}_{[ab]}\mathrm{al}_{ab}$,
   i.e., $\dfrac{\partial}{\partial u_g}\mathrm{al}_{ab} = \kappa^{\circ}_{[ab]}\mathrm{al}_{ab}$,
4. $\dfrac{\partial^2}{\partial u_g^2}\mathrm{al}_{ab} = \left(\dfrac{\partial}{\partial u_g}\phi^{[b]} + \dfrac{\partial}{\partial u_g}\phi^{[a]}\right)\mathrm{al}_a\mathrm{al}_b$, *i.e*,
   $\dfrac{\partial^2}{\partial u_g^2}\mathrm{al}_{ab} = (b_b - b_a)\left(\dfrac{(\phi^{[b]\prime} + \phi^{[a]\prime})}{\phi^{[b]\prime} - \phi^{[a]\prime}}\right)\mathrm{al}_{ab}$, *or* $\dfrac{\partial^2}{\partial u_g^2}\mathrm{al}_{ab} = \Delta^{\circ}_{ab}\mathrm{al}_{ab}$.

Proof Combinations among the relations in Theorem 4.46 and Lemma 4.48 show these relations. ∎

These relations could be simpler if we introduce $\phi^{\pm}_{[ab]} = \phi_{[a]} \pm \phi_{[b]}$. Further, as a corollary, we also have the following relations:

Proposition 4.50 *1.* $\Delta^{\circ}_{ab} = \kappa^{\circ 2}_{[ab]} + \dfrac{\partial}{\partial u_g}\kappa^{\circ}_{[ab]}$,

2. $\dfrac{\partial^2}{\partial u_g^2}\kappa^{\circ}_{[ab]} = 2\kappa^{\circ}_{[ab]}(\kappa^{\circ 2}_{[ab]} - \Delta^{\circ}_{ab}) + \dfrac{\partial}{\partial u_g}\Delta^{\circ}_{[ab]}$,
3. $\kappa^{\circ 2}_{[ab]} - \Delta^{\circ}_{ab} = \kappa^{\circ 2}_{[ab]}\left(\dfrac{(b_2 - b_1) - (\phi_b^{\prime 2} - \phi_a^{\prime 2})}{b_2 - b_1}\right)$,
4. $\dfrac{\partial}{\partial u_g}\Delta^{\circ}_{[ab]} = \Delta^{\circ\prime}_{[ab]} = 2\kappa^{\circ}_{[ab]}\dfrac{\phi^{[b]\prime}\phi^{[a]\prime\prime} - \phi^{[a]\prime}\phi^{[b]\prime\prime}}{(b_b - b_a)(\phi^{[b]\prime} - \phi^{[a]\prime})}$.

4.2.2.3 The Baker–Akhiezer Functions

We show the explicit expression of the *Baker function* or the *Baker–Akhiezer function* (3.116) for the hyperelliptic curves X, whose essence is given in [Ba1897, p. 421].

First, we recall that $\psi(u; v) = \dfrac{\mathrm{e}^{-\zeta(v)u}\sigma(u+v)}{\sigma(u)\sigma(v)}$ for $g = 1$ curve, $y = x^3 + \lambda_2 x^2 + \lambda_1 x + \lambda_0$, satisfies

$$\left[\frac{d^2}{du^2} - 2\wp(u)\right]\psi(u; v) = (\wp(v) + \lambda_2)\psi(u; v). \tag{4.40}$$

It is naturally obtained from (3.38)–(3.42), and the relation, $d\psi(u; v)/du = (\zeta(u + v) - \zeta(u) - \zeta(v))\psi(u; v)$, i.e.,

$$\left[\frac{d}{du} - (\zeta(u+v) - \zeta(u))\right]\psi(u; v) = \zeta(v)\psi(u; v). \tag{4.41}$$

It is easy to find that $\mu := \dfrac{d}{du}\log\psi(u; v)$ satisfies the MKdV equation.[3]

Let us consider the hyperelliptic analogue of (4.40) for the hyperelliptic curve X : $y^2 = f(x)$ of genus g. We basically follow the expression (3.116) by Nakayashiki [Nak2010b, Lemma 3] and Buchstaber, Enolskii, and Leykin [BucEL2012, Chap. 12].[4] However, due to Lemma 3.81, we introduce $\widetilde{\zeta}_i(P) := -\int_\infty^P \nu_i^{\mathrm{II}}$, and employ the definition

$$\psi(u; v) = \frac{\mathrm{e}^{-\sum_i \widetilde{\zeta}_i(P)u_i}\sigma(u+v)}{\sigma(u)\sigma_{\natural_1}(v)} \tag{4.42}$$

for $u \in \mathbb{C}^g$ and $v = w(P)$ of $P \in X$, which agrees with the al function when $v = \omega_r$. The factor $\sigma_{\natural_1}(v)$ in the denominator of (4.42) is required for the agreement, which is identical to the prime form $\widetilde{\mathscr{E}}(\infty, Q)$ in (3.67) and (3.112).

We show that $\psi(u; v)$ in (4.42) satisfies a generalization of (4.40) as follows.

First, we show the generalization of (3.42):

Proposition 4.51 *For $u = \widetilde{w}(\iota_X(x_1, y_1), \ldots, \iota_X(x_g, y_g))$ and $v = \widetilde{w}(\iota_X(\widetilde{x}, \widetilde{y}))$,*

$$\wp_{gg}(u+v) + \wp_{gg}(u-v) + 2\wp_{gg}(u) =$$
$$2\left[\sum_r \frac{y_r}{(\widetilde{x} - x_r)F'(x_r)}\right]^2 + 2\frac{f(\widetilde{x})}{F(\widetilde{x})^2} - 2\lambda_{2g} - 2\widetilde{x}. \tag{4.43}$$

---

[3] $\mu = \dfrac{y - \widetilde{y}}{x - \widetilde{x}}$ holds the Miura relation, $2x + \lambda_2 + \widetilde{x} = \mu^2 + \mu_u$, and the MKdV equation,

$$2(\lambda_2 + 3\widetilde{x})\mu_u + 6\mu^2\mu_u + \mu_{uuu} = 0,$$

as mentioned in Remark 4.43. It is remarked that $\psi(u, v)$ is defined on $\mathbb{C}$ but not on the Jacobian J_E in general. It is in contrast that $\mathrm{al}_r(u) = \psi(u, \omega_r)$ is a function of $\widehat{J}_{E,r}$. Thus we consider only the special case that $\widetilde{x}$ is a branch point.

[4] Note that $u_{w_{g-i+1}}$ in (3.116) corresponds to $-u_i$ in our convention.

Proof We differentiate Lemma 3.81 2 by u:

$$\wp_{gg}(u+v)+\wp_{gg}(u-v)-2\wp_{gg}(u)=\sum_{\ell}\frac{2y_\ell}{F'(x_\ell)}\frac{\partial}{\partial x_\ell}\sum_r\frac{2y_r}{(\widetilde{x}-x_r)F'(x_r)}$$

$$=2\left[\sum_r\frac{y_r}{(\widetilde{x}-x_r)F'(x_r)}\right]^2+\sum_r\frac{2}{F'(x_r)}\frac{\partial}{\partial x}\left[\frac{f(x)}{(\widetilde{x}-x)F'(x)}\right]\Bigg|_{x=x_r}$$

$$=2\left[\sum_r\frac{y_r}{(\widetilde{x}-x_r)F'(x_r)}\right]^2+2\left[\frac{f(\widetilde{x})}{F(\widetilde{x})^2}-\lambda_{2g}-\widetilde{x}-2\wp_{gg}(u)\right].$$

Here we use the strategy introduced in (4.33)–(4.36). ■

Lemma 4.52 *For* $u=\widetilde{w}(\iota_X(x_1,y_1),\dots,\iota_X(x_g,y_g))$ *and* $v=\widetilde{w}(\iota_X(\widetilde{x},\widetilde{y}))$,

$$[\zeta_g(u+v)-\zeta_g(u-v)-2\widetilde{\zeta}_g(v)]=\frac{2\widetilde{y}}{F(\widetilde{x})}. \tag{4.44}$$

Proof It can be obtained via (3.35) [BucEL2012, Chap. 12], but we show it directly, because of its sign. By Lemma 3.53, we find that the left-hand side is a meromorphic function on $S^gX\times X$. We have the expression,

$$\zeta_i(u+v)-\zeta_i(u-v)=-2\sum_{ij}\wp_{ij}(u)v_j-\frac{2}{3!}\sum_{ij}\wp_{ijk\ell}(u)v_jv_kv_\ell+d_{>3}(v_i).$$

whereas $v_g^{\mathrm{II}}=-\widetilde{t}^{-2}d\widetilde{t}+d_{>g+1}(\widetilde{t})$ for the arithmetic parameter $\widetilde{t}=\widetilde{x}^{g+1}/\widetilde{y}$. Thus we have

$$\zeta_g(u+v)-\zeta_g(u-v)-2\widetilde{\zeta}_g(v)=2(x_1+\cdots+x_g)\widetilde{t}+\frac{2}{\widetilde{t}}+d_{>0}(\widetilde{t}).$$

$$\frac{2\widetilde{y}}{F(\widetilde{x})}=\frac{2}{\widetilde{t}}\left(1+\frac{(x_1+\cdots+x_g)}{x}+d_{>0}(\widetilde{t})\right).$$

Thus we have the results. ■

Though we don't have its proof, Lemma 4.52 might be generalized to the following:

$$\begin{aligned}
[\zeta_{g-1}(u+v)-\zeta_{g-1}(u-v)+2\widetilde{\zeta}_{g-1}(v)]&=\frac{2\widetilde{y}}{F(\widetilde{x})}(\widetilde{x}-\wp_{gg}(u)),\\
[\zeta_{g-2}(u+v)-\zeta_{g-2}(u-v)+2\widetilde{\zeta}_{g-2}(v)]&=\frac{2\widetilde{y}}{F(\widetilde{x})}(\widetilde{x}^2-\wp_{gg}(u)\widetilde{x}+\wp_{g,g-1}(u)),\\
&\vdots
\end{aligned} \tag{4.45}$$

Further, by noting Remark 3.82, it is conjectured that $\widetilde{\zeta}_i$ might equal to $\zeta_i^{\natural_j}$ in Lemma 3.81.

Then the following are straightforward:

Lemma 4.53 *For $u = \widetilde{w}(\iota_X(x_1, y_1), \dots, \iota_X(x_g, y_g))$ and $v = \widetilde{w}(\iota_X(\widetilde{x}, \widetilde{y}))$, we have*

$$\zeta_g(u+v) - \zeta_g(u) - \widetilde{\zeta}_g(v) = \frac{\widetilde{y}}{F(\widetilde{x})} - \sum_{r=1}^{g} \frac{y_r}{(\widetilde{x} - x_r)F'(x_r)},$$

$$\wp_{gg}(u+v) - \wp_{gg}(u-v) = -\sum_{r=1}^{g} \frac{4\widetilde{y}y_r}{(\widetilde{x} - x_r)F'(x_r)F(\widetilde{x})},$$

and

$$\wp_{gg}(u+v) + \wp_{gg}(u) + \widetilde{x} = \left[\sum_r \frac{y_r}{(\widetilde{x} - x_r)F'(x_r)} - \frac{\widetilde{y}}{F(\widetilde{x})}\right]^2 - \lambda_{2g}.$$

Finally, we have the generalization of (4.40) and (4.41):

Proposition 4.54 *For $u = \widetilde{w}(\iota_X(x_1, y_1), \dots, \iota_X(x_g, y_g))$ and $v = \widetilde{w}(\iota_X(\widetilde{x}, \widetilde{y}))$, we have the relations*

$$\left[\frac{\partial^2}{\partial u_g^2} - 2\wp_{gg}(u)\right]\psi(u; v) = (\widetilde{x} + \lambda_{2g})\psi(u; v), \tag{4.46}$$

$$\left[\frac{\partial}{\partial u_g} - (\zeta_g(u+v) - \zeta_g(u))\right]\psi(u; v) = \widetilde{\zeta}_g(v)\psi(u; v). \tag{4.47}$$

Proof It is the same as the case of $g = 1$. ■

We note that we can find more general solutions to the modified KdV equations than the ones in Sect. 4.2.1 in terms of the above $\psi(u; v)$.

These relations might be connected with the nonlinear Schrödinger equations [Pr1985] and the complex modified Korteweg–de Vries equations. We recall that Eilbeck, Enol'skii, and Kostov investigated a similar problem in [EilEK2000].

4.2.3 Sine-Gordon Equations in Weierstrass' 1854 Paper and Their Extension

We show the sine-Gordon equation and its relation to a classical analysis by Weierstrass in this subsection.

As mentioned in Sect. 1.1.1, Gauss studied the differential equation (1.11), and Weierstrass followed Gauss' approach, which he learned from Gudermann probably, and obtained the expansion of his elliptic Al function [Wei1840]. Further, when Weierstrass investigated the hyperelliptic curves, he also took the approach of Gauss as a model to obtain his theta series Al's and completely obtained the Jacobi inversion formula of the hyperelliptic curves in 1854 [Wei1854]. In other words, Weierstrass generalized (1.11) to settle the Jacobi inversion problem for the hyperelliptic curves [Wei1854], and thus he essentially considered the sine-Gordon equation,

$$\partial_{v_1}\partial_{v_2}\varphi = \sin\varphi,$$

to obtain his theta series, Al functions, which were later redefined as the hyperelliptic σ function by Klein [Kl1886, Kl1888].

Modern expressions of the sine-Gordon equation in terms of the Riemann theta function were given by Mumford [Mum1984, 3.241] and Previato [Pr1986].

We follow Weierstrass' arguments and show that the al function satisfies the sine-Gordon equation. The hyperelliptic curve given in (2.65) with the parametrization (2.66), $f_X(x, y) = y^2 - f(x) = 0$. In terms of the al functions, Weierstrass implicitly considered the relation [Mat2002d, Mat2005],

$$\frac{\partial^2}{\partial u_1^{(r)} \partial u_g^{(r)}} \log \mathrm{al}_r^2 = \mathrm{al}_r^2 - \frac{f'(b_r)}{\mathrm{al}_r^2}, \tag{4.48}$$

$$\frac{\partial^2}{\partial v_r^{(g)} \partial v_s^{(g)}} \log \frac{\mathrm{al}_r}{\mathrm{al}_s} = \frac{1}{(b_r - b_s)} \left(f'(b_s) \left(\frac{\mathrm{al}_r}{\mathrm{al}_s}\right)^2 + f'(b_r) \left(\frac{\mathrm{al}_s}{\mathrm{al}_r}\right)^2 \right), \tag{4.49}$$

where $u_j^{(r)} = \sum_{i=1}^{g} \int_\infty^{(x_i, y_i)} \frac{(x - b_r)^{a-1} dx}{2y}$, $f'(x) := df(x)/dx$ and $v^{(g)}$'s are the parameter in $\mathbb{C}^g$ defined by the holomorphic one-form $\widehat{\nu}_i^{\mathrm{I}}$ (4.52); the expansion of $\widehat{\nu}_i^{\mathrm{I}}$ at the ramification point $(b_i, 0)$ in terms of the local parameter t ($t(b_i, 0) = 0$) is given by $\widehat{\nu}_i^{\mathrm{I}} = dt(1 + d_{>0}(t))/f'(b_i)$ as in Lemma 4.57. Weierstrass defined them to investigate the al functions and find his theta series Al functions. We call v_i *twisted W-normalization* and (4.49) the *Weierstrass relation* in this subsection. Baker investigated the hyperelliptic al functions in terms of $v_i^{(g)}$ in 1898 [Ba1898a] to clarify the structure of the sigma function.

In this subsection, we focus on (4.49) and generalize the flows in Remark 4.1 to the subvariety in J_X as in Sect. 2.5.3. We use the equivalence between $\widehat{W}_X^m = [-1]\widehat{W}_X^m$ in Lemma 4.56 and $S^m X$ in a certain sense for $m \le g$. The differential equations in W_X^m can also be regarded as algebraic relations in $S^m R_X$, and we concentrate on the later ones because we can treat them algebraically. Here we recall Remark 3.23 and use the latter one mentioned there. We show the flow.

1. Following Weierstrass' approach, we employ a twisted W-normalized holomorphic one-form, v_a, which locally agrees with the local parameter at ramification point $(B_a, \dots, B_a) \in S^m X$ [Ba1898a].
2. Our aim is to obtain "interesting" relations in the differential ring $\mathscr{D}_{\widehat{W}_X^m} := \mathscr{M}_{\widehat{W}_X^m}\left[\frac{\partial}{\partial v_1^{(m)}}, \dots, \frac{\partial}{\partial v_m^{(m)}}\right]$, where $\mathscr{M}_{\widehat{W}_X^m}$ is the ring of the meromorphic function over $\widehat{W}_X^m$ given by $\sigma_{\natural_m^{(i)}}$ $(i = 1, \dots, m+1)$ as mentioned in Sect. 3.3.2.
3. By using the identity $\dfrac{\partial}{\partial v_r^{(m)}} = \displaystyle\sum_{i=1}^{m} \frac{\partial x_i}{\partial v_r} \frac{\partial}{\partial x_i} =: \frac{\widetilde{\partial}}{\partial v_r^{(m)}, x}$ and the Jacobi inversion formula in Sect. 3.3.2, we consider some relations as ones in a ring $\mathscr{D}_{X^m}$ induced from $\mathscr{D}_X$,

$$\mathscr{D}_{X^m} := \mathscr{Q}(S^m R_X)\left[\frac{\widetilde{\partial}}{\partial v_1^{(m)}, x}, \dots, \frac{\widetilde{\partial}}{\partial v_m^{(m)}, x}\right],$$

where $S^m R_X$ is the m-th symmetric tensor product of R_X over $\mathbb{C}$.
4. We translate these relations in $\mathscr{D}_{X^m}$ into ones in $\widehat{W}_X^m$, which are the desired relations. Further, we consider some stratum of $\widehat{X}$ and $\widehat{J}_X$ instead of X and J_X as mentioned at the beginning of this section (see also Sect. 4.5.1).

We remark on the two types of differential operators as in Remark 3.23. In this subsection, we use the latter one in there.

By obtaining the relations among the meromorphic functions on $\widehat{W}_X^m$, we could find several properties of the sigma function on $\kappa_J^{-1}\widehat{W}_X^m$ and of $\widehat{W}_X^m$ itself, as Gauss and Weierstrass found the elliptic theta series by considering the static sine-Gordon equation (1.11) as mentioned in Chap. 1, though such discussion of them is beyond the scope of this book and is omitted.

4.2.3.1 Differentials on Hyperelliptic Curves

We handle the hyperelliptic curves X given by $f_X(x, y) = y^2 - f(x) = 0$ in (2.65) and (2.66). Here we use the further decomposition $f_{\rm I}(x) = f_{\rm I}^{(1)}(x) f_{\rm I}^{(2)}(x)$,

$$\begin{aligned} f_{\rm I}^{(1)}(x) &:= (x - a_1)(x - a_2)\cdots(x - a_m), \\ f_{\rm I}^{(2)}(x) &:= (x - a_{m+1})(x - a_{m+2})\cdots(x - a_g), \\ f_{\rm II}(x) &:= (x - c_1)(x - c_2)\cdots(x - c_g)(x - c), \end{aligned} \tag{4.50}$$

where $a_k = b_{2k-1}$, $c_k = b_{2k}$, $(k = 1, \cdots, g)$, $c = c_0 = b_0$.

Definition 4.55 [Ba1897, Ba1898a] For points $(x_i, y_i) \in X$, $(i = 1, \dots, g)$ and a non-negative integer $m \le g$, we define the following quantities.

1. The twisted W-normalized differentials of the first kind at $P = (x, y)$ are defined by

$$\widehat{\nu}^{\mathrm{I}}_k(P) := \frac{f_{\mathrm{I}}(x)dx}{2(x-a_k)f_{\mathrm{I}}'(a_k)y}, \quad (k=1,\cdots,m). \tag{4.51}$$

2. For the twisted Abel–Jacobi map,

$$\widehat{w} : S^g X \longrightarrow J_X = \mathbb{C}^g/\Gamma_X, \left(\widehat{w}((x_1,y_1),\cdots,(x_g,y_g)) := \sum_{i=1}^{g}\int_{\infty}^{(x_i,y_i)} \widehat{\nu}^{\mathrm{I}}_k\right), \tag{4.52}$$

we define the subspace

$$\widehat{W}^m_X := \widehat{w}(S^m X \times (a_{m+1},0)\times\cdots\times(a_g,0))/\Gamma_X. \tag{4.53}$$

Lemma 4.56 $[-1]\widehat{W}^m_X = \widehat{W}^m_X$.

Proof Using the style in Proposition 2.172 and μ_g Definition 2.156, we have the relation

$$\sum_{i=1}^{m}(x_i,y_i) + \sum_{i=m+1}^{g}(a_i,0) - g\infty \sim -\left[\sum_{i=1}^{m}(x_i,y_i) + \sum_{i=m+1}^{g}(a_i,0) - g\infty\right].$$

■

The Jacobi variety J_X is defined as a complex torus as $J_X := \widehat{W}^g_X$. As $\widehat{W}^m_X$ $(m<g)$ is embedded in J_X whose dimension as subvariety is m, the differential forms $(d\nu_k)_{k=1,\cdots,g}$ are not linearly independent. We find the basis as $\nu^{(m)}_k := \widehat{w}((x_1,y_1),\cdots,(x_m,y_m),(a_{m+1},0),\cdots,(a_g,0))$, $(k=1,\cdots,m)$ at $\widehat{W}^m_X$, and a filter structure,

$$\widehat{W}^0_X \subset \widehat{W}^1_X \subset \widehat{W}^2_X \subset\cdots\subset \widehat{W}^{g-1}_X \subset \widehat{W}^g_X = J_X.$$

The following shows the characterization of the twisted W-normalization differential $\widehat{\nu}^{\mathrm{I}}_i$, which corresponds to the property of the al function in Theorem 2.205 3:

Lemma 4.57 *Let the local parameter t at $(a_i,0)\in X$ such that $t^2=(x-a_i)$,*

$$\widehat{\nu}^{\mathrm{I}}_i = \frac{1}{f'(a_i)}dt(1+d_{>0}(t)).$$

For $((x_1,y_1),\cdots,(x_m,y_m))\in S^m X$, we consider

$$F_m(x) = (x-x_1)\cdots(x-x_m), \tag{4.54}$$

in Corollary 2.157, and in terms of $F_m(x)$, a hyperelliptic *al* function over a nonsingular locus of $\widehat{W}^m_X$ [Ba1898a, p. 340] [Wei1854],

$$\mathrm{al}_r^{(m)}(v^{(m)}) = \sqrt{F_m(b_r)}, \tag{4.55}$$

as in (2.128) and Theorem 2.205. Further, we introduce $m \times m$-matrices,

$$\mathscr{M}_m := \begin{pmatrix} \dfrac{1}{x_1 - a_1} & \dfrac{1}{x_2 - a_1} & \cdots & \dfrac{1}{x_m - a_1} \\ \dfrac{1}{x_1 - a_2} & \dfrac{1}{x_2 - a_2} & \cdots & \dfrac{1}{x_m - a_2} \\ \vdots & \vdots & \ddots & \vdots \\ \dfrac{1}{x_1 - a_m} & \dfrac{1}{x_2 - a_m} & \cdots & \dfrac{1}{x_m - a_m} \end{pmatrix},$$

$$\mathscr{Q}_m = \begin{pmatrix} \sqrt{\frac{f_{\mathrm{I}}(x_1)}{f_{\mathrm{II}}(x_1)}} & & & \\ & \sqrt{\frac{f_{\mathrm{I}}(x_2)}{f_{\mathrm{II}}(x_2)}} & & \\ & & \ddots & \\ & & & \sqrt{\frac{f_{\mathrm{I}}(x_m)}{f_{\mathrm{II}}(x_m)}} \end{pmatrix}, \quad \mathscr{A}_m = \begin{pmatrix} Q'(a_1) & & & \\ & Q'(a_2) & & \\ & & \ddots & \\ & & & Q'(a_m) \end{pmatrix}.$$

Lemma 4.58 *1.* $\det \mathscr{M}_m = \dfrac{(-1)^{m(m-1)/2} V(x_1, \cdots, x_m) V(a_1, \cdots, a_m)}{\prod_{k,l}(x_k - a_l)}$, *where* $V(z_1, \cdots, z_m) := \prod_{i<j}(z_i - z_j)$.

2. $\mathscr{M}_m^{-1} = \left[\left(\dfrac{F_m(a_j) f_{\mathrm{I}}^{(1)}(x_i)}{F_m'(x_i) f_{\mathrm{I}}^{(1)\prime}(a_j)(a_j - x_i)}\right)_{i,j}\right]$, *where* $F_m'(x) := dF_m(x)/dx$ *and* $f_{\mathrm{I}}^{(1)\prime}(x) = df_{\mathrm{I}}^{(1)}(x)/dx$.

3. $(\mathscr{M}\mathscr{Q})^{-1}\mathscr{A} = \left[\left(\dfrac{2y_i F_m(a_j)}{F_m'(x_i) f_{\mathrm{I}}^{(2)}(x_i)(a_j - x_i)}\right)_{i,j}\right]$.

Proof *1* is a well-known relation [Tak1930]. Since the divisor in the left-hand side gives the right-hand side as

$$C f_{\mathrm{II}}(x_1, \cdots, x_m) f_{\mathrm{II}}(a_1, \cdots, a_m) / \prod_{k,l}(x_k - a_l),$$

for a certain constant C. In order to determine C, we multiply both sides by $\prod_{k,l}(x_k - a_l)$ and let $x_1 = a_1$, $x_2 = a_2$, $\cdots$, and $x_m = a_m$. Then C is determined as above. *2* is obtained by the Laplace formula and the minor determinant for the inverse matrix. On *3*, we note that $f_{\mathrm{I}}^{(1)} f_{\mathrm{I}}^{(2)} = f_{\mathrm{I}}(x)$ in (4.50) and thus $f_{\mathrm{I}}^{(1)}(x) \times \sqrt{f_{\mathrm{II}}(x)/f_{\mathrm{I}}(x)} = y/f_{\mathrm{I}}^{(2)}$ except for the sign. Then we obtain *3*. ∎

Corollary 4.59 *Let* $\partial_{v_i}^{(r)} := \partial/\partial v_i^{(r)}$, *and* $\partial_{x_i} := \partial/\partial x_i$. *Then we formally have*

$$\begin{pmatrix} \partial_{v_1} \\ \partial_{v_2} \\ \vdots \\ \partial_{v_m} \end{pmatrix} = 2(\mathscr{M}\mathscr{Q}_m)^{-1}\mathscr{A}_m \begin{pmatrix} \partial_{x_1} \\ \partial_{x_2} \\ \vdots \\ \partial_{x_m} \end{pmatrix}. \tag{4.56}$$

Here we recall Remark 3.23 again. $\partial_{v_i}^{(r)}$ corresponds to the latter one mentioned there.

4.2.3.2 Weierstrass Relations on $\widehat{W}_X^m$

The hyperelliptic solution of the sine-Gordon equation over J_X related to ramified points $(a_1, 0)$ and $(a_2, 0)$ is obtained as (4.49) by Mumford [Mum1984], whose expression in an old fashion is the Weierstrass relation (4.49). We show an extension of the Weierstrass relation (4.49) over $\widehat{W}_X^m$ as our main theorem in this subsection.

Theorem 4.60 $\mathrm{al}_r^{(m)}$ *and* $\mathrm{al}_s^{(m)}$ $(r, s \in \{1, 2, \cdots, m\})$ *over a non-singular locus of* W_X^m *in (4.53) obey the relation, i.e., the Weierstrass relation for* $\widehat{W}_X^m$,

$$\begin{aligned} &\frac{\partial}{\partial v_r^{(m)}}\frac{\partial}{\partial v_s^{(m)}} \log \frac{\mathrm{al}_r^{(m)}(v^{(m)})}{\mathrm{al}_s^{(m)}(v^{(m)})} \\ &= \frac{1}{(a_r - a_s)}\left(\frac{f'(a_r)}{(f_{\mathrm{I}}^{(2)}(a_r))^2}\left(\frac{\mathrm{al}_s^{(m)}(v^{(m)})}{\mathrm{al}_r^{(m)}(v^{(m)})}\right)^2 + \frac{f'(a_s)}{(f_{\mathrm{I}}^{(2)}(a_s))^2}\left(\frac{\mathrm{al}_r^{(m)}(v^{(m)})}{\mathrm{al}_s^{(m)}(v^{(m)})}\right)^2\right) \\ &+ \frac{f'(a_{m+1})(\mathrm{al}_r^{(m)}(v^{(m)}))^2(\mathrm{al}_s^{(m)}(v^{(m)}))^2(a_r - a_s)}{(a_{m+1} - a_r)(a_{m+1} - a_s)(\mathrm{al}_{m+1}^{(m)}(v^{(m)}))^4(f_{\mathrm{I}}^{(2)\prime}(a_{m+1}))^2} \\ &+ \cdots\cdots + \frac{f'(a_g)(\mathrm{al}_r^{(m)}(v^{(m)}))^2(\mathrm{al}_s^{(m)}(v^{(m)}))^2(a_r - a_s)}{(a_g - a_r)(a_g - a_s)(\mathrm{al}_g^{(m)}(v^{(m)}))^4(f_{\mathrm{I}}^{(2)}(a_g)')^2}. \end{aligned} \tag{4.57}$$

Proof From (4.55), we consider the following formula instead of (4.57) without loss of generality,

$$\begin{aligned} \frac{\partial}{\partial v_1^{(m)}}\frac{\partial}{\partial v_2^{(m)}} \log \frac{F_m(a_1)}{F_m(a_2)} &= \frac{2F_m(a_1)F_m(a_2)}{(a_1 - a_2)}\Bigg[\frac{f'(a_1)}{[F_m(a_1)f_{\mathrm{I}}^{(2)}(a_1)]^2} + \frac{f'(a_2)}{[F_m(a_2)f_{\mathrm{I}}^{(2)}(a_1)]^2} \\ &+ \frac{f'(a_{m+1})(a_1 - a_2)^2}{(a_{m+1} - a_1)(a_{m+1} - a_2)F_m(a_{m+1})^2(Q_m^{(2)\prime}(a_{m+1}))^2} \\ &+ \cdots + \frac{f'(a_g)(a_1 - a_2)^2}{(a_g - a_1)(a_g - a_2)F_m(a_g)^2(Q_m^{(2)\prime}(a_g))^2}\Bigg]. \end{aligned} \tag{4.58}$$

As mentioned in the beginning of this section, we translate the descriptions on the Jacobian into those on the curves. We rewrite the differentials $v_{(r)}^{(m)}$'s in terms of the differentials over curves as in (4.59). We count the residue of integration and use a combinatorial trick as in Lemma 4.61. Then we obtain (4.58).

From Lemma 4.58 3 and (4.56), the derivative v's over $\widehat{W}_X^m$ in (4.53) are expressed by the affine coordinate x_i's,

$$\frac{\partial}{\partial v_i^{(m)}} = F_m(a_i) f_{\mathrm{I}}^{(2)}(a_i) \sum_{j=1}^{m} \frac{2y_j}{F_m'(x_j) f_{\mathrm{I}}^{(2)}(x_j)(x_j - a_i)} \frac{\partial}{\partial x_j}. \tag{4.59}$$

The left-hand side of (4.58) becomes

$$\frac{\partial^2}{\partial v_1 \partial v_2} \log \frac{F_m(a_1)}{F_m(a_2)} = F_m(a_1) f_{\mathrm{I}}^{(2)}(a_1)$$
$$\sum_{j=1 i=1}^{m} \frac{2y_j}{(x_i - a_1) F_m'(x_j) f_{\mathrm{I}}^{(2)}(x_j)} \frac{\partial}{\partial x_j} \frac{2y_i F_m(a_2) f_{\mathrm{I}}^{(2)}(a_1)}{F_m'(x_i) f_{\mathrm{I}}^{(2)}(x_i)(x_i - a_2)} \frac{(a_1 - a_2)}{(x_i - a_1)(x_i - a_2)}.$$

The left-hand side is

$$F_m(a_1)F_m(a_2)\Bigg[\sum_{i=1}^{m} \frac{1}{F_m'(x_i)} \left[\frac{\partial}{\partial x}\left(\frac{f(x)(a_2 - a_1)}{(x - a_1)^2 (x - a_2)^2 (f_{\mathrm{I}}^{(2)}(x))^2 F_m'(x)}\right)\right]_{x = x_i}$$
$$-\sum_{k,l,k\neq l} \frac{2y_k y_l (a_2 - a_1)}{F_m'(x_k) F_m'(x_l)(x_l - a_1)(x_l - a_2) f_{\mathrm{I}}^{(2)}(x_l)(x_k - a_1)(x_k - a_2) f_{\mathrm{I}}^{(2)}(x_k)(x_l - x_k)}\Bigg].$$

Then the proof of Theorem 4.60 is completely done due to the following lemma. ■

Lemma 4.61 *1.* $\displaystyle\sum_{i=1}^{m} \frac{1}{F_m'(x_i)} \left[\frac{\partial}{\partial x}\left(\frac{f(x)}{(x - a_1)^2 (x - a_2)^2 (f_{\mathrm{I}}^{(2)}(x)^2 F_m'(x)}\right)\right]_{x = x_i}$ *is equal to*

$$\frac{2}{(a_1 - a_2)^2}\Bigg(\frac{f'(a_1)}{F_m(a_1)^2 (f_{\mathrm{I}}^{(2)}(a_1))^2} + \frac{f'(a_2)}{F_m(a_2)^2 (f_{\mathrm{I}}^{(2)}(a_1))^2}$$
$$+ \frac{f'(a_{m+1})(a_1 - a_2)^2}{(a_{m+1} - a_1)(a_{m+1} - a_2) F_m(a_{m+1})^2 (Q_m^{(2)\prime}(a_{m+1}))^2}$$
$$+ \cdots + \frac{f'(a_g)(a_1 - a_2)^2}{(a_g - a_1)(a_g - a_2) F_m(a_g)^2 (Q_m^{(2)\prime}(a_g))^2}\Bigg).$$

2. $$\sum_{k,l,k\neq l}\left(\frac{2y_k y_l (a_2 - a_1)}{F_m'(x_k) F_m'(x_l)(x_l - a_1)(x_l - a_2) f_{\mathrm{I}}^{(2)}(x_l)(x_k - a_1)(x_k - a_2) f_{\mathrm{I}}^{(2)}(x_k)(x_l - x_k)}\right) = 0.$$

Proof *1* is proved by the following residual computations. We recall that ∂D_X is the boundary of a polygon representation D_X of X as illustrated in Fig. 2.3 and consider

$$\oint_{\partial D_X} \frac{f(x)}{((x-a_1)(x-a_2)F_m(x)f_{\mathrm{I}}^{(2)}(x))^2}dx = 0. \tag{4.60}$$

The divisor of the integrand of (4.60) is

$$\sum_{i=1}^{2g+1}(b_i,0) - 4\sum_{i=1,2,m+1,m+2,\cdots,g}(a_i,0) - 2\sum_{i=1}^{m}(x_i,y_i) - 2\sum_{i=1}^{m}(x_i,-y_i) + 3\infty.$$

We check these poles: First, we consider the contribution at ∞.

$$\begin{aligned}\mathrm{res}_{(x_k,\pm y_k)}&\frac{f(x)dx}{((x-a_1)(x-a_2)F_m(x)f_{\mathrm{I}}^{(2)}(x))^2}\\ &=\frac{1}{F_m'(x_k)}\left[\frac{\partial}{\partial x}\left(\frac{f(x)}{((x-a_1)(x-a_2)f_{\mathrm{I}}^{(2)}(x))^2F_m'(x)}\right)\right]_{x=x_k}.\end{aligned}$$

At the point $(a_1,0)$, noting that the local parameter t is given by $t^2=(x-a_1)$ there, we have

$$\mathrm{res}_{(a_1,0)}\frac{f(x)dx}{((x-a_1)(x-a_2)F_m(x)f_{\mathrm{I}}^{(2)}(x))^2}=\frac{2f'(a_1)}{((a_1-a_2)F_m(a_1)f_{\mathrm{I}}^{(2)}(a_1))^2}.$$

The residue at $(a_2,0)$ is similarly obtained. For the points $(a_k,0)$ $(g\ge k>m)$, we have

$$\mathrm{res}_{(a_k,0)}\frac{f(x)dx}{((x-a_1)(x-a_2)F_m(x)f_{\mathrm{I}}^{(2)}(x))^2}=\frac{2f'(a_k)}{((a_k-a_1)(a_k-a_2)F_m(a_k)f_{\mathrm{I}}^{(2)}(a_k))^2}.$$

By arranging them, we obtain *1*. *2* is obvious. ■

We obtain the Weierstrass relation (4.49) or the sine-Gordon equation as a corollary [Mat2005]. For the case $g=m$, the following is much more noticeable:

Corollary 4.62 *For a curve satisfying the relations $c_j=a_j$ for $j=m+1,\cdots,g$, $\mathrm{al}_r^{(m)}$ and $\mathrm{al}_s^{(m)}$ $(r,s\in\{1,2,\cdots,m\})$ over $\widehat{W}_X^m$ in (4.53) obey the relation,*

$$\frac{\partial}{\partial v_r^{(m)}}\frac{\partial}{\partial v_s^{(m)}}\log\frac{\mathrm{al}_r^{(m)}}{\mathrm{al}_s^{(m)}}=\frac{1}{a_r-a_s}\left[\frac{f'(a_r)}{(f_{\mathrm{I}}^{(2)}(a_r))^2}\left(\frac{\mathrm{al}_s^{(m)}}{\mathrm{al}_r^{(m)}}\right)^2+\frac{f'(a_s)}{(f_{\mathrm{I}}^{(2)}(a_s))^2}\left(\frac{\mathrm{al}_r^{(m)}}{\mathrm{al}_s^{(m)}}\right)^2\right]. \tag{4.61}$$

For $g=m$ case, we have (4.49).

Proof Since the condition $c_j = a_j$ for $j = m+1, \cdots, g$ means $f'(a_j) = 0$ for $j = m+1, \cdots, g$, Theorem 4.60 reduces to this one. ■

Assuming the same as in Corollary 4.62, let $A := \dfrac{\sqrt{f'(a_r)f'(a_s)}}{(a_r - a_s)f_{\mathrm{I}}^{(2)}(a_r)f_{\mathrm{I}}^{(2)}(a_s)}$ and

$$\phi_m^{(r,s)}(u) := \frac{1}{\sqrt{-1}} \log \sqrt{\frac{f'(a_r)}{f'(a_s} \frac{f_{\mathrm{I}}^{(2)}(a_r)}{f_{\mathrm{I}}^{(2)}(a_s)} \frac{F_m(a_r)}{F_m(a_s)}}.$$

We rephrase (4.62) so that $\phi_m^{(r,s)}$ obeys the sine-Gordon equation

$$\frac{\partial}{\partial v_r^{(m)}} \frac{\partial}{\partial v_s^{(m)}} \phi_m^{(r,s)} = A \cos(\phi_m^{(r,s)}). \tag{4.62}$$

Similarly, (4.48) can be also proved, though we don't mention it in this book [Mat2002d].

We remark that since the constant Gauss curvature surfaces S immersed in the three euclidean spaces are governed by the sine-Gordon equation due to Hilbert [Hil1901, TanD1979] and the general hyperelliptic solutions of the sine-Gordon equation given by the flat plane in the Jacobian J_X as mentioned above, various classes of the surfaces S might be embedded in the Jacobian.

4.2.4 Neumann System and Hyperelliptic al Function

In this subsection, we show the hyperelliptic solutions of the Neumann system in terms of the hyperelliptic al functions. It also means that a system obeying the Neumann system is described by the hyperelliptic sigma function. As the solution is described well in Mumford's book [Mum1984], we provide it along the line of the approach in Remark 4.1.

4.2.4.1 Neumann System

We shortly review the Neumann system $(q, \dot{q}) \in \mathbb{R}^{2g+2}$; $\dot{q} := \frac{dq}{dt}$, $\ddot{q} := \frac{d^2q}{dt^2}$ for the time parameter t. The Neumann dynamical system is a well-known integrable nonlinear dynamical system, whose Lagrangian for $(q, \dot{q}) \in \mathbb{R}^{2g+2}$ is given by

$$L = \frac{1}{2}\sum_{i=0}^{g} \dot{q}_i^2 - \frac{1}{2}\sum_{i=0}^{g} c_i q_i^2 \tag{4.63}$$

with a holonomic constraint,

$$\Phi(q) = 0, \quad \Phi(q) := \sum_{i=0}^{g} q_i^2 - 1, \tag{4.64}$$

which was proposed by C. Neumann in 1859 for the case of $g = 2$ [Neu1859]. This system is studied well in frameworks of the dynamical system [Mos1980], of the symplectic geometry [GuiS1984], of the algebraic geometry [Mum1984], and of the representation of the infinite Lie algebra [AdaHP1988, Sch1987].

The constraint (4.64) means $\dot{\Phi}(q) = 0$,

$$\sum_{i=0}^{g} \dot{q}_i q_i = 0. \tag{4.65}$$

The canonical momentum p_i to q_i is given as

$$p_i = \frac{\partial L}{\partial \dot{q}_i} = \dot{q}_i.$$

Purely kinematic investigations lead to the following proposition [Mum1984].

Proposition 4.63 *The Hamiltonian of this system is given by*

$$H := \frac{1}{2}\sum_{i=0}^{g} \dot{q}_i^2 + \frac{1}{2}\sum_{i=0}^{g} c_i q_i^2, \tag{4.66}$$

and the Hamiltonian vector field is given by

$$D_H = \sum_{i=0}^{g} \dot{q}_i \frac{\partial}{\partial q_i} - \sum_{i=0}^{g} c_i q_i \frac{\partial}{\partial \dot{q}_i} + \left(\sum_{i=0}^{g} \left[c_i q_i^2 - \dot{q}_i^2\right]\right) \sum_{i=0}^{g} q_i \frac{\partial}{\partial \dot{q}_i}. \tag{4.67}$$

The equation of motion is described as

$$\ddot{q}_i = -(2L + c_i) q_i. \tag{4.68}$$

Weierstrass al functions in Sects. 2.8.3 and 3.6.1 give natural descriptions of the Neumann dynamical system. As in Theorem 4.68, the configuration q^i of i-th particle (or coordinate) is directly given by the al function of X given by (2.66), $f_X(x, y) = y^2 - f(x) = 0$, $f(x) = \prod_r (x - b_r)$,

$$q_i(t) = \mathfrak{al}_i(t), \quad \mathfrak{al}_i = \frac{\sqrt{F(c_i)}}{\sqrt{f_{\mathrm{II}}'(c_i)}}.$$

Here $\mathfrak{al}_i(t)$ is defined in Definition 2.206.

4.2.4.2 UVW Expression and al Function

In this subsubsection, we consider a hyperelliptic curve X given by (2.66).

As Mumford studied the Neumann system using the UVW-expression of the hyperelliptic functions [Mum1984], we define U, V and W functions [Mum1984] as follows: Let $((x_1, y_1), \cdots, (x_g, y_g))$ be a point in $S^g X$. U is the same as F in Corollary 2.157, i.e., $U(x) := (x - x_1)\cdots(x - x_g) = F(x)$. We define V and W by

$$V(x) := \sum_{a=1}^{g} \frac{y_a F(x)}{F'(x_a)(x - x_a)} = F(x)\frac{\partial}{\partial u_g}\wp_{gg} = F(x)\Delta_x,$$

$$W(x) := \frac{f(x) - V(x)^2}{F(x)},$$

where we recall Δ_x in Definition 4.4.

In this section, we express the system in terms of the hyperelliptic $\wp$ functions and the al functions, which are written only in terms of the data of the curve. Let us introduce these functions as follows.

We recall the fact that the hyperelliptic $\wp_{gi}$ $(i = 1, 2, \cdots, g)$ functions are related to $F(x)$ via $F(x) = x^g + \sum_{i=1}^{g}(-1)^i \wp_{gi} x^{g-i}$, e.g., $\wp_{gg} := x_1 + \cdots + x_g$.

The Weierstrass al functions al_i and $\mathfrak{al}_i$ are defined in Definitions 2.202 and 2.206, which satisfy Theorem 2.205 [BucEL1997b, Ba1898a, Wei1854]:

Definition 4.64 We define

$$\mathrm{al}_r^{[i]}(u) := \frac{\partial}{\partial u_i}\mathrm{al}_r(u), \quad \mathfrak{al}_r^{[i]}(u) := \frac{\partial}{\partial u_i}\mathfrak{al}_r(u).$$

Furthermore, we prepare the primitive relations between differentials of the al functions and the UVW expressions:

Lemma 4.65 *1.* $\mathrm{al}_i^{[g]}(u) = -\dfrac{V(c_i)(u)}{\mathrm{al}_i(u)}, \quad \mathfrak{al}_i^{[g]}(u) = -\dfrac{V(c_i)(u)}{\mathfrak{al}_i(u) f_{\mathrm{II}}'(c_i)}.$

2. $$\frac{F(x)}{f_{\mathrm{II}}(x)} = \sum_{i=0}^{g} \frac{\mathfrak{al}_i(u)^2}{x - c_i}, \quad \frac{V(x)}{f_{\mathrm{II}}(x)} = -\sum_{i=0}^{g} \frac{\mathfrak{al}_i(u)\mathfrak{al}_i^{[g]}(u)}{x - c_i}, \quad \frac{W(x)}{f_{\mathrm{II}}(x)} = \sum_{i=0}^{g} \frac{\mathfrak{al}_i^{[g]}(u)^2}{x - c_i}.$$

Proof Noting $\dfrac{\partial}{\partial u_g} = \sum_{a=1}^{g} \dfrac{2y_a}{F'(x_a)}\dfrac{\partial}{\partial x_a}$ in Lemma 4.8 [Ba1903] and $\dfrac{\partial}{\partial x_a}F(x) = -\dfrac{F(x)}{(x - x_a)}$, we find that $\dfrac{1}{2}\dfrac{\partial}{\partial u_g}F(x) = -V(x)$, which directly gives the relations in *1*. The relations in *2* are obtained from the definition of $\mathfrak{al}_r$, the relations in *1* and the fact that $f(c_i)$ vanishes. ■

We have the following proposition including Theorem 2.208:

Proposition 4.66

$$\sum_{i=0}^{g} \mathfrak{al}_i^2(u) = 1, \quad \sum_{i=0}^{g} \frac{1}{c_i^2}[\mathfrak{al}_i^{[g]}]^2(u) = 0.$$

Proof As in the proof of Theorem 2.208, we also obtain the relations for the $\mathfrak{al}_r^{[g]}$ by evaluating $W(x)/x^2 f_{\rm II}(x)$ in Lemma 4.65 2. ■

There is a natural relation between the al function and the $\wp_{gg}$ function due to Theorem 4.46 2 (Miura transformation).

Proposition 4.67

$$\frac{\partial^2}{\partial u_g^2}\mathfrak{al}_i(u) = \frac{\partial}{\partial u_g}\mathfrak{al}_i^{[g]}(u) = \left(\sum_{j=0, b_i \neq c_i}^{2g} b_j - 2\wp_{gg}(u)\right)\mathfrak{al}_i,$$

$$\frac{1}{2}\frac{\partial}{\partial u_g}V(c_i) = F(c_i)\left(\sum_{j=0}^{2g} b_j - c_i - 2\sum_{a=1}^{g} x_a\right) - \frac{1}{F(c_i)}V(c_i)^2.$$

4.2.4.3 Neumann System and Hyperelliptic al Function

This section gives our main theorem as follows.

Theorem 4.68 *Assume that configurations of $(x_1, \cdots, x_g) \in S^g X$ are given so that* (al$_i$) *belongs to $\mathbb{R}^{g+1}$ and $(u_g, u_-) \in \mathbb{R}^g$, where $u_g \in \mathbb{R}$ by fixing $u_- \in \mathbb{R}^{g-1}$.*

1. *$\mathfrak{al}_i$ obeys the Neumann system, i.e.,*

$$q_i(t) = \mathfrak{al}_i(u_-, t), \quad \dot{q}_i = -\mathfrak{al}_i^{[g]}(u_-, t), \tag{4.69}$$

where the time t of the system is identified with $-u_g$, and thus the Hamiltonian vector field is given by

$$D_H := \frac{d}{dt} = -\frac{\partial}{\partial u_g}.$$

2. *The Hamiltonian and the Lagrangian or this system are given by*

$$H = \frac{1}{2}\left(\sum_{i=0}^{g} c_i - \sum_{j=1}^{g} a_j\right), \quad L = \frac{1}{2}\left(2\wp_{gg} - \sum_{i=0}^{2g} b_i\right).$$

3. *The conserved quantities are* a_i $(i = 1, \cdots, g)$ *and*

$$m_i := q_i^2 + \sum_{i=0}^{g} \sum_{j=0,\neq i}^{g} \frac{(q_i \dot{q}_j - q_j \dot{q}_i)^2}{c_i - c_j}, \quad (i = 0, \cdots, g),$$

which obey the relations

$$m_i = \frac{f_{\mathrm{I}}(c_i)}{f'_{\mathrm{II}}(c_i)}, \quad \sum_{i=0}^{g} m_i = 1, \quad \sum_{i=0}^{g} c_i m_i = H.$$

These relations were essentially proved in [Mum1984] by using the UVW expression and the properties of the theta function without al functions. However, by following the method [Ba1898a, Wei1854], we show them directly using the nature of the al functions without theta function following the flow in Remark 4.1.

Proof Assumptions are asserted by Lemma 4.8 for certain $\{b_r\}_r$. We consider *1*. Due to Proposition 4.66, $\mathfrak{al}_r$'s obviously obey the constraint condition $\Phi(\mathfrak{al}) = 0$, and $\dot{\Phi}(\mathfrak{al}) = 0$ by differentiating both sides of the identity in u_g. We should check whether they obey the equation of motion (4.68), which is proved in Proposition 4.67 if we assume the form of the Lagrangian L in *2*. *2* is directly obtained by using the relations in Lemma 4.70. Finally, *3* is proved in Remark 4.72. ■

Remark 4.69 1. The equation of motion (4.68) is directly related to Proposition 4.67, which is connected with the Miura transformation. Further, the constraint (4.64) satisfies due to the identity of $\mathfrak{al}$ functions as mentioned in Proposition 4.66. These exhibit the essentials of al functions. Hence the Neumann system should be expressed by the al functions as some dynamical systems are expressed by Jacobi sn, cn, dn functions.
2. We remark that the Hamiltonian depends only upon b_r's (c_i's and a_i's), which determine the hyperelliptic curve X. Thus it is manifest that it is invariant for the time u_g development of the system.
3. There are $2g$ degrees of freedom as a kinematic system because the constraints Φ and $\dot{\Phi}$ reduce them from $(2g + 2)$. The independent conserved quantities m_i are $g = g + 1 - 1$; "-1" comes from $\sum m_i = 1$. Since the sum of m_i gives the Hamiltonian H, H is not a linearly independent conserved quantity. Since there are other g conserved quantities a_i but their sum gives the Hamiltonian $\sum m_i$, the number of independent a_i is $g - 1$. However, $\sum_{i=0}^{g} \dot{q}_i^2 / c_i^2 = 0$ compensates for what is lacking. Hence the degrees of freedom of this system are equal to the number of the conserved quantities.

Let us give a lemma and remarks as follows, which are parts of the proofs of the Theorem 4.68.

Lemma 4.70 *1.* $\displaystyle\sum_{i=0}^{g} [\mathfrak{al}_i^{[g]}(u)]^2 = \wp_{gg}(u) - \sum_{j=1}^{g} a_j.$

2. $\sum_{i=0}^{g} c_i \mathfrak{al}_i(u)^2 = \sum_{i=0}^{g} c_i - \wp_{gg}(u)$.

Proof We consider the relation in *1*. Due to Lemma 4.65, we compute the integral $\oint_{\partial D_X} \frac{V(x)^2}{F(x) f_{\mathrm{II}}(x)} dx = 0$ for ∂D_X as illustrated in Fig. 2.3 to provide

$$2\sum_{i=0}^{g} \frac{V(c_i)^2}{F(c_i) f'_{\mathrm{II}}(c_i)} + \sum_{a=1,\epsilon=\pm}^{g} \mathrm{res}_{(x_a,\epsilon y_a)} \frac{V(x)^2}{F(x) f_{\mathrm{II}}(x)} dx + \mathrm{res}_\infty \frac{V(x)^2}{F(x) f_{\mathrm{II}}(x)} dx = 0,$$

whereas the third term vanishes, and each element in the second term is given by

$$\mathrm{res}_{(x_a,\pm y_a)} \frac{V^2(x)}{F(x) f_{\mathrm{II}}(x)} dx = \frac{f_{\mathrm{I}}(x_a)}{F'(x_a)}.$$

Further, we also evaluate the integral $\oint_{\partial D_X} \frac{f_{\mathrm{I}}(x)}{F(x)} dx = 0$. The integrand has singularities at $(x_a, \pm y_a)$ and infinity. Similar consideration leads us to the identities

$$\sum_{a=1,\epsilon=\pm}^{g} \frac{f_{\mathrm{I}}(x_a)}{F'(x_a)} = 2(a_1 + \cdots + a_g) - 2(x_1 + \cdots + x_g).$$

Due to these relations, we have the relation in *1*.

We consider the second relation by evaluating the integral $\oint_{\partial D_X} x \frac{F(x)}{f_{\mathrm{II}}(x)} dx = 0$. A residual computation gives $\sum_{i=0}^{g} c_i \frac{F(c_i)}{f'_{\mathrm{II}}(c_i)} = -\mathrm{res}_\infty x \frac{F(x)}{f_{\mathrm{II}}(x)} dx$. The contribution at infinity is $2((x_1 + \cdots x_g) - (c_1 + \cdots c_{g+1}))$. Hence we also have the relation in 2. ■

Remark 4.71 Using the fact $\frac{\partial x_a}{\partial u_g} = \frac{2 y_a}{F'(x_a)}$, we take a different expression of Lemma 4.70 [DulRVW2001],

$$\sum_{i=0}^{g} [\mathfrak{al}_i^{[g]}]^2 = \sum_{a,b=1}^{g} g(x)_{a,b} \frac{\partial x_a}{\partial u_g} \frac{\partial x_b}{\partial u_g},$$

where $g(x)_{a,b} := -\sum_{i=0}^{g} \frac{F(c_i)}{(c_i - x_a)(c_i - x_b) f'_{\mathrm{II}}(c_i)}$ whose off-diagonal part does not vanish for the case genus $g > 2$ in general.

Remark 4.72 (Proof of Theorem 4.68) Here we give the conserved quantities of the Neumann system. Let us consider

$$m_i(x) = q_i^2 + \sum_{i=1}^{g+1} \sum_{j=1,\neq i}^{g+1} \frac{(q_i\dot{q}_j - q_j\dot{q}_i)^2}{x - c_j}.$$

Then we have the identities

$$\frac{f(x)}{f_{\mathrm{II}}(x)^2} = \frac{F(x)W(x) + V(x)^2}{f_{\mathrm{II}}(x)^2} = \sum_{i=0}^{g} \frac{m_i(x)}{x - c_i},$$

$$m_i = \mathrm{res}_{c_i} \frac{m_i(x)}{x - c_i} = q_i^2 + \sum_{i=0}^{g} \sum_{j=0,\neq i}^{g} \frac{(q_i\dot{q}_j - q_j\dot{q}_i)^2}{c_i - c_j}.$$

The direct computation gives the relations in the theorem when we consider the integrals of differentials $\frac{f_{\mathrm{I}}(x)}{f_{\mathrm{II}}(x)}dx$, and $\frac{xf_{\mathrm{I}}(x)}{f_{\mathrm{II}}(x)}dx$.

4.3 Differential Relations from Addition Theorem

The addition theorems described in Sect. 3.4 also play a crucial role in the integrable system. After we recall the relation between the Toda lattice equation and the addition theorem of genus one, we show the quasi-periodic and periodic hyperelliptic solutions of the Toda lattice equations and their extension to the trigonal Toda equation as applications of the addition theorems to the integrable systems.

4.3.1 Toda Lattice: g = 1 Case

In this subsection, we demonstrate how one gets an elliptic solution of the Toda lattice [To1967, Mat2003b]. Let E be an elliptic curve given by

$$\begin{aligned} E: \frac{1}{4}\widehat{y}^2 = y^2 &= x^3 + \lambda_2 x^2 + \lambda_1 x + \lambda_0 \\ &= (x - e_1)(x - e_2)(x - e_3), \end{aligned} \tag{4.70}$$

where the e's are complex numbers and $\lambda_2 = -(e_1 + e_2 + e_3) = 0$.

The Weierstrass elliptic σ function associated with the curve E is connected with the Weierstrass $\wp$ and ζ functions by

$$\wp(u) = -\frac{d^2}{du^2} \log \sigma(u), \quad \zeta(u) = \frac{d}{du} \log \sigma(u).$$

Here the coordinate u in the Jacobian J_E associated with E is given by

$$u = \int_\infty^{(x,y)} \frac{dx}{2y},$$

with $x(u) = \wp(u)$, $\widehat{y}(u) = \wp'(u)$ and ∞ is the infinity point of E. The Jacobian is given by $J_E = \mathbb{C}/(\mathbb{Z}\omega' + \mathbb{Z}\omega'')$ and the double period (ω', ω'').

The key to obtain a $\wp$ function solution of the Toda lattice is the addition formula

$$\wp(u) - \wp(v) = -\frac{\sigma(u+v)\sigma(u-v)}{[\sigma(v)\sigma(u)]^2}. \tag{4.71}$$

By differentiating the logarithm of (4.71) with respect to u twice, we have

$$-\frac{d^2}{du^2}\log[\wp(u) - \wp(v)] = \wp(u+v) - 2\wp(u) + \wp(u-v).$$

For a constant number u_0, by letting $u = nu_0 + t + t_0$, $v = u_0$, we have

$$\begin{aligned} -\frac{d^2}{dt^2}\log[\wp(nu_0+t+t_0) - \wp(u_0)] &= [\wp((n+1)u_0+t+t_0) - \wp(u_0)] \\ &\quad - 2[\wp(nu_0+t) - \wp(u_0)] + [\wp((n-1)u_0+t) - \wp(u_0)]. \end{aligned}$$

Further, by letting $V_n(t) := -\wp(nu_0+t+t_0)$, $V_c := -\wp(u_0)$ and $q_n := -\log[V_n(t) - V_c]$, we have

$$-\frac{d^2}{dt^2}q_n = \mathrm{e}^{-q_{n+1}} - 2\mathrm{e}^{-q_n} + \mathrm{e}^{-q_{n-1}} \qquad (n = 1, 2, \ldots), \tag{4.72}$$

which is reduced to the *Toda lattice equation*. In fact, by letting $q_n = Q_n - Q_{n-1}$, $(Q_n = \sum_{i=1}^n q_n + Q_0)$, Q_n obeys the differential-difference equation of a nonlinear lattice [To1967],

$$\frac{d^2}{dt^2}Q_n = \mathrm{e}^{Q_n - Q_{n+1}} - \mathrm{e}^{Q_{n-1} - Q_n} \qquad (n = 1, 2, \ldots). \tag{4.73}$$

Here we assume $Q_0 = 0$ which corresponds to the base point of the oscillation.In this book, we also refer to (4.72) as the Toda lattice equation, though (4.73) is usually called that.

It is clear that this elliptic solution comes from the addition formula (4.71). By letting $t_0 = -\omega''$, we have the relation

$$\wp(t + nu_0 - \omega'') = (e_1 - e_3)\,\mathrm{dn}((e_1 - e_3)^{1/2}(u + nu_0))^2 + e_1,$$

which provides the connection with Toda's original solution in [To1967] for appropriate $e_1, e_2, e_3 \in \mathbb{R}$.

4.3.2 Toda Lattice: General g Case

Let us consider the hyperelliptic curves $X = X_g$ given by $f_X(x, y) = y^2 - f(x) = 0$ in (2.65) and (2.66) in this subsection.

We deal with the special case of Proposition 3.87, $v^{[1]} = v^{[2]}$, which leads us to the Toda equation as follows:

$$\begin{aligned}\frac{\sigma(u + 2v^{[1]})\sigma(u - 2v^{[1]})}{\sigma(u)^2\sigma_{\natural_2}(2v^{[1]})^2} &= f_{1,2}(x_1') - \sum_{i=1}^{g}\sum_{j=1}^{g} \wp_{ij}(u){x_1'}^{i+j-2}\\ &= f_{1,2}(x_1') + \sum_{i=1}^{g}\sum_{j=1}^{g} {x_1'}^{i-1}\frac{\partial}{\partial u_i}{x_1'}^{j-1}\frac{\partial}{\partial u_j}\log\sigma(u),\end{aligned} \tag{4.74}$$

where

$$f_{1,2}(x) := \frac{\partial_x^2 f(x)}{2f(x)} - f_{1,2}^I(x), \quad f_{1,2}^I(x) := \sum_{i=0}^{g}(i^2\lambda_{2i+1}x^{2i-1} + i(i-1)\lambda_{2i}x^{2i}).$$

In the derivation of (4.74), we compute the right-hand side in the equation in Proposition 3.87 by taking the limit as the l'Hôpital theorem,

$$\lim_{(x_2,y_2)\to(x_1,y_1)} \frac{G_1((x_1, y_1), (x_2, y_2))}{G_2((x_1, y_1), (x_2, y_2))} = \lim_{(x_2,y_2)\to(x_1,y_1)} \frac{\partial G_1((x_1, y_1), (x_2, y_2))/\partial x_1}{\partial G_2((x_1, y_1), (x_2, y_2))/\partial x_1}.$$

For $(x_i, y_i) \in X_g$ $(i = 1, \ldots, g)$, $(x_1', y_1') \in X_g$, we handle the Abel–Jacobi maps $u = w((x_1, y_1), \ldots, (x_g, y_g))$ and $v^{[1]} = w((x_1', y_1'))$. Let the point $c := 2v^{[1]}$ in the Jacobian, the differential operator

$$D_j := \sum_{i=1}^{g} {x_j'}^{i-1}\frac{\partial}{\partial u_i}$$

and the quantity

$$\mathscr{V}(u) := \mathscr{V}(u, v^{[1]}) := \sum_{i=1}^{g}\sum_{j=1}^{g} \wp_{ij}(u){x_1'}^{i+j-2} = -D_1^2\log\sigma(u), \quad \mathscr{V}_c(c) := f_{1,2}(x_1').$$

The addition theorem (4.74) turns out to be

$$\mathscr{V}(u) - \mathscr{V}_c(c) = -D_1^2 \log \sigma(u) - \mathscr{V}_c(c) = -\frac{\sigma(u+c)\sigma(u-c)}{\sigma(u)^2\sigma(c)^2}.$$

Let the operator $-D_1^2 \log$ act on both sides and thus we have

$$\begin{aligned}-D_1^2 \log(\mathscr{V}(u) - \mathscr{V}_c(c)) &= \mathscr{V}(u+c) - 2\mathscr{V}(u) + \mathscr{V}(u-c)\\ &= (\mathscr{V}(u+c) - \mathscr{V}_c(c)) - 2(\mathscr{V}(u) - \mathscr{V}_c(c))\\ &\quad + (\mathscr{V}(u-c) - \mathscr{V}_c(c)).\end{aligned}$$

By fixing $c \in J_X$, the direction of D_1 is determined in $\mathbb{C}^g$. Thus we decompose the vector space; let $t \in \mathbb{C}D_1 \subset \mathbb{C}^g$, i.e., $\langle dt, D_1\rangle = 1$, and let $t^\perp \in \mathbb{C}^g$ such that $\langle dt^\perp, D_1\rangle = 0$. We can let $u = t + nc + t^\perp$ and then we have

$$\begin{aligned}&\frac{d^2}{dt^2} \log\left(\mathscr{V}(t+nc+t^\perp) - \mathscr{V}_c(c)\right)\\ &= \mathscr{V}(t+(n+1)c+t^\perp) - 2\mathscr{V}(t+nc+t^\perp) + \mathscr{V}(t+(n-1)c+t^\perp).\end{aligned} \tag{4.75}$$

Thus we introduce a function $q_n(t; t^\perp, c) := \log\left(cV(t+nc+t^\perp) - \mathscr{V}_c(c)\right)$ on $\mathbb{C}^g$. Then it satisfies the Toda lattice equation:

Proposition 4.73

$$-\frac{d^2}{dt^2} q_n(t; t^\perp, c) = \mathrm{e}^{q_{n+1}(t;t^\perp,c)} - 2\mathrm{e}^{q_n(t;t^\perp,c)} + \mathrm{e}^{q_{n-1}(t;t^\perp,c)}. \tag{4.76}$$

This is a straightforward generalization of the genus one case (4.72), which is the quasi-periodic solution of the Toda lattice equation since in general, we cannot find an integer ℓ such that ℓc is in the lattice Γ_X. (When $g = 1$, for given c, we have $Q \in E$ satisfying $c = w(Q)$.)

4.3.3 *Periodic Solutions of Toda Lattice Equation and n-division Points*

In this subsection, we study the periodic solution of the Toda lattice equation in Proposition 4.73 [KodMP2013]. From Lemma 3.98 and Theorem 2.186, we consider the solution space of the $n(> g)$-division equation. We define

$$\Phi_n := \{P \in X_g \mid P \text{ is a zero of } \psi_n\}.$$

We recall the division points in Sect. 2.6.2, especially Corollary 2.187:

Lemma 4.74 [Can1994, MatP2009a] *For a positive integer $n(> g)$ and the set*

$$\Xi_n := \Phi_{n-g+1} \cap \ldots \cap \Phi_{n-1} \cap \Phi_n \cap \Phi_{n+1} \cap \ldots \cap \Phi_{n+g-1},$$

$P \in \Xi_n \subset J_X$ satisfies $nP \in \Gamma$.

For $2N(> g)$ satisfying $P \in \Xi_{2N}$, and $c = 2w(P)$, we have

$$\mathscr{V}(u) = \mathscr{V}(u + Nc).$$

Proposition 4.75 *For $P \in \Xi_{2N}$ and $c = 2w(P)$, the quasi-periodic solution $q_n(t; t^\perp, c)$ in Proposition 4.73 satisfies the periodic condition $q_{n+N}(t; t^\perp, c) = q_n(t; t^\perp, c)$ and is a periodic solution.*

Following Adler and van Moerbeke, and Kac and van Moerbeke [AdvM1991, KacM1975], we define

$$\sigma^{(n)}(t; t^\perp) := \sigma(t + nc + t^\perp), \quad \sigma^{(c)} := \sigma_{\natural_2}(c),$$

$$\zeta^{(n)}(t; t^\perp) := \sum_{i=1}^{g} {x'_1}^{i-1} \zeta_i(t + nc + t^\perp), \quad \zeta^{(c)} := \frac{1}{2} D_{1'} \log \sigma_{\natural_2}(c),$$

$$\wp^{(n)}(t; t^\perp) := \sum_{i,j=1}^{g} {x'_1}^{i+j-2} \wp_{ij}(t + nc + t^\perp), \quad \wp^{(c)}(t^\perp) := f_{1,2}(x'_1),$$

$$a_n = \wp^{(n)}(t; t^\perp) - \wp^{(c)}(t^\perp) = \frac{\sigma^{(n+1)}(t; t^\perp)\sigma^{(n-1)}(t; t^\perp)}{\sigma^{(n)}(t; t^\perp)^2 {\sigma^{(c)}}^2},$$

$$b_{n-1} = \frac{d}{dt} \log \frac{\sigma^{(n)}(t; t^\perp)}{\sigma^{(n-1)}(t; t^\perp)} - \zeta_c = \zeta^{(n)}(t; t^\perp) - \zeta^{(n-1)}(t; t^\perp) - \zeta^{(c)}.$$

Then we have the lemmas:

Lemma 4.76 [AdvM1991, KacM1975] *a_n and b_n satisfy*

$$\begin{cases} \dfrac{d}{dt} a_k = a_k(b_{k+1} - b_k), \\ \dfrac{d}{dt} b_k = a_k - a_{k-1}, \end{cases} \qquad (k = 1, 2, \ldots, N).$$

and its Lax matrix

$$\mathscr{L} := \begin{pmatrix} b_1 & 1 & 0 & \cdots & a_N \hat{w}^{-1} \\ a_1 & b_2 & 1 & \cdots & 0 \\ \vdots & \ddots & \ddots & \ddots & \vdots \\ 0 & \cdots & a_{N-2} & b_{N-1} & 1 \\ \hat{w} & \cdots & \cdots & a_{N-1} & b_N \end{pmatrix}$$

provides the hyperelliptic curve with the genus $N-1$ given by

$$\det(\mathscr{L}-z) = -\left(\hat{w} + \frac{\prod_{i=1}^{N} a_i}{\hat{w}} - \mathscr{P}(z)\right) = 0.$$

Lemma 4.77 *Let the hyperelliptic curve $\widehat{X}_{g,N-1}$ given in Lemma 4.76,*

$$\hat{X}_{g,N-1} := \{(z,\hat{w}) \mid \hat{w}^2 - \mathscr{P}(z)\hat{w} + \prod_{i=1}^{N} a_i = 0\} \bigcup \{\infty\}.$$

Then the polynomial $\mathscr{P}$ is given by

$$\mathscr{P}(z) := \tilde{\Delta}_{1,N}(z) - \tilde{\Delta}_{2,N-1}(z),$$

$$\tilde{\Delta}_{n,m} := \begin{vmatrix} b_m - z & 1 & 0 & \cdots & 0 \\ a_m & b_{m+1} - z & 1 & \cdots & 0 \\ \vdots & \ddots & \ddots & \ddots & \vdots \\ 0 & \cdots & a_{n-2} & b_{n-1} - z & 1 \\ 0 & \cdots & \cdots & a_{n-1} & b_n - z \end{vmatrix},$$

$$\mathscr{P}(z) := (-1)^N z^N + \sum_{k=1}^{N} (-1)^{N+k} \mathscr{I}_k z^{N-k},$$

where the coefficients in $\mathscr{P}(z)$ with respect to z are given by

$$\mathscr{I}_1 = \sum_{i=1}^{N} b_i, \quad \mathscr{I}_2 = \sum_{i>j}^{N} b_i b_j - \sum_{i=1}^{N} a_i, \quad \ldots\ldots,$$

$$\mathscr{I}_N = \tilde{\Delta}_{1,N}(0) - a_N \tilde{\Delta}_{2,N-1}(0), \quad \mathscr{I}_{N+1} = \prod_{i=1}^{N} a_i.$$

4.3.4 *Periodic and Quasi-periodic Solutions in Toda Equation*

This subsection summarizes the periodic and quasi-periodic solutions in the Toda equation. We have natural problems:

1. Show the explicit relation between X_g of genus g, and $\widehat{X}_{g,N-1}$ of genus N.
2. Show the explicit equation of $\widehat{X}_{g,N-1}$ in terms of R_X and zero points of ψ_{2N}.

As Hashimoto and Sakai investigated the relation between a hyperelliptic curve of genus two and an elliptic curve as a solution to the Poncelet problem [HasSa2009], it can be regarded as a case of $N = 2$ and $g = 1$ in the above problem since the elliptic Toda lattice solutions (4.72) have a direct transformation to the solutions of the Poncelet problem [KodMP2013, Appendix]. In other words, our problem is a generalization of the Poncelet problem [GriH1978].

For the problems, we have the following primitive observations.

1. The ramified points in $\widehat{X}_{g,N-1}$ are given by meromorphic functions in terms of the zeros of ψ_{2N}.
2. For $t^\perp \in \mathbb{C}^g$, and $u_0 \in \Xi_{2N}$ ($c = 2u_0$), the orbit in the Toda lattice equation,

$$M_{t^\perp,u_0} := \bigcup_{n=0}^{N-1} M_{n,t^\perp,u_0}$$

   has the cyclic group $\mathfrak{C}_N$ action of order N, where $M_{n,t^\perp,u_0} := \{tD_1 + 2nu_0 + t^\perp \in J_X \mid t \in \mathbb{C}\}$. There is a projection $\pi : M_{t^\perp,u_0} \to M_{0,t^\perp,u_0}$, and thus $\widehat{X}_{g,N-1}$ might be associated with the covering of X_g.
3. (a_n, b_n) in Lemma 4.76 or $(z, \widehat{w})$ in Lemma 4.77 should be regarded as sections of a line bundle on X_g.

4.3.5 *Trigonal Toda Equation*

In this subsection, we consider the trigonal extension of the Toda lattice equation by using the trigonal symmetry for the case of the elliptic curves [Mat2020].

4.3.5.1 Properties of Equiharmonic Elliptic Curves

Let us consider an elliptic curve E given by the affine equation

$$y(y - s) = x^3, \tag{4.77}$$

which is mentioned in Sects. 2.8.4 and 3.6.3.

The image of the incomplete elliptic integrals is acted on by $\mathrm{SL}(2, \mathbb{Z})$ and the cyclic group ζ_3. For $u, v \in \mathbb{C}$ ($v \neq 0$), we define a lattice

$$\mathscr{L}_{v,u} := \mathbb{Z}[\zeta_3]v + u := \{\ell_1 v + \ell_2 \zeta_3 v + u \mid \ell_1, \ell_2 \in \mathbb{Z}\}. \tag{4.78}$$

The fact that $\zeta_6 = \zeta_3 + 1$ for $\zeta_6 := \mathrm{e}^{2\pi\sqrt{-1}/6}$ implies $\mathbb{Z}[\zeta_3] = \mathbb{Z}[\zeta_6]$. Since $\mathscr{L}_{2\omega',0}$ agrees with the lattice of the periodicity, i.e., $x(u + L) = x(u)$, $y(u + L) = y(u)$ for $L \in \mathscr{L}_{2\omega',0} = \Gamma_E$, the Jacobian J_E of the curve E is given by

$$J_E = \mathbb{C}/\mathscr{L}_{2\omega',0}, \quad \kappa : \mathbb{C} \to J_E.$$

These points of the integrals for the branch points of the curve E are illustrated in Fig. 4.1. We also regard x and y as meromorphic functions on J_E due to their periodicity.

Further, Lemma 2.209 shows that ω_s and ω_0 belong to $\frac{1}{3}\mathscr{Z}_{2\omega',0}$.

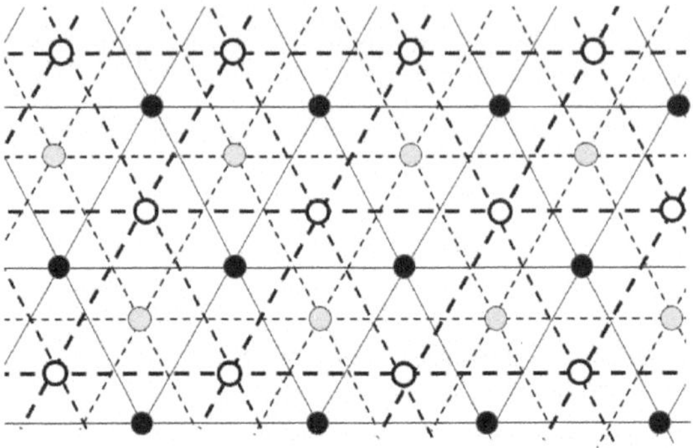

Fig. 4.1 The lattice points of $\mathscr{Z}_{2\omega',0}$ are denoted by the black dots, ω_0 and ω_s with $\mathscr{Z}_{2\omega',0}$ translations are denoted by gray dots and white dots respectively. This figure is the same as Fig. 2.12

4.3.5.2 Trigonal Toda Lattice Equation

The addition formula (3.75) gives the following lemma:

Lemma 4.78 *The quantity*

$$q(u,v) := \log(y(u) - y(v)), \quad \left(y(u) - y(v) = \mathrm{e}^{q(u,v)}\right)$$

satisfies the relation

$$-\frac{1}{2}\frac{d^3}{du^3}q(u,v) = \mathrm{e}^{q(u-v,v)} + \mathrm{e}^{q(u-\zeta_3 v,v)} + \mathrm{e}^{q(u-\zeta_3^2 v,v)} - 3\mathrm{e}^{q(u,v)}.$$

Proof We consider the logarithm of both sides of (3.75) and differentiate both sides three times with respect to u. Then we obtain

$$-\frac{d^3}{du^3}\log(y(u) - y(v)) = 2(y(u-v) + y(u-\zeta_3 v) + y(u-\zeta_3^2 v)) - 6(y(u)).$$

■

The right-hand side in Lemma 4.78 should be expressed by a difference operator. In order to express it, we prepare the geometry associated with Lemma 4.78.

We fix the complex numbers u_0 and v_0. We regard $\mathscr{Z}_{v_0,u_0}$ as the set of nodes $\mathscr{N}_{v_0,u_0}$ of an infinite directed (oriented) 6-regular graph $\mathscr{G}_{v_0,u_0}$ whose incoming degree and outgoing degree at each node are three, $\mathscr{N}_{v_0,u_0} = \mathscr{Z}_{v_0,u_0}$ [GodR2001]; $\mathscr{G}_{v_0,u_0}$ is

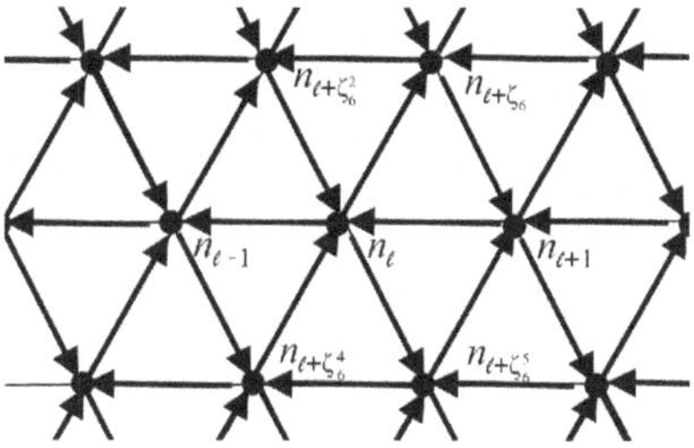

Fig. 4.2 The directed graph $\mathscr{G}_{v_0,u_0}$

illustrated in Fig. 4.2. Every node n_ℓ in $\mathscr{N}_{v_0,u_0}$ is labeled by an Eisenstein integer $\ell \in \mathbb{Z}[\zeta_3]$.

It is noted that every $v_0 \in \mathbb{C}$ is decomposed to $v_0 = v_0'\omega' + v_0''\omega''$ by the facts that $v_0', v_0'' \in \mathbb{R}$. Further, the quotient set of the lattice points modulo $\mathscr{Z}_{2\omega',0}$ is denoted by $\mathscr{N}_{v_0,u_0}/\mathscr{Z}_{2\omega',0} = \kappa(\mathscr{N}_{v_0,u_0})$. The following are obvious:

Lemma 4.79 *1. For $v_0 = v_0'\omega' + v_0''\omega''$ of $v_0', v_0'' \in \mathbb{Q}\bigcap[0, 2]$, the cardinality $|\mathscr{N}_{v_0,u_0}/\mathscr{Z}_{2\omega',0}|$ is finite for every $u_0 \in \mathbb{C}$, and*
2. for $v_0 = v_0'\omega' + v_0''\omega''$ of $v_0', v_0'' \in (\mathbb{R} \setminus \mathbb{Q})\bigcap[0, 2]$, $\mathscr{N}_{v_0,u_0}/\mathscr{Z}_{2\omega',0}$ is dense in J_E for every $u_0 \in \mathbb{C}$.

Let us introduce the function spaces, Ω and $\log\Omega$,

$$\Omega := \{Q : \mathbb{C} \times \mathscr{N}_{v_0,u_0} \to \mathbb{P} \mid \text{meromorphic}\}, \quad \log\Omega := \{q : \mathbb{C} \times \mathscr{N}_{v_0,u_0} \to \mathbb{P} \mid \mathrm{e}^q \in \Omega\}.$$

For an Eisenstein integer $\ell \in \mathbb{Z}[\zeta_3]$ or $n_\ell \in \mathscr{N}_{v_0,u_0}$, $t \in \mathbb{C}$ and fixed $u_0, v_0 \in \mathbb{C}$, let us consider an element in $\log\Omega$,

$$q_\ell(t; u_0, v_0) := q(t + u_0 - \ell v_0, v_0) = \log(y(t + u_0 - \ell v_0) - y(v_0)), \qquad (4.79)$$

which is denoted by $q_\ell(t)$ for brevity.

Lemma 4.78 gives the nonlinear relation on $\log\Omega$:

Proposition 4.80 *For $n_\ell \in \mathscr{N}_{v_0,u_0}$, $t \in \mathbb{C}$ and fixed $u_0, v_0 \in \mathbb{C}$, $q_\ell(t) = q_\ell(t; u_0, v_0)$ satisfies the relation*

$$-\frac{1}{2}\frac{d^3}{dt^3}q_\ell(t) = \mathrm{e}^{q_{\ell+1}(t)} + \mathrm{e}^{q_{\ell+\zeta_3}(t)} + \mathrm{e}^{q_{\ell+\zeta_3^2}(t)} - 3\mathrm{e}^{q_\ell(t)}. \qquad (4.80)$$

It is emphasized that (4.80) can be regarded as a differential-difference non-linear equation and its special solution is given by (4.79) for the elliptic curve (4.77). Its derivation is basically the same as Toda's original derivation of the Toda lattice equation [To1967] and that in [AdvM1991, KacM1975]. Further, it is related to the infinite graph $\mathscr{G}_{v_0,u_0}$. Thus we call this relation *the trigonal Toda lattice equation*.

We recall $\zeta_6 = \zeta_3 + 1$. For a given $n_\ell \in \mathscr{N}_{v_0,u_0}$, let us consider subgraph $\mathscr{G}_\ell \subset \mathscr{G}_{v_0,u_0}$ given by its nodes $\mathscr{N}_\ell := \{n_\ell, n_{\ell+1}, n_{\ell+\zeta_6}, n_{\ell+\zeta_6^2}, \ldots, n_{\ell+\zeta_6^5}\}$ $(\subset \mathscr{N}_{v_0,u_0})$; $\mathscr{N}_\ell$

consists of the center point n_ℓ with a hexagon $n_{\ell+\zeta_6^i}$ $(i = 0, 1, \ldots, 5)$. The submatrix of the incoming adjacency matrix $\mathscr{A}_{\mathrm{in}}$ for $\mathscr{G}_\ell$ is given by

$$\mathscr{A}_{\mathrm{in}}|_{\mathscr{G}_\ell} = \begin{pmatrix} 0&1&0&1&0&1&0\\ 0&0&1&0&0&0&1\\ 1&0&0&0&0&0&0\\ 0&0&1&0&1&0&0\\ 1&0&0&0&0&0&0\\ 0&0&0&0&1&0&1\\ 1&0&0&0&0&0&0 \end{pmatrix}.$$

The diagonal matrix gives the incoming degree matrix $\mathscr{D}_{\mathrm{in}}$ whose diagonal element is three. Thus we define the incoming Laplacian [GodR2001],

$$\Delta_{\mathrm{in}} := \mathscr{D}_{\mathrm{in}} - \mathscr{A}_{\mathrm{in}}.$$

Let us consider the functions $q \in \log \Omega$ and $e^q \in \Omega$ whose components at n_ℓ are given by $q_\ell(t)$ and $\mathrm{e}^{q_\ell(t)}$. We regard them as column vectors for each $\ell \in \mathbb{Z}[\zeta_3]$. Then the Laplacian acts on the vector spaces.

The incoming Laplacian shows that Proposition 4.80 is reduced to the following formula.

Proposition 4.81 *Using the above notations, (4.80) is written as*

$$\frac{d^3}{dt^3} q(t) = -\Delta_{\mathrm{in}} e^q(t).$$

It turns out that the trigonal Toda lattice equation consists of the third-order differential operators and trigonal graph Laplacian, which is a natural generalization of the original Toda lattice equation (4.72) [To1967].

Remark 4.82 We apply the above method to the trigonal curve (2.67).

By letting

$$D_t = \frac{1}{F(x_1', x_1, x_2, x_3)} \sum_{i=1}^{3} \phi_{i-1}(x_1', y_1') \frac{\partial}{\partial u_i}, \quad D_3 = \frac{\partial}{\partial u_3}$$

and

$$V(u + \hat{\zeta}^i v) := \frac{\mu_4((x_1', \zeta^i y_1'; (x_1, y_1), \ldots, (x_3, y_2))}{F(x_1', x_1, x_2, x_3)},$$

we have

$$\sum_{i=0}^{2} D_t D_3 \log V(u + \hat{\zeta}^i v) = 3V(u) - \sum_{i=0}^{2} V(u + \hat{\zeta}^i v),$$

which is the generalized Toda lattice equation of the non-hyperelliptic trigonal curve.

4.4 Difference Relations from Addition Theorem

In this section, we investigate the addition structures of curves of genus one and two as the difference equations following [Mat2002a]:[5] after [Mat2002a], Braden, Enolskii and Hone studied similar systems [BraEH2005, Hon2007].

The difference equations and ultra-discrete equations have been studied in integrable system for three decades, and play a crucial role in the integrable system [HirT2003, HieJN2016]. Thus in this section, we show a connection between discrete equations and algebraic curves. Since we study them based on the recursion relations in Proposition 3.102, we show a sketch of the proof as follows and show their behavior as the discrete equations.

4.4.1 *Proof of the Recursion Relation of $\widetilde{\psi}_n$ of $g = 2$*

In this subsection, we demonstrate the proof of the second relation in Proposition 3.102, i.e., for positive integers $m, n > 3$,

$$\widetilde{\psi}_2^2\widetilde{\psi}_m\widetilde{\psi}_n\widetilde{\psi}_{m+n}\widetilde{\psi}_{m-n} = \begin{vmatrix} \widetilde{\psi}_{m-2}\widetilde{\psi}_n & \widetilde{\psi}_{m-1}\widetilde{\psi}_{n+1} & \widetilde{\psi}_m\widetilde{\psi}_{n+2} \\ \widetilde{\psi}_{m-1}\widetilde{\psi}_{n-1} & \widetilde{\psi}_m\widetilde{\psi}_n & \widetilde{\psi}_{m+1}\widetilde{\psi}_{n+1} \\ \widetilde{\psi}_m\widetilde{\psi}_{n-2} & \widetilde{\psi}_{m+1}\widetilde{\psi}_{n-1} & \widetilde{\psi}_{m+2}\widetilde{\psi}_n \end{vmatrix}. \tag{4.81}$$

We assume that m and n are greater than three in this subsection. For $u \in w(X)$, we introduce the following quantities,

$$\xi_0(m, n, u) := \frac{\widetilde{\psi}_{m-n}(u)\widetilde{\psi}_{m+n}(u)}{\widetilde{\psi}_m(u)^2\widetilde{\psi}_n(u)^2}, \quad \xi_1(m, u) := \frac{\widetilde{\psi}_{m-2}(u)\widetilde{\psi}_{m+2}(u)}{\widetilde{\psi}_2(u)^2\widetilde{\psi}_m(u)^2},$$

$$\xi_2(m, n, u) := \frac{\widetilde{\psi}_{m-1}(u)\widetilde{\psi}_{m+1}(u)}{\widetilde{\psi}_m(u)^2\widetilde{\psi}_2(u)^2}\left[\frac{\widetilde{\psi}_{n-1}(u)^2\widetilde{\psi}_{n+2}(u)}{\widetilde{\psi}_n(u)^3} + \frac{\widetilde{\psi}_{n+1}(u)^2\widetilde{\psi}_{n-2}(u)}{\widetilde{\psi}_n(u)^3}\right].$$

The following is obvious:

Lemma 4.83 *The recursion relation (4.81) is reduced to*

$$\xi_0(m, n, u) + \xi_1(m, u) - \xi_1(n, u) + \xi_2(m, n, u) - \xi_2(n, m, u) = 0.$$

The following is the key lemma in this subsection:

Lemma 4.84 *For $u = w((x, y))$, we have the following representations:*

1. $\xi_0(m, n, u) = -(\wp_{11}(mu) - \wp_{11}(nu) + \wp_{12}(mu)\wp_{22}(nu) - \wp_{22}(mu)\wp_{12}(nu))$.

---

[5] In the paper [Mat2002a], “the discrete Painlevé equation” should read “the q-discrete Painlevé equation”.

2. $\xi_1(m,u) = -(\wp_{11}(mu) - \wp_{11}(2u) + 2\wp_{12}(mu)x + \wp_{22}(mu)x^2)$.
3. $\xi_2(m,n,u) - \xi_2(n,m,u) = (\wp_{22}(mu)x^2 + 2\wp_{12}(mu)x - \wp_{12}(nu)\wp_{22}(mu))$
$-(\wp_{22}(nu)x^2 + 2\wp_{12}(nu)x - \wp_{12}(mu)\wp_{22}(nu))$.

Clearly the right-hand sides of these equations in Lemma 4.84 identically satisfy the relation in Lemma 4.83. Thus the aim of this subsection is to prove Lemma 4.84, and we show the proof.

We first show the expressions of ξ's.

Lemma 4.85

$$\xi_0(m.n.u) = \frac{\sigma((m+n)u)\sigma((m-n)u)}{\sigma(mu)^2\sigma(nu)^2}, \quad \xi_1(m,u) = \frac{\sigma((m-2)u)\sigma((m+2)u)}{\sigma(mu)^2\sigma(2u)^2},$$

$$\begin{aligned}\xi_2(m,n,u) =& \frac{\sigma_2(u)^8}{\sigma(2u)^2}\frac{\sigma((m-1)u)\sigma((m+1)u)}{\sigma(mu)^2\sigma_2(u)^2}\left[\frac{\sigma((n+1)u)\sigma((n-1)u)}{\sigma(nu)^2\sigma_2(u)^2}\right]^2\\ &\times\left[\frac{\sigma(nu)\sigma((n+2)u)}{\sigma((n+1)u)^2\sigma_2(u)^2} + \frac{\sigma(nu)\sigma((n-2)u)}{\sigma((n-1)u)^2\sigma_2(u)^2}\right].\end{aligned}$$

Proof

$$\xi_0(m.n.u) = \frac{\sigma((m+n)u)\sigma((m-n)u)}{\sigma_2(u)^{(m+n)^2}\sigma_2(u)^{(m-n)^2}}\frac{\sigma_2(u)^{2m^2}\sigma_2(u)^{2n^2}}{\sigma(mu)^2\sigma(nu)^2},$$

$$\xi_1(m,u) = \frac{\sigma((m-2)u)\sigma((m+2)u)}{\sigma_2(u)^{(m-2)^2+(m+2)^2}}\frac{\sigma_2(u)^{2m^2}}{\sigma(mu)^2}\frac{\sigma_2(u)^4}{\sigma(2u)^2},$$

$$\begin{aligned}\xi_2(m,n,u) =& \frac{\sigma_2(u)^8}{\sigma(2u)^2}\frac{\sigma((m-1)u)\sigma((m+1)u)}{\sigma_2(u)^{(m-1)^2+(m+1)^2}}\frac{\sigma_2(u)^{2m^2+3n^2}}{\sigma(mu)^2\sigma(nu)^3}\\ &\times\left[\frac{\sigma((n-1)u)^2\sigma((n+2)u)}{\sigma_2(u)^{2(n-1)^2+(n+2)^2}} + \frac{\sigma((n+1)u)^2\sigma((n-2)u)}{\sigma_2(u)^{2(n+1)^2+(n-2)^2}}\right]\\ =& \frac{\sigma((m-1)u)\sigma((m+1)u)}{\sigma(mu)^2\sigma(2u)^2}\\ &\times\left[\frac{\sigma((n-1)u)^2\sigma((n+2)u)}{\sigma(nu)^3} + \frac{\sigma((n+1)u)^2\sigma((n-2)u)}{\sigma(nu)^3}\right].\end{aligned}$$

We recall the relations for hyperelliptic curve X of $g = 2$, which show the explicit representations of $\wp$'s:

Lemma 4.86 *For* $u = w((x_1, y_1), (x_2, y_2)) \in J_X$,

1. $\wp_{11} = \dfrac{f_\Omega(x_1, x_2) - 2y_1y_2}{(x_1 - x_2)^2}$, $\wp_{12} = -x_1x_2$, $\wp_{22} = x_1 + x_2$,
2. $x_a^2 - \wp_{22}(u)x_a - \wp_{21} = 0$, $a = 1, 2$,

3. $\wp_{222} = 2\dfrac{y_1 - y_2}{(x_1 - x_2)}$, $\wp_{221} = 2\dfrac{x_1 y_2 - x_2 y_1}{(x_1 - x_2)}$, $\wp_{211} = -2\dfrac{x_1^2 y_2 - x_2^2 y_1}{(x_1 - x_2)}$,
$\wp_{111} = 2\dfrac{y_2 \widetilde{f_\Omega}(x_1, x_2) - y_1 \widetilde{f_\Omega}(x_2, x_1)}{(x_1 - x_2)}$, *where* $\widetilde{f_\Omega}(x_1, x_2) = 4\lambda_0 + \lambda_1(3x_1 + x_2) + 2\lambda_2 x_1(x_1 + x_2) + \lambda_3 x_1^2(x_1 + 3x_2) + 4\lambda_4 x_1^3 x_2 + x_1^3 x_2(3x_1 + x_2)$, *and*
4. $2y_a = \wp_{222}(u)x_a + \wp_{221}(u)$, $a = 1, 2$.

Proof *1* and *2* are obvious. On *3*, we use the relation,

$$\begin{pmatrix} \partial_{u_1} \\ \partial_{u_2} \end{pmatrix} = \frac{1}{x_2 - x_1} \begin{pmatrix} 2x_2 y_1 & -2x_1 y_2 \\ -2y_1 & 2y_2 \end{pmatrix} \begin{pmatrix} \partial_{x_1} \\ \partial_{x_2} \end{pmatrix},$$

and then we obtain these relations directly. ■

The addition formula (3.43) in Corollary 3.89 is rewritten as follows.

Lemma 4.87 *Let u, v be points in $\mathbb{C}^2$.*

1.

$$\frac{\sigma(u+v)\sigma(u-v)}{\sigma(u)^2\sigma(v)^2} = Q(u, v), \tag{4.82}$$

$$Q(u, v) := -(\wp_{11}(u) - \wp_{11}(v) + \wp_{12}(u)\wp_{22}(v) - \wp_{12}(v)\wp_{22}(u)). \tag{4.83}$$

2.

$$\wp_{ij}(u+v) + \wp_{ij}(u-v) = 2\wp_{ij}(u) - \frac{Q_{ij}Q - Q_i Q_j}{Q^2}, \tag{4.84}$$

where $Q_i(u, v) := \partial_{u_i} Q$ and $Q_{ij}(u, v) := \partial_{u_j} Q_i$.

Let us introduce a lifted function defined as

$$\widetilde{\Psi}_n(u) = \frac{\sigma(nu)}{\sigma_2(u)^{n^2}}, \tag{4.85}$$

for $u \in J_X$, which is limit as $\widetilde{\psi}_n(u) = \lim_{u \to w(X)} \widetilde{\Psi}_n(u)$.[6] After considering relations over J_X, we will restrict some quantities into $w(X)$. We also note that even if the local parameter u is in $w(X)$, the point nu $(n > 1)$ is in J_X but not in $w(X)$ in general. (4.85) shows that the addition formula (4.82) is expressed by

$$\frac{\widetilde{\Psi}_{m+n}(u)\widetilde{\Psi}_{m-n}(u)}{\widetilde{\Psi}_n(u)^2\widetilde{\Psi}_m(u)^2} = Q(mu, nu). \tag{4.86}$$

[6] The first half of the following is similar to the relational expression that Uchida's division polynomial has [U2011].

We introduce a quantity $\Xi_3(nu, u)$ as

$$\Xi_3(nu, u) := \frac{1}{\widetilde{\Psi}_n(u)^3}\left(\widetilde{\Psi}_{n-1}(u)^2\widetilde{\Psi}_{n+2}(u) + \widetilde{\Psi}_{n+1}(u)^2\widetilde{\Psi}_{n-2}(u)\right) \tag{4.87}$$

for $u \in J_X$, whose restriction to $w(X)$ is a part of ξ_2 in Lemma 4.85 as in (4.88). Ξ_3 has a natural representation:

Lemma 4.88 $\Xi_3(mu, u)$ *is expressed by*

$$\begin{aligned}
\widetilde{\Psi}_1(u)^6\Big(&Q(mu, u)\big[2Q(mu, u)^2 + Q_{11}(mu, u) - \wp_{22}(u)Q_{12}(mu, u)\\
&\qquad + \wp_{12}(u)Q_{22}(mu, u)\big]\\
&- Q_1(mu, u)^2 - \wp_{22}(u)Q_1(mu, u)Q_2(mu, u)) + \wp_{12}(u)Q_2(mu, u)^2\Big).
\end{aligned}$$

Proof First, we rewrite the right-hand side of (4.87) by

$$\widetilde{\Psi}_1(u)^6\frac{\widetilde{\Psi}_{n-1}(u)^2\widetilde{\Psi}_{n+1}(u)^2}{\widetilde{\Psi}_n(u)^4\widetilde{\Psi}_1(u)^4}\left(\frac{\widetilde{\Psi}_{n+2}(u)\widetilde{\Psi}_n(u)}{\widetilde{\Psi}_{n+1}(u)^2\widetilde{\Psi}_1(u)^2} + \frac{\widetilde{\Psi}_{n-2}(u)\widetilde{\Psi}_n(u)}{\widetilde{\Psi}_{n-1}(u)^2\widetilde{\Psi}_1(u)^2}\right).$$

Lemma 4.87 shows the following:

$$\begin{aligned}
&\widetilde{\Psi}_1(u)^6 Q(nu, u)^2(Q((n+1)u, u) + Q((n-1)u, u))\\
&= \widetilde{\Psi}_1(u)^6 Q(nu, u)^2\Big[2\wp_{11}(u) - (\wp_{11}((n+1)u) + \wp_{11}((n-1)u))\\
&\quad - (\wp_{12}((n+1)u) + \wp_{12}((n-1)u))\wp_{22}(u)\\
&\quad + (\wp_{22}((n+1)u) + \wp_{22}((n-1)u))\wp_{12}(u)\Big]\\
&= \widetilde{\Psi}_1(u)^6 Q(nu, u)^2\Big[2\wp_{11}(u) - 2\wp_{11}(nu) + \frac{Q_{11}(nu, u)Q(nu, u) - Q_1(nu, u)^2}{Q(nu, u)^2}\\
&\quad - \wp_{22}(u)\Big(2\wp_{12}(nu) - \frac{Q_{12}(nu, u)Q(nu, u) - Q_1(nu, u)Q_2(nu, u)}{Q(nu, u)^2}\Big)\\
&\quad + \wp_{12}(u)\Big(2\wp_{22}(nu) - \frac{Q_{22}(nu, u)Q(nu, u) - Q_2(nu, u)^2}{Q(nu, u)^2}\Big)\Big]\\
&= \widetilde{\Psi}_1(u)^6\Big[2Q(nu, u)^2(\wp_{11}(u) - \wp_{11}(nu) + \wp_{22}(u)\wp_{12}(nu) - \wp_{12}(u)\wp_{22}(nu))\\
&\quad + Q_{11}(nu, u)Q(nu, u) - Q_1(nu, u)^2\\
&\quad - \wp_{22}(u)(Q_{12}(nu, u)Q(nu, u) - Q_1(nu, u)Q_2(nu, u))\\
&\quad + \wp_{12}(u)(Q_{22}(nu, u)Q(nu, u) - Q_2(nu, u)^2)\Big].
\end{aligned}$$

By arranging these terms, we obtain the formula. ■

Hereafter, let us consider the restriction of these quantities to the embedded curve $w(X) \subset J_X$.

Lemma 4.89 *1. For $u = w(x, y)$, $\wp_{22}(2u) = 2x$, $\wp_{12}(2u) = -x^2$, and*

$$\wp_{11}(2u) = -6x^3 - 4\lambda_4 x^2 - 2\lambda_3 x - \lambda_4 + \frac{f(x)'^2}{4f(x)},$$

where $f(x)' := df(x)/dx$.

2. *For $u \in w(x, y)$, $\widetilde{\psi}_2(u) = 2y$.*
3. *For $u \in J_X$, $\lim_{u\to w(x,y)} \left(\widetilde{\Psi}_1(u)^2 \wp_{11}((x, y))\right) = x^2$, $\lim_{u\to w(X)} \left(\widetilde{\Psi}_1(u)^2 \wp_{12}(u)\right) = -x$, and $\lim_{u\to w(x,y)} \left(\widetilde{\Psi}_1(u)^2 \wp_{22}(u)\right) = 1$.*

Proof Lemma 4.86 leads to *1* directly. On *2*, $\widetilde{\psi}_2(u) = \dfrac{\sigma(2u)}{\sigma_2(u)^4}$ is given in Lemma 3.70. From Corollary 3.62, we have

$$x = \lim_{u\to w(X)} \frac{\wp_{12}(u)}{\wp_{22}(u)} = \lim_{u\to w(X)} \frac{\sigma_1(u)}{\sigma_2(u)}.$$

Noting the definition of $\wp$ and using the relation

$$\lim_{u\to w(X)} \left(\widetilde{\Psi}_1(u)^2 \wp_{ij}(u)\right) = \lim_{u\to w(X)} \left(\frac{\sigma(u)^2}{\sigma_2(u)^2} \frac{\sigma_i(u)\sigma_j(u) - \sigma_{ij}(u)\sigma(u)}{\sigma(u)^2}\right),$$

we obtain *3*. ∎

Lemma 4.89 leads to the following lemma:

Lemma 4.90

1. $\lim_{u\to w(X)} Q(nu, 2u) = \wp_{11}(nu) - \wp_{11}(2u) - 2\wp_{12}(nu)x - \wp_{22}(nu)x^2$.
2. *The quantities,*

$$q(nu, u) := \lim_{u\to w(X)} \widetilde{\Psi}_1(u)^2 Q(nu, u), \quad q_i(nu, u) := \lim_{u\to w(X)} \widetilde{\Psi}_1(u)^2 Q_i(nu, u),$$

$$q_{ij}(nu, u) := \lim_{u\to w(X)} \widetilde{\Psi}_1(u)^2 Q_{ij}(nu, u),$$

become

$$\begin{aligned} q(nu, u) &= -x^2 + \wp_{22}(nu)x + \wp_{12}(nu), \\ q_{ij}(nu, u) &= \wp_{12ij}(nu) + \wp_{22ij}(nu)x, \\ q_i(nu, u) &= \wp_{12i}(nu) + \wp_{22i}(nu)x. \end{aligned}$$

3. $\xi_3(nu, u) := \lim_{u\in w(X)} \Xi_3(nu, u)$ *satisfies*

$$\xi_3(nu,u) = q(nu,u)\Big(2q(nu,u)^2 - q_{12}(nu,u) - xq_{22}(nu,u)\Big)$$
$$- q_2(nu,u)\Big(q_1(nu,u)) + xq_2(nu,u))\Big). \tag{4.88}$$

Proof $Q(nu, 2u)$ in (4.83) is restricted to $w(X)$ due to Lemma 4.89 *3*, and then *1* is obtained. Similarly *2* and *3* are proved. In *3*, we note that

$$\lim_{u \to w(X)} \widetilde{\Psi}_1(u)^6 Q_{11}(nu,u)^2 = 0, \quad \lim_{u \to w(X)} \widetilde{\Psi}_1(u)^6 Q(nu,u) Q_{11}(nu,u) = 0,$$

due to the order of the zero at $w(X)$. ■

Finally, we show a sketch of the proof of Lemma 4.84.

4.4.1.1 Proof of Lemma 4.84

1 and *2* are obvious by (4.86) and Lemma 4.90. *3* requires careful calculations. $\xi_2(m,n,u)$ consists of $\xi_3(nu,u)$ due to the definition in Lemma 4.88.

The paper [Mat2003a] precisely shows the proof by using the identities in Theorem 4.25 for genus two, and relations among $\wp_{ijk}$ and $\wp_{mn}$ for Lemma 4.86. However, it is quite complicated and requires tedious computations. Thus, we demonstrate the expressions of $\xi_3(nu,u)$ by considering the particular case $(n-1)u = w(x_{\mathrm{n}}, y_{\mathrm{n}})$ for a certain point $(x_{\mathrm{n}}, y_{\mathrm{n}}) \in X$, though in general, $(n-1)u$ belongs to $J_X \setminus w(X)$. For the case u, $(n-1)u \in w(X) \subset J_X$, we have the following relations:

1. $q(nu,u) = -(x^2 - \wp_{22}(nu)x - \wp_{21}(nu)) = 0$ due to Lemma 4.86 *2*,
2. $q_1(nu,u) = -2\dfrac{x^2 y_{\mathrm{n}} - x_{\mathrm{n}}^2 y}{x - x_{\mathrm{n}}} + 2\dfrac{x^2 y_{\mathrm{n}} - x x_{\mathrm{n}} y}{x - x_{\mathrm{n}}} = 2x_{\mathrm{n}} y = -2\dfrac{\wp_{21}(nu)}{x} y$ due to Lemma 4.86 *1*,
3. $q_2(nu,u) = 2y$ due to Lemma 4.86 *4*, and
4. $(q_1(nu,u) + xq_2(nu,u)) = 2y\left[x - \dfrac{\wp_{21}(nu)}{x}\right]$.

Thus we conclude that $\xi_3(nu,u) = 4y^2\left[x - \dfrac{\wp_{21}(nu)}{x}\right]$. (This formula holds for a general $u \in J_X$.) Hence in terms of this expression of ξ_3, we have the expression of ξ_2

$$\begin{aligned}\xi_2(m,n,u) &= -\frac{1}{4y^2}(x^2 - \wp_{22}(mu)x - \wp_{21}(mu))\xi_3(nu,u)\\ &= -[x^3 - \wp_{21}(nu) - \wp_{22}(mu)x + \wp_{22}(mu)\wp_{21}(nu)\\ &\quad + \wp_{21}(mu)x - \wp_{21}(mu)\wp_{21}(nu)].\end{aligned}$$

Finally, we obtain $\xi_2(m,n,u) - \xi_2(n,m,u)$ as in Lemma 4.84. ■

4.4.2 Difference Relations via Recursion Relations of $g = 1$

This subsection mentions the difference relations via recursion relations of $g = 1$. The division polynomial $\widetilde{\psi}$ of genus one (3.65) obeys the recursion relation in Proposition 3.102 for the $g = 1$ case. For the $m = 2$ case, we have a bilinear difference equation,

$$\widetilde{\psi}_{n+2}\widetilde{\psi}_{n-2} - \widetilde{\psi}_2^2\widetilde{\psi}_{n-1}\widetilde{\psi}_{n+1} + \widetilde{\psi}_3\widetilde{\psi}_1\widetilde{\psi}_n\widetilde{\psi}_n = 0. \tag{4.89}$$

By introducing the quantity $\beta_n = \widetilde{\psi}_{n+1}\widetilde{\psi}_{n-1}/\widetilde{\psi}_n^2$, (4.89) turns out to be the difference equation

$$\beta_{n+1}\beta_{n-1} = \frac{\widetilde{\psi}_2^2}{\beta_n} - \frac{\widetilde{\psi}_3\widetilde{\psi}_1}{\beta_n^2}. \tag{4.90}$$

This equation is identified with the q-discrete Painlevé I,

$$\beta_{n+1}\beta_{n-1} = \frac{z}{\beta_n} + \frac{a}{\beta_n^2}, \tag{4.91}$$

[GraTRCT2002, (2.16)], [OhtRGT1996]. We have a special solution of (4.91) as

$$\beta_n = \frac{\sigma((n+1)u)\sigma((n-1)u)}{\sigma(u)^2\sigma(ru)^2}.$$

4.4.3 Difference Relations via Recursion Relations of $g = 2$

This subsection shows the difference relations via recursion relations of $g = 2$. The division polynomial $\widetilde{\psi}$ of genus two (3.65) obeys the recursion relation in Proposition 3.102 for the $g = 2$ case. Noting $\widetilde{\psi}_0 = 0$, $\widetilde{\psi}_1 = 0$, $\widetilde{\psi}_2 = 2y$ and $\widetilde{\psi}_3 = 8y^3$, we have the $n = 3$ case of (4.92),

$$\begin{aligned}\widetilde{\psi}_3\widetilde{\psi}_2^2\widetilde{\psi}_{m+3}\widetilde{\psi}_{m-3}\widetilde{\psi}_m - \widetilde{\psi}_3^3\widetilde{\psi}_{m+2}\widetilde{\psi}_m\widetilde{\psi}_{m-2} - \widetilde{\psi}_5\widetilde{\psi}_2^2\widetilde{\psi}_{m+1}\widetilde{\psi}_m\widetilde{\psi}_{m-1}\\ + \widetilde{\psi}_4\widetilde{\psi}_3\widetilde{\psi}_2(\widetilde{\psi}_{m-2}\widetilde{\psi}_{m+1}^2 + \widetilde{\psi}_{m+2}\widetilde{\psi}_{m-1}^2) = 0,\end{aligned} \tag{4.92}$$

and the $n = 4$ case,

$$\begin{aligned}\widetilde{\psi}_4\widetilde{\psi}_2^2\widetilde{\psi}_{m+4}\widetilde{\psi}_{m-4}\widetilde{\psi}_m - \widetilde{\psi}_4^3\widetilde{\psi}_{m+2}\widetilde{\psi}_m\widetilde{\psi}_{m-2} + \widetilde{\psi}_6\widetilde{\psi}_4\widetilde{\psi}_2\widetilde{\psi}_m^3\\ - (\widetilde{\psi}_5^2\widetilde{\psi}_2 + \widetilde{\psi}_6\widetilde{\psi}_3^2)\widetilde{\psi}_{m+1}\widetilde{\psi}_m\widetilde{\psi}_{m-1}\\ + \widetilde{\psi}_5\widetilde{\psi}_4\widetilde{\psi}_3(\widetilde{\psi}_{m-2}\widetilde{\psi}_{m+1}^2 + \widetilde{\psi}_{m+2}\widetilde{\psi}_{m-1}^2) = 0.\end{aligned} \tag{4.93}$$

Due to Cantor [Can1994], (4.92) and (4.93) are reduced to

$$\begin{aligned}\widetilde{\psi}_4\widetilde{\psi}_2^3\widetilde{\psi}_{n+4}\widetilde{\psi}_{n-4} - \widetilde{\psi}_5\widetilde{\psi}_3\widetilde{\psi}_2^2\widetilde{\psi}_{n+3}\widetilde{\psi}_{n-3} + (\widetilde{\psi}_5\widetilde{\psi}_3^3 - \widetilde{\psi}_4^3\widetilde{\psi}_2)\widetilde{\psi}_{n+2}\widetilde{\psi}_{n-2} \\ + \widetilde{\psi}_6\widetilde{\psi}_3^2\widetilde{\psi}_2\widetilde{\psi}_{n+1}\widetilde{\psi}_{n-1} - \widetilde{\psi}_6\widetilde{\psi}_4\widetilde{\psi}_2^2\widetilde{\psi}_n\widetilde{\psi}_n = 0.\end{aligned} \tag{4.94}$$

This bilinear difference equation (4.94) is a genus two analogue of the bilinear equation (4.89) and also characterizes the $\widetilde{\psi}$ function over a genus two curve. We emphasize that from the point of view of bilinear difference equations, (4.94) is a very natural extension of (4.89) and cannot be reduced to a three-term relation in general.

From here, we consider modifications of (4.92)–(4.94) to find difference equations which have the form

$$x_{n+m}x_{n-m'} = \frac{f_1(x_{n+m-1}, \cdots, x_{n-m'+1})}{f_2(x_{n+m-1}, \cdots, x_{n-m'+1})}, \tag{4.95}$$

where f's are polynomials of $x_{n+m-1}, \cdots, x_{n-m'+1}$ and $m, m' = 1, 2, 3, \cdots$, by deforming (4.94) as a generalization of (4.91).

By setting $b_n = \widetilde{\psi}_{n+2}\widetilde{\psi}_{n-2}/\widetilde{\psi}_n^2$, (4.94) becomes a sixth-order ordinary difference equation

$$\begin{aligned}\widetilde{\psi}_4\widetilde{\psi}_2^3 b_{n+3}b_{n-3} - \frac{1}{b_{n-2}^2 b_{n-1}^3 b_n^4 b_{n+1}^3 b_{n+2}^2}\Big(\widetilde{\psi}_5\widetilde{\psi}_3\widetilde{\psi}_2^2 b_{n-2}b_{n-1}^2 b_n^3 b_{n+1}^2 b_{n+2} \\ - (\widetilde{\psi}_5\widetilde{\psi}_3^3 - \widetilde{\psi}_4^3\widetilde{\psi}_2)b_{n-1}b_n^2 b_{n+1} + \widetilde{\psi}_6\widetilde{\psi}_3^2\widetilde{\psi}_2 b_n - \widetilde{\psi}_6\widetilde{\psi}_4\widetilde{\psi}_2^2\Big) = 0\end{aligned} \tag{4.96}$$

for a generic point in $\widetilde{w} \circ \iota_X(X)$. (4.96) can be regarded as a generalization of (4.91).

Further, we consider other difference equations by dealing with special points. First, we consider points $(x, 0)$, $y = 0$ or the ramification point $B_r := (b_r, 0)$, $(r = 1, 2, 3, 4, 5)$. We recall $\alpha_n \in \mathbb{C}[x]$ in Corollary 2.193 and $\psi_3 = \psi_2^2$.

$$\widetilde{\psi}_n = \begin{cases} \widetilde{\psi}_3\alpha_n(x) & \text{for } n \text{ odd}, \\ \widetilde{\psi}_2\alpha_n(x) & \text{for } n \text{ even}. \end{cases}$$

For the point, (4.94) with even n becomes

$$\alpha_{n+4}\alpha_{n-4} - (\alpha_4^2)\alpha_{n+2}\alpha_{n-2} + \alpha_6\alpha_n\alpha_n = 0. \tag{4.97}$$

When we define $c_n = \alpha_{n+2}\alpha_{n-2}/\alpha_n^2$, (4.97) turns out to be the q-discrete Painlevé I (4.91) [GraTRCT2002, (2.16)],

$$c_{n+2}c_{n-2} = \frac{\alpha_4^2}{c_n} - \frac{\alpha_6}{c_n^2}.$$

In other words, we have a hyperelliptic solution of the q-discrete Painlevé I,

$$c_n(u) = \frac{\sigma((n+2)u)\sigma((n-2)u)}{\sigma(nu)^2\sigma_2(u)^8},$$

where u is $\widetilde{w} \circ \iota_X(B_r)$.

Further, using (4.97), we have that for a point $u \in \widetilde{w} \circ \iota_X(\{B_r\})$,

$$d_m(u) := \frac{\widetilde{\psi}_{m+1}(u)\widetilde{\psi}_{m-2}(u)}{\widetilde{\psi}_m(u)\widetilde{\psi}_{m-1}(u)} = \frac{\sigma((m+1)u)\sigma((m-2)u)}{\sigma(mu)\sigma((m-1)u)\sigma_2(u)^4} \tag{4.98}$$

obeys the third-order difference equation

$$d_{m+2}d_{m-1} - \frac{\alpha_5}{d_{m+1}d_m} + \alpha_4\left(\frac{1}{d_{m+1}} + \frac{1}{d_m}\right) = 0. \tag{4.99}$$

We note that although $\widetilde{\psi}_n$ vanishes for a point $\widetilde{w} \circ \iota_X(B_r)$, d_m has non-trivial values because it can be expressed by

$$d_m(u) := \frac{\alpha_{m+1}(u)\alpha_{m-2}(u)}{\alpha_m(u)\alpha_{m-1}(u)}. \tag{4.100}$$

In other words, the third-order difference equation (4.100) is well-defined, and its solution is associated with an algebraic curve. (4.100) can be also regarded as a generalization of (4.91).

We consider a point satisfying $\widetilde{\psi}_4 = 0$. Then we apparently obtain a fourth-order ordinary difference equation

$$\widetilde{\psi}_5\widetilde{\psi}_3\widetilde{\psi}_2^2 b_{n+2}b_{n-2} + \frac{1}{b_{n-1}^2 b_n^3 b_{n+1}^2}\left(\widetilde{\psi}_5\widetilde{\psi}_3^3 b_{n-1}b_n^2 b_{n+1} - \widetilde{\psi}_6\widetilde{\psi}_3^2\widetilde{\psi}_2 b_n\right) = 0 \tag{4.101}$$

at the point. However, if $\widetilde{\psi}_4$ vanishes at a point P, P is a torsion point. It means that $\widetilde{\psi}_2 = \widetilde{\psi}_3 = \widetilde{\psi}_4 = 0$, $\widetilde{\psi}_3 = \widetilde{\psi}_4 = \widetilde{\psi}_5 = 0$ or $\widetilde{\psi}_4 = \widetilde{\psi}_5 = \widetilde{\psi}_6 = 0$ from the definition of $\widetilde{\psi}$ function of genus two [Can1994, On2002].

4.5 Burgers–Hopf and Dispersionless KP Hierarchies

In this section, we give solutions to the dispersionless KP equations for plane W-curves X or (r, s)-curves mentioned in Sect. 2.3.1 along the line of the flow in Remark 4.1 following [MatP2009b].

Dispersionless hierarchies have been of interest ever since integrable hierarchies were first identified in the late 1960s; one reason for their importance in applied mathematics is to approximate the solution of an integrable equation by a finite-gap solution (defined over a Riemann surface of genus g) by "slowly modulating" g-phase wavetrains [FlFM1980]; the disadvantage is that they may develop singularities of shock-wave type [KodK2002, KodK2003, To1989]. We do not survey the abundant previous results (nor do we know of a comprehensive survey of them), except for stressing that—to our knowledge—the algebro-geometric solutions that have been constructed thus far are defined over a family of curves. In contrast, the investigation that we show in this subsection involves one fixed curve.

We use the following definition of the dispersionless Kadomtsev–Petviashvili equation (dKP):

$$\partial_1\partial_3 v + \partial_1(v\partial_1 v) + \partial_2^2 v = 0,$$

which can be obtained from the KP equation:

$$\partial_1\partial_3 u = \frac{1}{4}\partial_1^4 u + 3\partial_1(u\partial_1 u) + \frac{3}{4}\partial_2^2 u = 0$$

by rescaling each variable so that $\partial_n := \epsilon\partial_n$, expanding $v = u + O(\epsilon)$ and taking a limit $\epsilon \to 0$. In particular for the KdV reduction, we take the dispersionless Korteweg–de Vries (dKdV) equation to be

$$\partial_2 v = v\partial_1 v.$$

and similarly to the equations of the Burgers–Hopf hierarchy,

$$\partial_n v = v^n\partial_1 v,$$

which is obtained as a dispersionless limit of the KdV hierarchy [KodK2002].

We found our solution rather surprising, for the following reason. So far, the Riemann-surface solutions to dispersionless hierarchies (again, there are many sources, and we cannot list them all; in [Z2001], a clear account of the geometry is given) have used analytic functions on the Riemann surface which are not meromorphic (uniformization), and the variables (moments) are coefficients of expansions near a point in such a way that the solutions u, v depend on a moving family of Riemann surfaces. This picture arises quite naturally in the slow modulation we mentioned [FlFM1980], where the $\epsilon \to 0$ limit clearly causes the t_n part of the slow variables ϵt_n in the wave function to survive, contrary to what happens for KP.

The ϵ-expansions in these slow variables are worked out by an analogue of inverse scattering in [CarK1995]. Our variables are instead Abel variables on the Jacobian of a fixed curve and the solutions u, v are meromorphic functions on this curve. In order to shed some light on our type of variables, we have worked out the case of genus 1 (cf. Sect. 4.5.2), which also provided the 1-soliton solution to KdV, by rescaling time t along the same direction as x. We intend to develop a theory of wave functions in our context and eventually relate the two approaches to evolution in the Sato–Grassmannian manifold.

4.5.1 Functions on W_X^1

This subsection shows the differential relations among meromorphic functions on W_X^1. The stratum W_X^1 is smooth since it is isomorphic to the curve by the Abel theorem. We identify it with the curve for simplicity. Note now that the vector $u := w(x, y) \in \mathbb{C}^g$ is only defined modulo periods; nevertheless, it makes sense locally to view any of its coordinates u_i as functions on the curve W_X^1 [On2005]; moreover, none of them is trivial on the curve, again by the Abel theorem, and in fact it is invertible in the sense of the implicit function theorem of advanced calculus (the algebraic coordinates x and y of the point (x, y) are functions of u_i), unless one of the u_i (but not all) has a multiplicity due to a "branchpoint" of the projection of the Abel–Jacobi map—for example, on a hyperelliptic curve of genus two, $y^2 = \prod_{i=1}^5 (x - b_i)$, the projection to $u_1 = \int dx/y$ has a branchpoint at $\infty \in X$, and u_1 is not invertible at that point; equivalently, this means that at the point the differential ν_i^{I} has order of zero more than at the generic point. If we exclude this finite number of points from our construction, then we can define the maximal open subset U_i of the curve where the i-th coordinate is invertible:

$$x = x(u_i), \quad y = y(u_i).$$

On the overlaps $U_{ij} := U_i \bigcap U_j$, the transition map Ψ_{ij} between u_i and u_j is given by the Jacobi matrix,

$$\frac{\partial u_i}{\partial u_j} = \frac{\phi_{i-1}}{\phi_{j-1}}.$$

On U_{ij}, we use the notation $\partial/\partial u_k$ $(k = i, j)$ for the derivative with respect to u_k, cf. also [MatP2008, Remark 5.4 (3)]. The map Ψ_{ij} shows the following lemma:

Lemma 4.91 *On U_{ij}, the following relation holds:*

$$\frac{\partial}{\partial u_j} = \frac{\phi_{i-1}}{\phi_{j-1}} \frac{\partial}{\partial u_i}.$$

Here we recall Remark 3.23. $\partial/\partial u_i$ corresponds to the latter one mentioned there.

Given the choice of holomorphic differentials, for the arithmetic parameter t at ∞, the following relation holds:

$$u_{g-i} = t^{2g-N(i-1)-1}(1 + d_{>0}(t_\infty)).$$

4.5.2 On dKdV and dKP Equations of Genus One

In this subsection, we exhibit a class of solutions of the dKdV and the dKP equations. The construction uses polynomials in two variables viewed as functions on a plane curve with poles at infinity. The differential identities that enable the construction (Lemmas 4.91 and 4.94) follow from the transition map Ψ_{ij} between the Abelian variables defined on two charts of (the Abel–Jacobi image of) the curve.

Since the genus-one curve has a single chart, we don't have transition maps or the given type of solution. However, for the elliptic curve $y^2 = x^3 - g_2' x - g_3'$ ($\widehat{y} = 2y$, $g_a' = g_a/4$ ($a = 2, 3$)), we have the relations

$$-\frac{d\zeta(u)}{du} = \wp(u) = x, \quad \frac{d\wp(u)}{du} = 2y,$$

and

$$du = \frac{dx}{2y} = \frac{dx}{\widehat{y}}.$$

Hence we have

$$\frac{dx}{d\zeta} = -\frac{du}{d\zeta}\frac{dx}{du} = \frac{1}{x}\widehat{y},$$

and deduce the dKdV equation,

$$\frac{dx(u)}{du} = x\frac{dx}{d\zeta}.$$

Thus, we obtained this equation by a kind of generalization of our method: one of the evolution variables is still an Abelian integral, the other is not merely a rescaling (as for the KdV soliton in the genus-1 case) but an antiderivative, which can still be regarded as a quasi-periodic function on the same torus.

Finally, it should be noted that our equations can be extended to the general (r, s)-curve, (2.8), again, the dKP variables being Abelian variables implemented on the curve embedded in its Jacobian by Jacobi inversion. As stated in the introduction, the next project is to find a relation between these classes of functions and the tau function-type solutions [Z2001, Sect. 2] whose variables are moments of a uniformizing parameter at ∞ and the curve varies with the motion.

4.5.3 Burgers–Hopf Hierarchy in Plane W-curves

This subsection mentions the Burgers–Hopf hierarchy in plane W-curves X. For the symmetric W-curve, we recall the basis $S_X = \widehat{S}_X$ of $R_X^{\mathfrak{c}} = R_X$ in Definitions 2.18 and 2.108, $S_X = \{\phi_i\}$, $\phi_0 = 1$ and $\phi_1 = x$, or $N(0) = 0$ and $N(1) = r$.

We consider plane W-curves. We have the subset of the basis of the holomorphic one-forms in

$$d\mathfrak{s}_i := \nu^{\mathrm{I}}_{\chi(i)} = \frac{x^i dx}{f_{X,y}(x, y)}. \tag{4.102}$$

We have the following theorem [MatP2008, Remark 5.4 (3)].

Theorem 4.92 *We realize the solution of the Burgers–Hopf hierarchy,*

$$\frac{\partial}{\partial \mathfrak{s}_i} x = x^{j-i} \frac{\partial}{\partial \mathfrak{s}_j} x$$

in terms of R_X.

Proof $\dfrac{\partial}{\partial \mathfrak{s}_i} = \dfrac{f_{X,y}}{x^i} \dfrac{d}{dx}$. ∎

4.5.4 Dispersionless KP Hierarchy

This subsection shows the dispersionless KP hierarchy. We also consider the plane (r, s) W-curves. We assume that $r > 3$.

Since there are Weierstrass gaps from 1 to $r - 1$ at ∞, by Serre duality or Riemann–Roch the following is a basis of holomorphic differentials:

$$\nu^{\mathrm{I}}_g = \frac{\phi_{g-1} dx}{f_{X,y}(x, y)}, \quad \nu^{\mathrm{I}}_{g-1} = \frac{\phi_{g-2} dx}{f_{X,y}(x, y)}, \quad \nu^{\mathrm{I}}_{g-2} = \frac{\phi_{g-3} dx}{f_{X,y}(x, y)}, \tag{4.103}$$

whose Sato–Weierstrass weights are given by $\mathrm{wt}(\nu^{\mathrm{I}}_g) = 0$, $\mathrm{wt}(\nu^{\mathrm{I}}_{g-1}) = 1$, and $\mathrm{wt}(\nu^{\mathrm{I}}_{g-2}) = 2$.

Lemma 4.93 *We find* $\widetilde{\phi}_i \in R_X$ *such that*

$$\widetilde{\phi}_i = \phi_i + \sum_{i=0}^{i-1} a_{i,j} \phi_j, \quad a_{i,j} \in \mathbb{C}$$

satisfying $\widetilde{\phi}_{g-1} \widetilde{\phi}_{g-3} = \widetilde{\phi}^2_{g-2}$.

Lemma 4.94

$$\frac{d}{dx}(\widetilde{\phi}_{g-1} \widetilde{\phi}_{g-3}) = 2 \widetilde{\phi}_{g-2} \frac{d}{dx} \widetilde{\phi}_{g-2}.$$

We define

$$d\mathfrak{t}_1 := \frac{\widetilde{\phi}_{g-1} dx}{f_{X,y}(x, y)}, \quad d\mathfrak{t}_2 := \frac{\widetilde{\phi}_{g-2} dx}{f_{X,y}(x, y)}, \quad d\mathfrak{t}_3 := \frac{\widetilde{\phi}_{g-3} dx}{f_{X,y}(x, y)}. \tag{4.104}$$

Then we have

$$\mathfrak{t}_1 = t(1 + d_{>0}(t)), \quad \mathfrak{t}_2 = t^2(1 + d_{>0}(t)), \quad \mathfrak{t}_3 = t^3(1 + d_{>0}(t)),$$

where t is the arithmetic local parameter in Definition 2.16. We introduce

$$v := \frac{\widetilde{\phi}_{g-3}}{\widetilde{\phi}_{g-1}},$$

which has the properties

$$v = t_\infty^{-2}(1 + d_{>0}(t_\infty)).$$

For any element of the semigroup generated by r and s, there are meromorphic functions with exactly that order of pole at ∞ and regular elsewhere, and the orders of pole yield the following relations:

Proposition 4.95 *On $U_{g-2,g-1,g} := U_g \bigcap U_{g-1} \bigcap U_{g-2}$ of $W^1_{r,s}$, the transition maps $\Psi_{g-1,g-2}$, $\Psi_{g-1,g}$, $\Psi_{g-2,g}$ induce the following relation:*

$$\frac{\partial}{\partial \mathrm{t}_3}\frac{\partial}{\partial \mathrm{t}_1} + \frac{\partial}{\partial \mathrm{t}_1} v \frac{\partial}{\partial \mathrm{t}_1} - 2\frac{\partial}{\partial \mathrm{t}_2}\frac{\partial}{\partial \mathrm{t}_2} = 0.$$

Proof (4.104) gives:

$$\frac{\partial}{\partial \mathrm{t}_3} = v\frac{\partial}{\partial \mathrm{t}_1}.$$

Straightforward computation gives the relations

$$\frac{\partial}{\partial \mathrm{t}_1} v \frac{\partial}{\partial \mathrm{t}_1} = \frac{\partial}{\partial \mathrm{t}_1}\frac{\partial}{\partial \mathrm{t}_3} = \frac{(f_{X,y})^2}{\widetilde{\phi}_{g-1}\widetilde{\phi}_{g-3}}\left[\frac{1}{f_{X,y}}\frac{d}{dx} f_{X,y} - \frac{d}{dx}\log\widetilde{\phi}_{g-3} + \frac{d}{dx}\right]\frac{d}{dx},$$

$$\frac{\partial}{\partial \mathrm{t}_3}\frac{\partial}{\partial \mathrm{t}_1} = \frac{(f_{X,y})^2}{\widetilde{\phi}_{g-1}\widetilde{\phi}_{g-3}}\left[\frac{1}{f_{X,y}}\frac{d}{dx} f_{X,y} - \frac{d}{dx}\log\widetilde{\phi}_{g-1} + \frac{d}{dx}\right]\frac{d}{dx},$$

$$\frac{\partial}{\partial \mathrm{t}_2}\frac{\partial}{\partial \mathrm{t}_2} = \frac{(f_{X,y})^2}{\widetilde{\phi}_{g-1}\widetilde{\phi}_{g-3}}\left[\frac{1}{f_{X,y}}\frac{d}{dx} f_{X,y} - \frac{d}{dx}\log\widetilde{\phi}_{g-2} + \frac{d}{dx}\right]\frac{d}{dx}.$$

We conclude by Lemma 4.94. ∎

Proposition 4.95 gives a solution of the dKP equation as follows:

Theorem 4.96 *For the local chart $U_g \bigcap U_{g-1} \bigcap U_{g-2}$ of $W^1_{r,r+1}$, we have a solution of the dKP equation,*

$$\frac{\partial}{\partial \mathrm{t}_3}\frac{\partial}{\partial \mathrm{t}_1} v + \frac{\partial}{\partial \mathrm{t}_1}\left(v\frac{\partial}{\partial \mathrm{t}_1} v\right) - 2\frac{\partial}{\partial \mathrm{t}_2}\frac{\partial}{\partial \mathrm{t}_2} v = 0.$$

Remark 4.97 Lemma 4.93 could be extended for general W-curves with $r > 3$ due to Sect. 2.4.2, which means that we can find the solutions of the dKP equation in Theorem 4.96 for more general W-curves.

Remark 4.98 As mentioned in Sects. 2.2.3 and 3.9.4, the monomial curves are crucial in the Weierstrass Abelian function theory and the properties of the sigma functions.

The above relations are given by the embedding of the curve X into $\mathbb{C}^g$, and their essential parts are determined by the numerical semigroup and the monomial curves.

There are three approaches to the deformations of the algebraic curves in this book, the algebraic approaches mentioned in Appendix 2.11, the approach by means of the heat equation in Appendix 3.11 and the approaches by using the dKP hierarchy. As Pinkham studied the deformation of the monomial curves to the non-singular curves in [Pin1974], the monomial curves also appear in the heat equations and this dKP hierarchy, and play a crucial role. Thus these three approaches should be investigated from a unified standpoint.

Further, we find the dKP hierarchy even for the curve associated with $\langle 6, 13, 14, 15, 16\rangle$ mentioned in Sect. 2.3.4. The Faber polynomials related to $\{1, 2, 3, 4, 5, 7, 8, 9, 10, 11, 17, 23\}$ appear due to the results by Carroll and Kodama [CarK1995] [KomMP2013]. It is a contrast that the replicable functions associated with Monster group are related to the Faber polynomials with Norton numbers $\{1, 2, 3, 4, 5, 7, 8, 9, 11, 17, 19, 23\}$ [McH2022]. The monomial algebra associated with the $\mathbb{G}_m$ action at a singularity is related to simple Lie group [Sl1980], which might be connected with McKay's another discovery.

4.6 Generalized Elastica in Statistical Mechanics of Elastica

This section is the final section of this book.[7] As mentioned in Chap. 1, Euler and the Bernoullis studied the elastica problem, which critically contributed to the development of the Abelian function theory, especially the elliptic function theory. In this section, we consider the modernized elastica problem. The real problem concerning the elastica is quite crucial and sometimes enhances the development of the abstract theory. In terms of the theory of Abelian function theory, we consider the generalized elastica problem to produce the shapes of the supercoiled DNA.

As in several books on biophysics, the DNA has a large-scale geometrical structure, which is called a supercoil, besides the coiled structure. Their typical size is several sub-micrometers. These shapes are illustrated in Fig. 4.3. These shapes look to be governed by a particular weak elastic force. However, these shapes cannot be reproduced by the minimal points of the energy functional, including the Euler–Bernoulli energy functional, even if we consider the three-dimensional effects.[8] These minimal points are given by the elliptic functions, but it is obvious that the double periodic structure cannot express the complicated shapes. Thus it is natural to consider the

[7] This generalized elastica might be related to the replicable functions for the representation of the Monster group [Mat2008b, MatP2016, McH2022].

[8] It should be acknowledged that the author of this book learned this problem from Hideo Tsuru based on his works with Wadati [Ts1987, TsW1986].

excited states of the elastica due to the thermal effect to reproduce the shapes of the supercoiled DNA. When we consider the shape of DNA in the thermal bath, it is important to consider the statistical mechanics of isometric loops with elasticity in two-dimensional and three-dimensional spaces [Mat1998, Mat1999]. The observed shapes of DNA in laboratories are two-dimensional. Furthermore, the treatment of such an object is not straightforward, and its essentials are reduced to the two-dimensional case. Thus we have considered the two-dimensional case.

It turns out that they are expressed by the hyperelliptic functions rather than the elliptic functions. Thus we show our attempt to attack this problem in this section, even though our problem is not solved well at this stage. However, we illustrate curves associated with the hyperelliptic curves of genus two, which partially reproduce the shapes in Fig. 4.3.

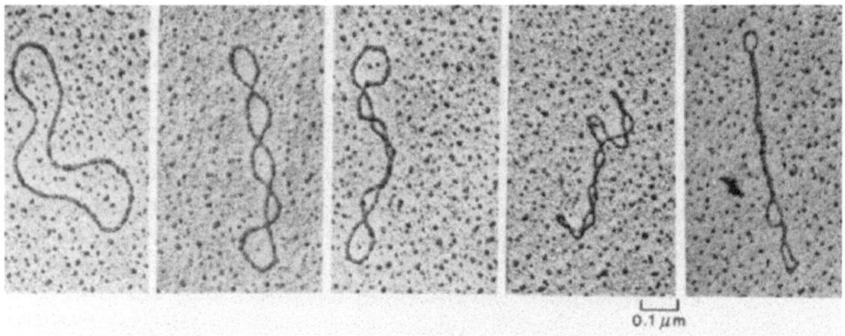

Fig. 4.3 Electron micrographs of DNA (mini ColE1 plasmid dimer, 5 kb) by Laurien Polder [KorB1992, Fig. 1–24 p. 36], Reproduced with permission from Roger Kornberg

Let us consider an analytic and isometric immersion Z of a loop S^1 in the complex plane $\mathbb{C}$ [MatP2016],

$$\mathscr{M} := \{Z : S^1 \hookrightarrow \mathbb{C} \mid \text{analytic, isometric}\},$$

and its arclength and curvature are denoted by s and $k := \frac{1}{\sqrt{-1}} \partial_s \log \partial_s Z$ respectively for $\partial_s := \partial/\partial s$. The elastica is given as the minimal point of the Euler–Bernoulli energy

$$\mathscr{E}[Z] := \frac{1}{2} \oint_{S^1} k^2(s) ds = \oint_{S^1} \{Z, s\}_{\text{SD}} ds, \tag{4.105}$$

where $\{Z, s\}_{\text{SD}} = \frac{\partial}{\partial s} \left(\frac{\partial^2 Z/\partial s^2}{\partial Z/\partial s} \right) - \frac{1}{2} \left(\frac{\partial^2 Z/\partial s^2}{\partial Z/\partial s} \right)^2$ is the Schwarzian derivative.[9]

The condition $|\partial_s Z| = 1$, namely $\partial_s Z \in \mathrm{U}(1)$, allows us to define an action $g_{s_0} Z(s) = Z(s - s_0)$ (where we identified S^1 with $\mathbb{R}$ modulo the integers); we call

[9] As partially mentioned in [MatP2016] it is closely related to Tjurin's investigation [Tj1978] and associated with the Faber polynomials studied by Schiffer [Sch1966]. The generalized elastica problem might be connected with McKay's replicable function theory [McH2022].

a transformation with this property "Euclidean motion". We set

$$\mathbf{M} := \mathcal{M}/\sim, \quad \mathfrak{M} := \mathbf{M}/\mathrm{U}(1),$$

where $\sim$ means the Euclidean motion.

We let the natural projections be denoted by $pr_1 : \mathcal{M} \to \mathbf{M}$ and $pr_2 : \mathbf{M} \to \mathfrak{M}$.

When we consider a tiny loop, such as DNA, in the thermal bath, it is vital to consider the statistical mechanics of elastica, which are formally given by [Mat1998]

$$\mathcal{Z}[\beta] = \int_{\mathcal{M}} DZ\mathrm{e}^{-\beta\mathcal{E}[Z]},$$

where β is a positive parameter, and DZ might be an appropriate measure in $\mathcal{M}$. It is noted that from a physical viewpoint, at zero temperature ($\beta \to \infty$), the problem is reduced to the minimal energy problem or the Euler–Bernoulli elastica problem. In other words, the statistical mechanics of elastica naturally contain the elastica problem.

However, it is a challenging problem because one cannot pose the statistical mechanics of elastica as a mathematical problem. However, it should be solved in "mathematics" as Euler and the Bernoullis did for their elastica problem. Thus the author of this book with Emma Previato proposed a generalized elastica, which can be treated in mathematics to approach the final goal, i.e., the generalized elastica problem.

The generalized elastica problem beyond Euler's elastica is given as follows:

1. to explicitly express the points in $\mathcal{M}$ (in terms of hyperelliptic functions),
2. to classify the moduli space $\mathcal{M}$ with respect to the Euler–Bernoulli energy,
3. to assign topology in $\mathcal{M}$ and to define an appropriate measure with respect to the Euler–Bernoulli energy, and
4. to analyze $\mathcal{M}$ for a given β.

The author of this book with Previato investigated the geometrical structure of $\mathcal{M}$ as a generalized elastica problem [Mat2010, MatO2003a, MatP2016, Pr1993]. It has the orbital decomposition in terms of the loop soliton equation whose curvature obeys the modified KdV equation; Since the MKdV equation preserves the Euler–Bernoulli energy of the deformation of curves. $\mathcal{M}$ is decomposed to

$$\mathcal{M} = \bigsqcup_{E\in\mathbb{R}} \mathcal{M}_E, \quad \mathcal{M}_E := \{Z \in \mathcal{M} \mid \mathcal{E}[Z] = E\},$$

which naturally includes Euler's elastica as its minimal point.

Physical investigation [Mat1998] predicts that $\mathcal{Z}[\beta]$ could be evaluated like

$$\mathcal{Z}[\beta] = \int dE \frac{\partial \mathrm{Vol}(\mathcal{M}_E)}{\partial E} \mathrm{e}^{-\beta E},$$

where $\mathrm{Vol}(\mathscr{M}_E)$ is the volume of $\mathscr{M}_E$.[10]

As we show below, it turns out that $\mathscr{M}_E$ is given by the orbit spaces satisfying the MKdV hierarchy, and it can be classified by finite-dimensional space, denoted by g, related to the algebraic solutions of the MKdV equation. The volume might consist of the volume of the real $(g-1)$-dimensional subspace of the hyperelliptic Jacobi variety.

In other words, the generalized elastica could have been solved in a decade if there had been a nice tool like the Weierstrass elliptic function theory. However, this did not exist in 2000, and thus, the author of this book with Previato has constructed the tools to express the problem for two decades, i.e., the Weierstrass sigma function theory in the higher genus, our main theme in this book.

As a tool, it has not yet reached a sufficient level of perfection for this problem. However, we stand at a level where we have made some findings in the generalized elastica problem, and also we can show them as the theory of the Weierstrass sigma functions in this book. On the generalized elastica problem, in particular, in 2021, the author of this book with Previato was able to compute, albeit incompletely, a shape that closely resembles the shape of DNA by treating a hyperelliptic function with a genus two. In this section, we give an overview of these in a mathematical framework based on [MatP2016, MatP2022].

Further, we note that Brylinski studied a similar problem on the moduli of the embedding of a loop into three-dimensional space in [Bry1993]; he showed that the moduli could be classified by the nonlinear Schrödinger equations, which is consistent with the result in three-dimensional extension of the generalized elastica problem [Mat1999].

4.6.1 Notations and Preliminaries

We let $\mathscr{A}^p_{S_1}(\mathbb{R})$ denote the real-analytic p-forms on S^1 with $\mathbb{R}$-valued coefficients, and by $\mathscr{A}^p_{S_1}(\mathbb{C})$ the real-analytic p-forms on S^1 with $\mathbb{C}$-valued coefficients.

Lemma 4.99 *The kernel of the $\mathbb{R}$-linear map* $\oint : \mathscr{A}^1_{S_1}(\mathbb{R}) \to \mathbb{R}$ *is equal to* $d\mathscr{A}^0_{S_1}(\mathbb{R})$.

Proof For $\eta \in \mathscr{A}^1_{S_1}(\mathbb{R})$, there exists $F \in \mathscr{C}^\omega(\mathbb{R}, \mathbb{R})$ such that $\eta := dF$ since $S^1 = \mathbb{R}/2\pi\mathbb{Z}$. When η is in the kernel of $\oint$, $F(2\pi) = F(0)$, so F belongs to $\mathscr{A}^0_{S_1}(\mathbb{R})$. ■

[10] If we replace β with $\sqrt{-1}/\hbar$, the problem is reduced to a functional integration representation of a quantum system related to a geometrical object. In theoretical physics, the method to sum a weight with respect to an invariant functional over possible states is known as the Feynman path integral (functional integral) method for statistical mechanics and quantum mechanics. Thus the author of this book called it quantized elastica. However, since mathematicians, in general, are not familiar with the method but only know mathematical models related to quantum algebra, the name might confuse them. Thus we now call it a generalized elastica problem.

For a given $Z \in \mathbf{M}$, the tangential angle $\phi \in \mathscr{C}^\omega(\mathbb{R}, \mathbb{R})$ and the curvature $k \in \mathscr{A}^0_{S_1}(\mathbb{R})$ are defined as

$$\phi(s) := \phi[Z](s) := \frac{1}{\sqrt{-1}} \log \partial_s Z(s) \in \mathscr{C}^\omega(\mathbb{R}, \mathbb{R}),$$

$$k(s) := k[Z](s) := \partial_s \phi[Z](s),$$

consistent with the terminology at the beginning of this section. Since $|\partial_s Z(s)| = 1$, for fixed parameters t_i, we view the curvature k as an element of $\mathscr{A}^0_{S_1}(\mathbb{R})$ and kds as an element of $\mathscr{A}^1_{S_1}(\mathbb{R})$.

4.6.2 Isometric Deformations

In this subsection, we consider the tangent space of the isometric deformations in $\mathbf{M}$, namely a fiber space $T_Z\mathbf{M}$ over the tangent space $T\mathbf{M}$ of $\mathbf{M}$ at Z in $\mathbf{M}$. By regarding the parameter s in Z as fixed, the fiber is given by the isometric deformations of $\mathbf{M}$. We let $\mathfrak{IDiff}(\mathbf{M})$ be the group acting on $\mathbf{M}$ that gives isometric deformations, and we consider the action of its Lie algebra $\mathfrak{idiff}(\mathbf{M})$ at a point Z in $\mathbf{M}$. We identify the infinitesimal orbit of $\mathfrak{idiff}(\mathbf{M})$, or the infinitesimal deformations, and the elements of $T_Z\mathbf{M}$ at $Z \in \mathbf{M}$. We do not need to prove the existence of this, say, analytic Lie group, because it will consist of the flows that we construct concretely. We take the limit of the flow parameterized by $t \in [0, \varepsilon)$ at $Z \in \mathbf{M}$ as ε approaches zero, in order to investigate the fiber $T_Z\mathbf{M}$. As a result, this limit gives us tangent vectors and differential operators belonging to the fiber, without a need to introduce a connection: for our purposes, we only need local calculations for given coordinates.

A deformation is defined by

$$\partial_t Z(s) = v(s)\partial_s Z(s), \quad s \in S^1, \tag{4.106}$$

where v belongs to $\mathscr{A}^0_{S_1}(\mathbb{C})$ and

$$v(s) = v^{(r)}(s) + \sqrt{-1}v^{(i)}(s),$$

where $v^{(r)}$ and $v^{(i)}$ denote real and imaginary parts respectively, so that

$$v^{(r)}, v^{(i)} \in \mathscr{A}^0_{S_1}(\mathbb{R}) := \mathscr{C}^\omega(S^1, \mathbb{R}).$$

Recalling that the "isometric" condition means that s, t are independent variables, namely $\partial_t\partial_s Z = \partial_s\partial_t Z$, we check that:

Proposition 4.100 *For a point $Z \in \mathbf{M}$, the isometric condition $\partial_t\partial_s Z = \partial_s\partial_t Z$ is reduced to*

$$\partial_t \partial_s \phi(s) = \partial_s \partial_t \phi(s) \quad s \in S^1,$$

where $\phi = \phi[Z]$.

Proof The statement follows from the calculations, $\partial_t \dfrac{\partial_s^2 Z}{\partial_s Z} = \dfrac{\partial_t \partial_s^2 Z}{\partial_s Z} - \dfrac{\partial_s^2 Z \partial_s \partial_t Z}{(\partial_s Z)^2}$ and $\partial_s \dfrac{\partial_t \partial_s Z}{\partial_s Z} = \dfrac{\partial_s \partial_t \partial_s Z}{\partial_s Z} - \dfrac{\partial_s^2 Z \partial_t \partial_s Z}{(\partial_s Z)^2}$. ■

An isometric deformation of $\partial_s Z = \mathrm{e}^{\sqrt{-1}\phi}$ belongs to $\mathscr{C}^\omega(S^1, \mathrm{U}(1))$, and its derivative along $t \in [0, \varepsilon)$ is given by

$$\partial_t \partial_s Z = \sqrt{-1}\partial_t \phi(s) \partial_s Z.$$

By applying ∂_s to both sides of (4.106), we have

$$\partial_s \partial_t Z = \partial_s (v \partial_s Z).$$

$$\sqrt{-1}(\partial_t \phi)\partial_s Z = (\partial_s v + v\sqrt{-1}k)\partial_s Z. \tag{4.107}$$

Since $\partial_t \phi$ is real-valued, we obtain two conditions by equating the real and imaginary parts:

Proposition 4.101 *For a point* $Z \in \mathbf{M}$, *the isometric condition* $[\partial_s, \partial_t]Z = 0$ *is reduced to two conditions* [GolP1991, GolP1993]*:*

$$\begin{aligned} \partial_t k &= \partial_s (\partial_s \frac{1}{k} \partial_s + k) v^{(r)} = \partial_s (\partial_s + k \partial_s^{-1} k) v^{(i)}, \\ &= \Omega^{\mathrm{I}} v^{(r)} = \Omega^{\mathrm{II}} v^{(i)}, \qquad (4.108) \\ k v^{(i)} &= \partial_s v^{(r)}, \qquad (4.109) \end{aligned}$$

where the differential operators, $\Omega^{\mathrm{I}} := \partial_s (\partial_s \frac{1}{k} \partial_s + k)$, *and* $\Omega^{\mathrm{II}} := \partial_s^2 + \partial_s (k \partial_s^{-1} k)$.

Here Ω^{II} is seen to agree with the recursion operator for the MKdV hierarchy.

Remark 4.102 (4.107) means that we have

$$\partial_t \phi \, \partial_s Z = (\partial_s \frac{1}{k} \partial_s + k) v^{(r)} \partial_s Z = (\partial_s + k \partial_s^{-1} k) v^{(i)} \partial_s Z.$$

Further, instead of the deformation parameter t, we can consider the deformation parameter $t = \sqrt{-1}\tau$ because $\sqrt{-1}\partial_\tau Z$, is expressed by the rotated velocity; $\sqrt{-1}\partial_\tau Z = v \partial_s Z$ means $\partial_\tau Z = -\sqrt{-1} v \partial_s Z = (v^{(i)} - \sqrt{-1} v^{(e)})\partial_s Z$. Then we have $\partial_\tau k = \Omega^{\mathrm{I}} v^{(i)} = -\Omega^{\mathrm{II}} v^{(r)}$ as

$$\sqrt{-1}\partial_\tau \phi\, \partial_s Z = \sqrt{-1}(\partial_s \frac{1}{k}\partial_s + k) v^{(i)} \partial_s Z = \sqrt{-1}(\partial_s + k\partial_s^{-1} k) v^{(r)} \partial_s Z.$$

The imaginary deformation means the replacement between $v^{(i)}$ and $v^{(r)}$.

In the limit as ε approaches zero, these conditions hold on the fiber of the tangent space. Therefore we can define the following length-preserving linear transformation on the tangent space, which we identified with $\mathfrak{i}\mathfrak{d}\mathfrak{i}\mathfrak{f}\mathfrak{f}(\mathbf{M})$,

$$\partial_s \begin{pmatrix} v^{(i)} \\ v^{(r)} \end{pmatrix} = \begin{pmatrix} 0 & \partial_s \frac{1}{k} \partial_s \\ k & 0 \end{pmatrix} \begin{pmatrix} v^{(i)} \\ v^{(r)} \end{pmatrix}, \quad \begin{pmatrix} \partial_s & -\partial_s \frac{1}{k} \partial_s \\ -k & \partial_s \end{pmatrix} \begin{pmatrix} v^{(i)} \\ v^{(r)} \end{pmatrix} = 0.$$

We derive some consequences from the second condition, (4.109). Let us consider the map

$$\ell_d : \mathscr{A}^0_{S_1}(\mathbb{R}) \to \mathscr{A}^1_{S_1}(\mathbb{R}), \quad \ell_d(v^{(i)}) = k v^{(i)} ds,$$

and denote the inverse image of $d\mathscr{A}^0_{S_1}(\mathbb{R}) \subset \mathscr{A}^1_{S_1}(\mathbb{R})$ by $\widehat{\mathscr{A}^0_{S^1}}(\mathbb{R}) := \ell_d^{-1} d\mathscr{A}^0_{S_1}(\mathbb{R})$, together with the map

$$\widehat{\ell_d} : \widehat{\mathscr{A}^0_{S^1}}(\mathbb{R}) \to d\mathscr{A}^0_{S_1}(\mathbb{R}) \subset \mathscr{A}^1_{S_1}(\mathbb{R}), \quad \widehat{\ell_d}(v^{(i)}) = k v^{(i)} ds = \partial_s v^{(r)} ds = dv^{(r)}.$$

We define the bijection $\ell_r : \widehat{\mathscr{A}^0_{S^1}}(\mathbb{R}) \to \mathscr{A}^0_{S_1}(\mathbb{R})/\mathbb{R}$ as a $\mathbb{R}$-vector space where $\ell_r(v^{(i)}) = \int_0^s k v^{(i)} ds = \int_0^s \partial_s v^{(r)} ds$. We will need (cf. Lemma 4.112) only deformations whose real part $v^{(r)}$ is unique up to multiplicative constant, so we fix a representative for the image in the quotient $\ell_r^0 : \widehat{\mathscr{A}^0_{S^1}}(\mathbb{R}) \to \mathscr{A}^0_{S_1}(\mathbb{R})$ of ℓ_r so that $\ell_r^0(v^{(i)}) = 0$, by requiring

$$\oint ds \ell_r^0(f) = 0. \tag{4.110}$$

We have an induced map,

$$\ell : \widehat{\mathscr{A}^0_{S^1}}(\mathbb{R}) \to \mathscr{A}^0_{S_1}(\mathbb{C}), \quad \ell(f) = \ell_r^0(f) + \sqrt{-1} f.$$

Lemma 4.103 *The euclidean move is parameterized by $\mathbb{C} \times \mathbb{R}$, i.e.:*

1. *For real parameters C and ϕ_0, the translation $\partial_t Z = C e^{\sqrt{-1}\phi_0}$ is an isometric deformation whose $v^{(i)} = -C \sin(\phi_0 - \phi)$.*
2. *For real parameters c, $\partial_t Z = c\sqrt{-1} Z$ is an isometric deformation whose $v^{(i)} = c\sqrt{-1}(X\partial_s X + Y \partial_s Y)$ by decomposing Z to real-valued functions of S^1 such that $Z = X + \sqrt{-1} Y$.*

Proof 1 is obtained from the facts $\partial_t Z = C e^{\sqrt{-1}(\phi_0 - \phi)} \partial_s Z$ and $\ell(v^{(i)}) = C e^{\sqrt{-1}(\phi_0 - \phi)}$. 2: Similarly we have $c\sqrt{-1} Z = \dfrac{c\sqrt{-1} Z}{\partial_s Z} \partial_s Z = c\sqrt{-1} Z \overline{\partial_s Z} \partial_s Z$.

■

For given $Z \in \mathscr{M}$, and

$$\partial_t Z|_{s=0} = \ell(v^{(i)})\partial_s Z|_{s=0} =: C,$$

we can arrange it as

$$\partial_t (Z - Ct)|_{s=0} = C - C = 0;$$

similarly, we can add a constant to $\partial_t \phi(s)$ by using rotation as the Euclidean motion. We will use this choice of representative in $\mathbf{M}$ for an element $Z \in \mathscr{M}$.

The following proposition is essentially the same as Proposition 3.3.4 (i) in [Bry1993], following from (4.109) for the map ℓ.

Proposition 4.104 *For $v \in \widehat{\mathscr{A}^0_{S^1}}(\mathbb{R})$ and $\tilde{Z} \in \mathscr{M}$, ℓ induces the bijection $\ell^\sharp$ and the surjection $\ell^\flat$,*

$$\begin{array}{ccc} \widehat{\mathscr{A}^0_{S^1}}(\mathbb{R}) & \xrightarrow{\ \ell^\sharp\ } & T_{\tilde{Z}}(\mathscr{M}) \\ & \searrow^{\ell^\flat} & \downarrow{\scriptstyle \mathrm{pr}_{1*}} \\ & & T_{\mathrm{pr}_1(\tilde{Z})}(\mathbf{M}) \end{array}$$

in the sense that, for every element v in the vector space $T_Z(\mathbf{M})$, where Z has the representative $\tilde{Z}$ in $\mathscr{M}$, we have an element $v^{(i)} \in \widehat{\mathscr{A}^0_{S^1}}(\mathbb{R})$ such that $v = \ell(v^{(i)})$.

We have the isomorphism $T_{\mathrm{pr}_1(\tilde{Z})}(\mathbf{M}) \cong \widehat{\mathscr{A}^0_{S^1}}(\mathbb{R})/(\mathbb{C} \oplus \mathbb{R}) \cong \mathscr{A}^0_{S^1}(\mathbb{R})/(\mathbb{C} \oplus \mathbb{R} \oplus \mathbb{R})$.

Given this result, it will be more convenient and sufficient for our purposes to handle elements of $\mathbf{M}$ rather than $\mathscr{M}$.

Lemma 4.105 *The infinitesimal isometric deformation (4.106) is reduced to*

$$\partial_t k = -\Omega^{\mathrm{II}} v^{(i)}$$

as an element in the fiber of $T_Z\mathbf{M}$.

4.6.3 Isoenergy Deformations

This subsection mentions the isoenergy deformations associated with the Euler–Bernoulli energy functional. We now impose, in addition to the isometric condition, that the Euler–Bernoulli energy functional $\mathscr{E}[Z]$ is preserved by the deformation. We call the Lie group of isometric and isoenergy motions acting on $\mathbf{M}$ $\mathfrak{IIDiff}(\mathbf{M}) \subset \mathfrak{IDiff}(\mathbf{M})$, and its Lie algebra acting at a point Z in $\mathbf{M}$, $\mathfrak{iidiff}(\mathbf{M})$. We identify the infinitesimal orbit of an element of $\mathfrak{iidiff}(\mathbf{M})$ with the infinitesimal deformation itself and with the corresponding element of $T_Z\mathbf{M}$ of $Z \in \mathbf{M}$.

Definition 4.106 We define

$$\mathbf{M}_E := \{Z \in \mathbf{M} \mid \mathscr{E}[Z] = E \text{ is preserved}\},$$

and let pr_E denote the natural projection $\mathbf{M}_E \to \mathfrak{M}_E := \mathbf{M}_E/\mathrm{U}(1)$.

From Lemma 4.99, we have the following result:

Proposition 4.107 *For $Z \in \mathbf{M}$, $\partial_t \mathscr{E}(Z)$ vanishes if and only if $k\partial_t k ds \in d\mathscr{A}^0_{S_1}(\mathbb{R})$, i.e., there exists a function $f \in \mathscr{A}^0_{S_1}(\mathbb{R})$ satisfying $k\partial_t k = \partial_s f$.*

Note that (4.109), one of the isometric conditions, says that $k\partial_t k$ is an exact differential, akin to the isoenergy condition.

We now construct an isometric deformation: let $v'^{(i)} := \partial_t k$ and $v'^{(r)} := \ell_r(v'^{(i)})$, which satisfies $k\partial_t k = kv'^{(i)} = \partial_s v'^{(r)}$. This yields the isometric condition

$$\partial_{t'} Z = (\ell(v'^{(i)}))\partial_s Z,$$

or

$$\partial_{t'} k = \Omega^{\mathrm{I}} v'^{(i)}, \quad kv'^{(i)} = \partial_s v'^{(r)}.$$

The following proposition and corollary use this construction:

Proposition 4.108 *If two isometric deformations, $v^{(i)}, v'^{(i)} \in \widehat{\mathscr{A}^0_{S^1}}(\mathbb{R})$,*

$$\partial_t Z = (\ell(v^{(i)}))\partial_s Z \quad \text{or} \quad \partial_t k = \Omega^{\mathrm{II}} v^{(i)}, \quad \text{and}$$
$$\partial_{t'} Z = (\ell(v'^{(i)}))\partial_s Z \quad \text{or} \quad \partial_{t'} k = \Omega^{\mathrm{II}} v'^{(i)},$$

satisfy the relation $\partial_t k = v'^{(i)}$, then:

1. *the deformation $\partial_t Z$ is isoenergy, i.e., $\partial_t \mathscr{E}[Z] = 0$, and*
2. $\partial_{t'} k = \Omega^{\mathrm{II}} v'^{(i)} = \Omega^{\mathrm{II}^2} v^{(i)}$.

The converse also holds:

Corollary 4.109 *If the isometric deformation,*

$$\partial_t Z = (\ell(v^{(i)}))\partial_s Z \quad \text{for which} \quad \partial_t k = \Omega^{\mathrm{II}} v^{(i)},$$

is isoenergy, then there is another isometric flow given by $\ell(v'^{(i)})$ satisfying $\partial_t k = v'^{(i)}$.

Remark 4.110 Proposition 4.108 (2) gives rise to a recursive construction: indeed, if $k\partial_{t'} k$ belongs to $d\mathscr{A}^0_{S_1}(\mathbb{R})$, there is another isometric deformation $\partial_{t''} k = \Omega^{\mathrm{II}} v''^{(i)}$, satisfying $\partial_{t'} k = v''^{(i)}$; then the deformation $\partial_{t'} Z$ is also isoenergy, and

$$\partial_{t''} k = \Omega^{\mathrm{II}^3} v^{(i)}$$

holds.

It follows from this remark that:

Proposition 4.111 *If* $v^{(i)} \in \mathscr{A}^0_{S^1}(\mathbb{R})$ *is given such that each element of the sequence* $\{\Omega^{\mathrm{II}^n} v^{(i)}\}_{n=0,1,2,\ldots}$ *belongs to* $\widehat{\mathscr{A}^0_{S^1}}(\mathbb{R})$*, by introducing parameters* $(\widetilde{w}_1, \widetilde{w}_2, \ldots) \in [0, \varepsilon)$*, we obtain infinitesimal isometric and isoenergy deformations satisfying*

$$\partial_{\widetilde{w}_1} k = \Omega^{\mathrm{II}} v^{(i)},$$

$$\partial_{\widetilde{w}_2} k = \Omega^{\mathrm{II}} \partial_{\widetilde{w}_1} k = \Omega^{\mathrm{II}^2} v^{(i)},$$

$$\partial_{\widetilde{w}_3} k = \Omega^{\mathrm{II}} \partial_{\widetilde{w}_2} k = \Omega^{\mathrm{II}^2} \partial_{\widetilde{w}_1} k = \Omega^{\mathrm{II}^3} v^{(i)},$$

$$\partial_{\widetilde{w}_4} k = \Omega^{\mathrm{II}} \partial_{\widetilde{w}_3} k = \Omega^{\mathrm{II}^2} \partial_{\widetilde{w}_2} k = \Omega^{\mathrm{II}^3} \partial_{\widetilde{w}_1} k = \Omega^{\mathrm{II}^4} v^{(i)},$$

$$\vdots$$

However, the problem remains of finding an initial isometric and isoenergy deformation such that each element of the sequence $\{\Omega^{\mathrm{II}^n} v^{(i)}\}_{n=0,1,2,\ldots}$ belongs to $\widehat{\mathscr{A}^0_{S^1}}(\mathbb{R})$. We will construct a natural one in the next subsection.

4.6.4 Stationary Deformation

In this subsection, we consider the stationary deformation. Since $Z(s)$ and $Z(s - s_0)$ have the same image loop, we call this the "stationary" deformation, in accordance with the terminology for the MKdV equation, when the solution $u(x, t_2, \ldots, t_n, \ldots)$ is viewed as a function of x only (N.B. In previous work by the author of this book in [Mat1998, Sect. 3], this is called the "trivial" deformation).

Lemma 4.112 *For any given real number* $c \in \mathbb{R}$ *and* $Z \in \mathbf{M}$*,* $Z(s + ct)$*, the stationary deformation, is a solution of*

$$\partial_t Z = cZ.$$

Proposition 4.113 *The stationary deformation is compatible with the map* $\mathrm{pr}_2 : \mathbf{M} \to \mathfrak{M}$ *and with its lifting map* $\mathrm{id} : \mathfrak{M}_E \to \mathfrak{M}_E$*, the identity.*

Remark 4.114 The stationary deformation is isometric and isoenergy, according to the equation

$$\partial_{t_1} k = \Omega^{\mathrm{II}^0} \partial_s k = \partial_s k.$$

Proposition 4.115 *For $Z \in \mathbf{M}$ and $k := k[Z]$, starting with the stationary deformation:*

$$\partial_{t_1} k = \partial_s k,$$

each element of the sequence $\{\Omega^{\mathrm{II}^n} \partial_s k\}_{n=0,1,2,\dots}$ belongs to $\widehat{\mathscr{A}}^0_{S^1}(\mathbb{R})$, yielding a sequence of isometric and isoenergy relations,

$$\partial_{t_2} k = \Omega^{\mathrm{II}} \partial_{t_1} k = \Omega^{\mathrm{II}} \partial_s k,$$

$$\partial_{t_3} k = \Omega^{\mathrm{II}} \partial_{t_2} k = \Omega^{\mathrm{II}^2} \partial_{t_1} k = \Omega^{\mathrm{II}^2} \partial_s k,$$

$$\partial_{t_4} k = \Omega^{\mathrm{II}} \partial_{t_2} k = \Omega^{\mathrm{II}^2} \partial_{t_2} k = \Omega^{\mathrm{II}^3} \partial_{t_1} k = \Omega^{\mathrm{II}^3} \partial_s k,$$

$$\vdots$$

This recursive sequence is known as the (focusing) modified KdV hierarchy, including the (focusing) modified KdV equation over $\mathbb{R}$,

$$\partial_{t_2} k = \frac{1}{2} k^2 \partial_s k + \partial_s^3 k, \quad \partial_{t_2} \phi = \frac{1}{8} (\partial_s \phi)^3 + \partial_s^3 \phi, \tag{4.111}$$

Proof We let $v^{(i)}$ in Proposition 4.111 correspond to the stationary deformation $v^{(i)} = \partial_s k$, and $\widetilde{w}_i = t_{i-1}$ for $i = 2, 3, \dots$. The equalities hold because the sequence of t_i are the higher MKdV flows, since Ω^{II} is the MKdV recursion operator. ∎

Remark 4.116 For the stationary deformation, we can write

$$\partial_s k = \Omega^{\mathrm{II}} 0 = (\partial_s^2 + \partial_s k \partial_s^{-1} k) \cdot 0 = \partial_s k \partial_s^{-1} k \cdot 0$$

by choosing the factor, $\partial_s^{-1} 0 = 1$, and thus write

$$\partial_{t_1} k = \Omega^{\mathrm{II}} 0.$$

Then we can take $v^{(i)}$ in Proposition 4.111 equal to 0. Then, with $v^{(i)} = 0$, $\widetilde{w}_i = t_i$ for $i = 1, 2, 3, \dots$.

By letting $w_j = \partial_{t_j} \phi$, we also have a matrix format:

$$\partial_s \begin{pmatrix} 1 \\ w_1 \\ w_2 \\ \vdots \\ w_\ell \\ \vdots \end{pmatrix} = \begin{pmatrix} \Omega^{\mathrm{II}} & & & & \\ & \Omega^{\mathrm{II}} & & & \\ & & \ddots & & \\ & & & \Omega^{\mathrm{II}} & \\ & & & & \ddots \end{pmatrix} \partial_s \begin{pmatrix} 1 \\ w_1 \\ w_2 \\ \vdots \\ w_\ell \\ \vdots \end{pmatrix}.$$

It is well-known that the MKdV hierarchy is integrable, and its orbits are well-defined, thus, the Lie-group and Lie-algebra actions we used for isometric and isoenergy diffeomorphism $\mathfrak{IIDiff}(\mathbf{M})$ on $\mathbf{M}$, make sense.

If the curvature k is given, ϕ is determined modulo a constant of integration. If ϕ is given, Z is determined modulo Euclidean motion. Therefore, the MKdV hierarchy acts on the space $\mathfrak{M}$ rather than $\mathbf{M}$:

Proposition 4.117 *There exists an action $\mathfrak{IIDiff}(\mathfrak{M})$ on $\mathfrak{M}$ given by the MKdV hierarchy.*

4.6.5 Finite-Dimensional Orbits

In this subsection, we investigate the finite-dimensional orbits. The (disjoint) orbits of the action $\mathfrak{IIDiff}(\mathfrak{M})$ on $\mathfrak{M}$ give a decomposition of $\mathfrak{M}$.

The orbits are finite-dimensional when the hierarchy defined recursively contains a finite number of linearly-independent flows. This happens, in particular, when the deformations ∂_{t_i} are stationary for $i > g+1$, although this is not the only case: as an analogy, the flows of a finite-dimensional orbit of the KdV equation span the Jacobian of a spectral curve, of given genus g; they are coordinatized by the coefficients of the expansion of a basis of holomorphic differentials in (the inverse of) a local parameter t on the curve, $\mathrm{t}_i = \sum_{j=0}^{\infty} c_{ij} t^j$, $i = 1, ..., g$, and of course all vectors $(c_{1j}, ..., c_{gj})$ could be nonzero. We call the union of the particular finite-dimension orbits for which the flows become stationary exactly at the g stage $\mathbf{M}_g$, although the orbit, in fact, could have any smaller dimension, equal to the dimension of the span of the first $g-1$ flows:

$$\mathbf{M}_g := \{Z \in \mathbf{M} \mid \text{for } k = k[Z], \Omega^{\mathrm{II}g} k = 0\},$$

$$\mathfrak{M}_g := \{Z \in \mathfrak{M} \mid \text{for } k = k[Z], \Omega^{\mathrm{II}g} k = 0\} = \mathrm{pr}_2(\mathbf{M}_g),$$

giving a filtration:

$$\mathbf{M}_g \subset \mathbf{M}_{g+1}, \quad \text{and} \quad \mathfrak{M}_g \subset \mathfrak{M}_{g+1}.$$

Definition 4.118 [KonIW1981, Ic1981] A one-parameter family of curves $\{C_t\}$ for real parameter $t \in \mathbb{R}$ is called a loop soliton, if its curvature obeys the MKdV equation; for $q := k/2$, (4.111).

It is noted that when we consider the submanifold Dirac operator on this system, the real curves which obey the MKdV hierarchy provide the geometrical realization of the inverse scattering method [Mat2008c, MatT1992]. Further, the above arguments have also applied a curve in three-euclidean space as in [Mat1999].

4.6.6 Finite-Dimensional Orbits and Hyperelliptic Functions

In this subsection, we express the finite-dimensional orbits in terms of the hyperelliptic functions. Since we study the hyperelliptic solutions in Sect. 4.2.1, we apply the results to this generalized elastica problem. As a first step, we apply them formally. The meaning of formal will be clarified in the following subsection.

We consider the hyperelliptic curve (2.65), $X_H : y^2 = f(x) = \prod_{i=0}^{2g}(x - b_i)$ with $F(x) = \prod_{i=1}^{g}(x - x_i)$ for a point $((x_1, y_1), \dots, (x_g, y_g)) \in S^g X_H$ and $u = w(\iota_X((x_1, y_1), \dots, (x_g, y_g))) \in \mathbb{C}^g$.

Theorem 4.119 [Mat2002b, Mat2007] *Fix a branch point b_a, $(a = 1, 2, \dots, 2g + 1)$. By letting*

$$\mu^{(a)} := \frac{1}{2}\partial_{u_g}\phi^{(a)}, \quad \phi^{(a)}(u) := \frac{1}{\sqrt{-1}}\log F(b_a), \quad F(b_a) = e^{\sqrt{-1}\phi^{(a)}},$$

$\mu^{(a)}$ and $\phi^{(a)}$ satisfy the relations,

$$(\partial_{u_{g-1}} - \frac{1}{2}(\lambda_{2g} + 3b_a)\partial_{u_g})\mu^{(r)} + 6\mu^{(r)2}\partial_{u_g}\mu^{(r)} + \partial_{u_g}^3\mu^{(r)} = 0, \tag{4.112}$$

$$(\partial_{u_{g-1}} - \frac{1}{2}(\lambda_{2g} + 3b_a)\partial_{u_g})\phi^{(a)} + \frac{1}{8}\left(\partial_{u_g}\phi^{(a)}\right)^3 + \frac{1}{4}\partial_{u_g}^3\phi^{(a)} = 0, \tag{4.113}$$

as differential identities in $S^g X_H$ and $\mathbb{C}^g$, where $\partial_{u_i} := \partial/\partial u_i$.

We call (4.112) and (4.113) modified KdV equations over $\mathbb{C}$. We can apply them formally to the loop soliton.

Corollary 4.120 [Mat2002b] *Let the configuration of the x-components $(x_1, \cdots, x_g)$ of the affine coordinates of the hyperelliptic curves $S^g X_H$ satisfy*

$$|F(b_a)| = 1,$$

for a unit circle. For such $(x_1, y_1), \cdots, (x_g, y_g)$, we have the image of its Abel–Jacobi map $u := w((x_1, y_1), \cdots, (x_g, y_g))$.

1. *By setting $s := u_g + (\lambda_{2g} + 3b_a)u_{g-1}/2$ and $t := u_{g-1}/4$,*

$$\partial_{u_g}Z^{(a)} := F(b_a), \quad or \quad |\partial_s Z^{(a)}| = 1, \tag{4.114}$$

completely characterizes the loop soliton over $\mathbb{C}$.

2. *The loop soliton over* $\mathbb{C}$ *is given by*

$$Z^{(a)} = \frac{1}{r_0}\left(b_a^g u_g + \sum_{i=1}^{g} b_a^i \zeta_{i-1}\right).$$

3. *The Euler–Bernoulli energy over* $\mathbb{C}$ *is given by*

$$\oint_{\alpha_a} k^2 ds = \frac{1}{r_0}(-4\eta'_{ag} + 2(\lambda_{2g} + b_a)\omega'_{ag}).$$

Proof Let us consider the linear relation among ∂_{u_i} and ∂_{t_i} in terms of the matrix $M := (\partial_{t_{2i+1}} u_{3-j})$,

$$\begin{pmatrix}\partial_s\\ \partial_t\end{pmatrix} = \begin{pmatrix}\partial_s u_g & \partial_s u_{g-1}\\ \partial_t u_g & \partial_t u_{g-1}\end{pmatrix}\begin{pmatrix}\partial_{u_g}\\ \partial_{u_{g-1}}\end{pmatrix} = \begin{pmatrix}1 & 0\\ -2(\lambda_{2g} + 3b_a) & 4\end{pmatrix}\begin{pmatrix}\partial_{u_g}\\ \partial_{u_{g-1}}\end{pmatrix}.$$

By considering ${}^tM$ and its inverse, we have the linear relations among one-forms,

$$\begin{pmatrix}du_g\\ du_{g-1}\end{pmatrix} = \begin{pmatrix}1 & -2(\lambda_{2g} + 3b_a)\\ 0 & 4\end{pmatrix}\begin{pmatrix}ds\\ dt\end{pmatrix}, \quad \begin{pmatrix}ds\\ dt\end{pmatrix} = \begin{pmatrix}1 & (\lambda_{2g} + 3b_a)/2\\ 0 & 1/4\end{pmatrix}\begin{pmatrix}du_g\\ du_{g-1}\end{pmatrix}.$$

The other is directly obtained. ∎

Remark 4.121 $\zeta_j(u)$ in $Z^{(r)}$ is expressed by Proposition 3.67

$$-\zeta_i(u) = \sum_{k=1}^{g} \int_{\infty}^{x_i} dr_i - \frac{1}{2}\det A_{g-i}.$$

4.6.7 *MKdV Equation over* $\mathbb{R}$

In this subsection, we investigate the MKdV equation over $\mathbb{R}$ to show what "formal" means in the previous subsection. Though we obtain the hyperelliptic solutions of the modified KdV equation and loop soliton equation, it is over $\mathbb{C}$, not a real geometrical solution. In the study of the algebro-geometric solutions of the integrable system, the real solutions are obtained by the natural embedding $\mathbb{R}$ to $\mathbb{C}$ ($x \mapsto x \in \mathbb{C}$). For example, we consider $\mathbb{R}[x, y]/(f_X(x, y))$ by restricting the coefficients in f_X to real values. However, it cannot bring us to periodic solutions in the case of the hyperelliptic solutions of the (focusing) MKdV equation over $\mathbb{R}$. By investigating the geometric structure of hyperelliptic solutions (4.114), which asserts that ϕ must be real, the real solution of the generalized elastica should be obtained in the framework of the hyperelliptic curves over $\mathbb{C}$.

Thus we clarify the difference between the MKdV equations (4.111) over $\mathbb{R}$ and (4.113) over $\mathbb{C}$. By introducing real and imaginary parts, $u_b = u_{b\,\mathrm{r}} + \sqrt{-1}u_{b\,\mathrm{i}}$ and $\phi^{(a)} = \phi_{\mathrm{r}}^{(a)} + \sqrt{-1}\phi_{\mathrm{i}}^{(a)}$, we have the Cauchy–Riemann relations, $\dfrac{\partial}{\partial u_{b\,\mathrm{r}}}\phi_{\mathrm{r}}^{(a)} = \dfrac{\partial}{\partial u_{b\,\mathrm{i}}}\phi_{\mathrm{i}}^{(a)}$ and $\dfrac{\partial}{\partial u_{b\,\mathrm{r}}}\phi_{\mathrm{i}}^{(a)} = -\dfrac{\partial}{\partial u_{b\,\mathrm{i}}}\phi_{\mathrm{r}}^{(a)}$.

Lemma 4.122 $\dfrac{\partial}{\partial u_b}\phi^{(a)} = \dfrac{\partial}{\partial u_{b\,\mathrm{r}}}\phi^{(a)} = 2\dfrac{\partial}{\partial u_b}\phi_{\mathrm{r}}^{(a)}$.

Proof Since $\dfrac{\partial}{\partial u_b} = \dfrac{1}{2}\left(\dfrac{\partial}{\partial u_{b\,\mathrm{r}}} - \sqrt{-1}\dfrac{\partial}{\partial u_{b\,\mathrm{i}}}\right)$, we have

$$\begin{aligned}
\frac{\partial}{\partial u_b}\phi^{(a)} &= \frac{1}{2}\left(\frac{\partial}{\partial u_{b\,\mathrm{r}}} - \sqrt{-1}\frac{\partial}{\partial u_{b\,\mathrm{i}}}\right)(\phi_{\mathrm{r}}^{(a)} + \sqrt{-1}\phi_{\mathrm{i}}^{(a)})\\
&= \frac{\partial}{\partial u_{b\,\mathrm{r}}}(\phi_{\mathrm{r}}^{(a)} + \sqrt{-1}\phi_{\mathrm{i}}^{(a)})\\
&= \left(\frac{\partial}{\partial u_{b\,\mathrm{r}}} - \sqrt{-1}\frac{\partial}{\partial u_{b\,\mathrm{i}}}\right)\phi_{\mathrm{r}}^{(a)}\\
&= 2\frac{\partial}{\partial u_b}\phi_{\mathrm{r}}^{(a)}.
\end{aligned}$$

■

Lemma 4.123 *The hyperelliptic solution (4.113) of MKdV equation over $\mathbb{C}$ means*

$$\begin{aligned}
&\left(\frac{\partial}{\partial u_{g-1\,\mathrm{r}}} - \frac{1}{2}(\lambda_{2g} + 3b_a)\frac{\partial}{\partial u_{g\,\mathrm{r}}}\right)\phi_{\mathrm{r}}^{(a)}\\
&\quad + \frac{1}{8}\left[\left(\frac{\partial}{\partial u_{g\,\mathrm{r}}}\phi_{\mathrm{r}}^{(a)}\right)^3 - 3\left(\frac{\partial}{\partial u_{g\,\mathrm{r}}}\phi_{\mathrm{i}}^{(a)}\right)^2\left(\frac{\partial}{\partial u_{g\,\mathrm{r}}}\phi_{\mathrm{r}}^{(a)}\right)\right] + \frac{1}{4}\frac{\partial^3}{\partial u_{g\,\mathrm{r}}^3}\phi_{\mathrm{r}}^{(a)} = 0,
\end{aligned}$$

$$\begin{aligned}
&\left(\frac{\partial}{\partial u_{g-1\,\mathrm{r}}} - \frac{1}{2}(\lambda_{2g} + 3b_a)\frac{\partial}{\partial u_{g\,\mathrm{r}}}\right)\phi_{\mathrm{i}}^{(a)}\\
&\quad - \frac{1}{8}\left[\left(\frac{\partial}{\partial u_{g\,\mathrm{r}}}\phi_{\mathrm{i}}^{(a)}\right)^3 - 3\left(\frac{\partial}{\partial u_{g\,\mathrm{r}}}\phi_{\mathrm{r}}^{(a)}\right)^2\left(\frac{\partial}{\partial u_{g\,\mathrm{r}}}\phi_{\mathrm{i}}^{(a)}\right)\right] + \frac{1}{4}\frac{\partial^3}{\partial u_{g\,\mathrm{r}}^3}\phi_{\mathrm{i}}^{(a)} = 0.
\end{aligned}$$

The real part of (4.113) is interpreted as a gauged MKdV equation as follows:

Theorem 4.124 *The real part of (4.113) is reduced to the gauged MKdV equation with gauge field $A(u) = (\lambda_{2g} + 3b_a - \frac{3}{4}(\partial_{u_g\,\mathrm{r}}\phi_{\mathrm{i}}^{(a)})^2)/2$,*

$$-(\partial_{u_{g-1}\,\mathrm{r}} - A(u)\partial_{u_g\,\mathrm{r}})\phi_{\mathrm{r}}^{(a)} + \frac{1}{8}\left(\partial_{u_g\,\mathrm{r}}\phi_{\mathrm{r}}^{(a)}\right)^3 + \frac{1}{4}\partial_{u_g\,\mathrm{r}}^3\phi_{\mathrm{r}}^{(a)} = 0. \tag{4.115}$$

Remark 4.125 One might regard that the problem is simple because we guess that we could make $\phi_{\mathrm{i}}^{(a)}$ zero. However, it is noted that the Abelian coordinate u_g and

u_{g-1} are determined by the configurations of $(x_i, y_i) \in S^g X_H$. The restriction of the configuration in $S^g X_H$ affects the Abelian coordinates u_g and u_{g-1} in Jacobian. The reality condition is to control both the Abelian integral image and preimage to satisfy some conditions as follows. It is an essential problem in the generalized elastica problem. We emphasize that in order to deal with them simultaneously, the reconstruction of the Abelian function theory we provide in this book is required. (Moreover, it is imperative that this issue be addressed with greater sophistication and improvement.)

In order to obtain a real solution of (4.111) over $\mathbb{R}$ or a generalized elastica in terms of the data in Theorem 4.119, the following conditions must be satisfied [Mat2007]:

CI $\prod_i |x_i - b_a| = \text{a constant} = \gamma > 0$ in Theorem 4.119,
CII $du_{g\mathrm{i}} = du_{g-1\mathrm{i}} = 0$ in Theorem 4.119,
CIII $A(u)$ is a real constant: if $A(u) = \varepsilon$ constant, (4.115) is reduced to (4.111).

We consider the solutions of (4.115) which satisfy the above conditions in Remark 4.125 as follows.

4.6.8 Genus One: Elastica

We handle the elastica of genus one in this subsection. First, we investigate the original elastica problem from a modern viewpoint. The genus one case is studied by Euler and Bernoulli. Using the isometric deformation $Z(\delta t) = Z(0) + \delta t \partial_t Z(0)$, $(\partial_t Z(0) = (v^{(r)} + \sqrt{-1}v^{(i)})\partial_s Z)$, we consider the minimal problem of the Euler–Bernoulli energy, (4.105). For the deformation, we evaluate $\dfrac{\mathscr{E}[Z(\delta t)] - \mathscr{E}[Z(0)]}{v^{(r)}\delta t} = 0$ modulo δt. It yields the static MKdV equation,

$$ak + \frac{1}{2}k^3 + \partial_s^2 k = 0,$$

which is associated with the elliptic curve $(\partial_s k)^2 + \frac{1}{4}k^4 + ak^2 + b = 0$, where a and b are integral constants.

We introduce affine parameters

$$\begin{aligned} x(s) &:= \frac{\sqrt{-1}}{4\alpha}\partial_s k + \frac{1}{8}k^2 + \frac{1}{12}a, \\ y(s) &:= \frac{1}{2\alpha}\partial_s x = \frac{1}{2}\left(\sqrt{-1}\left(\frac{1}{8}k^3 + \frac{1}{4}ak + \frac{\sqrt{-1}}{4\alpha}k\partial_s k\right)\right), \end{aligned} \tag{4.116}$$

so that they satisfy

$$\frac{\hat{y}^2}{4} = y^2 = \left(x + \frac{1}{6}a\right)\left(x - \frac{1}{12}a - \frac{1}{4}\sqrt{b}\right)\left(x - \frac{1}{12}a + \frac{1}{4}\sqrt{b}\right) \tag{4.117}$$
$$= (x - e_1)(x - e_2)(x - e_3),$$

where $e_1 = \frac{1}{6}a$, $e_2 = -\frac{1}{12}a + \frac{1}{4}\sqrt{b}$, and $e_3 = -\frac{1}{12}a - \frac{1}{4}\sqrt{b}$; $(b_{i-1} = e_i$ in $X_H)$. For later convenience, we let $a^2 - b = 16$.

The elliptic curve E is given by

$$E = \{(x, y) \in \mathbb{C}^2 \mid y^2 = (x - e_1)(x - e_2)(x - e_3)\} \cup \{\infty\}.$$

This means that $a = 2(e_2 + e_3 - 2e_1)$ and $b = -(e_2 - e_3)^2$. $(x, \widehat{y} = 2y)$ corresponds to the Weierstrass standard form [WhW1927].

The coordinate u of the complex plane $\mathbb{C}$ is given by

$$\int^{(x,y)} du, \quad du = \frac{dx}{2y}. \tag{4.118}$$

It is known that a shape of the (classical) elastica, i.e., a loop soliton with genus one, $Z : \mathbb{R} \hookrightarrow \mathbb{C}$ $(u \mapsto Z(u) = X^1(u) + \sqrt{-1}X^2(u))$ with $\partial_u Z = e^{\sqrt{-1}\phi}$ satisfies the differential equation

$$a\partial_u(\phi) + \frac{1}{2}(\partial_u \phi)^3 + \partial_u^3 \phi = 0, \tag{4.119}$$

where $\partial_u := d/du$.

The following is evident.

Proposition 4.126 (Euler [Eu1744]) *A solution of (4.119) is given by*

$$\partial_u Z^{(a)} = (x - e_a), \quad Z^{(a)} = \zeta(u) - e_a u + Z_0$$

for an elliptic curve given by the form (4.117). If it is a loop soliton if and only if it satisfies the reality condition:

1. $|\partial_u Z^{(a)}| = 1$.
2. $u \in \mathbb{R}$.

Proof The reality condition is proved after Proposition 4.128. ■

Let $(a, b, c) = (1, 2, 3)$, and $\phi^{(a)} = 2\varphi_a = 2\varphi$.

Lemma 4.127 *Let* $e^{2\sqrt{-1}\varphi} := (x - e_a)/c_{cba}$, $e_{ab} := e_a - e_b$ *and* $c_{cba} := \sqrt{e_{ca}e_{ba}}$. *The elliptic differential of the first kind (4.118) up to sign is*

$$du = \frac{d\varphi}{\sqrt{(\sqrt{e_{ba}} - \sqrt{e_{ca}})^2 + 4\sqrt{e_{ba}e_{ca}}\sin^2\varphi}},$$

where $\gamma = 1$, $(\sqrt{e_{ba}} - \sqrt{e_{ca}})^2 = (a - 4)/2$.

Proof Direct computations give

$$dx = 2c_{cba}\sqrt{-1}\mathrm{e}^{2\sqrt{-1}\varphi}d\varphi,$$

$$\begin{aligned} y &= c_{cba}\sqrt{-1}\mathrm{e}^{2\sqrt{-1}\varphi}\sqrt{e_{ba}(\mathrm{e}^{-2\sqrt{-1}\varphi} - c_{cba}e_{ba}^{-1})(\mathrm{e}^{2\sqrt{-1}\varphi} - c_{cba}^{-1}e_{ca})} \\ &= c_{cba}\sqrt{-1}\mathrm{e}^{2\sqrt{-1}\varphi}\sqrt{e_{ba}(\mathrm{e}^{-2\sqrt{-1}\varphi} - \sqrt{e_{ca}/e_{ba}})(\mathrm{e}^{2\sqrt{-1}\varphi} - \sqrt{e_{ba}/e_{ca}})} \\ &= c_{cba}\sqrt{-1}\mathrm{e}^{2\sqrt{-1}\varphi}\sqrt{e_{ba} + e_{ca} - 2\sqrt{e_{ba}e_{ca}}\cos 2\varphi}, \end{aligned}$$

up to sign. The addition formula $\cos(2\varphi) = 1 - 2\sin^2\varphi$ leads to the result. $\gamma^2 = a_1 a_2 = \dfrac{1}{16}(a^2 - b) = 1$ and $(\sqrt{e_{ba}} - \sqrt{e_{ca}})^2 = (e_{ba} + e_{ca} - 2\sqrt{e_{ba}e_{ca}}) = \dfrac{1}{2}(a - 4)$. ■

Let us use the standard representations

$$\widetilde{k} := \frac{2\sqrt{-1}\sqrt[4]{e_{ba}e_{ca}}}{\sqrt{e_{ba}} - \sqrt{e_{ca}}},$$

and then du is expressed by

$$du = \frac{2\widetilde{k}\,d\varphi}{\sqrt{1 - \widetilde{k}^2\sin^2\varphi}}. \tag{4.120}$$

By letting $w := \sin(\varphi)$, (4.120) becomes

$$\begin{aligned} du &= \frac{dw}{\sqrt{(1 - w^2)((\sqrt{e_{ba}} - \sqrt{e_{ca}})^2 + 4\sqrt{e_{ba}e_{ca}}w^2)}} \\ &= \frac{dw}{(\sqrt{e_{ba}} - \sqrt{e_{ca}})\sqrt{(1 - w^2)(1 - \widetilde{k}^2 w^2)}}. \end{aligned} \tag{4.121}$$

Proposition 4.128 *Let $\varphi = \varphi_{\mathrm{r}} + \sqrt{-1}\varphi_{\mathrm{i}}$ and $u = u_{\mathrm{r}} + \sqrt{-1}u_{\mathrm{i}}$ as real representations. For the case of genus one, the Abelian integral image of the real part of $d\varphi_{\mathrm{r}}$ corresponds to real part du_{r} of du.*

The equation $\partial_u\varphi + 2(\partial_u\varphi)^3 + \partial_u^3\varphi = 0$ (4.119) is reduced to the static MKdV equation over $\mathbb{R}$, $\partial_{u_{\mathrm{r}}}\varphi_{\mathrm{r}} + +2(\partial_u\varphi_{\mathrm{r}})^3 + \partial_u^3\varphi_{\mathrm{r}} = 0$ by assuming that $\varphi_{\mathrm{i}}(u) = \partial_{u_{\mathrm{r}}}\varphi_{\mathrm{i}}(u) = 0$ at $\{u_{\mathrm{i}} = 0\}$.

The condition CIII in Remark 4.125 is trivially satisfied.

Proof The equation (4.120) asserts the first statements in the proposition. The reality condition for both $\varphi_{\mathrm{i}} = 0$ and $u_{\mathrm{i}} = 0$ simultaneously satisfies. The Cauchy–Riemann relations in Lemma 4.122 show the statements. ■

The complex analytic equation is reduced to the real analytic equation for the $g = 1$ case.

4.6.8.1 Moduli Space of Elastica

Euler implicitly considered the moduli space of the elastica, whereas Mumford explicitly showed it by using the data of the theta function and the Jacobian. In this subsubsection, we consider the moduli space of elastica by using the algebraic data of the elliptic curves.

Proposition 4.129 (Mumford)[Mum1993] *The moduli* Λ *of elastica or loop soliton of genus one is given by the following subspace in the upper half plane* $\mathbb{H} := \{z \in \mathbb{C} \mid \Im z > 0\}$ *modulo* $\mathrm{PSL}(2, \mathbb{Z})$,

$$\Lambda := \sqrt{-1}\mathbb{R}_{>0} \cup \left(\frac{1}{2} + \sqrt{-1}\mathbb{R}_{>0}\right) \cup \infty \quad \text{modulo} \quad \mathrm{PSL}(2, \mathbb{Z}).$$

Here $\mathbb{R}_{>0}$ *is* $\{x \in \mathbb{R} \mid x > 0\}$.

Though Mumford led this result using the geometry of the Abelian variety of genus one [Mum1993], we provide alternative proof only by using the algebraic data of the elliptic curves. Due to the reality condition, φ is a subregion of a real number. For any φ in a certain region $[\varphi_l, \varphi_u]$, the reality condition of the loop soliton $Z^{(a)}$ requires that the denominator in (4.121) should be real and thus that $\sqrt{e_{ba}e_{ca}}$ and $(\sqrt{e_{ba}} - \sqrt{e_{ca}})^2$ should be real.

$$\Im\sqrt{e_{ba}} = \Im\sqrt{e_{ca}}, \quad \arg(e_{ba}) = -\arg(e_{ca}),$$

where $\arg(a) := \Im \log(a)$ for $a \in \mathbb{C}$. Accordingly we introduce an expression $e_{ba} =: \beta_{ba} e^{\sqrt{-1}\alpha_{ba}}$, using $\alpha_{ba} \in [0, \pi)$ and $\beta_{ba} \in \mathbb{R}$.[11] The reality condition of the loop soliton $Z^{(a)}$ requires alternative cases:

1. α_{ba} and α_{ca} vanish, i.e., e_{ba} and e_{ca} belong to $\mathbb{R}$, or
2. $\alpha_{ba} = -\alpha_{ca}$ and $\beta_{ba} = \beta_{ca}$.

However, the second case means that $(\sqrt{e_{ba}} - \sqrt{e_{ca}})^2$ is equal to zero and corresponds to $\widetilde{k} = \infty$. Thus we obtain the following lemma.

Lemma 4.130 *The reality condition of the loop soliton* $Z^{(a)}$ *is reduced to two alternative cases:*

I-1 $e_{ba} > 0$ *and* $e_{ca} > 0$, *i.e.*, $\widetilde{k} \in \sqrt{-1}\mathbb{R}_{\geq 0}$, $w = \sin\varphi \in [-1, 1]$.
I-2 $e_{ba} \leq 0$ *and* $e_{ca} \leq 0$, *i.e.*, $\widetilde{k} > 1$ *and* $w = \sin\varphi \in [1/\widetilde{k}, 1]$ *or* $w = \sin\varphi \in [-1, -1/\widetilde{k}]$.

Proof For a certain $\varphi \in \mathbb{R}$, u must be real. Hence the candidates for the e_{ba}'s are the following: (I-0) $e_{ba} < 0$ and $e_{ca} > 0$, or $e_{ba} > 0$ and $e_{ca} < 0$, (I-1) $e_{ba} > 0$ and $e_{ca} > 0$, and (I-2) $e_{ba} \leq 0$ and $e_{ca} \leq 0$.

---

[11] Here we defined $\beta_{ba} \in \mathbb{R}$ rather than $\beta_{ba} \in \mathbb{R}_{\geq 0}$ due to the domain of α_{ba}.

Since in the (I-0) case $(\sqrt{e_{ba}} - \sqrt{e_{ca}})$ belongs to $\mathbb{C} \setminus (\mathbb{R} \cup \mathbb{R}\sqrt{-1})$, the angle in the complex plane cannot be canceled by the other factors so that du is real. We cannot employ the (I-0) case. (I-1) is naturally selected for $\varphi \in \mathbb{R}$.

Let us consider the case (I-2). We note that 1) the region of $\sin \phi_a$ must be a subset of $[-1, 1]$, and 2) the prefactor $1/(\sqrt{e_{ba}} - \sqrt{e_{ca}})$ generates the factor $\sqrt{-1}$. Hence we conclude that for $\widetilde{k} > 1$ and $\sin \phi_a \in [1/\widetilde{k}, 1]$ or $\sin \phi_a \in [-1, -1/\widetilde{k}]$, du is real. ■

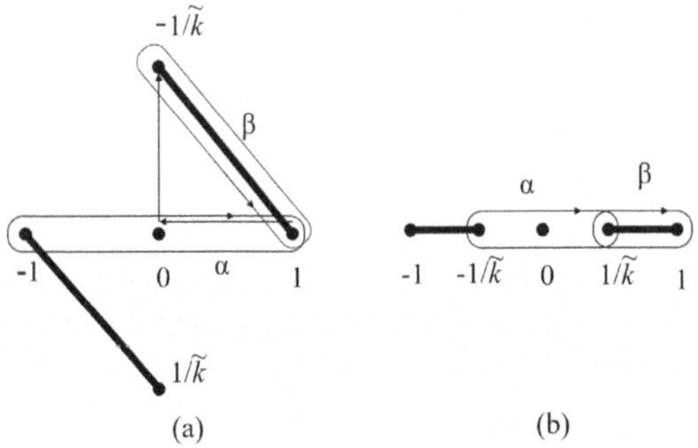

Fig. 4.4 Geometry of contours: α and β are the homology basis of the elliptic curves

Proof of Proposition 4.129: Let us consider the geometry of the integration. Figure 4.4 displays our situations; Fig. 4.4 (a) corresponds to the I-1 case and (b) to the I-2 case in Lemma 4.130.

1. The periodicity $(4\omega, 2\omega')$ of $\sqrt{(x - e_a)}$ for the I-1 case is given by

$$\omega' = \int_0^1 \frac{dw}{\sqrt{(1-w^2)((\sqrt{e_2} - \sqrt{e_1})^2 + 4\sqrt{e_2 e_1} w^2)}},$$

$$\omega'' = \left(\int_1^0 + \int_0^{\sqrt{-1}/|\widetilde{k}|}\right) \frac{dw}{\sqrt{(1-w^2)((\sqrt{e_2} - \sqrt{e_1})^2 + 4\sqrt{e_2 e_1} w^2)}}.$$

   Thus ω'' is equal to $\omega' + \sqrt{-1}L[\widetilde{k}]$ for general $\widetilde{k}$ with a certain real-valued function L. On the other hand, it is obvious that for $\widetilde{k} \to 0$, $L \to \infty$ and for $\widetilde{k} \to \infty$, $L \to 0$. Further, $L[\widetilde{k}]$ is a continuous function of $\widetilde{k}$ and its range is $\mathbb{R}_{>0}$. Hence $\tau = \omega''/\omega' \in (1/2 + \sqrt{-1}\mathbb{R}_{>0})$.
2. The periodicity $(4\omega', 2\omega'')$ of $\sqrt{(x - e_1)}$ for the I-2 case is given by

$$\omega' = 2\int_0^{1/\widetilde{k}} \frac{dw}{\sqrt{(1-w^2)((\sqrt{e_2} - \sqrt{e_1})^2 + 4\sqrt{e_2 e_1} w^2)}},$$

$$\omega'' = \int_{1/\widetilde{k}}^1 \frac{dw}{\sqrt{(1-w^2)((\sqrt{e_2} - \sqrt{e_1})^2 + 4\sqrt{e_2 e_1} w^2)}}.$$

It is obvious that for $\widetilde{k} \to 0$, $\omega' \to \infty$ and for $\widetilde{k} \to \infty$, $\omega' \to 0$, whereas ω'' is finite. Further, $\omega'[\widetilde{k}]$ and $\omega'[\widetilde{k}]$ are continuous in $\widetilde{k}$. Hence $\tau = \omega''/\omega' \in \sqrt{-1}\mathbb{R}_{>0}$.
Since the theory of the Jacobi elliptic functions gives the fact that $\widetilde{k}' := \sqrt{1-\widetilde{k}^2}$ gives the inversion of moduli $\tau \to -1/\tau$, the constraint $\widetilde{k} > 1$ in Lemma 4.130 is less important.

We note that the periodicity of $\sqrt{(x-e_1)}$ differs from $\partial_u Z^{(a)}$ by a factor of two but the difference is not so significant. We have a complete proof of Proposition 4.129 based upon the algebraic data of the elliptic curves. ■

Remark 4.131 We list its special cases for $a = 1$:

1. $\widetilde{k} = 0$ in I-1: its shape is a circle and its related curve is $y^2 = (x-e_1)^2(x-e_2)$.
2. $\widetilde{k} = \infty$ in I-2: its shape is a loop soliton solution, and its related curve is $y^2 = (x-e_1)(x-e_2)^2$.

For the map $\partial_u Z : S^1 \to S^1$, we can find the index as a winding number as shown in Fig. 4.5. We call it the index $(\partial_u Z)$.

Corollary 4.132 *The* index$(\partial_u Z)$ *is given as follows:*
I-1 index$(\partial_u Z) = \pm 1$.
I-2 index$(\partial_u Z) = 0$.

Proof In the case I-1, since the contour $w = \sin\varphi$ is $[-1, 1]$ which is identified with the range of sine function, φ becomes a monotonic increasing function of u. In fact, passing by $w = \pm 1$ changes the sign of $\sqrt{1-w^2}$ or $\cos\varphi$. By paying attention to the orientation of the contour, we have the sign of the index. On the other hand, in the case I-2, φ does not wind around S^1 like Fig. 4.5b. The branch point $(1/\widetilde{k}, 0)$ does not have an effect of the sign of $\sqrt{1-w^2}$. ■

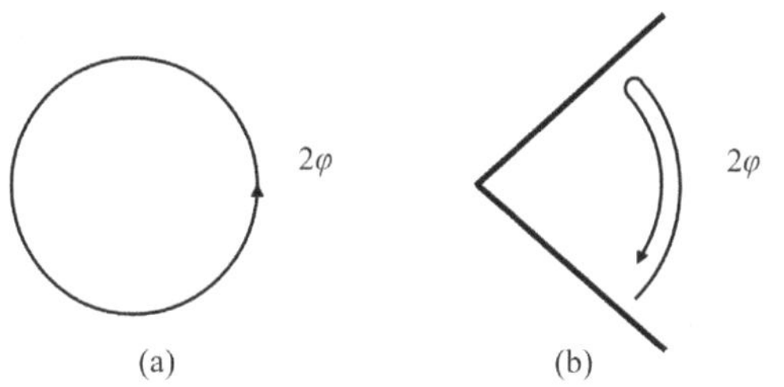

Fig. 4.5 The behavior of φ

4.6.8.2 Computations of Elastica by Euler's Method

In the cases of genera zero and one, it is easy to impose the conditions CI–III, i.e., $\partial_{ur}\psi_{\rm i} = 0, du_{\rm i} = 0$ by $|x - e_1| = \gamma$, and therefore we are concerned with $\phi = \psi_{\rm r} = 2\varphi_{\rm r}$, which obeys (4.25) and (4.115).

It is well-known that the solutions are expressed by the Jacobi elliptic functions e.g. [Lo1892], and can instead be written in terms of the Weierstrass elliptic functions [Mat2010] [Mat2002b, Mat2010]. Here in order to evaluate these solutions numerically, we show that Euler's figure eight is directly expressed by means of Euler's numerical quadrature method, though current mathematical software easily reproduces Euler's original results; our computation in Fig. 4.6 shows a simple numerical method suffices.

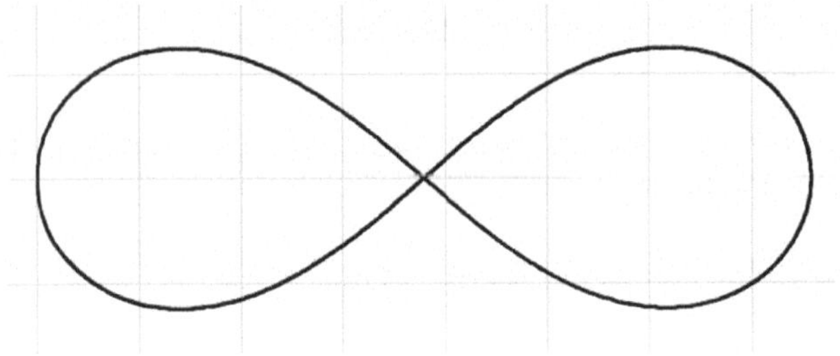

Fig. 4.6 Euler's elastica of figure-eight by Euler's method

1. $g = 0$ case: We consider the limit of $\widetilde{k} \to 0$ with $2kd\varphi = ds$. By identifying du with ds, we have the circle as the minimal path of the Euler–Bernoulli energy.
2. $g = 1$ case: We let $du = ds, \widetilde{k} = 2.398107502$ for $a = 2.608918126$. We numerically evaluate the figure-eight curve of in Fig. 4.6, as illustrated in Euler's book [Eu1744]: here we plot $Z(s) = (X + \sqrt{-1}Y)(s)$ as a function of s. We used Euler's numerical quadrature method, $\varphi(s + \delta) \approx \varphi(s) + \frac{\partial\varphi}{\partial s}\delta$ and $Z(s + \delta) = Z(s) + e^{2\sqrt{-1}\varphi}\delta$ for a small number δ. In the computations, φ moves back and forth in the interval $[\varphi_s, \varphi_e]$ where $\varphi_s = \sin^{-1}(\widetilde{k}^{-2}) \in [0, \pi/2)$ and $\varphi_e = \pi - \varphi_s$.

4.6.9 *Genus Two: Generalized Elastica*

In this subsection, we investigate the conditions CI–III for hyperelliptic curves X_H of genus $g = 2$.

For general genus g, due to condition CI, $d\varphi_{1\rm i} = -\sum_{j=2}^{g} d\varphi_{j\rm i}$, where φ_j is defied by $e^{2\sqrt{-1}\varphi_j} = (x_j - b_a)$ for $x_j \in X_H$. The condition CII should be extended to $2g - 1$ linear independent conditions, $du_{g\rm i} = 0, du_{j\rm r} = du_{j\rm i} = 0, j = 1, \ldots, g - 1$, though we do not know whether CIII hold or not. However, since the dimension of J_{X_H} is

$2g$, we cannot find the embedding of S^1 into J_{X_H} if the conditions are linearly independent. Thus our problem is to find the curve X_H and the configuration of $x_i \in \mathbb{C}$ for $i = 1, \dots, g$ so that they satisfy these conditions. However, it is too difficult to solve the problem at this stage. Accordingly, we consider these conditions for $g = 2$ to embed S^1 into J_{X_H} as follows, even though du_{2i} is not satisfied completely. Then we show that even data in $S^2 X_H$ give trajectories for generalization of elastica, cf. Fig. 4.8. We choose coordinates $u = {}^t(u_1, u_2)$ in $\mathbb{C}^2$; $u_i = u_i^{(1)} + u_i^{(2)}$ where $u_i^{(j)} = v_i((x_j, y_j))$. We let $a = 0$ in Theorem 4.119, $b_0 = -\gamma = -1$, and $e_c := b_c - b_0$ $(c = 1, 2, 3, 4)$. Let $\phi^{(0)} = 2\varphi_1 + 2\varphi_2$.

We restrict the moduli (rather, parameter) space of the curve X_H by the following:

Conditions: $\sqrt{e_{2a-1}} = \alpha_a + \sqrt{-1}\beta_a$, $\sqrt{e_{2a}} = \alpha_a - \sqrt{-1}\beta_a$ where $\alpha_a, \beta_a \in \mathbb{R}$, $a, b = 1, 2$, satisfying $\alpha_a^2 + \beta_a^2 = \gamma$.

Under this assumption, we have the natural extension of genus-one elastica; indeed, direct computation shows the following:

Lemma 4.133 *Let* $\gamma e^{2\sqrt{-1}\varphi} := (x - b_0)$, *(2.65)* $g = 2$ *is*

$$y^2 = 16\frac{\gamma^5 e^{6\sqrt{-1}\varphi}}{k_1^2 k_2^2}(1 - k_1^2 \sin^2\varphi)(1 - k_2^2 \sin^2\varphi), \tag{4.122}$$

where $k_a = \dfrac{2\sqrt{-1}\sqrt[4]{e_{2a-1}e_{2a}}}{\sqrt{e_{2a-1}} - \sqrt{e_{2a}}} = \dfrac{\sqrt{\gamma}}{\beta_a}$, $(a = 1, 2)$.

We consider a point $((x_1, y_1), (x_2, y_2))$ in $S^2 X_H$ under the condition CI. We define the variable φ_a by $x_c = \gamma e^{\sqrt{-1}\varphi_c}(e^{\sqrt{-1}\varphi_c} + (b_0/\gamma)e^{-\sqrt{-1}\varphi_c})$ $(c = 1, 2)$. Noting $dx_c = 2\sqrt{-1}\gamma e^{2\sqrt{-1}\varphi_c} d\varphi_c$ and $x_c dx_c = -4\gamma e^{3\sqrt{-1}\varphi_c} \sin\varphi_c \, d\varphi_c$, we have the holomorphic one-forms $(du_1^{(c)}, du_2^{(c)})$ $(c = 1, 2)$,

$$\left(\frac{(\sin\varphi_c + \sqrt{-1}\cos\varphi_c)\, d\varphi_c}{2\gamma K(\varphi_c)}, -\frac{\sin\varphi_c \, d\varphi_c}{K(\varphi_c)}\right), \tag{4.123}$$

where $K(\varphi) := \dfrac{\sqrt{\gamma(1 - k_1^2\sin^2\varphi)(1 - k_2^2\sin^2\varphi)}}{k_1 k_2}$.

Lemma 4.134 *Let* $K_c := K(\varphi_c)$. *The following holds:*

1. $$\begin{pmatrix} du_1 \\ du_2 \end{pmatrix} = \begin{pmatrix} \frac{\sqrt{-1}\exp(-\sqrt{-1}\varphi_1)}{2\gamma K_1} & \frac{\sqrt{-1}\exp(-\sqrt{-1}\varphi_2)}{2\gamma K_2} \\ \frac{-\sin(\varphi_1)}{K_1} & \frac{-\sin(\varphi_2)}{K_2} \end{pmatrix} \begin{pmatrix} d\varphi_1 \\ d\varphi_2 \end{pmatrix},$$

2. $$\begin{pmatrix} \partial_{u_1} \\ \partial_{u_2} \end{pmatrix} = \frac{2\gamma K_1 K_2}{\sqrt{-1}\sin(\varphi_2 - \varphi_1)} \begin{pmatrix} -\frac{\sin(\varphi_2)}{K_2} & \frac{\sin(\varphi_1)}{K_1} \\ -\frac{\sqrt{-1}\exp(-\sqrt{-1}\varphi_2)}{2\gamma K_2} & \frac{\sqrt{-1}\exp(-\sqrt{-1}\varphi_1)}{2\gamma K_1} \end{pmatrix} \begin{pmatrix} \partial_{\varphi_1} \\ \partial_{\varphi_2} \end{pmatrix}.$$

Due to Lemma 4.134 and the expression $\varphi_c = \varphi_{cr} + \sqrt{-1}\varphi_{ci}$, we have

$$\frac{\partial \varphi_{1\mathrm{i}}}{\partial u_{2\mathrm{r}}} = \frac{K_1 \sin \varphi_{2\mathrm{r}}}{\sin(\varphi_{1\mathrm{r}} - \varphi_{2\mathrm{r}})}, \quad \frac{\partial \varphi_{2\mathrm{i}}}{\partial u_{2\mathrm{r}}} = \frac{K_2 \sin \varphi_{1\mathrm{r}}}{\sin(\varphi_{2\mathrm{r}} - \varphi_{1\mathrm{r}})} \tag{4.124}$$

and $\phi^{(0)} = 2(\varphi_1 + \varphi_2)$ satisfies (4.113) and (4.115).

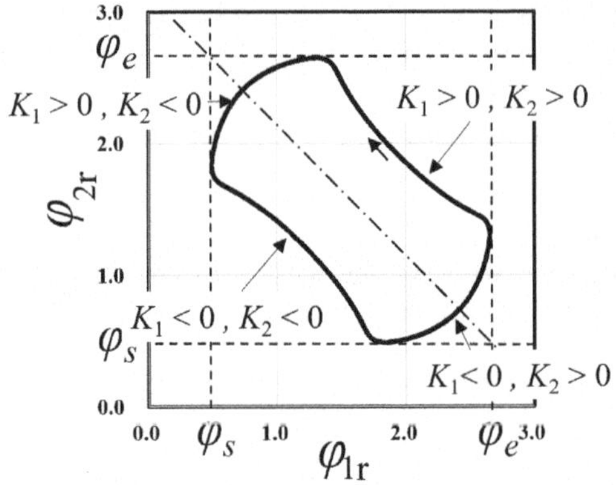

Fig. 4.7 The orbit C of $\partial_{u_2 \mathrm{r}} \phi_{\mathrm{i}}^{(0)} =$ constant: $\varphi_{1\mathrm{r}}$ depends on $\varphi_{2\mathrm{r}}$ for Fig. 4.8c (The arrow shows the direction of $d\eta$)

Due to condition CI, $d\varphi_{1\mathrm{i}} = -d\varphi_{2\mathrm{i}}$.

Let us investigate the conditions CII and CIII. Assume $A(u) = \varepsilon\gamma$; then, $\begin{pmatrix} \partial_t \\ \partial_s \end{pmatrix} = \begin{pmatrix} -4 & -4\varepsilon\gamma \\ 0 & 1 \end{pmatrix} \begin{pmatrix} \partial_{u_1} \\ \partial_{u_2} \end{pmatrix}$ in (4.115) shows the correspondence between the Abelian parameters (u_1, u_2) and (t, s) in the MKdV equation (4.111) via

$$\begin{pmatrix} dt \\ ds \end{pmatrix} = \begin{pmatrix} -1/4 & 0 \\ \varepsilon\gamma & 1 \end{pmatrix} \begin{pmatrix} du_1 \\ du_2 \end{pmatrix}.$$

Though we should evaluate Lemma 4.134 1 for a general point $(\varphi_{c\mathrm{r}}, \varphi_{c\mathrm{i}}), (c = 1, 2)$ in [MatP2022], it is computed at $\varphi_{1\mathrm{i}} = \varphi_{2\mathrm{i}} = 0$ under $|x_c - b_a| = \gamma$ $(c = 1, 2)$ instead of CI for simplicity. We refer to the condition as CI'.

Using this expression, $dt = dt_\mathrm{r} + \sqrt{-1}dt_\mathrm{i}$ and $ds = ds_\mathrm{r} + \sqrt{-1}ds_\mathrm{i}$, we have the following lemma:

Lemma 4.135 *At* $\varphi_{1\mathrm{i}} = \varphi_{2\mathrm{i}} = 0$, *we have* $\begin{pmatrix} dt_\mathrm{r} \\ dt_\mathrm{i} \\ ds_\mathrm{r} \\ ds_\mathrm{i} \end{pmatrix} = M \begin{pmatrix} d\varphi_{1\mathrm{r}} \\ d\varphi_{1\mathrm{i}} \\ d\varphi_{2\mathrm{r}} \\ d\varphi_{2\mathrm{i}} \end{pmatrix}$ *where*

$$M := \begin{pmatrix} -\frac{\sin\varphi_1}{8\gamma K_1} & \frac{\cos\varphi_1}{8\gamma K_1} & -\frac{\sin\varphi_2}{8\gamma K_2} & \frac{\cos\varphi_2}{8\gamma K_2} \\ -\frac{\cos\varphi_1}{8\gamma K_1} & -\frac{\sin\varphi_1}{8\gamma K_1} & -\frac{\cos\varphi_2}{8\gamma K_2} & -\frac{\sin\varphi_2}{8\gamma K_2} \\ \frac{(2+\varepsilon)\sin\varphi_1}{2K_1} & -\frac{\varepsilon\cos\varphi_1}{2K_1} & \frac{(2+\varepsilon)\sin\varphi_2}{2K_2} & -\frac{\varepsilon\cos\varphi_2}{2K_2} \\ \frac{\varepsilon\cos\varphi_1}{2K_1} & \frac{(2+\varepsilon)\sin\varphi_1}{2K_1} & \frac{\varepsilon\cos\varphi_2}{2K_2} & \frac{(2+\varepsilon)\sin\varphi_2}{2K_2} \end{pmatrix}. \tag{4.125}$$

In [MatP2022], the condition CI': $|x_c - b_a| = \gamma$ $(c = 1, 2)$ is employed. As mentioned in the beginning of this subsection, our problem is to find the curve X_H and the configuration of $x_i \in \mathbb{C}$ for $i = 1, \dots, g$ so that the condition CI, $d\varphi_{1\mathrm{i}} = -\sum_{j=2}^{g} d\varphi_{j\mathrm{i}}$ is expressed by $du_{g\mathrm{i}} = 0, du_{j\mathrm{r}} = du_{j\mathrm{i}} = 0, j = 1, \dots, g-1$. The number of the conditions is $2g$, whereas the degree of freedom is $2g$ due to the dimension of the Jacobian J_{X_H}. However, we require one-dimensional for a curve. Thus one of the $2g$ conditions must be determined by the others.

Accordingly, we show a result of a $g = 2$ analogue in [MatP2022] as follows:

Assume that CI' is satisfied. Then $dt_\mathrm{r} \propto ds_\mathrm{r}$. Under CI', we consider condition CIII, i.e., an orbit C of constant $\partial_{u_2\mathrm{r}}\phi_\mathrm{i}^{(0)}$ in the $\varphi_{1\mathrm{r}}$-$\varphi_{2\mathrm{r}}$ plane by requiring

$$d\left(\frac{\partial\phi_\mathrm{i}^{(0)}}{\partial u_{2\mathrm{r}}}\right) = \frac{\partial(\partial_{u_2\mathrm{r}}\phi_\mathrm{i}^{(0)})}{\partial\varphi_{1\mathrm{r}}}d\varphi_{1\mathrm{r}} + \frac{\partial(\partial_{u_2\mathrm{r}}\phi_\mathrm{i}^{(0)})}{\partial\varphi_{2\mathrm{r}}}d\varphi_{2\mathrm{r}} = 0, \tag{4.126}$$

due to (4.124). By parameterizing C by $\eta \in \mathbb{R}$, we have the curve equation

$$d\varphi_{1\mathrm{r}} = \frac{\partial(\partial_{u_2\mathrm{r}}\phi_\mathrm{i}^{(0)})}{\partial\varphi_{2\mathrm{r}}}d\eta, \quad d\varphi_{2\mathrm{r}} = -\frac{\partial(\partial_{u_2\mathrm{r}}\phi_\mathrm{i}^{(0)})}{\partial\varphi_{1\mathrm{r}}}d\eta, \tag{4.127}$$

On C, $\phi^{(0)}$, s and t are functions of $\eta \in \mathbb{R}$.

We obtain C by numerically solving (4.127) as in Fig. 4.7 using Euler's numerical quadrature method. In the computations, φ_a moves back and forth in the interval $[\varphi_s, \varphi_e]$ where $\varphi_s = \sin^{-1}(k_1^{-2}) \in [0, \pi/2)$ and $\varphi_e = \pi - \varphi_s$. Since φ_s and φ_e are branch points in X_H (the interval $[\varphi_s, \varphi_e]$ corresponds to one of the homology basis of X_H), the sign of K_a changes at the turns.

Due to multiplication by M in (4.125), ds_r vanishes on the line $L : \varphi_{1\mathrm{r}} = \pi - \varphi_{2\mathrm{r}}$ in the $\varphi_{1\mathrm{r}}$-$\varphi_{2\mathrm{r}}$ plane. For our $g = 2$ case, there exist the crossing points between the line L and the orbit C (Fig. 4.7), and at the points ds_r vanishes, whereas ds_i (or $du_{2\mathrm{i}}$) does not. Thus $\frac{d\phi_\mathrm{r}^{(0)}}{ds_\mathrm{r}} = \frac{\partial\phi_\mathrm{r}^{(0)}}{\partial s_\mathrm{r}} - \frac{\partial\phi_\mathrm{r}^{(0)}}{\partial s_\mathrm{i}}\frac{ds_\mathrm{i}}{ds_\mathrm{r}}$ diverges at the points. Since due to $g = 2$, the crossing occurs for any parameters k_c $(c = 1, 2)$, the divergence shows that the conditions CII and CIII are contradictory, and thus we cannot find the solution of (4.111) based on Theorem 4.119 in genus two. To satisfy the conditions, we need higher-genus hyperelliptic curves X_H $(g > 2)$.

However, the pole behavior of $d\phi_\mathrm{r}^{(0)}/ds_\mathrm{r}$ is of type $1/\sqrt{s_\mathrm{r} - s_{\mathrm{r}0}}$ at $s_\mathrm{r} = s_{\mathrm{r}0}$ up to a constant factor. Therefore $\phi_\mathrm{r}^{(0)}$ can be defined as a hyperelliptic (2-branch) function of s_r by using the data in Theorem 4.119 along C. Since for finite $\delta\eta$, we obtain $\delta\varphi_{c\mathrm{r}}$ $(c = 1, 2)$ by (4.127), we compute δs_r using M in (4.125), and approximate

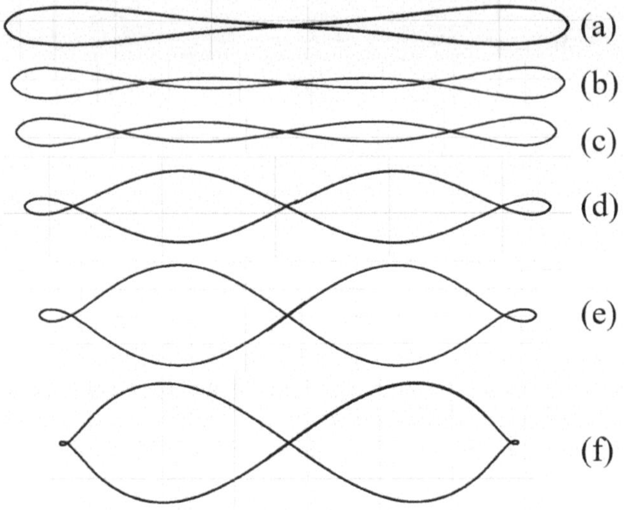

Fig. 4.8 Generalized elastica of genus two, where (k_1, k_2) are as follows: **a**: (1.60, 2.20), **b**: (1.90, 2.15), **c**: (2.08, 2.32), **d**: (3.08, 3.32), **e**: (3.80.4.20), **f**: (4.00, 10.00)

$Z(s_\mathrm{r} + \delta s_\mathrm{r}) \approx Z(s_\mathrm{r}) + \mathrm{e}^{2(\varphi_{1\mathrm{r}}+\varphi_{2\mathrm{r}})\sqrt{-1}}\delta s_\mathrm{r}$ using Euler's numerical quadrature method for appropriate initial data $\varphi_a(s = 0, t = 0)$ so that Z is periodic in s_r; since Z is in general not periodic, we determine appropriate initial states by the shooting method with the bisection process.

Finally, we numerically obtain the shapes of a generalization of the elastica of genus two in Fig. 4.8. Though they are not solutions of (4.111), they identically satisfies (4.113) and are a natural extension of Euler's results. We also numerically compute and display their $\phi_\mathrm{r}^{(0)}$ in Fig. 4.9 and the energy density $k^2/2$ in Fig. 4.10. We remark that the $\phi_\mathrm{r}^{(0)}$ in Fig. 4.9 exhibits 1) periodicity and 2) mixing of so-called kink (anti-kink) type (or $\pm$ tanh-shape type) mode and breather (wavy) type mode [AblS1981]; due to the kink type mode, the energy density (or elastic force) is localized (cf. Fig. 4.10) and has a singularity; the local extrema correspond to the two points of maximal curvature of the loops in Fig. 4.8. Such curves cannot be expressed by a function with a double period, such as elliptic functions, since these properties are complicated and require geometrical structures like the Jacobian J_{X_H} for the higher genus. Further, we emphasize that Fig. 4.8 partially reproduces the geometrical properties of DNA in Fig. 4.3. Figure 4.11 shows the contribution of $\varphi_{1,\mathrm{r}}$ and $\varphi_{2,\mathrm{r}}$ to $\phi_\mathrm{r}^{(0)}$.

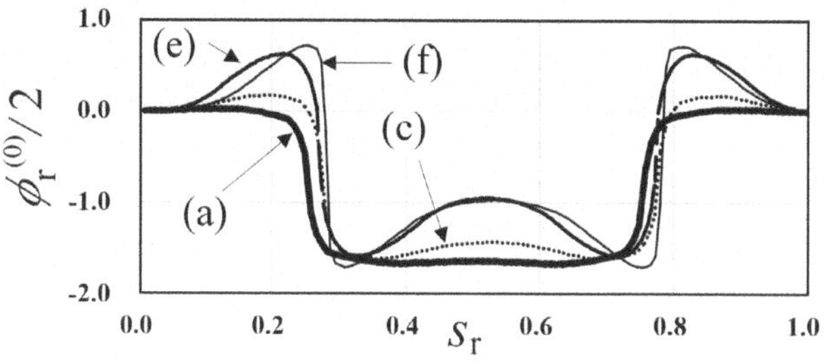

Fig. 4.9 $\phi_{\mathrm{r}}^{(0)}/2$ of Fig. 4.8a, c, e, and f

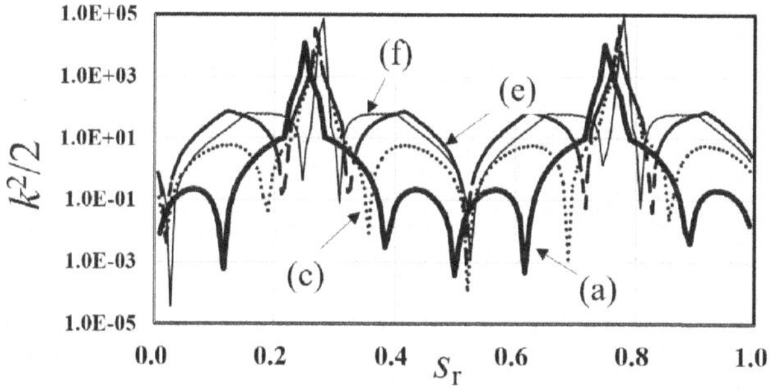

Fig. 4.10 $k^2/2$ of Fig. 4.8a, c, e, and f

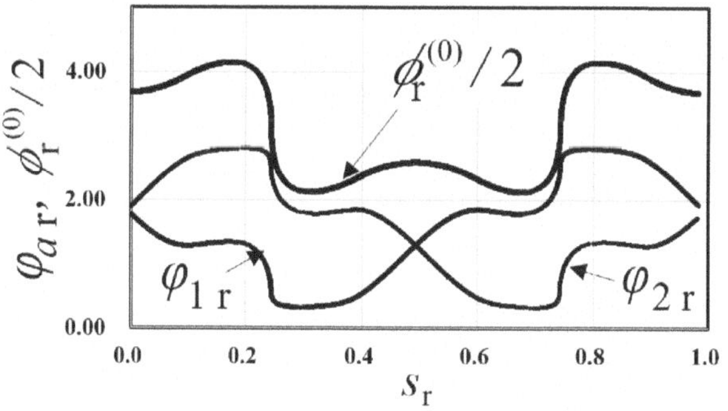

Fig. 4.11 The decomposition of $\phi_{\mathrm{r}}^{(0)}/2$ to $\varphi_{1\mathrm{r}}$ and $\varphi_{2\mathrm{r}}$ of Fig. 4.8d

After finishing writing this book, the author of this book improved this method and applied it to genus three to obtain plane curves related to the focusing MKdV equation over $\mathbb{R}$, which reproduced the properties of supercoiled DNA. [Mat2024a].[12] It means that our result can be regarded as a first step beyond the study of elastica by Euler, Bernoulli, Lagrange, and Gauss mentioned in Chap. 1.

[12] It should be noted that Emma Previato and the author planned to attack the genus three case immediately after [MatP2022].

References

[Abl1828] N. Abel, *Remarques sur quelques propriétés générales d'une certaine sorte de fonctions transcendantes*, Crelle's J., **3** (1828) 313–323.

[Abl1829] N. Abel, *Démonstration d'une propriété générale d'une certaine classe de de fonctions transcendentes*, Crelle's J., **4** (1828) 200–201 (written Jan. 1829).

[Abl1830] N. Abel, *Fernere mathematische Bruchstücke aus Herrn N. H. Abel's Briefen, (Fortsetzuni; von No. 28 Bd. V. Heft 4. S. 336), Schreibe11 des Herrn N. H. Abel an Herrn Legendre Zlt Paris*, Crelle's J., **6** (1830) 73–80.

[Abl1841] N. Abel, *Mémoire sur une propriété générale d'une classe trés étendue des fonctions transcendantes*, Paris, 1841.

[AblS1981] M. J. Ablowitz and H. Segur, Solitons and the inverse scattering transform, SIAM, 1981.

[AdaHP1988] M. R. Adams, J. Harnad, and E. Previato, *Isospectral Hamiltonian flows in finite and infinite dimensions*, Comm. Math. Phys., **117** (1988) 451–500.

[AdvM1991] M. Adler, and P. van Moerbeke, *The Toda lattice, Dynkin diagrams, singularities and Abelian varieties*, Invent. Math., **103** (1991) 223–278.

[ArCGH1985] E. Arbarello, M. Cornalba, P. A. Griffiths and J. Harris, Geometry of Algebraic Curves Vol. I, Springer-Verlag, New York, 1985.

[ArDNB2017] C. Armanini, F. Dal Corso, D. Misseroni and D. Bigoni, *From the elastica compass to the elastica catapult: an essay on the mechanics of soft robot arm*, Proc. R. Soc. A, **473** (2017) 20160870.

[Ape1946] R. Apéry, *Surles branches superlineaires des courbes algebriques*, C. R. Acad. Sci. Paris, **222** (1946) 1198–1200.

[ADGS2016] A. Assi, M. D'Anna, and P. García-Sánchez, Numerical semigroups and applications (RSME Springer Series 3), Springer-Verlag, New York, 2020.

[Ath2008] C. Athorne, *Identities for hyperelliptic $\wp$-functions of genus one, two and three in covariant form*, J. Phys. A: Math. Theor., **41** (2008) 415202 (20pp).

[AthEE2003] C. Athorne, J. C. Eilbeck and V. Z. Enolskii, *Identities for classical $\wp$ functions*, J. Geom. Phys., **48** (2003) 354–368.

[AthEE2004] C. Athorne, J. C. Eilbeck and V. Z. Enolskii, *A* SL(2) *covariant theory of genus 2 hyperelliptic functions*, Math. Proc. Camb. Phil. Soc., **136** (2004) 269–286.

[At1971] M. F. Atiyah, *Riemann surfaces and spin structure*, Ann. sci. l'É.N.S., **4** (1971) 47–62.

S. Matsutani, *The Weierstrass Sigma Function in Higher Genus and Applications to Integrable Equations*, Springer Monographs in Mathematics,
https://doi.org/10.1007/978-981-97-8163-8

[At1976] M. F. Atiyah, *William Vallance Douglas Hodge. 17 June 1903 – 7 July 1975*, Biographical Memoirs of Fellows of the Royal Society, **22** (1976) 169–192.

[Ay2014] T. Ayano, *Sigma functions for telescopic curves*, Osaka J. Math., **51** (2014) 459–481.

[Ay2016] T. Ayano, *On Jacobi inversion formulae for telescopic curves*, SIGMA, **12** (2016) 086, 21 pages.

[Ay2022] T. Ayano, *Hurwitz integrality of the power series expansion of the sigma function for a telescopic curve*, to appear in J. Math. Soc. Japan, arXiv:2208.10079.

[AyB2019] T. Ayano and V. M. Buchstaber, *Construction of two parametric deformation of KdV-hierarchy and solution in terms of meromorphic functions on the sigma divisor of a hyperelliptic curve of genus 3*, SIGMA, **15** (2019) 032, 15 pages.

[AyB2022] T. Ayano and V. M. Buchstaber, *Relationships between hyperelliptic functions of genus 2 and elliptic functions*, SIGMA, **18** (2022) 010, 30 pages.

[AyN2013] T. Ayano and A. Nakayashiki, *On addition formulae for sigma functions of telescopic curves*, SIGMA, **9** (2013) 046, 14 pages.

[Ba1897] H. F. Baker, Abelian functions: Abel's theorem and the allied theory of theta functions, Cambridge University Press, Cambridge, 1995, Reprint of the 1897 original.

[Ba1898a] H. F. Baker, *On the hyperelliptic sigma functions*, Amer. J. Math., **20** (1898) 301–384.

[Ba1898b] H. F. Baker, *On a certain system of differential equations determining periodic functions*, Proc. Camb. Phil. Soc., **9** (1898) 513–520.

[Ba1903] H. F. Baker, *On a system of differential equations leading to periodic functions*, Acta Math., **27** (1903) 135–156.

[Ba1907] H. F. Baker, An introduction to the theory of multiply-periodic functions, Cambridge University Press, Cambridge, 1907.

[Ba1928] H. F. Baker, *Note on the foregoing paper, "Commutative ordinary differential operators," by J. L. Burchnall and J. W. Chaundy*, Proc. Royal Soc. London (A), **118** No. 780 (1928) 584–593.

[BalEGO2008] S. Baldwin, J. C. Eilbeck, J. Gibbons, Y. Ônishi, *Abelian functions for cyclic trigonal curves of genus 4*, J. Geom. Phys., **58** (2008) 450–467.

[BelBEIM1994] E. D. Belokolos, A. I. Bobenko, V. Z. Enolskii, A. R. Its, and V. B. Matveev, Algebro-geometric approach to nonlinear integrable equations, Springer-Verlag, New York, 1994.

[BelE2001] E. D. Belokolos, V. Z. Enolskii, *Reduction of Abelian functions and algebraically integrable systems I*, J. Math. Sci., **106** (2001) 3395–3486.

[BelE2002] E. D. Belokolos, V. Z. Enolskii, *Reduction of Abelian functions and algebraically integrable systems II*, J. Math. Sci., **108** (2002) 295–374.

[BelES2005] E. D. Belokolos, V. Z. Enolskii, M. Salerno, *Wannier functions for quasi-periodic potentials*, Teor. Mat. Fiz., **144** (2005) 1081–1099.

[Ber2009] J. Bernatska, *General derivative Thomae formula for singular half-periods*, Lett. Math. Phys., **110** (2020) 2983–3014.

[BerL2019] J. Bernatska and D. Leykin, *On degenerate sigma-functions of genus two*, Glasgow Math. J., **61** (2019) 169–193.

[BerL2022] J. Bernatska and D. Leykin, *Solution of the Jacobi inversion problem on non-hyperelliptic curves*, Lett. Math. Phys., **20** (2023) 110, 38

[BerLN2020] J. Bernatska, V. Enolski, A. Nakayashiki, *Sato Grassmannian and degenerate sigma function*, Comm. Math. Phys., **374** (2020) 627–660.

[Ber1694a] Jacob Bernoulli, *Curvatura laminae elasticae*, Acta Eruditorum (June 1694), 262–280.

[Ber1694b] Jacob Bernoulli, *Constructio curvae accessus et recessus aquabilis, ope rectificationis curvae cujusdam algebraicae, addenda nuperae solutioni mensis Junii*, Acta Eruditorum (Sept. 1694), 336–338.

[BiV2003] Ch. Birkenhake and P. Vanhaecke, *The vanishing of the theta function in the KP direction: a geometric approach*, Compositio Math., **135** (2003) 323–330.

[Bi2023] S. R. Bistafa, *Euler's Variational approach to the elastica*, Euleriana, **3** (2023) 156–175.

[Bj1992] J-E. Björk, *Analytic D -modules and applications (Math. its Appl., 247)*, Kluwer Netherlands, 1993.

[Bol1887] O. Bolza, *Üeber die Reduction hyperelliptischer Integrale erster Ordnung und erster Gattung auf elliptische durch eine Transformation vierten Grades*, Math. Ann., **28** (1887) 447–456.

[Bol1895] O. Bolza, *On the first and second logarithmic derivatives of hyperelliptic sigma functions*, Amer. J. Math., **17** (1895) 11–36.

[Bol1899a] O. Bolza, *The partial differential equation for the hyper elliptic* Θ *and* σ*-functions*, Amer. J. Math., **21** (1899) 107–125.

[Bol1899b] O. Bolza, *Proof of Brioschifs recursion formula for the expansion of the even* σ*-functions of two variables*, Amer. J. Math., **21** (1899) 175–190.

[Bos1993] H. J. M. Bos, Lectures in the history of mathematics, AMS, 1993.

[BotT1993] R. Bott and L. W. Tu, Differential forms in algebraic topology, GTM82, Springer-Verlag, New York, 1982.

[Bot1986] U. Bottazzini, The higher calculus: a history of real and complex analysis from Euler to Weierstrass (English trans. by Van Egmond Warren), Springer-Verlag, New York, 1986.

[Bot2003] U. Bottazzini, *"Algebraic truths" vs. "geometric fantasies": Weierstrass' response to Riemann*, Proceedings of the International Congress of Mathematicians, Vol. III (Beijing, 2002), Higher Ed. Press (2002), pp. 923-934.

[BraDE2011] H. W. Braden, A, D'Avanzo, and V. Z. Enolski, *On charge-3 cyclic monopoles*, Nonlinearity, **24** (2011) 643–675.

[BraEH2005] H. W. Braden, V. Z. Enolskii and A. N. W. Hone, *Bilinear recurrences and addition formulae for hyperelliptic sigma functions*, J. Nonlinear Math. Phys., **12** (2005) 46–62.

[Bra1942] A. Brauer, *On a problem of partitions*, Amer. J. Math., **64** (1942) 299–312.

[BraS1962] A. Brauer and J. E. Shockley, *On a problem of Frobenius*, J. Reine Angew. Math., **211** (1962) 215–220.

[Bre1975] H. Bresinsky, *On prime ideals with generic zero* $x_i = t^{n_i}$, Proc. Amer. Math. Soc., **47** (1975) 329–332.

[BriH1864] F. Brioschi and M. Hermite, *Sur quelques formules pour la multiplication des fonctions elliptiques*, C. R. Acad. Sci. Paris, **59** (1864) 769–775.

[BryG1986] R. Bryant and P. Griffiths, *Reduction for Constrained Variational Problems and* $\int \kappa^2/2ds$, Amer. J. Math., **108** (1986) 525–570.

[Bry1993] J. -L. Brylinski, Loop spaces, characteristic classes and geometric quantization, Progress in Mathematics 107, Birkhäuser, Boston, 1993.

[BucB2020] V. M. Buchstaber and E. Y. Bunkova, *Lie algebras of heat operators in a nonholonomic frame*, Math. Notes., **108** (2020) 15–28.

[BucEL1997a] V. M. Buchstaber, V. Z. Enolskiĭ and D. V. Leĭkin, *Hyperelliptic Kleinian functions and applications* in Solitons, geometry, and topology: on the crossroad 1–33, Amer. Math. Soc. Transl. Ser. 2, 179, Adv. Math. Sci., 33, Amer. Math. Soc., Providence, RI, 1997.

[BucEL1997b] V. M. Buchstaber, V. Z. Enolskiĭ and D. V. Leĭkin, *Kleinian functions, hyperelliptic Jacobians and applications*, Reviews in Mathematics and Mathematical Physics, **10** (1997) 1–103 (and, unpublished revised version 2002).

[BucEL1999] V. M. Bukhshtaber, V. Z. Ènol'skiĭ and D. V. Leĭkin, *Rational analogues of Abelian functions*, Funct. Anal. Appl., **33** (1999) 83–94.

[BucEL2012] V. M. Buchstaber, V. Z. Enolski, and D. V. Leykin, *Multi-dimensional sigma-functions*, arXiv:1208.0990,2012, 267 pages.

[BucEL2020] V. M. Buchstaber, V. Z. Enolski, and D. V. Leykin, *σ functions: old and new results* in Integrable systems and algebraic geometry Vol. 2 edited by R. Donagi and T. Shaska, London Math Soc. Lect. Note Series **459** (2020) 175–214.

[BucL2002] V. M. Buchstaber and D. V. Leykin, *Polynomial Lie Algebras*, Funct. Anal. Appl., **36** (2002) 267–280.

[BucL2004] V. M. Buchstaber and D. V. Leykin, *Heat equations in a nonholonomic frame* Funct. Anal. Appl., **38** (2004) 88–101.

[BucL2005] V. M. Buchstaber and D. V. Leykin, *Addition laws on Jacobian varieties of plane algebraic curves*, Proc. Steklov Inst. Math., **251** (2005) 1–72.

[BucL2008] V. M. Buchstaber and D. V. Leykin, *Solution of the problem of differentiation of abelian functions over parameters for families of* (n, s)*-curves*, Func. Anal. Its Appl., **42** (2008) 268–278.

[BucLE1999] V. M. Buchstaber, D. V. Leĭkin and V. Z. Enolskiĭ, σ*-functions of* (n, s)*-curves* Uspekhi Mat. Nauk, **327** (1999) 155–156; translation in Russian Math. Surveys, **54** (1999) 628–629.

[BucM2017] V. M. Buchstaber and A. V. Mikhailov, *Infinite-dimensional Lie algebras determined by the space of symmetric squares of hyperelliptic curves*, Funct. Anal. Its Appl., **51** (2017) 2–21.

[BucM2021] V. M. Buchstaber and A. V. Mikhailov, *Integrable polynomial Hamiltonian systems and symmetric powers of plane algebraic curves*, Russian Math. Surveys, **76** (2021) 587–652.

[BucG1979] R-O. Buchweitz and G-M. Greuel, *Le nombre de Milnor equisingularite et deformations de singularites des courbes reduites*, C. R. Acad. Sci. Paris Seri. A et B, **288** (1979) A35–A38.

[BucG1980] R-O. Buchweitz and G-M. Greuel, *The Milnor number and deformations of complex curve singularities*, Invent. Math., **58** (1980) 241–281.

[BunB2009] E. Y. Bunkova and V. M. Buchstaber, *Heat equations and families of two-dimensional sigma functions*, Proc. of the Steklov Inst. Math., **266** (2009) 1–28.

[BurC1922] J. L. Burchnall and T. W. Chaundy, *Commutative ordinary differential operators*, Proc. Lond. Math. Soc., **21** (1922) 420–440.

[BurC1928] J. L. Burchnall and T. W. Chaundy, *Commutative ordinary differential operators*, Proc. R. Soc. Lond. (A), **118** (1928) 557–583.

[BurC1931] J. L. Burchnall and T. W. Chaundy, *Commutative ordinary differential operators II*, Proc. R. Soc. Lond. (A), **134** (1931) 471–485.

[Can1994] D. G. Cantor, *On the analogue of the division polynomials for hyperelliptic curves*, J. reine angew. Math., **447** (1994) 91–145.

[CarK1995] R. Carroll and Y. Kodama, *Solution of the dispersionless Hirota equations*, J. Phys. A: Math. Gen., **28** (1995) 6373–6387.

[Ch1951] C. Chevalley, Introduction to the theory of algebraic functions of one variable, Math. Surv. VI Amer. Math. Soc., 1951.

[CleG1866] A. Clebsch and P. Gordan, Theorie der abelschen funktionen, Teubner, 1866.

[CliP1967] A. H. Clifford and G. B. Preston, The algebraic theory of semigroups I, II A.M.S., 1961 and 1967.

[Cop1988] M. Coppens, *Weierstrass points with two prescribed non-gaps*, Pacific J. Math., **131** (1988) 71–10.

[CopK2014] M. Coppens and T. Kato, *Weierstrass points with first two non-gaps equal to* n *and* $n + 2$, Kyushu J. Math., **68** (2014) 139–147.

[Cor1989] A. Córdoba, *Dirac combs*, Lett. Mat. Phys., **17** (1989) 191–196.

[Cor1987] M. Cornalba, *On the locus of curves with automorphisms*, Ann. Mat. Pura Appl., **149** (1987) 135–151.

[CouH1953] R. Courant and D. Hilbert, Methods of mathematical physics, Vol. 1 [English edition], Wiley, 1953.

[Cox2012] D. A. Cox, Galois theory, 2nd edition, Wiley, 2012.

[CrCS1993] B. Crespi, S. -J. Chang, and K. -J. Shi, *Elliptical billiards and hyperelliptic functions*, J. Math. Phys., **34** (1993) 2257–2289.

[dV1888] L. Da Vinci, The complete botebooks of Leonardo Da Vinci, Volume 1, translated by J. P. Richter, London, 1888.

[DedW1882] R. Dedekind and H. Weber, Theory of algebraic functions of one variable (History of mathematics), translated into English by J. Stillwell, Amer. Math. Soc. 2012; *Theorie der algebraischen Functionen einer Veränderlichen*, Crelle's J., **92** (1882) 181–290.

[Del2008] A. Del Centina, *Weierstrass points and their impact in the study of algebraic curves: a historical account from the "Lückensatz" to the 1970s*, Ann Univ Ferrara, **54** (2008) 37–59.

[Del1989] P. Deligne (Notes by J. Milne), *Hodge cycles on abelian varieties*, Hodge cycles, motives, and Shimura varieties, ed. by P. Deligne, J. S. Milne, A. Ogus and K. Shih, Corrected Printing, Springer-Verlag, Berlin, 1989, 9–100.

[Des1637] R. Descartes, Discours de la méthode pour bien conduire sa raison, et chercher la verité dans les sciences (La Dioptrique, Les Météores, La Géométrie), 1637.

[Di1972] J. A. E. Dieudonné, *The Historical development of algebraic geometry*, A.M.S. Monthly, **79** (1972) 827–866.

[DAGJMMRT2023] R. Donagi, M. R. Adams, J. Glazebrook, L. Jeffrey, S. Matsutani, D. Mumford, V. Roubtsov, and B. Thompson, *Emma Previato: Memorial tribute*, Notices of the AMS Sept. (2023) 1243–1251.

[DoS2020] R. Donagi and T. Shaska, Integrable systems and algebraic geometry Vol. 1, Vol. 2: London Math Soc. Lect. Note Series 458, 459, Cambridge Univ. Press, 2020.

[Dub1975] B. Dubrovin, *Periodic problem for the Korteweg–de Vries equation in the class of finite-zone potentials*, Funct. Anal. Appl., **9** (1975) 215–223.

[Dub1981] B. Dubrovin, *Theta functions and non-linear equations*, Uspekhi Mat. Nauk, **36** (1981) 11–80, Russian Math. Surveys **36** (1981) 11–92.

[DulRVW2001] H. R. Dullin, P. H. Richter, A. P. Veselov, and H. Waalkens, *Actions of the Neumann systems via Picard–Fuchs equations*, Physica D, **155** (2001) 159–183.

[EaK1986] C. J. Earle and I. Kra, *Half-canonical divisors on variable Riemann surfaces* J. Math. Kyoto Univ., **26** (1986) 39–64.

[EilEO2011] J. C. Eilbeck, M. England and Y. Ônishi, *Abelian functions associated with genus three algebraic curves*, LMS J. Comput. Math., **14** (2011) 291–326.

[EilEG2010] J. C. Eilbeck, V. Z. Enolskii and J. Gibbons, *Sigma, tau and Abelian functions of algebraic curves*, J. Phys. A: Math. Theor., **43** (2010) 455216 (20pp).

[EilEK2000] J. C. Eilbeck, V. Z. Enol'skii and N. A. Kostov, *Quasiperiodic and periodic solutions for vector nonlinear Schrödinger equations*, J. Math. Phys., **41** (2000) 8236–8248.

[EilEL2000] J. C. Eilbeck, V. Z. Enolskii and D. V. Leykin, *On the Kleinian construction of abelian functions of canonical algebraic curves*, SIDE III—symmetries and integrability of difference equations (Sabaudia, 1998), CRM Proc. Lecture Notes, 25, Amer. Math. Soc., 2000 121–138.

[EilEMOP2007] J. C. Eilbeck, V. Z. Enol'skii, S. Matsutani, Y. Ônishi, and E. Previato, *Abelian functions for trigonal curves of genus three*, Int. Math. Res. Notices, **2007** (2007) 1–38,

[EilEMOP2008] J. C. Eilbeck, V. Z. Enol'skii, S. Matsutani, Y. Ônishi, and E. Previato, *Addition formulae over the Jacobian pre-image of hyperelliptic Wirtinger varieties*, J. Reine Angew Math., **619** (2008) 37–48.

[EilEP2001] J. C. Eilbeck, V. Z. Enolskii, and E. Previato, *Varieties of elliptic solitons*, J. Phys. A: Math. Gen., **34** (2001) 2215–2228.

[EilEP2003] J. C. Eilbeck, V. Z. Enolskii, and E. Previato, *On a generalized Frobenius–Stickelberger addition formula*, Lett. Math. Phys., **65** (2003) 5–17.

[EilGOY2017] J. C. Eilbeck, J. Gibbons, Y. Ônishi and S. Yasuda, *Theory of heat equations for sigma functions*, arXiv:1711.08395.

[EilMO2011] J. C. Eilbeck, S. Matsutani and Y. Ônishi, *Addition formulae for Abelian functions associated with specialized curves*, Phil. Trans. R. Soc. A, **369** (2011) 1245–1263.

[EilO2020] J. C. Eilbeck and Y. Ônishi, *Recursion relations on the power series expansion of the universal Weierstrass sigma function*, RIMS Kokyuroku Bessatsu, **78** (2020) 77–98.

[EisH1987] D. Eisenbud and J. Harris, *Existence, decomposition, and limits of certain Weierstrass points*, Invent. Math., **87** (1987) 495–515.

[EnE2009] M. England and J. C. Eilbeck, *Abelian functions associated with a cyclic tetragonal curve of genus six*, J. Phys. A: Math. and Theor. Phys., **42** (2009) 095210.

[EnG2003] V. Z. Enolskii and J. Gibbons, *Addition theorems on the strata of the theta divisor of genus three hyperelliptic curves*, Imperial College London, London (preprint), (2003).

[EnHKKLS2012] V. Enolski, B. Hartmann, V. Kagramanova, J. Kunz, C. Lämmerzahl and P. Sirimachan, *Inversion of a general hyperelliptic integral and particle motion in Horava–Lifshitz black hole space-times*, J. Math. Phys., **53** (2012) 012504.

[EnMO2008] V. Z. Enolskii, S. Matsutani and Y. Ônishi, *The addition law attached to a stratification for a hyperelliptic Jacobian variety*, Tokyo J. Math., **31** (2008) 27–38.

[EnR2008] V. Z. Enolskii and P. Richter, *Periods of hyperelliptic integrals expressed in terms of θ-constants by means of thomae formulae*, Phil. Trans. Math. Phys. Eng. Sci., **366** (2008) 1005–1024.

[EnPR2003] V. Z. Enolskii, M. Pronine and P. Richter, *Double pendulum and θ-divisor*, J. Nonlinear Sci., **13** (2003) 157–174.

[Eu1744] L. Euler, Methodus Inveniendi Lineas Curvas Maximi Minimive Proprietate Gaudentes 1744, Leonhardi Euleri Opera Omnia Ser. I, Vol. 14.

[FarK1991] H. M. Farkas and I. Kra, Riemann surfaces (GTM 71), Springer-Verlag, New York, 1991.

[FarZ2011] H. M. Farkas, S. Zemel, *Generalizations of Thomae's formula for Z_n curves*, Springer-Verlag, New York, 2011.

[Fay1973] J. D. Fay, Theta functions on Riemann surfaces, LNM 352, Springer-Verlag, New York, 1973.

[Fay1984] J. D. Fay, *On the even-order vanishing of Jacobian theta functions*, Duke Math. J., **51** (1984) 109–132.

[FeKMPA2022] Y. Fedorov, J. Komeda, S. Matsutani, E. Previato, K. Aomoto, *The sigma function over a family of cyclic trigonal curves with a singular fiber*, Isr. J. Math., **250** (2022) 345–402.

[FePU1955] E. Fermi, J. Pasta and S. Ulam, *Studies of nonlinear problems I*, Los Alamos Report, LA-1940 (1955).

[FlFM1980] H. Flaschka, M. G. Forest and D. W. McLaughlin, *Multiphase averaging and the inverse spectral solution of the Korteweg–de Vries equation*, Comm. Pure Appl. Math., **33** (1980) 739–784.

[Fo1981] O. Forster, Lectures on Riemann surfaces, translated by B. Gilligan, GTM 81, Springer-Verlag, Berlin–New York, 1981.

[Fra1991] C. G. Fraser, *Mathematical technique and physical conception in Euler's investigation of the elastica*, Centaurus, **34** (1991) 211–246.

[FroS1877] F. G. Frobenius and L. Stickelberger, *Zur Theorie der Elliptischen Functionen*, J. Reine Angew. Math., **83** (1877) 175–179.

[Fu1969] W. Fulton, Algebraic curves: an introduction to algebraic geometry, W. A Benjamin Inc., New York, 1969.

[FuH1991] W. Fulton and J. Harris, Representation theory, GTM 129, Springer-Verlag, Berlin–New York, 1991.

[GaPS2000] S. D. Galbraith. S. M. Paulus, and N. P. Smart, *Arithmetic on superelliptic curves*, Math. Comp., **71** (2000) 393–405.

[Ga1846] E. Galois, *OEuvres Mathématiques d'Évariste Galois*, J. de math. pures et appliquées, **11** (1846) 381–444.

[Ga1984] F. R. Gantmacher, Introduction to algebraic curves, Vol. 1, Chelsea, New York, 1984.

[Ga1929] C. F. Gauss, Carl Friedrich Gauss, Werke, 1–12, 1863–1929, Göttingen.

[GeH2003] F. Gesztesy and H. Holden, Soliton equations and their algebro-geometric solutions: volume 1, (1+1)-dimensional continuous models, CSAM 79, Cambridge Univ. Press, Cambridge, 2003.

[GeHMT2008] F. Gesztesy and H. Holden, J. Michor and G. Teschl, Soliton equations and their algebro-geometric solutions: volume 2,(1+1)-dimensional discrete models, CSAM 114, Cambridge Univ. Press, Cambridge, 2003.

[GiMO2013] J. Gibbons, S. Matsutani and Y. Ônishi, *Prime form and sigma function*, J. Phys. A: Math. Theor., **46** (2013) 175203.

[GodR2001] C. Godsil and G. Royle, Algebraic graph theory, Springer, Berlin, 2001.

[GolP1991] R. E. Goldstein and D. M. Petrich, *The Korteweg–de Vries hierarchy as dynamics of closed curves in the plane*, Phys. Rev. Lett., **67** (1991) 3203–3206.

[GolP1993] R. E. Goldstein and D. M. Petrich, *Solitons, Euler's equation, and the geometry of curve motion* in Singularities in fluids, plasmas and optics (Heraklion, 1992), 93-109, NATO Adv. Sci. Inst. Ser. C Math. Phys. Sci., 404 Kluwer Acad. Publ., Dordrecht, 1993.

[Goo1974] D. C. Goodman, Science and Belief – From Copernicus to Darwin: Towards a Mechanistic Philosophy Unit 4–5 (Course AMST283), Open UP, Milton Keynes, 1974.

[GraTRCT2002] B. Grammaticos, T. Tamizhmani, A. Ramani, A. S. Carstea, and K. M. Tamizhmani, *A bilinear approach to the discrete Painlevé I equations*, J. Phys. Soc. Jpn, **71** (2002) 443–447.

[Gra1990] D. Grant, *Formal groups in genus two*, J. reine angew. Math., **411** (1990) 96–121.

[Gri1989] P. A. Griffiths, Introduction to algebraic curves, AMS Translations of Math. Monographs **76**, A.M.S., 1989.

[GriH1978] P. Griffiths and J. Harris, *On Cayley's explicit solution to Poncelet's porism* L'Enseignement Mathématique. Revue Internationale. 2e Série, **24** (1978) 31–40.

[GuiS1984] V. Guillemin and S. Sternberg, Symplectic techniques in physics, Cambridge Univ. Press, Cambridge, 1984.

[Gun1984] R. C. Gunning, *Riemann surfaces and their associated Wirtinger varieties*, Bull. Amer. Math. Soc., **11** (1984) 287–316.

[Hag1970] Y. Hagihara, Celestial mechanics, Vol. 1: dynamical principles and transformation theory, The MIT Press, 1970.

[HanB1980] J. H. Hannay and M. V. Berry, *Quantization of linear maps on a torus – Fresnel diffraction by periodic grating*, Physica D, **1** (1980) 267–290.

[Han1989] V. L. Hansen, Braids and coverings: selected topics (LMAAT 18) (English edition), Cambridge University Press, Cambridge, 1989.

[Har2003] D. Harbater, *Patching and Galois theory*, Galois group and fundamental groups, MSRI Publ., **41** (2003) 315–426.

[Har1977] R. Hartshorne, Algebraic geometry, GTM52, Springer, New York, 1977.

[HasSa2009] K. Hashimoto and Y. Sakai, *General form of Humbert's modular equation for curves with real multiplication of* $\Delta = 5$, Proc. Japan Acad., **85** (2009) 171–176.

[Haw2013] T. Hawkins, The mathematics of Frobenius in context: A journey through 18th to 20th century mathematics (SHMP), Springer, New York, 2013.

[Hej1972] D. A. Hejhal, Theta functions, kernel, functions and Abelian integrals, Memories of the AMS **129** AMS, 1972.

[Her1970] J. Herzog, *Generators and relations of Abelian semigroup and semigroup ring*, Manuscripta Math., **3** (1970) 175–193.

[HieJN2016] J. Hietarinta, N. Joshi and F. W. Nijhoff, Discrete systems and integrability (CTAM 54), Cambridge University Press, Cambridge, 2016.

[Hil1901] D. Hilbert, *Ueber Flächen von constanter Gausscher Krümmung*, Trans. Amer. Math. Soc., **2** (1901) 87–99.

[Hir1971] R. Hirota, *Exact solutions of Korteweg–de Vries equation for multiple collisions of solitons*, Phys. Rev. Lett., **27** (1971) 1192–1194.

[Hir1976] R. Hirota, *Direct method of finding exact solutions of nonlinear evolution equations*, Bäcklund Transformations, LNM 515, eds by R. M. Miura, Springer, New York, 1976.

[HirT2003] R. Hirota and D. Takahashi, Sabun-to-Chosabun (Difference and ultra-discrete), [in Japanese] Kyoritsu, Tokyo, 2003.

[Hod1956] W. V. D. Hodge, *Henry Frederick Baker. 1866–1956*, Biographical Memoirs of Fellows of the Royal Society, **2** (1956) 49–68.

[Hon2007] A. N. W. Hone, *Sigma function solution of the initial value problem for Somos 5 sequence*, Trans. Amer. Math. Soc., **359** (2007) 5019–5034.

[HonJJPP2007] S.-K. Hong, J.-G. Jee, Y. H. Jang, N.-C. Park, and Y.-P. Park, *Dynamic analysis of flexible media by dynamic elastica theory with aerodynamic effect*, Microsyst. Technol., **13** (2007) 1141–1151.

[Hou1978] C. Houzel, *Fonctions elliptiques et intégrales abéliennes*, in Abrégé D'histoire des Mathématiques, Vol. 2, ed. by J. Dieudonné, Hermann, Paris, 1978.

[Hou2002] C. Houzel, *The Work of Niels Henrik Abel*, in The legacy of Niels Henrik Abel, ed. by O. A. Laudal and R. Piene, Springer-Verlag, New York, 2002.

[HuaR1949] L. K. Hua and I. Reiner, *On the generators of the symplectic modular group*, Trans. of A. M. S. **65** (1949) 415–426.

[Hur1888] A. Hurwitz, *Üeber diejenigen algebraischen Gebilde, welche eindeutige Transformationen in sich zulassen*, Math. Ann., **32** (1888) 290–308.

[Hur1891] A. Hurwitz, *Über Riemann'sche Flachen mit gegebenen Verzweigungspunkten*, Math. Ann., **39** (1891) 1–61.

[Hur1893] A. Hurwitz, *Über algebraischer Gebilde mit eindeutigen Transformationen in sich*, Math. Ann., **41** (1893) 403–442.

[HurC1893] A. Hurwitz and R. Courant, *Vorlesungen Ueber allgemeine Funktionentheorie und elliptische Funktionen (Grundlehren der mathematischen Wissenschaften, 3)*, Springer-Verlag, New York, 1893.

[Hus1936] E. Husserl, *The origin of geometry, 1936*, in Introduction to Husserl's the origin of geometry, by Derrida, Jacques, 1976 (English).

[Ic1981] Y. Ishimori, *On the modified Korteweg–de Vries soliton and the loop soliton*, J. Phys. Soc. Jpn, **50** (1981) 2471–2472.

[Ig1956] J. Igusa, *Fibre systems of Jacobian varieties*, Amer. J. Math., **78** (1956) 171–199.

[Ig1972] J. Igusa, Theta Functions (GMW (194)), Springer-Verlag, New York, 1972.

[It1986] A. R. Its, *On connections between solitons and finite-gap solutions of the nonlinear Schrödinger equation*, Sel. Math. Sov., **5** (1986) 29–43.

[ItM1976] A. R. Its and V. B. Matveev, *On a class of solutions of the Korteweg–de Vries equation*, Problemy Mat. Fiz., **8** (1976) 70–92 (Russian).

[Iw1993] K. Iwasawa, Algebraic functions, trans. by G. Kato, AMS, 1993.

[Jac1829] C. G. J. Jacobi, *Fundamenta Nova Theoriae Functionum Ellipticarum*, Werke Vol. 1. 1829.

[Jac1832a] C. G. J. Jacobi, *De theoremate Abeliano observatio*, Crelle's J., **9** (1832) 99.

[Jac1832b] C. G. J. Jacobi, *Considerationes generales de transcendentibus Abelianis*, Crelle's J., **9** (1832) 394–403.

[Jac1835] C. G. J. Jacobi, *De functionibus duarum variabilium quadrupliciter periodics quibus theoria transcendentium Abelianarum innititur*, Crelle's J., **13** (1835) 55–78.

[Jac1841] C. G. J. Jacobi, *De functionibus alternantibus earumque divisione per productum e differentiis elementorum conflatum*, Crelle's J., **22** (1835) 360–371.

[Jac1846] C. G. J. Jacobi, *Note sur les fonctions Abeliennes*, Crelle's J., **30** (1846) 183–184.

[Jah2003] H. N. Jahnke, A history of analysis (history of mathematics), Amer. Math. Soc., 2003.

[Jo1992] J. Jorgenson, *On directional derivatives of the theta function along its divisor*, Israel J. Math., **77** (1992) 273–284.

[KacM1975] M. Kac and P. van Moerbeke, *On some periodic Toda lattices*, Proc. Nat. Acad. Sci. U.S.A., **72** (1975) 1627–1629. *A complete solution of the periodic Toda problem, ibid.* **72** (1975) 2879–2880.

[KadP1970] B. B. Kadomtsev and V. I. Petviashvili, *On the stability of solitary waves in weakly dispersive media*, Sov. Phys. Dokl., **15** (1970) 539–541.

[Kat1979] T. Kato, *On Weierstraß points whose first non-gaps are three*, Crelle J, **312** (1979) 99–109.

[Kat1980] T. Kato, *Weierstrass normal form of a Riemann surface and its applications [in Japanese]*, Sûgaku, **32** (1980) 73–75.

[KawNTY1988] N. Kawamoto, Y. Namikawa, A. Tsuchiya, and Y. Yamada, *Geometric realization of conformal field theory on Riemann surafces*, Comm. Math. Phys., **116** (1988) 247–308.

[Ke1973] G. Kempf, *On the geometry of a theorem of Riemann*, Ann. Math., **98** (1973) 178–185.

[Ki1873] L. Kiepert, *Wirkliche Ausführung der ganzzahligen Multiplication der elliptischen Functionen*, J. reine angew. Math., **76** (1873) 21–33.

[Kl1886] F. Klein, *Ueber hyperelliptische Sigmafunctionen*, Math. Ann., **27** (1886) 431–464.

[Kl1888] F. Klein, *Ueber hyperelliptische Sigmafunctionen (Zweite Abhandlung)*, Math. Ann., **32** (1888) 351–380.

[Kl1890] F. Klein, *Zur theorie der Abel'schen functionen*, Math. Ann., **36** (1890) 1–83.

[Kl1903] F. Klein, *Gauß' wissenschaftliches Tagebuch 1796–1814*, Math. Ann. **57** (1903) 1–32.

[Kl1926] F. Klein, Vorlesungen ueber die Entwicklung der Mathematik im 19 Jahrhundert: Teil I, Springer-Verlag, Berlin, 1926.

[Kod1963] K. Kodaira, *On compact complex analytic structure II*, Ann. Math., **77** (1963) 563–626.

[Kod2017] Y. Kodama, KP Solitons and the Grassmannians: combinatorics and geometry of two-dimensional wave patterns, Springer, Berlin, 2017.

[KodK2002] Y. Kodama and B. G. Konopelchenko, *Singular sector of the Burgers–Hopf hierarchy and deformations of hyperelliptic curves*, J. Phys. A: Gen. Math., **35** (2002), L489–L500.

[KodK2003] Y. Kodama and B. G. Konopelchenko, *Deformations of plane algebraic curves and integrable systems of hydrodynamic type*, Nonlinear Physics: Theory and Experiment II, Proc. of the Workshop, Gallipoli, Italy 2002 World Scientific, 2003, 234–240.

[KodMP2013] Y. Kodama, S. Matsutani and E. Previato, *Quasi-periodic and periodic solutions of the Toda lattice via the hyperelliptic sigma function*, Annales de l'institut Fourier, **63** (2013) 655–688.

[KodX2021] Y. Kodama and Y. Xie, *Space curves and solitons of the KP hierarchy I: The ℓ-th generalized KdV hierarchy*, SIGMA, **17** (2021) 024(43 pages).

[Kom1983] J. Komeda, *On Weierstrass points whose first non-gaps are four*, J. Reine. Angew. Math., **341** (1983) 68–86.

[Kom1993] J. Komeda, *On Weierstrass points whose semigroups are generated by two elements*, Res. Rep. Kanagawa Inst. Tech., **B-17** (1993) 211–217.

[Kom1999] J. Komeda, *Existence of the primitive Weierstrass gap sequences on curves of genus 9*, Bol. Soc. Bras. Mat., **30** (1999) 125–137.

[Kom2004] J. Komeda, *On 6-semigroups generated by 4-elements from which affine toric varieties can be constructed*, Res. Rep. Kanagawa Inst. Tech., **B-28** (2004) 79–85.

[KomM2020] J. Komeda and S. Matsutani, *Jacobi inversion formulae for a curve in Weierstrass normal form*, in Integrable systems and algebraic geometry Vol. 2, edited by R. Donagi and T. Shaska, London Math Soc. Lect. Note Series **459** (2020) 383–404

[KomMP2013] J. Komeda, S. Matsutani and E. Previato, *The sigma function for Weierstrass semigroups* $\langle 3, 7, 8\rangle$ *and* $\langle 6, 13, 14, 15, 16\rangle$, Internat. J. Math., **24** (2013) 1350085, 58 pp.

[KomMP2016] J. Komeda, S. Matsutani and E. Previato, *The Riemann constant for a non-symmetric Weierstrass semigroup*, Arch. Math. (Basel), **107** (2016) 499–509.

[KomMP2019] J. Komeda, S. Matsutani, and E. Previato, *The sigma function for trigonal cyclic curves*, Lett. Math. Phys., **109** (2019) 423–447.

[KomMP2022a] J. Komeda, S. Matsutani and E. Previato, *Complementary modules of Weierstrass canonical forms*, SIGMA, **18** (2022), 098, 39 pages.

[KomMP2022b] J. Komeda, S. Matsutani, and E. Previato, *Algebraic construction of the sigma functions for general Weierstrass curves*, Mathematics, MDPI, **10(16)** (2022) 10, 32 pages, with correction **11(9)** (2023) 2164.

[KomO2004] J. Komeda and A. Ohbuchi, *Weierstrass points with first non-gap four on a double covering of a hyperelliptic curve*, Serdica Math. J., **30** (2004) 43–54; Corrigendum **32** (2006) 375–378.

[KomO2008a] J. Komeda and A. Ohbuchi, *Weierstrass points with first non-gap four on a double covering of a hyperelliptic curve II*, Serdica Math. J., **34** (2008) 771–782.

[KomO2008b] J. Komeda and A. Ohbuchi, *Existence of the non-primitive Weierstrass gap sequences on curves of genus 8*, Bull. Braz. Math. Soc. News Series, **39** (2008) 109–121.

[KonIW1981] K. Konno, Y. Ichikawa and M. Wadati, *A Loop soliton propagating along a stretched rope*, J. Phys. Soc. Japan, **50** (1981) 1025–1026.

[Koo1999] J. K. Koo, *On holomorphic differentials of some algebraic function field of one variable over* $\mathbb{C}$, Bull. Austral. Math. Soc., **43** (1991) 399–405.

[KorB1992] A. Kornberg and T. A. Baker, DNA Replication, 2nd ed., W. H. Freeman & Co., 1992.

[Kor2020] D. Korotkin, *Bergman tau-function: from Einstein equations and Dubrovin–Frobenius manifolds to geometry of moduli space*, in Integrable systems and algebraic geometry Vol. 2, ed. by R. Donagi and T. Shaska, London Math. Soc. Lect. Note Series, **459** (2020) 215–287.

[KorS2012] D. Korotkin and V. Shramchenko, *On higher genus Weierstrass sigma-function*, Physica D, **241** (2012) 2086–2094.

[KorV1885] D. J. Korteweg and G. de Vries, *On the change of form of long waves advancing in a rectangular canal and on a new type of long stationary waves*, Phil. Mag. 5th series, **39** (1885) 422–443.

[Kr1997] J. M. Krichever, *Methods of algebraic geometry in the theory of nonlinear equations*, Russ. Math. Surv., **32** (1977) 185–213.

[Ku1970] E. Kunz, *The value-semigroup of a one-dimensional Gorenstein ring*, Proc. Amer. Math. Soc., **25** (1970) 748–751.

[Ku2004] E. Kunz, Introduction to plane algebraic curves, trans. by R. G. Belshoff, Birkhäuser, Boston, 2004.

[Lag1869] J. L. Lagrange, Oeuvres de Lagrange, 1869.

[Lam1980] G. L. Lamb, Jr., Elements of soliton theory (Pure & Appl. Math. S.), John Wiley & Sons Inc., 1980.

[Lan1978] S. Lang, Elliptic curves Diophantine analysis, Grundlehren der mathematischen Wissenschaften 231. Springer-Verlag, New York–Berlin, 1978.

[Lan1982] S. Lang, Introduction to algebraic and abelian functions, Second edition, GTM 89, Springer-Verlag, New York–Berlin, 1982.

[Le1964] J. Lewittes, *Riemann surfaces and the theta functions*, Acta Math., **111** (1964) 37–61.

[Lo1892] A. E. H. Love, A treatise on the mathematical theory of elasticity, Cambridge University Press, Cambridge, 1892.

[Mac1985] I. G. Macdonald, Symmetric functions and Hall polynomials, Oxford Mathematical Monographs (2nd ed.) 1985.

[Mac1742] C. Maclaurin, A treatise on fluxions, 1742.

[Man1964] Ju. I. Manin, *Algebraic curves over fields with differentiation*, A.M.S. Transl., Twenty two papers on algebra, number theory and differential geometry, **37** (1964) 59–78.

[Map2019] Maplesoft a division of Waterloo Maple Inc.,. 2019. Maple. Waterloo, Ontario. Retrieved from https://hadoop.apache.org.

[Mar1996] A. I. Markushevich, *Analytic function theory*, in Mathematics of the 19th century geometry, analytic function theory, ed. by A. N. Kolmogorov and A. P. Yushkevich, Birkhäuser, Boston, 1996, 119–272.

[Mat1986] H. Matsumura, Commutative ring theory, CSAM 8 trans. by M. Reid, Cambridge Univ. Press, Cambridge, 1986.

[Mat1995] S. Matsutani, *The relation of lemniscate and a loop soliton as 3/2 and 1 spin fields along the modified Korteweg–de Vries equation*, Mod. Phys. Lett. A, **10** (1995) 717–721.

[Mat1998] S. Matsutani, *Statistical mechanics of elastica on a plane: origin of the MKdV hierarchy*, J. Phys. A: Math. & Gen., **31** (1998) 2705–2725.

[Mat1999] S. Matsutani, *Statistical mechanics of non-stretching elastica in three dimensional space*, J. Geom. Phys., **29** (1999) 243–259.

[Mat2001a] S. Matsutani, *Hyperelliptic solutions of KdV and KP equations: reevaluation of Baker's study on hyperelliptic sigma functions*, J. Phys. A: Math. & Gen., **34** (2001) 4721–4732.

[Mat2001b] S. Matsutani, *Soliton solutions of Kortweg–de Vries equations and hyperelliptic sigma functions*, arXiv:math-ph0105032, (2001).

[Mat2002a] S. Matsutani, *Elliptic and hyperelliptic solutions of discrete Painleve I and its extensions to higher order difference equations*, Phys. Lett. A, **300** (2002) 233–242.

[Mat2002b] S. Matsutani, *Hyperelliptic loop solitons with genus g : investigation of a quantized elastica*, J. Geom. Phys., **43** (2002) 146–162.

[Mat2002c] S. Matsutani, *Hyperelliptic solutions of modified Kortweg–de Vries equation of genus g: essentials of Miura transformation*, J. Phys. A: Math. & Gen., **35** (2002) 4321–4333.

[Mat2002d] S. Matsutani, *On a relation of Weierstrass al-functions*, Int. J. Appl. Math., **11** (2002) 295–307.

[Mat2003a] S. Matsutani, *Recursion relation of hyperelliptic psi-functions of genus two*, Integral Transforms and Special Func., **14** (2003) 517–527.

[Mat2003b] S. Matsutani, *Toda Equation and σ-Function of Genera One and Two*, J. Nonlinear Math. Phys., **10** (2003) 555–561.

[Mat2005] S. Matsutani, *Relations of al functions over subvarieties in a hyperelliptic Jacobian*, CUBO A Math. J., **7** (2005) 75–85.

[Mat2007] S. Matsutani, *Reality conditions of loop solitons genus g*, Elec. J. Diff. Eqns., **2007** (2007) 1–12.

[Mat2008a] S. Matsutani, *Neumann system and hyperelliptic al functions*, Surveys Math. and its Appl., **2008** (2008) 13–25.

[Mat2008b] S. Matsutani, *Relations in a quantized elastica*, J. Phys. A: Math. Theor., **41** (2008) 075201(12pp).

[Mat2008c] S. Matsutani, *A generalized Weierstrass represetation for a submanifold S* in $\mathbb{E}^n$ arising from the submanifold Dirac operator, Adv. Stud. Pure Math. **51** (2008) 259–283.

[Mat2010] S. Matsutani, *Euler's elastica and beyond*, J. Geom. Symm. Phys., **17** (2010) 45–86.

[Mat2020] S. Matsutani, *Trigonal Toda lattice equation*, J. Nonlinear Math. Phys., **27** (2020) 697–704.

[Mat2024a] S. Matsutani, *Statistical mechanics of elastica for the shape of supercoiled DNA: hyperelliptic elastica of genus three*, Physica A, **643** (2024) 129799 (11pages).

[Mat2024b] S. Matsutani, *Math Libre: Tales of elastica I, [Japanese]*, Gendai-Sugaku, **7** (2024) 82–84.

[MatK2013] S. Matsutani and J. Komeda, *Sigma functions for a space curve of type (3,4,5)*, J. Geom. Symmetry Phys., **30** (2013) 75–91.

[MatO2003a] S. Matsutani and Y. Ônishi, *On the moduli of a quantized elastica in* $\mathbb{P}$ *and KdV flows: study of hyperelliptic curves as an extension of Euler's perspective of elastica I*, Rev. Math. Phys., **15** (2003) 559–628.

[MatO2003b] S. Matsutani and Y. Ônishi, *Wave-particle complementarity and reciprocity of Gauss sums on Talbot effects*, Found. Phys. Lett., **16** (2003) 325–341.

[MatP2008] S. Matsutani and E. Previato, *Jacobi inversion on strata of the Jacobian of the* C_{rs} *curve* $y^r = f(x)$, J. Math. Soc. Japan, **60** (2008) 1009–1044.

[MatP2009a] S. Matsutani and E. Previato, *A generalized Kiepert formula for* C_{ab} *curves*, Israel J. Math., **171** (2009) 305–323.

[MatP2009b] S. Matsutani and E. Previato, *A class of solutions of the dispersionless KP equation*, Phys. Lett. A, **373** (2009) 3001–3004.

[MatP2014] S. Matsutani and E. Previato, *Jacobi inversion on strata of the Jacobian of the* C_{rs} *curve* $y^r = f(x)$ *II*, J. Math. Soc. Japan, **66** (2014) 647–692.

[MatP2015] S. Matsutani and E. Previato, *The al function of a cyclic trigonal curve of genus three*, Collectanea Mathematica, **66** (2015) 311–349.

[MatP2016] S. Matsutani and E. Previato, *From Euler's elastica to the mKdV hierarchy, through the Faber polynomials*, J. Math. Phys., **57** (2016) 081519.

[MatP2022] S. Matsutani and E. Previato, *An algebro-geometric model for the shape of supercoiled DNA*, Physica D, **430** (2022) 133073.

[MatT1992] S. Matsutani and H. Tsuru, *Physical relation between quantum mechanics and soliton on a thin elastic rod*, Phys. Rev. A, **46** (1992) 1144–1147.

[McH2022] J. McKay and Y-H. He, *Kashiwa lectures on new approaches to the Monster*, Notices, Int. Cons. Chinese Math., **10** (2022) 71–88.

[Mir1995] R. Miranda, Algebraic curves and Riemann surfaces (GSM5), A.M.S. Rhode Island, 1995.

[Miu1998] S. Miura, *Linear codes on affine algebraic curves*, IEICE Trans., **J81-A** (1998), 1398–1421 [in Japanese].

[MorP1986] I. Morrison and H. Pinkham, *Galois Weierstrass points and Hurwitz characters*, Ann. Math., **124** (1986) 591–625.

[Mos1980] J. Moser, *Geometry of quadrics and spectral theory*, The Chern symposium pp. 147–188, Springer, Berlin, 1980.

[Mul1984] M. Mulase, *Cohomological structure in soliton equations and Jacobian varieties*, J. Diff. Geom., **19** (1984) 403–430.

[Mum1977] D. Mumford, *An algebro-geometric construction of commuting operators and of solutions to the Toda lattice equation, Korteweg–de Vries equation and related non-linear equations*, Proceedings Int. Symp. Algebraic Geometry, Kyoto (1977) 115–153. Kinokuniya Book Store, Tokyo, 1978. Birkhäuser, Boston, 1984.

[Mum1981] D. Mumford, Tata lectures on theta Vol. I, Birkhäuser, Boston, 1981.

[Mum1984] D. Mumford, Tata lectures on theta Vol. II, Birkhäuser, Boston, 1984.

[Mum1993] D. Mumford, *Elastica and computer vision*, in Algebraic geometry and its applications, ed. by C. Bajaj, Springer-Verlag, Berlin, 1993, 507–518.

[Mum1999] D. Mumford, The Red Book of Varieties and Schemes: Includes the Michigan Lectures (1974) on Curves and their Jacobians (LNM.1358), Springer, Berlin, 1999.
[Nak2010a] A. Nakayashiki, *On algebraic expansions of sigma functions for* (n, s) *curves*, Asian J. Math., **14** (2010) 175–212.
[Nak2010b] A. Nakayashiki, *Sigma function as a tau function*, Int. Math. Res. Not., **2010** (2010) 373–394.
[Nak2016] A. Nakayashiki, *Tau function approach to theta functions*, Int. Math. Res. Not., **2016** (2016) 5202–5248.
[Nak2018] A. Nakayashiki, *Degeneration of trigonal curves and solutions of the KP-hierarchy*, Nonlinearity, **31** (2018) 3567–3590.
[Nak2019] A. Nakayashiki, *On reducible degeneration of hyperelliptic curves and soliton solutions*, SIGMA, **15** (2019) 009 (18 pages).
[Nak2020] A. Nakayashiki, *One step degeneration of trigonal curves and mixing of solitons and quasi-periodic solutions of the KP Equation*, in Geometric Methods in Physics XXXVIII: Workshop, Bialowieza, Poland, 2019 edited by P. Kielanowski, A. Odzijewicz, and E. Previato, Birkhäuser, Boston, 2020.
[NakY2012] A. Nakayashiki and K. Yori, *Derivatives of Schur, tau and sigma functions*, On Abel–Jacobi images, in symmetries, integrable systems and representations, edited by K. Iohara, S. Morier-Genoud, and B. Remy, Springer-Verlag, New York, 2012, 429–462.
[Nam1979] M. Namba, Families of Meromorphic functions on compact Riemann surfaces, LNM 767, Springer-Verlag, New York, 1979.
[Neu1999] J. Neukirch, Algebraic number theory, trans. by N. Schappacher, Springer, Berlin, 1999.
[Neu1859] C. Neumann, *De problemate quodam mechanico, quod ad primam integralium ultraellipticorum classem revocatur*, Crelle's J., **56** (1859) 46–63.
[New1985] A. C. Newell, Solitons in mathematics and physics, SIAM, 1985.
[New1711] I. Newton, Analysis per quantitatum series, fluxiones, ac differentias : cum enumeratione linearum tertii ordinis, 1711.
[Noe1918] E. Noether, *Invariante Variationsprobleme*, Nachr. v. d. Ges. d. Wiss. zu Göttingen, (1918) 235–257.
[Nov1974] S. P. Novikov, *The periodic problem for the Korteweg–de Vries equation*, Funct. Anal. Appl., **8** (1974) 236–246.
[Ohs2017] T. Ohsawa, Gendai-Fukusokaiseki-heno-dohyo (Guideposts to Modern Complex Analysis: The Range of Legends), Gendai-sugaku, Kyoto, 2017, [in Japanese].
[Ohs2018a] T. Ohsawa, L^2 *Joku-ikou-no-Saikin-no-Yoso (Recent aspect of* L^2*-Joku-ikou)*, Suugaku, **70** (2018) 184–203, [in Japanese].
[Ohs2018b] T. Ohsawa, L^2 approaches in several complex variables: towards the Oka–Cartan theory with precise bounds, Springer-Verlag, New York, 2018.
[OhtRGT1996] Y. Ohta, A. Ramani, B. Grammaticos, K. M. Tamizhmani, *From discrete to continuous Painlevé equations: a bilinear approach*, Phys. Lett., **24** (1996) 255–261.
[On1998] Y. Ônishi, *Complex multiplication formulae for hyperelliptic curves of genus three*, Tokyo J. Math., **21** (1998) 381–431.
[On2002] Y. Ônishi, *Determinant expressions for Abelian functions in genus two*, Glasgow Math. J., **44** (2002) 353–364.
[On2005] Y. Ônishi, *Determinant expressions for hyperelliptic functions*, Proc. Edinburgh Math. Soc., **48** (2005) 705–742.
[On2009] Y. Ônishi, *Abelian functions for trigonal curves of dgree four and determinantal formulae in purely trigonal case*, Int. J. Math., **20** (2009) 427–441.
[On2010] Y. Ônishi, *Frobenius–Stickelberger type formulae for general curves*, (unpublished) in "The higher-genus sigma function and applications", ICMS at Edinburgh, 11–15, Oct. 2010.
[On2011] Y. Ônishi, *Determinant formulae in Abelian functions for a general trigonal curve of degree five*, Comp. Methods Func. Theory, **11** (2011) 547–574.
[On2013] Y. Ônishi, Abel Function Theory (in Japanese), Chuo university, 2013.
[On2018] Y. Ônishi, *Arithmetical power series expansion of the sigma function for a plane curve*, Proc. of the Edinburgh Math. Soc., **61** (2018) 995–1022.

[Os1912] W. F. Osgood, Lehrbuch der Funktionentheorie, Teubner, Berlin, 1907; second edition, 1912.

[Pic1883] E. Picard, *Sur les fonctions de deux variables independentes analogues aux fonctions modulaires*, Acta Math, **2** (1883) 114–135.

[Pin1974] H. C. Pinkham, *Deformation of algebraic varieties with G_m action*, Astérisque, **20** (1974) 1–131.

[Po1910] H. Poincaré, Savants et écrivains, Flammarion, Paris, 1910.

[Pr1985] E. Previato, *Hyperelliptic quasi-periodic and soliton solutions of the nonlinear Schrödinger equation*, Duke Math. J., **52** (1985) 329–377.

[Pr1986] E. Previato, *A particle-system model of the sine-Gordon hierarchy*, Physica D: Nonlinear Phenomena, **18** (1986) 312–314.

[Pr1987] E. Previato, *Generalized Weierstrass $\mathfrak{P}$-functions and KP flows in affine space*, Comment. Math. Helv., **62** (1987) 292–310.

[Pr1993] E. Previato, *Geometry of the modified KdV equation*, in Geometric and quantum aspects of integrable systems (Scheveningen, 1992), Lecture Notes in Phys., 424, Springer-Verlag, Berlin, 1993, 43–65.

[Pr1996] E. Previato, *Seventy years of spectral curves: 1923–1993*, Integrable systems and quantum groups (Montecatini Terme, 1993), Lecture Notes in Math. 1620, Springer, Berlin, 1996, 419–481.

[PrBMEBKL] E. Previato, V. M. Buchstaber, V. B. Matveev, J. C. Eilbeck, H. Braden, J. Kunz and C. Lämmerzahl, *Victor Enolski (1945–2019)*, Notices of A.M.S., (2020) 1755–1767.

[Ra1973] H. Rademacher, Topics in analytic number theory (GMW 169), Springer-Verlag, New York, 1973.

[Ric2008] A. Rice, *In search of the "birthday" of elliptic functions*, The Mathematical Intelligencer **30** (2008) 48–56.

[Rie1857] G. F. B. Riemann, *Theorie der Abel'schen Functionen*, J. reine angew. Math., **54** (1857).

[Rie1866] G. F. B. Riemann, *Ueber das Verschwinden der θ -Functionen*, J. reine angew. Math., **65** (1866) 161–172.

[Ros1986] M. Rosen, *Abelian varieties over* $\mathbb{C}$ Arithmetic geometry, edited by G. Cornell and J. H. Silverman, Springer-Verlag, New York, 1986, 79–101.

[RosGS2009] J. C. Rosales and P. A. García-Sánchez, Numerical semigroup, Springer-Verlag, New York, 2009.

[Sag2001] B. E. Sagan, The Symmetric group: representations, combinatorial algorithms, and symmetric functions (GTM 203), Springer-Verlag, New York, 2001.

[SatM1980] M. Sato and Y. Mori (Sato), *Hirota-shi-no bilear equations ni-tsuite (On Hirota's bilinear equations [in Japanese])*, RIMS Kokyuroku, **338** (1980) 183–204.

[SatN1984] M. Sato, and M. Noumi, *Soliton equation and universal Grassmann manifold [in Japanese]*, Sophia University Kokyuroku in Mathematics, **18** (1984) 1–131.

[SatS1981] M. Sato and Y. Sato, *Hirota-shi-no bilear equations ni-tsuite II (On Hirota's bilinear equations II [in Japanese])*, RIMS Kokyuroku, **414** (1981) 181–202.

[SatS1982] M. Sato and Y. Sato, *Soliton equations as dynamical systems on infinite dimensional Grassmann manifold*, in Nonlinear partial differential equations in applied sciences, edited by P. D. Lax, H. Fujita, and G. Strang, 259–271, Amsterdam: North-Holland, 1982.

[ScaR2011] G. M. Scarpello and D. Ritelli, *Exact solutions of nonlinear equation of rod deflections involving the Lauricella hypergeometric functions*, Int. J. Math. Math. Sci., **2011** (2011) 838924, 22 pages.

[Sch1966] M. Schiffer, *Half-order differentials on Riemann surfaces*, SIAM J. App. Math., **14** (1966) 922–934.

[SchS1951] M. Schiffer and D. C. Spencer, Functionals of finite Riemann surfaces, Dover, 1951.

[Sch1987] R. J. Schilling, *Generalizations of the Neumann system*, Comm. Pure Appl. Math., **XL** (1987) 455–522.

[Sch1968] M. Schlessinger, *Functors of Artin rings*, Trans. Amer. Math. Soc., **130** (1968) 208–222.

[Sch1971] M. Schlessinger, *Rigidity of quotient singularities*, Inv. Math., **14** (1971) 17–26.

[Sch1887] F. Schottky, *Ueber eine specielle Function, welche bei einer bestimmten linearen Transformation ihres Arguments unverändert bleibt*, Crelle J., **101** (1887) 227–272.

[Sch1901] I. Schur, *Uber eine Klasse von Matrizen, die sich einer gegebenen Matrix zuordnen lassen*, Dieterich, 1901.

[SebF2007] A. Sebbar and T. Falliero, *Equilibrium point of Green's function for the annulus and Eisenstein series*, Proc. Amer. Math. Soc., **135** (2007) 313–328.

[SeW1985] G. Segal and G. Wilson, *Loop groups and equations of KdV type*, IHES, **61** (1985) 5–65.

[Ser1989] J-P. Serre, Topics in Galois theory: Research Notes in Mathematics (English Edition), Dover, 1989.

[Shi1988] H. Shiga, *On the representation of the Picard modular function by 0 constants I–II*, Publ. RIMS, Kyoto Univ., **24** (1988) 311–360.

[Shi1986] T. Shiota, *Characterization of Jacobian varieties in terms of soliton equations*, Inv. Math., **83** (1986) 333–382.

[Si1972] S. L. Siegel, *Topics complex function theory Vol III*, trans. by E. Gotschling and M. Tretkoff, John Wiley & Sons Inc., New York, 1972.

[Sl1980] P. Slodowy, Simple singularities and simple algebraic groups, LNM 815, Springer-Verlag, New York, 1980.

[Sp1966] E. H. Spanier, Algebraic topology, Springer-Verlag, New York, 1966.

[Ste1986] J. Steenbrink, *Limits of Hodge structures*, Inv. Math., **31** (1976) 229–257.

[Sti2008] H. Stichtenoth, Algebraic function fields and codes (GTM 254) (English Edition), Springer-Verlag, New York, 2008.

[Sui1972] N. Suita, *Capacities and kernels on Riemann surfaces*, Arch. Rational Mech. Anal, **46** (1972) 212–217.

[Suz2017] J. Suzuki, *Klein's fundamental 2-form of second kind for the* C_{ab} *curves*, SIGMA, **13** (2017) 017, 13.

[Sy1884] J. J. Sylvester, *Question 7382*, Mathematical questions from the educational times, **37** (1884) 26.

[Tak1930] T. Takagi, Daisuu-Gaku-Kougi (Lecture of algebra [in Japanese]), Kyouritsu, Tokyo, 1930.

[Tak1971] T. Takagi, Daisuu-Seisuu-ron (Algebraic number theory) [in Japanese], Iwanami, Tokyo, 1971.

[Tak1998] M. Takase, Abel/Galois Daen-kansuuron (elliptic function theory) [in Japanese], Asakura, Tokyo, 1998.

[Tak2012] M. Takase, Jacobi Daenkansuuron (elliptic function theory) [in Japanese], Kodansha, Tokyo, 2012.

[Tak2016] M. Takase, Riemann-to-Daisu-kannsuuron (Riemann and algebraic function theory) [in Japanese], The Univ. Tokyo Press, Tokyo, 2016.

[TanD1979] S. Tanaka and E. Date, KdV houteishiki [in Japanese], Kinokuniya, 1979.

[Tj1978] A. N. Tjurin, *Periods of quadratic differentials*, Usp. Mat. Nauk, **33** (1978) 149–195.

[To1967] M. Toda, *Vibration of a chain with nonlinear interaction*, J. Phys. Soc. Japan, **22** (1967) 431–436; *Wave propagation in anharmonic lattices*, *ibid.* **23** (1967) 501–506.

[To1989] M. Toda, Hisenkei Hadō To Soriton KTK Scientific Publ., Tokyo, Kluwer, Boston, 1989.

[Tr1960] C. Truesdell, The Rational Mechanics of Flexible or Elastic Bodies 1638–1788, Leonhardi Euleri Opera Omnia Ser. II, Vol. 11 part 2, Orell Füssli, Zürich, 1960.

[Tr1968] C. Truesdell, Essays in the History of Mechanics, Springer-Verlag, New York, 1968.

[Tr1983] C. Truesdell, *The influence of elasticity on analysis: the classic heritage*, Bull. Amer. Math. Soc., **9** (1983) 293–310.

[Ts1987] H. Tsuru, *Equilibrium shapes and vibrations of thin elastic rod*, J. Phys. Soc. Jpn., **56** (1987) 2309–2324.

[TsW1986] H. Tsuru and M. Wadati, *Elastic model of highly supercoiled DNA*, Biopolymers, **25** (1986) 2083–2096.

[U2011] Y. Uchida, *Division polynomials and canonical local heights on hyperelliptic Jacobians*, Manuscripta Math., **134** (2011) 273–308.

[vH1983] M. van Hoeij, *An algorithm for computing the Weierstrass normal form*, Proceedings of the 1995 international symposium on Symbolic and algebraic computation, ACM, New York, 1995, 90–95.

[Web1908] H. Weber, Lehrbuch der Algebra III, Friedr. Vieweg & Sohn, 1908.

[Wei1983] A. Weil, Number Theory, an approach through history: From Hammurapi to Legendre, Birkhäuser, Boston, 1983.

[Wei1840] K. Weierstrass, *Über die Entwicklung der Modular-Functionen*, Westernkotten in Westfalen, Sommer 1840 in Mathematische Werke I, 1–49, Mayer & Müller, Berlin, 1895.

[Wei1854] K. Weierstrass, *Zur Theorie der Abelschen Functionen*, J. Reine Angew. Math., **47** (1854) 289–306.

[Wei1867] K. Weierstrass, *Uber Normalformen algebraischer Gebilde*, Mathematische Werke III, 297–307 Mayer & Müller, Berlin, 1894.

[Wei1875a] K. Weierstrass, *Theorie der Abelschen Transcendenten*, Mathematische Werke IV, 1875–6, Mayer & Müller, Berlin, 1902.

[Wei1875b] K. Weierstrass, *Theorie der Elliptischen Functionen*, Mathematische Werke V, 1875–6, Mayer & Müller, Berlin, 1915.

[Wei1882] K. Weierstrass, *Zur Theorie der Elliptischen Functionen (Sitzungsberichte der Akademie der Wissenschaften zu Berlin, 1882, S. 443–451)*, Mathematische Werke II, 245–255, Mayer, & Müller, Berlin, 1895.

[Wei1915] K. Weierstrass, Mathematische Werke I–VI, Mayer & Müller, Berlin, 1894, 1895, 1903, 1902, 1915, 1915.

[Wey1955] H. Weyl, The concept of a Riemann surface, translated by G. R. MacLane, Dover, 1955.

[WhW1927] E. T. Whittaker and G. N. Watson, A course of modern analysis, Cambridge University Press, Cambridge, 1927.

[Yo1970] K. Yosida, Functional analysis (GMW 123), Springer-Verlag, New York, 1970.

[Z2001] A. V. Zabrodin, *The dispersionless limit of the Hirota equations in some problems of complex analysis*, Teoret. Mat. Fiz., **129** (2001) no. 2, 239–257; translation in Theoret. and Math. Phys., **129** (2001) 1511–1525.

Index

Symbols

S. Matsutani, *The Weierstrass Sigma Function in Higher Genus and Applications to Integrable Equations*, Springer Monographs in Mathematics,
https://doi.org/10.1007/978-981-97-8163-8

A

The manufacturer's authorised representative in the EU is Springer Nature Customer Service Centre GmbH, Europaplatz 3, 69115 Heidelberg, Germany. If you have any concerns regarding our products, please contact ProductSafety@springernature.com

Printed and bound by CPI Group (UK) Ltd, Croydon, CR0 4YY
01/10/2026
02225938-0007